Oxidative Stress in Microbial Diseases

Sajal Chakraborti • Tapati Chakraborti
Dhrubajyoti Chattopadhyay
Chandrima Shaha
Editors

Oxidative Stress in Microbial Diseases

Springer

Editors
Sajal Chakraborti
Department of Biochemistry and Biophysics
University of Kalyani
Kalyani, West Bengal, India

Tapati Chakraborti
Department of Biochemistry and Biophysics
University of Kalyani
Kalyani, West Bengal, India

Dhrubajyoti Chattopadhyay
Amity University
Kolkata, West Bengal, India

Chandrima Shaha
National Institute of Immunology
New Delhi, India

ISBN 978-981-13-8762-3 ISBN 978-981-13-8763-0 (eBook)
https://doi.org/10.1007/978-981-13-8763-0

This Springer imprint is published by the registered company Springer Nature Singapore Pte Ltd.
The registered company address is: 152 Beach Road, #21-01/04 Gateway East, Singapore 189721, Singapore

Professor Avadhesha Surolia is an eminent Glycobiologist at the Indian Institute of Science, Bangalore. Presently, he is an SERB Distinguished Fellow and an Honorary Professor at the Molecular Biophysics Unit, Indian Institute of Science (IISc), Bangalore. Professor Surolia was born at Kishangarh, Rajasthan, on December 3, 1947. He received BSc in Chemistry and Biology from Jodhpur University (1970); MSc in Biochemistry from Maharaja Sayajirao (MS) University, Baroda (1972); and PhD from Madras University (1976). In 1978 he received DSc degree from the Madras University. In recognition of his contributions to medicine and science, he

was conferred an Honorary DSc degree in 2013 by Queen's University, Belfast, UK. Prof. Surolia worked as a Scientist (1976–1981) at the Indian Institute of Chemical Biology, Kolkata; Assistant Professor (1981–1986), Associate Professor (1986–1991), Professor (1991–2013) and Chairman of the Molecular Biophysics Unit (2000–2006), all at the Indian Institute of Science (IISc), Bangalore; and Director at the National Institute of Immunology, New Delhi (2006–2011). Prof. Surolia is known for his work on lectin structure and interactions, orientation and dynamics of cell surface carbohydrate receptors and protein folding, diabetes, anti-malarials, anti-mycobacterials and anti-cancer agents based on curcumin, flavonoids, etc. He has made seminal contributions on the molecular features of protein-carbohydrate recognition that eventually led to the foundation of macromolecular crystallography as it exists in the country today. In addition, neuropathic pain, neurodegenerative disorders and the link between immunity and Obsessive-Compulsive Disorder (OCD) are areas of his current research interest. Prof. Surolia has published over 400 scientific papers in different international journals of repute. He is the recipient of Shanti Swarup Bhatnagar Prize conferred by the Council of Scientific and Industrial Research, Govt. of India. Prof. Surolia is a member of the Third World Academy of Sciences (TWAS), Trieste, Italy, and International Molecular Biologists Network. He is a Fellow of all the science academies in the country—the National Science Academy, Indian National Science

Academy and Indian Academy of Sciences. He was a member of the Board of Trustees of the Human Frontier Scientific Program (HFSP, Strasbourg; 2006–2013). He is the only Indian member of the International Glycoconjugate Organization since 1998 and served as its President from 2001 to 2004. He is Member of the Executive Committee of the International Union of Biochemistry and Molecular Biology (2011–2017). He has been a visiting Scientist at Massachusetts Institute of Technology (1982, 2004 and 2005), University of Maryland (1998–2005) and University of Michigan, USA (1983). He is an Editorial Board Member of the Proceedings of the National Academy of Sciences *(India) and the* Indian Journal of Biotechnology, *IUBMB-Life. Professor Surolia undoubtedly is a legendary figure in Indian science as is evident from the recent bibliometric analysis conducted by Department of Science and Technology (DST), Government of India (http://mbu.iisc.ac.in/~aslab/Bibliometric_study_DST_2015.pdf).*

He has inspired and motivated countless young researchers. We feel honoured to dedicate this book to Prof. Avadhesha Surolia and wish him good health and success in his long fruitful activities.

Preface

> "When antibiotics first came out, nobody could have imagined we'd have the resistance problem we face today. We didn't give bacteria credit for being able to change and adapt so fast". Bonnie Bassler (an American Molecular Biologist)

Infectious diseases are a major threat to human as well as animals. Parasitic diseases such as malaria, leishmaniasis, trypanosomiasis, amoebiasis, helminthiasis, brucellosis, leprosy and tuberculosis are major human health problems, especially in developing countries. A large number of these and other parasitic diseases are transmitted by vectors, where ROS plays an important role. Therefore, understanding of the basic mechanisms in oxidant -mediated microbial diseases will eventually provide novel therapies to combat transmission of pathogens and thereby progression of microbial diseases.

At physiologically low levels, reactive oxygen species (ROS) provide the specificity in their interaction with effectors in signalling processes. Balancing the generation and elimination of ROS is important for proper function of redox sensitive normal signalling processes. To maintain redox homeostasis and to eliminate ROS, aerobes are equipped with enzymatic/nonenzymatic antioxidants and metal sequestering proteins either to prevent or to intercept the formation of pro-oxidants.

Besides coping with an increase in ROS levels generated from intrinsic sources, microbes have to deal with the oxidative stress imposed by the hosts' immune response. Since the redox system plays such an important role for parasites survival within their host, drugs that are involved in either promoting ROS production or inhibiting cellular antioxidant systems will cause redox imbalance by pushing ROS levels above a certain level, which could lead to parasites death.

Infectious diseases are a leading cause of human death. To prevent microbial diseases in the developing countries, it is important to make an effective approach to public health and hygiene system with the participation, conviction and dedication of common people. With his experience on how China substantially eradicated microbial diseases, Joshua S. Horn, an English surgeon once narrated: "The first concept rests on the conviction that the ordinary people possess great strength and wisdom and that when their initiative is given full play they can accomplish miracles; that the art of leadership is to learn from the masses, to refine and systematize their experience and, on this basis, to decide on policy" (Joshua S. Horn : Away

With All Pests). In order to provide effective treatments of microbial diseases, researchers have, in the recent past, considered the changing pattern of microbes in producing relevant diseases. The complexity of microbial infection has now become a bit understandable as researchers are able to discover mechanisms of atleast some microbial pathogenesis.

This book contains 29 chapters contributed by established scientists working in the field of oxidative stress biology. They provided comprehensive and updated review on their respective topics. The expert contributors have elucidated the mechanisms of ROS production and their regulation, balance between oxidative stress and antioxidant system, and oxidative stress-induced pathologies and strategies to ameliorate a variety of microbial diseases. The novel descriptions provided in the chapters of the book will enhance our understanding of oxidative stress biology, which will help the readers to gain an in-depth insight and the latest development in the field of research on oxidative stress-induced microbial diseases. Thus, better understanding of regulation of oxidative stress could be utilized for devising strategies for development of novel therapeutics for clinical intervention in oxidative stress-induced microbial diseases. Certainly, the readers including postgraduate students and biomedical researchers will find this book extremely informative, interacting and stimulating.

Thanks are due to all contributors of the book for their considerable energy, time and effort to accomplish complete chapters, which are informative and readable. Thanks are also due to Ms. Ishrath Ara, Dr. Madhurima Kahali and Mr. Daniel Ignatius Jagadisan (Springer-Nature) for their understanding, cooperation and support during the preparation of this book.

Kalyani, West Bengal, India Sajal Chakraborti

Contents

Nikhilesh Joardar Parasitology Laboratory, Department of Zoology, Visva-Bharati University, Santiniketan, West Bengal, India

Hasan Karsen Department of Infectious Diseases and Clinical Microbiology, School of Medicine, Harran University, Sanliurfa, Turkey

Ginpreet Kaur Department of Pharmacology, SPP School of Pharmacy & Technology Management, SVKM's NMIMS, Mumbai, India

Asad U. Khan Medical Microbiology and Molecular Biology Laboratory, Interdisciplinary Biotechnology Unit, Aligarh Muslim University, Aligarh, Uttar Pradesh, India

Saloni Khogta Department of Pharmacology, SPP School of Pharmacy & Technology Management, SVKM's NMIMS, Mumbai, India

Suhyun Kim School of Life Sciences, BK21 Plus KNU Creative Bio Research Group, Kyungpook National University, Daegu, South Korea

Amit Kumar Department of Immunology & Defence Mechanism, College of Biotechnology, Sardar Vallabhbhai Patel University of Agriculture & Technology, Meerut, Uttar Pradesh, India

Dong Gun Lee School of Life Sciences, BK21 Plus KNU Creative Bio Research Group, Kyungpook National University, Daegu, South Korea

Swetapadma Majhi Electron Microscopy Unit, Sophisticated Analytical Instrument Facility, CSIR-Central Drug Research Institute, Lucknow, India
Academy of Scientific and Innovative Research, Chennai, India

S. Martínez-Subiela Department of Animal Medicine and Surgery, Faculty of Veterinary Medicine, Regional Campus of International Excellence "Campus Mare Nostrum", University of Murcia, Murcia, Spain

Greanious Alfred Mavondo Departments of Pathology and Physiology, Faculty of Medicine, Mpilo Hospital NUST Complex, National University of Science and Technology (NUST), Bulawayo, Zimbabwe

Joy Mavondo Imagegate Diagnostics c/o Dalony Distributors (PVT) Ltd, Bulawayo, Zimbabwe

Rubem Figueiredo Sadok Menna-Barreto Laboratório de Biologia Celular, IOC, FIOCRUZ, Rio de Janeiro, Brazil

Catharine Mhlanga Departments of Pathology and Physiology, Faculty of Medicine, Mpilo Hospital NUST Complex, National University of Science and Technology (NUST), Bulawayo, Zimbabwe

Lama Misba Medical Microbiology and Molecular Biology Laboratory, Interdisciplinary Biotechnology Unit, Aligarh Muslim University, Aligarh, Uttar Pradesh, India

Arijit Misra Microbiology Laboratory, Department of Botany, Institute of Science, Visva-Bharati (A Central University), Santiniketan, West Bengal, India

Kalyan Mitra Electron Microscopy Unit, Sophisticated Analytical Instrument Facility, CSIR-Central Drug Research Institute, Lucknow, India
Academy of Scientific and Innovative Research, Chennai, India

Blessing Nkazimulo Mkhwanazi Dietetics and Human Nutrition, School of Agriculture, Earth and Environment Science, College of Agricultural, Engineering and Science, University of KwaZulu Natal, Durban, South Africa

Suprabhat Mukherjee Parasitology Laboratory, Department of Zoology, Visva-Bharati University, Santiniketan, West Bengal, India

Department of Animal Science, Kazi Nazrul University, Asansol, West Bengal, India

Nyasha Mukonowenzou Departments of Pathology and Physiology, Faculty of Medicine, Mpilo Hospital NUST Complex, National University of Science and Technology (NUST), Bulawayo, Zimbabwe

Mayibongwe Loius Mzingwane Departments of Pathology and Physiology, Faculty of Medicine, Mpilo Hospital NUST Complex, National University of Science and Technology (NUST), Bulawayo, Zimbabwe

Suman K. Nandy Bioinformatics Infrastructure Facility (BIF), North-Eastern Hill University (NEHU), Tura, Meghalaya, India

Tomoyoshi Nozaki Department of Biomedical Chemistry, Graduate School of Medicine, The University of Tokyo, Tokyo, Japan

Amit Pal Division of Pathophysiology, National Institute of Cholera and Enteric Diseases (NICED), Beliaghata, Kolkata, India

Santanu Palchaudhuri Amity Institute of Biotechnology, Amity University Kolkata, Kolkata, West Bengal, India

Pijush Kanti Pramanik Department of Biochemistry and Biophysics, University of Kalyani, Kalyani, West Bengal, India

Dibyendu Raj Vivekananda College, Kolkata, India
Division of Parasitology, ICMR-National Institute of Cholera and Enteric Diseases, Kolkata, India

Anu Rahal Goat Health Division, ICAR-Central Institute for Research on Goats (CIRG), Mathura, Uttar Pradesh, India

Dandugudumula Ramu Department of Life Sciences, School of Natural Sciences, Shiv Nadar University, Greater Noida, Uttar Pradesh, India

C. P. Rubio Department of Animal Medicine and Surgery, Faculty of Veterinary Medicine, Regional Campus of International Excellence "Campus Mare Nostrum", University of Murcia, Murcia, Spain

S. Scarcella Lab. Biología Celular y Molecular, Dpto. de Cs. Biológicas, FCV-UNICEN-CIVETAN-CONICET, Tandil, Argentina

Jessica Scoffield Department of Pediatric Dentistry, The University of Alabama at Birmingham, Birmingham, AL, USA
Department of Microbiology, The University of Alabama at Birmingham, Birmingham, AL, USA

Vijeta Sharma Special Centre for Molecular Medicine, Jawaharlal Nehru University, New Delhi, India

Shailja Singh Special Centre for Molecular Medicine, Jawaharlal Nehru University, New Delhi, India

Santi P. Sinha Babu Parasitology Laboratory, Department of Zoology, Visva-Bharati University, Santiniketan, West Bengal, India

Hugo Solana Lab. Biología Celular y Molecular, Dpto. de Cs. Biológicas, FCV-UNICEN-CIVETAN-CONICET, Tandil, Argentina

M. V. Solana Lab. Biología Celular y Molecular, Dpto. de Cs. Biológicas, FCV-UNICEN-CIVETAN-CONICET, Tandil, Argentina

Rima Tapader Division of Pathophysiology, National Institute of Cholera and Enteric Diseases (NICED), Beliaghata, Kolkata, India

Federica Turrini Department of Pharmacy, University of Genoa, Genoa, Italy

Francesco Michelangelo Turrini Department of Genetics, Biology and Biochemistry, University of Turin, Turin, Italy

A. Tvarijonaviciute Department of Animal Medicine and Surgery, Faculty of Veterinary Medicine, Regional Campus of International Excellence "Campus Mare Nostrum", University of Murcia, Murcia, Spain

Hui Wu Department of Pediatric Dentistry, The University of Alabama at Birmingham, Birmingham, AL, USA
Department of Microbiology, The University of Alabama at Birmingham, Birmingham, AL, USA

Peng Zhou Department of Pediatric Dentistry, The University of Alabama at Birmingham, Birmingham, AL, USA

Paola Zunin Department of Pharmacy, University of Genoa, Genoa, Italy

Part I

General Aspects of Oxidant Stress in Microbial Diseases

Role of Reactive Oxygen Species in the Correlation Between Host and Microbes

1

Suhyun Kim and Dong Gun Lee

Abstract

Free radicals have the nature of unpaired electrons in their outer orbital that render them as highly reactive species. Specifically, aerobic cells produce hydroxyl radicals, hydrogen peroxide, and superoxide anions known as reactive oxygen species (ROS). Host immune systems generate ROS as defense mechanisms to clear pathogens. It is plausible that microbes can thrive in oxidative environments by decreasing host responses mediated by ROS during infection or inducing several genes encoding antioxidant enzymes. However, ROS at high concentrations have an adverse effect on any cellular components. Based on the significant impact on antimicrobial actions, antimicrobials efficiently kill bacteria by inducing ROS and targeting production of ROS, thus strengthening antimicrobial activity. Therefore, the balance of ROS needs to be investigated to control microbial infections and associated diseases. This chapter is focused on the role of ROS between host and pathogens.

Keywords

Reactive oxygen species · Pathogenic microbes · Antioxidant enzyme · Oxidative stress

S. Kim · D. G. Lee (✉)
School of Life Sciences, BK21 Plus KNU Creative Bio Research Group,
Kyungpook National University, Daegu, South Korea
e-mail: dglee222@knu.ac.kr

S. Chakraborti et al. (eds.), *Oxidative Stress in Microbial Diseases*,
https://doi.org/10.1007/978-981-13-8763-0_1

1.1 Introduction

Emerging of diseases associated with microbial infection is a significant burden on global economies and public health [1]. Symbiosis with promising dangerous microorganisms and evolutionary adaptations for resistance may be a major reason for immune diseases to occur [1]. However, resistance to multiple antimicrobial agents including pathogenic bacteria, fungi, virus, and parasite have become a serious hazard. Over-reliance on antimicrobial drugs to treat infections has led to resistance in various strains of microorganism and even cancerous cells [2]. As a response to microbial disease, macrophages lead to programmed cell death [3]. In that point, understanding of reactive oxygen species (ROS) signaling for host responses to challenges presented by microbial invasion is important. ROS play a role in clearance of colonizing microorganisms, thus exploration of the mechanisms between oxygen radicals and innate immune responses are associated with pathogen clearance [4]. Both researchers and health professionals were globally interested in the mechanisms of chronic infections caused by opportunistic pathogens. Typically, chronic infection can cause an increase in the immune response, which can often lead to further destruction. ROS is involved in the harmful response, which results in self-damage. Excessive ROS production results in robust cellular death by acute infection or inflammation, while lower ROS levels play a critical physiological role to signal for growth, apoptosis, immune response, and microbial colonization.

ROS, intracellular chemical species, are formed by partial reduction of oxygen [5]. A variety of signaling roles of ROS are involved in a broad spectrum of organism from bacteria to mammalian cells [6]. High ROS levels may cause macromolecule damage, leading to promotion of aging and age-related diseases and progress intense research because of its importance in aging and age-related diseases [7]. ROS are also produced within the gastrointestinal (GI) tract [8]. Despite the barriers in mucosa for protection of infection, microbial pathogens can result in oxidative injury and GI inflammatory responses in the epithelium and immune/inflammatory cells. This pathogenesis involved in various GI diseases including peptic ulcers, gastrointestinal cancers, and inflammatory bowel disease is caused by oxidative stress [8]. In insects, restriction of glucose and respiration inhibition are activators of ROS and the life span extension. Glucose restriction and slower metabolic rate also occur during insect diapause, suggesting that ROS could be involved in regulation of diapause [7]. They may inactivate Gram-negative bacteria characterized by an impermeable outer cell membrane which contains endotoxins and blocks antibiotics, dyes, and detergents for protection of the sensitive inner membrane and cell wall [9].

ROS are by-product generated from cellular respiration, protein folding, and end products of a number of normal metabolic reactions [5] Additionally, ultraviolet radiation, cigarette smoking, alcohol, nonsteroidal anti-inflammatory drugs, ischemia-reperfusion injury, chronic infections, and inflammatory disorders can produce ROS [8]. Exceedingly high level of ROS over an organism's detoxification and repair capabilities indiscriminately reacts with DNA, RNA, proteins, and lipids, leading to oxidative stress [5, 10]. Superoxide anion ($O_2^{\bullet -}$), hydrogen peroxide

(H_2O_2), and the hydroxyl radical (•OH) are representative ROS. Among ROS, $O_2^{\bullet-}$ results from the one-electron reduction of molecular oxygen [11]. H_2O_2 is generated from superoxide produced by mitochondria and NADPH oxidases. Typically, •OH are generated from H_2O_2 through Fenton reaction in the presence of ferrous ions [11]. These ROS tightly regulate within cells [5]. Thus, this chapter focuses on the role and regulation of ROS in microbial infection.

1.2 Involvement of ROS in Cellular Signaling Network

ROS (e.g., $O_2^{\bullet-}$ and H_2O_2) act as secondary signaling molecules, control the action of several signaling pathways for inflammation and immune responses, and directly kill microbes that infect tissues [4, 12, 13]. ROS are then closely linked to cell homeostasis and metabolism. ROS are not stored in specific compartments and their levels are controlled by a subtle balance between production and cleaning as they are quickly generated or eliminated anywhere in the cell [14]. The local increase in ROS production may be specific to spatial control of ROS accumulation, confined to specific locations of cells such as specific membrane patches or cell organelles [6]. Thus, each cell carries ROS signals in an autonomous manner to activate its own ROS production [6, 15]. Signal transduction takes advantage of the unique biological properties of each oxidizing agent species, including chemical reactivity, stability, and lipid diffusion ability [5]. There are various types of ROS such as having charge (e.g., $O_2^{\bullet-}$) or not (e.g., H_2O_2) and being lipophilic (e.g., lipid peroxides). Therefore, they can be confined to specific intracellular compartments or can be easily moved through the membrane [14]. ROS is generally a signal which serves priming or activating cellular signaling networks. Moreover, other signals work with the ROS to deliver specificity [6]. These other signals are small peptides, hormones, lipids, cell wall fragments, and others. The calcium signal has a specific pattern of oscillation within a defined cell location. Likewise, the ROS signal is also delivered with the decoded message [6]. Since each individual cell or organelles has a set of ROS receptors, the generated ROS signal is decoded and transmitted by other networks, such as calcium or protein phosphorylation. Therefore, ROS can be incorporated with other signaling pathways [6, 15].

1.3 ROS Production in Host Cells in Response to Microbial Infection

Mammalian innate immune systems have elaborately evolved mechanisms to recognize and kill foreign invaders. It uses a combination of ROS, RNS, and enzymatic degradation as the main process of phagocytosis of macrophages and neutrophils [16]. ROS production rapidly rises during microbial infections (Fig. 1.1) to promote pathogen removal and contribute to signaling pathways involving inflammation, cell proliferation, and immune responses [17].

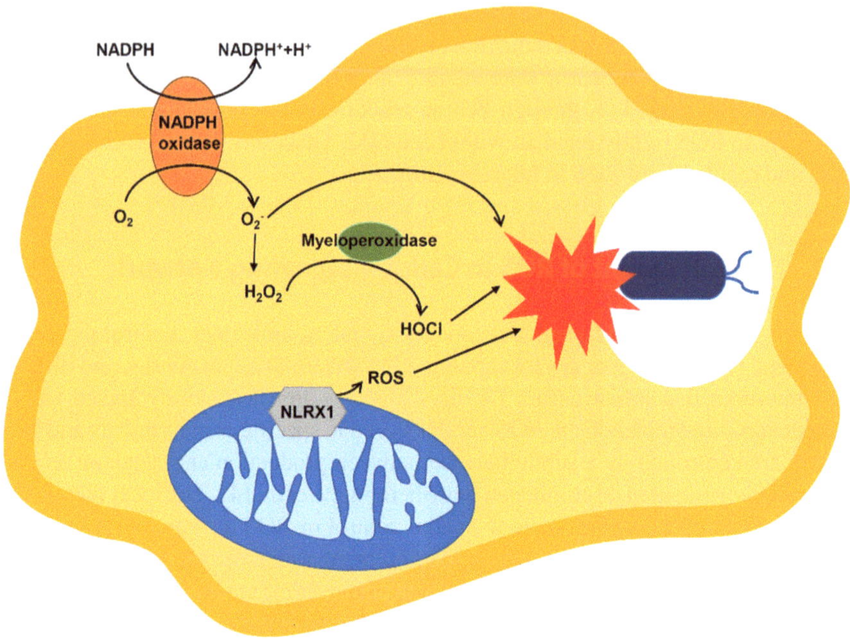

Fig. 1.1 ROS production in phagocyte against pathogenic microorganisms

A well-characterized ROS source in host-microbial interactions is the membrane-associated NADPH oxidase complex. Activated neutrophils and macrophages mainly produce ROS via NADPH oxidase [18]. The phagocytic NADPH oxidase complex is composed of cytochrome consisting of gp91phox, a phagocytic cell oxidase, and p22phox, which is embedded in the membrane. In addition, the cytoplasmic subunits p47phox, p67phox, and p40 phox and rac translocate to cytochrome when oxidase is activated. NADPH is oxidized to $NADP^+$, and electrons are transported to oxygen by a reduction potential gradient, forming a $O_2^{\bullet-}$ [19]. When a disorder occurs in NADPH oxidase, phagocytes are defective in the ROS production, resulting in repeated life-threatening microbial infections [20].

Myeloperoxidase (MPO) is an iron-containing heme protein present in neutrophilic granulocyte azurophilic granules and lysosomes of monocytes [21]. MPO has a catalytic activity to form HOCl from H_2O_2 and chloride ions, generating substances with strong antimicrobial activity. It plays a role in the killing of several microorganisms including bacteria, fungi, viruses, red cells, and malignant and nonmalignant nucleated cells [16, 22–26]. Although MPO seems to be not essential for host defense due to low frequency of infections in MPO-deficient individuals, hypohalous acids (HOCl)) dramatically improves microbicidal activity with H_2O_2 by increasing susceptibility to fungal species like *Candida* [27, 28].

Microbial infection and mitochondrial ROS have been also connected to innate immune responses. NLRX1, which contains a nucleotide-binding domain (NBD)

and a Leucine-rich repeat (LRR)-containing family member receptor (NLR), alerts the presence of intracellular intruders [29, 30]. NLRX1 stimulates the production of ROS, allowing ROS to interact directly with the pathogen as well as acting as a secondary signaling molecule [31–33]. ROS induced by NLRX1 is important in terms of infection, especially its signaling ability. The elevated ROS levels enhance the downstream signaling of NF-κB and Jun amino-terminal kinases-dependent (JNK), which demonstrate that ROS mediates pro-inflammatory gene transcription and cytokine-based signal responses caused by microbial invasion [34, 35]. It is important to understand the ROS signaling for the host response as a feature of innate immunity during infection. Colonizing microorganisms are efficiently prevented by ROS, thus the study of mechanism of correlation between oxygen radicals and innate immune responses is crucial for pathogen elimination [4].

1.4 Modulations of Host ROS Level by Pathogenic Microbes

ROS plays a large role in immune cells by killing phagocytosed microorganisms or involving in immune and inflammatory signaling cascades. However, a wide range of microorganisms modulate ROS and increase the persistent infectivity [36]. Control of ROS levels seems to be a conserved strategy utilized by various microorganisms. Through inducing or inhibiting oxygen radicals, pathogens reduce ROS-mediated host responses to infection and facilitate colonization. Prevention of NADPH oxidase assembly is the direct method to interfere with ROS accumulation in pathogens. *Francisella tularensis* directly prevented recruitment of gp91phox/p22phox and p47/p67phox to the phagosome during infection. Thereby the bacterium attenuates the assembly of NADPH oxidase and keeps neutrophils from oxidative burst mechanism [37]. *Anaplasma phagocytophilum* has the ability of attenuating NADPH oxidase activity by averting p22 and gp91phox incorporation into microbial-containing phagosomes [38, 39]. The transient increase in ROS within macrophages could be a way to suppress oxidatvie bursts [40]. *Chlamydia trachomatis* incorporate a subunit of NADPH oxidase, Rac, to the *Chlamydia*-containing vacuole to suppress oxidative bursts. *Chlamydia* adjusts the intracellular ROS level back to normal state, thus avoiding their immune sensors and inflammatory reactions and obscuring their presence [40].

Microorganisms present outside immune cell also have the ability to regulate intracellular ROS to persist infection. In neutrophils, these microorganisms primarily form exopolysaccharide, thereby inhibiting ROS production. Regardless of their intracellular origination, they directly scavenge free oxygen radicals by producing the exopolysaccharides [41]. Another case reported that phagocytic engulfment of apoptotic cells is interfered by production of pathogen's secondary metabolite. This secondary metabolite reduces intracellular Rho GTPase activities and ROS levels in host cells, enhancing inflammation at the infection site owing to failure of phagocytosis [42].

Unlike described organisms above, *Entamoeba histolytica* activates NADPH oxidase to increase ROS, leading to apoptosis in host cells [43]. This parasite utilizes

ROS molecules as a secondary signaling messenger, derived from NADPH oxidase. ROS stimulate extracellular signal-related kinase (ERK) 1/2 signaling to cause host cell death [44–46]. Other pathogens may induce host cell death using proliferating ROS and may contribute to pathogen spread through cell apoptosis. Japanese encephalitis virus (JEV) mediates intracellular ROS accumulation and activates pro-apoptotic signaling, resulting in apoptosis in host cells. Therefore, the virus promotes the spread of virions and persists the viral infection [47]. *Trypanosoma cruzi*, the causative agent of Chagas' disease, can infect humans and remains in the incubation period, but it causes chronic inflammation at the time of re-emergence and damages organs. *T. cruzi* optimally persist infection by elevating ROS generation owing to pathogen-induced mitochondrial membrane potential perturbation [48].

1.5 Enzymatic Antioxidant System in Microbes Against ROS

Bacteria, fungi, and parasites exhibit enzymatic activity to avoid oxidative protection of the host during the immune response to the infection process. Pathogens induce several genes encoding enzymes which serve maintenance of cellular redox environment and thus can endure the high levels of ROS [49]. Superoxide dismutase (SOD) is the first line of defense against $O_2^{\bullet -}$, which catalyze dismutation of $O_2^{\bullet -}$ in H_2O_2 [50]. Catalases (Cat) then degraded H_2O_2 into water and oxygen. Other enzymatic systems that detoxify hydroperoxide radicals are glutathione and thioredoxin systems. Glutathione peroxidases, glutathione reductases, and glutaredoxins belong to glutathione system [51–53]. Thioredoxin system is composed of peroxiredoxin, thioredoxin, and thioredoxin reductase [54–56]. There are other antioxidant enzymes, cytochrome *C* peroxidase and the chloroperoxidase that reduce organic hydroperoxide [57, 58].

Since SOD acts first as bacterial enzyme to detoxify ROS, these enzymes are important for microorganisms to combat oxidative stresses. *Staphylococcus aureus* encoded two genes of SOD [50, 59]. During stationary phase or superoxide anion generation, SodA gene was expressed to a Mn-SOD [53]. Phagocytic cells such as macrophage also increase the activity of SOD in bacteria [60, 61]. It has been demonstrated that Δ*SodA* and Δ*SodM* strains are more vulnerable to the oxidative stress [59, 62]. Likewise, Cat is also involved in ROS detoxification. Cat catalyzes the reduction of H_2O_2 to water and oxygen. Only one Cat (kat A) was found in *S. aureus* and it is activated when being phagocytized by macrophages. There is no effect on survival in kat A gene deletion strains, but the nasal colonization drastically decreased compared to the wild-type strain. Thus, kat A appears to be important in the colonization process, but it is not essential to the occurrence of the compensatory systems [50, 62]. The Cat of *Escherichia coli*, katE and katG, can detoxify the H_2O_2 [63, 64], but their significance for virulence is unclear due to compensatory mechanisms [63]. In *Streptococcus* spp., no Cat existed, so the catalysis of H_2O_2 must be dependent on other thioredoxin system enzymes. In *Streptococcus pneumoniae*, thiol peroxidase named TpxD is induced in response to oxidative stress. TpxD increases the survival rate of bacteria and also helps intranasal infection [65, 66]. Furthermore, the peroxiredoxin AhpC of *S. aureus* is involved in its resistance against external oxidative stress and is crucial for its survival, persistence, and nasal colonization [51, 67]. The effect on the virulence of AhpC in *E. coli* has not been proven, but H_2O_2 can be detoxified [60, 61, 68]. *S. aureus* does not encode glutathione-based enzymes but expresses genes corresponding to thioredoxin and thioredoxin reductase, which are induced in response to chemical oxidants [69].

Contrary to bacteria, the function of the enzyme antioxidant system in relation to the virulence of the fungi has not been clearly elucidated. ROS, produced by the host owing to fungi and thereafter in fungal virulence, can be detoxified by SODs. SOD1 and SOD5 are considered as virulent factors in *Candida albicans* because they protect yeast and hyphal cells from oxidative stress produced by macrophages [51, 67–69]. Cu/Zn-SOD (sod1) and Mn-SOD (sod2), the only SOD enzymes in *Cryptococcus neoformans*, are associated with virulence [70, 71]. Likewise, three Cat have been discovered in *C. albicans*. They are involved in the H_2O_2 detoxification and leukocyte resistance [72, 73]. *Aspergillus fumigatus* also has three Cat, which are all involved in the degradation of H_2O_2, but it was not known whether CatA is involved in virulence [74–76]. Recent studies indicate that glutathione systems could also participate in ROS detoxification. The glutaredoxin GRX2 which was shown to be associated with virulence in systemic infection participates in the defense to chemical oxidative stress and macrophages [51, 77]. Glutathione peroxidases GPX1 and GPX2 are responses to chemical oxidative stress and protect cells from ROS, but are not necessary for virulence in *C. neoformans* [52]. Another antioxidant enzyme, thioredoxin reductase, is capable of degrading ROS produced by neutrophils. Also, thioredoxin reductase is concerned with resistance in *C. albicans* strains isolated from recurrent candidiasis of vagina [54, 78, 79]. In *C. neoformans*, there are two thiol peroxidases called TSA1 and TSA3 that are overexpressed in the presence of H_2O_2. Further, TSA1 is considered as a virulence factor [80]. The thioredoxins TRX1 and TRX2 play a role in the oxidative and nitrosative stress defense expressed under stressful conditions, but TRX1 is not required for fungal virulence [55, 81]. Asp f 3 allergen in *A. fumigatus* is presumed to be a potential thioredoxin peroxidase [82] and has been shown to be an immunogenicity marker of allergic bronchial aspergillosis [83, 84]. It is found that H_2O_2 induces this protein strongly [82]. Other peroxidases also participate in antioxidant activity, particularly cytochrome *C* peroxidase, which is used to survive macrophages in response to oxidative stress and participates in the virulence of fungi [57, 85]. Likewise, the chloroperoxidase plays a role in oxidative stress response in various microbial strains including *A. flavus*, *Caldariomyces fumago*, *Sinirhizobium meliloti*, and *Streptomyces coelicolor* [58, 86–88].

1.6 Intracellular Response by Oxidative Stress

Although the product of the reaction is $O_2^{\bullet-}$ catalyzed by the phagocyte oxidase complex, subsequent reactions of $O_2^{\bullet-}$ can result in production of additional intermediates. Such intermediates like H_2O_2, •OH, singlet oxygen ($^1O_2$), and HOCl have different reactivity, stability, compartmentalization, and biological activity [89].

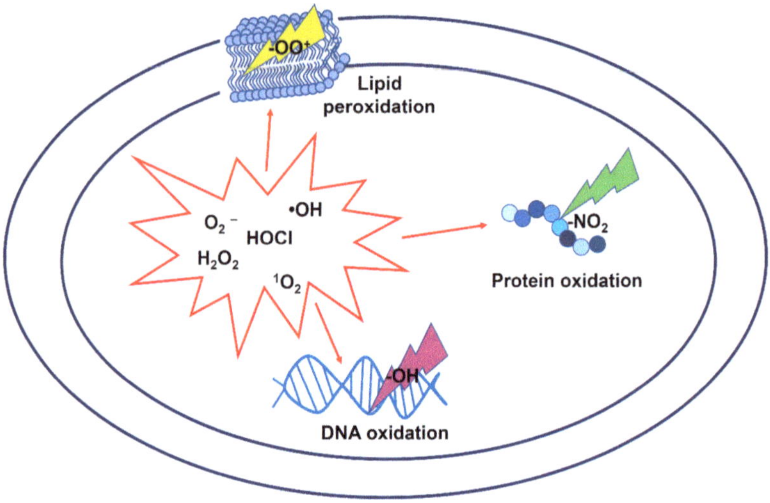

Fig. 1.2 Oxidative damage site induced by ROS in microorganisms

The generation and high concentrations of oxygen radicals can induce oxidative stress by causing imbalance between the production of oxidants and the antioxidant defenses. When ROS exceed those of antioxidants, cell responses such as apoptosis, tumorgenesis, and immune-response are potentiated [4, 90, 91]. The overproduction of ROS disturbs the prooxidant/antioxidant balance, resulting in damage of the cellular components, such as lipids, proteins, and DNA (Fig. 1.2). Many of the chemical modifications reflect the evolutionary transformation of these molecules from signaling mediators to cytotoxic species [92].

Lipid peroxidation is a universal impairment of ROS in polyunsaturated lipid-rich eukaryotes [93]. The membrane lipids present in the subcellular organelles are sensitive to free radical damage [12]. Lipids can cause direct and indirect effects by reacting with free radicals through producing highly damaging chain reaction of lipid peroxidation [12, 94]. The peroxidative sequence is initiated by abstracting a hydrogen atom from a methylene group, leaving behind an unpaired electron on the carbon atom of the cell [12]. It inhibits the membrane assembly and thus perturbs the membrane fluidity and permeability [94]. Lipid peroxidation of bacteria has been observed after ingestion by neutrophils [95]. However, it has not been proven that this affects bacterial killing. Chain-peroxidation reactions are limited by the saturated and monounsaturated fatty acids in bacterial membranes, but membrane lipids may be important part of oxidative damage only in eukaryotic pathogens. Yeast cells cannot synthesize unsaturated fatty acids, but when grown in a culture medium enriched with polyunsaturated fatty acids, they are easily incorporated into the membrane structure which eventually increase the possibility of lipid peroxidation to oxidative stress [96]. Indeed, yeast cells with polyunsaturated fatty acid-rich yeast membranes were hypersensitive to cadmium toxicity through lipid peroxidation [97]. Lipid peroxidation has also been reported in yeast cells undergoing autophagy induced by rapamycin and H_2O_2- and menadione-induced oxidative stress [98–100].

Oxidative damage of the protein adversely affects the homeostatic function of the cell and rapidly lowers the cell viability [101]. Most oxidized proteins that are functionally inactivated are rapidly eliminated, but some are accumulated slowly, contributing to a various diseases [12]. Protein carbonyl is formed through oxidation of arginine, proline, or lysine and is readily identified after oxidative injury [102]. Protein carbonyl is an irreversible product and is therefore susceptible to proteolytic degradation. When *E. coli* was treated with H_2O_2, proteins (e.g., alcohol dehydrogenase E, DnaK, OppA, F_0F_1 -ATPase) are oxidized [102]. As a tendency to form large protein aggregates that cannot be degraded by cells through normal protein degradation pathways, prompt removal of protein carbonyl is important. Without effectively destroying these aggregates, the carbonyl accumulation will be further strengthened and cell homeostasis will eventually be disturbed [103]. In addition, dysfunctional protein can be synthesized by ROS, such as mRNA mistranslation. For example, oxygen-dependent chromate ion Cr (VI) in *Saccharomyces cerevisiae* induces mistranslation of mRNA and causes accumulation of insoluble and toxic protein carbonyl aggregates containing inappropriately synthesized inactive proteins. This is regarded as a major means of toxicity to Cr (VI) in yeast [104].

Nucleic acid can cause mutations in various forms due to ROS [105]. DNA damage rely on the presence of iron, thus the hydroxyl or peryl radical is a toxic intermediate generated from the Fenton reaction [106]. Because of sensitivity to attack by •OH, the C4–C5 double bond of pyrimidine generates a spectrum of oxidative pyrimidine damage products, including thymine glycol, uracil glycol, urea residue, 5-hydroxydeoxyuridine, 5-hydroxydeoxycytidine, hydantoin, and others. Similarly, •OH oxidizes purines bases into 8-hydroxydeoxyguanosine, 8-hydroxydeoxyadenosine, and formamidopyrimidines, a well-characterized marker of oxidative DNA damage [12, 105]. ROS damage creates lesions that produce cross-linking of DNA to other molecules and block replication. It also induces single and double strand breaks in the backbone [105]. Studies in *E. coli* show that the major mechanism of ROS-dependent antimicrobial activity at low concentrations of H_2O_2 is DNA damage, while the mortality rate at high concentrations of H_2O_2 is due to ROS-mediated damage to multiple cellular targets [107, 108].

1.7 Natural Antimicrobial Agent for Controlling Microbial Infection

Secondary metabolites extracted from the edible, medicinal, and herbal plants and their derived essential oils could be natural antimicrobials [109]. Plant-produced secondary metabolites are promising healthy ingredients or human disease controlling agents. Secondary metabolites are of great interest because of their many benefits including antimicrobial properties against pathogenic microbes [110]. Many chemical antimicrobials still threaten our health; thus more attention has been given toward the potential antimicrobial activities of natural products. Natural

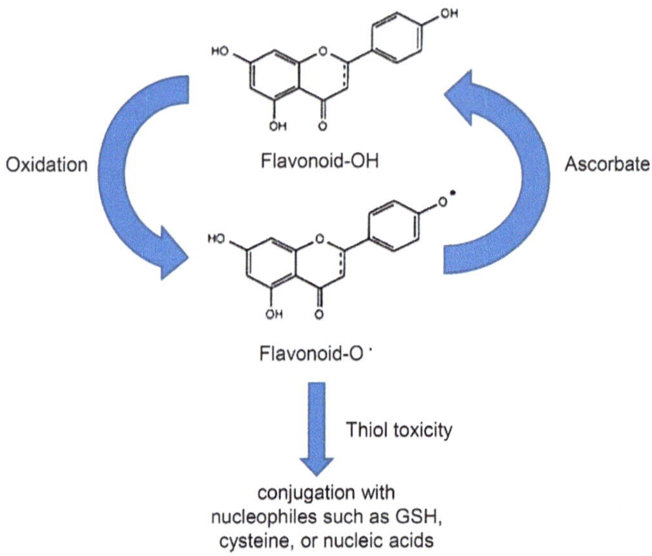

Fig. 1.3 An effect on thiol toxicity of prooxidant chemistry of flavonoid

products of dietary components such as medicinal plants and herbs are known to have potent antioxidant activity and free radical scavenging activity [111, 112]. Nevertheless, antioxidants also have prooxidant activity under certain conditions [111], leading to cause oxidative damage on biomolecules such as lipids, proteins, and DNA [112, 113]. The paradoxical role of antioxidants is also directly related to redox signaling of antioxidants [111]. Flavonoids, well-known phytochemicals, induce detoxifying enzymes by exerting prooxidant activity to enhance their ability to promote health and thus have a beneficial effect in toxic chemical reactions [112, 114]. Their prooxidant activity was related to the structural characteristics of the flavonoids in a concentration-dependent manner [112, 115]. The total number of hydroxyl groups in the flavonoid molecule is thought to be proportional to the prooxidant activity [112]. Flavonoid quinones can be stabilized by conjugation with nucleophiles such as GSH, cysteine, or nucleic acids which is responsible for one of the prooxidant effects of flavonoids (Fig. 1.3) [116]. Rapid NADH oxidization by phenoxyl radicals is shown to the flavonoids apigenin, naringenin, and naringin as prooxidant properties, resulting in extensive oxygen uptake and hydroxyl radical formation [112].

1.8 Conclusion

ROS is produced by a variety of physiological systems such as aerobic metabolism and plays an important role in host immune defense against pathogenic microorganisms as well as cellular signaling pathways. The role of phagocytes is to recognize

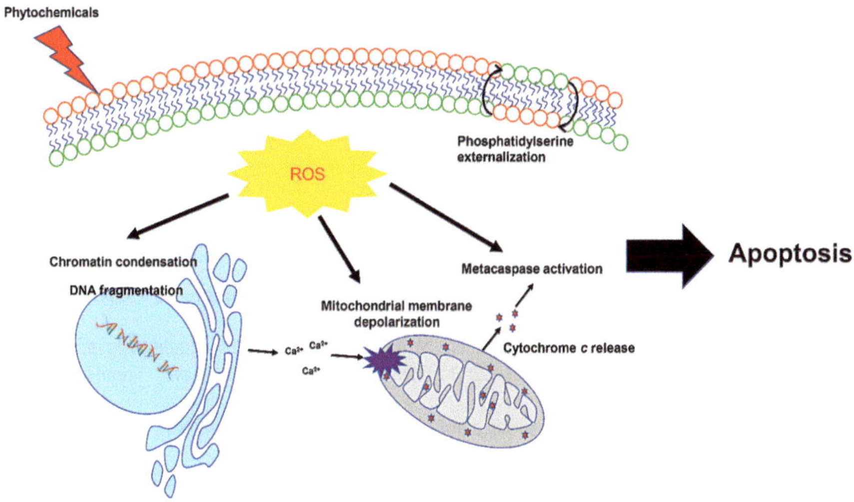

Fig. 1.4 Regulation of cell functions induced by prooxidant property of phytochemicals

microbes through their many molecular patterns and to engulf them. Once a microorganism is phagocyted, it determines the treatment that is enacted within the phagosome by the molecule recognized on the surface of the microorganism. NADPH oxidase generates ROS in response to pathogens and contributes to get rid of many pathogens. Microorganisms can avoid engulfment through a number of methods including avoidance of recognition by phagocytic receptors and destruction of the phagocytic signaling pathway. Bacteria and fungi also exhibit enzymatic activities to avoid host-induced oxidative defense. To maintain ROS levels at a non-toxic concentration, cellular antioxidants coordinate the balance between production and degradation. SOD, catalase, glutathione system, thioredoxin, peroxidase, and glutathione systems which represent the prominent enzymatic antioxidants are used to scavenge excess ROS. ROS, which is involved in cellular signaling or physiological processes at low concentrations, have a harmful effect when the concentration is too high. Excessive ROS level has deleterious effects due to destruction of cell structure, lipid, DNA, and protein, ultimately leading to cell death. Natural products have enormous potential as effective antimicrobial agents and can be employed to incorporate antimicrobial agents to generate ROS. These prooxidant properties are related to cell signaling and allow the flavonoids to contribute to the regulation of cell function (Fig. 1.4). To provide greater therapies under different conditions, unraveling the signaling events related to the oxidative free radicals at the cellular level as well as the physiological response to such stress needs to be conducted. Moreover, discovery of the mechanisms and roles of ROS during infection can also help to find new strategies to increase the antimicrobial potential of currently available antimicrobials.

References

1. Jones KE, Patel NG, Levy MA, Storeygard A, Balk D, Gittleman JL, Daszak P (2008) Global trends in emerging infectious diseases. Nature 451:990–993
2. Lakshmaiah Narayana J, Chen JY (2015) Antimicrobial peptides: possible anti-infective agents. Peptides 72:88–94
3. Chow SH, Deo P, Naderer T (2016) Macrophage cell death in microbial infections. Cell Microbiol 18:466–474
4. Spooner R, Yilmaz O (2011) The role of reactive-oxygen-species in microbial persistence and inflammation. Int J Mol Sci 12:334–352
5. Reczek CR, Chandel NS (2015) ROS-dependent signal transduction. Curr Opin Cell Biol 33:8–13
6. Mittler R, Vanderauwera S, Suzuki N, Miller G, Tognetti VB, Vandepoele K, Gollery M, Shulaev V, Van Breusegem F (2011) ROS signaling: the new wave? Trends Plant Sci 16:300–309
7. Palli SR (2017) New roles for old actors, ROS and PRMT1. Proc Natl Acad Sci U S A 114:10810–10812
8. Bhattacharyya A, Chattopadhyay R, Mitra S, Crowe SE (2014) Oxidative stress: an essential factor in the pathogenesis of gastrointestinal mucosal diseases. Physiol Rev 94:329–354
9. Sperandio FF, Huang YY, Hamblin MR (2013) Antimicrobial photodynamic therapy to kill gram-negative bacteria. Recent Pat Antiinfect Drug Discov 8:108–120
10. Brynildsen MP, Winkler JA, Spina CS, MacDonald IC, Collins JJ (2013) Potentiating antibacterial activity by predictably enhancing endogenous microbial ROS production. Nat Biotechnol 31:160–165
11. Schieber M, Chandel NS (2014) ROS function in redox signaling and oxidative stress. Curr Biol 24:R453–R462
12. Devasagayam TP, Tilak JC, Boloor KK, Sane KS, Ghaskadbi SS, Lele RD (2004) Free radicals and antioxidants in human health: current status and future prospects. J Assoc Physicians India 52:794–804
13. Terazawa S, Nakajima H, Shingo M, Niwano T, Imokawa G (2012) Astaxanthin attenuates the UVB-induced secretion of prostaglandin E2 and interleukin-8 in human keratinocytes by interrupting MSK1 phosphorylation in a ROS depletion-independent manner. Exp Dermatol 21(Suppl 1):11–17
14. Gilroy S, Suzuki N, Miller G, Choi WG, Toyota M, Devireddy AR, Mittler R (2014) A tidal wave of signals: calcium and ROS at the forefront of rapid systemic signaling. Trends Plant Sci 19:623–630
15. Zhang J, Wang X, Vikash V, Ye Q, Wu D, Liu Y, Dong W (2016) ROS and ROS-mediated cellular signaling. Oxidative Med Cell Longev 2016:4350965
16. Malle E, Furtmüller P, Sattler W, Obinger C (2007) Myeloperoxidase: a target for new drug development? Br J Pharmacol 152:838–854
17. Sareila O, Kelkka T, Pizzolla A, Hultqvist M, Holmdahl R (2011) NOX2 complex-derived ROS as immune regulators. Antioxid Redox Signal 15:2197–2208
18. Segal BH, Grimm MJ, Khan ANH, Han W, Blackwell TS (2012) Regulation of innate immunity by NADPH oxidase. Free Radic Biol Med 53:72–80
19. Bedard K, Lardy B, Krause K-H (2007) NOX family NADPH oxidases: not just in mammals. Biochimie 89:1107–1112
20. Kuhns DB, Alvord WG, Heller T, Feld JJ, Pike KM, Marciano BE, Uzel G, DeRavin SS, Priel DAL, Soule BP (2010) Residual NADPH oxidase and survival in chronic granulomatous disease. N Engl J Med 363:2600–2610
21. Winterbourn CC, Hampton MB, Livesey JH, Kettle AJ (2006) Modeling the reactions of superoxide and myeloperoxidase in the neutrophil phagosome: implications for microbial killing. J Biol Chem 281:39860

22. Stendahl O, Coble B, Dahlgren C, Hed J, Molin L (1984) Myeloperoxidase modulates the phagocytic activity of polymorphonuclear neutrophil leukocytes. Studies with cells from a myeloperoxidase-deficient patient. J Clin Invest 73:366–373
23. Nauseef WM (1986) Myeloperoxidase biosynthesis by a human promyelocytic leukemia cell line: insight into myeloperoxidase deficiency. Blood 67:865–872
24. Lanza F (1998) Clinical manifestation of myeloperoxidase deficiency. J Mol Med 76:676–681
25. Malle E, Buch T, Grone H-J (2003) Myeloperoxidase in kidney disease. Kidney Int 64:1956–1967
26. Van der Veen BS, de Winther MP, Heeringa P (2009) Myeloperoxidase: molecular mechanisms of action and their relevance to human health and disease. Antioxid Redox Signal 11:2899–2937
27. Prokopowicz Z, Marcinkiewicz J, Katz DR, Chain BM (2012) Neutrophil myeloperoxidase: soldier and statesman. Arch Immunol Ther Exp 60:43–54
28. Lehrer RI, Cline MJ (1969) Leukocyte myeloperoxidase deficiency and disseminated candidiasis: the role of myeloperoxidase in resistance to Candida infection. J Clin Invest 48:1478–1488
29. Arnoult D, Soares F, Tattoli I, Castanier C, Philpott DJ, Girardin SE (2009) An N-terminal addressing sequence targets NLRX1 to the mitochondrial matrix. J Cell Sci 122:3161–3168
30. Moore CB, Bergstralh DT, Duncan JA, Lei Y, Morrison TE, Zimmermann AG, Accavitti-Loper MA, Madden VJ, Sun L, Ye Z (2008) NLRX1 is a regulator of mitochondrial antiviral immunity. Nature 451:573
31. Groeger G, Quiney C, Cotter TG (2009) Hydrogen peroxide as a cell-survival signaling molecule. Antioxid Redox Signal 11:2655–2671
32. Liou G-Y, Storz P (2010) Reactive oxygen species in cancer. Free Radic Res 44:479–496
33. Dworakowski R, Alom-Ruiz SP, Shah AM (2008) NADPH oxidase-derived reactive oxygen species in the regulation of endothelial phenotype. Pharmacol Rep 60:21
34. Tattoli I, Carneiro LA, Jehanno M, Magalhaes JG, Shu Y, Philpott DJ, Arnoult D, Girardin SE (2008) NLRX1 is a mitochondrial NOD-like receptor that amplifies NF-κB and JNK pathways by inducing reactive oxygen species production. EMBO Rep 9:293–300
35. Abdul-Sater AA, Saïd-Sadier N, Lam VM, Singh B, Pettengill MA, Soares F, Tattoli I, Lipinski S, Girardin SE, Rosenstiel P (2010) Enhancement of reactive oxygen species production and chlamydial infection by the mitochondrial Nod-like family member, NLRX1. J Biol Chem 285(53):41637. https://doi.org/10.1074/jbc.M110.137885
36. Cirillo SL, Subbian S, Chen B, Weisbrod TR, Jacobs WR, Cirillo JD (2009) Protection of Mycobacterium tuberculosis from reactive oxygen species conferred by the mel2 locus impacts persistence and dissemination. Infect Immun 77:2557–2567
37. McCaffrey RL, Allen LAH (2006) Francisella tularensis LVS evades killing by human neutrophils via inhibition of the respiratory burst and phagosome escape. J Leukoc Biol 80:1224–1230
38. Carlyon JA, Latif DA, Pypaert M, Lacy P, Fikrig E (2004) Anaplasma phagocytophilum utilizes multiple host evasion mechanisms to thwart NADPH oxidase-mediated killing during neutrophil infection. Infect Immun 72:4772–4783
39. Woldehiwet Z (2008) Immune evasion and immunosuppression by Anaplasma phagocytophilum, the causative agent of tick-borne fever of ruminants and human granulocytic anaplasmosis. Vet J 175:37–44
40. Boncompain G, Schneider B, Delevoye C, Kellermann O, Dautry-Varsat A, Subtil A (2010) Production of reactive oxygen species is turned on and rapidly shut down in epithelial cells infected with chlamydia trachomatis. Infect Immun 78:80–87
41. Bylund J, Burgess L-A, Cescutti P, Ernst RK, Speert DP (2006) Exopolysaccharides from Burkholderia cenocepacia inhibit neutrophil chemotaxis and scavenge reactive oxygen species. J Biol Chem 281:2526–2532
42. Bianchi SM, Prince LR, McPhillips K, Allen L, Marriott HM, Taylor GW, Hellewell PG, Sabroe I, Dockrell DH, Henson PW (2008) Impairment of apoptotic cell engulfment by

pyocyanin, a toxic metabolite of Pseudomonas aeruginosa. Am J Respir Crit Care Med 177:35–43
43. Sim S, Yong T-S, Park S-J, Im K-i, Kong Y, Ryu J-S, Min D-Y, Shin MH (2005) NADPH oxidase-derived reactive oxygen species-mediated activation of ERK1/2 is required for apoptosis of human neutrophils induced by Entamoeba histolytica. J Immunol 174:4279–4288
44. Stanciu M, Wang Y, Kentor R, Burke N, Watkins S, Kress G, Reynolds I, Klann E, Angiolieri MR, Johnson JW (2000) Persistent activation of ERK contributes to glutamate-induced oxidative toxicity in a neuronal cell line and primary cortical neuron cultures. J Biol Chem 275:12200–12206
45. Wang X, Martindale JL, Holbrook NJ (2000) Requirement for ERK activation in cisplatin-induced apoptosis. J Biol Chem 275:39435–39443
46. Singh S, Upadhyay AK, Ajay AK, Bhat MK (2007) p53 regulates ERK activation in carboplatin induced apoptosis in cervical carcinoma: a novel target of p53 in apoptosis. FEBS Lett 581:289–295
47. Yang T-C, Lai C-C, Shiu S-L, Chuang P-H, Tzou B-C, Lin Y-Y, Tsai F-J, Lin C-W (2010) Japanese encephalitis virus down-regulates thioredoxin and induces ROS-mediated ASK1-ERK/p38 MAPK activation in human promonocyte cells. Microbes Infect 12:643–651
48. Gupta S, Bhatia V, Wen J-j, Wu Y, Huang M-H, Garg NJ (2009) Trypanosoma cruzi infection disturbs mitochondrial membrane potential and ROS production rate in cardiomyocytes. Free Radic Biol Med 47:1414–1421
49. Crawford DR, Davies K (1994) Adaptive response and oxidative stress. Environ Health Perspect 102:25
50. Gaupp R, Ledala N, Somerville GA (2012) Staphylococcal response to oxidative stress. Front Cell Infect Microbiol 2:33
51. Chaves GM, da Silva WP (2012) Superoxide dismutases and glutaredoxins have a distinct role in the response of *Candida albicans* to oxidative stress generated by the chemical compounds menadione and diamide. Memórias do Instituto Oswaldo Cruz 107:998–1005
52. Missall TA, Cherry-Harris JF, Lodge JK (2005) Two glutathione peroxidases in the fungal pathogen *Cryptococcus neoformans* are expressed in the presence of specific substrates. Microbiology 151:2573–2581
53. Dickinson DA, Forman HJ (2002) Glutathione in defense and signaling. Ann N Y Acad Sci 973:488–504
54. Godoy JSR, Kioshima ÉS, Abadio AKR, Felipe MSS, de Freitas SM, Svidzinski TIE (2016) Structural and functional characterization of the recombinant thioredoxin reductase from Candida albicans as a potential target for vaccine and drug design. Appl Microbiol Biotechnol 100:4015–4025
55. Steen B, Zuyderduyn S, Toffaletti D, Marra M, Jones S, Perfect J, Kronstad J (2003) Cryptococcus neoformans gene expression during experimental cryptococcal meningitis. Eukaryot Cell 2:1336–1349
56. Wood ZA, Schröder E, Harris JR, Poole LB (2003) Structure, mechanism and regulation of peroxiredoxins. Trends Biochem Sci 28:32–40
57. Giles SS, Perfect JR, Cox GM (2005) Cytochrome c peroxidase contributes to the antioxidant defense of Cryptococcus neoformans. Fungal Genet Biol 42:20–29
58. Jacks TJ, De Lucca AJ, Morris NM (1999) Effects of chloroperoxidase and hydrogen peroxide on the viabilities of Aspergillus flavus conidiospores. Mol Cell Biochem 195:169–172
59. Clements MO, Watson SP, Foster SJ (1999) Characterization of the major superoxide dismutase of Staphylococcus aureus and its role in starvation survival, stress resistance, and pathogenicity. J Bacteriol 181:3898–3903
60. Das D, Saha S, Bishayi B (2008) Intracellular survival of Staphylococcus aureus: correlating production of catalase and superoxide dismutase with levels of inflammatory cytokines. Inflamm Res 57:340–349
61. Das D, Bishayi B (2010) Contribution of catalase and superoxide dismutase to the intracellular survival of clinical isolates of Staphylococcus aureus in murine macrophages. Indian J Microbiol 50:375–384

62. Beavers WN, Skaar EP (2016) Neutrophil-generated oxidative stress and protein damage in *Staphylococcus aureus*. Pathogens Dis 74(6):pii:ftw060
63. Seaver LC, Imlay JA (2001) Alkyl hydroperoxide reductase is the primary scavenger of endogenous hydrogen peroxide in Escherichia coli. J Bacteriol 183:7173–7181
64. Park S, You X, Imlay JA (2005) Substantial DNA damage from submicromolar intracellular hydrogen peroxide detected in Hpx-mutants of Escherichia coli. Proc Natl Acad Sci 102:9317–9322
65. Yesilkaya H, Andisi VF, Andrew PW, Bijlsma JJ (2013) Streptococcus pneumoniae and reactive oxygen species: an unusual approach to living with radicals. Trends Microbiol 21:187–195
66. Hajaj B, Yesilkaya H, Benisty R, David M, Andrew PW, Porat N (2012) Thiol peroxidase is an important component of *Streptococcus pneumoniae* in oxygenated environments. Infect Immun 80(12):4333–4343. https://doi.org/10.1128/IAI.00126-12
67. Fradin C, De Groot P, MacCallum D, Schaller M, Klis F, Odds FC, Hube B (2005) Granulocytes govern the transcriptional response, morphology and proliferation of Candida albicans in human blood. Mol Microbiol 56:397–415
68. Hwang C-S, Rhie G-e, Oh J-H, Huh W-K, Yim H-S, Kang S-O (2002) Copper-and zinc-containing superoxide dismutase (cu/ZnSOD) is required for the protection of Candida albicans against oxidative stresses and the expression of its full virulence. Microbiology 148:3705–3713
69. Lamarre C, LeMay J-D, Deslauriers N, Bourbonnais Y (2001) Candida albicans expresses an unusual cytoplasmic manganese-containing superoxide dismutase (SOD3 gene product) upon the entry and during the stationary phase. J Biol Chem 276:43784–43791
70. Cox GM, Harrison TS, McDade HC, Taborda CP, Heinrich G, Casadevall A, Perfect JR (2003) Superoxide dismutase influences the virulence of Cryptococcus neoformans by affecting growth within macrophages. Infect Immun 71:173–180
71. Narasipura SD, Chaturvedi V, Chaturvedi S (2005) Characterization of Cryptococcus neoformans variety gattii SOD2 reveals distinct roles of the two superoxide dismutases in fungal biology and virulence. Mol Microbiol 55:1782–1800
72. Wysong DR, Christin L, Sugar AM, Robbins PW, Diamond RD (1998) Cloning and sequencing of a Candida albicans catalase gene and effects of disruption of this gene. Infect Immun 66:1953–1961
73. Komalapriya C, Kaloriti D, Tillmann AT, Yin Z, Herrero-de-Dios C, Jacobsen MD, Belmonte RC, Cameron G, Haynes K, Grebogi C (2015) Integrative model of oxidative stress adaptation in the fungal pathogen Candida albicans. PLoS One 10:e0137750
74. Shibuya K, Paris S, Ando T, Nakayama H, Hatori T, Latge J-P (2006) Catalases of Aspergillus fumigatus and inflammation in aspergillosis. Nippon Ishinkin Gakkai Zasshi 47:249–255
75. Paris S, Wysong D, Debeaupuis J-P, Shibuya K, Philippe B, Diamond RD, Latgé J-P (2003) Catalases of Aspergillus fumigatus. Infect Immun 71:3551–3562
76. Philippe B, Ibrahim-Granet O, Prevost M, Gougerot-Pocidalo M, Perez MS, Van der Meeren A, Latge J (2003) Killing of Aspergillus fumigatus by alveolar macrophages is mediated by reactive oxidant intermediates. Infect Immun 71:3034–3042
77. Chaves G, Bates S, Maccallum DM, Odds FC (2007) Candida albicans GRX2, encoding a putative glutaredoxin, is required for virulence in a murine model. Genet Mol Res 6:1051–1063
78. Kusch H, Engelmann S, Albrecht D, Morschhäuser J, Hecker M (2007) Proteomic analysis of the oxidative stress response in Candida albicans. Proteomics 7:686–697
79. Ratti BA, Godoy JSR, de Souza Bonfim Mendonça P, Bidóia DL, Nakamura TU, Nakamura CV, Consolaro MEL, Svidzinski TIE, de Oliveira Silva S (2015) Microbicidal activity of neutrophils is inhibited by isolates from recurrent vaginal candidiasis (RVVC) caused by *Candida albicans* through fungal thioredoxin reductase. Cell Immunol 293:22–29
80. Missall TA, Pusateri ME, Lodge JK (2004) Thiol peroxidase is critical for virulence and resistance to nitric oxide and peroxide in the fungal pathogen, Cryptococcus neoformans. Mol Microbiol 51:1447–1458

81. Missall TA, Lodge JK (2005) Function of the thioredoxin proteins in Cryptococcus neoformans during stress or virulence and regulation by putative transcriptional modulators. Mol Microbiol 57:847–858
82. Lessing F, Kniemeyer O, Wozniok I, Loeffler J, Kurzai O, Haertl A, Brakhage AA (2007) The Aspergillus fumigatus transcriptional regulator AfYap1 represents the major regulator for defense against reactive oxygen intermediates but is dispensable for pathogenicity in an intranasal mouse infection model. Eukaryot Cell 6:2290–2302
83. Asif AR, Oellerich M, Amstrong VW, Riemenschneider B, Monod M, Reichard U (2006) Proteome of conidial surface associated proteins of Aspergillus f umigatus reflecting potential vaccine candidates and allergens. J Proteome Res 5:954–962
84. Hemmann S, Menz G, Ismail C, Blaser K, Crameri R (1999) Skin test reactivity to 2 recombinant Aspergillus fumigatus allergens in a fumigatus-sensitized asthmatic subjects allows diagnostic separation of allergic bronchopulmonary aspergillosis from fungal sensitization. J Allergy Clin Immunol 104:601–607
85. Upadhya R, Campbell LT, Donlin MJ, Aurora R, Lodge JK (2013) Global transcriptome profile of Cryptococcus neoformans during exposure to hydrogen peroxide induced oxidative stress. PLoS One 8:e55110
86. Buchhaupt M, Ehrich K, Hüttmann S, Guder J, Schrader J (2011) Over-expression of chloroperoxidase in Caldariomyces fumago. Biotechnol Lett 33:2225
87. De Mot R, Schoofs G, Nagy I (2007) Proteome analysis of Streptomyces coelicolor mutants affected in the proteasome system reveals changes in stress-responsive proteins. Arch Microbiol 188:257–271
88. Barloy-Hubler F, Cheron A, Hellegouarch A, Galibert F (2004) Smc01944, a secreted peroxidase induced by oxidative stresses in Sinorhizobium meliloti 1021. Microbiology 150:657–664
89. Nathan C, Shiloh MU (2000) Reactive oxygen and nitrogen intermediates in the relationship between mammalian hosts and microbial pathogens. Proc Natl Acad Sci 97:8841–8848
90. Peralta MA, da Silva MA, Ortega MG, Cabrera JL, Paraje MG (2015) Antifungal activity of a prenylated flavonoid from Dalea elegans against Candida albicans biofilms. Phytomedicine 22:975–980
91. Abedelahi A, Salehnia M, Allameh AA, Davoodi D (2010) Sodium selenite improves the in vitro follicular development by reducing the reactive oxygen species level and increasing the total antioxidant capacity and glutathione peroxide activity. Hum Reprod 25:977–985
92. Stamler JS, Lamas S, Fang FC (2001) Nitrosylation: the prototypic redox-based signaling mechanism. Cell 106:675–683
93. Catalá A (2009) Lipid peroxidation of membrane phospholipids generates hydroxy-alkenals and oxidized phospholipids active in physiological and/or pathological conditions. Chem Phys Lipids 157:1–11
94. Catala A (2009) Lipid peroxidation of membrane phospholipids generates hydroxy-alkenals and oxidized phospholipids active in physiological and/or pathological conditions. Chem Phys Lipids 157:1–11
95. Shohet S, Pitt J, Baehner R, Poplack D (1974) Lipid peroxidation in the killing of phagocytized pneumococci. Infect Immun 10:1321–1328
96. Biliński T, Litwińska J, Błszczyński M, Bajus A (1989) Superoxide dismutase deficiency and the toxicity of the products of autooxidation of polyunsaturated fatty acids in yeast. Biochimica et Biophysica Acta (BBA) Lipids Lipid Metabol 1001:102–106
97. Howlett N, Avery S (1997) Relationship between cadmium sensitivity and degree of plasma membrane fatty acid unsaturation in Saccharomyces cerevisiae. Appl Microbiol Biotechnol 48:539–545
98. Kiššová I, Deffieu M, Samokhvalov V, Velours G, Bessoule J-J, Manon S, Camougrand N (2006) Lipid oxidation and autophagy in yeast. Free Radic Biol Med 41:1655–1661
99. Reekmans R, Smet KD, Chen C, Hummelen PV, Contreras R (2005) Old yellow enzyme interferes with Bax-induced NADPH loss and lipid peroxidation in yeast. FEMS Yeast Res 5:711–725

100. Kim I-S, Sohn H-Y, Jin I (2011) Adaptive stress response to menadione-induced oxidative stress in Saccharomyces cerevisiae KNU5377. J Microbiol 49:816–823
101. Berlett BS, Stadtman ER (1997) Protein oxidation in aging, disease, and oxidative stress. J Biol Chem 272:20313–20316
102. Tamarit J, Cabiscol E, Ros J (1998) Identification of the major oxidatively damaged proteins in Escherichia coli cells exposed to oxidative stress. J Biol Chem 273:3027–3032
103. Cecarini V, Gee J, Fioretti E, Amici M, Angeletti M, Eleuteri AM, Keller JN (2007) Protein oxidation and cellular homeostasis: emphasis on metabolism. Biochimica et Biophysica Acta (BBA) Mol Cell Res 1773:93–104
104. Holland S, Lodwig E, Sideri T, Reader T, Clarke I, Gkargkas K, Hoyle DC, Delneri D, Oliver SG, Avery SV (2007) Application of the comprehensive set of heterozygous yeast deletion mutants to elucidate the molecular basis of cellular chromium toxicity. Genome Biol 8:R268
105. Cabiscol E, Tamarit J, Ros J (2000) Oxidative stress in bacteria and protein damage by reactive oxygen species. Int Microbiol 3:3–8
106. McCormick ML, Buettner GR, Britigan BE (1998) Endogenous superoxide dismutase levels regulate iron-dependent hydroxyl radical formation in Escherichia coli exposed to hydrogen peroxide. J Bacteriol 180:622–625
107. Imlay JA, Linn S (1986) Bimodal pattern of killing of DNA-repair-defective or anoxically grown Escherichia coli by hydrogen peroxide. J Bacteriol 166:519–527
108. Imlay JA, Linn S (1988) DNA damage and oxygen radical toxicity. Science 240:1302–1309
109. Tiwari BK, Valdramidis VP, O'Donnell CP, Muthukumarappan K, Bourke P, Cullen P (2009) Application of natural antimicrobials for food preservation. J Agric Food Chem 57:5987–6000
110. Ngwoke KG, Odimegwu DC, Esimone C (2011) Antimicrobial natural products, science against microbial pathogens: communicating current research and technology advances. FORMATEX, Badajoz, pp 1011–1026
111. Devasagayam T, Tilak J, Boloor K, Sane KS, Ghaskadbi SS, Lele R (2004) Free radicals and antioxidants in human health: current status and future prospects. J Assoc Phys India 52:4
112. Prochazkova D, Bousova I, Wilhelmova N (2011) Antioxidant and prooxidant properties of flavonoids. Fitoterapia 82:513–523
113. Kim HI, Kim JA, Choi EJ, Harris JB, Jeong SY, Son SJ, Kim Y, Shin OS (2015) In vitro and in vivo antimicrobial efficacy of natural plant-derived compounds against Vibrio cholerae of O1 El Tor Inaba serotype. Biosci Biotechnol Biochem 79:475–483
114. Manach C, Scalbert A, Morand C, Rémésy C, Jiménez L (2004) Polyphenols: food sources and bioavailability. Am J Clin Nutr 79:727–747
115. Yen G-C, Duh P-D, Tsai H-L, Huang S-L (2003) Pro-oxidative properties of flavonoids in human lymphocytes. Biosci Biotechnol Biochem 67:1215–1222
116. Schnelldorfer T, Gansauge S, Gansauge F, Schlosser S, Beger HG, Nussler AK (2000) Glutathione depletion causes cell growth inhibition and enhanced apoptosis in pancreatic cancer cells. Cancer Interdisc Intl J Am Cancer Soc 89:1440–1447

Blood Biomarkers of Oxidative Stress in Human and Canine Leishmaniosis

2

C. P. Rubio, D. Escribano, J. Hernández, J. J. Cerón, A. Tvarijonaviciute, and S. Martínez-Subiela

Abstract

Leishmaniosis includes a set of diseases produced by protozoan parasites of the genus *Leishmania* by invasion of the mononuclear phagocyte system of mammalian hosts. Different studies performed in human and canine leishmaniosis have reported that there is an increase in the biomarkers of oxidant status and a decrease in the biomarkers of antioxidant defense, causing an imbalance among oxidants and antioxidants and thus showing that the oxidative damage plays an important role in both diseases. In this chapter the main changes in blood biomarkers of oxidative stress occurring in human and canine leishmaniosis are discussed.

Keywords

Leishmaniosis · Human · Dog · Oxidative stress · Biomarker

C. P. Rubio · J. J. Cerón · A. Tvarijonaviciute · S. Martínez-Subiela (✉)
Department of Animal Medicine and Surgery, Faculty of Veterinary Medicine, Regional Campus of International Excellence "Campus Mare Nostrum", University of Murcia, Murcia, Spain
e-mail: silviams@um.es

D. Escribano
Department of Animal and Food Science, Faculty of Veterinary Medicine, Universitat Autonoma de Barcelona, Bellaterra, Spain

J. Hernández
Department of Plant Biology, Faculty of Biology, University of Murcia, Murcia, Spain

S. Chakraborti et al. (eds.), *Oxidative Stress in Microbial Diseases*,
https://doi.org/10.1007/978-981-13-8763-0_2

2.1 General Importance of Leishmaniosis

Leishmaniosis includes a set of diseases produced by protozoan parasites of the genus *Leishmania* by invasion of the mononuclear phagocyte system of mammalian hosts. They are transmitted primarily by female phlebotomine sand flies belonging to the genera *Lutzomyia* (new world) and *Phlebotomus* (old world); hence sand flies are responsible for spreading the disease [1]. These neglected diseases are prevalent in at least 98 countries and 3 territories on 5 continents, of which the majority are underdeveloped countries [2, 3].

Most of the leishmaniases are zoonoses, though anthroponotic forms also exist, and in these cases humans are the sole reservoir. The reservoir hosts are several mammal species (around 70 species), including canines, rodents, and marsupials, and they are responsible for the long-term conservation of *Leishmania* in nature [4, 5]. The majority of the reservoir hosts suffer subclinical or only mild infections that may keep for years. However, the dog is an important exception, since it may develop a generalized and fatal disease [6].

The diversity of clinical manifestations, epidemiology, and immunopathology makes leishmaniosis a complex disease to study [7]. Human leishmaniosis (HuL) is a compound disease causing an extensive variety of clinical forms ranging in severity from self-healing cutaneous leishmaniosis (CL) to fatal disseminated visceral leishmaniosis (VL) [8]. Among the recognized clinical forms of the disease in humans, kala-azar, or VL, is the most severe and progressive form, as it is almost always fatal if untreated [9]. Canine leishmaniosis (CaL) is a systemic disease that leads to a broad range of clinical signs that may culminate to death [10, 11].

2.2 Oxidative Stress in the Pathogenesis of the Leishmaniosis

When *Leishmania* spp. is recognized, macrophages are activated and become so-called effector cells that can phagocytose and destroy the unwanted guest. After the activation of the macrophage, different cellular processes start, such as production of phagolysosomal degradation enzymes (e.g., proteases, nucleases, phosphatases, lipases, and esterases), oxidative burst, and nitric oxide (•NO) production [12]. The generation of these cytokines, reactive oxygen species (ROS), and •NO derivatives usually eliminates the phagocytosed microorganism; however, *Leishmania* spp. can survive and propagate in this hostile environment [13]. *Leishmania* parasite, after entering the macrophage, can protect itself against the macrophage oxidative outbreak. This can be achieved by stimulating the synthesis of antioxidants and at the same time inhibiting superoxide anion radical ($O_2^{\bullet-}$) and •NO generation in the macrophage. Different forms of *Leishmania* during its cycle have various suppressing effects, and in general, the effects of the parasite on the macrophage increase with time [12]. Parasite persistence inside the macrophages depends on the balance between the capacity of the immune response to adequately activate *Leishmania*-infected macrophages and the ability of the parasite to combat cytotoxic mechanisms of macrophage activation [14].

On the other hand, the higher levels of reactive species (ROS, and •NO derivatives), that are produced primarily to attack the invading parasite, may destabilize the oxidant-antioxidant balance; leading to a potential damage to the tissues or host cells and contributing to the progression of the disease [15].

In this chapter, we aimed to show and describe the main changes in blood biomarkers of oxidative stress that have been reported to occur in human and canine leishmaniosis. In each biomarker the changes occurring will be indicated; as well, when this information is available, the possible explanation of its physiopathological mechanism and their possible clinical applications.

2.3 Biomarkers of Oxidative Stress in Human Leishmaniosis

The presence of intracellular *Leishmania* amastigotes in mononuclear phagocytes may impair the oxidant-antioxidant balance [15]. Therefore, to evaluate the impact of the host-parasite interaction on the host oxidants and antioxidant defense system, different indicators have been investigated in human cutaneous leishmaniosis.

2.3.1 Biomarkers of Oxidant Status

2.3.1.1 Erythrocyte/Plasma Lipid Peroxidation (LPO)

The assay most frequently used to evaluate products of lipid peroxidation is the malondialdehyde (MDA) formation, while other products such as lipid hydroperoxides have been measured in human CL.

Increased erythrocyte and serum/plasma LPO levels (1.5-fold) according to MDA formation have been reported in patients with active cutaneous leishmaniosis with no effect on age and sex (Table 2.1) [15–21]. These increases could be attributed to different causes, like the respiratory burst that occurs due to the presence of amastigotes in the phagocytes which originate excess of free radicals, and, as a consequence, lead to peroxidation of polyunsaturated fatty acids of the membrane lipids [15]. In addition, LPO levels decreased in healed patients, reaching similar values to those of healthy controls [15, 18].

Interestingly, a close relationship has been found between MDA and DNA damage, indicating that oxidants generated during CL cause a DNA damage in these patients [19].

A significant increase in erythrocyte MDA has also been reported in kala-azar patients in comparison with controls [27, 28]. In addition, an inverse relationship of MDA with hemoglobin content that suggested a lipid peroxidation increases when the anemia progressed in VL patients [27].

2.3.1.2 Plasma Total Peroxides (TP)

Levels of TP were significantly increased in CL patients when measured with the FOX2 method [19].

Table 2.1 LPO levels according to malondialdehyde (MDA) formation reported in patients with active cutaneous leishmaniosis

Authors	Method	Region	Sample	Mean in control group	Mean in diseased group
Kocyigit et al. [16]	Jain et al. [22]	Anatolia, Turkey	Erythrocytes	4.16 nmol/g Hb	8.68 nmol/g Hb
Vural et al. [15]	Jain et al. [23]	Anatolia, Turkey	Erythrocytes	1.56 nmol/g Hb	2.44 nmol/g Hb
Ozbilge et al. [17]	Jain [24]	Anatolia, Turkey	Serum	0.25 nmol/mL	0.46 nmol/mL
Serarslan et al. [18]	Esterbauer et al. [25]	Iskenderun, Turkey	Serum	1.3 nmol/mL	1.7 nmol/mL
Kocyigit et al. [19]	Kocyigit et al. [16]	Anatolia, Turkey	Plasma	3.06 nmol/mL	4.85 nmol/mL
AbdulGhani et al. [20]	Fong et al. [26]	Taiz, Yemen	Serum	1.38 nmol/mL	3.4 nmol/mL
Asmaa et al. [21]	Fong et al. [26]	Taiz, Yemen	Serum	1.38 nmol/mL	3.4 nmol/mL

2.3.1.3 Plasma Carbonyl Content (PCC)

No significant differences could be seen in PCC in patients with CL comparing with healthy controls [19].

2.3.1.4 Nitrite and Nitrate Concentrations

Nitrite and nitrate concentrations were found to be higher in CL patients when compared to healthy controls [18, 29, 30]. Both analytes reflect the NO production, which has been reported to have antimicrobial and leishmanicidal activity; and their increase showed that reactive nitrogen intermediates metabolism is altered in patients with CL [29]. In addition, a significant decrease in NO levels was found after the treatment [18]. Moreover, Cabrera et al. [30] showed that chronic forms of the disease (intermediate or chronic cutaneous leishmaniosis) had significantly lower concentrations of nitrate and nitrite than localized or diffuse forms; therefore, NO may have a different role in different forms of the disease.

2.3.2 Biomarkers of Antioxidant Status

2.3.2.1 Total Antioxidant Status (TAS)

A significant decrease in TAS of CL patients have been reported indicating that the antioxidant defense of these patients was impaired [19] probably due to an excess amount of reactive species.

2.3.2.2 Superoxide Dismutase Activity (SOD)

Contradictory results have been found in SOD in different studies performed in erythrocytes versus serum samples. Erel et al. [29] and Kocyigit et al. [16] found increased SOD activity in erythrocytes of patients suffering CL in comparison with controls. However, Serarslan et al. [18] reported that SOD activity in serum of CL patients was lower than controls. Biswas et al. have also observed significantly reduced values of SOD in kala-azar patients [27]. Upon treatment of CL patients, significantly higher SOD activities were observed [18]. Most isoforms of SOD are located intracellularly which could be the source of the different activity observed by the previous studies.

Increased SOD activity results in increased conversion of superoxide anion into hydrogen peroxide (H_2O_2). Other studies have also demonstrated that SOD can effectively accentuate vascular relaxation and the half-life of NO [31, 32], and that the addition of SOD to macrophages raised leishmanicidal activity [32]; hence it could be a mechanism of defense induced by cytokines like IL-1 and TNF alpha [16].

2.3.2.3 Reduced Glutathione (rGSH), Glutathione Peroxidase (GSH-Px), and Glutathione Reductase (GR)

Glutathione plays an essential role in protecting cells exposed to oxidative stress. The glutathione redox cycle is a major protection system for detoxification of ROS within the red cells [33].

Glutathione peroxidases catalyze the reduction of H_2O_2 in the presence of rGSH [34]. Increased rGSH concentrations were observed in erythrocytes of CL patients [15]. High rGSH concentrations were also found in erythrocytes of patients diagnosed with VL with a positive correlation with MDA [28]; however, the increase was not significant in comparison with control patients. Conversely, significantly decreased values of whole blood rGSH were found in patients with VL compared to controls, and it has been suggested that it could be due to oxidative stress and overutilization of GSH by the cells. In addition, its levels remain decreased after the treatment probably due to an involvement of GSH in drug metabolism [28].

Decreased GSH-Px activities have been found in CL patients compared with controls both in erythrocytes [15, 16, 29] and serum [18]. In addition, a decrease in rGSH and an increase in GSH-Px activity were observed after the treatment [15, 18]. A significant decrease in GSH-Px has been reported also in VL [27]. Significantly decreased values of GR were found in kala-azar patients compared with controls [27].

Results found in human CL showed that although the reducing potential is increased partially as a result of accumulation of rGSH, a marked decrease of the enzyme GSH-Px is observed during the active CL infection [15]. This could indicate a poor utilization of rGSH in erythrocytes and therefore, sparse efficacy and participation of the glutathione redox cycle in detoxification of radical-mediated damage [15]. Furthermore, it has been suggested that as a result of increased ROS and reactive nitrogen species (RNS) generation in CL patients, GSH-Px could be consumed during ROS scavenging, and thus the activity and/or levels of the enzyme decrease during CL [18].

2.3.2.4 Catalase

Similar to GSH-Px, catalase activity was reported to be decreased in erythrocytes of CL patients by Erel et al. [29] and Kocyigit et al. [16] as well as in VL patients by Biswas et al. [27]. Contrarily, other authors such as AbdulGhani et al. [20] and Asmaa et al. [21] did not find a significant decrease in serum catalase activity in their studies. Catalase, like GSH-Px, uses H_2O_2 as substrate; therefore decrease in catalase activity enables H_2O_2 to stay in the medium for a long time and at higher concentrations [29]. The nonsignificant decrease in catalase activity found by some authors could be due to either a protection mechanism of the parasite itself from the toxic oxygen metabolites or differences in the methods or type of sample used for the measurement [20, 21]. Effect of age or sex on catalase activity was not found [21].

2.3.2.5 Uric Acid

Higher levels of uric acid has been reported in CL patients compared to controls [20, 21]. Uric acid is an important contributor to total antioxidant capacity; it provides a significant antioxidant defense against nitration by nitroperoxynitrite; hence it could represent a mechanism of defense against oxidative stress caused by leishmaniosis infection [21, 35].

2.3.2.6 Vitamin C

Vitamin C was found to be significantly decreased in CL patients, and an increase was observed in healed patients [15]. This antioxidant is expected to be consumed in situations of increased formation of oxygen-derived free radicals and is often low in infections where it could be a direct inducer of the cellular immune response [36].

2.4 Biomarkers of Oxidative Stress in Canine Leishmaniosis

The interest on the biomarkers of oxidative stress in blood canine leishmaniosis has increased in the last years. Although there is variability in the findings of different authors, in general, in canine leishmaniosis, there is an increase in the biomarkers of oxidant status and a decrease in the biomarkers of antioxidant defense. This would indicate that the oxidative damage plays an important role also in this disease.

2.4.1 Biomarkers of Oxidant Status

2.4.1.1 Total Oxidant Status(TOS)

TOS has been reported to be increased in dogs with canine leishmaniosis, though contradictory results have been found about changes observed regarding the severity of the disease [37, 38]. In addition, TOS was not found to be a suitable biomarker to monitor treatment in canine leishmaniosis [39].

2.4.1.2 Erythrocyte/Plasma Lipid Peroxidation (LPO)

As indicated before, the product of lipid peroxidation frequently measured to evaluate oxidative damage to lipids is the MDA. MDA has been reported to be increased in canine leishmaniosis, indicating the presence of an oxidative damage in this disease. MDA and other products of lipid peroxidation are accompanied by degradation of cellular membranes and altered membrane permeability which produce cell damage and could be one of the causes of anemia that occurs in canine leishmaniosis [40]. In addition, increase in MDA in dogs seropositive to leishmaniosis has been associated with the presence of clinical signs and correlated positively with urea and creatinine that would indicate that MDA can be involved in the kidney damage associated to leishmaniosis [41].

2.4.2 Biomarkers of Antioxidant Status

2.4.2.1 Total Antioxidant Status (TAS)

Trolox equivalent antioxidant capacity (TEAC), ferric ion reducing ability of plasma (FRAP), and cupric ion reducing antioxidant capacity (CUPRAC) are different assays that can be used to determine TAS of a sample and have been reported in canine leishmaniosis by different authors. Changes in TAS were not detected in dogs with canine leishmaniosis with the TEAC method by Gultekin et al. [37]; however, Almeida et al. [38] detected significantly reduced values of TEAC. However, when TAS was evaluated by FRAP, dogs with canine leishmaniosis showed lower values of FRAP than healthy dogs [42], and when TAS was analyzed with a commercial assay, dogs seropositive to canine leishmaniosis showed lower values of TAS compared to healthy dogs, and the appearance of clinical signs was associated to lower TAC values [41].

On the other hand, TEAC and FRAP were not useful for monitoring leishmaniosis treatment [39]. However, CUPRAC did show a significant increase in dogs that showed a successful response to the treatment. These differences could be due to the analytes that evaluate each assay. For example, FRAP evaluates mainly uric acid [43] that could be not so important as a marker of oxidative stress in the serum of the dog compared to humans. However, CUPRAC evaluates ascorbic acid, carotene, thiol, and blood GSH that are decreased in dogs affected by leishmaniosis [38, 40]. This raises the importance of using an appropriate substrate and assay for the evaluation of the total antioxidant status in canine leishmaniosis.

2.4.2.2 Paraoxonase-1 (PON-1)

PON-1 is an enzyme that prevents low density lipoprotein (LDL) oxidation, and it has been reported to decrease in experimentally induced and naturally occurred canine leishmaniosis [44]. In the experimentally induced leishmaniosis, PON-1 reached its minimum values at 2 months postinfection, time that was 2 months before the appearance of clinical signs. Therefore, it could be considered as an early marker of leishmaniosis. In addition, lower values of PON-1 have been reported in dogs with canine leishmaniosis and proteinuria; therefore, it could also be a marker of disease severity.

The minimum value of PON-1 in experimental infection appeared at the same time when there was an increase in acute phase proteins such as C-reactive protein (CRP), thus showing a significant negative correlation with this marker. Therefore, the inflammation could be a cause of decrease in PON-1 in this disease. In addition, other reports have indicated that the increased oxidative stress could be another reason for the decrease in PON-1 in canine leishmaniosis [40, 41].

PON-1 showed increases and return to similar values of those observed in healthy dogs when the dogs were successfully treated. Therefore, it can be considered as a biomarker of treatment although its response is slow in comparison with acute phase proteins such as CRP or ferritin [44].

2.4.2.3 Thiols

Thiols are a class of organic compounds that contain a sulfhydryl group and have critical roles in protect cells from oxidative damage by the degradation of electrophilic groups of ROS [45, 46]. It has been demonstrated that thiols increase after a successful treatment for canine leishmaniosis and are correlated with PON-1 [39].

2.4.2.4 Superoxide Dismutase (SOD)

SOD in canine leishmaniosis has a very interesting role. It is involved in the decrease of the oxidant response in the leishmania-infected macrophages and therefore is found to be increased in these macrophages.

However, in serum it is increased in dogs with clinical signs of leishmaniosis, having a possible protecting role against the oxidant products produced in the inflammation associated with the process. When the dogs have a very high clinical score, the activity of this enzyme decreases, possibly due to previous consumption trying to compensate the increased oxidant concentrations during this disease [47].

2.5 Conclusions

The following statements can be concluded:

- In both human and canine leishmaniosis, there is an increase in oxidant biomarkers and a decrease in the biomarkers of the antioxidant defense.
- The biomarkers of oxidative stress could have a potential application mainly in the evaluation of the severity of the disease and also for monitoring the treatment.

These conclusions are limited because usually the studies in this topic involve a reduced number of biomarkers, and also employ different assays that can result in contradictory results. It would be desirable the development of studies in which a wide panel of various biomarkers both of the oxidant and antioxidant response will be measured in order to get a broad view of the process. In addition, the use of standardized and validated assays in order to get results that could be applied in different laboratories would be recommended.

References

1. Peters W, Killick-Kendrick R (1987) The Leishmaniases in biology and medicine. Academic, London/Orlando
2. WHO (2010) Control of the leishmaniases. World Health Organization, Geneva
3. Alvar J, Vélez ID, Bern C, Herrero M, Desjeux P, Cano J, Jannin J, den BM, Team the WLC (2012) Leishmaniasis worldwide and global estimates of its incidence. PLoS One 7:e35671
4. Magill AJ (2012) Leishmaniasis. In: Magill AJ, Strickland GT, Maguire JH, Ryan ET, Solomon T (eds) Hunter's tropical medicine and emerging infectious diseases. Elsevier Health Sciences, Edinburgh, p 1215
5. Ashford RW (1996) Leishmaniasis reservoirs and their significance in control. Clin Dermatol 14:523–532
6. Dedet JP, Pratlong F (2009) Protozoa infection. In: Cook GC, Gordon C, Zumla A, Manson P (eds) Manson's tropical diseases Saunders/Elsevier, Amsterdam, p 1830
7. Fátima Horta M, Mendes BP, Roma EH, Soares F, Noronha M, Pereira Macêdo J, Oliveira LS, Duarte MM, Vieira LQ (2012) Reactive oxygen species and nitric oxide in cutaneous Leishmaniasis. J Parasitol Res 2012:11
8. Dumonteila E, Ramirez-Sierra MJ, Escobedo-Ortegon J, García-Miss M del R (2003) DNA vaccines induce partial protection against Leishmania mexicana. Vaccine 21:2161–2168
9. Ribeiro RR, Michalick MSM, da Silva ME, dos Santos CCP, Frézard FJG, da Silva SM (2018) Canine Leishmaniasis: an overview of the current status and strategies for control. Biomed Res Int 2018:1–12
10. Solano-Gallego L, Miró G, Koutinas A et al (2011) LeishVet guidelines for the practical management of canine leishmaniosis. Parasit Vectors 4:86
11. Noli C, Saridomichelakis MN (2014) An update on the diagnosis and treatment of canine leishmaniosis caused by Leishmania infantum (syn. L. chagasi). Vet J 202:425–435
12. Van Assche T, Deschacht M, da Luz RAI, Maes L, Cos P (2011) Leishmania–macrophage interactions: insights into the redox biology. Free Radic Biol Med 51:337–351
13. Stafford JL, Neumann NF, Belosevic M (2002) Macrophage-mediated innate host defense against protozoan parasites. Crit Rev Microbiol 28:187–248
14. Mukbel RM, Patten C, Gibson K, Ghosh M, Petersen C, Jones DE (2007) Macrophage killing of Leishmania amazonensis amastigotes requires both nitric oxide and superoxide. Am J Trop Med Hyg 76:669–675
15. Vural H, Aksoy N, Ozbilge H (2004) Alterations of oxidative–antioxidative status in human cutaneous leishmaniasis. Cell Biochem Funct 22:153–156
16. Kocyigit A, Gurel M, Ulukanligil M, Kocyigit A, Gurel M, Ulukanligil M (2003) Erythrocyte antioxidative enzyme activities and lipid peroxidation levels in patients with cutaneous leishmaniasis. Parasite 10:277–281
17. Ozbilge H, Aksoy N, Kilic E, Saraymen R, Yazar S, Vural H (2005) Evaluation of oxidative stress in cutaneous leishmaniasis. J Dermatol 32:7–11
18. Serarslan G, Yilmaz HR, Sogut S (2005) Serum antioxidant activities, malondialdehyde and nitric oxide levels in human cutaneous leishmaniasis. Clin Exp Dermatol 30:267–271
19. Kocyigit A, Keles H, Selek S, Guzel S, Celik H, Erel O (2005) Increased DNA damage and oxidative stress in patients with cutaneous leishmaniasis. Mutat Res 585:71–78
20. AbdulGhani S, Hezam A (2014) Biochemical and hematological levels in patients with cutaneous Leishmaniasis in Yemen (Online)

21. Asmaa Q, Al-Shamerii S, Al-Tag M, Al-Shamerii A, Li Y, Osman BH (2017) Parasitological and biochemical studies on cutaneous leishmaniasis in Shara'b district, Taiz, Yemen. Ann Clin Microbiol Antimicrob 16:47
22. Jain SK, Levine SN, Duett J, Hollier B (1990) Elevated lipid peroxidation levels in red blood cells of streptozotocin-treated diabetic rats. Metabolism 39:971–975
23. Jain SK, McVie R, Duett J, Herbst JJ (1989) Erythrocyte membrane lipid peroxidation and glycosylated hemoglobin in diabetes. Diabetes 38:1539–1543
24. Jain SK (1988) Evidence for membrane lipid peroxidation during the in vivo aging of human erythrocytes. Biochim Biophys Acta 937:205–210
25. Esterbauer H, Cheeseman KH (1990) Determination of aldehydic lipid peroxidation products: malonaldehyde and 4-hydroxynonenal. Methods Enzymol 186:407–421
26. Fong KL, McCay PB, Poyer JL, Keele BB, Misra H (1973) Evidence that peroxidation of lysosomal membranes is initiated by hydroxyl free radicals produced during flavin enzyme activity. J Biol Chem 248:7792–7797
27. Biswas T, Ghosh DK, Mukherjee N, Ghosal J (1997) Lipid peroxidation of erythrocytes in visceral leishmaniasis. J Parasitol 83:151–152
28. Neupane DP, Majhi S, Chandra L, Rijal S, Baral N, Baral N (2008) Erythrocyte glutathione status in human visceral leishmaniasis. Indian J Clin Biochem 23:95–97
29. Erel O, Kocyigit A, Bulut V, Gurel S (1999) Reactive nitrogen and oxygen intermediates in patients with cutaneous Leishmaniasis. Mem Inst Oswaldo Cruz Rio Janeiro 179:179–179
30. Cabrera M, Rodriguez O, Monsalve I, Tovar R, Hagel I (2003) Variations in the serum levels of soluble CD23, nitric oxide and IgE across the spectrum of American cutaneous leishmaniasis. Acta Trop 88:145–151
31. Gryglewski RJ, Palmer RMJ, Moncada S (1986) Superoxide anion is involved in the breakdown of endothelium-derived vascular relaxing factor. Nature 320:454–456
32. Assreuy J, Cunha FQ, Epperlein M, Noronha-Dutra A, O'Donnell CA, Liew FY, Moncada S (1994) Production of nitric oxide and superoxide by activated macrophages and killing of Leishmania major. Eur J Immunol 24:672–676
33. Saltman P (1989) Oxidative stress: a radical view. Semin Hematol 26:249–256
34. Rush JWE, Sandiford SD (2003) Plasma glutathione peroxidase in healthy young adults: influence of gender and physical activity. Clin Biochem 36:345–351
35. Glantzounis GK, Tsimoyiannis EC, Kappas AM, Galaris DA (2005) Uric acid and oxidative stress. Curr Pharm Des 11:4145–4151
36. Irwin MI, Hutchins BK (1976) A conspectus of research on vitamin C requirements of man. J Nutr 106:821–879
37. Gultekin M, Pasa S, Ural K, Balikci C, Ekren Asici GS, Gultekin G (2018) Oxidative status and lipid profile among dogs at different stages of visceral Leishmaniasis. Turkish J Parasitol 41:183–187
38. Almeida BFM, Narciso LG, Melo LM, Preve PP, Bosco AM, Lima VMF, Ciarlini PC (2013) Leishmaniasis causes oxidative stress and alteration of oxidative metabolism and viability of neutrophils in dogs. Vet J 198:599–605
39. Rubio CPCP, Martinez-Subiela S, Tvarijonaviciute A, Hernández-Ruiz J, Pardo-Marin L, Segarra S, Ceron JJ (2016) Changes in serum biomarkers of oxidative stress after treatment for canine leishmaniosis in sick dogs. Comp Immunol Microbiol Infect Dis 49:51–57
40. Bildik A, Kargın F, Seyrek K, Pasa S, Özensoy S (2004) Oxidative stress and non-enzymatic antioxidative status in dogs with visceral Leishmaniasis. Res Vet Sci 77:63–66
41. Heidarpour M, Soltani S, Mohri M, Khoshnegah J (2012) Canine visceral leishmaniasis: relationships between oxidative stress, liver and kidney variables, trace elements, and clinical status. Parasitol Res 111:1491–1496
42. Rubio CP, Martinez-Subiela S, Hernández-Ruiz J, Tvarijonaviciute A, Ceron JJ (2017) Analytical validation of an automated assay for ferric-reducing ability of plasma in dog serum. J Vet Diagnostic Investig 29:574–578
43. Benzie IFF, Strain JJ (1996) The ferric reducing ability of plasma (FRAP) as a measure of "antioxidant power": the FRAP assay. Anal Biochem 239:70–76

44. Martinez-Subiela S, Cerón JJ, Strauss-Ayali D, Garcia-Martinez JD, Tecles F, Tvarijonaviciute A, Caldin M, Baneth G (2014) Serum ferritin and paraoxonase-1 in canine leishmaniosis. Comp Immunol Microbiol Infect Dis 37:23–29
45. Yardim-Akaydin S, Ozkan Y, Ozkan E, Torun M, Simşek B (2003) The role of plasma thiol compounds and antioxidant vitamins in patients with cardiovascular diseases. Clin Chim Acta 338:99–105
46. Dickinson DA, Forman HJ (2002) Cellular glutathione and thiols metabolism. Biochem Pharmacol 64:1019–1026
47. Solcà MS, Andrade BB, Abbehusen MMC et al (2016) Circulating biomarkers of immune activation, oxidative stress and inflammation characterize severe canine visceral Leishmaniasis. Sci Rep 6:32619

Reactive Oxygen and Nitrogen Species in the Oral Cavity

3

Peng Zhou, Jessica Scoffield, and Hui Wu

Abstract

In the oral cavity, reactive oxygen and nitrogen species are continuously generated by bacterial metabolism and host-mediated cellular factors. These reactive species facilitate a critical function that regulates the outcomes of both oral and systemic diseases, such as dental caries, periodontitis, and the maintenance of blood pressure. As a result, the protective effects of ROS and RNS are being explored for future therapeutic applications.

Keywords

Reactive oxygen species (ROS) · Reactive nitrogenous species (RNS) · Microbiome · Oral cavity

3.1 Introduction

Hundreds of microbial species reside in the oral cavity. For many species, the oral cavity can be a hostile environment due to the consistent production of reactive oxygen and nitrogen species. Exposure to oxidative and nitrosative stress impacts the survival of oral microbes, bacterial competition, and the development of

P. Zhou
Department of Pediatric Dentistry, The University of Alabama at Birmingham, Birmingham, AL, USA

J. Scoffield · H. Wu (✉)
Department of Pediatric Dentistry, The University of Alabama at Birmingham, Birmingham, AL, USA

Department of Microbiology, The University of Alabama at Birmingham, Birmingham, AL, USA
e-mail: hwu@uab.edu

S. Chakraborti et al. (eds.), *Oxidative Stress in Microbial Diseases*,
https://doi.org/10.1007/978-981-13-8763-0_3

bacteria-associated oral diseases. This chapter deals with four relevant topics: (1) sources of oxidative and nitrosative stress in the oral cavity and (2) their influence on oral diseases, (3) the interplay between oral bacteria and reactive oxygen and nitrogen intermediates in maintaining systemic health, and (4) development of oxygen and nitric oxide-based therapeutics.

3.2 Reactive Oxygen and Nitrogen Species in the Oral Cavity

3.2.1 Reactive Oxygen Species (ROS)

The generation of ROS in the oral cavity can be host- or bacteria-induced, or both, and functions as a source of oxidative stress or an antioxidant defense [1, 2]. There are several forms of ROS, which are reactive species containing the diatomic oxygen molecule O_2, whereas O_2 itself is relatively inert in contrast to its radical forms [1, 3]. Highly reactive O_2 molecules can attack numerous biological macromolecules, e.g., DNA, RNA, lipids, and proteins, thus leading to abnormal cellular processes such as tumorigenesis, aging, and the development of many diseases [1, 2]. Hydrogen peroxide (H_2O_2) is a form of ROS that can be readily reduced to hydroxyl

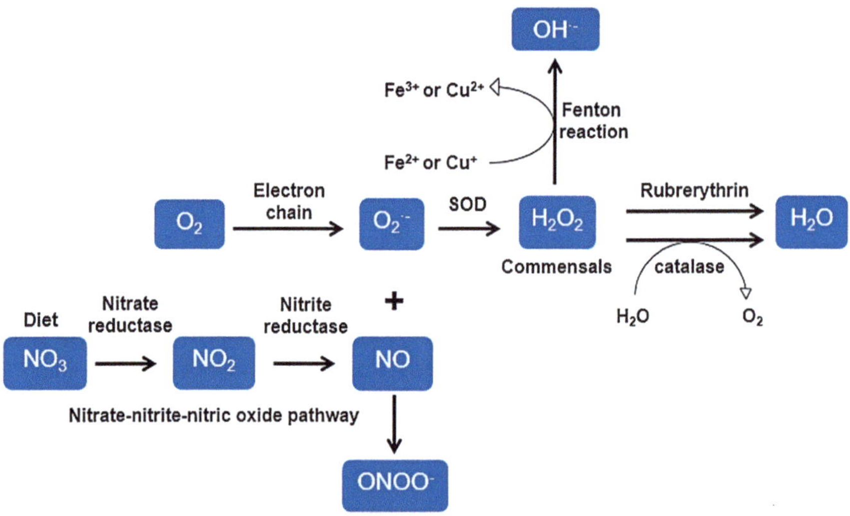

Fig. 3.1 Schematic diagram of generation of reactive oxygen species (ROS) and reactive nitrogen species (RNS) in the oral cavity. Superoxide ($O_2^{\cdot-}$), generated by the electron transport chain, can be dismutated by superoxide dismutase (SOD) to hydrogen peroxide (H_2O_2). Commensal bacteria also produce H_2O_2. In bacteria, H_2O_2 can be eliminated by two important antioxidant enzymes: (1) catalase decomposes H_2O_2 to O_2 and H_2O; (2) rubrerythrin reduces H_2O_2 to two water molecules. In the presence of transition metals, such as Fe^{2+} and Cu^+, H_2O_2 can be easily reduced to hydroxyl radicals ($OH^{\cdot-}$) by the Fenton reaction. Dietary nitrate (NO_3) is converted to nitrite (NO_2) by nitrate reductase, and NO_2 is converted to nitric oxide (NO) by nitrite reductase. NO can react with $O_2^{\cdot-}$ to generate peroxynitrite ($ONOO^-$)

radicals ($OH^{\cdot-}$) (Fig. 3.1), which nondiscriminatively damage all biological macromolecules, e.g., DNA, RNA, lipids, carbohydrates, and proteins [1, 2]. This Fenton reaction is catalyzed by the transition metals, ferrous and copper ions (Fig. 3.1) [1, 2]. Due to its high energy and reactivity, hydroxyl radicals can quickly oxidize and damage biological macromolecules. Moreover, the half-life of hydroxyl radicals in vivo is short lived (approximately 10^{-9} s), and no enzymatic reaction can efficiently capture and remove this dangerous compound [4]. Thus, this makes hydroxyl radicals a notorious and detrimental chemical to living organisms. Elimination of H_2O_2 and blockage of the Fenton reaction is a crucial antioxidant defense in most cells. In prokaryotes, H_2O_2 is usually converted to O_2 or H_2O by the catalytic activity of two important antioxidant enzymes, catalase and rubrerythrin (Fig. 3.1) [5, 6]. Catalase, present in most living organisms, can decompose H_2O_2 to oxygen and water, while rubrerythrin, normally found in anaerobic bacteria, is a non-heme iron enzyme responsible for reducing H_2O_2 to two water molecules (Fig. 3.1) [5, 6]. Both enzymes are crucial for protecting cells from oxidative damage by H_2O_2.

Superoxide ($O_2^{\cdot-}$) is another form of ROS generated from the electron chain in vivo and is considered a key radical leading to the oxidation of proteins [3]. Once produced, $O_2^{\cdot-}$ is quickly dismutated by superoxide dismutase (SOD) to generate H_2O_2 and O_2; thus SOD is recognized as an important enzyme in cellular antioxidant defenses [7]. Singlet oxygen ($^1O_2$) is an inorganic compound in gaseous state, kinetically unstable but with a low decay rate [8]. $^1O_2$ is generated via irradiation of oxygen in the presence of organic dye or photochemical materials. Due to its highly reactive olefins (also known as alkene, which is an unsaturated hydrocarbon with carbon-carbon double bond), singlet oxygen can readily peroxidize lipids [3] and becomes detrimental to living cells.

3.2.2 Reactive Nitrogen Species (RNS)

Ecological niche in the oral cavity is often dynamic and modulated by interactions between the host immune response, the microbial community, and dietary factors. The impact of oxidative stress (mainly H_2O_2) on the physiology of oral microbes has been well documented [9]; however, reactive nitrogen species (RNS) are not well characterized despite that they are generated at physiologically meaningful levels in the oral cavity and play a prominent role not only in oral health and dysbiosis but are highly relevant to overall systemic health.

Dietary nitrate is the primary precursor for intermediates that produce reactive nitrogen species in the oral cavity. Leafy green vegetables provide approximately 85% of dietary nitrate [10], which is accumulated and circulated by the salivary glands and reduced by oral commensal bacteria by the process of denitrification. Although the oral cavity harbors approximately 700 bacterial species [11], the process of denitrification requires specific enzymes that are unique to a select few genera of oral commensal bacteria, mainly *Prevotella*, *Veillonella*, and *Actinomyces* spp. [12]. During denitrification, nitrate (NO_3) is reduced to nitrite (NO_2) by nitrate reductase, and nitrite is reduced to nitric oxide (NO) by nitrite reductase, collectively known as the nitrate-nitrite-nitric oxide pathway (Fig. 3.1). NO is a free radical and signaling molecule produced by endogenous commensal bacteria and the host immune response (neutrophils) to defend against invasion by microbial pathogens [13]. In addition, NO can react with superoxide to form peroxynitrite ($ONOO^-$) (Fig. 3.1), a highly reactive nitrogen species that has antimicrobial activity and mediates the nitration of tyrosine residues, thus damaging proteins and cells [14]. Salivary nitrate and nitrite levels can exceed 1 mM under nitrate-rich diets [15], and these precursors can facilitate the generation of various RNS that modulate oral health and disease.

3.2.3 Oxidative and Nitrosative Stress in Oral Microbial Diseases

The human oral cavity typically maintains a homeostatic biofilm community; however, the disruption of this balance leads to the development of diseases [16]. Dental caries, endodontic infections, and periodontitis are three main infectious diseases that are initiated by changes in composition of bacterial communities in the oral cavity. Moreover, the ability of the microorganisms to trigger diseases is heavily dependent on their ability to deal with environmental stress, such as ROS and RNS.

Dental caries is the decay of the tooth surface due to acids produced by bacteria and is the most prevalent and costly oral health issue in children and adults [17]. Dental caries can lead to a number of symptoms including pain, tooth loss, and inflammation or infection of the surrounding tissues [17]. In the past five decades, studies have demonstrated that mutans streptococci, such as *Streptococcus mutans* and *Streptococcus sobrinus*, and lactobacilli are the most common pathogens associated with dental caries [18]. *S. mutans*, a gram-positive cocci bacterium, can produce abundant lactic acid by fermenting dietary carbohydrates to lower pH and demineralize teeth [19]. Lactic acid can also be used as a chemical weapon to outcompete with other oral bacterial species [19]. For instance, commensal streptococcal species, such as *Streptococcus oligofermentans*, can inhibit the growth of *S. mutans* by oxidizing lactate to generate H_2O_2, a lethal chemical to *S. mutans* [20]. In contrast to H_2O_2-generating streptococcal species, *S. mutans* does not produce H_2O_2 and is more sensitive to H_2O_2. Other oral microbes like *Mycoplasma salivarium* can scavenge ROS and reduce oxidative stress by employing different antioxidant systems, thus promoting the growth of cariogenic pathogens [21]. In addition, enzymatic strategies (catalase and superoxide dismutase) and nonenzymatic antioxidant systems, such as uric acid, can potentially reduce oxidative stress in the saliva of patients with dental caries [22].

Periodontitis affects about 11% of the population around the world and is a main cause of alveolar bone resorption, tooth loss, and halitosis [23]. The pathogenicity of periodontitis is multifactorial, and a large number of studies have shown that dysbiosis between the host and oral microbiota leads to the development of periodontitis [24]. The red complex pathogens (*Porphyromonas gingivalis*, *Tannerella forsythia*, and *Treponema denticola*) as well as *Fusobacterium nucleatum* have been implicated in the development of periodontal diseases [25]. Periodontal pathogens trigger an inflammatory response in host cells or tissues, which causes the destruction of the periodontium and induces bone resorption [26]. The presence of a dysbiotic biofilm is key to the pathogenesis. The development of human oral biofilms is a sequential process, which starts with initial colonizers adhering to the tooth surface, followed by the incorporation of early, middle, and late colonizers to form a mature biofilm [27]. All periodontal pathogens are considered middle to late bacterial colonizers and often colonize the subgingival, which is anaerobic and consistent with the fact that these pathogens are strict anaerobes and extremely susceptible to oxygen stress. On the other hand, initial colonizers, such as *Streptococcus sanguinis*, produce copious amounts of H_2O_2 in oral biofilms [28]. This oxygen stress should inhibit the growth of these strict anaerobic pathogens; however, *P. gingivalis* and *F. nucleatum* are often isolated and identified in the early biofilm community with streptococcus [29]. Antioxidant mechanisms have been reported in *P. gingivalis* and *F. nucleatum*. *P. gingivalis* uses rubrerythrin to reduce H_2O_2, which partially explains their increased tolerance to oxygen. In addition, *F. nucleatum* can support the growth of *P. gingivalis* in oxygenated environments [30]. Furthermore, catalase-positive bacteria, such as *V. parvula*, can remove H_2O_2 and reduce oxidative stress in their microniche which supports the growth of these anaerobic periodontal pathogens [31]. *Aggregatibacter actinomycetemcomitans* is another periopathogen that is associated with localized aggressive periodontitis. During coculture with the H_2O_2-producing commensal, *S. gordonii*, *A. actinomycetemcomitans* displays H_2O_2-dependent tolerance to human serum [32], suggesting that H_2O_2 production by commensal bacteria can protect this pathogen against host immunity. Interestingly, H_2O_2 production by the oral commensal, *Streptococcus parasanguinis*, restricts *A. actinomycetemcomitans* from integrating into dual species biofilms while promoting *S. parasanguinis*' biofilm and reducing *A. actinomycetemcomitans* survival [33]. Overall, these studies demonstrate that ROS produced by the oral biofilm community mediates dynamic bacterial interactions within polymicrobial communities and plays a crucial role in the development and pathogenicity of periodontitis.

3.2.4 Regulation of Systemic Homeostasis and Dysbiosis by Oral Reactive Oxygen and Nitrogen Species

Current evidence suggests that oral diseases can have a profound effect on systemic health by enhancing the circulation of ROS and RNS. Specifically, periodontitis has been shown to be a significant source of systemic ROS, especially during colonization with the periopathogens *P. gingivalis* and *A. actinomycetemcomitans*. *P. gingivalis* and *A. actinomycetemcomitans* induce a systemic inflammatory response that is associated with an increase of systemic ROS [34]. Systemic ROS induced by periopathogens is suspected to promote atherosclerosis, lipid oxidation, and organ damage [35]. In contrast, coordinated metabolic activities between nitrogenous intermediates and commensal bacteria have been demonstrated to have beneficial outcomes on overall systemic health. The generation of NO by oral nitrate-reducing bacteria has been shown to provide a systemic health benefit. In an animal model of infection, dietary nitrate supplementation has been shown to decrease blood pressure and heart rate but increase plasma NO and the abundance of bacterial nitrate reducers (*Haemophilus* spp.) [36]. Moreover, direct supplementation of NO improves blood pressure, endothelial function, and vascular compliance in hypertensive patients [37]. NO lowers blood pressure by inducing vasodilation [38]. However, the blood pressure lowering benefits of NO are abolished following the use of mouthwash because the antiseptic drastically lowers the number of nitrate-reducing bacteria [39]. Once considered toxic to human body or having no significant role in mammalian physiology, recent studies exploring the role of the nitrate-nitrite-nitric oxide pathway now point to new avenues of potential therapeutic benefits.

In addition to modulating cardiovascular health through NO production, there is evidence that oral commensal streptococci disseminate to lung infections in cystic fibrosis (CF) patients and stabilize lung infections by generating RNS. CF is a genetic disorder caused by a defect in the cystic fibrosis transmembrane conductance regulator (CFTR) protein. Defects in the CFTR protein inhibit proper mucociliary clearance of microbes from the lung, and as a result, the lung is chronically colonized with recalcitrant polymicrobial infections [40]. Historically, *Pseudomonas aeruginosa* has been recognized as the major contributor of morbidity and mortality in CF lung disease [40]. However, recent microbiome analysis of CF sputum reveals that oral commensals, particularly *S. parasanguinis* and *Streptococcus salivarius*, are abundantly present in clinically stable CF patients compared to patients who experienced acute lung exacerbations [41]. Studies that have explored mechanisms of how oral commensal streptococci promote CF lung stability have discovered that these streptococci are capable of inhibiting *P. aeruginosa*, the major CF pathogen. H_2O_2-producing oral streptococci can inhibit *P. aeruginosa* in a nitrite-dependent manner by generating peroxynitrite ($ONOO^-$) [42, 43]. Moreover, a functional nitrite reductase is required by *P. aeruginosa* to resist inhibition of oral streptococci H_2O_2 production [42]. Similar to RNS and ROS, peroxynitrite damages DNA, proteins, and lipids and can be formed from reactions with NO and superoxide or NO_2 and H_2O_2 at an acidic pH [14]. The ability of nitrite to work in concert with H_2O_2 to potentially produce peroxynitrite or other RNS in the CF lung suggests that RNS may be generated in the oral cavity as a consequence of denitrification and the presence of H_2O_2-producing oral commensals.

3.3 ROS- and RNS-Based Therapeutics

Due to a wide spectrum of antimicrobial activity, H_2O_2 mouthwashes have been used to control plaque and prevent periodontal disease in dentistry [44]. New strategies for controlling oral infections are based on the antimicrobial activity of ROS. Antimicrobial photodynamic therapy (aPDT) is a concept that utilizes visible light to activate a photosensitizer to generate ROS [45]. aPDT has been shown to be effective for inhibiting cariogenic, endodontic, and periodontal pathogens [45]. Recently, much attention has been focused on exploring the efficacy of nitrogenous intermediates as potent antimicrobials for oral infectious diseases. Both clinical and experimental observations highlight the possible utility of these compounds as therapeutic agents. For instance, elevated concentrations of salivary NO have been associated with a decreased prevalence of dental caries [46]. *S. mutans* establishes sticky biofilms on the tooth surface and produces acid that promotes the demineralization of tooth enamel. NO is a potent antimicrobial that is used as a host defense mechanism. Due to the association of increased NO with a lower incidence of dental caries, NO-releasing nanoparticles are being considered as potential therapeutics for caries prevention. At a low pH, NO-releasing nanoparticles are effective against *S. mutans* growth and biofilm formation [47]. Moreover, NO-releasing nanoparticles induce self-killing and inhibit biofilm formation by the oral fungal pathogen, *Candida albicans* [48]. Furthermore, dietary consumption of nitrate-rich beet juice correlates with increased salivary nitrite and nitric oxide, in addition to an increase in salivary pH, which provides an unfavorable environment for caries development [49]. Nitrogenous intermediates are also being used to target periodontal pathogens *A. actinomycetemcomitans* and *P. gingivalis* [50]. The occurrence of elevated salivary nitrates and nitrites in periodontal disease patients is hypothesized to be a mechanism of defense against periodontal microbes. Taken together, the development of NO-based therapeutics appears to be promising for treating bacterial-induced oral diseases.

References

1. Zhou P, Liu L, Tong H, Dong X (2012) Role of operon aaoSo-mutT in antioxidant defense in Streptococcus oligofermentans. PLoS One 7(5):e38133. https://doi.org/10.1371/journal.pone.0038133. PubMed PMID: 22666463; PubMed Central PMCID: PMC3364214
2. Birben E, Sahiner UM, Sackesen C, Erzurum S, Kalayci O (2012) Oxidative stress and antioxidant defense. World Allergy Organ J 5(1):9–19. https://doi.org/10.1097/WOX.0b013e3182439613. PubMed PMID: 23268465; PubMed Central PMCID: PMCPMC3488923
3. Halliwell B, Gutteridge JMC (2007) Free radicals in biology and medicine, 4th edn. University Press, Oxford/New York, xxxvi, 851 p
4. Sies H (1993) Strategies of antioxidant defense. Eur J Biochem 215(2):213–219. PubMed PMID: 7688300
5. Andrews SC (2010) The ferritin-like superfamily: evolution of the biological iron storeman from a rubrerythrin-like ancestor. Biochim Biophys Acta 1800(8):691–705. https://doi.org/10.1016/j.bbagen.2010.05.010. PubMed PMID: 20553812
6. Zamocky M, Furtmuller PG, Obinger C (2008) Evolution of catalases from bacteria to humans. Antioxid Redox Signal 10(9):1527–1548. https://doi.org/10.1089/ars.2008.2046. PubMed PMID: 18498226; PubMed Central PMCID: PMCPMC2959186
7. Touati D (1989) The molecular genetics of superoxide dismutase in E. coli. An approach to understanding the biological role and regulation of SODS in relation to other elements of the defence system against oxygen toxicity. Free Radic Res Commun 8(1):1–9. PubMed PMID: 2684801
8. Rosenthal I (1976) Recent developments in singlet molecular oxygen chemistry. Photochem Photobiol 24(6):641–645. PubMed PMID: 798214
9. Lamont RJ (2016) Hydrogen peroxide is a central determinant of oral polymicrobial synergy. Environ Microbiol. https://doi.org/10.1111/1462-2920.13537. PubMed PMID: 27654314

10. van Velzen AG, Sips AJ, Schothorst RC, Lambers AC, Meulenbelt J (2008) The oral bioavailability of nitrate from nitrate-rich vegetables in humans. Toxicol Lett 181(3):177–181. https://doi.org/10.1016/j.toxlet.2008.07.019. PubMed PMID: 18723086
11. Aas JA, Paster BJ, Stokes LN, Olsen I, Dewhirst FE (2005) Defining the normal bacterial flora of the oral cavity. J Clin Microbiol 43(11):5721–5732. https://doi.org/10.1128/JCM.43.11.5721-5732.2005. PubMed PMID: 16272510; PubMed Central PMCID: PMCPMC1287824
12. Burleigh MC, Liddle L, Monaghan C, Muggeridge DJ, Sculthorpe N, Butcher JP et al (2018) Salivary nitrite production is elevated in individuals with a higher abundance of oral nitrate-reducing bacteria. Free Radic Biol Med 120:80–88. https://doi.org/10.1016/j.freeradbiomed.2018.03.023. PubMed PMID: 29550328
13. Mariano FS, Campanelli AP, Nociti Jr FH, Mattos-Graner RO, Gonçalves RB (2012) Antimicrobial peptides and nitric oxide production by neutrophils from periodontitis subjects. Braz J Med Biol Res 45(11):1017–1024. https://doi.org/10.1590/s0100-879x2012007500123
14. Radi R (2013) Peroxynitrite, a stealthy biological oxidant. J Biol Chem 288(37):26464–26472. https://doi.org/10.1074/jbc.R113.472936. PubMed PMID: 23861390; PubMed Central PMCID: PMCPMC3772193
15. Pannala AS, Mani AR, Spencer JPE, Skinner V, Bruckdorfer KR, Moore KP et al (2003) The effect of dietary nitrate on salivary, plasma, and urinary nitrate metabolism in humans. Free Radic Biol Med 34(5):576–584. https://doi.org/10.1016/s0891-5849(02)01353-9
16. Marsh PD (2006) Dental plaque as a biofilm and a microbial community – implications for health and disease. BMC Oral Health 6(Suppl 1):S14. https://doi.org/10.1186/1472-6831-6-S1-S14. PubMed PMID: 16934115; PubMed Central PMCID: PMC2147593
17. Selwitz RH, Ismail AI, Pitts NB (2007) Dental caries. Lancet 369(9555):51–59. https://doi.org/10.1016/S0140-6736(07)60031-2. PubMed PMID: 17208642
18. Beighton D, Manji F, Baelum V, Fejerskov O, Johnson NW, Wilton JM (1989) Associations between salivary levels of Streptococcus mutans, Streptococcus sobrinus, lactobacilli, and caries experience in Kenyan adolescents. J Dent Res 68(8):1242–1246. https://doi.org/10.1177/00220345890680080601. PubMed PMID: 2632612
19. Loesche WJ (1986) Role of Streptococcus mutans in human dental decay. Microbiol Rev 50(4):353–380. PubMed PMID: 3540569; PubMed Central PMCID: PMCPMC373078
20. Tong H, Chen W, Merritt J, Qi F, Shi W, Dong X (2007) Streptococcus oligofermentans inhibits Streptococcus mutans through conversion of lactic acid into inhibitory H2O2: a possible counteroffensive strategy for interspecies competition. Mol Microbiol 63(3):872–880. https://doi.org/10.1111/j.1365-2958.2006.05546.x. PubMed PMID: 17302806
21. Kornspan JD, Ginsburg I, Rottem S (2013) The oxidant scavenging capacity of the oral mycoplasma salivarium. Arch Oral Biol 58(10):1378–1384. https://doi.org/10.1016/j.archoralbio.2013.05.001. PubMed PMID: 23735812
22. Silva PVD, Troiano JA, Nakamune A, Pessan JP, Antoniali C (2016) Increased activity of the antioxidants systems modulate the oxidative stress in saliva of toddlers with early childhood caries. Arch Oral Biol 70:62–66. https://doi.org/10.1016/j.archoralbio.2016.06.003. PubMed PMID: 27328152
23. Marcenes W, Kassebaum NJ, Bernabe E, Flaxman A, Naghavi M, Lopez A et al (2013) Global burden of oral conditions in 1990-2010: a systematic analysis. J Dent Res 92(7):592–597. https://doi.org/10.1177/0022034513490168. PubMed PMID: 23720570; PubMed Central PMCID: PMCPMC4484374
24. Darveau RP (2010) Periodontitis: a polymicrobial disruption of host homeostasis. Nat Rev Microbiol 8(7):481–490. https://doi.org/10.1038/nrmicro2337. PubMed PMID: 20514045
25. Hajishengallis G, Lamont RJ (2012) Beyond the red complex and into more complexity: the polymicrobial synergy and dysbiosis (PSD) model of periodontal disease etiology. Mol Oral Microbiol 27(6):409–419. https://doi.org/10.1111/j.2041-1014.2012.00663.x. PubMed PMID: 23134607; PubMed Central PMCID: PMCPMC3653317
26. Kinane DF, Preshaw PM, Loos BG (2011) Working group 2 of seventh European workshop on P. host-response: understanding the cellular and molecular mechanisms of host-microbial interactions--consensus of the seventh European workshop on periodontology. J Clin

Periodontol 38(Suppl 11):44–48. https://doi.org/10.1111/j.1600-051X.2010.01682.x. PubMed PMID: 21323703
27. Kolenbrander PE, Palmer RJ, Jr., Rickard AH, Jakubovics NS, Chalmers NI, Diaz PI. Bacterial interactions and successions during plaque development. Periodontol 2000 42:47–79. doi: https://doi.org/10.1111/j.1600-0757.2006.00187.x. PubMed PMID: 16930306
28. Chen L, Ge X, Dou Y, Wang X, Patel JR, Xu P (2011) Identification of hydrogen peroxide production-related genes in Streptococcus sanguinis and their functional relationship with pyruvate oxidase. Microbiology 157(Pt 1):13–20. https://doi.org/10.1099/mic.0.039669-0. PubMed PMID: 20847003; PubMed Central PMCID: PMC3069532
29. Periasamy S, Kolenbrander PE (2009) Mutualistic biofilm communities develop with Porphyromonas gingivalis and initial, early, and late colonizers of enamel. J Bacteriol 191(22):6804–6811. https://doi.org/10.1128/JB.01006-09. PubMed PMID: 19749049; PubMed Central PMCID: PMC2772475
30. Diaz PI, Zilm PS, Rogers AH (2002) Fusobacterium nucleatum supports the growth of Porphyromonas gingivalis in oxygenated and carbon-dioxide-depleted environments. Microbiology 148(Pt 2):467–472. https://doi.org/10.1099/00221287-148-2-467. PubMed PMID: 11832510
31. Zhou P, Li X, Huang IH, Qi F (2017) Veillonellae catalase protects the growth of fusobacterium nucleatum in microaerophilic and Streptococcus gordonii-present environments. Appl Environ Microbiol. https://doi.org/10.1128/AEM.01079-17. PubMed PMID: 28778894
32. Ramsey MM, Whiteley M (2009) Polymicrobial interactions stimulate resistance to host innate immunity through metabolite perception. Proc Natl Acad Sci U S A 106(5):1578–1583. https://doi.org/10.1073/pnas.0809533106. PubMed PMID: 19164580; PubMed Central PMCID: PMCPMC2629492
33. Duan D, Scoffield JA, Zhou X, Wu H (2016) Fine-tuned production of hydrogen peroxide promotes biofilm formation of Streptococcus parasanguinis by a pathogenic cohabitant Aggregatibacter actinomycetemcomitans. Environ Microbiol. https://doi.org/10.1111/1462-2920.13425. PubMed PMID: 27348605
34. Wang Y, Andrukhov O, Rausch-Fan X (2017) Oxidative stress and antioxidant system in periodontitis. Front Physiol 8:910. https://doi.org/10.3389/fphys.2017.00910. PubMed PMID: 29180965; PubMed Central PMCID: PMCPMC5693842
35. Kurita-Ochiai T, Jia R, Cai Y, Yamaguchi Y, Yamamoto M (2015) Periodontal disease-induced atherosclerosis and oxidative stress. Antioxidants (Basel) 4(3):577–590. https://doi.org/10.3390/antiox4030577. PubMed PMID: 26783845; PubMed Central PMCID: PMCPMC4665422
36. Hyde ER, Luk B, Cron S, Kusic L, McCue T, Bauch T et al (2014) Characterization of the rat oral microbiome and the effects of dietary nitrate. Free Radic Biol Med 77:249–257. https://doi.org/10.1016/j.freeradbiomed.2014.09.017. PubMed PMID: 25305639
37. Houston M, Hays L (2014) Acute effects of an oral nitric oxide supplement on blood pressure, endothelial function, and vascular compliance in hypertensive patients. J Clin Hypertens (Greenwich) 16(7):524–529. https://doi.org/10.1111/jch.12352. PubMed PMID: 24962851
38. Goto C, Nishioka K, Umemura T, Jitsuiki D, Sakagutchi A, Kawamura M et al (2007) Acute moderate-intensity exercise induces vasodilation through an increase in nitric oxide bioavailiability in humans. Am J Hypertens 20(8):825–830. https://doi.org/10.1016/j.amjhyper.2007.02.014. PubMed PMID: 17679027
39. Kapil V, Haydar SM, Pearl V, Lundberg JO, Weitzberg E, Ahluwalia A (2013) Physiological role for nitrate-reducing oral bacteria in blood pressure control. Free Radic Biol Med 55:93–100. https://doi.org/10.1016/j.freeradbiomed.2012.11.013. PubMed PMID: 23183324; PubMed Central PMCID: PMCPMC3605573
40. Filkins LM, O'Toole GA (2015) Cystic fibrosis lung infections: Polymicrobial, complex, and hard to treat. PLoS Pathog 11(12):e1005258. https://doi.org/10.1371/journal.ppat.1005258. PubMed PMID: 26719892; PubMed Central PMCID: PMCPMC4700991
41. Filkins LM, Hampton TH, Gifford AH, Gross MJ, Hogan DA, Sogin ML et al (2012) Prevalence of streptococci and increased polymicrobial diversity associated with cystic

fibrosis patient stability. J Bacteriol 194(17):4709–4717. https://doi.org/10.1128/JB.00566-12. PubMed PMID: 22753064; PubMed Central PMCID: PMCPMC3415522
42. Scoffield JA, Wu H (2016) Nitrite reductase is critical for Pseudomonas aeruginosa survival during co-infection with the oral commensal Streptococcus parasanguinis. Microbiology 162(2):376–383. https://doi.org/10.1099/mic.0.000226. PubMed PMID: 26673783; PubMed Central PMCID: PMCPMC4766596
43. Scoffield JA, Wu H (2015) Oral streptococci and nitrite-mediated interference of Pseudomonas aeruginosa. Infect Immun 83(1):101–107
44. Kobayashi Y, Hayashi M, Yoshino F, Tamura M, Yoshida A, Ibi H et al (2014) Passive ultrasonic irrigation in the presence of a low concentration of hydrogen peroxide enhances hydroxyl radical generation and bactericidal effect against Enterococcus faecalis. J Oral Sci 56(1):35–39. https://doi.org/10.2334/josnusd.56.35
45. Cieplik F, Tabenski L, Buchalla W, Maisch T (2014) Antimicrobial photodynamic therapy for inactivation of biofilms formed by oral key pathogens. Front Microbiol 5:405. https://doi.org/10.3389/fmicb.2014.00405. PubMed PMID: 25161649; PubMed Central PMCID: PMCPMC4130309
46. Syed M, Sachdev V, Chopra R (2016) Intercomparison of salivary nitric oxide as a biomarker of dental caries risk between caries-active and caries-free children. Eur Arch Paediatr Dent 17(4):239–243. https://doi.org/10.1007/s40368-016-0234-z. PubMed PMID: 27357363
47. Backlund CJ, Worley BV, Schoenfisch MH (2016) Anti-biofilm action of nitric oxide-releasing alkyl-modified poly(amidoamine) dendrimers against Streptococcus mutans. Acta Biomater 29:198–205. https://doi.org/10.1016/j.actbio.2015.10.021. PubMed PMID: 26478472; PubMed Central PMCID: PMCPMC4695967
48. Ahmadi MS, Lee HH, Sanchez DA, Friedman AJ, Tar MT, Davies KP et al (2016) Sustained nitric oxide-releasing nanoparticles induce cell death in *Candida albicans* yeast and hyphal cells, preventing biofilm formation in vitro and in a rodent central venous catheter model. Antimicrob Agents Chemother 60(4):2185–2194. https://doi.org/10.1128/AAC.02659-15. PubMed PMID: 26810653; PubMed Central PMCID: PMCPMC4808184
49. Hohensinn B, Haselgrubler R, Muller U, Stadlbauer V, Lanzerstorfer P, Lirk G et al (2016) Sustaining elevated levels of nitrite in the oral cavity through consumption of nitrate-rich beetroot juice in young healthy adults reduces salivary pH. Nitric Oxide 60:10–15. https://doi.org/10.1016/j.niox.2016.08.006. PubMed PMID: 27593618
50. Backlund CJ, Sergesketter AR, Offenbacher S, Schoenfisch MH (2014) Antibacterial efficacy of exogenous nitric oxide on periodontal pathogens. J Dent Res 93(11):1089–1094. https://doi.org/10.1177/0022034514529974. PubMed PMID: 25139363; PubMed Central PMCID: PMCPMC4293763

Role of Gut Microbiota in Combating Oxidative Stress

4

Bomba Dam, Arijit Misra, and Sohini Banerjee

Abstract

We are at a fascinating junction of medical microbiology, witnessing a paradigm shift in the basic understanding of diseases and their treatment strategies. Here, we have summarized different diseases that induce oxidative stress like pathophysiological, metabolic, neurodegenerative, and microbial infectious diseases and the role of gut microbiota correspondingly to alleviate the toxic state. The concepts of oxidative stress, gastrointestinal tract, and healthy gut microbiota are briefly introduced followed by an elucidated account of their relationships in different diseased conditions. Almost all diseases are linked to, or lead to gut dysbiosis, particularly characterized by an overall decline in gut microbial diversity; reduction in number of beneficial microbial members like *Lactobacillus*, *Bifidobacterium*, and anaerobic short-chain fatty acid producers (e.g., *Bacteroidetes* and *Faecalibacterium prausnitzii*); and altered proportions of *Firmicutes* and *Bacteroidetes* members. This is accompanied by an increase in aerotolerant or facultative anaerobic opportunistic pathogens like the Gram-negative proteobacterial members of *Enterobacteriaceae* and *Enterococcaceae*. Thus, future therapeutic interventions must be directed towards maintaining or restoring the gut microbiota composition to its healthy state to overcome the oxidative stress generated in different diseased conditions.

B. Dam (✉) · A. Misra
Microbiology Laboratory, Department of Botany, Institute of Science, Visva-Bharati (A Central University), Santiniketan, West Bengal, India
e-mail: bomba.dam@visva-bharati.ac.in

S. Banerjee
Microbiology Laboratory, Department of Botany, Institute of Science, Visva-Bharati (A Central University), Santiniketan, West Bengal, India

Department of Environmental Studies, Institute of Science, Visva-Bharati (A Central University), Santiniketan, West Bengal, India

S. Chakraborti et al. (eds.), *Oxidative Stress in Microbial Diseases*,
https://doi.org/10.1007/978-981-13-8763-0_4

Keywords
Oxidative stress · Gut microbiota · Dysbiosis · Short-chain fatty acid · Firmicutes-to-Bacteroidetes ratio · Next-generation sequencing

4.1 Introduction

Intracellular production of reactive oxygen species (ROS) and reactive nitrogen species (RNS) at low concentrations as a byproduct of normal oxidative cellular metabolism plays a fundamental role in different life processes [1]. They are extremely reactive with one or more unpaired electrons like nitric oxide (•NO), hydroxyl radical (•OH), superoxide ($O_2^{\bullet-}$), alkoxyl radicals (RO•), and hydrogen peroxide (H_2O_2). Elevated levels of ROS/RNS that host defense mechanisms cannot counterbalance lead to disturbances in cellular redox potential. This in turn could damage cellular components (like proteins, DNA, and membranes), alter enzyme functions, and also act as signaling molecules leading to a number of diseases [1, 2]. The gastrointestinal (GI) tract or gut is a major source of ROS [3]. Apart from the host itself, gut microorganisms, particularly in an altered diseased state produce ROS which facilitates disease development [2, 4]. Oxidative stress is responsible for various pathophysiological diseases of the gut like inflammatory bowel disease (IBD), gastro-duodenal ulcers, and GI malignancies [5, 6]. Recent studies extended this list to metabolic disorders and infectious, neurodegenerative, and immunodegenerative diseases as well [7–10].

Defense arsenal against oxidative stress includes enzymatic and non-enzymatic components (Fig. 4.1). The first line of indispensable defense is provided by endogenously produced antioxidant enzymes like superoxide dismutase (SOD), catalase, and glutathione peroxidase that help to remove free radicals [11]. Among non-enzymatic antioxidants, some are again produced by the host, like glutathione, coenzyme Q, melatonin, or metal-binding proteins like transferrin, ferritin, and albumin [12]. The remaining are exogenously acquired as nutrient supplements like vitamin E (tocopherols), vitamin C, polyphenols, vitamin A (carotenoids), and trace minerals such as manganese, copper, selenium, and zinc that function as cofactors in catalytic activities to remove ROS and reinstate the redox balance [13].

A large number of studies have been performed during the last few years to understand the relationship between altered gut microbiota or dysbiosis and diseased conditions (Fig. 4.2). Here, we have tried to bring together all major groups of diseases that induce oxidative stress and comprehend them based on a common parameter, i.e., dysbiosis as compared to healthy gut microbiota that include bacteria and other minor components like fungi, archaea, and viruses. In addition, we have tried to provide the recent developments in microbiome studies and the continued progress in this area that can enable more precision diagnostics and future medicine.

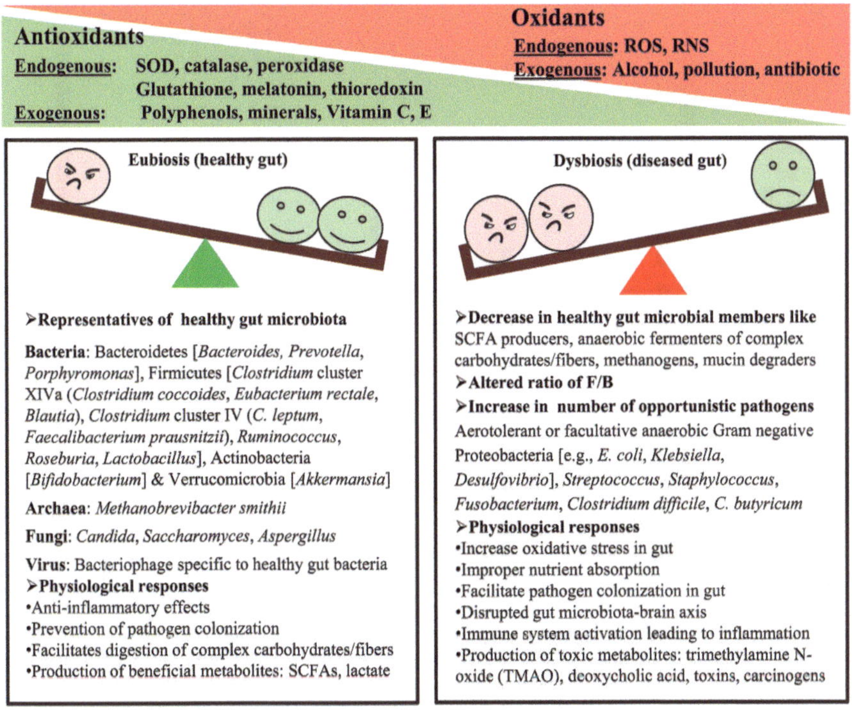

Fig. 4.1 The fine balance between oxidants and antioxidants and its relation to gut eubiosis or dysbiosis. In healthy individuals, a delicate balance is maintained between the endogenous and exogenous oxidants produced/accumulated in the gut and their detoxification by antioxidant systems. Any misbalance in this homeostatic condition increases the oxidative stress and leads to diseased state with an altered gut microbial composition called dysbiosis. Cartoons represent the eubiosis and dysbiosis condition, respectively, in healthy and diseased states, with the major microbial players and their physiological responses listed below them

4.2 Gastrointestinal Tract: Its Influence on Gut Microbiota

The GI tract is the large tubular organ that starts from mouth and terminates at anus, where muscle movements and release of hormones and enzymes facilitates proper digestion followed by removal of waste products. It has four distinct sections, namely, the esophagus, stomach, small intestine, and large intestine, which varies in their substrate availability, digestive enzymes, pH, and oxygen tension that result in contrasting microbial loads and composition [14]. Direct influence of pH was revealed in previous studies where slightly acidic pH (<5.5) maintained by adding short-chain fatty acids (SCFAs) in the media inhibited growth of pathogenic *Bacteroides* members but not butyrate-producing *Eubacterium rectale* [15, 16]. Lumen of healthy colon is strictly anoxic (Eh = ~250 mV) and thus any oxidative stress can shift the equilibrium of gut microflora from anaerobic to aerotolerant or facultative

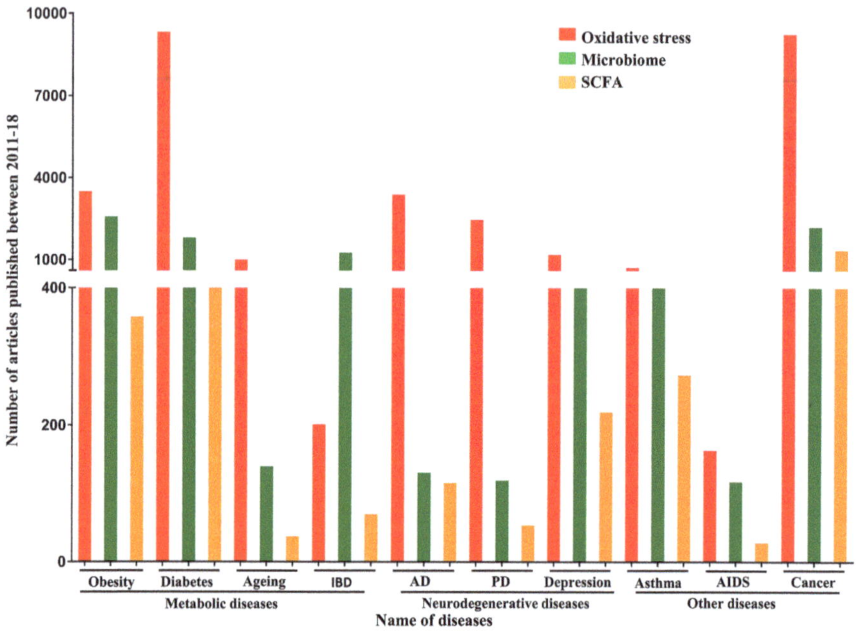

Fig. 4.2 Publication trends that refer the role of gut microbiota and SCFAs in diseases involving oxidative stress. The "Advance search tool" in PubMed with three field combinations was used, namely, "name of the disease," "year of publication from 01/01/2011 to 30/08/2018," and either of the key words "oxidative stress," "microbiome/microbiota," or "short-chain fatty acids/SCFA.". (Adapted from Refs. [14, 19, 25])

anaerobic groups [17]. Microbial load in gut segments increases from the anterior to posterior end. While secretion of bile salt and acidic pH in the stomach restricts the number to only 10^3 bacteria per ml, it increases a bit to 10^4–10^6 in the small intestine (with pancreatic secretions), and eventually goes up to 10^{12} in the large intestine. Near-neutral to slightly acidic pH, slow nutrient flow rates, and abundant protective antioxidant enzymes in the intestine selectively favors colonization of metabolically diverse obligate anaerobic groups like bacteroides, bifidobacteria, lactobacilli, sulfate reducers, methanogens, and SCFA producers [18–22].

4.3 Gut Microbiota: Health and Diseases

Humans have evolved to co-exist with microbes and in this process the latter have learnt to perform important functions in the body, eventually succeeding as a symbiont that colonize mostly in the gut mucosal surface [20, 23]. This "gut microbiota" as commonly referred comprises a dense and diverse community of bacteria and some members of fungi, archaea, protozoa, and even viruses. In fact, it is now acknowledged that the relationship between host and its gut microbiota is much deeper and the same was beautifully titled in a recent editorial preface as "we are married to our gut microbiota" [24]. Our body is made of around 10^{13} human cells, and recent enumerations predict that it also harbors at least 10^{14} microbial cells, an obvious outnumber of 10 to 1 [25]. In total, these microbial cells may weigh up to 1–2 kg, which is approximately the weight of human brain, thereby leading to a belief that gut microbiota is perhaps an extrasensory organ in the body [24].

4.3.1 Early Colonization

Until recently, newborns were presumed to be sterile but studies suggest that mother to fetus microbial transmission and even potential in utero colonization is common and is influenced by the nature of birth [26]. While *Lactobacillus* and *Prevotella*, which reside in the vagina, colonize normal born babies, C-section-delivered ones have skin representatives like *Staphylococcus*, *Corynebacterium*, and *Propionibacterium*. Feeding habit also determines the colonization pattern. Compared to infants bottle-fed with formula milk, breastfeeding favors colonization of *Lactobacillus*, *Bifidobacterium*, and *Enterococcus* members that help in digestion of sugars present in breast milk [27]. Subsequent introduction of solid food increases microbial richness until a stabilized core gut microflora is developed. The environment and diet during the first 3–5 years of the child have a long-lasting impact on its adult gut microbiota composition that in turn influences development of the immune and neurologic systems. For example, early colonization by *Lactobacilli* and *Bifidobacterium* reduces subsequent allergy development, while *Bacteroides fragilis* reduces lipopolysaccharide (LPS) responsiveness signifying induction of immune tolerance [28].

4.3.2 Healthy Gut Microbiota

Despite the huge diversity in gut microbial population, certain members are repeatedly detected in high numbers irrespective of their geographical, ethnic, or socioeconomic status. At the phylum level, *Firmicutes*, *Bacteroidetes*, and *Actinobacteria* are the most dominant followed by *Proteobacteria* and *Verrucomicrobia* [25, 29] (Fig. 4.1). *Firmicutes* are Gram-positive bacteria with low G + C content and commonly include the large class of *Clostridia* and lactic acid bacteria. *Actinobacteria* are also Gram-positive but with a high G + C content and include *Bifidobacterium*. *Bacteroidetes* members form the most diverse group comprising of Gram-negative, nonspore-forming, anaerobic or aerobic, rod-shaped bacteria and include three large heterogenous classes namely *Bacteroides*, *Prevotella*, and *Porphyromonas* [20, 23]. *Bacteroidetes* members can utilize diverse polysaccharides and their genomes encode a proportionally higher number of carbohydrate-active enzymes than any other gut phylum [30]. Moreover, they can adapt quickly to changes in ecological niches as evident from the highly plastic nature of their genome with frequent gene duplications, rearrangements, and horizontal gene transfers [31].

Healthy gut is comprised of more than 1000 different species distributed over 70 genera like *Bacteroides*, *Clostridium*, *Faecalibacterium*, *Bifidobacterium*, *Ruminococcus*, *Eubacterium*, *Peptococcus*, and *Lactobacillus* [18, 29, 32]. Among them, *Bacteroides* is the most predominant contributing ~25% of the total gut members and include several SCFA producers [33]. *Bifidobacteria* and *Lactobacillus* are the two major autochthonous healthy gut colonizers from birth [34].

These highly stable healthy gut microorganisms, acts as the first line of defense against foreign invaders by preventing their colonization on the intestinal mucosal layer and activate several immune functions [35, 36]. In fact, the collective genomes of all gut microbes, called "microbiome," also outnumber the host genomic content by at least 100–150 times and encode proteins that facilitate several metabolic functions not imparted by those encoded from the human genome [37]. This includes fermentation of complex dietary polysaccharides, vitamins, hormones, and neurotransmitter production, absorption of ions, and regulation of different host metabolic and immune functions [38, 39]. They also help in the overall brain health and control several of its function [40].

4.3.3 Leaky Gut and Dysbiosis

"Gut dysbiosis" and "leaky gut" are real physiological conditions associated with several chronic diseases. The relation between these two conditions in combating oxidative stress is illustrated in Fig. 4.3. The intestinal lumen is separated from the adjoining tissues by a single layer of cells called the mucosal barrier and helps in nutrient exchange with blood vessels. Sealing of the paracellular space of these cells by tight junctions allows selective permeability and prevents entry of most large molecules and pathogens [41]. Several pathophysiological conditions like psychological or oxidative stress can result in increased permeability of this epithelial barrier or make it "leaky," thereby facilitating translocation of intact pathogens, their endotoxins, and other antigens like LPS to the bloodstream, and provide them access to the enteric nerves and immune cells [42]. This eventually results in an imbalance of gut microbial equilibrium or "dysbiosis," particularly reflected by reduced microbial richness and decreased abundance of beneficial anaerobes along with increased aerotolerant or facultative anaerobic opportunistic pathogens [38, 43] (Fig. 4.1). The "leaky gut syndrome" leads to a number of diseased conditions like GI inflammations, autoimmunity, metabolic syndromes, and even neurodegenerative and infectious diseases [43].

4.3.4 F/B Ratio: An Important Determinant of Gut Health

Members of *Firmicutes* and *Bacteroidetes* are the two major bacterial domains commonly identified in gut environments and account for at least 50–70% of the total microbiota. While the colonic microbiota of the former domain mostly includes the *Clostridium* cluster XIVa (commonly termed as *Clostridium coccoides-Eubacterium rectale* group) and *Clostridium* cluster IV (*C. leptum* including *Faecalibacterium prausnitzii*), the latter is represented by the *Bacteroides-Prevotella-Porphyromonas* group. Diseased conditions like obesity, diabetes mellitus, Crohn's disease, autism, and depression have all been linked to the differential levels of these groups (Table 4.1). While the diversity of *Firmicutes* normally reduce in pathological states like IBD and Crohn's diseases, patients suffering from obesity and type 2 diabetes (T2D) have lower *Firmicutes* and/or higher *Bacteroidetes* members, which lead to an elevated F/B ratio (Table 4.1). Not only in humans, a higher F/B ratio can also be correlated to better feed conversion ratio, growth performance, and increased body weight in farm animals like poultry birds [44]. The ratio was low in infants and aged people as compared to the normal middle-aged healthy subjects [45]. In centenarians (>100 years), the ratio reduced even further [46]. This has resulted into a common saying by dieticians worldwide that "to be firm and cute, reduce *Firmicutes*, and increase number of *Bacteroidetes* in your gut."

4.3.5 Microbial SCFAs in Combating Oxidative Stress

SCFAs are straight-chain saturated fatty acids composed of two to six carbon atoms and named accordingly as acetic, propionic, butyric, valeric, and carpoic acids. Healthy gut microbiota produce up to 50–100 $mmol \cdot L^{-1}$ per day of these volatile organic acids with the first three being the most common and present at a molar ratio of approximately 60:20:20 [15, 47, 48]. They were detected as the major anions in the human colon, as early as in 1878, and subsequently identified in gut of both herbivore and omnivore animals like horse, kangaroo, rabbit, dog, and pig [49]. SCFAs have been directly used as a therapeutic agent for diseases like metabolic syndrome, bowel disorders, autism [50, 51], colitis [48, 52], and cancer [53–55].

SCFAs have a significant role on colonic health and its energy metabolism [48]. Butyrate, the most studied SCFA is transported by gut epithelial cells as a preferred energy source and influences several functions of the colonic mucosal layer such as enhanced visceral perception and appetite, inhibition of inflammation and carcinogenesis, reduced oxidative stress, and revival of intestinal barrier function [56]. In addition, SCFAs in colonic lumen lower the pH thereby determining microbial colonization pattern [15]. Several receptors for SCFAs have been recently discovered on the surface of immune cells, intestinal epithelial cells, and even on adipocytes, thereby making them the strongest contender for signaling biomolecules between host and gut microbiota [57] (Fig. 4.3). In fact they are associated with two major signaling mechanisms, namely, inhibition of histone deacetylases (HDACs) and activation of transmembrane G-protein-coupled receptors (GPR) like GPR41, GPR43, and GPR109A [58, 59]. Reduced production of SCFAs in the gut or a lower representation of the producing microorganism and/or their corresponding genes in metagenome sequences has been positively correlated to several diseased states like IBD, obesity, type 1 diabetes (T1D) and T2D [60], autism [61], major depression

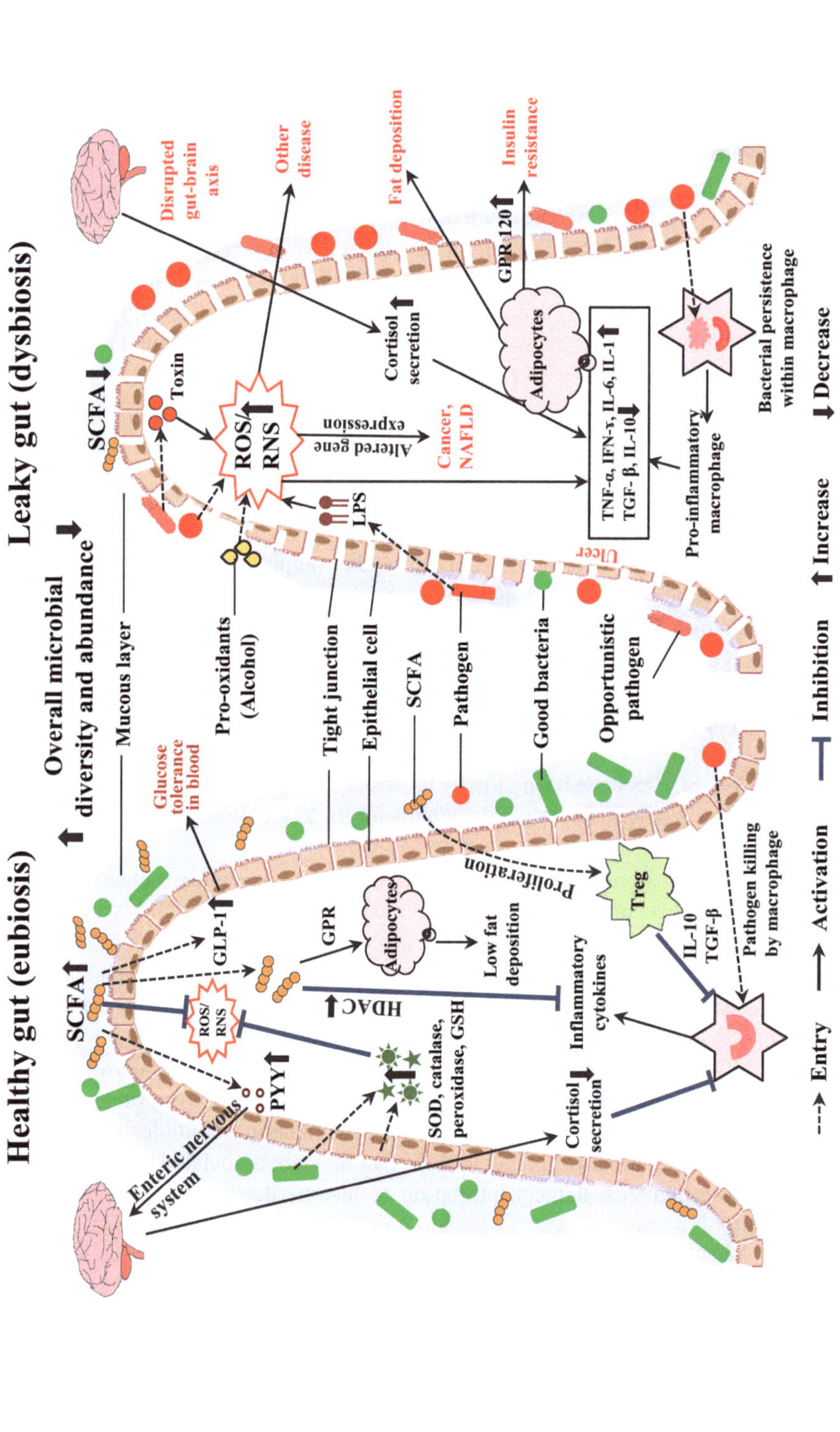
Healthy gut (eubiosis)
Leaky gut (dysbiosis)
Overall microbial diversity and abundance
Mucous layer
Glucose tolerance in blood
Pro-oxidants (Alcohol)
Tight junction
Epithelial cell
SCFA
Pathogen
Good bacteria
Opportunistic pathogen
SCFA
GLP-1
PYY
ROS/RNS
HDAC
GPR
Adipocytes
Low fat deposition
Inflammatory cytokines
SOD, catalase, peroxidase, GSH
Cortisol secretion
Enteric nervous system
Proliferation
Treg
IL-10 TGF-β
Pathogen killing by macrophage
Disrupted gut-brain axis
Other disease
Fat deposition
Insulin resistance
Toxin
LPS
Altered gene expression
Cancer, NAFLD
Cortisol secretion
Adipocytes
GPR120
TNF-α, IFN-ɤ, IL-6, IL-1
TGF- β, IL-10
Ulcer
Pro-inflammatory macrophage
Bacterial persistence within macrophage
Entry
Activation
Inhibition
Increase
Decrease

Fig. 4.3 Illustrative model representing function of the intestinal barrier, gut microbiota, and SCFAs in combating oxidative stress. In healthy intestine, epithelial cells are covered with thick mucous layer that acts as a physical and chemical barrier to invading microbes and help to maintain the mucosal surface by keeping it well hydrated and lubricated. In diseased state, pro-oxidants results in depletion of this layer and loosening of the tight junctions between epithelial cells that maintains gut integrity. This compromised intestinal permeability also called "leaky gut" allows passage of pro-oxidants, microbial toxins, lipopolysaccharide (LPS), or even intact microbes into the systemic blood. This leads to increased production of ROS and RNS and triggers onset of ulcers, cancer, non-alcoholic fatty liver disease (NAFLD), and others. In normal condition, ROS/RNS are effectively counterbalanced by the antioxidant enzymes or molecules produced by the epithelial cells and the residing bacteria. The overall microbial diversity and abundance is also higher as compared to dysbiosis. Among others, SCFA producers are present in high numbers in healthy condition and the produced volatile fatty acids act as signaling molecule to fight toxic effect of ROS/RNS in several diseases by triggering the function of intestinal macrophages that are non-inflammatory and engulfs foreign bacteria escaped into the lumen. SCFAs inhibit the conversion of macrophages to pro-inflammatory by stimulating proliferation of regulatory T-cells (Treg) to secrete interleukin 10 (IL-10) and transforming growth factor-beta (TGF-β) superfamily of cytokines. By inhibiting histone deacetylase (HDAC) function, they block production of other pro-inflammatory cytokines like tumor necrosis factor alpha (TNF-α) and interferon gamma (IFN-γ) by the macrophages. SCFA binds to G-protein receptors (like GPR41) and alters production of the hormone Glucagon-like peptide 1 (GLP-1) that improves glucose tolerance and triggers mucosal L-cells to overexpress the peptide tyrosine tyrosine (PYY). This short peptide also regulates the enteric nervous system. However, in dysbiosis, the pathogens use cellular receptors like GPR41 and GPR43 and toll-like receptors (TLRs) to get access into the cell. Their persistence within macrophages accompanied by reduced production of TGF-β and IL-10 by Treg cells induce them to convert into pro-inflammatory macrophages, which produce inflammatory cytokines (like TNF-α, IFN- γ, IL-6, IL-1) in adipocytes that can reduce GPR120 production leading to insulin resistance and increased fat deposition. Oxidative stress can also affect gut-brain axis by induced secretion of the stress hormone, cortisol that disrupts healthy microflora by altering acid secretion and gut motility. (Adapted from Refs. [190–196])

Table 4.1 Representative case studies of different diseases that are known to induce oxidative stress and are influenced by shifts in gut microbiota

Sl No	Disease	Study design (no. of individuals/age/sex/area/economic status, etc.)[a]; Sample source	Shift in major microbial groups/individuals in diseased condition as compared to healthy control				Method used for microbial population analysis[c]	Important conclusions	References
			Decrease	Increase	F/B ratio	SCFAs:gene/producers[c]			
Metabolic diseases									
1	Obesity	Human: 31 monozygotic, and 23 dizygotic female twin pairs (21–32 yr) and their mothers where available (n = 46), USA; feces	Overall diversity, *Bacteroidetes*	*Actinobacteria*	Increased	–	16S (full length), sanger and 16S (V2 and V6), 454	A core gut microbiome exists at the level of shared genes but not bacteria	[100]
2	Obesity	Human: 12 obese (21–65 yr) fed with either a fat- or carbohydrate-restricted low-calorie diet, USA; feces	*Bacteroidetes*	*Firmicutes*	Increased	–	16S (full length), clone libraries, sanger	Carbohydrate- or fat-restricted low-calorie diet restores normal gut microbiota	[101]
3	Obesity	Human: 9 obese and 12 lean; fed with varied calorie diet, USA; feces	*Bacteroidetes*	*Firmicutes* (*Clostridia*)	Increased	–	16S (V2), 454	Gut microbiota help in nutrient harvest and is influenced by its load	[197]
4	Obesity[b]	Human: 36 obese adolescents (13–15 yr) with calorie-restricted diet and increased physical activity, Spain; feces	Total bacteria, *Bacteroides fragilis* and *Lactobacillus* group	*Clostridium coccoides*, *Bifidobacterium longum*, *E. coli*	Increased	–	qPCR for total bacteria and different groups	Low-calorie diet and physical activity reduces body weight by regulating gut microbiota	[75]

5	Obesity[b]	Germ-free mice ingested with healthy human feces and fed with high-fat diet to induce obesity; 6 mice treated with oolong tea polyphenols and 6 controls; feces	*Bacteroidetes* (*Prevotellaceae*)	*Firmicutes*	Increased	Increased butyrate and acetate producers	16S (V3–V4), Illumina HiSeq	Tea polyphenols are good prebiotics supplements to prevent obesity	[93]
6	T1D	Caucasian children: 16 T1D patients and 16 controls (~7 yrs), USA; feces	Total bacteria, *Firmicutes*; *Lactobacillus*, *Prevotella*, *Bifidobacterium*, *Blautia coccoides-E. rectale* group	*Bacteroidetes*; *Bacteroides*, *Clostridium*, *Veillonella*	Decreased	Increased butyrate-producing bacteria	DGGE (16S, V2–V3) and qPCR	First study linking T1D to gut microbiota composition	[73]
7	T1D	Human: 4 T1D children with autoimmunity and 4 controls; feces	*Firmicutes* (human firmicute CO19)	*Bacteroidetes* (*Bacteroides ovatus*)	Decreased	Low SCFA producers	16S (V3), 454	There is an autoimmune microbiota for T1D and can help early diagnosis	[198]
8	T2D	Human: 345 for metagenome and 23 for T2D classification studies, China; feces	*Faecalibacterium prausnitzii*, *Roseburia*, *Eubacterium rectale*	*E. coli*, *Clostridium*, *Akkermansia*; *Desulfovibrio*	–	Less SCFA synthesis genes and producers	Complete metagenome; Illumina HiSeq	Gut microbial markers can help in classifying T2D	[95]

(continued)

Table 4.1 (continued)

Sl No	Disease	Study design (no. of individuals/age/sex/area/economic status, etc.)[a]; Sample source	Shift in major microbial groups/individuals in diseased condition as compared to healthy control				Method used for microbial population analysis[c]	Important conclusions	References
			Decrease	Increase	F/B ratio	SCFAs:gene/producers[c]			
9	T2D	Human: 18 patients and 18 controls (male adult), Denmark; feces	*Firmicutes*; *Clostridia*	*Betaproteobacteria*, *Bacteroidetes Bacilli*	Decreased	–	16S (V4), 454; qPCR	T2D is associated with changes in intestinal microbiota composition	[77]
10	T2D	Human: 16 patients and 12 controls (48–65 yrs), China; feces	*Bifidobacterium*, *Bacteroides vulgatus*	*Clostridium leptum*	–	–	16S (V3) DGGE; qPCR	Gut microbiota changes are associated with T2D occurrence and development	[199]
11	Aging	Human: 21 centenarians (~100 yrs), 22 elderly (~72 yrs), 20 young adults (~31 yrs), and 21 elderly offspring of centenarians (~67 yrs), Italy; feces	*F. prausnitzii*, total *Bifidobacteria*	*Eubacterium limosum*, *E. coli*, *Klebsiella*, *Haemophilus*, *Proteus*, *Vibrio*	Low in ~31 and 100 yrs, high in elderly	Low copy number of *F. prausnitzii*	Microarray chip; qPCR of bacterial and archaeal 16S rRNA gene	Maintenance of microbial homeostasis is essential for gut health during aging and proper diet can improve life expectancy	[46]
12	Malnutrition in child	Human: One healthy and one malnutrition female (16 months), India; feces	*Actinobacteria*; *Enterobacteriaceae*, *Streptococcaceae*, *Methanosarcinaceae*	*Campylobacteraceae*, *Helicobacteraceae*, *Bacteroidaceae*	Decreased	–	Metagenome sequence, 454	Enteric pathogens are abundant in malnourished child gut that cause intestinal inflammation	[200]

13	Metabolic syndrome (obesity and T2D)	Human: Caucasian obese male (9 each) infused with own feces (autologous) or from lean male donors (allogenic), Netherlands; feces and duodenal biopsies	Microbial diversity, *Clostridium* cluster XIVa, *Roseburia intestinalis*, *Eubacterium hallii*	*Bacteroidetes*	–	Low SCFA in feces and low butyrate producers	Microarray, qPCR for 16S rRNA gene of bacteria and archaea	Increased gut microbial diversity is associated with improved insulin resistance	[201]
14	Metabolic syndrome (obesity and T2D)	Rat: 12 in each group fed with high-fat diet (HFD), HFD and lipoic acid and control diet; colonic content	*Lactobacilli*	*E. coli*, *Enterococcus*	–	–	16S DGGE	HFD alter gut microbiota by changing the redox state that can be improved by antioxidants	[10]
Inflammatory disease									
15	IBD (Crohn's disease or CD)	Human: 447 CD diagnosed and 221 control group of children and adolescents (<17 yrs), North America; feces and biopsy (ileum, rectum)	*F. prausnitzii*, *Blautia*, several members of *Bacteroides*, *Bifidobacterium*, and *Ruminococcus*	*Enterobacteriaceae* (*E. coli*), *Pasteurellaceae*, *Veillonellaceae*, *Fusobacteriaceae*	–	Low *F. prausnitzii*	16S (V4) and metagenome; Illumina MiSeq	Early diagnosis of CD is possible by the rectal mucosal-associated microbiota; antibiotics amplify dysbiosis	[85]
16	IBD (CD)	Human: 49 CD patients and 54 healthy controls, China; feces	*F. prausnitzii*, *Bifidobacterium*	*E. coli*, *Klebsiella*, *pneumoniae*, *Clostridium*, *Streptococcus*	–	Low *but* genes and SCFA producers	Complete metagenome; Illumina HiSeq	CD gut microbiome shift toward enhanced pro-inflammatory capacity	[94]

(continued)

Table 4.1 (continued)

Sl No	Disease	Study design (no. of individuals/age/sex/area/economic status, etc.)[a]; Sample source	Shift in major microbial groups/individuals in diseased condition as compared to healthy control				Method used for microbial population analysis[c]	Important conclusions	References
			Decrease	Increase	F/B ratio	SCFAs:gene/producers[c]			
17	IBD (ulcerative colitis or UC)[b]	Colon cancer cell line treated with probiotic strains; Mice: Probiotic fed to dextran sodium sulfate-induced colitis mice; chow diet control; feces	*Lachnospiraceae*; *Lactobacillus animalis*, *Helicobacter ganmani*	*Moraxellaceae*; *Akkermansia*, *Ruminococcus*, *Erysipelatoclostridium*	–	–	16S (V3–V4); Illumina MiSeq	Treatment with probiotics (*Lactobacillus* and *Bifidobacterium*) decreases oxidative stress and is useful for UC	[34]
18	Colitis	Mice: *Citrobacter rodentium*-induced colitis and control; feces	Total *Bacteroidetes*	–	–	–	FISH and qPCR	Gut microbiota can help to overcome inherent genetic susceptibility	[70]
19	Coeliac disease	Children: 30 active, 18 non-active, and 30 controls, Spain; feces and duodenal biopsy	Total *Bifidobacterium* and *B. longum*	–	–	–	qPCR using *Bifidobacterium*-specific primers	Level of *Bifidobacterium* and *B. longum* in feces can be used as CD index	[78]
20	IBS	Human: 37 patients (~37 yrs) and 20 controls (~39 yrs), Sweden; feces	*Bacteroidetes*; *Alistipes*	*Firmicutes*, *Proteobacteria*, *Cyanobacteria*	Increased	Low SCFA producers	16S (V4); 454	First study to define IBS patient subgroups based on gut microbial profiles	[135]

Neurodegenerative diseases									
21	Alzheimer's disease	Human: 25 patients and 25 controls, USA; feces	Overall diversity, *Firmicutes*, (*Clostridium*, *Dialister*, *Turicibacter*, *Blautia*), *Bifidobacterium*	*Bacteroides*, *Akkermansia*	Decreased	–	16S (V4); Illumina MiSeq	Gut microbiota are potential target for therapeutic intervention of Alzheimer's disease	[141]
22	Alzheimer's disease[b]	AD-induced rats; 8 each fed with D-galactose, D-gal and OMO, D-gal and high OMO, and water (control); feces [OMO: Fructooligosaccharides from *Morinda officinalis*]	*Bacteroides*, *Prevotella*, *Parabacteroides*, *Coprococcus*, *Desulfovibrio*, *Bifidobacterium*, *Lactobacillus*	*Corynebacterium*, *Staphylococcus*, *Ruminococcus*, *Roseburia*, *Dorea*, *Sutterella*	–	Lack of SCFA-producing genes in metagenome	16S; Illumina MiSeq	Prebiotics like OMO can enhance antioxidative activities and regulate the synthesis and secretion of neurotransmitters	[87]
23	Parkinson's disease	Human: 38 patients and 34 healthy controls, USA; feces and mucosa	*Faecalibacterium*, *Blautia*, *Coprococcus*, and *Roseburia*	*Proteobacteria*; *Ralstonia*	No difference	Low SCFA producers	16S (V4); amplicon sequencing	Pro-inflammatory dysbiosis is present in PD patients	[202]
24	Stress-induced memory dysfunction[b]	Germ-free mice infected with *Citrobacter rodentium*, stressed with acute water avoidance and fed with probiotic; feces	*Bacteroides*	*Firmicutes*, *Enterobacteriaceae*, *Eubacteria rectale*	–	–	qPCR for eubacteria and bacterial groups	Intestinal microbiota can influence memory formation and probiotics control its dysfunction	[91]

(continued)

Table 4.1 (continued)

Sl No	Disease	Study design (no. of individuals/age/sex/area/ economic status, etc.)[a]; Sample source	Shift in major microbial groups/individuals in diseased condition as compared to healthy control				Method used for microbial population analysis[c]	Important conclusions	References
			Decrease	Increase	F/B ratio	SCFAs:gene/ producers[c]			
Other diseases									
25	Asthma	Children (1 yr): 87 atopic, 136 wheezing, 22 atopic and wheezing, and 74 controls, Canada; feces	*Faecalibacterium*, *Lachnospira*, *Veillonella*, *Rothia*	–	–	Low acetate in feces	16S (V3); Illumina MiSeq, qPCR	Microbial dysbiosis in first 100 days of life is linked to the risk of asthma and allergy	[182]
26	Myocardial infarction[b]	Mice exercised on treadmill for 4 weeks or caged (control) before coronary artery ligation; feces	*Butyricimonas*, *Akkermansia*, *Prevotella*, *Lactobacillus*	*Corynebacterium*, *Staphylococcus*, *Enterobacteriaceae*	–	–	16S (V4); Illumina HiSeq	Gut microbial alteration by exercise is effective to cure cardiac diseases	[91]
27	Symptomatic atherosclerosis	Human: 12 patients and 13 controls, Sweden; feces	*Methanobrevibacter smithii*; *Eubacterium*, *Roseburia*, *Bacteroides*	*Collinsella*	–	Low *but* gene	Complete metagenome; Illumina HiSeq	Inflammation in arterial wall is directly associated with gut microbiota	[96]
28	Colorectal cancer (CRC)	Human: 46 patients and 56 controls, China; feces	*Bacteroides*, *Roseburia*, *Alistipes*, *Eubacterium*, *Parasutterella*	*Enterococcus*, *Escherichia*, *Shigella*, *Klebsiella*, *Streptococcus*, *Porphyromonas*	–	Low copy no. of *but* genes and butyrate producers	16S (V3) 454; qPCR against 16S and *but* genes	Increased opportunistic pathogens and reduced butyrate producers is a significant feature of CRC	[203]

29	AIDS	Human: 31 patients (18 with antiviral treatment) and 27 controls, France; feces	Microbial diversity; *F. prausnitzii*, *Blautia*, *Ruminococcus*, *Bifidobacterium*	*Gammaproteobacteria* (*Citrobacter*, *Escherichia*, *Shigella*)	–	Low SCFA producers	16S (V3–V4); Illumina MiSeq	Oxidative stress-induced imbalance in aerobic and anaerobic bacteria in the gut is common in AIDS	[145]

[a]Control refers to healthy individuals. Individuals used in most of the case studies have not taken antibiotics for at least few months prior to the study
[b]Studies where results are interpreted as changes in microbial composition after treatment. These data were changed accordingly to fit the title of the table
[c]Abbreviations: *but* genes, butyrate (or other SCFA) biosynthesis gene; 16S, 16S rRNA gene with the hypervariable region mentioned in parentheses; 454, 454 pyrosequencing; DGGE, density gradient gel electrophoresis; FISH, fluorescence in situ hybridization

[62], colon cancer [53], Alzheimer's disease [63], kidney diseases, *Clostridium difficile* infections [64], and even altered circadian rhythm [65] (Table 4.1).

SCFAs are produced by anaerobic fermentation of indigestible fibers mostly by bacteria belonging to the phylum *Bacteroidetes* and *Firmicutes* with acetate and propionate being produced by the former, and butyrate by the latter members [66]. Some key SCFA-producing genera include the *Clostridium* cluster IV member, *F. prausnitzii*, and *Clostridium* cluster XIVa representatives like *E. rectale*, *E. ramulus*, and *Roseburia cecicola* [67]. *F. prausnitzii* has been consistently detected in amplicon libraries and also isolated as pure cultures particularly in healthy individuals but depleted in diseased conditions (Table 4.1) and thus is the most targeted biomarker to diagnose dysbiosis [68].

4.4 Understanding the Secret World of Microbes: Technological Advancements

The GI tract is a complex environment facilitating growth of microorganisms with several unique physiological and biochemical competences making it difficult to formulate a mimicking culture media to support growth of all the microbial components. In spite of that, several attempts have been made to grow them in pure cultures using different selective media [44, 69]. Limited success with culture-based studies has paved the way for several molecular techniques to understand gut microbial population dynamics and behavior in different model systems.

4.4.1 Molecular Techniques

To understand the overall community composition in any environment, several culture-independent molecular techniques are used which includes fluorescence in situ hybridization (FISH), density gradient gel electrophoresis (DGGE), or quantitative PCR (qPCR). Success of all these techniques relies on efficient identification of specific regions of the 16S rRNA gene or any other functional gene to design probes for FISH, or primers for DGGE or qPCR. FISH probes have been used to monitor total eubacteria or specific groups in gut samples of patients suffering from ulcerative colitis [70] and obesity [71, 72]. Genus specific primers have been effectively used in DGGE for monitoring T1D [73]. SYBR Green- or TaqMan probe-based real-time qPCR assays have been used frequently either as an independent tool or to validate amplicon sequence results in patients with obesity [71, 74, 75], T1D [76], T2D [77], coeliac disease [78], metabolic endotoxemia [79], and aging [45, 46]. Primer pairs designed against important gut microbial groups like total eubacteria, *Bacteroides-Prevotella-Porphyromonas*, *Clostridium* cluster XIVa and *Lactobacilli*, or specific genus/species like *Bifidobacterium*, *Lactobacillus acidophilus*, *Eubacterium rectale*, *F. prausnitzii*, and even the archaeal genera *Methanobrevibacter* have been efficiently used in qPCR to enumerate these members in fecal or gut samples (Table 4.1).

4.4.2 Sequence-Based Metagenomics

The advent of high-throughput next-generation sequencing (NSG) technologies during the last decade has revolutionized microbial population studies by providing an alternative approach to analyze the community composition and their gene content in any environment by metagenomics. Major technological advancement of these platforms like 454, Illumina, and Ion Torrent over the till date gold standard Sanger sequencing is that they facilitate massively parallel sequencing runs in a single tube, a concept referred as "polony sequencing" [80]. This is possible because of techniques like emulsion PCR or bridge amplification reactions carried out in nano- or pico-reaction centers either in tubes or on surfaces of flow cells as used, respectively, in Ion Torrent and Illumina platforms [81]. The basics of different sequencing platforms, like the chemistry or technology used, read length and major advantages are summarized in Table 4.2.

Presently, construction of 16S rRNA gene amplicon libraries and their high depth sequencing on NGS platforms is a faster, reliable, cost-effective, and thus the most preferred choice for microbial population studies. Several primer pairs designed against different hypervariable regions of 16S rRNA gene have been used by scientists worldwide to generate libraries with V3-V4 region being the most preferred (Table 4.1). However, different regions must be tried as few recent reports have shown that the amplicon libraries generated from the V1-V2 [44] or V1-V3 region [82] yield much higher resolution at least for gut samples. In addition, false amplification of eukaryotic 18S rDNA was detected using primers against the V3, V3-V5, and V6-V8 regions [83].

4.4.3 Choice of Samples and Model System

The most important factor that influences the outcome in any population-based study is sample source. For studying gut microbiota, fecal samples are the preferred choice due to its ease in collection and handling (Table 4.1). Several studies have claimed fecal microflora to be similar to that of the rumen and almost a true representation of the entire colon [49]. Also, reports suggest that fecal samples stored for up to 3 days at room temperature do not alter (only 10% deviation) the microbial community structure [84]. However, some studies claim otherwise, like in patients with onset of Crohn's disease, mucosal dysbiosis was feebly deciphered in stool, but the biopsy samples could accurately trace the shift in gut microbiota [85].

As model system, although humans remain the best choice, it cannot be used ethically in all situations. Instead, mouse, due to its phylogenetic relatedness to human and similar GI tract anatomy and microbial compositions [86], is the preferred organism to understand disease onset, progression and its response to different treatment regimens [10, 65, 70, 79, 87–91] (Table 4.1). Various strains of mice have been constructed, out of which the germ-free mice (e.g., Swiss-Webster) has really paved the way for gut population studies [91, 92]. Scientists can introduce the gut content of different group of human subjects in mice and study the fluctuation

Table 4.2 Summary of different sequencing techniques

Sequencing platform	Sequencing chemistry; detection system	Library amplification	Read length[a] (yield per run)	Major advantages	Shortcomings
Sanger, Applied Biosystems[b]	Dideoxy chain termination reaction during synthesis; fluorescence	Normal PCR with single primer	600–1000 bp (96 kb)	Gold standard of sequencing	Low throughput, expensive per run
454 Pyrosequencing, Roche	Pyrophosphate release during synthesis; series of enzymatic reactions eventually emitting light	Emulsion PCR, barcode- and adapter-linked primers	Up to 700 bp (700 Mbp)	Fast, longer reads, high accuracy (99.90%)	Runs are expensive, homopolymer errors, discontinued
ABI-SOLiD, Life Technologies[b]	Sequencing by ligation using DNA ligase; fluorescence detection of di-base probes	Emulsion PCR in agarose bead-bound DNA fragments	35–75 bp (2–4 Gb)	Low cost per base, each base identified twice, high accuracy (99.90%)	Longer time (1–2 weeks), short read length
Ion torrent, Thermo fisher scientific	Detects release of protons during synthesis; semiconductor-based detection of slight change in pH	Emulsion PCR, barcode- and adapter-linked primers	200–600 bp (chips) (1–8 Gb)	Less expensive equipment, fast; preferred for bacterial genome and amplicon sequencing	Homopolymer errors, accuracy (98%)
HiSeq/MiSeq/NovaSeq, Illumina	A proprietary DNA polymerase used that detects single nucleotide incorporation; fluorescence, reversible terminators	Bridge amplification on solid flow cells, barcode- and adapter-linked primers	2 x 300 bp (500–3000 Gb)	Best option for high-throughput outputs as required for human genome/exome sequencing	Expensive instrument, accuracy (98%)

Sequel system, PacBio	Single-molecule real-time synthesis; fluorescence, terminally phospholinked	No amplification required; single molecule can be directly sequenced	20 kb (10 Gb) for genome 40 kb (20 Gb) for amplicons	Long read length facilitates de novo assembly, fast, high accuracy (99.999%); low degree of bias, epigenetic studies	Expensive instrument
Helicos, Heliscope	Single-molecule synthesis; fluorescence; virtual terminator	No amplification required	30 bp (15 Gb)	Two flow cells can be run in parallel	Low read length
MiniION/GridION/ PromethION, Oxford Nanopore technologies	Changes in electric current density across nanopore surfaces	No amplification required; direct single-molecule sequencing	Up to 2 Mb (40 Gb to 15 Tb)	Portable and affordable instrument; true real-time, largest read length	Newly launched, not well tested

[a]All sequencing platforms generate single-end reads, while Illumina supports both single- and paired-end reads
[b]Presently, "Thermo Fisher Scientific"

with corresponding disease manifestation and observe the response to different treatments [70, 93].

4.4.4 Modern High-Throughput Techniques

Complete metagenomics or sequencing the entire gene content of gut microbiota is now used frequently to understand their overall metabolic capabilities and how they help in combating oxidative stress, as done for Crohn's disease (CD), IBD [85, 94], T2D [95], and symptomatic atherosclerosis [96]. However, use of direct metatranscriptomics, metaproteomics, or even metabolomics (particularly estimation of SCFAs) on a regular basis would be rewarding.

Combinatorial approaches including culture-dependent and culture-independent molecular techniques yield better results. A recent high-throughput culture-enriched molecular profiling study that combines microbial cultivation techniques with 16S rRNA gene sequencing suggests that contrary to our common understanding a large number of human gut microorganisms are in fact cultivable [97]. Even the recently introduced "culturomics" approach on human gut samples has successfully cultured and simultaneously identified a total of 1057 prokaryotic species, adding 531 new members to the existing human gut repertoire, of which 187 bacterial and 1 archaeal species had never been previously isolated from humans [98]. This new technical advancement combines high-throughput multiple culture conditions, matrix-assisted laser desorption/ionization-time of flight (MALDI-TOF), and 16S rRNA gene sequencing for identification [98]. Meticulous use of these modern techniques in coming days will surely pace the development in this arena.

4.5 Diseases Inducing Oxidative Stress: Role of Microbiota

Representative case studies of different classes of diseases that are known to induce oxidative stress and are influenced by gut microbial community structure and function are summarized in Table 4.1 and discussed in brief below.

4.5.1 Metabolic Diseases

Incidences of metabolic or lifestyle-induced diseases like obesity, diabetes, hypertension, and aging are increasing due to alterations in modern lifestyle patterns.

4.5.1.1 Obesity

Obesity was recognized as a health problem as early as the fifth century BC by the famous physician Hippocrates. The World Health Organization in its latest report (Feb 2018) has claimed 39% adults above 18 years to be obese worldwide with the prevalence tripling between 1975 and 2016. Obesity occurs due to excessive adipose tissue deposition, which leads to release of pro-inflammatory cytokines like tumor necrosis factor alpha (TNF-α), interleukin-1β (IL-1β), and IL-6 resulting in oxidative stress and production of ROS [99].

Obesity is controlled by shifts in gut microbial composition whereby microorganisms vary in their efficiency to harvest energy from the same diet resulting in energy imbalance and overweight. Studies show that germ-free mice have 40% less fat deposition in their body when compared to normal mice, keeping both diets constant [100]. An array of studies have demonstrated a positive correlation between weight gain and SCFA-producing bacteria, F/B ratio, and oxidative stress [100–102].

4.5.1.2 Diabetes

Diabetes is a group of metabolic disorder illustrated by high levels of blood sugar or hyperglycemia. It is presently the most prevalent diseased condition with its first mentions dating back to 1552 BC in the ancient Egyptian "Ebers Papyrus" [103]. Paul Langerhans in 1869 was the first to correlate the role of pancreas with diabetes [104]. The disease is now known to be caused by a myriad of factors like defects in insulin production and/or its action and reduced ability to metabolize carbohydrates, lipids, and proteins, resulting in malfunction of kidneys, eyes, nerves, heart, and blood vessels [105]. While the rare (5–10% individuals) and hereditary T1D results from autoimmune destruction of the pancreatic β-cells that secrete insulin, T2D is more widespread and is due to nonresponsiveness to insulin action and also lack of its production in later stages [106]. Several studies have revealed that progression of diabetes and its complications is due to oxidative stress imparted by changes in energy metabolism patterns resulting in elevated levels of glycoxidation and lipoxidation products in plasma and tissues [107, 108]. This is accompanied by decreased antioxidants like reduced glutathione, vitamin E, and activities of SOD and catalase [109, 110].

Progression of T1D and T2D has now been associated with oxidative stress and alterations in the GI tract microbiota and their metabolic functions [60, 95, 111–113]. Lower values of SCFAs in the diabetic gut due to decline in their producers like *E. rectale*, *F. prausnitzii*, and *Roseburia intestinalis* can reduce the production of glucagon-like peptide-1 (GLP-1) leading to impaired insulin sensitivity [114]. In addition, diabetic GI tracts have enriched Gram-negative opportunistic pathogens belonging to *Bacteroidetes* and *Proteobacteria* that leads to inflammatory response due to their endotoxins or LPS layers [115].

4.5.1.3 Aging

Harman in 1956 proposed that oxidative stress-induced free radicals are responsible for age-related damage and are still considered the most important cause of this metabolic downfall [116, 117]. ROS generated by mitochondria or from any other cellular or external sites damages the organelle and leads to the aging process [118]. A gut microbiota-dependent metabolite, trimethylamine N-oxide (TMAO), generated from dietary choline, betaine, and carnitine, by several families of bacteria like *Prevotellaceae*, *Deferribacteraceae*, *Anaeroplasmataceae*, and *Enterobacteriaceae*, has been associated with several geriatric diseases including endothelial cell senescence, vascular inflammation, atherosclerosis, multiple sclerosis, and other cardiovascular diseases [51, 119, 120].

The microbial composition in the intestine undergoes transformation from birth to adulthood that gets altered again with aging [45]. Aged person has higher abundance of *Proteobacteria* and *Bacteroidetes* and an overall decrease in total bacterial diversity, particularly that of *Firmicutes* members like *Bifidobacterium*, *Lactobacillus*, and *Clostridium* cluster IV [119, 121]. The deterioration of the gut microbial composition is even severe in centenarians (>100 years) with an enriched population of facultative anaerobes, mostly belonging to *Bacilli* and *Proteobacteria* like *E. coli*, *Klebsiella pneumoniae*, *Haemophilus*, *Proteus*, *Pseudomonas*, *Serratia*, *Vibrio*, and *Yersinia*. F/B ratio was also lower in them (3.6) as compared to the elderly people (5.1) along with a sharp decrease in number of *F. prausnitzii* [46].

Skeletal muscle that helps in locomotion and correct posture comprises 40% of total body mass, and any degenerative loss of its strength and function mostly in aged individuals leads to sarcopenia. Gut microbiota, particularly those capable of producing compounds like folate, vitamin B12, tryptophan, glycine betaine, and SCFA, have roles in determining muscle structure and function and improving its anabolism [122]. Decrease in SCFA-producing genera like *Faecalibacterium* and *Clostridium* has been associated with age-related loss of muscle mass and subsequent development of sarcopenia. Even the gut microbiota in patients of atherosclerotic cardiovascular disease had an increased abundance of *Enterobacteriaceae* members and a decrease in butyrate-producing bacteria like *Roseburia intestinalis* and *F. prausnitzii* [123].

Other metabolic syndromes that have been linked to oxidative stress and dysbiosis include high blood pressure, dyslipidemia, fibrosis, non-alcoholic fatty liver disease, cirrhosis, and even hepatocellular carcinoma [17, 117, 124–129].

4.5.2 Inflammatory Diseases

Inflammatory bowl (IBD) and irritable bowel syndrome (IBS) are common diseases associated with gut inflammation and bowel dysfunction respectively. In IBD, chronic inflammation and sores/ulcers occurs in the colon and small intestine as in ulcerative colitis, or in the digestive tract lining as in Crohn's disease. IBS on the other hand does not cause inflammation and thus is less severe but is characterized by alterations in bowel habits and abdominal or visceral pain, accompanied by anxiety and depression [130]. Although no definite pathogen has been associated with these diseases that would fulfill Koch's postulates, the involvement of oxidative stress has been highly advocated [5]. ROS, NOS, and pro-inflammatory cytokines are both the cause and effect of IBD [131, 132]. Similarly, lipid peroxidation and disturbance in the oxidant-antioxidant homeostasis are recognized to be associated with pathogenesis of IBS [6, 133].

IBD is also reported to be associated with reduced intestinal microbial diversity, particularly within the phylum *Firmicutes* and a relative increase in *Proteobacteria* [132]. Major depleted genera include members of *Clostridium* cluster XIVa, *F.*

prausnitzii, *Lactobacillus*, *Bacteroides*, and *Bifidobacterium* [134]. In new CD patients, oxidative stress in gut led to shifts from anaerobic to a pathobiont-like auxotrophy with higher proportions of aerotolerant or facultative anaerobic taxa [85]. The archaeal members also shift from methanogen to acetogen to reduce the accumulated hydrogen, and their metagenomes have higher representation of benzoate degradation pathway and glycerophospholipid and LPS metabolism [85]. The diseased state was exclusively associated with dysbiosis and an increase in F/B ratio [135].

4.5.3 Neurodegenerative Diseases

High metabolic rate in the brain contributes to an elevated oxidative stress resulting in neurodegenerative diseases like Parkinson's and Alzheimer's disease, mostly prevalent in aged individuals who develop dementia with distorted memory and loss of other cognitive functions. They are directly controlled by "leaky gut" condition [136] and depressive disorder signal induced by the altered gut microbiota which is transmitted to the brain via vagus nerve of the enteric nervous system, together known as gut microbiota-brain axis [130, 137]. Terminals of these widely distributed nerves in the gut also function as receptors for inflammatory cytokines like IL-1 and prostaglandins. Proper brain function by neurotransmitters like serotonin, epinephrine, dopamine, glutamate, and γ-aminobutyric acid are influenced by specific gut microbiota by either modulating their levels and availability or affecting immunological functions and hormonal signals [92, 138].

A number of factors contribute to the development of Parkinson's disease like low dopamine levels, accumulation of alpha-synuclein, or Lewy bodies in the brain. In addition, increased local inflammation in the gut leads to oxidative stress due to reduction in putative anti-inflammatory butyrate- and hydrogen-producing gut bacteria and an altered F/B ratio. Affected individuals have reduced number of *Lachnospiraceae*, *Prevotellaceae*, *Bacteroidetes*, *F. prausnitzii*, and *Clostridium*, but higher *Enterobacteriaceae* members and *Akkermansia* [139, 140].

Alzheimer's disease that involves irreversible progressive memory loss is caused either by genetic mutations or by amyloid cascade, oxidative stress, inflammation, and alterations in gut microbiota. In case-control studies, amplicon libraries from patients were found to have lower representation of *Firmicutes* particularly from the families, *Ruminococcaceae*, *Turicibacteraceae*, *Peptostreptococcaceae*, and *Clostridiaceae*, and higher abundance of *Bacteroidaceae* and *Rikenellaceae* members from *Bacteroidetes*, indicating tipping off of F/B ratio [141, 142].

4.5.4 Microbial Infectious Diseases

Apart from the pathophysiological, metabolic, and neurodegenerative diseases, several viral and bacterial infectious diseases also induce oxidative stress and have a direct correlation to gut microbiota, viz., infections by human immunodeficiency virus (HIV), hepatitis B virus (HBV), *Clostridium difficile*, or *Helicobacter pylori* [32].

HIV infection results in significant reduction of mature CD4 T-cells in the gut-associated lymphoid tissues that constitutes a crucial part of immunological network in the GI tract and can recognize the indigenous gut microbiota after being instructed by the T-regulatory cells (Treg) [143]. Interestingly, abundance of *Lactobacillales* members correlated negatively to CD4 cell count that gets reversed following anti-retroviral therapy [144]. Similarly, in another study on AIDS patients, gut microbial diversity was found to be considerably reduced with an increase in aerotolerant members that exerts higher oxidative stress, but the same was not restored even after antiviral treatments [145].

C. difficile is an anaerobic, Gram-positive, spore-forming bacillus and is the most common cause of antibiotic-associated diarrhea in adults. This opportunistic pathogen is a normal microflora of the human gut but at very low numbers. Treatments with wide-spectrum antibiotics like fluoroquinolones, cephalosporins, and penicillins disrupt gut microbial homeostasis and reduce overall diversity that favors rapid colonization of *C. difficile* [64]. The bacteria then produce toxins, like TcdA and TcdB, that disrupt the mucosal cell lining and increase oxidative stress leading to symptoms like watery diarrhea to life-threatening colon inflammation [64, 146]. The infection results in a decrease of butyrate producers and an increase in opportunistic pathogens [64].

H. pylori is an autochthonous member of the gastric niche at low numbers and its enhanced colonization induces a range of infections like peptic disease, gastritis, periodontitis, duodenal ulcers, and even gastric cancer with direct proofs of increased oxidative stress in these conditions along with reduced gut microbiota richness and altered composition [147–149].

4.6 Gut Microbiota: Rare Components

Gut microbiota "in toto" comprise not only bacteria but also members of fungi, archaea, and viruses albeit at much lower proportions and are also referred as the "rare biosphere" [86]. Several studies have highlighted the involvement of these minor populations in host health and disease.

4.6.1 Fungi

Existence of commensal fungal members has been known for years, but it was only during the start of this decade that attempts to characterize the mycobiome component consisting of a mere ~0.1% of the total gut population were made [150]. Although a core set of fungal species (similar to core microbiome) could not be demarcated in healthy guts, a few members have been frequently identified, viz., *Candida*, *Saccharomyces*, *Aspergillus*, *Penicillium*, *Rhodotorula*, *Torulopsis*, and *Blastomyces* [150], with the first two being most prevalent. Several members have been almost always reported in some diseases, like *Rhodotorula* in patients with IBD, and *Candida* with chronic hepatitis B [150, 151]. The microbiota and mycobiota are proposed to work in sync for disease manifestation. For example, in Crohn's disease, an elevated level of the fungus *C. tropicalis* is usually associated with higher proportions of *E. coli* and *Serratia marcescens*, and the three together are proposed to form strong biofilms that aggravate intestinal inflammation [152].

4.6.2 Archaea

Archaea, although constituting a separate domain, have been completely ignored as a component of gut microbiota. However, recent studies have identified them as important players, particularly the methanogenic group, that can produce methane by reducing hydrogen released from anaerobic fermentation of carbohydrates, thereby facilitating growth of the latter. These have been identified in feces of healthy individuals belonging to all age groups except infants, and include *Methanobrevibacter smithii*, *Methanosphaera stadtmanae*, and *Methanomassiliicoccus luminyensis*, with the first member present in incredibly high proportions (up to 94% of total Archaea) and also isolated as a pure culture from gut/fecal samples [153]. Absence of archaeal members in infants suggests their colonization from environmental exposures and as a proof *M. smithii* was identified in raw milk products suggesting its possible route of introduction in children [154]. In a recent survey on West African children, *M. smithii* was detected in 40–75% of healthy controls but not in any of the 20 stool samples of those with acute malnutrition, a diseased state with high oxidative stress, thereby making the gut lumen oxic and unfavorable for the growth of these anaerobic methanogens [22].

4.6.3 Virus

The virome community is now considered as decisive components in the gut ecosystem and includes bacteriophages (~90%) and very low proportion of eukaryotic viruses [155]. In fact, lysis of specific host bacteria by these viral communities can dictate the diversity of gut bacterial population [86]. A proper understanding of the dynamics of gut virome and the factors affecting them might help to fine-tune bacterial population toward a healthy state in the future [156]. In a recent study using fecal viral metagenome sequences of 12 individuals, a novel bacteriophage genome (named "crAssphage") was bioinformatically assembled, and its host was predicted to be a member of the genus *Bacteroides* based on a unique region present on the phage genome that can encode a carbohydrate-binding domain [157]. In fact, its unique ~97 kb genome has six-fold higher abundance in metagenome databases as compared to all other known phages taken together and surprisingly comprises almost 1.68% of sequence reads generated from human fecal metagenomic studies

[157]. Thus, explorations of gut microflora must include dynamics of these least studied microbial groups in addition to bacteria for deciphering their function in maintaining healthy state of an individual.

4.7 Retaining or Regaining Healthy Gut Microbiota: Possible Interventions

How best we can manage our gut microbiota will determine the ability to combat oxidative stress in diseased state and maintain a good health. Treatments aimed at restoration of healthy gut microbiota, instead of manipulating the host, are potentially attractive as microbes are relatively accessible and can be controlled with ease. They can be either directly administered through the safest passage, i.e., oral routes as probiotics, or modulated by food supplements like prebiotics, polyphenols, and antibiotics. Thus, diet is the most important factor that can efficiently modulate and help in restoration of gut microbial community structure and function to its healthy state as has been described in studies performed in human and other vertebrate animals, apes, and reptiles [71, 136, 158–163].

4.7.1 Probiotics

Probiotics are live bacteria used as food supplements that can help to restore the gut microbial composition to a healthy state after dysbiosis. Traditionally, man has understood the importance of fermented foods like yogurt, sauerkraut, kefir, tempeh, kimchi, and kombucha and has consumed them without knowing that the microbes therein helps to maintain the healthy state [164]. These classical food products are now known to be enriched with several healthy gut colonizers. Probiotic strains prescribed presently to cure different diseases mostly belong to the genera *Lactobacillus* and *Bifidobacterium* [34, 164–168]. However, they are either aerobic, aerotolerant, or facultative anaerobes. This suggest that we have not yet exploited the most beneficial and perfect probiotic strain(s), as healthy gut microbes are mostly strict anaerobes (Fig. 4.1, Table 4.1). Several of these anaerobic, SCFA producers are now targeted as potential new/next-generation probiotic strains [169]. *F. prausnitzii* is the most promising candidate with some successful attempts being made in this direction [68, 170]. This genus has been universally identified in healthy individuals, which diminishes in patients suffering from gastrointestinal disorders, obesity, diabetes, aging, Parkinson's disease, and even AIDS (Table 4.1). Archaeal strains like *M. smithii* have also been proposed as future probiotic [171], and fungal strains like *Saccharomyces boulardii* are already being effectively used against several gastrointestinal diseases [172].

4.7.2 Prebiotics

Prebiotics are specific food products like fibers, glycans, or nondigestible carbohydrates that our digestive system cannot utilize but have a stimulating effect on the growth of certain microbial groups like SCFA producers that use these complex substances as substrate [173]. They are present in plenty in whole grains, legumes, beans, bananas, asparagus, oats, and fruits like apple, pear, and kiwi. Their health benefits have long been acknowledged and recommended for several gastroenterological disorders without knowing its mode of action. However, recent studies have shown that dietary intakes of prebiotics hugely determine the gut microbial composition both in the short and long term and specifically promote healthy members like *Bifidobacteria* and inhibit colonization of opportunistic pathogens like *E. coli* [19, 87, 163]. In fact, in vitro production of SCFAs by pure cultures has been found to be strongly influenced by the prebiotic used, like starch is strongly butyrogenic, but pectin favors propionate and acetate production [47]. Prebiotic supplements have been successfully used to alleviate diseases like Crohn's disease [174], T2D [60], and Alzheimer's disease [87] and also for circadian clock adjustments [65].

On a usability scale, prebiotics are a better and easier option compared to the next-generation probiotics like *F. prausnitzii* that is very difficult to administer as a normal supplement due to its extreme oxygen sensitivity [170]. To resolve this problem, a New Zealand-based company (Anagenix) has recently launched a natural prebiotic supplement derived from gold kiwifruit that they claim to facilitate growth of *F. prausnitzii* and its restoration in the gut.

4.7.3 Polyphenols

Polyphenols, mostly the plant-derived ones with flavonoid structure, are known to have immense antioxidant activity and are present in plenty in red wine, dark chocolate, green tea, grape seed, and olive oil [175]. A diet rich in polyphenols can regulate the intestinal microbiota and help in reducing body weight [162]. Human intervention studies or those performed on rat models or in vitro fermentation experiments indicate that consumption of polyphenol-rich food(s) helps in either retaining or significantly increasing the number of good anaerobic commensal members, like *Bacteroides*, *Lactobacillus*, *Bifidobacterium*, *Prevotella*, *Blautia coccoides*, and *Eubacterium rectale*, while reducing or inhibiting the pathogenic genera [161, 175–179].

4.7.4 Antibiotics

Compared to the first three food supplements, antibiotics functions the other way round and is now categorized as an exogenous oxidant. Antibiotics were considered to be miracle drugs that can cure almost all diseases and were indiscriminately used not only as a human medicine but also as a growth-promoting agent in different poultry and farm animals [44]. Although they are prescribed to kill the bad bugs in our body, they also target many good bacteria and alter the normal gut microbial composition mostly resulting in weight gain and emergence of antibiotic resistance [180]. For example, antibiotic exposure amplified the microbial dysbiosis and increased the chance of Crohn's disease [85]. In another report, IBS-associated *Bifidobacterium* and *Bacteroides* spp. were found to be sensitive to antibiotic use [135]. Thus, it is recommended to follow a healthy lifestyle with regulated natural food and sleep, with minimum antibiotics to maintain friendly intestinal microbiota and normal rhythm.

4.8 Future Directions

We are presently at a very exciting juncture in our understanding of human health and medicine. A massive surge in the gut microbiome studies particularly due to recent advancements in sequencing techniques and efficient bioinformatic pipelines have enriched our knowledge about human body and its interaction with the microbial world. Capturing microbial diversity at much higher sequencing depth leads to better understanding of these functional interactions, particularly involving the minor taxa or those with relatively lesser shifts. Experimental designs that compare the diseased cases versus controls, using large cohorts of samples covering wide geographical locations, diverse age groups, and ethnicity, are getting popular and have been attempted for the healthy individuals [181], and patients of Crohn's disease [85], T2D [95], obesity [100], aging [46], and pediatric asthma [182] (Table 4.1). Results generated from such large multicenter studies yield databases that can augment the resolution and would be able to pinpoint specific group(s) or genera to a healthy or diseased state. With improved understanding of the gut microflora, future therapeutic strategies would surely be more personalized, accurate, and focused to restore the microbiota as closely as possible to its native state [183].

However, till date, there is no single dependable and reproducible method that specifically aims at enriching the gut microbes of interest. Monitoring disturbed gut microbiome in the normally sterile blood plasma is another major area that can help in quick detection of a diseased state as already done for a variety of inflammatory diseases [184]. One major treatment option that has been used infrequently for over 50 years and has immense hope to restore gut microbiota is fecal microbiota transplant (FMT) from tested healthy donors [185]. FMT for patients with multiple recurrences of *C. difficile* enterocolitis even after antibiotic treatments has been particularly promising and is now acknowledged as a safe and efficient (>90% success rates) alternative to standard drug therapy [186–188]. FMT can efficiently restore the normal colonic flora both in human and in mouse models and has been proposed to be effective in several other disorders like IBD, IBS, obesity, diabetes, neurodegenerative disorders, autoimmune diseases, and allergy [185]. If we can get over the uncomfortable thought of being fed with fecal material then perhaps FMT may prove useful as a single medicine for all diseased states, albeit its efficacy as a treatment option for different disorders need to be validated. Moreover, the microbial diversity present in the administered material and their potential long-term effects also needs to be properly evaluated and the process must be legislated like any other organ donation, scrutinizing the regulations to be imposed on both the donor and recipient before FMT is prescribed on a regular basis. However, the easiest approach to have a healthy gut, free of oxidative stress, is to maintain a regulated and healthy lifestyle and judicially use the three "P"s in our daily diet, i.e., probiotics, prebiotics, and polyphenols [189].

Acknowledgments Financial assistance from DST-SERB [No.SB/YS/LS-16/2013], CSIR [No. 38(1410)/15/ EMR-II)], UGC MRP [No. 43-462/2014(SR)] and WB-DST [No. ST/P/S&T/5G-18/2017] is thankfully acknowledged. Sohini Banerjee is grateful to CSIR [No. 38(1410)/15/EMR-II] for her fellowship.

References

1. Di Meo S, Reed TT, Venditti P, Victor VM (2016) Role of ROS and RNS sources in physiological and pathological conditions. Oxid Med Cell Longev 2016:44
2. Jones RM, Mercante JW, Neish AS (2012) Reactive oxygen production induced by the gut microbiota: pharmacotherapeutic implications. Curr Med Chem 19:1519–1529
3. Bhattacharyya A, Chattopadhyay R, Mitra S, Crowe SE (2014) Oxidative stress: an essential factor in the pathogenesis of gastrointestinal mucosal diseases. Physiol Rev 94:329–354
4. Neish AS (2013) Redox signaling mediated by the gut microbiota. Free Radic Res 47:950–957
5. Balmus IM, Ciobica A, Trifan A, Stanciu C (2016) The implications of oxidative stress and antioxidant therapies in inflammatory bowel disease: clinical aspects and animal models. Saudi J Gastroenterol 22:3–17
6. Choghakhori R, Abbasnezhad A, Hasanvand A, Amani R (2017) Inflammatory cytokines and oxidative stress biomarkers in irritable bowel syndrome: association with digestive symptoms and quality of life. Cytokine 93:34–43
7. Liu Z, Zhou T, Ziegler AC, Dimitrion P, Zuo L (2017) Oxidative stress in neurodegenerative diseases: from molecular mechanisms to clinical applications. Oxid Med Cell Longev 2017:11
8. Uttara B, Singh AV, Zamboni P, Mahajan RT (2009) Oxidative stress and neurodegenerative diseases: a review of upstream and downstream antioxidant therapeutic options. Curr Neuropharmacol 7:65–74
9. Maritim AC, Sanders R, Watkins Iii JB (2003) Diabetes, oxidative stress, and antioxidants: a review. J Biochem Mol Toxicol 17:24–38
10. Qiao Y, Sun J, Ding Y, Le G, Shi Y (2013) Alterations of the gut microbiota in high-fat diet mice is strongly linked to oxidative stress. Appl Microbiol Biotechnol 97:1689–1697
11. Ighodaro OM, Akinloye OA (2018) First line defence antioxidants-superoxide dismutase (SOD), catalase (CAT) and glutathione peroxidase (GPX): their fundamental role in the entire antioxidant defence grid. Alexandria Med J 4:287–293
12. Mironczuk-Chodakowska I, Witkowska AM, Zujko ME (2018) Endogenous non-enzymatic antioxidants in the human body. Adv Med Sci 63:68–78
13. Bouayed J, Bohn T (2010) Exogenous antioxidants--double-edged swords in cellular redox state: health beneficial effects at physiologic doses versus deleterious effects at high doses. Oxid Med Cell Longev 3:228–237
14. Miranda-Bautista J, Banares R, Vaquero J (2017) Chapter 1 – the gastrointestinal system: anatomy and sources of oxidative stress. In: Gracia-Sancho J, Salvado J (eds) Gastrointestinal tissue. Academic, Cambridge, MA, pp 3–20

15. Duncan SH, Louis P, Thomson JM, Flint HJ (2009) The role of pH in determining the species composition of the human colonic microbiota. Environ Microbiol 11:2112–2122
16. Walker AW, Duncan SH, Leitch ECM, Child MW, Flint HJ (2005) pH and peptide supply can radically alter bacterial populations and short chain fatty acid ratios within microbial communities from the human colon. Appl Environ Microbiol 71:3692–3700
17. Dapito DH, Mencin A, Gwak GY, Pradere JP, Jang MK et al (2012) Promotion of hepatocellular carcinoma by the intestinal microbiota and TLR4. Cancer Cell 21:504–516
18. Xu J, Gordon JI (2003) Honor thy symbionts. Proc Natl Acad Sci 100:10452–10459
19. Flint HJ, Scott KP, Louis P, Duncan SH (2012) The role of the gut microbiota in nutrition and health. Nat Rev Gastroenterol Hepatol 9:577–589
20. Backhed F, Ley RE, Sonnenburg JL, Peterson DA, Gordon JI (2005) Host-bacterial mutualism in the human intestine. Science 307:1915–1920
21. Louis P, Flint HJ (2009) Diversity, metabolism and microbial ecology of butyrate-producing bacteria from the human large intestine. FEMS Microbiol Lett 294:1–8
22. Million M, Tidjani Alou M, Khelaifia S, Bachar D, Lagier J-C et al (2016) Increased gut redox and depletion of anaerobic and methanogenic prokaryotes in severe acute malnutrition. Sci Rep 6:26051
23. Johnson EL, Heaver SL, Walters WA, Ley RE (2017) Microbiome and metabolic disease: revisiting the bacterial phylum Bacteroidetes. J Cell Mol Med 95:1–8
24. Ray K (2012) Gut microbiota: married to our gut microbiota. Nat Rev Gastroenterol Hepatol 9:555
25. Turnbaugh PJ, Ley RE, Hamady M, Fraser-Liggett CM, Knight R et al (2007) The human microbiome project. Nature 449:804
26. Kundu P, Blacher E, Elinav E, Pettersson S (2017) Our gut microbiome: the evolving inner self. Cell 171:1481–1493
27. Madan JC, Hoen AG, Lundgren SN, Farzan SF, Cottingham KL et al (2016) Association of cesarean delivery and formula supplementation with the intestinal microbiome of 6-week-old infants. JAMA Pediatr 170:212–219
28. Sjogren YM, Tomicic S, Lundberg A, Bottcher MF, Bjorksten B et al (2009) Influence of early gut microbiota on the maturation of childhood mucosal and systemic immune responses. Clin Exp Allergy 39:1842–1851
29. Khanna S, Tosh PK (2014) A clinician's primer on the role of the microbiome in human health and disease. Mayo Clin Proc 89:107–114
30. Kaoutari AE, Armougom F, Gordon JI, Raoult D, Henrissat B (2013) The abundance and variety of carbohydrate-active enzymes in the human gut microbiota. Nature Rev Microbiol 11:497
31. Thomas F, Hehemann JH, Rebuffet E, Czjzek M, Michel G (2011) Environmental and gut Bacteroidetes: the food connection. Front Microbiol 2:93
32. Wang B, Yao M, Lv L, Ling Z, Li L (2017) The human microbiota in health and disease. Engineering 3:71–82
33. Wexler HM (2007) *Bacteroides*: the good, the bad, and the nitty-gritty. Clin Microbiol Rev 20:593–621
34. Wang Y, Guo Y, Chen H, Wei H, Wan C (2018) Potential of *Lactobacillus plantarum* ZDY2013 and *Bifidobacterium bifidum* WBIN03 in relieving colitis by gut microbiota, immune, and anti-oxidative stress. Can J Microbiol 64:327–337
35. Faith JJ, Guruge JL, Charbonneau M, Subramanian S, Seedorf H et al (2013) The long-term stability of the human gut microbiota. Science 341:1237439
36. McDermott AJ, Huffnagle GB (2014) The microbiome and regulation of mucosal immunity. Immunology 142:24–31
37. Ursell LK, Haiser HJ, Van Treuren W, Garg N, Reddivari L et al (2014) The intestinal metabolome: an intersection between microbiota and host. Gastroenterology 146:1470–1476
38. Levy M, Kolodziejczyk AA, Thaiss CA, Elinav E (2017) Dysbiosis and the immune system. Nat Rev Immunol 17:219
39. Galland L (2014) The gut microbiome and the brain. J Med Food 17:1261–1272

40. Quigley EM (2017) Microbiota-brain-gut axis and neurodegenerative diseases. Curr Neurol Neurosci Rep 17:94
41. Odenwald MA, Turner JR (2013) Intestinal permeability defects: is it time to treat? Clin Gastroenterol Hepatol 11:1075–1083
42. Gareau MG, Silva MA, Perdue MH (2008) Pathophysiological mechanisms of stress-induced intestinal damage. Curr Mol Med 8:274–281
43. Houser MC, Tansey MG (2017) The gut-brain axis: is intestinal inflammation a silent driver of Parkinson's disease pathogenesis? NPJ Parkinsons Dis 3:3
44. Banerjee S, Sar A, Misra A, Pal S, Chakraborty A, Dam B (2018) Increased productivity in poultry birds by sub-lethal dose of antibiotics is arbitrated by selective enrichment of gut microbiota, particularly short chain fatty acid producers. Microbiology 164:142–153
45. Mariat D, Firmesse O, Levenez F, Guimaraes V, Sokol H et al (2009) The Firmicutes/Bacteroidetes ratio of the human microbiota changes with age. BMC Microbiol 9:123
46. Biagi E, Nylund L, Candela M, Ostan R, Bucci L et al (2010) Through ageing, and beyond: gut microbiota and inflammatory status in seniors and centenarians. PLoS One 5:e10667
47. Cummings JH, Pomare EW, Branch WJ, Naylor CP, Macfarlane GT (1987) Short chain fatty acids in human large intestine, portal, hepatic and venous blood. Gut 28:1221–1227
48. den Besten G, van Eunen K, Groen AK, Venema K, Reijngoud DJ et al (2013) The role of short chain fatty acids in the interplay between diet, gut microbiota, and host energy metabolism. J Lipid Res 54:2325–2340
49. Cummings JH (1981) Short chain fatty acids in the human colon. Gut 22:763–779
50. MacFabe DF, Cain DP, Rodriguez-Capote K, Franklin AE, Hoffman JE et al (2007) Neurobiological effects of intraventricular propionic acid in rats: possible role of short chain fatty acids on the pathogenesis and characteristics of autism spectrum disorders. Behav Brain Res 176:149–169
51. Burakoff R (1998) Less commonly used therapies for IBD or treatments on the fringe. Inflamm Bowel Dis 4:308–317
52. Tian Y, Xu Q, Sun L, Ye Y, Ji G (2018) Short chain fatty acids administration is protective in colitis-associated colorectal cancer development. J Nutr Biochem 57:103–109
53. Blouin JM, Penot G, Collinet M, Nacfer M, Forest C et al (2011) Butyrate elicits a metabolic switch in human colon cancer cells by targeting the pyruvate dehydrogenase complex. Int J Cancer 128:2591–2601
54. Brady LJ, Gallaher DD, Busta FF (2000) The role of probiotic cultures in the prevention of colon cancer. J Nutr 130:410S–414S
55. Fung KY, Cosgrove L, Lockett T, Head R, Topping DL (2012) A review of the potential mechanisms for the lowering of colorectal oncogenesis by butyrate. Br J Nutr 108:820–831
56. Hamer HM, Jonkers D, Venema K, Vanhoutvin S, Troost FJ et al (2008) Review article: the role of butyrate on colonic function. Aliment Pharmacol Ther 27:104–119
57. Morrison DJ, Preston T (2016) Formation of short chain fatty acids by the gut microbiota and their impact on human metabolism. Gut Microbes 7:189–200
58. Ohira H, Tsutsui W, Fujioka Y (2017) Are short chain fatty acids in gut microbiota defensive players for inflammation and atherosclerosis? J Atheroscler Thromb 24:660–672
59. Tan J, McKenzie C, Potamitis M, Thorburn AN, Mackay CR et al (2014) The role of short chain fatty acids in health and disease. Adv Immunol 121:91–119
60. Zhao L, Zhang F, Ding X, Wu G, Lam YY et al (2018) Gut bacteria selectively promoted by dietary fibers alleviate type 2 diabetes. Science 359:1151–1156
61. MacFabe DF (2012) Short chain fatty acid fermentation products of the gut microbiome: implications in autism spectrum disorders. Microb Ecol Health Dis 23:19260
62. Kelly JR, Borre Y, OBrien C, Patterson E, El Aidy S et al (2016) Transferring the blues: depression-associated gut microbiota induces neurobehavioural changes in the rat. J Psychiatr Res 82:109–118
63. Ho L, Ono K, Tsuji M, Mazzola P, Singh R et al (2017) Protective roles of intestinal microbiota derived short chain fatty acids in Alzheimers' disease-type beta-amyloid neuropathological mechanisms. Expert Rev Neurother 18:83–90

64. Gu S, Chen Y, Zhang X, Lu H, Lv T et al (2016) Identification of key taxa that favor intestinal colonization of *Clostridium difficile* in an adult Chinese population. Microbes Infect 18:30–38
65. Tahara Y, Yamazaki M, Sukigara H, Motohashi H, Sasaki H et al (2018) Gut microbiota-derived short chain fatty acids induce circadian clock entrainment in mouse peripheral tissue. Sci Rep 8:1395
66. Macfarlane S, Macfarlane GT (2003) Regulation of short chain fatty acid production. Proc Nutr Soc 62:67–72
67. Barcenilla A, Pryde SE, Martin JC, Duncan SH, Stewart CS et al (2000) Phylogenetic relationships of butyrate-producing bacteria from the human gut. Appl Environ Microbiol 66:1654–1661
68. Lopez-Siles M, Duncan SH, Garcia-Gil LJ, Martinez-Medina M (2017) *Faecalibacterium prausnitzii*: from microbiology to diagnostics and prognostics. ISME J 11:841
69. Madsen KL, Doyle JS, Jewell LD, Tavernini MM, Fedorak RN (1999) *Lactobacillus* species prevents colitis in interleukin 10 gene-deficient mice. Gastroenterology 116:1107–1114
70. Ghosh S, Dai C, Brown K, Rajendiran E, Makarenko S et al (2011) Colonic microbiota alters host susceptibility to infectious colitis by modulating inflammation, redox status, and ion transporter gene expression. Am J Physiol Gastrointest Liver Physiol 301:G39–G49
71. Duncan SH, Lobley GE, Holtrop G, Ince J, Johnstone AM et al (2008) Human colonic microbiota associated with diet, obesity and weight loss. Int J Obes 32:1720
72. Nadal I, Santacruz A, Marcos A, Warnberg J, Garagorri M et al (2009) Shifts in *Clostridia*, *Bacteroides* and immunoglobulin-coating fecal bacteria associated with weight loss in obese adolescents. Int J Obes 33:758
73. Murri M, Leiva I, Gomez-Zumaquero JM, Tinahones FJ, Cardona F et al (2013) Gut microbiota in children with type 1 diabetes differs from that in healthy children: a case-control study. BMC Med 11:46
74. Schwiertz A, Taras D, Schafer K, Beijer S, Bos NA et al (2009) Microbiota and SCFA in lean and overweight healthy subjects. Obesity 18:190–195
75. Santacruz A, Marcos A, Warnberg J, Marti A, Martin-Matillas M et al (2009) Interplay between weight loss and gut microbiota composition in overweight adolescents. Obesity 17:1906–1915
76. Roesch LF, Lorca GL, Casella G, Giongo A, Naranjo A et al (2009) Culture-independent identification of gut bacteria correlated with the onset of diabetes in a rat model. ISME J 3:536–548
77. Larsen N, Vogensen FK, van den Berg FWJ, Nielsen DS, Andreasen AS et al (2010) Gut microbiota in human adults with type 2 diabetes differs from non-diabetic adults. PLoS One 5:e9085
78. Collado MC, Donat E, Ribes-Koninckx C, Calabuig M, Sanz Y (2008) Imbalances in faecal and duodenal *Bifidobacterium* species composition in active and non-active coeliac disease. BMC Microbiol 8:232
79. Cani PD, Rodrigo B, Knauf C, Waget A, Neyrinck AM et al (2008) Changes in gut microbiota control metabolic endotoxemia-induced inflammation in high-fat diet-induced obesity and diabetes in mice. Diabetes 57:1470–1481
80. Buermans H, Den Dunnen J (2014) Next generation sequencing technology: advances and applications. Biochim Biophys Acta 1842:1932–1941
81. Quail MA, Smith M, Coupland P, Otto TD, Harris SR et al (2012) A tale of three next generation sequencing platforms: comparison of ion torrent, Pacific biosciences and Illumina MiSeq sequencers. BMC Genomics 13:341
82. Huse SM, Ye Y, Zhou Y, Fodor AA (2012) A core human microbiome as viewed through 16S rRNA sequence clusters. PLoS One 7:e34242
83. Huys G, Vanhoutte T, Joossens M, Mahious AS, De Brandt E et al (2008) Coamplification of eukaryotic DNA with 16S rRNA gene-based PCR primers: possible consequences for population fingerprinting of complex microbial communities. Curr Microbiol 56:553–557

84. Roesch LF, Casella G, Simell O, Krischer J, Wasserfall CH et al (2009) Influence of fecal sample storage on bacterial community diversity. Open Microbiol J 3:40–46
85. Gevers D, Kugathasan S, Denson Lee A, Vazquez-Baeza Y, Van Treuren W et al (2014) The treatment-naive microbiome in new-onset Crohn's disease. Cell Host Microbe 15:382–392
86. Hillman ET, Lu H, Yao T, Nakatsu CH (2017) Microbial ecology along the gastrointestinal tract. Microbes Environ 32:300–313
87. Chen D, Yang X, Yang J, Lai G, Yong T et al (2017) Prebiotic effect of fructooligosaccharides from *Morinda officinalis* on Alzheimer's disease in rodent models by targeting the microbiota-gut-brain axis. Front Aging Neurosci 9:403
88. Minter MR, Zhang C, Leone V, Ringus DL, Zhang X et al (2016) Antibiotic-induced perturbations in gut microbial diversity influences neuro-inflammation and amyloidosis in a murine model of Alzheimer's disease. Sci Rep 6:30028
89. Desbonnet L, Clarke G, Traplin A, OSullivan O, Crispie F et al (2015) Gut microbiota depletion from early adolescence in mice: implications for brain and behaviour. Brain Behav Immun 48:165–173
90. Liu Z, Liu HY, Zhou H, Zhan Q, Lai W et al (2017) Moderate-intensity exercise affects gut microbiome composition and influences cardiac function in myocardial infarction mice. Front Microbiol 8:1687
91. Gareau MG, Wine E, Rodrigues DM, Cho JH, Whary MT et al (2014) Bacterial infection causes stress-induced memory dysfunction in mice. Gut 60:307–317
92. Gareau MG (2014) Microbiota-gut-brain axis and cognitive function. Adv Exp Med Biol 817:357–371
93. Cheng M, Zhang X, Zhu J, Cheng L, Cao J et al (2018) A metagenomics approach to the intestinal microbiome structure and function in high fat diet-induced obesity mice fed with oolong tea polyphenols. Food Funct 9:1079–1087
94. He Q, Gao Y, Jie Z, Yu X, Laursen JM et al (2017) Two distinct metacommunities characterize the gut microbiota in Crohn's disease patients. GigaScience 6:1–11
95. Qin J, Li Y, Cai Z, Li S, Zhu J et al (2012) A metagenome-wide association study of gut microbiota in type 2 diabetes. Nature 490:55–60
96. Karlsson FH, Fak F, Nookaew I, Tremaroli V, Fagerberg B et al (2012) Symptomatic atherosclerosis is associated with an altered gut metagenome. Nat Commun 3:1245
97. Lau JT, Whelan FJ, Herath I, Lee CH, Collins SM et al (2016) Capturing the diversity of the human gut microbiota through culture-enriched molecular profiling. Genome Med 8:72
98. Lagier JC, Khelaifia S, Alou MT, Ndongo S, Dione N et al (2016) Culture of previously uncultured members of the human gut microbiota by culturomics. Nat Microbiol 1:16203
99. Marseglia L, Manti S, DAngelo G, Nicotera A, Parisi E et al (2014) Oxidative stress in obesity: a critical component in human diseases. Int J Mol Sci 16:378–400
100. Turnbaugh PJ, Hamady M, Yatsunenko T, Cantarel BL, Duncan A et al (2009) A core gut microbiome in obese and lean twins. Nature 457:480–484
101. Ley RE, Turnbaugh PJ, Klein S, Gordon JI (2006) Microbial ecology: human gut microbes associated with obesity. Nature 444:1022–1023
102. Abdallah Ismail N, Ragab SH, Abd Elbaky A, Shoeib AR, Alhosary Y et al (2011) Frequency of Firmicutes and Bacteroidetes in gut microbiota in obese and normal weight Egyptian children and adults. Arch Med Sci 7:501–507
103. Lakhtakia R (2013) The history of diabetes mellitus. Sultan Qaboos Univ Med J 13:368–370
104. Barach JH (1952) Paul Langerhans, 1847-1888. Diabetes 1:411–413
105. American Diabetes Association (2010) Diagnosis and classification of diabetes mellitus. Diabetes Care 33(Suppl 1):S62–S69
106. Asmat U, Abad K, Ismail K (2016) Diabetes mellitus and oxidative stress- a concise review. Saudi Pharm J 24:547–553
107. Baynes JW, Thorpe SR (1999) Role of oxidative stress in diabetic complications: a new perspective on an old paradigm. Diabetes 48:1–9

108. Vistoli G, De Maddis D, Cipak A, Zarkovic N, Carini M et al (2013) Advanced glycoxidation and lipoxidation end products (AGEs and ALEs): an overview of their mechanisms of formation. Free Radic Res 47:3–27
109. Haskins K, Bradley B, Powers K, Fadok V, Flores S et al (2003) Oxidative stress in type 1 diabetes. Ann N Y Acad Sci 1005:43–54
110. Giacco F, Brownlee M (2010) Oxidative stress and diabetic complications. Circ Res 107:1058–1070
111. Han H, Li Y, Fang J, Liu G, Yin J et al (2018) Gut microbiota and type 1 diabetes. Int J Mol Sci 19:995
112. Moreno-Indias I, Cardona F, Tinahones FJ, Queipo-Ortuno MI (2014) Impact of the gut microbiota on the development of obesity and type 2 diabetes mellitus. Front Microbiol 5:190
113. Walton EL (2017) Oxidative stress and diabetes: glucose response in the cROSsfire. Biomed J 40(5):241–244
114. Tolhurst G, Heffron H, Lam YS, Parker HE, Habib AM et al (2012) Short chain fatty acids stimulate glucagon-like peptide-1 secretion via the G-protein-coupled receptor FFAR2. Diabetes 61:364–371
115. Utzschneider KM, Kratz M, Damman CJ, Hullarg M (2016) Mechanisms linking the gut microbiome and glucose metabolism. J Clin Endocrinol Metab 101:1445–1454
116. Harman D (2006) Free radical theory of aging: an update. Ann N Y Acad Sci 1067:10–21
117. Aron-Wisnewsky J, Gaborit B, Dutour A, Clement K (2013) Gut microbiota and non-alcoholic fatty liver disease: new insights. Clin Microbiol Infect 19:338–348
118. Bonomini F, Rodella LF, Rezzani R (2015) Metabolic syndrome, aging and involvement of oxidative stress. Aging Dis 6:109–120
119. Ke Y, Li D, Zhao M, Liu C, Liu J et al (2018) Gut flora-dependent metabolite trimethylamine-N-oxide accelerates endothelial cell senescence and vascular aging through oxidative stress. Free Radical Bio Med 116:88–100
120. Kumar M, Babaei P, Ji B, Nielsen J (2016) Human gut microbiota and healthy aging: recent developments and future prospective. Nutr Healthy Aging 4:3–16
121. Anderson KE, Ricigliano VA, Mott BM, Copeland DC, Floyd AS et al (2018) The queen's gut refines with age: longevity phenotypes in a social insect model. Microbiome 6:108
122. Ticinesi A, Lauretani F, Milani C, Nouvenne A, Tana C et al (2017) Aging gut microbiota at the cross-road between nutrition, physical frailty, and sarcopenia: is there a gut-muscle axis? Nutrients 9
123. Jie Z, Xia H, Zhong SL, Feng Q, Li S et al (2017) The gut microbiome in atherosclerotic cardiovascular disease. Nat Commun 8:845
124. Ley RE, Backhed F, Turnbaugh P, Lozupone CA, Knight RD et al (2005) Obesity alters gut microbial ecology. Proc Natl Acad Sci U S A 102:11070–11075
125. Yang T, Santisteban MM, Rodriguez V, Li E, Ahmari N et al (2015) Gut dysbiosis is linked to hypertension. Hypertension 65:1331–1340
126. Davis CD (2016) The gut microbiome and its role in obesity. Nutr Today 51:167
127. Hoen AG, Li J, Moulton LA, Otoole GA, Housman ML et al (2015) Associations between gut microbial colonization in early life and respiratory outcomes in cystic fibrosis. J Pediatr 167:138–147.e133
128. Ridlon JM, Alves JM, Hylemon PB, Bajaj JS (2013) Cirrhosis, bile acids and gut microbiota: unraveling a complex relationship. Gut Microbes 4:382–387
129. Qin N, Yang F, Li A, Prifti E, Chen Y et al (2014) Alterations of the human gut microbiome in liver cirrhosis. Nature 513:59
130. Foster JA, Rinaman L, Cryan JF (2017) Stress & the gut-brain axis: regulation by the microbiome. Neurobiol Stress 7:124–136
131. Seril DN, Liao J, Yang GY, Yang CS (2003) Oxidative stress and ulcerative colitis-associated carcinogenesis: studies in humans and animal models. Carcinogenesis 24:353–362
132. Matsuoka K, Kanai T (2015) The gut microbiota and inflammatory bowel disease. Semin Immunopathol 37:47–55

133. Mete R, Tulubas F, Oran M, Yilmaz A, Avci BA et al (2013) The role of oxidants and reactive nitrogen species in irritable bowel syndrome: a potential etiological explanation. Med Sci Monit 19:762–766
134. Loh G, Blaut M (2012) Role of commensal gut bacteria in inflammatory bowel diseases. Gut Microbes 3:544–555
135. Jeffery IB, OToole PW, Ohman L, Claesson MJ, Deane J et al (2011) An irritable bowel syndrome subtype defined by species-specific alterations in faecal microbiota. Gut 61:997–1006
136. Caracciolo B, Xu W, Collins S, Fratiglioni L (2014) Cognitive decline, dietary factors and gut-brain interactions. Mech Ageing Dev 136-137:59–69
137. Oriach CS, Robertson RC, Stanton C, Cryan JF, Dinan TG (2016) Food for thought: the role of nutrition in the microbiota-gut-brain axis. Clin Nutr Exper 6:25–38
138. Dinan TG, Cryan JF (2017) Gut instincts: microbiota as a key regulator of brain development, ageing and neurodegeneration. J Physiol 595:489–503
139. Unger MM, Spiegel J, Dillmann KU, Grundmann D, Philippeit H et al (2016) Short chain fatty acids and gut microbiota differ between patients with Parkinson's disease and age-matched controls. Parkinsonism Relat Disord 32:66–72
140. Gerhardt S, Mohajeri M (2018) Changes of colonic bacterial composition in Parkinson's disease and other neurodegenerative diseases. Nutrients 10:708
141. Vogt NM, Kerby RL, Dill-McFarland KA, Harding SJ, Merluzzi AP et al (2017) Gut microbiome alterations in Alzheimer's disease. Sci Rep 7:13537
142. Kohler CA, Maes M, Slyepchenko A, Berk M, Solmi M et al (2016) The gut-brain axis, including the microbiome, leaky gut and bacterial translocation: mechanisms and pathophysiological role in Alzheimer's disease. Curr Pharm Des 22:6152–6166
143. Shulzhenko N, Morgun A, Hsiao W, Battle M, Yao M et al (2011) Crosstalk between B lymphocytes, microbiota and the intestinal epithelium governs immunity versus metabolism in the gut. Nat Med 17:1585–1593
144. Perez-Santiago J, Gianella S, Massanella M, Spina CA, Karris MY et al (2013) Gut *Lactobacillales* are associated with higher CD4 and less microbial translocation during HIV infection. AIDS 27:1921–1931
145. Dubourg G, Lagier JC, Hue S, Surenaud M, Bachar D et al (2016) Gut microbiota associated with HIV infection is significantly enriched in bacteria tolerant to oxygen. BMJ Open Gastroenterol:3
146. Heinlen L, Ballard JD (2010) *Clostridium difficile* infection. Am J Med Sci 340:247–252
147. Gao JJ, Zhang Y, Gerhard M, Mejias-Luque R, Zhang L et al (2018) Association between gut microbiota and *Helicobacter pylori* related gastric lesions in a high-risk population of gastric cancer. Front Cell Infect Mi 8:202
148. Sheh A, Fox JG (2013) The role of the gastrointestinal microbiome in *Helicobacter pylori* pathogenesis. Gut Microbes 4:505–531
149. Noto JM, Peek RM Jr (2017) The gastric microbiome, its interaction with *Helicobacter pylori*, and its potential role in the progression to stomach cancer. PLoS Pathog 13:e1006573
150. Cui L, Morris A, Ghedin E (2013) The human mycobiome in health and disease. Genome Med 5:63–63
151. Chen Y, Chen Z, Guo R, Chen N, Lu H et al (2011) Correlation between gastrointestinal fungi and varying degrees of chronic Hepatitis B virus infection. Diagn Microbiol Infect Dis 70:492–498
152. Hager CL, Ghannoum MA (2017) The mycobiome: role in health and disease, and as a potential probiotic target in gastrointestinal disease. Digest Liver Dis 49:1171–1176
153. Nkamga VD, Henrissat B, Drancourt M (2017) Archaea: essential inhabitants of the human digestive microbiota. Hum Microbiome J 3:1–8
154. van de Pol JA, van Best N, Mbakwa CA, Thijs C, Savelkoul PH et al (2017) Gut colonization by methanogenic archaea is associated with organic dairy consumption in children. Front Microbiol 8:355
155. Reyes A, Semenkovich NP, Whiteson K, Rohwer F, Gordon JI (2012) Going viral: next generation sequencing applied to human gut phage populations. Nat Rev Microbiol 10:607–617

156. Scarpellini E, Ianiro G, Attili F, Bassanelli C, De Santis A et al (2015) The human gut microbiota and virome: potential therapeutic implications. Digest Liver Dis 47:1007–1012
157. Dutilh BE, Cassman N, McNair K, Sanchez SE, Silva GGZ et al (2014) A highly abundant bacteriophage discovered in the unknown sequences of human faecal metagenomes. Nat Commun 5:4498
158. Vital M, Gao J, Rizzo M, Harrison T, Tiedje JM (2015) Diet is a major factor governing the fecal butyrate-producing community structure across mammalia, aves and reptilia. ISME J 9:832–843
159. Givens CE, Ransom B, Bano N, Hollibaugh JT (2015) Comparison of the gut microbiomes of 12 bony fish and 3 shark species. Mar Ecol Prog Ser 518:209–223
160. Bourassa MW, Alim I, Bultman SJ, Ratan RR (2016) Butyrate, neuroepigenetics and the gut microbiome: can a high fiber diet improve brain health? Neurosci Lett 625:56–63
161. Valdes L, Cuervo A, Salazar N, Ruas-Madiedo P, Gueimonde M et al (2015) The relationship between phenolic compounds from diet and microbiota: impact on human health. Food Funct 6:2424–2439
162. Rastmanesh R (2011) High polyphenol, low probiotic diet for weight loss because of intestinal microbiota interaction. Chem Biol Interact 189:1–8
163. David L, Maurice CF, Carmody RN, Gootenberg DB, Button JE et al (2014) Diet rapidly and reproducibly alters the human gut microbiome. Nature 505:559
164. Heller KJ (2001) Probiotic bacteria in fermented foods: product characteristics and starter organisms. Am J Clin Nutr 73:374s–379s
165. Sengul N, Isık S, Aslım B, Ucar G, Demirbag AE (2011) The effect of exopolysaccharide-producing probiotic strains on gut oxidative damage in experimental colitis. Digest Dis and Sci 56:707–714
166. Morrow LE, Kollef MH, Casale TB (2010) Probiotic prophylaxis of ventilator-associated pneumonia: a blinded, randomized, controlled trial. Am J Respir Crit Care Med 182:1058–1064
167. Amar J, Chabo C, Al W, Klopp P, Vachoux C et al (2011) Intestinal mucosal adherence and translocation of commensal bacteria at the early onset of type 2 diabetes: molecular mechanisms and probiotic treatment. EMBO Mol Med 3:559–572
168. Bordalo Tonucci L, Dos Santos KMO, De Luces Fortes Ferreira CL, Ribeiro SMR, De Oliveira LL et al (2015) Gut microbiota and probiotics: focus on diabetes mellitus. Crit Rev Food Sci Nutr 57:2296–2309
169. Endo H, Niioka M, Kobayashi N, Tanaka M, Watanabe T (2013) Butyrate-producing probiotics reduce nonalcoholic fatty liver disease progression in ats: new insight into the probiotics for the gut-liver axis. PLoS One 8:e63388
170. Martin R, Miquel S, Benevides L, Bridonneau C, Robert V et al (2017) Functional characterization of novel *Faecalibacterium prausnitzii* strains isolated from healthy volunteers: a step forward in the use of *F. prausnitzii* as a next-generation probiotic. Front Microbiol 8:1226
171. Brugere JF, Ben Hania W, Arnal ME, Ribiere C, Ballet N et al (2018) Archaea: microbial candidates in next-generation probiotics development. J Clin Gastroenterol 52:S71
172. McFarland LV, Bernasconi P (1993) *Saccharomyces boulardii:* a review of an innovative biotherapeutic agent. Microb Ecol Health Dis 6:157–171
173. Scott KP, Martin JC, Duncan SH, Flint HJ (2014) Prebiotic stimulation of human colonic butyrate-producing bacteria and *Bifidobacteria*. In Vitro FEMS Microbiol Ecol 87
174. Walters SS, Quiros A, Rolston M, Grishina I, Li J et al (2014) Analysis of gut microbiome and diet modification in patients with Crohn's disease. SOJ Microbiol Infect Dis 2:1–13
175. Dolara P, Luceri C, Filippo CD, Femia AP, Giovannelli L et al (2005) Red wine polyphenols influence carcinogenesis, intestinal microflora, oxidative damage and gene expression profiles of colonic mucosa in F344 rats. Mutat Res Fundam Mol Mech Mutagen 591:237–246
176. Cardona F, Andres-Lacueva C, Tulipani S, Tinahones FJ, Queipo-Ortuno MI (2013) Benefits of polyphenols on gut microbiota and implications in human health. J Nutr Biochem 24:1415–1422

177. Queipo-Ortuno MI, Boto-Ordonez M, Murri M, Gomez-Zumaquero JM, Clemente-Postigo M et al (2012) Influence of red wine polyphenols and ethanol on the gut microbiota ecology and biochemical biomarkers. Am J Clin Nutr 95:1323–1334
178. Cueva C, Sanchez-Patan F, Monagas M, Walton GE, Gibson GR et al (2013) In vitro fermentation of grape seed flavan-3-ol fractions by human faecal microbiota: changes in microbial groups and phenolic metabolites. FEMS Microbiol Ecol 83:792–805
179. Lee HC, Jenner AM, Low CS, Lee YK (2006) Effect of tea phenolics and their aromatic fecal bacterial metabolites on intestinal microbiota. Res Microbiol 157:876–884
180. Bailey LC, Forrest CB, Zhang P, Richards TM, Livshits A et al (2014) Association of antibiotics in infancy with early childhood obesity. JAMA Pediatr 168:1063–1069
181. Huttenhower C, Gevers D, Knight R, Abubucker S, Badger JH et al (2012) Structure, function and diversity of the healthy human microbiome. Nature 486:207
182. Arrieta MC, Stiemsma LT, Dimitriu PA, Thorson L, Russell S et al (2015) Early infancy microbial and metabolic alterations affect risk of childhood asthma. Sci Transl Med 7:307ra152–307ra152
183. Dubourg G, Surenaud M, Levy Y, Hue S, Raoult D (2017) Microbiome of HIV-infected people. Microb Pathogen 106:85–93
184. Potgieter M, Bester J, Kell DB, Pretorius E (2015) The dormant blood microbiome in chronic, inflammatory diseases. FEMS Microbiol Rev 39:567–591
185. Borody TJ, Khoruts A (2011) Fecal microbiota transplantation and emerging applications. Nat Rev Gastroenterol Hepatol 9:88–96
186. Brandt LJ (2013) Intestinal microbiota and the role of fecal microbiota transplant (FMT) in treatment of C. difficile infection. Am J Gastroenterol 108:177–185
187. Petrof EO, Gloor GB, Vanner SJ, Weese SJ, Carter D et al (2013) Stool substitute transplant therapy for the eradication of *Clostridium difficile* infection: 'RePOOPulating' the gut. Microbiome 1:3
188. Rao K, Young VB (2015) Fecal microbiota transplantation for the management of *Clostridium difficile* infection. Infect Dis Clin N Am 29:109–122
189. Marchesi JR, Adams DH, Fava F, Hermes GD, Hirschfield GM et al (2015) The gut microbiota and host health: a new clinical frontier. Gut 65:330–339
190. Lathrop SK, Bloom SM, Rao SM, Nutsch K, Lio CW et al (2011) Peripheral education of the immune system by colonic commensal microbiota. Nature 478:250
191. Blacher E, Levy M, Tatirovsky E, Elinav E (2017) Microbiome-modulated metabolites at the interface of host immunity. J Immunol Res 198:572–580
192. Mayer EA, Tillisch K, Gupta A (2015) Gut/brain axis and the microbiota. J Clin Invest 125:926–938
193. Ramezani A, Massy ZA, Meijers B, Evenepoel P, Vanholder R et al (2016) Role of the gut microbiome in uremia: a potential therapeutic target. Am J Kidney Dis 67:483–498
194. Sun M, Wu W, Liu Z, Cong Y (2017) Microbiota metabolite short chain fatty acids, GPCR, and inflammatory bowel diseases. J Gastroenterol Hepatol 52:1–8
195. Tremaroli V, Backhed F (2012) Functional interactions between the gut microbiota and host metabolism. Nature 489:242
196. Ahmad R, Sorrell MF, Batra SK, Dhawan P, Singh AB (2017) Gut permeability and mucosal inflammation: bad, good or context dependent. Mucosal Immunol 10:307
197. Jumpertz R, Le DS, Turnbaugh PJ, Trinidad C, Bogardus C et al (2011) Energy-balance studies reveal associations between gut microbes, caloric load, and nutrient absorption in humans. Am J Clin Nutr 94:58–65
198. Giongo A, Gano KA, Crabb DB, Mukherjee N, Novelo LL et al (2011) Toward defining the autoimmune microbiome for type 1 diabetes. ISME J 5:82–91
199. Wu X, Ma C, Han L, Nawaz M, Gao F et al (2010) Molecular characterisation of the faecal microbiota in patients with type II diabetes. Curr Microbiol 61:69–78
200. Gupta SS, Mohammed MH, Ghosh TS, Kanungo S, Nair GB et al (2011) Metagenome of the gut of a malnourished child. Gut Pathog 3: 7

201. Vrieze A, Van Nood E, Holleman F, Salojarvi J, Kootte RS et al (2012) Transfer of intestinal microbiota from lean donors increases insulin sensitivity in individuals with metabolic syndrome. Gastroenterology 143:913–916.e917
202. Keshavarzian A, Green SJ, Engen PA, Voigt RM, Naqib A et al (2015) Colonic bacterial composition in Parkinson's disease. Mov Disord 30:1351–1360
203. Wang T, Cai G, Qiu Y, Fei N, Zhang M et al (2011) Structural segregation of gut microbiota between colorectal cancer patients and healthy volunteers. ISME J 6:320

Role of ROS in *T. cruzi* Intracellular Development

5

Luciana O. Andrade and Patrícia Pereira Dias

Abstract

Chagas disease, caused by the protozoan *Trypanosoma cruzi*, presents a variable clinical course ranging from asymptomatic cases to more severe forms with cardiac, digestive, or cardio-digestive impairment. The factors involved in this clinical heterogeneity are not completely understood, but certainly both host and parasite genetic variability are important in this process. In the vertebrate host, the establishment of the infection depends on parasite host cell invasion and intracellular multiplication, as well as the host immune response to parasite colonization. *T. cruzi* is able to invade different cell types, but macrophages as a first defense cell and muscle cells (specially cardiomyocytes) are considered key during the establishment of infection in the host. Many factors regulate parasite invasion and intracellular development. Reactive oxygen species (ROS) have been shown to be important during parasite host cell infection. Although in many cases ROS is seen as detrimental to parasite development, recent evidences from the literature have shown that ROS may actually have a dual role during infection. While in some circumstances it could work in parasite control, in other scenarios, it may act to potentiate parasite intracellular multiplication. Here, we present a brief background of the disease and parasite genetic structure in order to discuss this dual role of ROS during parasite host cell colonization.

Keywords

T. cruzi · Oxidative stress · Intracellular development · Cardiomyocytes

L. O. Andrade (✉) · P. P. Dias
Departamento de Morfologia, Universidade Federal de Minas Gerais, Minas Gerais, Brazil
e-mail: lucianaandrade@ufmg.br

S. Chakraborti et al. (eds.), *Oxidative Stress in Microbial Diseases*,
https://doi.org/10.1007/978-981-13-8763-0_5

5.1 Introduction

Chagas disease or American trypanosomiasis is a chronic endemic malady, caused by the protozoan parasite *Trypanosoma cruzi*. According to the World Health Organization (WHO), about eight million people worldwide are infected with this parasite, with the majority of them living in Latin America [1, 2], due to the distribution of its insect vector. Under natural conditions, the parasite is transmitted to humans and other mammals, by insects belonging to the Reduviidae family (reviewed by [1, 3]), which are endemic of the American continent, mostly Central and South America. It is estimated that each year 10,000 people die as a result of the symptoms of this disease, and more than 25 million people are at risk of acquiring American trypanosomiasis [4]. There are also other forms of parasite transmission, such as blood transfusion, oral or congenital, through laboratory accidents and organ transplantation. Those alternative forms of transmission have become important forms of contamination with *T. cruzi* in recent years [3]. From those, outbreaks of oral transmission have been identified in several countries of Latin America in the last decade, the most significant in Caracas, Venezuela, with international repercussion [5–8]. These other alternative forms of transmission, together with the recent waves of people migration in a more globalized world, have allowed countries where the disease is not endemic, due to the absence of the transmitting insect, such as Canada, France, Spain, Switzerland, Japan, and Australia, to present cases of *T. cruzi* infection, making Chagas disease also a global health problem [9–13]. The migration of infected people to countries where Chagas disease is not endemic has been mainly due to sociocultural and political factors [14].

5.2 *T. cruzi* Life Cycle, Disease Development, and Parasite Population Structure

T. cruzi life cycle begins when the insect vector, a triatomine from Reduviidae family, while feeding on the blood from the host, releases in its feces and urine the metacyclic trypomastigote form of the *T. cruzi*, an elongated and flagellated form of the parasite, infective for the vertebrate host. These metacyclictrypomastigotes are capable of penetrating into the vertebrate host through the injured insect bite site or when in contact with mucous and connective membranes from the host [15].

Once inside the organism, the metacyclic trypomastigote form may invade a wide variety of nucleated cell types, including professional phagocytic cells, such as tissue macrophages, or nonprofessional phagocytic cells, such as striated and smooth muscle cells and fibroblasts, among others. Within the host cell, the trypomastigote form transforms into an aflagellate round-shaped form, called amastigote, which is able to replicate by binary fission inside the host cell. After successive divisions, the amastigote form transforms back into the flagellate trypomastigote form, which breaks the host cell due to its large numbers and intense movement. Once outside the cell, these forms can invade neighboring cells or fall into the bloodstream and infect other host cells, causing lesions that may induce the development of the disease. The cell cycle is closed when the insect vector ingests the trypomastigote forms present in the blood of the vertebrate host when doing the blood repast. Once in the digestive tract of the invertebrate host, the trypomastigote form transforms into the epimastigote form, a flagellated and infective form for the vector. Epimastigotes multiply and migrate to the insect's posterior gut, differentiating into the metacyclic trypomastigote form, which is eliminated in feces and urine during blood repast [15–17].

Infection in man is defined by two distinct clinical phases. The acute phase, corresponding to the initial period of infection, characterized by high parasitemia and tissue parasitism, followed by the chronic phase of the infection, which persists throughout the host's life and is characterized by lower tissue and blood parasitism [18]. While in the acute phase parasite is found in many tissues, upon the intervention of an immune response triggered during infection, not only parasitemia but also tissue parasitism is significantly reduced, and patients enter the chronic phase of infection [19]. Chronic infection has a variable clinical course. Initially, when patients enter the chronic phase of infection, they usually do not present any symptoms of the disease, which is called the asymptomatic or indeterminate form of the chronic infection. However, some of these patients, around 30–40% and usually 10–30 years after the initial infection, will develop clinical symptoms, characteristic of Chagas disease, which include heart (cardiac form) and/or digestive tract (megacolon or megaesophagus – digestive form) disorders. This is a consequence, not only of the persistence of the parasite in these organs but also of the inflammatory process generated upon parasite colonization [19].

It has been extensively shown in the literature that this clinical variability depends not only on genetic factors of the parasite, whose population structure is quite variable, but also on genetic factors of the host [20, 21]. Together, they would define a differential tissue distribution of parasite populations that would interfere with the development of the distinct clinical forms. In fact, previous studies performed by our group have shown that distinct populations of the parasite are found in different organs of infected patients [22], reinforcing data on the existence of a differential tissue tropism, related to the development of clinical forms [23].

Studies of *T. cruzi* population structure, through biochemical and most importantly at the genetic level, have revealed a very heterogeneous pattern, but parasite strains could be clustered into six groups or discrete typing units (DTUs), called TcI to TcVI. Into these groups, parasite populations are considered genetically more related to each other, allowing their identification by molecular or genetic markers [20]. TcI harbor populations of parasites usually related to the wild-type transmission cycle, as well as TcIII and TcIV [24]. The populations belonging to TcI exhibit a wide geographical distribution, which passes through the United States to Northern Chile and Argentina [24–26]. In Brazil, it is commonly found in the Amazon basin [27]. In some countries, such as Colombia, TcI is associated with chagasic cardiomyopathy, while in other countries, they show low pathogenesis [26, 28]. TcII, TcV, and TcVI groups were identified further south of South America, associated with human infections. TcII, in particular, is predominant in the central and southern regions of

South America. TcII representatives present high diversity among them and are found to be more involved in the domestic cycle and also in the more severe cases of Chagas disease [29–31].

5.3 *T. cruzi* Intracellular Life Cycle

Regardless from which *T. cruzi* group, the parasite subverts the mechanism that eukaryotic cells use to repair small insults in their plasma membrane in order to invade the host cell [32]. For this, infective trypomastigote forms adhere to the surface of the host cell and stimulate them through parasite surface proteins or secreted factors, which then lead to the release of calcium from host cells' intracellular stores [33–41]. It has been shown in the literature that trypomastigotes from different *T. cruzi* groups will differ in the expression of these factors and therefore induce different levels of calcium signaling in the host cell, which may influence their ability to invade the host cell [42, 43]. In addition to the calcium signals triggered by parasite factors, *T. cruzi* may also cause small ruptures in the plasma membrane while in contact with cells, provoking the influx of calcium from the extracellular to the intracellular milieu [32]. In both cases, the increase of intracellular calcium leads to recruitment and exocytosis of lysosomes, which in turn leads to a compensatory endocytosis process through which the parasite is carried into the host cell [32, 44]. The parasite is then internalized into a vacuole containing lysosomal and plasma membrane [35, 45]. More Lysosomes fuse with this initial vacuole until the whole vacuolar membrane is covered with lysosomal markers.

Subsequently, the trypomastigote escapes from the vacuole and falls into the cytoplasm of the cell, where it transforms into the replicative amastigote form and successively divides in the host cell cytoplasm until they differentiate back into the trypomastigote form [45–47]. Calcium signaling induced during parasite entry as well as changes through different environments during parasite cellular colonization process certainly can directly or indirectly influence *T. cruzi* host cell and parasite behavior.

5.4 Host Cell ROS Response to *T. cruzi* Infection

Infected cells are able to respond to the infection by activating several genes, which could interfere with the parasite's intracellular behavior [48–52]. It is important to note that some of the genes differentially expressed by the cell during the infection period may vary depending on the type of cell used for the infection, while other genes are common to different cell types [49]. Houston-Ludlam and coworkers (2016) analyzed HFF (human foreskin fibroblasts) transcripts 24 hours after infection with distinct *T. cruzi* populations belonging to TcI and TcII [53]. These cells showed increased expression of genes for interferon type I responses and also of a member of the tumor necrosis factor linker superfamily, associated with the induction of apoptosis in tumor cells. Other studies analyzing the transcriptome of HFF cells at the initial moments after infection with the *T. cruzi* Y strain showed that genes for the host cell cycle and immune response were also positively regulated [52]. Moraes and coworkers [54] also showed that upon infection with *T. cruzi*, H9C2, cardiomyocytes increase the production of interleukin 1β and cyclooxygenase 2, important inducers of inflammatory mediators [54].

Infection of cells with *T. cruzi* has also shown to induce the production of reactive oxygen species (ROS). Among the changes that occur during *T. cruzi* cell infection, there are important alterations in the metabolic activity of the cell that include increased fatty acid metabolism and β-oxidation that appear to be important for parasite growth [51]. It is known that increased fatty acid oxidation may be a contributing factor for increased production of reactive oxygen species (ROS) by mitochondria [55]. To the best of our knowledge, the first report of the induction of ROS by *T. cruzi* infection was in 1990 by Cardoni and coworkers [56]. In this work, the authors showed that 3–4 weeks after the infection of BALB/c mice with Tulahuen strain, an increase in ROS production was observed for spleen and peritoneal cells. Much later, it was shown that the oxidative stress generated upon *T. cruzi* infection led to oxidative modifications in the mitochondrial respiratory complexes in hearts of mice infected with the parasite, possibly contributing to disease outcome [57]. In a following work, they showed that the myocardium of mice was also able to produce ROS upon infection with *T. cruzi* as a consequence of mitochondrial electron leakage at complex III, inducing the formation of superoxide [58]. Additionally, in a related study, they showed that different organs also respond with ROS production upon *T. cruzi* infection, with high rates at the peripheral blood [59]. Those results led to the conclusion that host cell immune response would be contributing to this process. Later, they showed that *T. cruzi* infection of a lineage of mouse atrial cardiomyocytes (HL-1 cells), as well as primary cardiomyocytes of adult mice, could directly disturb mitochondrial function, inducing electron leakage and the production of superoxide anion by these cells, without the participation of the host immune system, although the latter was able to potentiate *T. cruzi*-induced cardiomyocyte ROS response [60]. Further, they showed that this induction of ROS in cardiomyocytes was dependent on the presence of intracellular parasites, since neither parasite lysate nor secreted proteins had the same effect [60]. On the other hand, Guiñazu and coworkers showed that for splenocytes, the exposure to *T. cruzi* cruzipain alone, a highly immunogenic glycoprotein, was able to induce ROS production by these cells [61]. Ba et al. (2010) also showed in human cardiomyocytes that the ROS produced by mitochondria contribute to the expression of cytokine genes in response to the infection of these cells with *T. cruzi* [62]. In the heart, it is common sense that the production of ROS is extremely detrimental to the organ, contributing to tissue damage and most certainly to the development of the clinical symptoms of the disease [63–65].

5.5 Dual Role of ROS in *T. cruzi* Infection of Macrophages

Although possibly cytotoxic to host cell and tissue components, ROS production has been long considered as a powerful resource against pathogen infection. ROS is one of the first responses produced by innate immune cells to pathogen invaders. Many examples are found in the literature. Macrophages' oxidative bursts, for example, have been shown to have a key role in controlling many bacterial infections (reviewed by [66]). ROS production by plants has also been shown to have an important role in pathogen resistance [67–69]. ROS controlling of parasitic infections has also been reported as an important resource for infection control [70, 71]. In this line, it has been described that many parasites use ROS scavenging systems as important resistance factors during host infections [71, 72].

In the case of *T. cruzi*, ROS has also been identified as a source of parasite control [60, 73, 74]. In the early stages of infections, it has been shown that macrophages play a key role in controlling parasite infection [75, 76]. Phagocytosis of the parasite by naive macrophages leads to the activation of NADPH oxidase that in turn generates superoxide radicals, which contribute to parasite control [77]. Macrophages primed with cytokines will then elicit a nitric oxide response, which will react with the superoxide radical, boosting the control of intraphagosomal *T. cruzi* parasites [78]. *T. cruzi*, on its turn, possesses a network of antioxidant enzymes, localized at different compartments (cytosol, mitochondria, and endoplasmic reticulum) and capable of dealing with and fighting the endogenous as well as reactive oxygen and nitrogen species produced during host cell infection. Among those antioxidant enzymes are trypanothione (TSH), tryparedoxin (TXN), ascorbate peroxidase (TcAPX), glutathione, mitochondrial peroxiredoxin (TcMPX), and superoxide dismutase (SOD), among others (reviewed by [79]). Some of them have been shown to be upregulated during metacyclogenesis, which could indicate a potential role to face the oxidative cell environment [80, 81]. These data altogether reinforce a role of ROS in controlling *T. cruzi* infection.

Recently though, an opposite effect of ROS has been described, suggesting that those molecules may also be present as positive regulators during *T. cruzi* infection. Paiva and coworkers (2012) demonstrated that the production of ROS contributes to the intracellular growth of the parasite in macrophages [82]. They showed that the endogenous expression of the nuclear factor, erythroid2-related factor 2 (NRF2), which orchestrates antioxidant defenses, such as expression of the enzyme heme-oxygenase-1 (HO-1), reduces parasitism in macrophages infected with *T. cruzi* Y strain. On the other hand, treatment of infected macrophages with pro-oxidant substances, such as tin protoporphyrin (SnPP), an inhibitor of HO-1 activity, promoted parasite growth [82]. They also showed that treatment of infected mice with this antioxidant components also increases parasite burden in infected tissues. Another example of a positive effect of ROS in *T. cruzi* intracellular development in macrophages came from the work of Goes and collaborators (2016) [83]. In this work, the authors showed a detriment in intracellular proliferation of *T. cruzi* Y strain in macrophages from mice deficient in the gp91phox subunit of NADPH oxidase (Phox KO), essential for the generation of the oxidative stress in these cells, when compared to wild-type mice [83]. They also evaluated the effect of gp91phox deficiency in infection of macrophages by a recombinant strain of *T. cruzi* that superexpresses MutT (an enzyme involved in DNA repair). DNA lesions are one of the main consequences of oxidative stress, and guanine is the most easily oxidized nucleotide, forming 8-oxo-GTP that, when incorporated to the DNA during replication, may cause DNA lesions [84]. This enzyme catalyzes the degradation of 8-oxo-GTP (8-oxo-2′-deoxy-guanosine −5′-triphosphate) in 8-oxo-GMP (8-oxo-2′-deoxy-guanosine-5′-monophosphate), preventing 8-oxo-GTP incorporation to the DNA [85, 86]. *T. cruzi* MutT superexpressors have been shown to be more efficient in cell colonization compared to wild parasites, meaning a higher parasite intracellular multiplication rate, suggesting that 8-oxo-GMP, generated by MutT, could serve as signaling molecule to produce parasites more adapted to the intracellular environment [87]. When Goes infected macrophages derived from Phox KO mice with *T. cruzi* MutT strain, parasite replication rate was much higher when compared to the wild-type strain, reinforcing a possible signaling role for role 8-oxo-GMP to boost parasite replication during oxidative stress [83].

5.6 Role of ROS in *T. cruzi* Intracellular Development in Cardiomyocytes

As mentioned previously, a couple of cells other than macrophages have also been shown to be able to induce a ROS response upon *T. cruzi* infection, including cardiomyocytes, considered a key cell during host infection [57, 88]. Although it has been suggested that ROS would be deleterious to parasite development in cardiomyocytes, most of the work was focused on the role of ROS in tissue damage during parasite infection [62, 89–91]. Our group has then investigated the role of ROS signaling in *T. cruzi* intracellular multiplication rates in cardiomyocytes. For this, we used two distinct *T. cruzi* monoclonal populations, JG strain (TcII) and a clone from Colombian strain, Col1.7G2 (TcI). In a former work, we had shown that for BALB/c mice in the chronic phase of infection with both JG and Col1.7G2, simultaneously, there was a predominance of JG in the heart, with very low levels of Col1.7G2 in this organ [92]. This result suggested that JG was more adapted to the heart of these mice. Histopathological analyses corroborated these data, since the inflammation profile found for heart from mice submitted to mixed infection was similar to the pattern seen for this same organ obtained from mice infected with JG only [92]. In vitro studies using primary cultures of cardiomyocytes, isolated from BALB/c mice embryos, when infected with Col1.7G2 or JG reproduced the in vivo data [93]. In this study, it was shown that although the number of infected cardiomyocytes for cultures exposed to JG strain was lower when compared to cultures exposed to Col1.7G2, the number of intracellular parasites per cell was much higher for cultures infected with JG, when compared to those infected with Col1.7G2 [93]. In addition, we showed that these isolated cardiomyocytes, when submitted to mixed infections (JG + Col1.7G2), showed a greater proportion of JG in relation to Col1.7G2 over time of infection. Thus, in this case, the intracellular multiplication rate was directly related to the tissue selection process. Moreover, it was shown that this behavioral profile was dependent on the cell type studied, reinforcing that not only the parasite but also host cell participates in the differential tissue tropism of *T. cruzi*. Based on these results, we tested whether ROS could be influencing these differences in intracellular development of JG and Col1.7G2 in cardiomyocytes. We showed that primary cardiomyocytes of BALB/c mice, as well as pluripotent human cells induced in cardiomyocytes, after infection with the *T. cruzi* JG strain, presented greater intracellular multiplication of the parasites over time of infection, concomitant to a higher production of reactive oxygen species when compared to the same conditions of infection with clone Col1.7G2 [94]. Inhibition of reactive oxygen species, by treatment with catalase, an enzyme that metabolizes hydrogen peroxide, was able to inhibit the growth of JG, but not Col1.7G2 [94]. In parallel, it was demonstrated that JG strain presented lower expression of enzymes of the antioxidant machinery in comparison to Col1.7G2, strongly suggesting that, as it was demonstrated for macrophages, ROS may also fuel parasite intracellular development in cardiomyocytes.

5.7 How Does ROS Signal to Parasite Intracellular Development?

How ROS may signal to parasite intracellular development is still not known. In our work we have demonstrated that JG exposure to ROS led to the production of a higher level of signaling molecules such as calcium and superoxide anion ($O_2^{\bullet-}$), which is not observed for Col1.7G2 [94]. Calcium has been shown to be important for multiplication and metacyclogenesis in epimastigotes [95]; therefore, it could be working as a signaling molecule for *T. cruzi* intracellular multiplication. Additionally, it has been demonstrated that low levels of ROS production induced by heme in *T. cruzi* epimastigotes favor parasite proliferation via a Ca^{2+} calmodulin kinase II (CaMKII)-like pathway [96]. With respect to $O_2^{\bullet-}$, although there is data in the literature showing that its increase may induce programmed cell death in *T. cruzi* [74], there are also reports showing that it could work as a signal for increased cell proliferation and as an inhibitor of apoptotic pathways [97, 98]. As mentioned previously, another signaling molecule could be the generation of 8-oxo-GMP during *T. cruzi* oxidative stress exposure. Boldogh and coworkers (2012) showed that excised 8-oxoG, during the DNA repair pathway, binds back to the excision enzyme OGG1 (outside the catalytic site) and interacts with the RasGTPase enzyme, acting as a guanine nucleotide exchange factor [99]. This interaction enables this GTPase to activate signaling pathways, including those involved in oxidative stress response.

5.8 Conclusion

What would determine whether ROS is responsible for death or proliferation would most likely be related to the amount to which parasites are exposed and the ability of the parasite to sense and trigger the intracellular signaling. In our studies with JG and Col1.7G2 *T. cruzi* infections in cardiomyocytes, it was not possible to determine whether only some strains are sensitive to ROS signaling or whether the ROS level produced is responsible for the effects of cell proliferation observed. Recently, Vilar-Pereira and colleagues studied the role of antioxidant agents in cardiac function during *T. cruzi* infection in mice [100]. For this, they evaluated the production of ROS in the hearts of BALB/c mice in the chronic phase of infection with the Colombian strain of *T. cruzi*. In this study, the authors showed that infection with the Colombian strain was able to induce high amounts of ROS in the hearts of mice in the chronic phase of infection, as well as an electrical and mechanical dysfunction of infected hearts. Treatment with different antioxidant agents was able to improve heart function, but only the treatment with resveratrol was able to reduce the burden of parasites. This may present at first, in contrast to our data and other data from the same group showing that the cardiac parasite load decreases in response to treatment with antioxidant agents [82]. However, in the work of 2012, the strain of *T. cruzi* used was Y strain. The Colombian strain and Y, as the clone of the Colombian strain (Col1.7G2) and JG used in our study, belong to two different DTUs of *T. cruzi*, TcI and II, respectively. It is possible that parasites from different DTUs respond differently to ROS. In this case, for TcII strains, ROS might not have an effect on control or signaling and that the effect of resveratrol could be on another signaling pathway, whereas TcI strains would respond to ROS. Our data support this hypothesis, since the treatment affects the population of TcI, but not TcII. Whether *T. cruzi* TcII strains do not really respond to ROS or whether this response depends on the amount or type of response triggered by ROS production is a question that still needs to be elucidated.

References

1. Araujo CA, Waniek PJ, Jansen AM (2009) An overview of Chagas disease and the role of triatomines on its distribution in Brazil. Vector Borne Zoonotic Dis 9(3):227–234
2. WHO, W.H.O (2018) Chagas disease (American trypanosomiasis) [cited 2018 25th July]
3. Coura JR, Dias JC (2009) Epidemiology, control and surveillance of Chagas disease – 100 years after its discovery. Mem Inst Oswaldo Cruz 104(4):31–40
4. WHO, W.H.O (2017) Chagas disease (American trypanosomiasis)
5. Barbosa-Ferreira JM et al (2010) Stroke in a chronic autochthonous chagasic patient from the Brazilian Amazon. Rev Soc Bras Med Trop 43:751–753
6. Alarcon de Noya B et al (2010) Large urban outbreak of orally acquired acute Chagas disease at a school in Caracas, Venezuela. J Infect Dis 201(9):1308–1315
7. Andrade SG et al (2011) Biological, biochemical and molecular features of Trypanosoma cruzi strains isolated from patients infected through oral transmission during a 2005 outbreak in the state of Santa Catarina, Brazil: its correspondence with the new *T. cruzi* Taxonomy Consensus (2009). Mem Inst Oswaldo Cruz 106(8):948–956
8. Souza-Lima Rde C et al (2013) Outbreak of acute Chagas disease associated with oral transmission in the Rio Negro region, Brazilian Amazon. Rev Soc Bras Med Trop 46(4):510–514
9. Florian Sanz F et al (2005) Chagasic cardiomyopathy in Spain: a diagnosis to bear in mind. An Med Interna 22(11):538–540
10. Schmunis GA (2007) Epidemiology of Chagas disease in non-endemic countries: the role of international migration. Mem Inst Oswaldo Cruz 102(Suppl 1):75–85
11. Lescure FX et al (2008) Chagas disease, France. Emerg Infect Dis 14(4):644–646

12. Munoz J et al (2009) Clinical profile of Trypanosoma cruzi infection in a non-endemic setting: immigration and Chagas disease in Barcelona (Spain). Acta Trop 111(1):51–55
13. Jackson Y et al (2010) Prevalence, clinical staging and risk for blood-borne transmission of Chagas disease among Latin American migrants in Geneva, Switzerland. PLoS Negl Trop Dis 4(2):e592
14. Dias JC et al (2016) Brazilian consensus on Chagas disease, 2015. Epidemiol Serv Saude 25(Special):7–86
15. Chagas C (1909) Nova Trypanozomiaze Humana. Memorias do Instituto Oswaldo Cruz 1:11–80
16. Tyler KM, Engman DM (2001) The life cycle of Trypanosoma cruzi revisited. Int J Parasitol 31(5–6):472–481
17. Rassi A Jr, Rassi A, Marin-Neto JA (2010) Chagas disease. Lancet 375(9723):1388–1402
18. Prata A (2001) Clinical and epidemiological aspects of Chagas disease. Lancet Infect Dis 1(2):92–100
19. Dutra WO, Gollob KJ (2008) Current concepts in immunoregulation and pathology of human Chagas disease. Curr Opin Infect Dis 21(3):287–292
20. Zingales B et al (2009) A new consensus for Trypanosoma cruzi intraspecific nomenclature: second revision meeting recommends TcI to TcVI. Mem Inst Oswaldo Cruz 104(7):1051–1054
21. Macedo AM, Oliveira RP, Pena SD (2002) Chagas disease: role of parasite genetic variation in pathogenesis. Expert Rev Mol Med 4(5):1–16
22. Vago AR et al (1996) Kinetoplast DNA signatures of Trypanosoma cruzi strains obtained directly from infected tissues. Am J Pathol 149(6):2153–2159
23. Franco DJ et al (2003) Trypanosoma cruzi: mixture of two populations can modify virulence and tissue tropism in rat. Exp Parasitol 104(1–2):54–61
24. Miles MA et al (2009) The molecular epidemiology and phylogeography of *Trypanosoma cruzi* and parallel research on Leishmania: looking back and to the future. Parasitology 136(12):1509–1528
25. Yeo M et al (2005) Origins of Chagas disease: Didelphis species are natural hosts of *Trypanosoma cruzi* I and armadillos hosts of Trypanosoma cruzi II, including hybrids. Int J Parasitol 35(2):225–233
26. Zingales B et al (2012) The revised Trypanosoma cruzi subspecific nomenclature: rationale, epidemiological relevance and research applications. Infect Genet Evol 12(2):240–253
27. Coura JR et al (2002) Emerging Chagas disease in Amazonian Brazil. Trends Parasitol 18(4):171–176
28. Ramirez JD et al (2013) Molecular epidemiology of human oral Chagas disease outbreaks in Colombia. PLoS Negl Trop Dis 7(2):e2041
29. Diosque P et al (2003) Multilocus enzyme electrophoresis analysis of Trypanosoma cruzi isolates from a geographically restricted endemic area for Chagas' disease in Argentina. Int J Parasitol 33(10):997–1003
30. Campbell DA, Westenberger SJ, Sturm NR (2004) The determinants of Chagas disease: connecting parasite and host genetics. Curr Mol Med 4(6):549–562
31. Lages-Silva E et al (2006) Variability of kinetoplast DNA gene signatures of Trypanosoma cruzi II strains from patients with different clinical forms of Chagas' disease in Brazil. J Clin Microbiol 44(6):2167–2171
32. Fernandes MC et al (2011) Trypanosoma cruzi subverts the sphingomyelinase-mediated plasma membrane repair pathway for cell invasion. J Exp Med 208(5):909–921
33. Tardieux I et al (1992) Oral susceptibility of *Aedes albopictus* to dengue type 2 virus: a study of infection kinetics, using the polymerase chain reaction for viral detection. Med Vet Entomol 6(4):311–317
34. Rodriguez A et al (1995) A trypanosome-soluble factor induces IP3 formation, intracellular Ca2+ mobilization and microfilament rearrangement in host cells. J Cell Biol 129:1263–1273
35. Rodriguez A et al (1996) Host cell invasion by trypanosomes requires lysosomes and microtubule/kinesin-mediated transport. J Cell Biol 134:349–362

36. Ruiz RC et al (1998) Infectivity of Trypanosoma cruzi strains is associated with differential expression of surface glycoproteins with differential Ca^{2+} signalling activity. Biochem J 330(Pt 1):505–511
37. Caler EV et al (1998) Oligopeptidase B-dependent signaling mediates host cell invasion by Trypanosoma cruzi. EMBO J 17(17):4975–4986
38. Scharfstein J et al (2000) Host cell invasion by Trypanosoma cruzi is potentiated by activation of bradykinin B(2) receptors. J Exp Med 192(9):1289–1300
39. Manque PM et al (2003) Cell adhesion and Ca2+ signaling activity in stably transfected Trypanosoma cruzi epimastigotes expressing the metacyclic stage-specific surface molecule gp82. Infect Immun 71(3):1561–1565
40. Maeda FY, Cortez C, Yoshida N (2012) Cell signaling during Trypanosoma cruzi invasion. Front Immunol 3:361
41. Martins NO et al (2015) Molecular characterization of a novel family of Trypanosoma cruzi surface membrane proteins (TcSMP) involved in mammalian host cell invasion. PLoS Negl Trop Dis 9(11):e0004216
42. Neira I, Ferreira AT, Yoshida N (2002) Activation of distinct signal transduction pathways in Trypanosoma cruzi isolates with differential capacity to invade host cells. Int J Parasitol 32(4):405–414
43. Seco-Hidalgo V, De Pablos LM, Osuna A (2015) Transcriptional and phenotypical heterogeneity of Trypanosoma cruzi cell populations. Open Biol 5(12):150190
44. Fernandes MC, Andrews NW (2012) Host cell invasion by Trypanosoma cruzi: a unique strategy that promotes persistence. FEMS Microbiol Rev 36(3):734–747
45. Andrews NW (1994) From lysosomes into the cytosol: the intracellular pathway of Trypanosoma cruzi. Braz J Med Biol Res 27(2):471–475
46. Dvorak JA, Hyde TP (1973) Trypanosoma cruzi: interaction with vertebrate cells in vitro. I. Individual interactions at the cellular and subcellular levels. Exp Parasitol 34:268–283
47. Andrade LO, Andrews NW (2005) The Trypanosoma cruzi-host-cell interplay: location, invasion, retention. Nat Rev Microbiol 3(10):819–823
48. Vaena de Avalos S et al (2002) Immediate/early response to Trypanosoma cruzi infection involves minimal modulation of host cell transcription. J Biol Chem 277(1):639–644
49. Costales JA, Daily JP, Burleigh BA (2009) Cytokine-dependent and-independent gene expression changes and cell cycle block revealed in Trypanosoma cruzi-infected host cells by comparative mRNA profiling. BMC Genomics 10:252
50. Manque PA et al (2011) Trypanosoma cruzi infection induces a global host cell response in cardiomyocytes. Infect Immun 79(5):1855–1862
51. Caradonna KL et al (2013) Host metabolism regulates intracellular growth of Trypanosoma cruzi. Cell Host Microbe 13(1):108–117
52. Li Y et al (2016) Transcriptome remodeling in Trypanosoma cruzi and human cells during intracellular infection. PLoS Pathog 12(4):e1005511
53. Houston-Ludlam GA, Belew AT, El-Sayed NM (2016) Comparative transcriptome profiling of human foreskin fibroblasts infected with the Sylvio and Y strains of Trypanosoma cruzi. PLoS One 11(8):e0159197
54. Moraes KC, Diniz LF, Bahia MT (2015) Role of cyclooxygenase-2 in Trypanosoma cruzi survival in the early stages of parasite host-cell interaction. Mem Inst Oswaldo Cruz 110(2):181–191
55. Rosca MG et al (2012) Oxidation of fatty acids is the source of increased mitochondrial reactive oxygen species production in kidney cortical tubules in early diabetes. Diabetes 61(8):2074–2083
56. Cardoni RL, Rottenberg ME, Segura EL (1990) Increased production of reactive oxygen species by cells from mice acutely infected with Trypanosoma cruzi. Cell Immunol 128(1):11–21
57. Wen JJ, Garg N (2004) Oxidative modification of mitochondrial respiratory complexes in response to the stress of Trypanosoma cruzi infection. Free Radic Biol Med 37(12):2072–2081

58. Wen JJ, Garg NJ (2008) Mitochondrial generation of reactive oxygen species is enhanced at the Q(o) site of the complex III in the myocardium of Trypanosoma cruzi-infected mice: beneficial effects of an antioxidant. J Bioenerg Biomembr 40(6):587–598
59. Dhiman M et al (2008) Enhanced nitrosative stress during Trypanosoma cruzi infection causes nitrotyrosine modification of host proteins: implications in Chagas' disease. Am J Pathol 173(3):728–740
60. Gupta S et al (2009) Trypanosoma cruzi infection disturbs mitochondrial membrane potential and ROS production rate in cardiomyocytes. Free Radic Biol Med 47(10):1414–1421
61. Guinazu N et al (2010) Induction of NADPH oxidase activity and reactive oxygen species production by a single Trypanosoma cruzi antigen. Int J Parasitol 40(13):1531–1538
62. Ba X et al (2010) Trypanosoma cruzi induces the reactive oxygen species-PARP-1-RelA pathway for up-regulation of cytokine expression in cardiomyocytes. J Biol Chem 285(15):11596–11606
63. Zacks MA et al (2005) An overview of chagasic cardiomyopathy: pathogenic importance of oxidative stress. An Acad Bras Cienc 77(4):695–715
64. Dhiman M, Garg NJ (2011) NADPH oxidase inhibition ameliorates Trypanosoma cruzi-induced myocarditis during Chagas disease. J Pathol 225(4):583–596
65. Paiva CN, Medei E, Bozza MT (2018) ROS and Trypanosoma cruzi: fuel to infection, poison to the heart. PLoS Pathog 14(4):e1006928
66. Fang FC (2004) Antimicrobial reactive oxygen and nitrogen species: concepts and controversies. Nat Rev Microbiol 2(10):820–832
67. Apostol I, Heinstein PF, Low PS (1989) Rapid stimulation of an oxidative burst during elicitation of cultured plant cells : role in defense and signal transduction. Plant Physiol 90(1):109–116
68. Levine A et al (1994) H2O2 from the oxidative burst orchestrates the plant hypersensitive disease resistance response. Cell 79(4):583–593
69. Simon-Plas F, Elmayan T, Blein JP (2002) The plasma membrane oxidase NtrbohD is responsible for AOS production in elicited tobacco cells. Plant J 31(2):137–147
70. Schirmer RH et al (1987) Oxidative stress as a defense mechanism against parasitic infections. Free Radic Res Commun 3(1–5):3–12
71. Bosch SS et al (2015) Oxidative stress control by apicomplexan parasites. Biomed Res Int 2015:351289
72. Staerck C et al (2017) Microbial antioxidant defense enzymes. Microb Pathog 110:56–65
73. Machado FS et al (2000) Trypanosoma cruzi-infected cardiomyocytes produce chemokines and cytokines that trigger potent nitric oxide-dependent trypanocidal activity. Circulation 102(24):3003–3008
74. Piacenza L et al (2007) Mitochondrial superoxide radicals mediate programmed cell death in Trypanosoma cruzi: cytoprotective action of mitochondrial iron superoxide dismutase overexpression. Biochem J 403(2):323–334
75. Munoz-Fernandez MA, Fernandez MA, Fresno M (1992) Activation of human macrophages for the killing of intracellular Trypanosoma cruzi by TNF-alpha and IFN-gamma through a nitric oxide-dependent mechanism. Immunol Lett 33(1):35–40
76. Kierszenbaum F et al (1974) Phagocytosis: a defense mechanism against infection with Trypanosoma cruzi. J Immunol 112(5):1839–1844
77. Alvarez MN et al (2004) Macrophage-derived peroxynitrite diffusion and toxicity to Trypanosoma cruzi. Arch Biochem Biophys 432(2):222–232
78. Ferrer-Sueta G, Radi R (2009) Chemical biology of peroxynitrite: kinetics, diffusion, and radicals. ACS Chem Biol 4(3):161–177
79. Piacenza L et al (2009) Enzymes of the antioxidant network as novel determiners of Trypanosoma cruzi virulence. Int J Parasitol 39(13):1455–1464
80. Parodi-Talice A et al (2007) Proteomic analysis of metacyclic trypomastigotes undergoing Trypanosoma cruzi metacyclogenesis. J Mass Spectrom 42(11):1422–1432
81. Piacenza L et al (2008) Peroxiredoxins play a major role in protecting Trypanosoma cruzi against macrophage- and endogenously-derived peroxynitrite. Biochem J 410(2):359–368

82. Paiva CN et al (2012) Oxidative stress fuels Trypanosoma cruzi infection in mice. J Clin Invest 122(7):2531–2542
83. Goes GR et al (2016) Trypanosoma cruzi needs a signal provided by reactive oxygen species to infect macrophages. PLoS Negl Trop Dis 10(4):e0004555
84. van Loon B, Markkanen E, Hubscher U (2010) Oxygen as a friend and enemy: how to combat the mutational potential of 8-oxo-guanine. DNA Repair (Amst) 9(6):604–616
85. Michaels ML, Miller JH (1992) The GO system protects organisms from the mutagenic effect of the spontaneous lesion 8-hydroxyguanine (7,8-dihydro-8-oxoguanine). J Bacteriol 174(20):6321–6325
86. Nakabeppu Y et al (2006) MTH1, an oxidized purine nucleoside triphosphatase, prevents the cytotoxicity and neurotoxicity of oxidized purine nucleotides. DNA Repair (Amst) 5(7):761–772
87. Aguiar PH et al (2013) Oxidative stress and DNA lesions: the role of 8-oxoguanine lesions in Trypanosoma cruzi cell viability. PLoS Negl Trop Dis 7(6):e2279
88. Wen JJ et al (2008) Tissue-specific oxidative imbalance and mitochondrial dysfunction during Trypanosoma cruzi infection in mice. Microbes Infect 10(10–11):1201–1209
89. Dhiman M et al (2012) Cardiac-oxidized antigens are targets of immune recognition by antibodies and potential molecular determinants in chagas disease pathogenesis. PLoS One 7(1):e28449
90. Wen JJ, Garg NJ (2018) Manganese superoxide dismutase deficiency exacerbates the mitochondrial ROS production and oxidative damage in Chagas disease. PLoS Negl Trop Dis 12(7):e0006687
91. Wen JJ, Yin YW, Garg NJ (2018) PARP1 depletion improves mitochondrial and heart function in Chagas disease: effects on POLG dependent mtDNA maintenance. PLoS Pathog 14(5):e1007065
92. Andrade LO et al (1999) Differential tissue distribution of diverse clones of Trypanosoma cruzi in infected mice. Mol Biochem Parasitol 100(2):163–172
93. Andrade LO et al (2010) Differential tissue tropism of Trypanosoma cruzi strains: an in vitro study. Mem Inst Oswaldo Cruz 105(6):834–837
94. Dias PP et al (2017) Cardiomyocyte oxidants production may signal to T. cruzi intracellular development. PLoS Negl Trop Dis 11(8):e0005852
95. Lammel EM et al (1996) Trypanosoma cruzi: involvement of intracellular calcium in multiplication and differentiation. Exp Parasitol 83(2):240–249
96. Nogueira NP et al (2017) Heme modulates Trypanosoma cruzi bioenergetics inducing mitochondrial ROS production. Free Radic Biol Med 108:183–191
97. Buetler TM, Krauskopf A, Ruegg UT (2004) Role of superoxide as a signaling molecule. News Physiol Sci 19:120–123
98. Pervaiz S, Clement MV (2002) A permissive apoptotic environment: function of a decrease in intracellular superoxide anion and cytosolic acidification. Biochem Biophys Res Commun 290(4):1145–1150
99. Boldogh I et al (2012) Activation of ras signaling pathway by 8-oxoguanine DNA glycosylase bound to its excision product, 8-oxoguanine. J Biol Chem 287(25):20769–20773
100. Vilar-Pereira G et al (2016) Resveratrol reverses functional Chagas heart disease in mice. PLoS Pathog 12(10):e1005947

Arginase: A Prospective Regulator of Oxidative Stress During Microbial Pathogenesis

6

Surajit Bhattacharjee

Abstract

Arginase is a manganese metalloenzyme that regulates the intracellular availability of L-arginine through its hydrolysis to urea and L-ornithine. Arginine is a common substrate for both nitric oxide synthase (NOS) and arginase. L-Arginine converted to nitric oxide (NO) and L-ornithine by nitric oxide synthase. Nitric oxide (NO) initiates the immunomodulatory effects within host and imparts defense by inducing microbicidal activity against invading pathogens within the cellular microenvironment. Therefore, the competition between arginase and nitric oxide synthase for arginine plays a determinant role in immune response against microbial pathogenesis. It was reported that depletion of arginine by myeloid cells expressing arginase 1 leads to suppression of T-cell immune response. Pathogens themselves were found to synthesize arginase to escape host immune response. Recent research has identified that reactive oxygen species (ROS) plays a significant role in the regulation of arginase activity and expression during microbial infection. Pathogen survival within the host system depends on the inclination to arginase pathway. Moreover, arginase participates in arginine metabolism which provides an essential route for sustainability of pathogen survival and reproduction within the host by negotiating host defense response. Researchers currently have shown that arginase modulates the host immune response and infection pathology in many ways during microbial, parasitic, and also viral infections. In this chapter, we have discussed about the involvement of arginase in the modulation of NO during inflammatory response coined by host upon encounter to pathogen. We have highlighted the involvement of arginase in the modulation of host immune response during bacterial,

S. Bhattacharjee (✉)
Department of Molecular Biology and Bioinformatics, Tripura University (A Central University), Suryamaninagar, Tripura, India
e-mail: surajit77@tripurauniv.in

S. Chakraborti et al. (eds.), *Oxidative Stress in Microbial Diseases*,
https://doi.org/10.1007/978-981-13-8763-0_6

viral, and parasitic infection. The current and potential therapeutic approaches against microbial diseases by targeting arginase function were discussed.

Keywords
Arginases · iNOS · Polyamines · Macrophages · Immune response · Host-pathogen interactions

6.1 Introduction

Arginase (EC 3.5.3.1) is a manganese-based ureohydrolase enzyme which converts L-arginine to L-ornithine and urea [1]. The arginase enzymes were found to maintain their existence throughout evolution starting from the early life forms like bacteria. It is present in almost all life forms like yeasts, plants, invertebrates, and vertebrates [2, 3]. There are two major isoforms of arginase, namely, arginase-1 (A1) and arginase-2 (A2). A1 is a cytosolic enzyme expressed mainly in the liver and to a lesser extent in the kidneys in the majority of animals to remove surplus nitrogen produced by the metabolism of amino acids and nucleotides through urea cycle [4, 5], whereas A2 is a mitochondrial protein found predominantly in the kidneys, small intestine, brain, monocytes, and macrophages. Interestingly, most of the microorganisms, plants, and invertebrates were found to possess only A2 [3–5]. Endothelium cells are found to possess both the isoforms of arginase. The isoforms are expressed differentially under specific stimulations. In human, the A1 isoform is composed of 322 amino acids, and A2 is composed of 354 amino acids [4, 5]. Each of these arginases is produced from distinct genes (ARG1 and ARG2) located on different chromosomes in humans. ARG1 is located in chromosome 6 in human and chromosome 10 in mouse. ARG2 is located in chromosome 14 in human and chromosome 12 in mouse. The regulation of ARG1 and ARG2 gene expression are independent of each other [4, 5]. Both the arginase isoforms exhibits similar pattern of action, as both of them requires manganese as a cofactor and arginine as substrate. The expression of these two isoforms of arginases in different cell types can be regulated by a wide range of mediators and conditions. Variability in tissue and species also leads to changes in arginase activity [6].

The expression of arginases is regulated by several factors including derivatives of oxidative stress [such as hydrogen peroxide, H_2O_2; superoxide (O_2^-)], reactive nitrogen species [such as nitric oxide (NO); peroxynitrite, ($ONOO^-$)], hypoxia, inflammatory mediators (such as lipopolysaccharides (LPS)), several cytokines [such as interleukins-(IL)-4, IL-10, IL-13, IL-6, tumor necrosis factor (TNF-α)], and angiotensin II [5, 7–12]. Several intracellular signaling pathways lead to stimulation of expression and enzyme activity of arginase. The increase in arginase activity under oxidative stress is reported to be mediated through PKC-activated RhoA/Rho kinase (ROCK) pathway. Induction of AII in LPS-stimulated macrophages is found to be dependent on intracellular signaling pathways associated with mitogen-activated protein kinases [such as extracellular signal-related kinases (ERK) and p38 kinases]. Inhibition of ERK in RAW 264.7 cells was found to suppress the expression of A2 in macrophages; on the contrary, it heightened the expression of iNOS. The expression of A1 is regulated by several transcription factors including (1) CCAAT/enhancer-binding protein β (C/EBPβ); (2) signal transducer and activator of transcription (STAT)-3/6; (3) Kruppel-like factor 4 (KLF4); (4) liver X receptor α (LXRα); (5) purine box factor 1 (PU.1); (6) c-Fos; and (7) peroxisome proliferator-activated receptors (PPARs γ and δ) [5, 7–12]. All of these transcription factors modulate the A1 expression by binding with distinct and specific sites of ARG1 gene. The expression of A2 was reported to be regulated by poly(ADP-ribose) polymerase 1 (PARP-1) in endothelial cells [5, 7–12].

Arginase plays two major functions: first, to detoxify the body from ammonia through its conversion to urea, and second, to form ornithine from arginine. The cellular concentration of L-ornithine is regulated by arginase in addition to other enzymes like ornithine decarboxylase, ornithine aminotransferase, and ornithine transcarbamylase. L-ornithine is a precursor of polyamines (e.g., spermine and spermidine) and proline. Spermine and spermidine are important for cell division, growth, differentiation, and survival, whereas proline, produced as a metabolic by product of L-ornithine, plays an important role in wound healing and tissue repairing as it is an important component of collagen and fibrin [5, 13, 14].

Arginine, the substrate of arginase, is also hydrolyzed by nitric oxide synthases. Nitric oxide synthases (NOSs) convert arginine to L-citrulline and nitric oxide. Nitric oxide synthases are present in three different isoforms, which are denoted as neuronal NOS (nNOS), inducible NOS (iNOS), and endothelial NOS (eNOS). Inducible NOS can be expressed in many cell types in response to LPS, cytokines, or other agents. Generation of NO by inducible NOS has cytostatic effects on target cells [4, 5, 15, 16]. Induction of NO in infected macrophages results in strand breaks and fragmentation of DNA. These result in NO-mediated cytotoxic and cytostatic effects on invading pathogens. Interestingly, NO produced by iNOS of nonimmune cells can also participate in generating the host-protective immune response. Inducible form of nitric oxide synthase, through the production of NO, is involved in healing of skin, killing of bacteria, regulation of T-cell proliferation and differentiation (TH1/TH2), etc. [4, 5, 15, 16].

The balance activity of arginase and nitric oxide synthases (NOS) regulates the cellular level of arginine. Increased concentration of extrahepatic arginase reduces the availability of L-arginine for NOS. Thus, reciprocal regulation of arginase and NOS appear to be responsible for the production of nitric oxide during immune response. The Km value of arginase is in the range of mM, whereas that of NOS is in μM range. Interestingly, the pattern of substrate usage by these two enzymes is found to be similar due to 1000 times higher V_{max} of arginase in comparison to NOS at body pH [6, 17]. Downregulation of NO synthesis by arginase leads to tissue regeneration and induction of fibrosis. Regulation of arginine metabolism under the influence of environmental factors and immune modulators results in the elimination of intracellular parasites through NO synthesis (in presence of NOS) or in the multiplication of parasite through synthesis of L-ornithine and polyamine (in presence of arginase). NO produced by nNOS is also involved in neuronal signaling and synaptic plasticity. Reaction of NO with superoxide results in the production of peroxynitrite ($ONOO^-$). This in turn decreases the levels of NO and increases the generation of a potent oxidant (i.e., $ONOO^-$). Both superoxide and peroxynitrite increase oxidative stress, which in turn induces the level of arginase expression [5, 6, 17, 18].

Arginase is also an important marker for the M1 and M2 type of macrophages. Macrophages play an important role in immune homeostasis during inflammation-associated diseases and infectious diseases. The enzymes associated with nitric oxide synthesis, i.e., NOS and arginase, were found to be related to the phenotypic responses of pro-inflammatory type (M1) and anti-inflammatory type (M2) of macrophages [19, 20]. In response to LPS stimulation, M1 macrophages generate NO from arginine through NOS, whereas M2 macrophages produce ornithine from arginine through arginase. NO produced in M1 macrophages results in the killing of intracellular pathogens. Production of ornithine and subsequently its downstream metabolites like polyamines and proline in M2 macrophages promotes cell proliferation and supports in repairing of tissue damage [4, 6, 19, 20]. In M1 and M2 macrophages, these types of responses are regulated by specific cytokines. PAMP (pathogen-associated molecular patterns) or IFNγ stimulates M1 macrophages to produce NO. In contrast, M2 activity is amplified by IL-4, IL-13, and TGFβ. In M2 macrophages, anti-inflammatory cytokines (i.e., IL-4, IL-13, and TGF-β), cAMP, and LPS are reported to enhance the expression of AI [4, 6, 19, 20]. Accumulated evidences reveal that the main function of M2 macrophages is tissue repair and wound healing by involvement of arginase activity. Alternatively, M1 macrophages are well suited to function as key effector cells for the eradication of intracellular pathogens and tumor cells due to the presence of nitric oxide. Enhanced arginase activity may also contribute to altered immune function in conditions of trauma, infection, transplantation, autoimmunity, and cancer due to impairment of NOS2 function [4, 6, 19, 20].

The present book chapter aims to summarize the involvement of arginase in oxidative stress response during microbial pathogenesis. Modulation of arginase by different pathogens during infection and its effect on host response are illustrated in detail. Finally, the therapeutic approach to combat the modulation of arginase pathway during dreadful pathogenic invasion and consequent generation of oxidative stress will also be summarized as they hold a great potentiality for the treatment of these deadly diseases.

6.2 Arginase in Immune Response

The expression and activity of arginases are regulated by a range of immunomodulators. Anti-inflammatory cytokines like IL-4, IL-10, TGF-β, and IL-13 were reported to induce the expression of A1 in myeloid cells. Among these cytokines IL-4 and IL-13 were found to stimulate transcription factors like STAT3, STAT6, and CCAAT/enhancer-binding protein β (C/EBPβ) [7–12]. Most of these transcription factors upregulated the expression of A1 by binding to an enhancer in the ARG1 locus. The expression of A1 also regulated by several other immunomodulators like granulocyte-macrophage colony-stimulating factor (GM-CSF), prostaglandin E2 (PGE2), cyclic adenosine monophosphate (cAMP), and agonists of toll-like receptor (TLR) [5, 6, 21]. The expression of arginase was found to be inhibited by nitric oxide (NO) in macrophages. NO is produced in macrophages as an antimicrobial and tumoricidal agent during immune response. In case of M1 macrophages, NO plays the key role in the eradication of intracellular pathogens. Conversely, generation of excessive nitric oxide results in undesirable cytotoxic effect to host. Macrophages use A1 to regulate the unwanted nitric oxide production by limiting the accessibility of arginine to iNOS. Moreover, it was found that in M2 macrophages, A1 plays a crucial role in generating immune response against pathogens. In M2 macrophages, A1 controls the immune responses by utilizing arginine away from iNOS and further metabolizing it through ornithine pathways [22]. It was reported that A1 in M2 macrophages plays a pivotal role in inhibiting inflammation and fibrosis following recovery from infection with *S. mansoni* [23]. Pesce et al. in 2009 showed that deletion of A1 in macrophages resulted in increased mortality in mice due to uncontrolled inflammatory response [24]. A1 was also found to modulate the T-cell response by suppressing the multiplication of T cell and augmentation of T regulatory (Treg) cells. Activation of A1 in macrophages was found to protect the host by inhibiting the generation of IL-12/IL-23p40 along with the suppression of Th17 and Th1 differentiation during schistosomiasis [24, 25]. Conversely, A1 in macrophages is not found to be crucial in murine models of asthma and lung inflammation. Therefore, it may be suggested that the regulation of inflammatory responses by macrophage A1 varies from one organ to another organ [6].

6.3 Arginase in Bacterial Infection

The pathogenic bacteria modulate arginases to maintain their survivability within the host. *H. pylori* is a common gastric pathogen responsible for the induction of prolonged inflammatory response in the stomach. These result in the occurrence of peptic ulcer and/or gastric adenocarcinoma [26]. It has been reported that arginase in *H. pylori* is encoded by rocF gene. *H. pylori* utilizes arginase to metabolize L-arginine into L-ornithine and urea. The latter acts as a substrate for synthesizing NH_3 for neutralization of the acidic pH in the stomach. Thus, the bacterium is shielded from the high acidic environment of the stomach and colonizes within the gastric niche without any obstacle [27]. *H. pylori* induces the expression of A2 in macrophages in order to induce the apoptosis. Along with its role in acid resistance in the stomach, *H. pylori* arginase can diminish the bactericidal property of macrophages by suppressing the generation of NO [28, 29]. L-arginine is reported to be an essential amino acid for the activation of T cell. *H. pylori* arginase alters the CD3ζ expression and T-cell multiplication by depleting L-arginine concentration. Therefore, these characteristics of *Helicobacter* arginase allow the bacteria to encounter the host immune response in order to maintain their survivability on the mucous layer. A2 knockout in mice resulted in enhanced Th1/Th2 response which in turn resulted in reduced bacterial load and augmented gastritis pathology in comparison to their wild-type counterpart [30]. The apoptosis of host macrophages is attributed to *H. pylori* arginase activity, which further supports them to escape the host immune response [31]. *H. pylori* arginase utilizes the host arginine to produce polyamines, and the oxidation of polyamines like spermine generates H_2O_2 by using the enzyme spermine oxidase. It may lead to the induction of apoptosis in macrophages. Spermine is also reported to reduce the expression of iNOS and pro-inflammatory cytokines in *H. pylori*-infected macrophages [32].

Staphylococcus aureus-associated infections resulted in serious health-care threat due to increased incidence of antibiotic tolerance. *S. aureus* biofilms in host were found to enhance the expression and activity of myeloid A1 resulting in anti-inflammatory and profibrotic activities [33]. Recruitment of myeloid-derived suppressor cells (MDSCs) during *S. aureus* infection promotes the anti-inflammatory characteristics of macrophage [33].

Mycobacterium tuberculosis (Mtb), the causative agent of tuberculosis, was found to evade the host immune response by altering the macrophage-mediated metabolism of L-arginine through increased arginase expression. L-arginine metabolism by arginase limits the generation of free radicals like nitric oxide (NO), which is one of the potent antimicrobial strategies conferred by host macrophages against Mtb [34]. Upregulation of A1 expression during Mtb infection requires the involvement of the TLR2-MyD88-mediated signaling pathway and the transcription factor C/EBPβ. Induction of IL-10, IL-6, and G-CSF (granulocyte-colony stimulatory factor) by TLR2-MyD88-C/EBPβ pathway in infected macrophages increases the expression of A1 through the transcription factor STAT3. In Mtb-infected patients, A1 produced by polymorphonuclear granulocytes (PMNs) was found to suppress the expression of CD3 in T cells [35]. A1 has also been reported to inhibit the proliferation of T cells and their ability to release the cytokines in Mtb patients. NO produced in granuloma macrophages is a key immune response to protect against *Mycobacterium* infection. Other reports suggest that the induction of arginase during *Mycobacterium* infection is dependent on IL-10 [6, 35]. A1 has been found to limit the availability of L-arginine and polyamine concentration in hypoxic granulomas during Mtb infection. A current report depicts that *Mycobacterium marinum* infection leads to M1 polarization through the upregulation of iNOS and inhibition of arginase. In arginase knocked out mice infected with BCG reported to exhibit heightened nitrotyrosine in the liver granuloma. It was found that the increased production of NO in the Arg-1-deficient mice resulted in the generation of peroxynitrite during infection with *M. tuberculosis* [6, 35, 36].

Another intracellular pathogen, *Salmonella typhimurium*, causes a range of infections in human. The types of infections caused by *Salmonella* include food poisoning, enteric fever, typhoid fever, and gastroenteritis. The host cell cationic amino acid transporter (mCAT1) was found to assist the *Salmonella*-containing vacuole to utilize cytosolic arginine reservoir during infection. *Salmonella* was found to induce the expression and activity of A1 and diminish the host arginine pool for iNOS which leads to significant reduction in the NO production in macrophages of the infected host [6, 37, 38]. It allows the bacteria to secure its pathogenesis in host by avoiding nitric oxide toxicity. It was found that intact LPS (lipopolysaccharide) or other surface pathogen-associated molecular patterns (PAMPs) which are present on *S. typhimurium* cell membrane stimulate the expression of A2 in host cells [37, 38].

Extracellular bacteria like *Streptococcus pneumoniae* also target host arginase for their survivability within host organism [39]. *S. pneumonia* infection of alveolar macrophages increases the expression of A1 in an IL-13-dependent manner, resulting in attenuated host defense against this pathogen during lung infection [39]. Likewise, an increase in arginase activity in *Pseudomonas pneumonia*-infected lungs was reported to be associated with a decrease in iNOS activity and inhibition in NO production [40].

6.4 Arginase in Viral Infection

Human immunodeficiency virus (HIV) the causative agent of acquired immune deficiency syndrome (AIDS) successfully escapes host immune response in numerous ways. In HIV-infected host, arginase was found to be upregulated in myeloid cells, and this in turn reduces the bioavailability of L-arginine in the microenvironment. Since L-arginine is essential for T-cell development, this reduction in the bioavailability of L-arginine causes impaired T-cell responses [6, 41]. Limiting the L-arginine was found to arrest the infected cell in the G0-G1 phase of the cell cycle [42]. The higher arginase activity during HIV infection also results in the lower expression of CD 3ζ in T cells which further decreases T-cell response. In addition to the impaired T-cell functions, arginase also favors HIV replication in macrophages because arginase catabolizes L-arginine into polyamines which are essential for HIV replication [6, 42].

In the case of hepatitis C virus (HCV) infection, the expression of A1 was found to be increased at both mRNA and protein level which helps in viral growth in host cells. It was reported that siRNA-mediated eradication of arginase restricts the hepatocellular growth during HCV infection. It was also observed that inhibition of arginase leads to the induction of cell (HCV-infected) death through heightened NO response. Therefore, A1 plays a determinant role in the maintenance of intracellular growth of HCV along with the regulation of hepatocellular growth [43].

6.5 Arginase in Parasitic Infections

Trypanosoma cruzi is a protozoan parasite and the causative organism responsible for Chagas disease. It has been reported that during *Trypanosoma cruzi* infections, growth of parasites is supported by the induction of A1 and A2 enzymes which compete with host iNOS for the utilization of common substrate L-arginine [44]. Stempin et al. (2002) reported that "Cruzipain" a parasitic antigen enhances the arginase activity through the induction of type 2 cytokines like IL-4, IL-10, IL-13, and TGF-β. These type 2 cytokines shift arginine metabolism in macrophages toward arginase pathway [45]. Subsequently, arginase induces the polyamine synthesis specifically spermidine, spermine, and putrescine from ornithine. These polyamines support the growth of *Trypanosoma* in infected macrophages since they provide essential growth factors for synthesizing DNA and the antioxidant trypanothione [46]. Chang et al. in 2000 reported that type 2 cytokines induce arginase activity through tyrosine kinase, protein kinase A, and p38MAPK pathways [47]. It is also found that A1 and A2 mRNA expression and their catalytic activity were higher in the susceptible mouse strain (i.e., BALB/c), compared to the resistant mouse strain C57BL/6. Thus, arginase might be considered as a host marker for susceptibility to *Trypanosoma* infection.

Malaria is a spectrum of disease caused by protozoan parasite plasmodium. Malarial parasite severely modifies the host metabolism for its own survival. It was reported that during *Plasmodium* infection, parasite arginases play an important role in depleting extracellular L-arginine pool, thereby restricting the production NO by iNOS. However, it has been observed that in vivo growth of plasmodium does not depend on the parasite A1, which is unlikely to other parasites like *Trypanosoma* and *Leishmania* [6, 46].

Another protozoan parasite leishmania resides within mammalian macrophages by inhibiting the host-protective immune responses including the escaping of phagolysosomal fusion [6]. It is well documented that macrophages regulate the intracellular growth of *Leishmania* by induction of pro-inflammatory response. NO is the major leishmanicidal agent produced during heightened inflammatory response. Numerous studies have depicted that the pathogenesis of *Leishmania* within the host organisms can be attributed to the activity of arginase. It is interesting to mention that restricting the availability of L-arginine at the site of pathology results in immunosuppression and leads to noncurative forms of cutaneous and visceral leishmaniasis. *Leishmania* parasite was also found to alter the host arginase pathway for the synthesis of polyamines by ornithine decarboxylase (ODC) from ornithine [6, 46]. Polyamines provide essential growth support for the proliferation of parasite within the host microenvironment. It is reported that the polyamines like spermidine and spermine suppress host pro-inflammatory cytokine response [46]. Anti-inflammatory cytokines like IL-10, IL-4, and TGF-β play a major role in inducing the expression of A1 during leishmanial infection. It is found that *L. donovani* transports L-arginine through a high affinity amino acid permease called LdAAP3 which binds selectively to L-arginine. Interference of arginine metabolism in infected macrophages due to the presence of LdAAP3 in parasites allows the multiplication of leishmania promastigote within the hostile environment [6, 48]. Suppression of *L. major* arginase reported to restrict the intracellular growth of parasite by reducing the accessibility of ornithine for the generation of polyamines. Interestingly, it is found that concurrent suppression of both host A1 and *L. major* arginase are unable to enhance the NO level or inflammatory responses during infection. It was found that *Leishmania* devoid of arginase enzyme can survive within the hostile environment by utilizing the polyamines produced by the host [6, 41, 46].

Arginase was found to play an important role in *Schistosoma* infection. Macrophages isolated from *S. mansoni*-infected mice showed an increased expression of A1. A1 was shown to enhance the availability of proline for collagen synthesis to help the parasite survivability within host macrophages [49].

6.6 Arginase in Fungal Infection

Candida albicans is a human commensal fungus which is a natural component of human skin, gastrointestinal, and genitourinary flora. It causes lethal systemic infection if the microorganism reaches out to bloodstream in the case of immunocompromised patients. It is the fourth leading cause of nosocomial bloodstream infection. Upon ingestion within macrophages, the survival of the pathogen and further disease progression were found to be dependent on arginine metabolism where intracellular arginase plays a pivotal role [6, 50].

Internalization of *C. albicans* inside the Mϕ was found to be associated with the upregulation of the genes responsible for L-arginine biosynthesis. The rate of metabolism of L-arginine was reported to be induced during *C. albicans* infection. Moreover, Sims (1986) and Bahn and Muhlschlegel in 2006 showed that elevated level of CO_2 triggered hyphal growth. It was found that urea was produced from arginine by the action A1 in *C. albicans*-infected macrophages. Urea amidolyase metabolized urea to CO_2 and NH_3. Further, bicarbonate produced from CO_2 activates cAMP-dependent protein kinase A which triggers the formation of yeast-hypha inside Mϕs [51–53]. Yeast-hypha switching was found to be dependent on the biosynthesis of arginine. The germ tube formation acts as a virulence factor which is required to escape from macrophages. It was found that *C. albicans* express three putative arginases which are (encoded by CAR1, ORF19.3418, and ORF 19.5862) strongly induced inside macrophages [51]. Among them, CAR supports successful escape from macrophages, whereas the other two arginases are excreted. The excreted arginases utilize L-arginine of host macrophage and limit the availability of substrate for iNOS and abrogate NO production which is another survival strategy of *C. albicans*. It was found that ROS (reactive oxygen species) produced, during oxidative burst of macrophages following phagocytosis of *C. albicans*, induces arginine biosynthesis to promote filamentous growth inside macrophages for survival [51].

6.7 Therapeutic Strategies by Targeting Arginase

Arginase can be targeted to develop therapeutic strategies against microbial pathogenesis. Inhibition of arginase and its downstream polyamine synthesis provides protection from infectious diseases like leishmaniasis and African sleeping sickness [54]. The availability of proline for cell proliferation and collagen deposition is regulated by arginase isoforms in diseases like asthma and cancer. Hence, disease progression in such cases can be delayed by inhibiting arginase by N(omega)-hydroxy-nor-l-arginine (nor-NOHA) [55, 56]. It was found that A1-specific siRNA can prevent the HCV virus-associated liver carcinoma [43]. DFMO (α-difluoromethylornithine), an inhibitor of ODC (ornithine decarboxylase), and cyclohexylamine, an inhibitor of spermidine biosynthesis, were reported to be used successfully in the treatment of trypanosomiasis and *H. pylori* infection. Polyamine synthesis inhibitor has been proved to be useful in the treatment of leishmaniasis and sleeping sickness in human [57, 58]. Since arginase is a crucial factor for the survival of many pathogens, attempts of making new and more specific arginase inhibitors can be an emerging field. Glucocorticoids were reported to inhibit the activity of arginase in the presence of LPS stimulation and therefore contribute in the treatment of inflammatory airborne diseases [59].

However, stimulation of arginase expression and enzyme activation were found to aid the host by interfering with the detrimental effect of NO. Increased activity of arginase was found to support the cellular multiplication by contributing polyamines as downstream metabolites. The upregulation of arginase expression enhanced the conversion of arginine to proline for collagen synthesis. Arginase appeared as an important determinant in the functioning of mammalian immune system as the enzyme is involved in different aspects of immune functions like activation of T lymphocytes and inflammation [5, 6, 55]. Therefore, intervention with different aspects of arginase activity specifically L-arginine metabolism holds great potential for the therapy of immunosuppressive infectious diseases.

6.8 Conclusions

The equilibrium between arginase and iNOS expression and enzyme activity controls the pro-inflammatory and anti-inflammatory responses. Different pathogens employ strategies to neutralize host immune responses by interfering with arginine metabolism. Therefore, targeting specific components of the arginase/NOS pathway is a promising therapeutic approach against infectious diseases through the regulation of NO levels. Arginase is also involved in normal physiological responses like cell growth, collagen synthesis, tissue repairing, etc. Therefore, inhibition of arginase activity may lead to interference with these physiological processes. More specific studies are needed to characterize the involvement and regulation of different isoforms of arginase in normal physiological processes and during disease condition. This will help us to design more effective therapeutic strategies by targeting arginase activity.

Acknowledgments The author acknowledges the support of Ms. Trishna Mahanty and Dr. Avik Sarkar in the preparation of the manuscript. Indian government extramural research funds from Indian Council of Medical Research and SERB-Department of Science and Technology to author's laboratory are sincerely acknowledged.

References

1. Romagnoli G, Verhoeven MD, Mans R et al (2014) An alternative, arginase-independent pathway for arginine metabolism in *Kluyveromyces lactis* involves guanidinobutyrase as a key enzyme. Mol Microbiol 93(2):369–389
2. Mielczarek-Puta M, Chrzanowska A, Barańczyk-Kuźma A (2008) New insights into arginase. Part I. Structure and characteristics. Postepy Hig Med Dosw (Online) 62:206–213
3. Jenkinson CP, Grody WW, Cederbaum SD (1996) Comparative properties of arginases. Comp Biochem Physiol B Biochem Mol Biol 114(1):107–132
4. Caldwell RB, Toque HA, Narayanan SP et al (2015) Arginase: an old enzyme with new tricks. Trends Pharmacol Sci 36(6):395–405
5. Caldwell RW, Rodriguez PC, Toque HA et al (2018) Arginase: a multifaceted enzyme important in health and disease. Physiol Rev 98(2):641–665
6. Das P, Lahiri A, Lahiri A et al (2010) Modulation of the arginase pathway in the context of microbial pathogenesis: a metabolic enzyme moonlighting as an immune modulator. PLoS Pathog 6(6):e1000899
7. Chandra S, Romero MJ, Shatanawi A et al (2012) Oxidative species increase arginase activity in endothelial cells through the RhoA/Rho kinase pathway. Br J Pharmacol 165:506–519
8. Gao X, Xu X, Belmadani S et al (2007) TNF-alpha contributes to endothelial dysfunction by upregulating arginase in ischemia/reperfusion injury. Arterioscler Thromb Vasc Biol 27:1269–1275
9. Pandey D, Bhunia A, Oh YJ et al (2014) OxLDL triggers retrograde translocation of arginase2 in aortic endothelial cells via ROCK and mitochondrial processing peptidase. Circ Res 115:450–459
10. Wang L, Bhatta A, Toque HA et al (2015) Arginase inhibition enhances angiogenesis in endothelial cells exposed to hypoxia. Microvasc Res 98(1–8):2015
11. Shatanawi A, Lemtalsi T, Yao L et al (2015) Angiotensin II limits NO production by upregulating arginase through a p38 MAPK-ATF-2 pathway. Eur J Pharmacol 746:106–114
12. Sheldon KE, Shandilya H, Kepka-Lenhart D et al (2013) Shaping the murine macrophage phenotype: IL-4 and cyclic AMP synergistically activate the arginase I promoter. J Immunol 191:2290–2298
13. Morris SM (2007) Arginine metabolism: boundaries of our knowledge. J Nutr 137(6):1602S–1609S
14. Morris S Jr (2002) Regulation of enzymes of the urea cycle and arginine metabolism. Annu Rev Nutr 22:87–105. https://doi.org/10.1146/annurev.nutr.22.110801.140547
15. Förstermann U, Sessa WC (2011) Nitric oxide synthases: regulation and function. Eur Heart J 33(7):829–837, 837a–837d
16. Xue Q, Yan Y, Zhang R et al (2018) Regulation of iNOS on immune cells and its role in diseases. Int J Mol Sci 19(12):3805
17. Wu G, Morris SM Jr (1998) Arginine metabolism: nitric oxide and beyond. Biochem J 336(Pt1):1–17
18. Maarsingh H, Zaagsma J, Meurs H (2009) Arginase: a key enzyme in the pathophysiology of allergic asthma opening novel therapeutic perspectives. Br J Pharmacol 158(3):652–664
19. Rath M, Müller I, Kropf P et al (2014) Metabolism via arginase or nitric oxide synthase: two competing arginine pathways in macrophages. Front Immunol 5:532
20. Yang Z, Ming XF (2014) Functions of arginase isoforms in macrophage inflammatory responses: impact on cardiovascular diseases and metabolic disorders. Front Immunol 5:533
21. Rodriguez PC, Ochoa AC, Al-Khami AA (2017) Arginine metabolism in myeloid cells shapes innate and adaptive immunity. Front Immunol 8:93
22. Stunault MI, Bories G, Guinamard RR et al (2018) Metabolism plays a key role during macrophage activation. Mediat Inflamm 2018:Article ID 2426138, 10 pages
23. Barron L, Wynn TA (2011) Macrophage activation governs schistosomiasis-induced inflammation and fibrosis. Eur J Immunol 41(9):2509–2514

24. Pesce JT, Ramalingam TR, Mentink-Kane MM et al (2009) Arginase-1-expressing macrophages suppress Th2 cytokine-driven inflammation and fibrosis. PLoS Pathog 5(4):e1000371
25. Hesse M, Modolell M, La Flamme AC et al (2001) Differential regulation of nitric oxide synthase-2 and arginase-1 by type 1/type 2 cytokines in vivo: granulomatous pathology is shaped by the pattern of L-arginine metabolism. J Immunol 67(11):6533–6544
26. Vogiatzi P, Cassone M, Luzzi I et al (2007) Helicobacter pylori as a class I carcinogen: physiopathology and management strategies. J Cell Biochem 102:264–273
27. McGee DJ, Radcliff FJ, Mendz GL et al (1999) Helicobacter pylori rocF is required for arginase activity and acid protection in vitro but is not essential for colonization of mice or for urease activity. J Bacteriol 181:7314–7322
28. Gobert AP, McGee DJ, Akhtar M et al (2001) Helicobacter pylori arginase inhibits nitric oxide production by eukaryotic cells: a strategy for bacterial survival. Proc Natl Acad Sci U S A 98:13844–13849
29. Chaturvedi R, Asim M, Lewis ND et al (2007) L arginine availability regulates inducible nitric oxide synthase-dependent host defense against helicobacter pylori. Infect Immun 75:4305–4315
30. Zabaleta J, McGee DJ, Zea AH et al (2004) Helicobacter pylori arginase inhibits T cell proliferation and reduces the expression of the TCR zeta-chain (CD3zeta). J Immunol 173:586–593
31. Gobert AP, Cheng Y, Wang JY et al (2002) Helicobacter pylori induces macrophage apoptosis by activation of arginase II. J Immunol 168:4692–4700
32. Bussiere FI, Chaturvedi R, Cheng Y et al (2005) Spermine causes loss of innate immune response to *Helicobacter pylori* by inhibition of inducible nitric-oxide synthase translation. J Biol Chem 280:2409–2412
33. Yamada KJ, Heim CE, Aldrich AL et al (2018) Arginase-1 expression in myeloid cells regulates *Staphylococcus aureus* planktonic but not biofilm infection. Infect Immun 86(7):pii:e00206-18
34. Duque-Correa MA, Kühl AA, Rodriguez PC et al (2014) Macrophage arginase-1 controls bacterial growth and pathology in hypoxic tuberculosis granulomas. Proc Natl Acad Sci U S A 111(38):E4024–E4032
35. Qualls JE, Neale G, Smith AM et al (2010) Arginine usage in mycobacteria-infected macrophages depends on autocrine-paracrine cytokine signaling. Sci Signal 3(135):ra62
36. Jamaati H, Mortaz E, Pajouhi Z (2017) Nitric oxide in the pathogenesis and treatment of tuberculosis. Front Microbiol 8:2008
37. Ruby T, McLaughlin L, Gopinath S et al (2012) Salmonella's long-term relationship with its host. FEMS Microbiol Rev 36(3):600–615
38. Das P, Lahiri A, Lahiri A et al (2010) Cationic amino acid transporters and Salmonella Typhimurium ArgT collectively regulate arginine availability towards intracellular Salmonella growth. PLoS One 5(12):e15466
39. Knippenberg S, Brumshagen C, Aschenbrenner F et al (2015) Arginase 1 activity worsens lung-protective immunity against *Streptococcus pneumoniae* infection. Eur J Immunol 45(6):1716–1726
40. Mehl A, Ghorbani P, Douda D (2014) Effect of arginase inhibition on pulmonary L-arginine metabolism in murine *Pseudomonas pneumonia*. PLoS One 9(3):e90232
41. Takele Y, Abebe T, Weldegebreal T et al (2013) Arginase activity in the blood of patients with visceral leishmaniasis and HIV infection. PLoS Negl Trop Dis 7(1):e1977
42. Cloke TE, Garvey L, Choi BS et al (2010) Increased level of Arginase activity correlates with disease severity in HIV-seropositive patients. J Infect Dis 202(3):374–385
43. Cao W, Sun B, Feitelson MA et al (2009) Hepatitis C virus targets over-expression of arginase I in hepatocarcinogenesis. Int J Cancer 124(12):2886–2892
44. Felizardo AA, Caldas IS, Mendonça AAS (2018) Impact of Trypanosoma cruzi infection on nitric oxide synthase and arginase expression and activity in young and elderly mice. Free Radic Biol Med 129:227–236
45. Aoki MP, Guiñazú NL, Pellegrini AV (2004) Cruzipain, a major Trypanosoma cruzi antigen, promotes arginase-2 expression and survival of neonatal mouse cardiomyocytes. Am J Physiol Cell Physiol 286(2):C206–C212

46. Balaña-Fouce R, Calvo-Álvarez E, Álvarez-Velilla R et al (2012) Role of trypanosomatid's arginase in polyamine biosynthesis and pathogenesis. Mol Biochem Parasitol 181(2):85–93
47. Chang CI, Zoghi B, Liao JC et al (2000) The involvement of tyrosine kinases, cyclic AMP/protein kinase A, and p38 mitogen-activated protein kinase in IL-13-mediated Arginase I induction in macrophages: its implications in IL-13-inhibited nitric oxide production. J Immunol 165:2134–2141
48. Atri C, Guerfali FZ, Laouini D (2018) Role of human macrophage polarization in inflammation during infectious diseases. Int J Mol Sci 19(6):pii:E1801
49. Rodrigues JPF, Caldas IS, Gonçalves RV (2017) S. mansoni-T. cruzi co-infection modulates arginase-1/iNOS expression, liver and heart disease in mice. Nitric Oxide 66:43–52
50. Achkar JM, Fries BC (2010) Candida infections of the genitourinary tract. Clin Microbiol Rev 23(2):253–273
51. Ghosh S, Navarathna DH, Roberts DD et al (2009) Arginine-induced germ tube formation in *Candida albicans* is essential for escape from murine macrophage line RAW 264.7. Infect Immun 77:1596–1605
52. Mogensen E, Mühlschlegel FA (2008) CO_2 sensing and virulence of Candida albicans. In: Brakhage AA, Zipfel PF (eds) Human and animal relationships. The Mycota (A comprehensive treatise on fungi as experimental systems for basic and applied research), vol 6. Springer, Berlin/Heidelberg
53. Mock RC, Pollack JH, Hashimoto T (1990) Carbon dioxide induces endotrophic germ tube formation in *Candida albicans*. Can J Microbiol 36(4):249–253
54. Heby O, Persson L, Rentala M (2007) Targeting the polyamine biosynthetic enzymes: a promising approach to therapy of African sleeping sickness, Chagas' disease, and leishmaniasis. Amino Acids 33:359–366
55. Bansal V, Ochoa JB (2003) Arginine availability, arginase, and the immune response. Curr Opin Clin Nutr Metab Care 6:223–228
56. Ng KP, Manjeri A, Lee LM (2018) The arginase inhibitor Nω-hydroxy-nor-arginine (nor-NOHA) induces apoptosis in leukemic cells specifically under hypoxic conditions but CRISPR/Cas9 excludes arginase 2 (ARG2) as the functional target. PLoS One 13(10):e0205254
57. Wang CC (1991) A novel suicide inhibitor strategy for antiparasitic drug development. J Cell Biochem 45:49–53
58. Gonzalez NS, Huber A, Algranati ID (2001) Spermidine is essential for normal proliferation of trypanosomatid protozoa. FEBS Lett 508:323–326
59. Klasen S, Hammermann R, Fuhrmann METAL (2001) Glucocorticoids inhibit lipopolysaccharide-induced up-regulation of arginase in rat alveolar macrophages. Br J Pharmacol 132(6):1349–1357

Oxidative Stress as a Determinant of Antimicrobial Action, Resistance, and Treatment

7

Satabdi Banerjee, Suman K. Nandy, and Sajal Chakraborti

Abstract

Multidrug resistance of bacterial strains due to inappropriate use of antibiotics is a great concern of current healthcare management. In the last century, the antibiotic research was mainly focused toward synthesis of new drugs for the traditional targets such as cell wall, protein, and DNA synthesis as well as understanding the bacterial counter strategies, namely, target modification, inactivation of drug, alteration of membrane permeability, active efflux of drug, etc. The shift of the paradigm of antibiotics research toward bacterial response to antibiotic action has recognized a common pathway of bactericidal antibiotic-induced oxidative stress-mediated cell killing. All bactericidal antibiotics, but not the bacteriostatic one, hyperactivate the citric acid cycle and electron transport chain and open the flood gate of reactive oxygen species (ROS) generation including the formation of highly toxic hydroxyl radical through Fenton reaction. ROS causes severe cellular damage to the biomolecules by lipid peroxidation, carbonylation of protein, and DNA strand breakage eventually leading to cell death. Bacterial stress response mechanisms counter the oxidative stress through active drug efflux, metabolic pathway modifications, synthesis of antioxidants, and SOS repair network, which additionally help microorganisms to

Authors Satabdi Banerjee and Suman K. Nandy have been contributed equally to this chapter.

S. Banerjee
Department of Environmental Management, William Carey University,
Shillong, Meghalaya, India

S. K. Nandy (✉)
Bioinformatics Infrastructure Facility (BIF), North-Eastern Hill University (NEHU),
Tura, Meghalaya, India
e-mail: sumanknandy@nehu.ac.in

S. Chakraborti
Department of Biochemistry and Biophysics, University of Kalyani,
Kalyani, West Bengal, India

S. Chakraborti et al. (eds.), *Oxidative Stress in Microbial Diseases*,
https://doi.org/10.1007/978-981-13-8763-0_7

acquire and propagate resistance. New treatment strategies are devised on pathogen-to-pathogen basis to overturn the resistance and potentiate the bactericidal action of antibiotics by targeting the SOS response regulatory proteins as well as increasing the ROS level at the site of infection by delivery of localized species-specific metabolites and antimicrobial agents.

Keywords
Drug resistance · Antibiotic · Bactericidal · Oxidative stress · ROS

7.1 Introduction

Throughout our life, we remain associated symbiotically to numerous bacteria – harmless, beneficial, neutral, or pathogenic. In terms of number of cells present in human body, only 10% are of human and 90% are microbes. Pathogens represent an extremely small part of this microbiota responsible for a number of diseases in human. Discovery of modern antibiotics in the early twentieth century has revolutionized the treatment of pathogenic infection and becomes an important part of contemporary healthcare management. After mere 17 years of the discovery of penicillin in 1945, while receiving the Nobel Prize in Medicine, Sir Alexander Fleming cautioned mankind from the threat of antibiotic resistance of bacteria caused by the misuse of antibiotics [1, 2]. Nowadays, the exponential growth of multiple antibiotic-resistant strains of bacteria poses an increasingly significant risk to public health.

Bacterial stress response mechanism enables bacteria to adapt in the adverse and fluctuating environmental conditions. Antibiotics elicit a huge selection pressure in bacterial population, while the plasticity of prokaryotic genetic material triggers diverse but specific responses leading to a rapid genetic evolution through adaptive mutations, horizontal gene transfer, modification of gene expression pattern, and alteration of cellular physiology ultimately producing bacterial population with heterozygous level of antibiotic resistance. In the past 60 years, most efforts are centered to determine the advancement of the antibiotics' direct effect on microorganisms such as modification of membrane permeability, inhibition of protein, DNA synthesis, etc. and manifestation of bacterial countermeasures, to wit modification of antibiotic structure, cellular target, and drug-target interaction, variation in membrane permeability, and active efflux of drug across bacterial membrane [3, 4]. The bacterial physiological response to antibiotic treatments leading to cell death occurs via oxidative stress (OS) response [4, 5], SOS DNA stress response [4, 6], heat-shock protein stress response [4, 7], and their role in drug resistance, which recently grabbed the attention of researchers. Understanding the rationale behind the antibiotic resistance will surely lay foundation for the new therapeutic avenues and boost the scope of development of new drugs.

In this review, we primarily concentrate on the antibiotic resistance in response to OS. Exposure of antibiotics induces different kinds of stress in bacterium and host, where oxidative immune response is one of the most prominent aspects in this scenario. Various studies have suggested that treatment with bactericidal antibiotics may result an increase in OS. Thus, increased antioxidant capacities may, therefore, protect a bacterium from both the host immune response and antibiotic therapy. So, to understand the tripartite relationship between OS, antibiotic action, and bacterial resistance mechanisms, one should start with how the antibiotics work and will gradually move to answer the question why they have stopped working and look for the possible remedies to get them back on track.

7.2 Antibiotics: Mode of Action

In general, antibiotics are used either to kill bacteria, i.e., the bactericidal effect, or to stop their growth, the bacteriostatic strategy, by selectively interfering with the essential biological pathways of the pathogens without bothering the host. Mostly, the goal is achieved by exploiting the relative biochemical differences between the prokaryotic and the eukaryotic molecular machineries, i.e., regulating or inhibiting enzymes of microbial protein synthesis or DNA replication and repair, or through inhibition of any nonhomologous enzyme of essential microbial metabolic pathways or by perturbing cell wall/membrane organization [8]. To disrupt the first three machineries, antibiotics are needed to enter into cell through one or more membranes, i.e., required to be capable of travel both hydrophilic and hydrophobic space.

7.2.1 Cell Wall/Membrane Organization

Bacterial cell wall biosynthesis was the first well-proved target utilized historically. Bacterial cell walls are composed of peptidoglycan scaffolds, covalently cross-linked meshworks of muramyl pentapeptide and glycans, that provide the mechanical strength to withstand osmotic lysis and harsh environmental conditions. Antibiotics utilize the advantageous absence of peptidoglycan in mammalian cells and selectively interfere at different levels of peptidoglycan synthesis (e.g., liposidomycins, tunicamycin, vancomycin), delivery (e.g., bacitracin), and cross-linking (e.g., β-lactam antibiotics/penicillin) in bacteria, resulting in weak cell walls that fail to survive the osmotic pressure and eventually burst [8–10]. Further antimicrobial peptides are used to perturb the integrity of cell membranes. Polymyxins are used as topical antibiotics as they disrupt integrity of the inner and outer cell membranes of both bacterial and eukaryotic cells by binding to the lipopolysaccharides. Ionophore antibiotics (e.g., daptomycin, monensin, and valinomycin) cause rapid depolarization of membrane potential by leaking ions, and that eventually destabilizes catabolic and anabolic pathways [11].

7.2.2 Protein Biosynthesis

Bacterial protein synthesis inhibitors form the broadest class of antibiotics, functions by interfering any of the three primary stages of bacterial protein translation, to wit (1) constitution of initiation complex comprises of small 30S ribosomal subunit (30S), mRNA, tRNA and initiation factors; (2) formation of 70S ribosome by the initiation complex and large 50S ribosomal subunit (50S); and (3) elongation and termination of polypeptide chains.

These antibiotics are mainly divided into two distinct categories considering their site of action, i.e., either as the 30S inhibitors or as the 50S inhibitors. Aminoglycosides (e.g., streptomycin, hygromycin) often show better result when used in combination with β-lactams, interact with 16S rRNA of 30S, and cause a conformational change leading to either misreading of tRNA or early termination of protein chain extension. Mistranslated proteins are known to be incorporated in cell membrane and subsequently compromise the membrane permeability and contribute to increase in uptake of aminoglycosides [12]. Tobramycin and kanamycin, on the other hand, inhibit the of 70S ribosome formation by blocking the binding site of 30S. Tetracyclines are also able to block the tRNA-binding site in both prokaryotic and eukaryotic ribosomes, but they show higher affinity toward the prokaryotic counterpart. Amphenicols (e.g., chloramphenicol), macrolides (e.g., erythromycin, azithromycin), lincosamides (e.g., lincomycin, clindamycin), and oxazolidinones (e.g., linezolid, radezolid) bind at conserved regions of peptidyl transferase center of 23S rRNA of 50S and either stop the initiation of translation or block the access of tRNA and/or terminate the elongation process prematurely by blocking the polypeptide export tunnel in 50S. All these antibiotics are generally bacteriostatic except aminoglycosides, although they are able to show bactericidal effects on treatment in species-specific manner [13, 14].

7.2.3 Nucleic Acid Synthesis

Quinolone class of antimicrobials, derivatives of nalidixic acid, is known to intervene with the preservation of chromosomal topology by binding to type II topoisomerases, e.g., bacterial DNA gyrases in Gram-negative bacteria, and topoisomerase IV in Gram-positive bacteria. Both enzymes share a highly similar structure and functional mechanism, principally introduce transient double-strand nick, relax the supercoiling, and again ligate them. Quinolones, viz., ciprofloxacin, and fluoroquinolones arrest bacterial topoisomerases between DNA strand breakage and joining i.e. at DNA relegation stage. Accumulation of permanent double-strand breaks blocks DNA replication, mRNA transcription, protein translation, and cell division, ultimately leading to the initiation of SOS repair and cell death [15]. Prevention of SOS response not only improves the bactericidal effects of quinolones but also nullifies the development of antibiotic resistance [16, 17]. Therapeutic strategies were also developed to stop the bacterial RNA synthesis by utilizing the conserved structure of DNA-dependent bacterial RNA polymerase. Rifamycins are the most popular member of the family rifampin/rifampicin that bind to RNA polymerase and impose a physical blockage in the way of nascent oligonucleotide elongation. Therefore if the transcription process is already executed beyond a certain number of steps, rifamycins fail to inhibit. The broad-spectrum antibiotic rifamycins efficaciously work against Gram-negative, Gram-positive, and obligate intracellular bacteria including mycobacterial infections such as leprosy and tuberculosis but show distinctly lower sensitivity for mammalian RNA polymerase [18].

7.2.4 Metabolic Pathway

Targeting the central metabolic pathways in bacteria still remains as a prospective but largely unexplored area [19]. Folic acid is essential for nucleic acid synthesis and metabolism of amino acids. While human manage the uptake of folic acid from dietary supplements, microorganisms synthesize it on their own. Thus, disruption of bacterial folic acid synthesis pathway will lead to cessation of growth and cell division of bacteria. A combination of sulfonamides (such as sulfamethoxazole, sulfacetamide, sulfisomidine, sulfathiazole) and diaminopyrimidines (e.g., trimethoprim, tetroxoprim, brodimoprim, iclaprim) is often used successfully against Gram-negative and Gram-positive microorganisms. Sulfonamides competitively inhibit the early enzyme dihydropteroate synthase by mimicking its substrate para-aminobenzoic acid, whereas trimethoprim inhibits the enzyme dihydrofolate reductase at the final step of the pathway [20].

7.3 Contribution of Reactive Oxygen Species (ROS) in Antibiotic Action

Apart from the well-documented mechanism of actions of antibiotics against microorganisms, recent studies on the microbial response to bactericidal antibiotics point toward a possible ROS-mediated common mechanism involving metabolic pathways that further boost the antibiotic action. The antibiotic activity of gramicidin A is generally attributed to the alteration of membrane permeability and composition; the new findings with *Staphylococcus aureus* point toward hydroxyl radical formation coupled with the disruption of TCA cycle and momentary reduction of NADH level [21]. Similar increase in ROS production is observed in wild-type *Staphylococcus epidermidis* in response to β-lactams, while the tolerant strains depicted TCA cycle dysfunction and cell surface modifications [22]. Duan et al. [23] have evidenced that in the presence of exogenous bacterial growth inhibitor l-serine, even the fluoroquinolone-resistant *E. coli* strains became susceptible to fluoroquinolones. The combinatorial treatment showed markedly higher efficacy against the resistant strains along with increased production of NADH, disruption of iron-sulfur clusters, and stimulation of bacterial ROS production. Further antibiotic-mediated ROS generation was directly documented by single-cell real-time fluorescence assay, chemiluminescence, and electron paramagnetic resonance in various studies [22, 24, 25]. Mutational and inhibitory studies involving triggered dysfunction of electron transport chain (ETC) and aerobic respiration have suggested reduced aminoglycoside susceptibility in *E. coli* and *Pseudomonas aeruginosa* [26, 27]. The varying antibiotic tolerance of superoxide dismutase (SOD)-deficient *Enterococcus faecalis* with the variation of energy resources provides additional support for the reliance of antibiotic action on metabolic pathways. Utilization of hexoses has shown the highest killing, followed by pentoses, whereas glycerol has failed to pose any threat for enterococcus survival. The above portrayed the importance of glycolysis pathway [28].

7.4 The Common Pathway of ROS-Mediated Antibiotic Killing

Endogenous ROS generation is the inevitable fallout of oxidative respiration in living organisms. Molecular oxygen (O_2) diffuses through the microorganisms and undergoes single-electron reduction by ETC flavoprotein complexes and generates superoxide radical ($O_2^{\bullet-}$). In steady state, bacterial Mn-SOD or Fe-SOD provides protection from $O_2^{\bullet-}$-based damages ($2O_2^{\bullet-} + 2H^+ \rightarrow H_2O_2 + O_2$) but also generates another toxic product, hydrogen peroxide (H_2O_2), and that reduced to water by

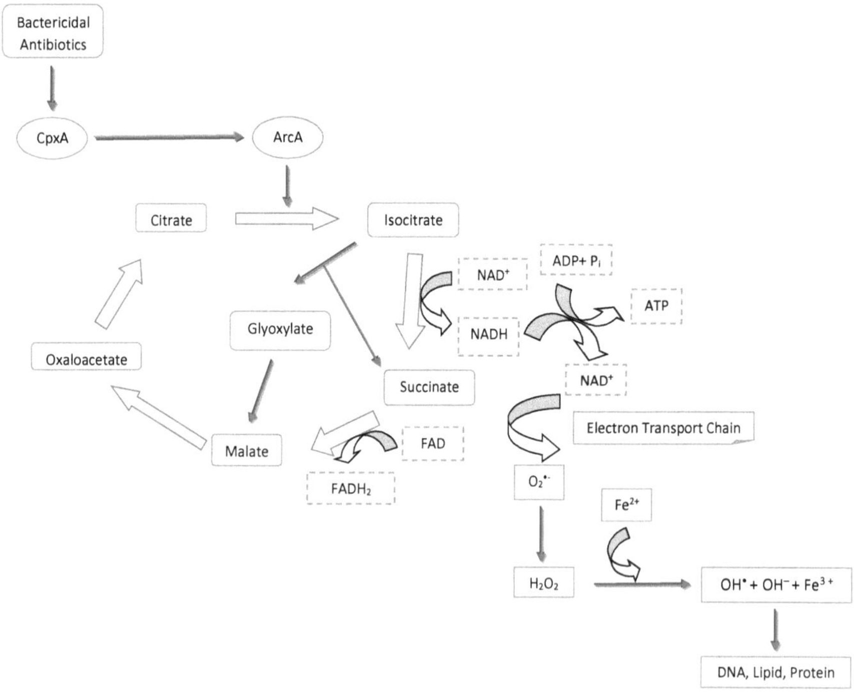

Fig. 7.1 Common mechanism of antibiotic induced ROS mediated cell killing [5]

catalases/peroxidase [29]. On the other hand, exposure of all major classes of bactericidal antibiotics in lethal doses, irrespective of their specific mode of action, induces CpxAR two-component envelope stress response system and successively redox responsive two-component regulatory system ArcA. ArcA, being a potent transcriptional regulator of energy metabolism and respiratory systems, hyperactivates the TCA cycle and ETC, thereby increasing both the ROS generation and demand of reducing equivalents, namely, NADH and $FADH_2$ (Fig. 7.1) [30, 31]. The increase in the levels of $O_2^{\bullet-}$ and H_2O_2 overcomes the dismutase/peroxidase defense mechanism and generates lethal hydroxyl radicals ($OH^{\bullet}$) via Fenton reaction ($H_2O_2 + Fe^{2+} \rightarrow OH^- + OH^{\bullet} + Fe^{3+}$). Unlike $O_2^{\bullet-}$ and H_2O_2, there is no known mechanism for in vivo $OH^{\bullet}$ detoxification. Rather $OH^{\bullet}$ can induce cytotoxic and mutagenic damage to biomolecules, namely, lipid peroxidation, loss of function, or structural integrity of protein and DNA breakage either by itself or via generation of other ROS. ROS can induce nick in single-strand or double-strand DNA, physically damage the nucleotide bases and phosphodiester backbone, as well oxidize the deoxynucleotide pool causing further damage to nucleic acids that subsequently leads to antibiotic-induced cell death [5, 32, 33]. The supply of ferrous iron (Fe^{2+}) in Fenton reaction is maintained by $O_2^{\bullet-}$ either through oxidation of available free ferric ions (Fe^{3+}) or by destabilization of iron-sulfur clusters of bacterial redox proteins causing disruption of iron regulation through Haber-Weiss reaction ($Fe^{3+} + O_2^{\bullet-} \rightarrow Fe^{2+} + O_2$). Thus, during oxidative stress, Haber-Weiss and Fenton reactions can be coupled to start a vicious cycle of $OH^{\bullet}$ and other ROS production and thereby damage various components of the cell [5, 29].

7.5 Counter Evidences of ROS-Mediated Antibiotic Killing

The evidence discussed in Sect. 7.3 points toward the involvement of TCA cycle and other metabolic pathways in antibiotic-mediated ROS formation. But, there are some counter reports questioning the quantification of antibiotic contribution in bacterial ROS generation. Wild-type and OS mutant strains of *Listeria monocytogenes,* a human pathogen with acyclic TCA cycle lacking α-ketoglutarate dehydrogenase, upon exposure to bactericidal antibiotics did not show any difference toward antibiotic sensitivity. The absence of α-ketoglutarate dehydrogenase ceases the formation of NADH [34]. But bactericidal antibiotic-mediated redox alterations as well as ROS-induced increase in the efficacy of antibiotic is acknowledged in a later study with *Streptococcus pneumoniae*, a bacterium deficient of ETC – the speculative source of ROS, indicate toward the contribution from some alternative source of ROS generation [35].

Antibiotics work with twin strategy: firstly, the important bacterial functional components that are also the major energy consumer are targeted and simultaneously the metabolic state of the microbes is modified. While the growth inhibition by bacteriostatic antibiotics is associated with downregulation of cellular respiration and metabolic pathways, the killing effect of bactericidal antibiotics is connected with the acceleration of respiration. Hence, the combinatorial application or use of growth/metabolic activity inhibitors such as $OH^{\bullet}$ scavenger and/or antioxidants (e.g., thiourea) or iron chelators (e.g., dipyridyl) decreases the efficacy of bactericidal antibiotics as reflected in many studies that question ROS contribution [36, 37] Conversely, addition of specific metabolites boosts the bactericidal cell killing in both aerobic and anaerobic conditions. Further, the physiological state of microorganisms is another important factor to decide their response to antibiotics. For example, nonavailability of nutrients may induce stringent response and illustrate antibiotic tolerance [38]. Therefore, the early reports of indifferent survival of *E. coli* in presence of a variety of antibiotics in lethal concentrations under aerobic and anaerobic conditions as well as ineffectiveness of thiourea to produce any protective effect on bacterial endurance [36, 37] are now in contrast to several later reports [21–23, 32, 39, 40]. However, the direct real-time evidences of antibiotic-mediated ROS generation are simply beyond any doubt [22, 24, 25].

Antibiotic-mediated ROS formation might not be a universal effect for bactericidal antibiotics. For example, the bactericidal effects of colistin, an antibacterial peptide, on *P. aeruginosa* are independent of $OH^{\bullet}$ generation [41]. Additionally, the genetic background of the strains also plays an important role in the antibiotic-induced ROS generation as well as the antibiotic concentration as seen in different strains of *S. aureus* and *S. pneumoniae*. As a result, antibiotics that have shown bactericidal effect in certain strains may act as a bacteriostatic in other strains of the same species [42, 43].

7.6 Bacterial Management of Antibiotic-Mediated OS

Microorganisms have portrayed diverse protective responses to all known bactericidal antibiotic-mediated ROS generation. Bacterial genetic response to the antibiotic exposure has formulated several different strategies to overcome antibiotic-induced OS. Overexpression of resistance determinants such as multidrug efflux pumps and their regulatory proteins is widely documented among bacteria. Continuous exposure of increasing concentrations of bactericidal antibiotics has shown indole production, which consecutively turn on the efflux pumps in the resistant strains of *E. coli* [44]. Additionally, the SoxR-controlled induction of acrAB multidrug efflux in *E. coli*, MgrA- and SarZ-mediated regulation of efflux genes in *S. aureus*, and mexAB-oprM drug efflux operon regulation by MexR in *P. aeruginosa* are also known to be sensitive to OS and boost bacterial survival against antibiotics [45–48].

Sub-lethal concentrations of antimicrobials are known to affect bacterial physiology and metabolism in more than one way resulting in the alteration in antibiotic sensitivity. Lower concentrations of antibiotics induce low level of ROS, which is unable to produce enough cellular damage to kill the bacteria, but stimulate the protective responses and become beneficial for bacteria in acquiring resistance. Pre-treatment with sub-optimal inhibitory concentrations of metabolic generators of superoxide induces SoxR-mediated multidrug efflux systems in *E. coli* and reduces aminoglycoside susceptibility [49]. Treatment with sub-lethal concentrations of tobramycin and gentamicin is found to decrease ROS generation in *P. aeruginosa* by partially bypassing the TCA cycle and downregulating Fe uptake to avoid the Fenton reaction [50]. Similar modifications of TCA cycle resulting in OS tolerance against β-lactum antibiotics through alteration of cell surface charge distribution and prevention of autolysis are also reported in clinical isolates of *S. epidermidis*, a normal inhabitant of human skin [22]. *Burkholderia cepacia* and *P. aeruginosa* also opt for an alternative pathway, the glyoxylate shunt, to control the antibiotic-induced ROS generation bypassing the decarboxylation steps involving reducing equivalents (NADH, $FADH_2$) production (Fig. 7.1) and thus decoupling ETC form TCA cycle [51, 52]. Sub-inhibitory concentrations of antibiotics such as tetracycline, ampicillin, cotrimoxazole, and gentamycin are known to induce differential gene expression in *Listeria monocytogenes* ensuring a shift toward anaerobic metabolism to avoid ROS formation [53]. Production of toxic molecules like hydrogen sulfide also exerts protective effect against the OS induced by wide range of bactericidal antibiotics in *Bacillus anthracis*, *S. aureus*, *E. coli*, and *P. aeruginosa* [54]. Sub-lethal doses of bactericidal antibiotics can also produce multidrug resistant micro-organisms through ROS-mediated mutagenesis by multiple mechanisms such as clash between replication and transcription machinery for highly transcribed genes, downregulation of mismatch repair system, and direct DNA damage as elaborated in Sect. 7.4 [33, 55, 56].

DNA damage caused by ROS eventually activates the SOS co-regulator RecA, which successively inactivates the SOS steric repressor protein LexA and initiates SOS gene expression. Depending upon the nature and frequency of lesions, the repair process may involve nucleotide base excision, recombination, and error-prone DNA synthesis by translesion DNA polymerases, e.g., DNA polymerases II, IV, and V. DNA polymerase V is mainly engaged in the repair of ROS-inflicted DNA damages. The lack of proofreading activity in translesion polymerases further contributes to its lower sequence fidelity resulting in increased mutation rate during SOS repair [57, 58]. ROS-mediated oxidation of most susceptible nucleotide guanine produced extremely mutagenic 8-oxo-deoxyguanine, which can form base pair with adenine and promotes mutagenesis through error-prone SOS repair system [33]. SOS response has been shown to be stimulated in *E. coli* by β-lactum antibiotics ensuing brief cessation of the cell division as a protective measure since the β-lactums are only active in growing cells [59]. In *S. aureus,* β-lactum-induced SOS repair even promotes horizontal transfer of virulence factors [60]. Activation of conserved SOS regulatory network by β-lactum-like nongenotoxic antibiotics can accentuate antibiotic-induced ROS generation. Further treatment with an antibiotic has been illustrated to gain multidrug resistance through SOS repair-mediated mutations. Ceftazidime resistance has been reported in *P. aeruginosa*, which results from metronidazole-mediated SOS response causing genetic rearrangement in integron gene cassette [61]. The importance of SOS response in resistance development and the special role of RecA have been recognized in *E. coli*, although the magnitude of the effect is highly dependent on the particular antibiotic class and associated optimal or stressed growth conditions [62].

7.7 Futuristic Treatment Strategies

Metabolism plays a guiding role in bactericidal antibiotic-mediated killing and inducing antibiotic resistance. Hyperactivation of TCA cycle and ETC generates ROS and boosts the antibiotic action, while low ROS concentrations are advantageous to microorganisms in gaining resistance through endogenous ROS-protective response. Hence, boosting bactericidal antibiotic action with appropriate supplements and perturbing the ROS defense mechanism of bacteria arise in two much-speculated avenues for treatment [63].

Metabolic models can be used to predict the ROS generation hotspots and to design species-specific therapeutic strategies to counter antibiotic tolerance. Increase in the susceptibility of *S. aureus* and *P. aeruginosa* is observed against aminoglycosides with fructose and fumarate supplementation, respectively, owing to the amplified electro chemical gradient across the bacterial membrane inducing enhancement of antibiotic uptake. Quinolone lethality could be improved against *E. coli* bacterial load by stimulating bacterial respiration through the addition of glucose and fumarate as metabolites [64, 65]. Site-specific delivery of metabolites at the site of infection may further boost the antibiotic activity. Cooperative action of interferon-gamma-induced ROS formation and application of ceftazidime antibiotic illustrated increased susceptibility of *Burkholderia pseudomallei* in infected macrophages [66]. L-serine has been reported to enhance ofloxacin activity against *E. coli* by increasing the ROS level through Fenton reaction [23]. Combinatorial treatment of isoniazid and artemisinin proved successful against dormant *M. tuberculosis* strains [65].

Hypothetical speculation of regulating the SOS gene network or inhibiting the key gene products to improve the efficacy of all bactericidal antibiotics is a long-standing postulate in the world of antibiotic design. Recently, reversal of quinolone activity has been demonstrated in a series of *E. coli* phenotypes of varying antibiotic tolerance through the suppression of SOS response [67]. Suppression of SOS response by targeting the major regulators, RecA and LexA, is a well-accepted strategy of restoring bacterial vulnerability and minimizing the emergence of resistance. Considering the presence of human homologues of RecA, the Rad51 family can form active quaternary structure with RecA, i.e., the RecA oligomerization interface comes out as a potent target. Bacteriophage-mediated delivery and expression of uncleavable LexA variants provide promising results, while combinatorial use of quinolones further increases the bacterial sensitivity [5, 68, 69]. Targeting auto-proteolysis of LexA is considered a difficult target until recently, but Mo et al. [70] came up with a solution through academia-industry partnership. Among the alternative approaches, bacteriophage-mediated species-specific delivery of antimicrobial peptides, gene networks promoting antibiotic effectiveness, or repressing repair mechanisms could be the next-generation treatment without affecting the human microbiota [6, 71]. Targeted and combination therapies with nanoantibiotics, nanomaterials loaded with antibiotics, will probably explore the advantage of both worlds and might be another promising approach toward future treatment [72].

Acknowledgments Thanks are due to the Department of Biotechnology, Ministry of Science and Technology, Government of India, for the funding and Bioinformatics Centre (BIF), NEHU, Tura Campus, for providing the infrastructure facility.

References

1. Fleming A (1929) On antibacterial action of culture of penicillium, with special reference to their use in isolation of B. influenzae. Br J Exp Pathol 10:226–236
2. Fleming A (1964) Penicillin. In: Nobel lectures, physiology or medicine 1942–1962. Elsevier Publishing Company, Amsterdam
3. Munita JM, Arias CA (2016) Mechanisms of antibiotic resistance. Microbiol Spectr 4(2). https://doi.org/10.1128/microbiolspec.VMBF-0016-2015
4. Poole K (2012) Bacterial stress responses as determinants of antimicrobial resistance. J Antimicrob Chemother 67(9):2069–2089
5. Kohanski MA, Dwyer DJ, Hayete B, Lawrence CA, Collins JJ (2007) A common mechanism of cellular death induced by bactericidal antibiotics. Cell 130:797–810
6. Kohanski MA, Dwyer DJ, Collins JJ (2010) How antibiotics kill bacteria: from targets to networks. Nat Rev Microbiol 8(6):423
7. Tran TD, Kwon HY, Kim EH, Kim KW, Briles DE, Pyo S, Rhee DK (2011) Decrease in penicillin susceptibility by heat shock protein ClpL in *Streptococcus pneumoniae*. Antimicrob Agents Chemother. https://doi.org/10.1128/AAC.01383-10
8. Walsh C (2000) Molecular mechanisms that confer antibacterial drug resistance. Nature 406(6797):775–781
9. Kimura KI, Ikeda Y, Kagami S, Yoshihama M, Suzuki K, Osada H, Isono K (1998) Selective inhibition of the bacterial peptidoglycan biosynthesis by the new types of liposidomycins. J Antibiot (Tokyo) 51(12):1099–1104
10. Storm DR (1974) Mechanism of bacitracin action: a specific lipid-peptide interaction. Ann N Y Acad Sci 235(1):387–398
11. Nelson ML, Grier MC, Barbaro SE, Ismail MY (2009) Polyfunctional antibiotics affecting bacterial membrane dynamics. Antiinfect Agents Med Chem 8(1):3–16
12. Davis BD, Chen LL, Tai PC (1986) Misread protein creates membrane channels: an essential step in the bactericidal action of aminoglycosides. Proc Natl Acad Sci U S A 83:6164–6168
13. McCoy LS, Xie Y, Tor Y (2011) Antibiotics that target protein synthesis. Wiley Interdiscip Rev RNA 2(2):209–232
14. Bhattacharjee MK (2016) Antibiotics that inhibit protein synthesis. In: Chemistry of antibiotics and related drugs. Springer, pp 129–151
15. Fàbrega A, Madurga S, Giralt E, Vila J (2009) Mechanism of action of and resistance to quinolones. Microb Biotechnol 2(1):40–61
16. Howard BM, Pinney RJ, Smith JT (1993) Function of the SOS process in repair of DNA damage induced by modern 4-quinolones. J Pharm Pharmacol 45:658–662
17. Cirz RT, Chin JK, Andes DR, de Crécy-Lagard V, Craig WA, Romesberg FE (2005) Inhibition of mutation and combating the evolution of antibiotic resistance. PLoS Biol 3:e176
18. Floss HG, Yu TW (2005) Rifamycin mode of action, resistance, and biosynthesis. Chem Rev 105(2):621–632
19. Murima P, McKinney JD, Pethe K (2014) Targeting bacterial central metabolism for drug development. Cell Chem Biol 21(11):1423–1432
20. Prescott JF (2013) Sulfonamides, diaminopyrimidines, and their combinations. In: Giguère S, Prescott JF, Dowling PM (eds) Antimicrobial therapy in veterinary medicine. Wiley, Hoboken, pp 279–294
21. Liou JW, Hung YJ, Yang CH, Chen YC (2015) The antimicrobial activity of gramicidin a is associated with hydroxyl radical formation. PLoS One 10(1):e0117065

22. Thomas VC, Kinkead LC, Janssen A, Schaeffer CR, Woods KM, Lindgren JK, Peaster JM, Chaudhari SS, Sadykov M, Jones J, AbdelGhani SM (2013) A dysfunctional tricarboxylic acid cycle enhances fitness of Staphylococcus epidermidis during β-lactam stress. MBio 4(4):e00437–e00413
23. Duan X, Huang X, Wang X, Yan S, Guo S, Abdalla AE, Huang C, Xie J (2016) L-serine potentiates fluoroquinolone activity against Escherichia coli by enhancing endogenous reactive oxygen species production. J Antimicrob Chemother 71(8):2192–2199
24. Choi H, Yang Z, Weisshaar JC (2015) Single-cell, real-time detection of oxidative stress induced in Escherichia coli by the antimicrobial peptide CM15. Proc Natl Acad Sci U S A 112(3):E303–E310
25. Páez PL, Becerra MC, Albesa I (2010) Antioxidative mechanisms protect resistant strains of Staphylococcus aureus against ciprofloxacin oxidative damage. Fundam Clin Pharmacol 24(6):771–776
26. Schurek KN, Marr AK, Taylor PK, Wiegand I, Semenec L, Khaira BK, Hancock RE (2008) Novel genetic determinants of low-level aminoglycoside resistance in Pseudomonas aeruginosa. Antimicrob Agents Chemother 52(12):4213–4219
27. Girgis HS, Hottes AK, Tavazoie S (2009) Genetic architecture of intrinsic antibiotic susceptibility. PLoS One 4(5):e5629
28. Ladjouzi R, Bizzini A, van Schaik W, Zhang X, Rincé A, Benachour A, Hartke A (2015) Loss of antibiotic tolerance in sod-deficient mutants is dependent on the energy source and arginine catabolism in enterococci. J Bacteriol:JB-00389. https://doi.org/10.1128/JB.00389-15
29. Imlay JA (2003) Pathways of oxidative damage. Annu Rev Microbiol 57(1):395–418
30. Kohanski MA, Dwyer DJ, Wierzbowski J, Cottarel G, Collins JJ (2008) Mistranslation of membrane proteins and two-component system activation trigger antibiotic-mediated cell death. Cell 135(4):679–690
31. Liu X, De Wulf P (2004) Probing the ArcA-P modulon of Escherichia coli by whole genome transcriptional analysis and sequence recognition profiling. J Biol Chem 279(13):12588–12597
32. Belenky P, Jonathan DY, Porter CB, Cohen NR, Lobritz MA, Ferrante T, Jain S, Korry BJ, Schwarz EG, Walker GC, Collins JJ (2015) Bactericidal antibiotics induce toxic metabolic perturbations that lead to cellular damage. Cell Rep 13(5):968–980
33. Neeley WL, Essigmann JM (2006) Mechanisms of formation, genotoxicity, and mutation of guanine oxidation products. Chem Res Toxicol 19(4):491–505
34. Feld L, Knudsen GM, Gram L (2012) Bactericidal antibiotics do not appear to cause oxidative stress in Listeria monocytogenes. Appl Environ Microbiol:AEM-00324. https://doi.org/10.1128/AEM.00324-12
35. Ferrándiz MJ, Martín-Galiano AJ, Arnanz C, Zimmerman T, De La Campa AG (2015) Reactive oxygen species contribute to the bactericidal effects of the fluoroquinolone moxifloxacin in Streptococcus pneumoniae. Antimicrob Agents Chemother:AAC-02299. https://doi.org/10.1128/AAC.02299-15
36. Liu Y, Imlay JA (2013) Cell death from antibiotics without the involvement of reactive oxygen species. Science 339(6124):1210–1213
37. Keren I, Wu Y, Inocencio J, Mulcahy LR, Lewis K (2013) Killing by bactericidal antibiotics does not depend on reactive oxygen species. Science 339(6124):1213–1216
38. Nguyen D, Joshi-Datar A, Lepine F, Bauerle E, Olakanmi O, Beer K, McKay G, Siehnel R, Schafhauser J, Wang Y, Britigan BE (2011) Active starvation responses mediate antibiotic tolerance in biofilms and nutrient-limited bacteria. Science 334(6058):982–986
39. Allison KR, Brynildsen MP, Collins JJ (2011) Metabolite-enabled eradication of bacterial persisters by aminoglycosides. Nature 473(7346):216
40. Lobritz MA, Belenky P, Porter CB, Gutierrez A, Yang JH, Schwarz EG, Dwyer DJ, Khalil AS, Collins JJ (2015) Antibiotic efficacy is linked to bacterial cellular respiration. Proc Natl Acad Sci U S A 201509743
41. Brochmann RP, Toft A, Ciofu O, Briales A, Kolpen M, Hempel C, Bjarnsholt T, Høiby N, Jensen PØ (2014) Bactericidal effect of colistin on planktonic Pseudomonas aeruginosa is independent of hydroxyl radical formation. Int J Antimicrob Agents 43(2):140–147

42. Liu Y, Liu X, Qu Y, Wang X, Li L, Zhao X (2012) Inhibitors of ROS accumulation delay and/or reduce lethality of several anti-staphylococcus agents. Antimicrob Agents Chemother:AAC-00754
43. Dridi B, Lupien A, Bergeron MG, Leprohon P, Ouellette M (2015) Antibiotic-induced oxidative stress responses between laboratory and clinical isolates of Streptococcus pneumoniae. Antimicrob Agents Chemother:AAC-00316
44. Lee HH, Molla MN, Cantor CR, Collins JJ (2010) Bacterial charity work leads to population-wide resistance. Nature 467(7311):82
45. Gu M, Imlay JA (2011) The SoxRS response of Escherichia coli is directly activated by redox-cycling drugs rather than by superoxide. Mol Microbiol 79(5):1136–1150
46. Chen PR, Bae T, Williams WA, Duguid EM, Rice PA, Schneewind O, He C (2006) An oxidation-sensing mechanism is used by the global regulator MgrA in Staphylococcus aureus. Nat Chem Biol 2(11):591
47. Chen PR, Nishida S, Poor CB, Cheng A, Bae T, Kuechenmeister L, Dunman PM, Missiakas D, He C (2009) A new oxidative sensing and regulation pathway mediated by the MgrA homologue SarZ in Staphylococcus aureus. Mol Microbiol 71(1):198–211
48. Chen H, Hu J, Chen PR, Lan L, Li Z, Hicks LM, Dinner AR, He C (2008) The Pseudomonas aeruginosa multidrug efflux regulator MexR uses an oxidation-sensing mechanism. Proc Natl Acad Sci U S A 105(36):13586–13591
49. Mosel M, Li L, Drlica K, Zhao X (2013) Superoxide-mediated protection of Escherichia coli from antimicrobials. Antimicrob Agents Chemother:AAC-00754
50. Mikkelsen H, Swatton JE, Lilley KS, Welch M (2010) Proteomic analysis of the adaptive responses of *Pseudomonas aeruginosa* to aminoglycoside antibiotics. FEMS Microbiol Lett. https://doi.org/10.1111/j.1574-6968.2009.01729.x
51. Van Acker H, Sass A, Bazzini S, De Roy K, Udine C, Messiaen T, Riccardi G, Boon N, Nelis HJ, Mahenthiralingam E, Coenye T (2013) Biofilm-grown Burkholderia cepacia complex cells survive antibiotic treatment by avoiding production of reactive oxygen species. PLoS One 8(3):e58943
52. Ahn S, Jung J, Jang IA, Madsen EL, Park W (2016) Role of glyoxylate shunt in oxidative stress response. J Biol Chem:jbc–M115
53. Knudsen GM, Fromberg A, Ng Y, Gram L (2016) Sublethal concentrations of antibiotics cause shift to anaerobic metabolism in listeria monocytogenes and induce phenotypes linked to antibiotic tolerance. Front Microbiol 7:1091
54. Shatalin K, Shatalina E, Mironov A, Nudler E (2011) H2S: a universal defense against antibiotics in bacteria. Science 334(6058):986–990
55. Kohanski MA, DePristo MA, Collins JJ (2010) Sublethal antibiotic treatment leads to multidrug resistance via radical-induced mutagenesis. Mol Cell 37(3):311–320
56. Jee J, Rasouly A, Shamovsky I, Akivis Y, Steinman SR, Mishra B, Nudler E (2016) Rates and mechanisms of bacterial mutagenesis from maximum-depth sequencing. Nature 534(7609):693
57. Yang W, Woodgate R (2007) What a difference a decade makes: insights into translesion DNA synthesis. Proc Natl Acad Sci U S A 104(40):15591–15598
58. Tang M, Pham P, Shen X, Taylor JS, O'donnell M, Woodgate R, Goodman MF (2000) Roles of E. coli DNA polymerases IV and V in lesion-targeted and untargeted SOS mutagenesis. Nature 404(6781):1014
59. Miller C, Thomsen LE, Gaggero C, Mosseri R, Ingmer H, Cohen SN (2004) SOS response induction by ß-lactams and bacterial defense against antibiotic lethality. Science 305(5690):1629–1631
60. Maiques E, Úbeda C, Campoy S, Salvador N, Lasa Í, Novick RP, Barbé J, Penadés JR (2006) β-lactam antibiotics induce the SOS response and horizontal transfer of virulence factors in Staphylococcus aureus. J Bacteriol 188(7):2726–2729
61. Hocquet D, Llanes C, Thouverez M, Kulasekara HD, Bertrand X, Plésiat P, Mazel D, Miller SI (2012) Evidence for induction of integron-based antibiotic resistance by the SOS response in a clinical setting. PLoS Pathog 8(6):e1002778

62. Händel N, Hoeksema M, Mata MF, Brul S, Ter Kuile BH (2016) Effects of stress, reactive oxygen species, and the SOS response on de novo acquisition of antibiotic resistance in Escherichia coli. Antimicrob Agents Chemother 60(3):1319–1327
63. Van Acker H, Coenye T (2017) The role of reactive oxygen species in antibiotic-mediated killing of bacteria. Trends Microbiol 25(6):456–466
64. Brynildsen MP, Winkler JA, Spina CS, MacDonald IC, Collins JJ (2013) Potentiating antibacterial activity by predictably enhancing endogenous microbial ROS production. Nat Biotechnol 31(2):160
65. Meylan S, Andrews IW, Collins JJ (2018) Targeting antibiotic tolerance, pathogen by pathogen. Cell 172(6):1228–1238
66. Mosovsky K, Silva E, Troyer R, Propst-Graham K, Dow S (2014) Interaction of IFN-γ induced reactive oxygen species with ceftazidime leads to synergistic killing of intracellular Burkholderia. Antimicrob Agents Chemother:AAC-02781
67. Recacha E, Machuca J, de Alba PD, Ramos-Güelfo M, Docobo-Pérez F, Rodriguez-Beltrán J, Blázquez J, Pascual A, Rodríguez-Martínez JM (2017) Quinolone resistance reversion by targeting the SOS response. MBio 8(5):e00971–e00917
68. Mo CY, Manning SA, Roggiani M, Culyba MJ, Samuels AN, Sniegowski PD, Goulian M, Kohli RM (2017) Systematically altering bacterial SOS activity under stress reveals therapeutic strategies for potentiating antibiotics. mSphere 1(4):e00163–e00116
69. Lu TK, Collins JJ (2009) Engineered bacteriophage targeting gene networks as adjuvants for antibiotic therapy. Proc Natl Acad Sci U S A 106(12):4629–4634
70. Mo CY, Culyba MJ, Selwood T, Kubiak JM, Hostetler ZM, Jurewicz AJ, Keller PM, Pope AJ, Quinn A, Schneck J, Widdowson KL (2017) Inhibitors of LexA autoproteolysis and the bacterial SOS response discovered by an academic–industry partnership. ACS Infect Dis 4(3):349–359
71. Braff D, Shis D, Collins JJ (2016) Synthetic biology platform technologies for antimicrobial applications. Adv Drug Deliv Rev 105:35–43
72. Jijie R, Barras A, Teodorescu F, Boukherroub R, Szunerits S (2017) Advancements on the molecular design of nanoantibiotics: current level of development and future challenges. Mol Syst Des Eng 2(4):349–369

Part II

Parasitic Diseases

The Biological Impact of Oxidative Metabolism in Trypanosomatid Parasites: What Is the Perfect Balance Between Reactive Species Production and Antioxidant Defenses?

8

Ana Cristina Souza Bombaça, Luiza Gervazoni Ferreira de Oliveira, Elmo Eduardo Almeida-Amaral, and Rubem Figueiredo Sadok Menna-Barreto

Abstract

Diseases caused by trypanosomatids include leishmaniasis (*Leishmania* spp.), Chagas disease (*Trypanosoma cruzi*), and sleeping sickness (*Trypanosoma brucei*) that affect millions of people, especially low-income populations, being classified as neglected tropical diseases. Limitations in the clinical treatment, associated with the huge number of cases, make these infections a health and socioeconomic problem worldwide. To complete their life cycle, trypanosomatids survive to environmental changes in different hosts, including oxidative stress. A paradoxal role of reactive oxygen species (ROS) has been proposed, such as signaling as a proliferation regulator or even presenting cytotoxic activity, depending on the concentration. Mitochondrial electron transport chain, especially complex III, is figured as one of the most important ROS resources in trypanosomatids. In relation to antioxidant defenses, trypanothione pathway plays a crucial role, being a peculiar thiol-redox system responsible for the maintenance of protozoa functions mediated by thiol-dependent processes. In this chapter, we discuss the biological aspects of oxidative stress in trypanosomatids and its implications for the success of the infection. The possible ROS resources in these protozoa and their consequent antioxidant machinery involved in detoxification were also focused in this review, including alternative strategies for the development of new drugs for these diseases based on oxidative stress modulation.

A. C. S. Bombaça · L. G. F. de Oliveira · E. E. Almeida-Amaral · R. F. S. Menna-Barreto (✉)
Laboratório de Biologia Celular, IOC, FIOCRUZ, Rio de Janeiro, Brazil
e-mail: rubemsadok@gmail.com

S. Chakraborti et al. (eds.), *Oxidative Stress in Microbial Diseases*,
https://doi.org/10.1007/978-981-13-8763-0_8

Keywords
Protozoa, Trypanosomatids · Parasitic diseases · Reactive species · Antioxidant defenses

8.1 Introduction

The World Health Organization (WHO) defined neglected tropical diseases (NTDs) as a group of illnesses that affects low-income populations in tropical and subtropical areas, without the adequate sanitary conditions, reaching over than a billion people worldwide. These conditions which are far from the ideal, together with severe limitations in the current chemotherapy, led to high mortality and morbidity rates of these diseases in developing countries [1]. Trypanosomatid parasites are responsible for some of the most important NTDs. *Leishmania* species, *Trypanosoma cruzi*, and *Trypanosoma brucei* cause leishmaniasis, Chagas disease, and sleeping sickness, respectively [2], and the implication of oxidative metabolism in the success of these infections is the focus of this review.

8.1.1 *Leishmania* spp. and Leishmaniasis

Leishmaniasis is a complex of diseases caused by over 20 species of parasites of the genus *Leishmania* [3]. This disease is endemic in 98 countries, with more than 12 million people infected and 350 million in risk areas [4]. The parasite life cycle is digenetic, with two hosts: an invertebrate host (sandfly), being also the transmitter of the disease, and the vertebrate host (mammalian), including humans and animals such as rodents, canines, marsupials, primates, and others [5–8]. *Leishmania* is responsible for two major clinical manifestations, tegumentary and visceral leishmaniasis. The tegumentary form can be divided into cutaneous, diffuse, and mucocutaneous; however, it is the visceral form which can lead to death by affecting internal organs, such as the liver and spleen [9]. Leishmaniasis chemotherapy has been based over seven decades on pentavalent antimonials; however, even being the first choice of treatment in several countries, its side effects and longstanding therapy added to all the resistance reports led to alternative drug choices, such as amphotericin B, pentamidine, paramomycin, and miltefosine [10, 11]. Amphotericin B and its lipid formulation, which reduces side effects but has a higher cost, is the first-line therapy in some countries and an alternative in cases of antimonial failure. Nevertheless, along with pentamidine and paramomycin, there is a vast report of failure cases, with side effects, longstanding treatment, and resistance reports [9]. Miltefosine, the first oral drug for leishmaniasis, has been emerging in leishmaniasis therapeutic scenario, but the resistance case reports and teratogenic effects limit its use [12].

Leishmania biological cycle presents two major hosts and two forms. When the hematophagous sandfly host bites the infected mammal, it ingests a mixture of blood and phagocytic cells infected with amastigote forms, which are released upon cellular rupture. Inside the insect gut (midgut or hindgut, depending on the *Leishmania* subgenus), amastigotes differentiate as procyclic promastigotes and develop in the sandfly. Promastigotes migrate to the proboscis and differentiate into metacyclic promastigotes, the mammalian infective form. When the infected sandfly bites a mammalian host, it regurgitates promastigotes, which are phagocytized by mononuclear phagocytic cells. Once, inside these cells, promastigotes transform into amastigotes, that proliferates up to host cell rupture, being phagocytized by other phagocytic cells, disseminating the infection [13].

8.1.2 *Trypanosoma cruzi* and Chagas Disease

Chagas disease, which is caused by the hemoflagellate protozoa *T. cruzi*, is typically a Latin American endemic illness. With the reduction in the transmission by triatominae vector, other transmission routes emerge, such as contaminated food or liquid ingestion in Brazil and transfusional transmission in Europe and North America, due to an increase in the migratory flux of infected people [14, 15]. In relation to clinical manifestations, Chagas disease presents an acute phase defined by high parasitemia in the patient bloodstream and a chronic phase where severe cardiac and/or digestive alterations are observed in 30–40% of the infected individuals decades after the acute infection [16, 17]. Up to now, benznidazole and nifurtimox are the clinical options for Chagas disease chemotherapy, being strongly effective on acute cases. On the other hand, both nitroderivatives show severe side effects as well as important activity limitations in the symptomatic chronic patients, particularly. Differences in the susceptibility of the parasite stocks isolated from distinct areas to these drugs also emphasize the necessity of the development of alternative compounds [18, 19].

The complexity of *T. cruzi* biological cycle, with different parasite evolutive stages and two hosts, also contributes for the delay in the development of novel active trypanocidal compounds. During the bloodmeal, triatomines ingest trypomastigotes with blood of the infected host. Reaching the midgut, bloodstream forms differentiate into proliferative epimastigotes that colonizes the insect digestive tract. The migration of epimastigotes to the low nutrient and acid environment of the triatomine posterior rectum triggers a new differentiation process in the parasite, and nonreplicative metacyclic trypomastigotes present in the insect feces will infect the vertebrate, invading host cells. In the intracellular environment, metacyclic forms differentiate into amastigotes that proliferates quickly. A new intracellular differentiation occurs to trypomastigotes that disrupt host cell and disseminate the infection by bloodstream [20, 21].

8.1.3 *Trypanosoma brucei* and Sleeping Sickness

Sleeping sickness, which is caused by the trypanosomatid *T. brucei rhodesiense* or *T. brucei gambiense*, is a disease restricted to sub-Saharan Africa, where the vector tsetse flies (*Glossina* spp.) is present. The transmission occurs by the insect bite, being *T. b. gambiense* the most abundant infection (98% of all cases), which is especially distributed in the Democratic Republic of Congo [22]. An estimation of WHO pointed to 70 million people under risk of infection and about 20,000 new cases per year [23–25]. Clinical manifestations of sleeping sickness mainly involve cognitive impairment, including mental confusion, personality alterations, and seizures, among others, deriving from the direct injury caused by the parasite presence in central nervous system of the host, being present for many years or months in *T. b. gambiense* or *T. b. rhodesiense* infections, respectively [26]. Unfortunately, this disease progression can lead to the patient death, if untreated. Due to the variety of illness stages and parasite subspecies, adaptations in chemotherapy approach are critical. Suramin and pentamine are the first-line drugs for early phase of *T. b. rhodesiense* and *T. b. gambiense* infections, respectively. On the other hand, the treatment of later stages is more complicated, once the compounds need to cross the blood-brain barrier. Melarsoprol and eflornithine (only for *T. b. gambiense*) are the current drugs for this stage, sometimes in combination with nifurtimox; however, their high toxicity associated with difficulties in the administration encourages the search for alternative therapies [27–29].

The evasion of the host immune system by the parasite represents a crucial challenge for the efficacy of novel drugs. Variant surface glycoproteins (VSGs) of *T. brucei* mammalian stages are produced constantly, varying the composition of the protozoa surface coat and hampering their recognition by host phagocytic cells [30]. The biological cycle of *T. brucei* starts during the tsetse fly foraging, when metacyclic trypomastigotes are inoculated after the insect bite, reaching the mammalian bloodstream. The first differentiation step takes place, and long slender forms proliferate, maintaining the infection. Central nervous system, as well as different other tissues, are infected when this stage crosses the endothelia. In order to guarantee the parasite survival in the tsetse environment, long slender forms differentiate into short stumpy forms that will be ingested by the insect. In the midgut, a new differentiation process occurs, and procyclic trypomastigotes will colonize the digestive tract of the flies. The migration of procyclic forms to the salivary glands takes place, tissue where the last parasite differentiation step will occur. Infective metacyclic trypomastigotes presented in the insect saliva will reach vertebrate host during tsetse bloodmeal [31].

8.2 Oxidative Stress

The harmful consequences of free radicals production for biological systems and its implications in aging and diseases were proposed only half century after the first description of these species by Moses Gomberg [32]. Curiously, many years later, it was demonstrated that free radicals can also present a beneficial role for the cells and tissues, and it was postulated their involvement in the killing of pathogens, and help in the immune system [33, 34]. So, the participation of these molecules has been demonstrated in a great variety of biochemical pathways, acting as regulators [35].

Even today, the terms "reactive species" and "free radicals" are employed as synonyms, generating a little confusion. Free radicals are instable and represent reactive molecules with an unpaired electron in their orbital [36, 37]. Some reactive species does not present an unpaired electron, being not considered a free radical [38]. Reactive nitrogen species (RNS) comprehend nitric oxide (NO), nitrogen dioxide (NO_2), peroxynitrite ($ONOO^-$), and nitroxyl (HNO) that participate in many crucial cellular processes [39], while also reactive oxygen species (ROS) include superoxide anion ($O_2^{\bullet-}$), hydrogen peroxide (H_2O_2), and hydroxyl radical (HO^-). Here, we only focused on the mechanisms of oxidative stress and their detoxification in pathogenic trypanosomatids and also its implications for the infection success.

In oxidative conditions, antioxidant machinery is highly expressed in order to downregulate the reactive species to physiological levels. Indeed, when such balance is broken, the concentration of ROS and/or RNS becomes higher, defining the oxidative stress condition that presents different biological implications, depending on molecular targets oxidized and the efficacy of antioxidant defenses, among other factors [40, 41]. The increase in ROS generation usually promotes this disbalance; however, the decrease in levels of antioxidant enzymes, due to reduction in expression or inactivation, also contributed for the oxidative regulation [42]. Regardless of the causes, the availability of reactive species scavengers strongly modulates almost all of biochemical pathways [43].

In the oxidative context, among all organelles and cellular structures, the mitochondrion stands out. Nobel Prize winner, Peter Mitchel, in 1961, described the co-dependency of cellular respiration and oxidative phosphorylation, being the oxygen metabolism and the ATP production causes of each other [44–46]. In the mitochondrial cristae, oxidative phosphorylation takes place, being the proton electrochemical gradient generated by electron transport through the complexes, leading to ATP synthesis [47, 48]. In aerobic conditions, the great majority of the oxygen directly reduced to water by complex IV (cytochrome c oxidase) in an electron transport chain (ETC) [49]. The partial reduction of oxygen by ETC electrons leakage represents one of the main ROS sources in eukaryotic cells, usually occurring in complexes I and III, as well as in coenzyme Q [50]. Alternatively to ETC, other enzymatic reactions, such as those catalyzed by oxidases, also can represent important ROS resource. In ETC, coenzyme Q shows an oxidized (UQ) or a reduced state (partially reduced UQ^- or fully reduced UQH_2), becoming one of the largest ROS generators in the mitochondrion. It is directly associated with its role in oxidative phosphorylation, where coenzyme Q reduction by complexes I and II allows the electrons targeting to complex III, in a step dependent of the full reduction of ubiquinone. This process includes oxidized ubiquinone that is partially reduced to semiquinone (transference of only one electron). During the return to stable state, coenzyme Q needs to be reduced again, and such reduction occurs with the semiquinone formation. This reaction is not so quick; many molecules including oxygen could be also reduced by semiquinone, producing mitochondrial ROS [51–53]. Chronologically, $O_2^{\bullet-}$ is the first ROS generated during ETC electron leakage, derived from oxygen reduction by only one electron. Due to its unstability, superoxide anion is detected in low concentrations in cells under physiological conditions, situation where superoxide dismutase (SOD) and other specific antioxidant enzymes are expressed in order to detoxify these free radicals [41, 52].

After superoxide anion production, the next ROS generated is H_2O_2 by the $O_2^{\bullet-}$ reduction with two protons accepted, concomitantly. H_2O_2 does not present an unpaired electron, being more stable and less reactive than $O_2^{\bullet-}$. Once such molecule is still more reactive than molecular oxygen, it is called ROS but it is not considered a free radical. Peroxidases and catalases are specific antioxidant enzymes involved in H_2O_2 removal [41]. In a new reduction step, HO^- is produced from H_2O_2, a highly reactive free radical that interacts with more protons and electrons, generating water as a product from Fenton reaction [54]. For sure, HO^- is the most damaging ROS, impairing a great variety of biological processes; however, due to its high instability, this molecule presents low half-life in cells compared to other ROS [41, 55, 56]. In summary, cellular physiology is directly affected by the consequences of these species production, damaging macromolecules, organelles, and structures, which can lead to phospholipids peroxidation, including in the plasma membrane, and causing subsequent cell rupture [57] (Fig. 8.1).

8.2.1 Oxidative Metabolism in Pathogenic Trypanosomatids

The mitochondrial metabolic processes that occur in the majority of the organisms are also present in trypanosomatids, despite some specific peculiarities of this organelle [58] (Fig. 8.1). ETC exhibits unique features when comparing their enzymatic complexes to the canonical system. In higher eukaryotes, the complex I (NADH: ubiquinone oxidoreductase) contains up to 30 accessory subunits whose function remains largely unknown [59–61]. It catalyzes the transfer of electrons from NADH to ubiquinone, restoring the NAD^+, with the concomitant translocation of four protons across the mitochondrial inner membrane [62]. In trypanosomatids, this complex consists in 19 subunits which were determined by proteomic analysis or deduced from genome sequence searches [63]. However, the functionality of complex I in these protozoa has been debated. Among the described subunits, all molecules known to participate in electron transfer are present, but four membrane subunits supposedly involved in proton translocation are missing. Indeed, NADH-dependent substrates are not able to stimulate ATP production in isolated mitochondrion [64]. Natural *T. cruzi* mutants which showed deletions in ND4, ND5, and ND7 genes coding for complex I subunits presented no significant differences in oxygen consumption, respiratory control ratio, and mitochondrial membrane potential in the presence of NADH-dependent substrates or $FADH_2$-generating succinate. In mammals, the complex I is also a site of ROS production; however, H_2O_2 formation induced by different substrates was not associated to complex I subunit deletions, demonstrating that these mutations are not important for the control of oxidative burst in trypanosomatids [65]. In *T. brucei*, NADH-induced respiration is sensitive to the complex I inhibitor rotenone, at a higher concentration than that dose required for inhibiting this complex in other models [66, 67], which is suggestive of the

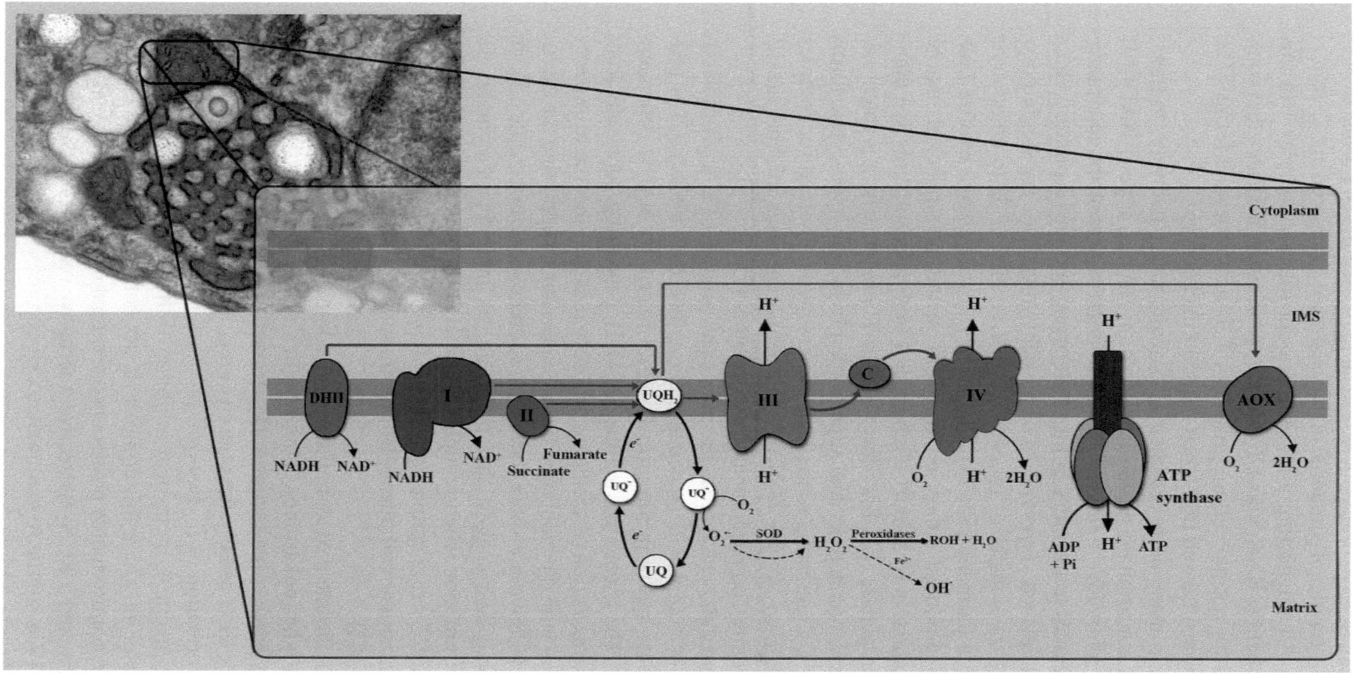

Fig. 8.1 During catabolic pathways, reduced equivalents are produced and enter in the respiratory chain at complexes I and II and at type-II NADH dehydrogenases (DHII), reducing coenzyme Q (UQH_2) pool in the inner membrane. Electrons from UQH_2 are successively transported to complex III, to cytochrome C (C), and to complex IV, where they reduce molecular O_2 to H_2O. Electron transference through complexes III and IV is coupled to H^+ translocation from the mitochondrial matrix to intermembrane space (IMS). This process induces the formation of an electrochemical gradient between both sides of the inner mitochondrial membrane, which is used to produce ATP from ADP and inorganic phosphate (Pi) by the ATP synthase. In *T. brucei*, the respiratory chain possesses an alternative oxidase (AOX) that catalyzes the reduction of O_2 to H_2O by UQH_2. The electron transference between coenzyme Q and complex III or AOX is a process dependent to the full reduction of ubiquinone (UQH_2). The partial reduction of this molecule generates a semiquinone (UQ^-), an unstable molecule that can reduce O_2, forming $O_2^{\bullet-}$. Naturally (dashed arrows) or by the action of superoxide dismutase (SOD), $O_2^{\bullet-}$ is transformed into H_2O_2, which can be reduced by the action of peroxidases or can react with metal ions, such as Fe^{2+} in a Fenton reaction, producing HO^-

inhibition of other electron carriers [68] or even a characteristic of an atypical complex I. Besides that, previous findings pointed to an increased mRNA levels of several kDNA encoded complex I subunits in bloodstream compared to *T. brucei* procyclic forms [69, 70], suggesting a more important role for complex I in this parasite stage. Nevertheless, RNAi knockdown of three subunits indicates that complex I is not required for normal culture growth of *T. brucei* procyclic forms [71], the same phenotype was observed in *T. cruzi* natural knockouts [65].

Trypanosomatids, as many other organisms, have alternative pathways to transfer electrons without concomitant proton translocation, such as type-II NAD(P)H dehydrogenases and alternative oxidases [57]. The contribution of the alternative NADH oxidizing enzymes to the entry point of electrons into the respiratory chain is not completely established. Type-II NAD(P)H dehydrogenases are single polypeptides that catalyze the transfer of two electrons from NAD(P)H to ubiquinone without coupled proton pumping [72, 73]. These enzymes are able to oxidize both NADH and NADPH produced either in cytosol or in the mitochondrial matrix, depending on the orientation of NADH binding site. In trypanosomatids, the alternative NADH dehydrogenase was first described in *T. brucei*, where a rotenone-insensitive NADH oxidation activity and superoxide production in procyclic forms were reported [74, 75]. Contrasting to the lack of phenotype observed for complex I mutants, the knockdown of type-II NAD(P)H dehydrogenases affects parasite growth and decreases mitochondrial membrane potential [71, 76, 77].

In most trypanosomatids, succinate is the major end product of glucose metabolism. This substrate of complex II (succinate:ubiquinone oxidoreductase) is produced in the glycosome and/or in the mitochondrial matrix by NADH-dependent fumarate reductases. These enzymes oxidize NADH generating succinate from fumarate [78–81]. The excretion of succinate probably indicates that the respiratory chain is not able to deal with the input of reducing equivalents, supporting the exclusive occurrence of oxidative phosphorylation from complex II to complex IV in trypanosomatids. Some data suggest that malonate, an inhibitor of complex II, impaired oxygen consumption when complex I and II substrates were added [64, 65, 82]. Complex II of *T. cruzi* epimastigotes and *L. donovani* promastigotes produce ROS after the treatment with thenoyltrifluoroacetone (TTFA), another inhibitor of complex II [83, 84].

The ubiquinol:cytochrome c oxidoreductase, or complex III, works similarly in trypanosomatids and other eukaryotes. This complex carries the electrons flow to cytochrome c, which reduces the complex IV. As well as in other organisms, trypanosomatid complex III is usually the major source of mitochondrial ROS, being responsible for $O_2^{\bullet-}$ formation in *L. donovani* promastigotes and *T. cruzi* epimastigotes induced by antimycin A, a classical inhibitor of this enzyme [83]. The only exception was *T. brucei*, in which this complex is not considered a potential ROS producer site [75]. Trypanosomatids possess two terminal oxidases: a classic complex IV KCN-sensitive and a KCN-insensitive alternative oxidase (AOX) [85–87]. The classic complex IV is similar to mammalian cytochrome c oxidase, transferring electrons from cytochrome c to the final acceptor, an oxygen molecule. This complex is important to mitochondrial functions, causing a decrease in this organelle membrane potential, reduced ATP production via oxidative phosphorylation, and redirected oxygen consumption to AOX when their indispensable subunits were repressed in *T. brucei* procyclic stage [88]. The impairment of cytochrome c oxidase also resulted in other severe mitochondrial phenotypes. Without active complex IV, the electrons flow through complex III is not able to completely reduce the oxygen. Therefore, the electrons flow from cytochrome-mediated pathway can be deviated to AOX. Gnipová et al. [88] also showed that knockdown of *T. brucei* complex IV subunits affects the complex III activity, suggesting that these subunits are responsible for the signaling mechanism that allows communication between these two sequential respiratory complexes. It can contribute to ROS production detected when cytochrome c oxidase was repressed in these parasites.

AOX, the second terminal oxidase in trypanosomatids, is restricted to the inner mitochondrial membrane of both *T. brucei* bloodstream and procyclic forms, not exhibiting proton translocation capacity and subsequent contribution to proton gradient that drives ATP formation. In bloodstream forms, the complex IV can be completely replaced by AOX, while in the insect stage, the enzyme coexists with this complex. Interestingly, KCN, a complex IV inhibitor, does not completely abolish *T. cruzi* and *L. donovani* respiratory rates, suggesting the existence of an AOX as an alternative to cytochrome c oxidase. However, the activity of salicylhydroxamic acid, an AOX inhibitor, was observed only in *T. cruzi*, suggesting a role of this enzyme in this parasite energetic metabolism [89, 90]. In *L. donovani*, the respiration KCN-insensitive still remains unclear, needing further investigation [91]. Fang and Beattie [72] showed that the inhibition of AOX by salicylhydroxamic acid stimulates ROS formation in *T. brucei*, resulting in an increase in oxidation of cellular proteins. Besides that, AOX activity increased when parasites were incubated in the presence of H_2O_2 and antimycin A, which leads to high ROS levels. These data suggested that the excess of reducing equivalents was removed by AOX in *T. brucei* transferring these equivalents to oxygen, preventing ROS production.

Kinetoplastid parasites have a complex life cycle in which they transit between invertebrate and vertebrate hosts. During the cycle, the protozoa change their morphology and metabolic profile, adaptating to diverse environmental conditions [92, 93]. In *T. cruzi*, the comparison between the energetic and oxidative metabolisms of bloodstream trypomastigotes and epimastigotes showed more active complex II–III and a restriction in electron flux to complex IV, reducing oxygen consumption and resulting in increased H_2O_2 generation in bloodstream forms [94]. These findings can be explained by the access to glucose at constant concentration that trypomastigotes have in vertebrate bloodstream. Similar results were found in *T. brucei* bloodstream stage that is essentially glycolytic, also living in an environment that presents high glucose levels. Besides that, these protozoa also lack many tricarboxylic acid cycle enzymes and cytochromes, affecting energy production [95–97]. In this regard, the invertebrate environment is glucose poor but is rich in amino acids released from intense digestion of blood proteins [92, 98, 99], resulting in high hemolymphatic levels of histidine [100, 101]. Several groups have discussed the possibility of oxidative environment as a stimulus to trypanosomatid growth. In *T. cruzi* epimastigotes, an increase in parasite proliferation was observed in response to H_2O_2 incubation, mediated by calmodulin kinase II activation. The exposure of epimastigotes to different redox-state molecules including heme, a pro-oxidant molecule derived from the insect blood digestion, increased mitochondrial ROS production and parasite replication, whereas mitochondrion-targeted antioxidant reduces ROS generation, impairing protozoa proliferation and increasing metacyclogenesis [102–104].

8.2.2 Oxidative Metabolism in the Hosts

Both invertebrate and vertebrate hosts of trypanosomatids share a common machinery of oxidative metabolism, which occurs in the mitochondrion. After decades of research in this field, critical proteins and molecules involved in the oxidative metabolism were described. The sequenced mitochondrial genome and uncoupling proteins (UCP) assays were extremely important for the overall comprehension of this process [105–107]. As described above for trypanosomatids, ETC is presented in the mitochondrial inner membrane, being the complexes I, II, III, IV, and ATP synthase functional [46]. The electrons entry in ETC occurs in complexes I or II, passing to complex III through coenzyme Q. Having received the electrons from complexes I or II, complex III uses cytochrome c to pass the electrons to complex IV, responsible for the reduction reaction that generates H_2O from O_2 [108]. As in trypanosomatids, hosts mitochondria are the main source of cellular ROS. To date, 11 sites of superoxide and/or hydrogen peroxide in mammalian mitochondria have been described, depending on the substrate metabolism, electron transport, and oxidative phosphorylation. The majority of these site-specific mitochondrial ROS production has been studied, measuring the maximum capacities of these sites under optimal conditions [52, 109]. As described above, complexes II and IV are not important ROS sources in mammals. Here, these two complexes are the main ROS-generating enzymes.

Differently of trypanosomatids, in mammals, complex I is the only entry point of electrons from NADH into the respiratory chain. This enzyme presents two domains: a hydrophilic portion located into mitochondrial matrix and a hydrophobic one, embedded in the inner mitochondrial membrane. All the known redox centers of complex I, the flavin mononucleotide cofactor (FMN), and eight FeS clusters are located in the hydrophilic domain [110]. The complex I has been recognized for a long time as one of the main sources of ROS production by the mammalian mitochondrial respiratory chain. This process was previously demonstrated, where the reduction of coenzyme Q pool and the generation of a large $\Delta\Psi m$ by succinate led to H_2O_2 production [111]. Subsequently, other authors showed that isolated complex I, in the presence of NADH, produces $O_2^{\bullet-}$ and that this production is enhanced by rotenone [112]. The mechanism of $O_2^{\bullet-}$ generation by isolated complex I is well understood, by the reaction of O_2 with the fully reduced FMN (set by $NADH/NAD^+$ ratio), which explains the enzyme inhibition by rotenone [113, 114]. During stress conditions, as ETC inhibition by damage, loss of cytochrome c, or low ATP demand and consequent low cellular respiration, the ratio of $NADH/NAD^+$ increases, leading to $O_2^{\bullet-}$ production [52]. The overexpression of a yeast NADH dehydrogenase in mammalian mitochondria reduces $O_2^{\bullet-}$ generation through the NADH/NAD$^+$ decrease [115]. As described previously to trypanosomatids, in mammals, the ROS production by complex III is dependent of coenzyme Q-cycle and semiquinone formation [116].

The oxidation of energetic substrates generates reducing cofactors, as NADH and $FADH_2$, which donate electrons to ETC. During the electron flow between the mitochondrial complexes, the dissipated energy is used by complexes I, III, and IV to translocate protons to intermembrane space, generating an $\Delta\Psi m$ across the inner membrane. Protons return to the mitochondrial matrix through ATP synthase, decrease the electrochemical gradient, and promote ATP synthesis [44]. However, the oxidative phosphorylation is partially coupled since protons can return to the mitochondrial matrix independently of ATP synthase and thereby without ATP synthesis [44, 117]. The energy-dissipating process (proton translocation to intermembrane space followed by the proton re-entry in mitochondrial matrix) apparently is present in all eukaryotic cells in a high proportion of cellular metabolic rates (up to 25% of the basal metabolic rate in the rats) and could prevent the oversupply of electrons to ETC, minimizing the probability of electron leak and $O_2^{\bullet-}$ production [118, 119]. Some authors showed the close relationship between the proton leakage and ROS production: uncoupler molecules and ADP, which increase respiration rate, stimulate ATP synthesis, and decrease $\Delta\Psi m$, are known for impaired ROS production in isolated mitochondria [120, 121].

There are at least two types of proton leakage: a basal and an inducible. The basal proton leak is unregulated and depends only on the presence but not on the activity of carrier proteins. In this case, the proton return to mitochondrial matrix occurs through the lipid bilayer and has low impact. The inducible proton leakage is a protein-mediated process that is regulated and could be reversibly activated and/or inhibited. Among the inner mitochondrial membrane carriers, the UCPs, proteins belonging to mitochondrial anion carrier protein (MACP) family, are the mitochondrial carriers whose participation in ROS production is better understood [122, 123]. In the late 1970s, the first UCP was described in mammalian brown adipose tissue and was designated UCP1; afterwards, UCP1 homologues were found in mammalian tissues (UCP1-5) [124–127]. While proton leakage mediated by UCP1 is crucial for adaptative thermogenesis in the cold [124, 128], the function of these protein homologues is not yet fully elucidated. One of the differences between UCP1 and their homologues is the abundance in individual cells and nonthermogenic tissues, influencing the oxidative phosphorylation and ROS production [129]. The role of UCP in H_2O_2 generation was first demonstrated in 1997, showing that the inhibition of UCP2 by GDP results in higher $\Delta\Psi m$ and ROS production [130]. Studies with UCP2 knockout mice demonstrated an oxidative burst in macrophages and liver and also improved resistance to *Toxoplasma gondii*. In contrast, UCP2 overexpression decreased ROS generation. In addition, $O_2^{\bullet-}$ and lipid peroxidation products (4-hydroxy-2-nonenal (HNE)) have been described to activate UCPs [131, 132]. Chemical uncouplers, such as 2,4-dinitrophenol (DNP) and FCCP (carbonyl cyanide p-tri-fluoromethoxyphenylhydrazone), also had a protector effect, where they contributed to a decrease in oxidative stress in the skeletal muscle, heart, and brain [133–135].

8.3 Antioxidant Machinery

As discussed above, mitochondrion is one of the major ROS sources. It is well-known that the reactions involving ATP synthesis are able to release toxic products such as ROS and RNS. ROS accumulation led to severe biological consequences, and a machinery to regulate oxidative stress is essential to minimize its deleterious effects [136]; therefore, eukaryotic cells, among them, trypanosomatids and their hosts, possess enzymatic and nonenzymatic antioxidant defenses. Antioxidants are molecules responsible for the prevention of substrates oxidation (or delay), decreasing the intensity of the characteristic oxidative phenotype that includes genotoxicity and injury in crucial molecules such as proteins and lipids, among others [137]. Eukaryotic cells use two distinct strategies against the oxidative stress described up to now: the blockage of radical formation by antioxidant molecules (enzymes or not) or even the increase in the expression of specific enzymes that remove oxidized biomolecules [138, 139].

Considering the molecular mass, antioxidants are classified into (i) high molecular mass antioxidants (≥10 KDa), which include antioxidant enzymes, such as transferrin, albumin, and ferritin, which strongly bind to metal ions, among other pro-oxidant molecules with high oxidizing potential [43], and (ii) low molecular mass antioxidants (≤1 KDa), such as tocopherol (vitamin E), ascorbic acid (vitamin C), anthocyanins, carotenoids, uric acid, and polyphenols, which are obtained during the alimentation of almost all organisms [43, 138].

8.3.1 Antioxidant Defenses in Pathogenic Trypanosomatids

8.3.1.1 Trypanothione/Trypanothione Reductase

The antioxidant system of trypanosomatids is based on trypanothione/trypanothione reductase, an alternative system to glutathione/glutathione reductase, absent in these protozoa [140] (Fig. 8.2). The trypanothione [N1, N8-bis (glutathionyl) spermidine] is well-characterized, being formed by the binding of two glutathiones (GSH) and one spermidine (SPD) molecule in the cytosol [141–145]. Trypanothione biosynthesis is divided into three different steps, including GSH and SPD synthesis, similar to the pathway in other organisms.

GSH synthesis begins with the formation of γ-glutamylcysteine from the binding of L-glutamate and L-cysteine, catalyzed by γ-glutamylcysteine synthase (GCS). Such step is a limiter of the reaction in mammals and trypanosomatids. Previous studies showed that low GCS levels in *L. infantum* reduce the resistance to oxidative stress and consequently parasite survival in activated macrophages [146–149]. In *T. brucei*, the reduction in GCS levels resulted in a decrease in GSH and trypanothione concentrations; however, GCS knockdown led to an increase in GSH uptake,

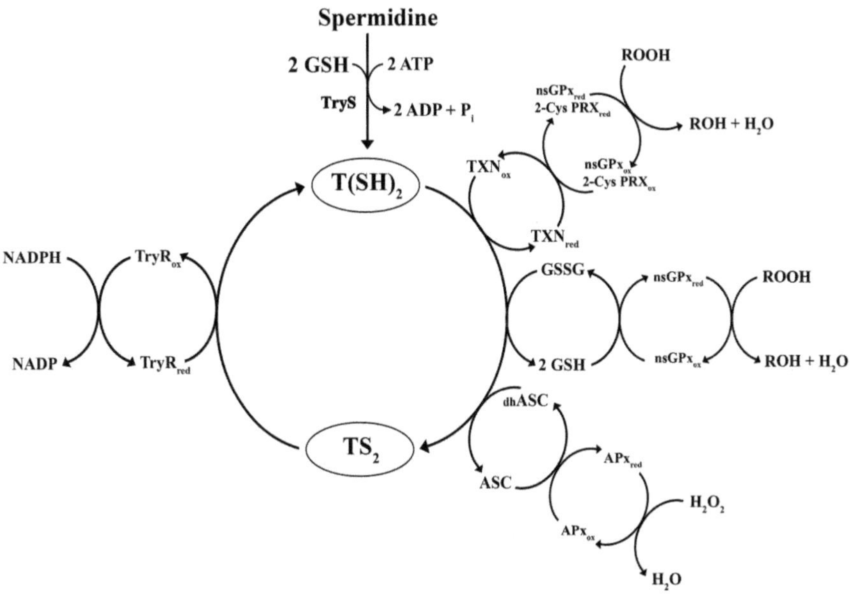

Fig. 8.2 The trypanothione is formed by the binding of two glutathione (GSH) molecules and one spermidine molecule in a reaction catalyzed by trypanothione synthetase (TryS). In this system, the maintenance of reduced trypanothione [dihydrotrypanothione – $T(SH)_2$] is dependent on trypanothione reductase (TryR) action by NADPH consumption. $T(SH)_2$ directly reduces tryparedoxin (TXN), dehydroascorbate (dhASC) to ascorbate (ASC), and glutathione disulfide (GSSG) to GSH. Trypanothione disulfide (TS_2); 2-Cys peroxiredoxin (2-Cys PRX); non-selenium glutathione peroxidase-like enzymes (nsGPx); ascorbate peroxidase (APx); hydroperoxides (ROOH)

reversing trypanothione levels in this parasite [145, 150]. Once human blood has low GSH levels, the GSH uptake is not the primary source of these molecules in *T. brucei* bloodstream forms [151]. Curiously, GSH transporters were not described in *Leishmania* spp. and *T. cruzi* [145]. For GSH synthesis, it is necessary the presence of cysteine, an amino acid with a thiol group, that confers its redox capacity. In trypanosomatids, cysteine can be taken and/or biosynthesized, although, despite *Leishmania* spp. and *T. cruzi* express cysteine carriers, the uptake levels are much lower than in *T. brucei*, which expressed highly efficient transporters [152–155]. Alternatively, *Leishmania* spp. and *T. cruzi* have two biosynthetic pathways, also present in other eukaryotes, de novo or cysteine assimilatory pathway (CAP) and trans-sulfuration pathway (TSP). Increased levels of protein expression and activity of cysteine synthase and cystathionine β-synthase in *L. braziliensis*, key enzymes of these pathways, led to elevated thiol concentrations in response to oxidative and nitrosative stresses, confirming the association between cysteine biosynthesis and stress response. Promastigotes and amastigotes expressed differently these pathways, with TSP pathway increased in insect form and higher de novo synthesis in mammalian form [156]. *T. brucei* possesses gene sequences only of TSP pathway; nevertheless, this parasite does not synthesize cysteine [145, 152–155]. The stage/

species-specific regulation of cysteine biosynthetic pathways may be due to complex life cycle and exogenous nutrients, which differ considerably between invertebrate and mammalian hosts. The de novo pathway occurs predominantly in the mammalian intracellular form, while the TSP is present in insect form. These observations are consistent with the fact that *T. brucei* and *Trypanosoma rangeli*, parasites without an intracellular stage, possess only TSP pathway [155–159].

Polyamines are simple aliphatic compounds that are found in all mammalian tissues and also in microorganisms. These molecules are essential for cell growth and differentiation and have several biological roles. In trypanosomatids, SPD is one of the most important polyamines, being involved in crucial cellular processes including synthesis of trypanothione. Polyamines could be obtained by de novo synthesis from ornithine and in some cases from arginine, or by the uptake from the extracellular medium [160]. *T. brucei* synthesize SPD through the ornithine decarboxylation mediated by ornithine decarboxylase, which generates putrescine, a substrate for spermidine synthase, while *Leishmania* spp., beyond de novo synthesis, can also uptake this molecule [161, 162]. In *T. cruzi*, enzymes of putrescine biosynthetic pathway are absent, being this parasite auxotrophic for polyamines [163, 164]. An increase in *T. cruzi* polyamine transport improves the resistance to oxidative stress in this protozoa, which can be generated by incubation with H_2O_2 or trypanocidal drugs, such nifurtimox and benznidazole [165].

The last step in trypanothione biosynthesis is the binding of GSH and SPD. This process is exclusive of few organisms, including trypanosomatids [145]. Trypanothione biosynthesis consists in an ATP-dependent addition of two GSH molecules to SPD amino groups by different pathways, depending on the protozoa. In pathogenic trypanosomatids, trypanothione is synthesized by trypanothione synthetase (TryS) in two steps: first this enzyme catalyzes the binding of one GSH molecule and SPD, forming glutathionylspermidine (intermediate product). After this, the same enzyme added a second GSH, forming the trypanothione molecule. In addition, TryS presents amidase function, hydrolyzing trypanothione and glutathionylspermidine to form GSH and SPD, suggesting an involvement in polyamine levels [145, 166–169].

In trypanosomatids, trypanothione can be found in reduced form dihydrotrypanothione ($T(SH)_2$) and/or in the oxidized form trypanothione disulfide (TS_2). $T(SH)_2$ is more reactive than GSH, being a dithiol, which favors the reduction of disulfides (different pKa values: 7.4 and 8.4 to trypanothione and GSH, respectively) [170–172]. Depending on the reactive species, $T(SH)_2$ can suffer one- or two-electron oxidation, forming thiyl radicals (RS^-) or sulfenic acid (RSOH), respectively. The thiyl radicals are formed in the reactions with peroxyl and hydroxyl radicals, nitrogen dioxide, and others, whereas the sulfenic acid is formed during the reactions with H_2O_2 and peroxynitrite [145]. Moreover, $T(SH)_2$ participate in many antioxidant pathways, reducing intermediate molecules which transfer electrons to peroxidases [173].

The $T(SH)_2$ levels are maintained by NADPH-dependent flavoenzyme trypanothione reductase (TryR). TryR has 40% identity with mammalian glutathione reductase and shares several physical and chemical characteristics with host enzyme, being their specificity of disulfides the main difference. The enzyme reduces positively charged oxidized forms of glutathionyl-polyamine conjugates, such as trypanothione, glutathionylspermidine, and others [174], whereas glutathione reductase only accepts negatively charged oxidized glutathione as a substrate. This characteristic is determined by the presence of five amino acid residues in the catalytic site, make this part wider, more hydrophobic and negatively charged when compared with glutathione reductase [175]. TryR distribution is still discussed, but some previous data suggested cytosolic and mitochondrial locations, while other reports proposed a glycosome localization [176–178]. *T. cruzi* and *T. brucei* have a carboxy-terminal extension with a tripeptide segment that would direct the enzyme to glycosome [179–181]. TryR is a potential drug target due to its importance for the survival of trypanosomatids. *L. donovani* mutants that presents only 15% of the wild-type TryR activity presented a normal growth in culture, but also showed an increased susceptibility to oxidative stress and a reduced viability inside macrophages. Similarly, the depletion of this enzyme in *T. brucei* led to an increase in sensitivity to H_2O_2 of the parasite, which then cannot successfully infect mice [182–184].

8.3.1.2 Tryparedoxin, a Trypanothione-Dependent Peroxidase System

Catalase and glutathione peroxidase (GPx) are absent in pathogenic trypanosomatids, making these parasites more susceptible to high concentrations of hydroperoxides [185, 186]. For many years, the elimination of low H_2O_2 concentrations was attributed to trypanothione [187, 188]. Recent studies showed the presence of three different classes of peroxidases; however, trypanothione remains essential for hydroperoxide removal (Fig. 8.2). Among these peroxidases, tryparedoxin (TXN) is a member of thioredoxin superfamily (Trx) that transfers reducing equivalents to different thiol proteins, which are found exclusively on Kinetoplastida [189, 190]. In contrast to classical thioredoxins, TXN is not directly reduced by a NADPH-dependent flavoprotein; however, these molecules are reduced by $T(SH)_2$/TryR system at the expenses of NADPH in the parasite [191–193]. Experimental analyses pointed cytosol as the localization of TXN in different trypanosomatids, but *in silico* approaches also suggested the mitochondrion and endoplasmic reticulum as possible target organelles to the enzyme in *Leishmania* spp. and *T. cruzi* [173, 177, 191–195]. TXN depletion in *L. infantum* and *T. brucei* promoted the impairment of the antioxidant metabolism, compromising the parasite survival [195, 196]. Studies with *L. infantum* and *L. donovani* showed that TXN is essential for promastigotes and amastigotes stages, particularly during the establishment of infection [195, 197, 198].

Tryparedoxin peroxidase (TXNPx) is a class of enzymes that use TXN as electron donor. This process occurs in two steps: (1) $T(SH)_2$ reduces TXN, being considered a regulatory reaction in the pathway; and (2) TXNPx is reduced by TXN [199]. The TXNPx includes two types of peroxidases, the peroxiredoxins (PRXs) and non-selenium glutathione-like enzymes (nsGPx). PRXs are a family of antioxidant enzymes that are present in several organisms, detoxifying hydroperoxides and peroxynitrite through cysteine (Cys) residues. These enzymes are divided in two categories, depending on the quantity of Cys residues involved in the reaction. The enzyme of trypanosomatids is a typical 2-Cys peroxiredoxin (2-Cys PRX) distributed in cytosol and mitochondrion, although, in *Leishmania* spp., one of the genes that encode this protein has a glycosomal signal sequence [190, 200]. The 2-Cys PRX has two identical reactions centers, being the substrate reduced by one of the Cys residues present in these centers, forming a cysteine sulfenic acid (Cys-SOH). Then, this residue is attacked by a Cys residue, forming a stable disulfide bond that is removed by TXN oxidoreductase [201]. 2-Cys PRX overexpression in *Leishmania* spp. and *T. cruzi* increased the resistance to hydroperoxides and peroxynitrite; however, the excess of substrate can reduce the enzyme activity in *T. cruzi* [202–205].

The second class of TNXPx, the nsGPxs, is structurally similar to glutathione peroxidase, but the selenocysteine residue in the active site of this enzyme was replaced for Cys residue [206]. These enzymes are classified as nsGPx-AI-III and nsGPx-B with different localizations. In *T. cruzi*, nsGPx-A1 is found in glycosome and cytosol, while nsGPX-AII-III are restricted to the cytoplasm. Interestingly, nsGPx-B is exclusively located in the endoplasmic reticulum in this parasite [177, 207]. In *T. brucei* and *Leishmania* spp., nsGPX-AI-II exhibit cytosolic localization, but on the other hand, nsGPx-AIII shows glycosomal and mitochondrial signal peptides [181, 208]. The nsGPx mechanism of action is similar to 2-Cys PRX, with the two Cys residues in the active site being responsible for substrate oxidation. The affinity of nsGPxs to substrates is a peculiarity of these enzymes. The depletion and/or mutation of amino acid residues in the catalytic site is responsible for a decrease in glutathione-binding capacity [207]. In this way, nsGPx-A uses TXN during substrate reduction, while nsGPx-B shows low affinity to both molecules [177, 209, 210].

8.3.1.3 Ascorbate Peroxidase

Despite the first description by Clark et al. [211], the biological relevance of ascorbic acid (vitamin C) in trypanosomatids was demonstrated almost a decade later [212]. This molecule acts as a cofactor for a wide range of enzymes involved in distinct metabolic processes, among them the antioxidant system [213]. The ascorbate peroxidase (APx), an antioxidant enzyme that uses ascorbic acid as a reductor agent, is a heme-containing peroxidase that catalyzes H_2O_2 reduction [209]. This enzyme has an endoplasmic reticulum localization in *T. cruzi* and *Leishmania* spp., whereas in *T. brucei*, there are no reports about this existence [185, 214, 215]. APx acts by cleaving O-O bound in H_2O_2 through the reaction between heme and reactive species, producing water. This process generates an intermediate that is reduced by ascorbate in two sequential electron transfer, restoring APx. Dehydroascorbate, the oxidized molecule, returns to the reduced form in two nonenzymatic reactions that may include $T(SH)_2$ and TXN [212, 216, 217]. Taylor et al. [218] showed that APx overexpression in *T. cruzi* confers protection against H_2O_2 exposure, whereas its depletion results in enhanced sensitivity. The enzyme activity is not required to parasite fundamental processes, such as replication and virulence.

8.3.1.4 Fe-Superoxide Dismutase

The SOD is an antioxidant enzyme described in most eukaryotes, being responsible for $O_2^{\bullet-}$ dismutation in H_2O_2 and O_2 [219]. This enzyme action is dependent on a metallic cofactor, such as Cu/ZnSOD and MnSOD, which are found in mammalian cytosol and mitochondria, respectively. In trypanosomatids, *in silico* analyses showed four SOD genes that use iron as cofactor. These isoforms are expressed in different organelles; FeSOD-B is distributed between cytoplasm and glycosomes of *T. cruzi* and *T. brucei*, whereas FeSOD-A and FeSOD-C are found in *T. brucei* mitochondrial matrix and intermembrane space, while only the first isoform was described in *T. cruzi* [220–223]. FeSOD-D was described in the three pathogenic trypanosomatids, being its activity related with apoptotic-like death in *T. cruzi* and *L. donovani*, while in *T. brucei*, it is considered nonessential [57, 224, 225]. The downregulation of FeSOD-A in *Leishmania* spp. increased the parasite sensitivity to menadione, a $O_2^{\bullet-}$ producer, whereas this enzyme overexpression protects the parasite from the oxidative stress. Furthermore, the SOD expression is related to differentiation and replication processes in *Leishmania* spp. Promastigotes in stationary phase showed increased resistance to oxidative stress and higher SOD activity during amastigote differentiation. In contrast, low SOD activity and ROS accumulation were found during promastigote logarithmic phase [226, 227]. In *T. cruzi*, the enzyme also has an important role in vertebrate host adaptation, with the enzymatic levels increasing during metacyclogenesis [92].

8.3.2 Antioxidant Defenses in the Hosts

8.3.2.1 Glutathione/Glutathione Reductase System

GSH (L-glutamyl-L-cysteine-L-glycine) is the most abundant low molecular mass thiol in eukaryotic cells [228, 229]. Its reduced form is also the active form (GSH) that is oxidized to glutathione disulfide (GSSG) in the oxidative stress. GSH is synthesized in cytosol and is transported into mitochondrion by dicarboxylate carrier protein and 2-oxoglutarate carrier protein [230]. Glutathione reductase (GR) is the enzyme responsible for maintain glutathione in its reduced form since the abundance of its oxidized form leads to a decrease in GSH/GSSG ratio, which serves as a warning of oxidative stress [229]. GR has been found in all organisms analyzed thus far, being a highly conserved enzyme in highly divergent organisms, such as *Homo sapiens* and *E. coli* [231]. GR has two cysteines in the catalytic domain and two other domains that bind to FAD and NADPH. Glutathione is found in cytoplasm, nucleus, mitochondria, and endoplasmic reticulum, as well as it seems to be present in lysosomes [232]. The formation of disulfides between GSH and protein cysteine residues constitutes a protective mechanism for thiols, which prevents their further oxidation, protecting cells from oxidative stress [233]. However, the formation of protein disulfides can alter the function of thiol-based proteins, such as receptors, protein kinases, and transcription factors, impairing cell signaling. In this context, GSSG is able to play a role in a nonspecific cell signalization [230]. GSH is considered the most important redox molecule present in organisms, above all, mammals. This molecule is able to play several roles, such as a redox buffer, acting as cofactor scavenger for antioxidant enzymes such as GPx. This enzyme is able to detoxify H_2O_2 and lipid peroxidation products by the reaction of selenocysteines (presents in its active site) with peroxide group, forming GSSG and H_2O [228, 230, 233, 234].

GSH is not only found in cytosol but also in endoplasmic reticulum, nucleus, and intermembrane space. The transport of this molecule to nucleus is thought to bcl-2 facilitating passive diffusion via nuclear pores [235], and nuclear pool of GSH is responsible for regulating the redox state of protein sulfhydryls, in order to prevent DNA oxidative damage [229, 230]. The regulation of cytosolic GSH transport to the nucleus appears to follow cell cycle progression, balancing the GSH cytosolic/GSH nuclear according to cellular proliferation [236]. Reaching the endoplasmic reticulum, GSH is responsible for maintaining the homeostasis and the thiol levels in catalytic sites for PDI (protein disulfide isomerase) protein folding. Besides endoplasmic reticulum capacity to produce ROS, other oxidative inducers can lead to an unbalanced environment, impairing natural protein folding [237, 238]. Inside mitochondria, GSH is able to control mitochondria ROS generation by ETC and preserve mitochondrial proteins and lipids integrity [239, 240]. GSH also prevents toxic effects of free intracellular metals, such as iron, preventing its reaction with O_2 and Fenton reaction [228].

8.3.2.2 Other Thiol-Dependent Enzymes

Trx is part of a major system called TRX system, which includes, besides the Trx, the thioredoxin reductase (TrxR) and NADPH. Trx is a ubiquitous protein with a redox-active dithiol/disulfide site [241]. As GSH/GSSG, Trx appears in a reduced form [$Trx(SH)_2$], with a dithiol group, and in an oxidized form ($TrxS_2$), with a disulfide bound. Trx contains a conserved site Cys-Gly-Pro-Cys found in all organisms. In mammalian cells, Trx is described by having two isoforms, known as Trx-1 and Trx-2. Trx-1 is localized in cytosol, being transported to the nucleus during oxidative stress [229], while Trx-2 is the exclusive mitochondrial Trx isoform, regulating mitochondrial homeostasis. As in GSH/GSSG system, the maintenance of reduced and oxidized Trx ratio is extremely important in TRX system. TrxR is an oxidoreductase, responsible for regulating $Trx(SH)_2/TrxS_2$ [241, 242].

PRXs are ubiquitous thiol-dependent enzymes, known as an ancient family of proteins, being evolutionarily conserved and present in all kinds of organisms [243]. This peroxidase was first observed in *Saccharomyces cerevisiae*, demonstrating an antioxidant activity [244]. Unlike GSH and Trx, PRXs are selenium and heme-free molecules, with a peroxidatic cysteine (Cp) conserved in N-terminal domain. PRX mechanism of action consists in Cp attack to O–O bonds present in peroxide (reaction 3), which is oxidized into cysteine sulfenic acid [245]. PRX uses Trx as a hydrogen donor, creating an electron flow. Besides the H_2O_2, PRXs are able to act as a scavenger for peroxynitrite and lipid peroxidation products. The ability to scavenge peroxides protects prokaryotic and eukaryotic cells from DNA, lipids, and proteins oxidative damages, caused by ROS and RNS [243, 245]. Some groups have been demonstrating the relation of H_2O_2 signaling with PRX inactivation, describing new roles for this protein. Low concentrations of H_2O_2 appear to generate PRX inactivation by hyperoxidation, a reversible reaction promoted by sulfiredoxin [246, 247]. This reaction appears to follow a circadian rhythm, whereas the H_2O_2 signaling is required for different cell functions [248]. In structural terms, PRXs are divided in two subcategories: 1-Cys and 2-Cys peroxiredoxins, depending on the quantity of Cys residues involved in the reaction. There are six PRX isoforms present in mammalian: PRX I, II, III, and IV (2-Cys PRX), PRX V (atypical 2-Cys PRX), and PRX VI (1-Cys PRX) [242, 243, 245, 249]. PRX I and II are found in cytosol and nucleus. PRX III has an affinity to mitochondria, while PRX IV is found at endoplasmic reticulum. PRX V has been already detected at cytosol, mitochondria, and peroxisomes [250].

Glutaredoxin (GRX) is a small dithiol protein, also known as thioltransferase, required at the redox system. GSH is able to reduce GRX, and it has been described the potential role of its sensing changes in GSH/GSSG ratio [251]. GRX has the Trx family motif Cys-X-X-Cys, and it is able to form a disulfide bond within GSH. GRXs are also under two forms, an oxidized and a reduced form, and its reduction reaction is done by GSH, GR, and NADPH [249]. GRX is also found inside the intermembrane space, endoplasmic reticulum, and cytosol. It has been described that GRX can also display a sensing role at GSH-dependent glucose deprivation, as its interactions result in mediated cell death [251]. GRX-GSH is extremely important when TRX system decreases its activity under circumstances such as lack of selenium or Trx/TrxR inhibition, acting as a backup redox system [252, 253]. It has been described that mitochondrial GRX serves as TrxR substrate, maintaining mitochondrial redox homeostasis.

8.3.2.3 Superoxide Dismutase and Catalase

In the enzymatic redox system, SOD and catalase appear as the main enzymes in cellular detoxification. SOD is the first defense against superoxide, converting $O_2^{\bullet-}$ in H_2O_2 and O_2. This enzyme is dependent on metal as a cofactor, and in mammals, there are three different isoforms: SOD1, a Cu/ZnSOD found in cytoplasm, nucleus, and plasma membrane; SOD2, a MnSOD found in mitochondrial matrix; and SOD3, a Cu/ZnSOD found in extracellular compartments, scavenging $O_2^{\bullet-}$ released in inflammatory cascades [254, 255]. High H_2O_2 concentrations are very dangerous for the cells, being necessary its elimination. Catalase is an enzyme that works directly connected with SOD, in order to complete $O_2^{\bullet-}$ detoxification, converting H_2O_2 into H_2O and O_2 [256]. The catalase types include Fe (heme)/Mn dependent [257]. These enzymes are largely found in all mammalian tissues, especially in red blood cells, and have been described as a cardiac and neural aging protectors [255, 258, 259].

8.4 Role of Oxidative Metabolism in Hosts/Trypanosomatid Infection

These parasites must thrive endogenous toxic metabolites produced by its aerobic metabolism and deal with the oxidative burst derived from the host immune system, which include ROS production. Once some antioxidant machinery in trypanosomatids such as catalase and classical GSH/GPx system is lacking, many authors suggested that ROS production by the hosts is a defense against parasite infection. However, it is well described the ability of trypanosomatids to overcome this situation, using ROS and RNS as important signaling molecules for their survival [260–262].

Both *Leishmania* spp. and *T. cruzi* have an intracellular stage inside the vertebrates. Macrophages and neutrophils are phagocytic cells, responsible for recognizing, internalizing, and destroying pathogens, being the first contact of infective metacyclic forms, performing a key role in infection control [263, 264]. For a successful infection, *T. cruzi* metacyclic trypomastigotes must invade macrophages and survive to oxidative burst found inside the phagosome. Previous data showed that cruzipain, an immunogenic glycoprotein, induces an increase in ROS production in murine cells during parasite invasion [265]. $O_2^{\bullet -}$ is produced by an associated-membrane NADPH oxidase, contributing to the formation of an oxidative environment during phagocytosis [266]. Besides that, the increase in H_2O_2 formation is also related with NADPH oxidase activation, once that spontaneous reactions (enzymatic or not) can convert $O_2^{\bullet -}$ in this molecule [267]. When the parasite is phagocytosed, a signaling cascade is triggered, culminating in the oxidative burst. The complex NADPH oxidase, known as NOX family, is a transmembrane multimeric protein able to transfer electrons to O_2 by NADPH oxidation [268, 269]. Cytochemical data demonstrated $O_2^{\bullet -}$ production when the parasite is attached to the macrophage surface, due to NADPH oxidase activation [270]. These enzymes need calcium or cytosolic proteins to be activated and thus lead to ROS production. In early stages of infection, NADPH oxidase is activated and its subunits are directed to phagosome membrane. Seven NADPH oxidase homologues have been identified in several cells types, being the NADPH oxidase 2 (NOX 2) the homologue present in phagocytic cells [269, 271].

Other important reactive species found during macrophages infection is NO. This RNS is a highly reactive free-radical produced by the oxidation of L-arginine by nitric oxide synthase (NOS) [272], participating in trypanosomatids killing, both directly or through the interaction with other free radicals, such as $O_2^{\bullet -}$, forming $ONOO^-$ [273, 274]. NO is not a strong reactive species by itself, being unlikely to account for a direct damage to the parasite. The NO can inhibit the respiratory chain in mammalian via interactions with cytochrome c oxidase, which is accompanied by a steady-state level of reduced respiratory complexes, which favors intramitochondrial $O_2^{\bullet -}$ formation [275–277]. Proinflammatory cytokine production by macrophages, such as IL-12, INF-γ, and TNF-α, activates the inducible nitric oxide synthase (iNOS), generating high amounts of NO, which is maintained by 24 h during infection [264, 278]. The upregulation of iNOS expression in BALB/c mice is protected from infection by *Leishmania major*. Furthermore, *T. cruzi* infection-resistant C57BL/6 mice produced larger amounts of NO, which is correlated with a better control of acute-phase parasitemia [279, 280]. In addition to NADPH oxidases and iNOS, myeloperoxidase (MPO) is another enzyme present in phagocytic cells, mainly neutrophils, which is pointed out as responsible for oxidative burst. MPO is able to catalyze the reaction of hypochlorite production, one of the major neutrophil antimicrobial responses [269]. This enzyme is a lysosomal hemoprotein, member of cyclooxygenase superfamily, stored in neutrophils azurophilic granules [281, 282]. When neutrophils are activated, the MPO is released in cytoplasm during degranulation, which also releases H_2O_2. The MPO-H_2O_2 system is able to convert halide ions such as Cl^-, Br^-, and I^-, producing their oxidized forms of hypohalous acids (HOX), potent antimicrobial molecules [283, 284].

The levels of parasite antioxidant defenses during macrophage invasion may improve pathogen survival [285]. Chronic chagasic cardiomyopathy is characterized by the presence of pseudo cysts of amastigotes nests in the cardiac fiber. During *T. cruzi* invasion in cardiomyocytes, there is an increase in the production of inflammatory mediators, such as TNF-α and IL-1β, and the induction of iNOS, with subsequent NO generation [286, 287]. The biochemical and genetic heterogeneity among *T. cruzi* strains is, in part, responsible for the diverse clinical manifestations of the disease, controlling several pathogenesis aspects [288]. Taking into account the establishment of a nitroxidative stress during the acute and chronic stages of Chagas disease, *T. cruzi* antioxidant enzymes are important virulence factors and become decisive for the infection success. In this context, several proteomic analyses showed an overexpression of *T. cruzi* antioxidant machinery in the infective metacyclic trypomastigotes compared with noninfective epimastigotes [92]. Such increase, found during metacyclogenesis of different *T. cruzi* strains, may act as a general preadaptation process to allow the parasite survival in the nitroxidative environment found in the vertebrate host. Piacenza et al. [289] showed a positive association between parasite virulence and levels of antioxidant enzymes *in vivo*.

In *Leishmania* spp. infection, some strategies to escape the oxidative burst present in phagocytic cells were demonstrated. Lipophosphoglycan (LPG), a molecule widely distributed in promastigotes surface, plays an important role in intracellular survival of these parasites. The protective effect of LPG is restricted to the establishment of infection during differentiation of promastigotes into amastigotes. In vitro experiments showed that LPG decreased oxidative burst in activated monocytes through inhibition of p67phox and p47phox recruitment to NADPH oxidase complex in phagosomes [290, 291]. LPG is also able to reduce $NO^\bullet$ production, regulating the iNOS expression in macrophages [292]. Additionally, $O_2^{\bullet-}$ generation after the infection by promastigotes and amastigotes is substantially different [272]. The hypothesis for this difference is the deficiency of NADPH oxidase activity after amastigote infection. Monocytes of patients with visceral leishmaniasis produce lower levels of $O_2^{\bullet-}$ and H_2O_2 and have a decreased NADPH activity when compared with healthy controls [293, 294]. To accomplish the successful NADPH oxidase complex assembly, the maturation of gp91phox is necessary, which is dependent on heme availability. During the infection, *Leishmania pifanoi* amastigotes induces the heme oxygenase-1 expression, an enzyme responsible for heme degradation, thereby blocking gp91phox maturation and preventing NADPH activity [295, 296]. Amastigotes also induced lower levels of p47phox phosphorylation mediated by protein kinase C, decreasing the phagosomal recruitment of p67phox and p47phox to NADPH oxidase complex. Interestingly, unlike promastigotes, such effect is not attributed to LPG in amastigotes and remains unclear [291, 297, 298].

Although ROS are expected to be responsible for pathogen elimination during oxidative burst as described above, evidences suggest that ROS production could also have a beneficial role to *T. cruzi* macrophage infection. Oxidative stress mobilizes iron from host intracellular storages, which is an essential for amastigote replication. Such process occurs by the regulation of elF2α kinase. In the absence of heme, elF2α kinase is active and promotes cell growth arrest, leading to the differentiation of proliferative amastigotes into nonproliferative trypomastigotes [299]. ROS also participate in *Leishmania* spp. differentiation in a process dependent on iron availability. In this case, changes in intracellular iron levels activate a ROS-dependent signaling pathway that induces promastigotes differentiation into infective amastigotes [300]. Trypanosomatids also have contact with ROS and RNS inside the invertebrate host, in which the blood digestion, derived from the hematophagic behavior, causes an increase of heme concentration and an oxidative burst [260, 301]. Heme-induced and mitochondrial ROS stimulates epimastigote proliferation, being the contribution of mitochondrial ROS in epimastigote growth confirmed by the use of mitochondrial-targeted antioxidant that impairs parasite proliferation [103, 104]. The contribution of mitochondrial ROS to epimastigote growth was confirmed using the mitochondrion-targeted antioxidant MitoTEMPO that decreased ROS and ATP production induced by heme, strongly impairing the cell proliferation.

8.5 Oxidative Mechanisms of Action of Anti-trypanosomatid Drugs

Considering all mechanistic studies of preclinical compounds performed in trypanosomatids, mitochondrion stands out among the most recurrent targets in these parasites, and mitochondrial damage has been described as part of the mode of action of distinct drugs classes [83, 84, 149, 302–311]. Surprisingly, the components and molecular events involved in the mitochondrial susceptibility to drugs, usually detected by phenotypes such as swelling and/or depolarization of the organelle, are completely undescribed [312]. Such scarcity of molecular information about the mechanisms of action of the great majority of the compounds makes hard to prove the mitochondrial direct effect. The specificity of this organelle as a target is very controversial, representing its injury a secondary target, derived from primary effects on another biochemical processes in different cellular structures. Independently of the origin or intensity, the mitochondrial damage caused by drugs regularly promotes a calcium homeostasis and/or ROS production [313].

Additionally, the activity of mitochondrial-specific inhibitors has been evaluated in trypanosomatid parasites, essentially targeting ETC complexes. Due to the fact that the biological activity of complex I had not been demonstrated in trypanosomatids up to now, the effect of the classical inhibitor of NADH dehydrogenase, rotenone, is very debatable [68, 77, 83, 314]. On the other hand, ROS generation was stimulated in parasites treated with complexes II and III inhibitors (noyltrifluoroacetone and antimycin A, respectively), being correlated to the mitochondrial depolarization and an apoptotic-like events in some cases [66, 75, 83, 94, 315]. Interestingly, the inhibition of complex II promoted a potentialization of leishmanicidal effect of the clinical drug pentamidine in vitro [83]. In *L. donovani*, 4,4′-bis((tri-n-pentylphosphonium)methyl)benzophenone dibromide and sitamaquine, complex II inhibitors, led to a ROS production and deep mitochondrial alterations, including reduction in oxygen consumption and ATP levels as well as the remarkable swelling of the organelle and consequent cell cycle arrestment [316, 317]. Similar alterations could also be observed after treatment of *L. donovani* with tafenoquine and miltefosine, which inhibited complexes III and IV, respectively [318, 319]. Furthermore, in almost all pathogenic trypanosomatids, complex IV activity was, at least partially, inhibited by KCN, a classical cytochrome c oxidase inhibitor in higher eukaryotes [84, 91, 94, 320–322]. The absence of AOX in humans makes this oxidase another interesting mitochondrial drug target [323]. In *T. brucei*, it was demonstrated that ascofuranone affects ubiquinol oxidase activity in vitro, producing an apoptotic-like phenotype [323–326]. The alkyl lysophospholipid analogue edelfosine interferes in ATP synthase activity in *Leishmania*, being suggested a correlation between this biological effect to the leishmanicidal mechanism of action [327].

It is well-established that some chemical characteristics confer high redox potential to some compounds, leading to ROS generation [328, 329]. In this context, structural differences presented in the quinoidal nucleus directly influenced the oxidative capacity of quinones [330]. The effect of naturally occurring quinones and derivatives has been investigated in different *Leishmania* spp., *T. cruzi*, and *T. brucei*, and a promising activity was observed [331–339]. The first description of the oxidative effect of a quinone in trypanosomatids was performed in the late 1970s, demonstrating that beta-lapachone induced ROS generation in *T. cruzi* epimastigotes [340, 341]. More recently, other naphthoquinones also showed similar mode of action, producing considerable amounts of reactive species in this parasite [321, 342–344]. Almost a decade ago, a mechanistic proposal was raised by our research group, in order to explain the anti-*T. cruzi* effect of naphthofuranquinones. In 2009, our data pointed to the mitochondrial depolarization, derived from electron flow impairment, probably due to the electrons deviation from ubiquinone to the compounds. It compromises electron flux, producing ROS, leading to the impairment of the mitochondrial function, reflected by the reduction in respiratory rates, complexes I–III activity, and the dilation of the organelle [321]. Natural products also promoted mitochondrial damage/ROS production on different species of *Leishmania*. Quercetin, apigenin, and epigallocatequin-3-gallate are flavonoids working as potent ROS inductors in both *Leishmania* forms, causing mitochondrial dysfunction such as a decrease in ATP levels by altering mitochondrial membrane potential [345–349]. Thiosemicarbazones, 1,3,4-thiadiazole, and triazoles derivatives, or even other classes of drugs such as LQB-118, Flau-A, clioquinol, and pyrazyl/pyridylhydrazones derivatives, also induced morphological injury in the mitochondrion and ROS production, suggesting the trigger of parasite cell death [350–356]. Interestingly, a trypanothione reductase inhibitor, cyclobenzaprine also increased ROS levels, which may be the mechanism of the antileishmanial effect caused by this compound [357].

After half century of the development of nifurtimox and benznidazole, the available clinical options for Chagas disease, their mode of action is still debatable. The first mechanistic hypothesis for the trypanocidal activity of both compounds was proposed in early 1980s, indicating $O_2^{\bullet-}$ and H_2O_2 generation induced by nifurtimox; however, such production was not observed after the treatment with benznidazole [358–361]. In this way, the effect of nifurtimox depends on the type-II nitroreductases activity that transforms nitroanion radical, producing ROS and their subsequent biological consequences such as lipids peroxidation [359, 362, 363]. Unfortunately, Boiani et al. [364] showed no correlation between anti-*T. cruzi* effect of nifurtimox and ROS production, which was demonstrated by the absence of redox cycling at trypanocidal concentrations together with low molecular weight thiol reduction. Recently, the trypanocidal activities of nifurtimox and also benznidazole were associated with type-I nitroreductase in oxygen-independent way, and nitroso and hydroxylamine intermediates would generate amine, using NADH as a cofactor [365–367]. Based on the nitroderivative, different reactions of these intermediates would take place. For example, furane ring would be cleaved, producing a highly reactive unsaturated open chain nitrile in case of nifurtimox [367, 368]. For benznidazole, such cleavage led to glyoxal and other metabolites production, which directly interacts with DNA [367].

Moreover, despite the current clinical drugs for sleeping sickness or leishmaniasis have not been associated with oxidative stress, a great variety of compounds leads to the mitochondrial functional impairment, increasing ROS production during the treatment. The molecular targets of anti-trypanosomatid drugs involved in these protozoa oxidative stress events reported experimentally were described in Fig. 8.3.

8.6 Conclusions

During the macrophage/parasite interaction, the host cell triggers a signaling cascade, recruiting immune cells to combat the infection. Macrophages induce NO production via iNOS and $O_2^{\cdot}$, which is also required in initial phagocytic steps; therefore, an oxidative burst against the parasites is developed [262, 369, 370]. Both superoxide and NO can also generate peroxynitrite, a toxic-free radical for pathogens [371]. Neutrophils, the first-line defense at pathogen infection, are also able to induce NO production in order to kill parasites, but in a smaller scale than macrophages [371, 372]. However, both leishmania and trypanosome parasites developed several evasion systems, being able to fool host oxidative burst mechanisms.

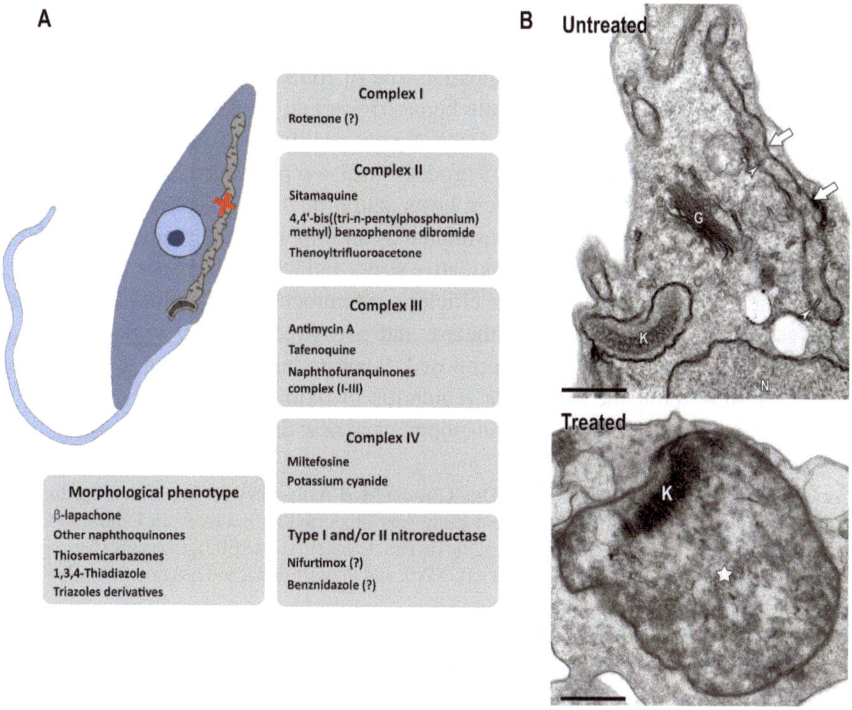

Fig. 8.3 Main mitochondrial molecular targets of anti-trypanosomatid drugs (**a**) and a recurrent morphological phenotype found in treated trypanosomatids are the mitochondrial swelling (**b**)

Antioxidant machinery of trypanosomatids, especially trypanothione/trypanothione reductase pathway, is considered an interesting drug target, and many efforts have been made to the design of novel-specific inhibitors that do not interfere with mammalian systems [184]. Trypanothione is the most characterized antioxidant system in these parasites, once glutathione/glutathione peroxidase pathway and classical enzymes such as catalase are absent [185, 373]. Unfortunately, up to now, no promising inhibitors of any antioxidant enzyme of pathogenic trypanosomatids were found, despite all the efforts employed [374]. The high susceptibility of these protozoa to ROS in relation to their hosts is an old-fashioned concept, due to the presence of efficient scavengers in trypanosomatids [51, 57, 375].

Several preclinical studies associated oxidative stress to the mode of action of anti-trypanosomatid compounds, being the parasites mitochondrion, the main ROS source [98]. Curiously, the participation of this organelle in pathogenic trypanosomatids treated with different drugs has been extensively demonstrated, but the molecular mechanisms involved are still unknown in a large number of cases. The mitochondrial injury could be derived from a randomic outcome of an indirect effect (probably in the great majority of the cases) or even resulted from ETC-specific inhibition that generates a redox imbalance [312]. Previous data published showed high host toxicity of anti-trypanosomatid compounds with redox mode of action, suggestive of low drug specificity to these protozoa.

As it was mentioned, the significative role of ROS in cell signaling cannot be neglected. The pro-oxidant molecule heme triggers an oxidative stress, leading to calmodulin kinase II activation and consequent proliferation of *T. cruzi* epimastigotes [103]. On the other hand, ROS production also represents a crucial step for the success of these parasites in different hosts, and mitochondrial plasticity (morphological and molecular) has been postulated as an important adaptation, being ETC impairment directly associated to oxidative stress and loss of the redox balance [50]. The clinical correlation between the efficiency of mitochondrial antioxidant machinery (especially trypanothione synthetase and peroxiredoxins) of trypanosomatids and these parasites virulence was proposed, favoring the progression of the disease [277]. Studies about the molecular regulation of oxidative stress processes could base promising strategies for the development of new drugs.

Acknowledgments The authors thank Dra. Claudia M. d´Ávila-Levy for her critical reading and English grammar corrections. The present study was supported by grants from Fundação Carlos Chagas Filho de Amparo a Pesquisa do Estado do Rio de Janeiro (FAPERJ), Conselho Nacional de Desenvolvimento Científico e Tecnológico (CNPq), and Papes/Fundação Oswaldo Cruz (Fiocruz).

References

1. World Health Organization (2016) Neglected tropical diseases. http://www.who.int/trypanosomiasis_african/en/index.html. Accessed 10 July 2018
2. World Health Organization (2008) The global burden of disease: 2004 update. http://www.who.int/healthinfo/global_burden_disease/2004_report_update/en/. Accessed 10 July 2018
3. World Health Organization (2018) What is leishmaniasis? http://www.who.int/leishmaniasis/disease/en/. Accessed 10 July 2018
4. Alvar J, Vélez ID, Bern C, Herrero M, Desjeux P, Cano J, Jannin J, den Boer M, Team WLC (2012) Leishmaniasis worldwide and global estimates of its incidence. PLoS One 7(5):e35671. https://doi.org/10.1371/journal.pone.0035671
5. Handman E (2001) Leishmaniasis: current status of vaccine development. Clin Microbiol Rev 14(2):229–243. https://doi.org/10.1128/cmr.14.2.229-243.2001
6. Anversa L, Tiburcio MGS, Richini-Pereira VB, Ramirez LE (2018) Human leishmaniasis in Brazil: a general review. Rev Assoc Med Bras (1992) 64(3):281–289. https://doi.org/10.1590/1806-9282.64.03.281
7. Ponte-Sucre A, Gamarro F, Dujardin JC, Barrett MP, López-Vélez R, García-Hernández R, Pountain AW, Mwenechanya R, Papadopoulou B (2017) Drug resistance and treatment failure in leishmaniasis: a 21st century challenge. PLoS Negl Trop Dis 11(12):e0006052. https://doi.org/10.1371/journal.pntd.0006052
8. Croft SL, Olliaro P (2011) Leishmaniasis chemotherapy – challenges and opportunities. Clin Microbiol Infect 17(10):1478–1483. https://doi.org/10.1111/j.1469-0691.2011.03630.x
9. Sundar S, Singh B (2018) Emerging therapeutic targets for treatment of leishmaniasis. Expert Opin Ther Targets 22(6):467–486. https://doi.org/10.1080/14728222.2018.1472241
10. Tiuman TS, Santos AO, Ueda-Nakamura T, Filho BP, Nakamura CV (2011) Recent advances in leishmaniasis treatment. Int J Infect Dis 15(8):e525–e532. https://doi.org/10.1016/j.ijid.2011.03.021

11. Ndjonka D, Rapado LN, Silber AM, Liebau E, Wrenger C (2013) Natural products as a source for treating neglected parasitic diseases. Int J Mol Sci 14(2):3395–3439. https://doi.org/10.3390/ijms14023395
12. Khalil NM, de Mattos AC, Carraro TC, Ludwig DB, Mainardes RM (2013) Nanotechnological strategies for the treatment of neglected diseases. Curr Pharm Des 19(41):7316–7329
13. Centers for Disease Control and Prevention (2018) Parasites – Leishmaniasis. https://www.cdc.gov/parasites/leishmaniasis/biology.html. Accessed 10 July 2018
14. Schmunis GA, Yadon ZE (2010) Chagas disease: a Latin American health problem becoming a world health problem. Acta Trop 115(1–2):14–21. https://doi.org/10.1016/j.actatropica.2009.11.003
15. Schofield CJ, Jannin J, Salvatella R (2006) The future of Chagas disease control. Trends Parasitol 22(12):583–588. https://doi.org/10.1016/j.pt.2006.09.011
16. Rassi A, Marin-Neto JA (2010) Chagas disease. Lancet 375(9723):1388–1402. https://doi.org/10.1016/s0140-6736(10)60061-x
17. Rassi A, Marcondes de Rezende J (2012) American trypanosomiasis (Chagas disease). Infect Dis Clin North Am 26(2):275–291. https://doi.org/10.1016/j.idc.2012.03.002
18. Urbina JA (2010) Specific chemotherapy of Chagas disease: relevance, current limitations and new approaches. Acta Trop 115(1–2):55–68. https://doi.org/10.1016/j.actatropica.2009.10.023
19. Soeiro MN, de Castro SL (2011) Screening of potential anti-*Trypanosoma cruzi* candidates: *in vitro* and *in vivo* studies. Open Med Chem J 5:21–30. https://doi.org/10.2174/1874104501105010021
20. Tyler KM, Engman DM (2001) The life cycle of *Trypanosoma cruzi* revisited. Int J Parasitol 31(5–6):472–481
21. Clayton J (2010) Chagas disease: pushing through the pipeline. Nature 465(7301):S12–S15. https://doi.org/10.1038/nature09224
22. Welburn SC, Molyneux DH, Maudlin I (2016) Beyond tsetse-implications for research and control of human african Trypanosomiasis epidemics. Trends Parasitol 32(3):230–241. https://doi.org/10.1016/j.pt.2015.11.008
23. Lumsden WH (1970) Trypanosomiasis. Trop Dis Bull 67(5):465–481
24. Malvy D, Chappuis F (2011) Sleeping sickness. Clin Microbiol Infect 17(7):986–995. https://doi.org/10.1111/j.1469-0691.2011.03536.x
25. World Health Organization (2016) Human African trypanosomiasis. http://www.who.int/trypanosomiasis_african/en/. Accessed 10 July 2018
26. Kennedy PG (2013) Clinical features, diagnosis, and treatment of human African trypanosomiasis (sleeping sickness). Lancet Neurol 12(2):186–194. https://doi.org/10.1016/s1474-4422(12)70296-x
27. Steverding D (2010) The development of drugs for treatment of sleeping sickness: a historical review. Parasit Vectors 3(1):15. https://doi.org/10.1186/1756-3305-3-15
28. Brun R, Don R, Jacobs RT, Wang MZ, Barrett MP (2011) Development of novel drugs for human African trypanosomiasis. Fut Microbiol 6(6):677–691. https://doi.org/10.2217/fmb.11.44
29. Simarro PP, Franco J, Diarra A, Postigo JA, Jannin J (2012) Update on field use of the available drugs for the chemotherapy of human African trypanosomiasis. Parasitology 139(7):842–846. https://doi.org/10.1017/s0031182012000169
30. Horn D (2014) Antigenic variation in African trypanosomes. Mol Biochem Parasitol 195(2):123–129. https://doi.org/10.1016/j.molbiopara.2014.05.001
31. Langousis G, Hill KL (2014) Motility and more: the flagellum of *Trypanosoma brucei*. Nat Rev Microbiol 12(7):505–518. https://doi.org/10.1038/nrmicro3274
32. Harman D (2009) Origin and evolution of the free radical theory of aging: a brief personal history, 1954–2009. Biogerontology 10(6):773–781. https://doi.org/10.1007/s10522-009-9234-2

33. Babior BM, Kipnes RS, Curnutte JT (1973) Biological defense mechanisms. The production by leukocytes of superoxide, a potential bactericidal agent. J Clin Invest 52(3):741–744. https://doi.org/10.1172/jci107236
34. Rossi F, Della Bianca V, de Togni P (1985) Mechanisms and functions of the oxygen radicals producing respiration of phagocytes. Comp Immunol Microbiol Infect Dis 8(2):187–204
35. Shaikhali J, Heiber I, Seidel T, Ströher E, Hiltscher H, Birkmann S, Dietz KJ, Baier M (2008) The redox-sensitive transcription factor Rap2.4a controls nuclear expression of 2-Cys peroxiredoxin A and other chloroplast antioxidant enzymes. BMC Plant Biol 8:48. https://doi.org/10.1186/1471-2229-8-48
36. Riley PA (1994) Free radicals in biology: oxidative stress and the effects of ionizing radiation. Int J Radiat Biol 65(1):27–33
37. Poljsak B, Šuput D, Milisav I (2013) Achieving the balance between ROS and antioxidants: when to use the synthetic antioxidants. Oxid Med Cell Longev 2013:956792. https://doi.org/10.1155/2013/956792
38. Sies H (2015) Oxidative stress: a concept in redox biology and medicine. Redox Biol 4:180–183. https://doi.org/10.1016/j.redox.2015.01.002
39. Chen X, Wang F, Hyun JY, Wei T, Qiang J, Ren X, Shin I, Yoon J (2016) Recent progress in the development of fluorescent, luminescent and colorimetric probes for detection of reactive oxygen and nitrogen species. Chem Soc Rev 45(10):2976–3016. https://doi.org/10.1039/c6cs00192k
40. Sies H (1991) Role of reactive oxygen species in biological processes. Klin Wochenschr 69(21–23):965–968
41. Pisoschi AM, Pop A (2015) The role of antioxidants in the chemistry of oxidative stress: A review. Eur J Med Chem 97:55–74. https://doi.org/10.1016/j.ejmech.2015.04.040
42. Xing J, Wang G, Zhang Q, Liu X, Gu Z, Zhang H, Chen YQ, Chen W (2015) Determining antioxidant activities of lactobacilli cell-free supernatants by cellular antioxidant assay: a comparison with traditional methods. PLoS One 10(3):e0119058. https://doi.org/10.1371/journal.pone.0119058
43. Lushchak VI (2014) Free radicals, reactive oxygen species, oxidative stress and its classification. Chem Biol Interact 224:164–175. https://doi.org/10.1016/j.cbi.2014.10.016
44. Mitchell P (1961) Coupling of phosphorylation to electron and hydrogen transfer by a chemi-osmotic type of mechanism. Nature 191:144–148
45. Vercesi AE, Oliveira HCF (2018) Contribution to mitochondrial research in Brazil: 10th anniversary of the mitomeeting. Cell Biol Int 42(6):626–629. https://doi.org/10.1002/cbin.10898
46. Guo R, Zong S, Wu M, Gu J, Yang M (2017) Architecture of Human Mitochondrial Respiratory Megacomplex I. Cell 170(6):1247–1257.e1212. https://doi.org/10.1016/j.cell.2017.07.050
47. Chance B, Williams GR (1956) The respiratory chain and oxidative phosphorylation. Adv Enzymol Relat Subj Biochem 17:65–134
48. Schägger H (2001) Respiratory chain supercomplexes. IUBMB Life 52(3–5):119–128. https://doi.org/10.1080/15216540152845911
49. Ott M, Gogvadze V, Orrenius S, Zhivotovsky B (2007) Mitochondria, oxidative stress and cell death. Apoptosis 12(5):913–922. https://doi.org/10.1007/s10495-007-0756-2
50. Venditti P, Di Stefano L, Di Meo S (2013) Mitochondrial metabolism of reactive oxygen species. Mitochondrion 13(2):71–82. https://doi.org/10.1016/j.mito.2013.01.008
51. Boveris A, Stoppani AO (1977) Hydrogen peroxide generation in *Trypanosoma cruzi*. Experientia 33(10):1306–1308
52. Murphy MP (2009) How mitochondria produce reactive oxygen species. Biochem J 417(1):1–13. https://doi.org/10.1042/bj20081386
53. Wang Y, Hekimi S (2016) Understanding Ubiquinone. Trends Cell Biol 26(5):367–378. https://doi.org/10.1016/j.tcb.2015.12.007
54. Gutteridge JM (1994) Biological origin of free radicals, and mechanisms of antioxidant protection. Chem Biol Interact 91(2–3):133–140

55. Gutteridge JM, Halliwell B (1992) Comments on review of free radicals in biology and medicine, second edition, by Barry Halliwell and John M. C. Gutteridge. Free Radic Biol Med 12(1):93–95
56. Kohen R, Nyska A (2002) Oxidation of biological systems: oxidative stress phenomena, antioxidants, redox reactions, and methods for their quantification. Toxicol Pathol 30(6):620–650. https://doi.org/10.1080/01926230290166724
57. Tomás AM, Castro H (2013) Redox metabolism in mitochondria of trypanosomatids. Antioxid Redox Signal 19(7):696–707. https://doi.org/10.1089/ars.2012.4948
58. Tielens AG, van Hellemond JJ (2009) Surprising variety in energy metabolism within Trypanosomatidae. Trends Parasitol 25(10):482–490. https://doi.org/10.1016/j.pt.2009.07.007
59. Marques I, Duarte M, Assunção J, Ushakova AV, Videira A (2005) Composition of complex I from *Neurospora crassa* and disruption of two "accessory" subunits. Biochim Biophys Acta 1707(2-3):211–220. https://doi.org/10.1016/j.bbabio.2004.12.003
60. Carroll J, Fearnley IM, Walker JE (2006) Definition of the mitochondrial proteome by measurement of molecular masses of membrane proteins. Proc Natl Acad Sci U S A 103(44):16170–16175. https://doi.org/10.1073/pnas.0607719103
61. Galkin A, Moncada S (2017) Modulation of the conformational state of mitochondrial complex I as a target for therapeutic intervention. Interface Focus 7(2):20160104. https://doi.org/10.1098/rsfs.2016.0104
62. Hatefi Y (1985) The mitochondrial electron transport and oxidative phosphorylation system. Annu Rev Biochem 54:1015–1069. https://doi.org/10.1146/annurev.bi.54.070185.005055
63. Opperdoes FR, Michels PA (2008) Complex I of Trypanosomatidae: does it exist? Trends Parasitol 24(7):310–317. https://doi.org/10.1016/j.pt.2008.03.013
64. Turrens JF (1989) The role of succinate in the respiratory chain of *Trypanosoma brucei* procyclic trypomastigotes. Biochem J 259(2):363–368
65. Carranza JC, Kowaltowski AJ, Mendonça MA, de Oliveira TC, Gadelha FR, Zingales B (2009) Mitochondrial bioenergetics and redox state are unaltered in *Trypanosoma cruzi* isolates with compromised mitochondrial complex I subunit genes. J Bioenerg Biomembr 41(3):299–308. https://doi.org/10.1007/s10863-009-9228-4
66. Beattie DS, Obungu VH, Kiaira JK (1994) Oxidation of NADH by a rotenone and antimycin-sensitive pathway in the mitochondrion of procyclic *Trypanosoma brucei brucei*. Mol Biochem Parasitol 64(1):87–94
67. Beattie DS, Howton MM (1996) The presence of rotenone-sensitive NADH dehydrogenase in the long slender bloodstream and the procyclic forms of *Trypanosoma brucei brucei*. Eur J Biochem 241(3):888–894
68. Hernandez FR, Turrens JF (1998) Rotenone at high concentrations inhibits NADH-fumarate reductase and the mitochondrial respiratory chain of *Trypanosoma brucei* and *T. cruzi*. Mol Biochem Parasitol 93(1):135–137
69. Jasmer DP, Feagin JE, Stuart K (1985) Diverse patterns of expression of the cytochrome c oxidase subunit I gene and unassigned reading frames 4 and 5 during the life cycle of *Trypanosoma brucei*. Mol Cell Biol 5(11):3041–3047
70. Souza AE, Myler PJ, Stuart K (1992) Maxicircle CR1 transcripts of *Trypanosoma brucei* are edited and developmentally regulated and encode a putative iron-sulfur protein homologous to an NADH dehydrogenase subunit. Mol Cell Biol 12(5):2100–2107
71. Verner Z, Cermáková P, Skodová I, Kriegová E, Horváth A, Lukes J (2011) Complex I (NADH:ubiquinone oxidoreductase) is active in but non-essential for procyclic *Trypanosoma brucei*. Mol Biochem Parasitol 175(2):196–200. https://doi.org/10.1016/j.molbiopara.2010.11.003
72. Fang J, Beattie DS (2003) Alternative oxidase present in procyclic T*rypanosoma brucei* may act to lower the mitochondrial production of superoxide. Arch Biochem Biophys 414(2):294–302

73. Fisher N, Bray PG, Ward SA, Biagini GA (2007) The malaria parasite type II NADH:quinone oxidoreductase: an alternative enzyme for an alternative lifestyle. Trends Parasitol 23(7):305–310. https://doi.org/10.1016/j.pt.2007.04.014
74. Fang J, Beattie DS (2003) Identification of a gene encoding a 54 kDa alternative NADH dehydrogenase in *Trypanosoma brucei*. Mol Biochem Parasitol 127(1):73–77
75. Fang J, Beattie DS (2002) Rotenone-insensitive NADH dehydrogenase is a potential source of superoxide in procyclic *Trypanosoma brucei* mitochondria. Mol Biochem Parasitol 123(2):135–142
76. Verner Z, Skodová I, Poláková S, Durišová-Benkovičová V, Horváth A, Lukeš J (2013) Alternative NADH dehydrogenase (NDH2): intermembrane-space-facing counterpart of mitochondrial complex I in the procyclic *Trypanosoma brucei*. Parasitology 140(3):328–337. https://doi.org/10.1017/s003118201200162x
77. Surve S, Heestand M, Panicucci B, Schnaufer A, Parsons M (2012) Enigmatic presence of mitochondrial complex I in *Trypanosoma brucei* bloodstream forms. Eukaryot Cell 11(2):183–193. https://doi.org/10.1128/ec.05282-11
78. Boveris A, Hertig CM, Turrens JF (1986) Fumarate reductase and other mitochondrial activities in *Trypanosoma cruzi*. Mol Biochem Parasitol 19(2):163–169
79. Besteiro S, Biran M, Biteau N, Coustou V, Baltz T, Canioni P, Bringaud F (2002) Succinate secreted by *Trypanosoma brucei* is produced by a novel and unique glycosomal enzyme, NADH-dependent fumarate reductase. J Biol Chem 277(41):38001–38012. https://doi.org/10.1074/jbc.M201759200
80. Coustou V, Biran M, Besteiro S, Rivière L, Baltz T, Franconi JM, Bringaud F (2006) Fumarate is an essential intermediary metabolite produced by the procyclic *Trypanosoma brucei*. J Biol Chem 281(37):26832–26846. https://doi.org/10.1074/jbc.M601377200
81. Coustou V, Besteiro S, Rivière L, Biran M, Biteau N, Franconi JM, Boshart M, Baltz T, Bringaud F (2005) A mitochondrial NADH-dependent fumarate reductase involved in the production of succinate excreted by procyclic *Trypanosoma brucei*. J Biol Chem 280(17):16559–16570. https://doi.org/10.1074/jbc.M500343200
82. Coustou V, Biran M, Breton M, Guegan F, Rivière L, Plazolles N, Nolan D, Barrett MP, Franconi JM, Bringaud F (2008) Glucose-induced remodeling of intermediary and energy metabolism in procyclic *Trypanosoma brucei*. J Biol Chem 283(24):16342–16354. https://doi.org/10.1074/jbc.M709592200
83. Mehta A, Shaha C (2004) Apoptotic death in *Leishmania donovani* promastigotes in response to respiratory chain inhibition: complex II inhibition results in increased pentamidine cytotoxicity. J Biol Chem 279(12):11798–11813. https://doi.org/10.1074/jbc.M309341200
84. Silva TM, Peloso EF, Vitor SC, Ribeiro LH, Gadelha FR (2011) O_2 consumption rates along the growth curve: new insights into *Trypanosoma cruzi* mitochondrial respiratory chain. J Bioenerg Biomembr 43(4):409–417. https://doi.org/10.1007/s10863-011-9369-0
85. Evans DA, Brown RC (1973) The inhibitory effects of aromatic hydroxamic acids on the cyanide-insensitive terminal oxidase of *Trypanosoma brucei*. Trans R Soc Trop Med Hyg 67(2):258
86. Edwards C, Chance B (1982) Evidence for the presence of two terminal oxidases in the trypanosomatid *Crithidia oncopelti*. J Gen Microbiol 128(7):1409–1414. https://doi.org/10.1099/00221287-128-7-1409
87. Opperdoes FR, Butenko A, Flegontov P, Yurchenko V, Lukeš J (2016) Comparative metabolism of free-living *Bodo saltans* and parasitic trypanosomatids. J Eukaryot Microbiol 63(5):657–678. https://doi.org/10.1111/jeu.12315
88. Gnipová A, Panicucci B, Paris Z, Verner Z, Horváth A, Lukeš J, Zíková A (2012) Disparate phenotypic effects from the knockdown of various *Trypanosoma brucei* cytochrome c oxidase subunits. Mol Biochem Parasitol 184(2):90–98. https://doi.org/10.1016/j.molbiopara.2012.04.013
89. Chaudhuri M, Ott RD, Hill GC (2006) Trypanosome alternative oxidase: from molecule to function. Trends Parasitol 22(10):484–491. https://doi.org/10.1016/j.pt.2006.08.007

90. Horváth A, Horáková E, Dunajcíková P, Verner Z, Pravdová E, Slapetová I, Cuninková L, Lukes J (2005) Downregulation of the nuclear-encoded subunits of the complexes III and IV disrupts their respective complexes but not complex I in procyclic *Trypanosoma brucei*. Mol Microbiol 58(1):116–130. https://doi.org/10.1111/j.1365-2958.2005.04813.x
91. Santhamma KR, Bhaduri A (1995) Characterization of the respiratory chain of *Leishmania donovani* promastigotes. Mol Biochem Parasitol 75(1):43–53
92. Atwood JA, Weatherly DB, Minning TA, Bundy B, Cavola C, Opperdoes FR, Orlando R, Tarleton RL (2005) The *Trypanosoma cruzi* proteome. Science 309(5733):473–476. https://doi.org/10.1126/science.1110289
93. Shah-Simpson S, Pereira CF, Dumoulin PC, Caradonna KL, Burleigh BA (2016) Bioenergetic profiling of *Trypanosoma cruzi* life stages using Seahorse extracellular flux technology. Mol Biochem Parasitol 208(2):91–95. https://doi.org/10.1016/j.molbiopara.2016.07.001
94. Gonçalves RL, Barreto RF, Polycarpo CR, Gadelha FR, Castro SL, Oliveira MF (2011) A comparative assessment of mitochondrial function in epimastigotes and bloodstream trypomastigotes of *Trypanosoma cruzi*. J Bioenerg Biomembr 43(6):651–661. https://doi.org/10.1007/s10863-011-9398-8
95. Tielens AG, Van Hellemond JJ (1998) Differences in energy metabolism between trypanosomatidae. Parasitol Today 14(7):265–272
96. Clayton CE, Michels P (1996) Metabolic compartmentation in African trypanosomes. Parasitol Today 12(12):465–471
97. Priest JW, Hajduk SL (1994) Developmental regulation of *Trypanosoma brucei* cytochrome c reductase during bloodstream to procyclic differentiation. Mol Biochem Parasitol 65(2):291–304
98. Bringaud F, Rivière L, Coustou V (2006) Energy metabolism of trypanosomatids: adaptation to available carbon sources. Mol Biochem Parasitol 149(1):1–9. https://doi.org/10.1016/j.molbiopara.2006.03.017
99. Silber AM, Tonelli RR, Lopes CG, Cunha-e-Silva N, Torrecilhas AC, Schumacher RI, Colli W, Alves MJ (2009) Glucose uptake in the mammalian stages of *Trypanosoma cruzi*. Mol Biochem Parasitol 168(1):102–108. https://doi.org/10.1016/j.molbiopara.2009.07.006
100. Harington JS (1956) Histamine and histidine in excreta of the blood-sucking bug *Rhodnius prolixus*. Nature 178(4527):268
101. Harington JS (1961) Studies of the amino acids of *Rhodnius prolixus* I. Analysis of the haemolymph. Parasitology 51:309–318
102. Finzi JK, Chiavegatto CW, Corat KF, Lopez JA, Cabrera OG, Mielniczki-Pereira AA, Colli W, Alves MJ, Gadelha FR (2004) *Trypanosoma cruzi* response to the oxidative stress generated by hydrogen peroxide. Mol Biochem Parasitol 133(1):37–43
103. Nogueira NP, de Souza CF, Saraiva FM, Sultano PE, Dalmau SR, Bruno RE, Gonçalves RL, Laranja GA, Leal LH, Coelho MG, Masuda CA, Oliveira MF, Paes MC (2011) Heme-induced ROS in *Trypanosoma cruzi* activates CaMKII-like that triggers epimastigote proliferation. One helpful effect of ROS. PLoS One 6(10):e25935. https://doi.org/10.1371/journal.pone.0025935
104. Nogueira NP, Saraiva FMS, Oliveira MP, Mendonça APM, Inacio JDF, Almeida-Amaral EE, Menna-Barreto RF, Laranja GAT, Torres EJL, Oliveira MF, Paes MC (2017) Heme modulates *Trypanosoma cruzi* bioenergetics inducing mitochondrial ROS production. Free Radic Biol Med 108:183–191. https://doi.org/10.1016/j.freeradbiomed.2017.03.027
105. Anderson S, Bankier AT, Barrell BG, de Bruijn MH, Coulson AR, Drouin J, Eperon IC, Nierlich DP, Roe BA, Sanger F, Schreier PH, Smith AJ, Staden R, Young IG (1981) Sequence and organization of the human mitochondrial genome. Nature 290(5806):457–465
106. Vercesi AE, Martins LS, MAP S, HMF L, Cuccovia IM, Chaimovich H (1995) PUMPing plants. Nature 375:24. https://doi.org/10.1038/375024a0
107. Vercesi AE, Borecký J, Maia IG, Arruda P, Cuccovia IM, Chaimovich H (2006) Plant uncoupling mitochondrial proteins. Annu Rev Plant Biol 57:383–404. https://doi.org/10.1146/annurev.arplant.57.032905.105335

108. Hassanpour SH, Dehghani MA, Karami SZ (2018) Study of respiratory chain dysfunction in heart disease. J Cardiovasc Thorac Res 10 (1):1–13. https://doi.org/10.15171/jcvtr.2018.01
109. Wong HS, Dighe PA, Mezera V, Monternier PA, Brand MD (2017) Production of superoxide and hydrogen peroxide from specific mitochondrial sites under different bioenergetic conditions. J Biol Chem 292(41):16804–16809. https://doi.org/10.1074/jbc.R117.789271
110. Sazanov LA, Hinchliffe P (2006) Structure of the hydrophilic domain of respiratory complex I from *Thermus thermophilus*. Science 311(5766):1430–1436. https://doi.org/10.1126/science.1123809
111. Hinkle PC, Butow RA, Racker E, Chance B (1967) Partial resolution of the enzymes catalyzing oxidative phosphorylation. XV. Reverse electron transfer in the flavin-cytochrome beta region of the respiratory chain of beef heart submitochondrial particles. J Biol Chem 242(22):5169–5173
112. Cadenas E, Boveris A, Ragan CI, Stoppani AO (1977) Production of superoxide radicals and hydrogen peroxide by NADH-ubiquinone reductase and ubiquinol-cytochrome c reductase from beef-heart mitochondria. Arch Biochem Biophys 180(2):248–257
113. Kussmaul L, Hirst J (2006) The mechanism of superoxide production by NADH:ubiquinone oxidoreductase (complex I) from bovine heart mitochondria. Proc Natl Acad Sci U S A 103(20):7607–7612. https://doi.org/10.1073/pnas.0510977103
114. Hirst J, King MS, Pryde KR (2008) The production of reactive oxygen species by complex I. Biochem Soc Trans 36(Pt 5):976–980. https://doi.org/10.1042/bst0360976
115. Seo BB, Marella M, Yagi T, Matsuno-Yagi A (2006) The single subunit NADH dehydrogenase reduces generation of reactive oxygen species from complex I. FEBS Lett 580(26):6105–6108. https://doi.org/10.1016/j.febslet.2006.10.008
116. Bleier L, Dröse S (2013) Superoxide generation by complex III: from mechanistic rationales to functional consequences. Biochim Biophys Acta 1827(11–12):1320–1331. https://doi.org/10.1016/j.bbabio.2012.12.002
117. Cadenas S (2018) Mitochondrial uncoupling, ROS generation and cardioprotection. Biochim Biophys Acta. https://doi.org/10.1016/j.bbabio.2018.05.019
118. Rolfe DF, Brand MD (1996) Contribution of mitochondrial proton leak to skeletal muscle respiration and to standard metabolic rate. Am J Physiol 271(4 Pt 1):C1380–C1389. https://doi.org/10.1152/ajpcell.1996.271.4.C1380
119. Stuart JA, Cadenas S, Jekabsons MB, Roussel D, Brand MD (2001) Mitochondrial proton leak and the uncoupling protein 1 homologues. Biochim Biophys Acta 1504(1):144–158
120. Boveris A, Chance B (1973) The mitochondrial generation of hydrogen peroxide. General properties and effect of hyperbaric oxygen. Biochem J 134(3):707–716
121. Korshunov SS, Skulachev VP, Starkov AA (1997) High protonic potential actuates a mechanism of production of reactive oxygen species in mitochondria. FEBS Lett 416(1):15–18
122. Jastroch M, Divakaruni AS, Mookerjee S, Treberg JR, Brand MD (2010) Mitochondrial proton and electron leaks. Essays Biochem 47:53–67. https://doi.org/10.1042/bse0470053
123. Woyda-Ploszczyca AM, Jarmuszkiewicz W (2017) The conserved regulation of mitochondrial uncoupling proteins: From unicellular eukaryotes to mammals. Biochim Biophys Acta 1858(1):21–33. https://doi.org/10.1016/j.bbabio.2016.10.003
124. Nicholls DG, Bernson VS, Heaton GM (1978) The identification of the component in the inner membrane of brown adipose tissue mitochondria responsible for regulating energy dissipation. Experientia Suppl 32:89–93
125. Fleury C, Neverova M, Collins S, Raimbault S, Champigny O, Levi-Meyrueis C, Bouillaud F, Seldin MF, Surwit RS, Ricquier D, Warden CH (1997) Uncoupling protein-2: a novel gene linked to obesity and hyperinsulinemia. Nat Genet 15(3):269–272. https://doi.org/10.1038/ng0397-269

126. Boss O, Samec S, Paoloni-Giacobino A, Rossier C, Dulloo A, Seydoux J, Muzzin P, Giacobino JP (1997) Uncoupling protein-3: a new member of the mitochondrial carrier family with tissue-specific expression. FEBS Lett 408(1):39–42
127. Mao W, Yu XX, Zhong A, Li W, Brush J, Sherwood SW, Adams SH, Pan G (1999) UCP4, a novel brain-specific mitochondrial protein that reduces membrane potential in mammalian cells. FEBS Lett 443(3):326–330
128. Klingenberg M (1990) Mechanism and evolution of the uncoupling protein of brown adipose tissue. Trends Biochem Sci 15(3):108–112
129. Sluse FE (2012) Uncoupling proteins: molecular, functional, regulatory, physiological and pathological aspects. Adv Exp Med Biol 942:137–156. https://doi.org/10.1007/978-94-007-2869-1_6
130. Nègre-Salvayre A, Hirtz C, Carrera G, Cazenave R, Troly M, Salvayre R, Pénicaud L, Casteilla L (1997) A role for uncoupling protein-2 as a regulator of mitochondrial hydrogen peroxide generation. FASEB J 11(10):809–815
131. Azzu V, Jastroch M, Divakaruni AS, Brand MD (2010) The regulation and turnover of mitochondrial uncoupling proteins. Biochim Biophys Acta 1797(6-7):785–791. https://doi.org/10.1016/j.bbabio.2010.02.035
132. Mailloux RJ, Harper ME (2011) Uncoupling proteins and the control of mitochondrial reactive oxygen species production. Free Radic Biol Med 51(6):1106–1115. https://doi.org/10.1016/j.freeradbiomed.2011.06.022
133. MacLellan JD, Gerrits MF, Gowing A, Smith PJ, Wheeler MB, Harper ME (2005) Physiological increases in uncoupling protein 3 augment fatty acid oxidation and decrease reactive oxygen species production without uncoupling respiration in muscle cells. Diabetes 54(8):2343–2350
134. Tahara EB, Navarete FD, Kowaltowski AJ (2009) Tissue-, substrate-, and site-specific characteristics of mitochondrial reactive oxygen species generation. Free Radic Biol Med 46(9):1283–1297. https://doi.org/10.1016/j.freeradbiomed.2009.02.008
135. Toime LJ, Brand MD (2010) Uncoupling protein-3 lowers reactive oxygen species production in isolated mitochondria. Free Radic Biol Med 49(4):606–611. https://doi.org/10.1016/j.freeradbiomed.2010.05.010
136. Brigelius-Flohé R, Flohé L (2011) Basic principles and emerging concepts in the redox control of transcription factors. Antioxid Redox Signal 15(8):2335–2381. https://doi.org/10.1089/ars.2010.3534
137. Scandalios JG (2005) Oxidative stress: molecular perception and transduction of signals triggering antioxidant gene defenses. Braz J Med Biol Res 38 (7):995–1014. doi:S0100-879x2005000700003
138. Babior BM, Curnutte JT, Kipnes RS (1975) Biological defense mechanisms. Evidence for the participation of superoxide in bacterial killing by xanthine oxidase. J Lab Clin Med 85(2):235–244
139. Griffith OW (1980) Determination of glutathione and glutathione disulfide using glutathione reductase and 2-vinylpyridine. Anal Biochem 106(1):207–212
140. Fairlamb AH, Blackburn P, Ulrich P, Chait BT, Cerami A (1985) Trypanothione: a novel bis(glutathionyl)spermidine cofactor for glutathione reductase in trypanosomatids. Science 227(4693):1485–1487
141. Shames SL, Fairlamb AH, Cerami A, Walsh CT (1986) Purification and characterization of trypanothione reductase from *Crithidia fasciculata*, a newly discovered member of the family of disulfide-containing flavoprotein reductases. Biochemistry 25(12):3519–3526
142. Krauth-Siegel RL, Enders B, Henderson GB, Fairlamb AH, Schirmer RH (1987) Trypanothione reductase from *Trypanosoma cruzi*. Purification and characterization of the crystalline enzyme. Eur J Biochem 164(1):123–128

143. Sullivan FX, Walsh CT (1991) Cloning, sequencing, overproduction and purification of trypanothione reductase from *Trypanosoma cruzi*. Mol Biochem Parasitol 44(1):145–147
144. Aboagye-Kwarteng T, Smith K, Fairlamb AH (1992) Molecular characterization of the trypanothione reductase gene from *Crithidia fasciculata* and *Trypanosoma brucei*: comparison with other flavoprotein disulphide oxidoreductases with respect to substrate specificity and catalytic mechanism. Mol Microbiol 6(21):3089–3099
145. Manta B, Comini M, Medeiros A, Hugo M, Trujillo M, Radi R (2013) Trypanothione: a unique bis-glutathionyl derivative in trypanosomatids. Biochim Biophys Acta 1830(5):3199–3216. https://doi.org/10.1016/j.bbagen.2013.01.013
146. Lueder DV, Phillips MA (1996) Characterization of *Trypanosoma brucei* gamma-glutamylcysteine synthetase, an essential enzyme in the biosynthesis of trypanothione (diglutathionylspermidine). J Biol Chem 271(29):17485–17490
147. Grondin K, Haimeur A, Mukhopadhyay R, Rosen BP, Ouellette M (1997) Co-amplification of the gamma-glutamylcysteine synthetase gene gsh1 and of the ABC transporter gene pgpA in arsenite-resistant *Leishmania tarentolae*. EMBO J 16(11):3057–3065. https://doi.org/10.1093/emboj/16.11.3057
148. Olin-Sandoval V, González-Chávez Z, Berzunza-Cruz M, Martínez I, Jasso-Chávez R, Becker I, Espinoza B, Moreno-Sánchez R, Saavedra E (2012) Drug target validation of the trypanothione pathway enzymes through metabolic modelling. FEBS J 279(10):1811–1833. https://doi.org/10.1111/j.1742-4658.2012.08557.x
149. Mukherjee P, Majee SB, Ghosh S, Hazra B (2009) Apoptosis-like death in *Leishmania donovani* promastigotes induced by diospyrin and its ethanolamine derivative. Int J Antimicrob Agents 34(6):596–601. https://doi.org/10.1016/j.ijantimicag.2009.08.007
150. Huynh TT, Huynh VT, Harmon MA, Phillips MA (2003) Gene knockdown of gamma-glutamylcysteine synthetase by RNAi in the parasitic protozoa *Trypanosoma brucei* demonstrates that it is an essential enzyme. J Biol Chem 278(41):39794–39800. https://doi.org/10.1074/jbc.M306306200
151. Jones DP, Carlson JL, Samiec PS, Sternberg P, Mody VC, Reed RL, Brown LA (1998) Glutathione measurement in human plasma. Evaluation of sample collection, storage and derivatization conditions for analysis of dansyl derivatives by HPLC. Clin Chim Acta 275(2):175–184
152. Duszenko M, Ferguson MA, Lamont GS, Rifkin MR, Cross GA (1985) Cysteine eliminates the feeder cell requirement for cultivation of *Trypanosoma brucei* bloodstream forms in vitro. J Exp Med 162(4):1256–1263
153. Duszenko M, Mühlstädt K, Broder A (1992) Cysteine is an essential growth factor for *Trypanosoma brucei* bloodstream forms. Mol Biochem Parasitol 50(2):269–273
154. Hesse F, Selzer PM, Mühlstädt K, Duszenko M (1995) A novel cultivation technique for long-term maintenance of bloodstream form trypanosomes *in vitro*. Mol Biochem Parasitol 70(1-2):157–166
155. Williams RA, Westrop GD, Coombs GH (2009) Two pathways for cysteine biosynthesis in *Leishmania major*. Biochem J 420(3):451–462. https://doi.org/10.1042/bj20082441
156. Romero I, Téllez J, Romanha AJ, Steindel M, Grisard EC (2015) Upregulation of Cysteine Synthase and Cystathionine β-Synthase Contributes to *Leishmania braziliensis* Survival under Oxidative Stress. Antimicrob Agents Chemother 59(8):4770–4781. https://doi.org/10.1128/aac.04880-14
157. Romero I, Téllez J, Yamanaka LE, Steindel M, Romanha AJ, Grisard EC (2014) Transsulfuration is an active pathway for cysteine biosynthesis in *Trypanosoma rangeli*. Parasit Vectors 7:197. https://doi.org/10.1186/1756-3305-7-197
158. Marciano D, Santana M, Nowicki C (2012) Functional characterization of enzymes involved in cysteine biosynthesis and H(2)S production in *Trypanosoma cruzi*. Mol Biochem Parasitol 185(2):114–120. https://doi.org/10.1016/j.molbiopara.2012.07.009

159. Okalang U, Nanteza A, Matovu E, Lubega GW (2013) Identification of coding sequences from a freshly prepared *Trypanosoma brucei brucei* expression library by polymerase chain reaction. Int J Biochem Mol Biol 4(2):73–82
160. Colotti G, Ilari A (2011) Polyamine metabolism in Leishmania: from arginine to trypanothione. Amino Acids 40(2):269–285. https://doi.org/10.1007/s00726-010-0630-3
161. Hasne MP, Ullman B (2005) Identification and characterization of a polyamine permease from the protozoan parasite *Leishmania major*. J Biol Chem 280(15):15188–15194. https://doi.org/10.1074/jbc.M411331200
162. Hasne MP, Ullman B (2011) Genetic and biochemical analysis of protozoal polyamine transporters. Methods Mol Biol 720:309–326. https://doi.org/10.1007/978-1-61779-034-8_19
163. Carrillo C, Cejas S, González NS, Algranati ID (1999) *Trypanosoma cruzi* epimastigotes lack ornithine decarboxylase but can express a foreign gene encoding this enzyme. FEBS Lett 454(3):192–196
164. Carrillo C, Cejas S, Huber A, González NS, Algranati ID (2003) Lack of arginine decarboxylase in *Trypanosoma cruzi* epimastigotes. J Eukaryot Microbiol 50(5):312–316
165. Reigada C, Sayé M, Vera EV, Balcazar D, Fraccaroli L, Carrillo C, Miranda MR, Pereira CA (2016) *Trypanosoma cruzi* Polyamine Transporter: Its Role on Parasite Growth and Survival Under Stress Conditions. J Membr Biol 249(4):475–481. https://doi.org/10.1007/s00232-016-9888-z
166. Smith K, Nadeau K, Bradley M, Walsh C, Fairlamb AH (1992) Purification of glutathionylspermidine and trypanothione synthetases from *Crithidia fasciculata*. Protein Sci 1(7):874–883. https://doi.org/10.1002/pro.5560010705
167. Oza SL, Tetaud E, Ariyanayagam MR, Warnon SS, Fairlamb AH (2002) A single enzyme catalyses formation of Trypanothione from glutathione and spermidine in *Trypanosoma cruzi*. J Biol Chem 277(39):35853–35861. https://doi.org/10.1074/jbc.M204403200
168. Oza SL, Ariyanayagam MR, Aitcheson N, Fairlamb AH (2003) Properties of trypanothione synthetase from *Trypanosoma brucei*. Mol Biochem Parasitol 131(1):25–33
169. Comini M, Menge U, Wissing J, Flohé L (2005) Trypanothione synthesis in crithidia revisited. J Biol Chem 280(8):6850–6860. https://doi.org/10.1074/jbc.M404486200
170. Gilbert HF (1990) Molecular and cellular aspects of thiol-disulfide exchange. Adv Enzymol Relat Areas Mol Biol 63:69–172
171. Moutiez M, Aumercier M, Schöneck R, Meziane-Cherif D, Lucas V, Aumercier P, Ouaissi A, Sergheraert C, Tartar A (1995) Purification and characterization of a trypanothione-glutathione thioltransferase from *Trypanosoma cruzi*. Biochem J 310(Pt 2):433–437
172. Fraser-L'Hostis C, Defrise-Quertain F, Coral D, Deshusses J (1997) Regulation of the intracellular pH in the protozoan parasite *Trypanosoma brucei brucei*. Biol Chem 378(9):1039–1046
173. Irigoín F, Cibils L, Comini MA, Wilkinson SR, Flohé L, Radi R (2008) Insights into the redox biology of *Trypanosoma cruzi*: Trypanothione metabolism and oxidant detoxification. Free Radic Biol Med 45(6):733–742. https://doi.org/10.1016/j.freeradbiomed.2008.05.028
174. Ariyanayagam MR, Oza SL, Mehlert A, Fairlamb AH (2003) Bis(glutathionyl)spermine and other novel trypanothione analogues in *Trypanosoma cruzi*. J Biol Chem 278(30):27612–27619. https://doi.org/10.1074/jbc.M302750200
175. Stoll VS, Simpson SJ, Krauth-Siegel RL, Walsh CT, Pai EF (1997) Glutathione reductase turned into trypanothione reductase: structural analysis of an engineered change in substrate specificity. Biochemistry 36(21):6437–6447. https://doi.org/10.1021/bi963074p
176. Meziane-Cherif D, Aumercier M, Kora I, Sergheraert C, Tartar A, Dubremetz JF, Ouaissi MA (1994) *Trypanosoma cruzi*: immunolocalization of trypanothione reductase. Exp Parasitol 79(4):536–541
177. Wilkinson SR, Meyer DJ, Taylor MC, Bromley EV, Miles MA, Kelly JM (2002) The *Trypanosoma cruzi* enzyme TcGPXI is a glycosomal peroxidase and can be linked to trypanothione reduction by glutathione or tryparedoxin. J Biol Chem 277(19):17062–17071. https://doi.org/10.1074/jbc.M111126200

178. Wilkinson SR, Prathalingam SR, Taylor MC, Ahmed A, Horn D, Kelly JM (2006) Functional characterisation of the iron superoxide dismutase gene repertoire in *Trypanosoma brucei*. Free Radic Biol Med 40(2):198–209. https://doi.org/10.1016/j.freeradbiomed.2005.06.022
179. Smith K, Opperdoes FR, Fairlamb AH (1991) Subcellular distribution of trypanothione reductase in bloodstream and procyclic forms of *Trypanosoma brucei*. Mol Biochem Parasitol 48(1):109–112
180. Sommer JM, Wang CC (1994) Targeting proteins to the glycosomes of African trypanosomes. Annu Rev Microbiol 48:105–138. https://doi.org/10.1146/annurev.mi.48.100194.000541
181. Schlecker T, Schmidt A, Dirdjaja N, Voncken F, Clayton C, Krauth-Siegel RL (2005) Substrate specificity, localization, and essential role of the glutathione peroxidase-type tryparedoxin peroxidases in *Trypanosoma brucei*. J Biol Chem 280(15):14385–14394. https://doi.org/10.1074/jbc.M413338200
182. Tovar J, Cunningham ML, Smith AC, Croft SL, Fairlamb AH (1998) Down-regulation of *Leishmania donovani* trypanothione reductase by heterologous expression of a trans-dominant mutant homologue: effect on parasite intracellular survival. Proc Natl Acad Sci U S A 95(9):5311–5316
183. Krieger S, Schwarz W, Ariyanayagam MR, Fairlamb AH, Krauth-Siegel RL, Clayton C (2000) Trypanosomes lacking trypanothione reductase are avirulent and show increased sensitivity to oxidative stress. Mol Microbiol 35(3):542–552
184. Leroux AE, Krauth-Siegel RL (2016) Thiol redox biology of trypanosomatids and potential targets for chemotherapy. Mol Biochem Parasitol 206(1–2):67–74. https://doi.org/10.1016/j.molbiopara.2015.11.003
185. Boveris A, Sies H, Martino EE, Docampo R, Turrens JF, Stoppani AO (1980) Deficient metabolic utilization of hydrogen peroxide in *Trypanosoma cruzi*. Biochem J 188(3):643–648
186. Kraeva N, Horáková E, Kostygov AY, Kořený L, Butenko A, Yurchenko V, Lukeš J (2017) Catalase in Leishmaniinae: With me or against me? Infect Genet Evol 50:121–127. https://doi.org/10.1016/j.meegid.2016.06.054
187. Penketh PG, Kennedy WP, Patton CL, Sartorelli AC (1987) Trypanosomatid hydrogen peroxide [corrected] metabolism. FEBS Lett 221(2):427–431
188. Carnieri EG, Moreno SN, Docampo R (1993) Trypanothione-dependent peroxide metabolism in *Trypanosoma cruzi* different stages. Mol Biochem Parasitol 61(1):79–86
189. Nogoceke E, Gommel DU, Kiess M, Kalisz HM, Flohé L (1997) A unique cascade of oxidoreductases catalyses trypanothione-mediated peroxide metabolism in *Crithidia fasciculata*. Biol Chem 378(8):827–836
190. Castro H, Tomás AM (2008) Peroxidases of trypanosomatids. Antioxid Redox Signal 10(9):1593–1606. https://doi.org/10.1089/ars.2008.2050
191. Castro H, Sousa C, Novais M, Santos M, Budde H, Cordeiro-da-Silva A, Flohé L, Tomás AM (2004) Two linked genes of *Leishmania infantum* encode tryparedoxins localised to cytosol and mitochondrion. Mol Biochem Parasitol 136(2):137–147
192. Castro H, Romao S, Gadelha FR, Tomás AM (2008) *Leishmania infantum*: provision of reducing equivalents to the mitochondrial tryparedoxin/tryparedoxin peroxidase system. Exp Parasitol 120(4):421–423. https://doi.org/10.1016/j.exppara.2008.09.002
193. Castro H, Romao S, Carvalho S, Teixeira F, Sousa C, Tomás AM (2010) Mitochondrial redox metabolism in trypanosomatids is independent of tryparedoxin activity. PLoS One 5(9):e12607. https://doi.org/10.1371/journal.pone.0012607
194. El-Sayed NM, Myler PJ, Blandin G, Berriman M, Crabtree J, Aggarwal G, Caler E, Renauld H, Worthey EA, Hertz-Fowler C, Ghedin E, Peacock C, Bartholomeu DC, Haas BJ, Tran AN, Wortman JR, Alsmark UC, Angiuoli S, Anupama A, Badger J, Bringaud F, Cadag E, Carlton JM, Cerqueira GC, Creasy T, Delcher AL, Djikeng A, Embley TM, Hauser C, Ivens AC, Kummerfeld SK, Pereira-Leal JB, Nilsson D, Peterson J, Salzberg SL, Shallom J, Silva JC, Sundaram J, Westenberger S, White O, Melville SE, Donelson JE, Andersson B, Stuart KD, Hall N (2005) Comparative genomics of trypanosomatid parasitic protozoa. Science 309(5733):404–409. https://doi.org/10.1126/science.1112181

195. Romao S, Castro H, Sousa C, Carvalho S, Tomás AM (2009) The cytosolic tryparedoxin of *Leishmania infantum* is essential for parasite survival. Int J Parasitol 39(6):703–711. https://doi.org/10.1016/j.ijpara.2008.11.009
196. Comini MA, Krauth-Siegel RL, Flohé L (2007) Depletion of the thioredoxin homologue tryparedoxin impairs antioxidative defence in African trypanosomes. Biochem J 402(1):43–49. https://doi.org/10.1042/bj20061341
197. Suman SS, Equbal A, Zaidi A, Ansari MY, Singh KP, Singh K, Purkait B, Sahoo GC, Bimal S, Das P, Ali V (2016) Up-regulation of cytosolic tryparedoxin in Amp B resistant isolates of *Leishmania donovani* and its interaction with cytosolic tryparedoxin peroxidase. Biochimie 121:312–325. https://doi.org/10.1016/j.biochi.2015.12.017
198. Suman SS, Amit A, Singh KP, Gupta P, Equbal A, Kumari A, Topno RK, Ravidas V, Pandey K, Bimal S, Das P, Ali V (2018) Cytosolic tryparedoxin of *Leishmania donovani* modulates host immune response in visceral leishmaniasis. Cytokine 108:1–8. https://doi.org/10.1016/j.cyto.2018.03.010
199. Gommel DU, Nogoceke E, Morr M, Kiess M, Kalisz HM, Flohé L (1997) Catalytic characteristics of tryparedoxin. Eur J Biochem 248(3):913–918
200. Barr SD, Gedamu L (2001) Cloning and characterization of three differentially expressed peroxidoxin genes from *Leishmania chagasi*. Evidence for an enzymatic detoxification of hydroxyl radicals. J Biol Chem 276(36):34279–34287. https://doi.org/10.1074/jbc.M104406200
201. Poole LB (2007) The catalytic mechanism of peroxiredoxins. Subcell Biochem 44:61–81
202. Levick MP, Tetaud E, Fairlamb AH, Blackwell JM (1998) Identification and characterisation of a functional peroxidoxin from *Leishmania major*. Mol Biochem Parasitol 96(1-2):125–137
203. Castro H, Budde H, Flohé L, Hofmann B, Lünsdorf H, Wissing J, Tomás AM (2002) Specificity and kinetics of a mitochondrial peroxiredoxin of *Leishmania infantum*. Free Radic Biol Med 33(11):1563–1574
204. Flohé L, Budde H, Bruns K, Castro H, Clos J, Hofmann B, Kansal-Kalavar S, Krumme D, Menge U, Plank-Schumacher K, Sztajer H, Wissing J, Wylegalla C, Hecht HJ (2002) Tryparedoxin peroxidase of *Leishmania donovani*: molecular cloning, heterologous expression, specificity, and catalytic mechanism. Arch Biochem Biophys 397(2):324–335. https://doi.org/10.1006/abbi.2001.2688
205. Budde H, Flohé L, Hecht HJ, Hofmann B, Stehr M, Wissing J, Lünsdorf H (2003) Kinetics and redox-sensitive oligomerisation reveal negative subunit cooperativity in tryparedoxin peroxidase of *Trypanosoma brucei brucei*. Biol Chem 384(4):619–633. https://doi.org/10.1515/bc.2003.069
206. Wilkinson SR, Meyer DJ, Kelly JM (2000) Biochemical characterization of a trypanosome enzyme with glutathione-dependent peroxidase activity. Biochem J 352(Pt 3):755–761
207. Wilkinson SR, Taylor MC, Touitha S, Mauricio IL, Meyer DJ, Kelly JM (2002) TcGPXII, a glutathione-dependent *Trypanosoma cruzi* peroxidase with substrate specificity restricted to fatty acid and phospholipid hydroperoxides, is localized to the endoplasmic reticulum. Biochem J 364(Pt 3):787–794. https://doi.org/10.1042/bj20020038
208. Colasante C, Ellis M, Ruppert T, Voncken F (2006) Comparative proteomics of glycosomes from bloodstream form and procyclic culture form *Trypanosoma brucei brucei*. Proteomics 6(11):3275–3293. https://doi.org/10.1002/pmic.200500668
209. Herbette S, Roeckel-Drevet P, Drevet JR (2007) Seleno-independent glutathione peroxidases. More than simple antioxidant scavengers. FEBS J 274(9):2163–2180. https://doi.org/10.1111/j.1742-4658.2007.05774.x
210. Hillebrand H, Schmidt A, Krauth-Siegel RL (2003) A second class of peroxidases linked to the trypanothione metabolism. J Biol Chem 278(9):6809–6815. https://doi.org/10.1074/jbc.M210392200
211. Clark D, Albrecht M, Arévalo J (1994) Ascorbate variations and dehydroascorbate reductase activity in *Trypanosoma cruzi* epimastigotes and trypomastigotes. Mol Biochem Parasitol 66(1):143–145

212. Wilkinson SR, Obado SO, Mauricio IL, Kelly JM (2002) *Trypanosoma cruzi* expresses a plant-like ascorbate-dependent hemoperoxidase localized to the endoplasmic reticulum. Proc Natl Acad Sci U S A 99(21):13453–13458. https://doi.org/10.1073/pnas.202422899
213. Halliwell B (2001) Vitamin C and genomic stability. Mutat Res 475(1-2):29–35
214. Docampo R, de Boiso JF, Boveris A, Stoppani AO (1976) Localization of peroxidase activity in *Trypanosoma cruzi* microbodies. Experientia 32(8):972–975
215. Adak S, Datta AK (2005) *Leishmania major* encodes an unusual peroxidase that is a close homologue of plant ascorbate peroxidase: a novel role of the transmembrane domain. Biochem J 390(Pt 2):465–474. https://doi.org/10.1042/bj20050311
216. Krauth-Siegel RL, Lüdemann H (1996) Reduction of dehydroascorbate by trypanothione. Mol Biochem Parasitol 80(2):203–208
217. Reckenfelderbäumer N, Krauth-Siegel RL (2002) Catalytic properties, thiol pK value, and redox potential of *Trypanosoma brucei* tryparedoxin. J Biol Chem 277(20):17548–17555. https://doi.org/10.1074/jbc.M112115200
218. Taylor MC, Lewis MD, Fortes Francisco A, Wilkinson SR, Kelly JM (2015) The *Trypanosoma cruzi* vitamin C dependent peroxidase confers protection against oxidative stress but is not a determinant of virulence. PLoS Negl Trop Dis 9(4):e0003707. https://doi.org/10.1371/journal.pntd.0003707
219. McCord JM, Fridovich I (1969) The utility of superoxide dismutase in studying free radical reactions. I. Radicals generated by the interaction of sulfite, dimethyl sulfoxide, and oxygen. J Biol Chem 244(22):6056–6063
220. Dufernez F, Yernaux C, Gerbod D, Noël C, Chauvenet M, Wintjens R, Edgcomb VP, Capron M, Opperdoes FR, Viscogliosi E (2006) The presence of four iron-containing superoxide dismutase isozymes in trypanosomatidae: characterization, subcellular localization, and phylogenetic origin in *Trypanosoma brucei*. Free Radic Biol Med 40(2):210–225. https://doi.org/10.1016/j.freeradbiomed.2005.06.021
221. Ismail SO, Paramchuk W, Skeiky YA, Reed SG, Bhatia A, Gedamu L (1997) Molecular cloning and characterization of two iron superoxide dismutase cDNAs from *Trypanosoma cruzi*. Mol Biochem Parasitol 86(2):187–197
222. Temperton NJ, Wilkinson SR, Meyer DJ, Kelly JM (1998) Overexpression of superoxide dismutase in *Trypanosoma cruzi* results in increased sensitivity to the trypanocidal agents gentian violet and benznidazole. Mol Biochem Parasitol 96(1–2):167–176
223. Taylor MC, Kelly JM (2006) pTcINDEX: a stable tetracycline-regulated expression vector for *Trypanosoma cruzi*. BMC Biotechnol 6:32. https://doi.org/10.1186/1472-6750-6-32
224. Piacenza L, Irigoín F, Alvarez MN, Peluffo G, Taylor MC, Kelly JM, Wilkinson SR, Radi R (2007) Mitochondrial superoxide radicals mediate programmed cell death in *Trypanosoma cruzi*: cytoprotective action of mitochondrial iron superoxide dismutase overexpression. Biochem J 403(2):323–334. https://doi.org/10.1042/bj20061281
225. Getachew F, Gedamu L (2012) *Leishmania donovani* mitochondrial iron superoxide dismutase A is released into the cytosol during miltefosine induced programmed cell death. Mol Biochem Parasitol 183(1):42–51. https://doi.org/10.1016/j.molbiopara.2012.01.005
226. Alzate JF, Arias AA, Moreno-Mateos D, Alvarez-Barrientos A, Jiménez-Ruiz A (2007) Mitochondrial superoxide mediates heat-induced apoptotic-like death in *Leishmania infantum*. Mol Biochem Parasitol 152(2):192–202. https://doi.org/10.1016/j.molbiopara.2007.01.006
227. Mittra B, Laranjeira-Silva MF, Miguel DC, Perrone Bezerra de Menezes J, Andrews NW (2017) The iron-dependent mitochondrial superoxide dismutase SODA promotes. J Biol Chem 292(29):12324–12338. https://doi.org/10.1074/jbc.M116.772624
228. Filomeni G, Rotilio G, Ciriolo MR (2002) Cell signalling and the glutathione redox system. Biochem Pharmacol 64(5-6):1057–1064
229. Circu ML, Aw TY (2010) Reactive oxygen species, cellular redox systems, and apoptosis. Free Radic Biol Med 48(6):749–762. https://doi.org/10.1016/j.freeradbiomed.2009.12.022

230. Valko M, Leibfritz D, Moncol J, Cronin MT, Mazur M, Telser J (2007) Free radicals and antioxidants in normal physiological functions and human disease. Int J Biochem Cell Biol 39(1):44–84. https://doi.org/10.1016/j.biocel.2006.07.001
231. Couto N, Wood J, Barber J (2016) The role of glutathione reductase and related enzymes on cellular redox homoeostasis network. Free Radic Biol Med 95:27–42. https://doi.org/10.1016/j.freeradbiomed.2016.02.028
232. Chiang HS, Maric M (2011) Lysosomal thiol reductase negatively regulates autophagy by altering glutathione synthesis and oxidation. Free Radic Biol Med 51(3):688–699. https://doi.org/10.1016/j.freeradbiomed.2011.05.015
233. Bindoli A, Fukuto JM, Forman HJ (2008) Thiol chemistry in peroxidase catalysis and redox signaling. Antioxid Redox Signal 10(9):1549–1564. https://doi.org/10.1089/ars.2008.2063
234. Flohé L (2010) Changing paradigms in thiology from antioxidant defense toward redox regulation. Methods Enzymol 473:1–39. https://doi.org/10.1016/s0076-6879(10)73001-9
235. Voehringer DW, McConkey DJ, McDonnell TJ, Brisbay S, Meyn RE (1998) Bcl-2 expression causes redistribution of glutathione to the nucleus. Proc Natl Acad Sci U S A 95(6):2956–2960
236. Markovic J, Borrás C, Ortega A, Sastre J, Viña J, Pallardó FV (2007) Glutathione is recruited into the nucleus in early phases of cell proliferation. J Biol Chem 282(28):20416–20424. https://doi.org/10.1074/jbc.M609582200
237. Chakravarthi S, Jessop CE, Bulleid NJ (2006) The role of glutathione in disulphide bond formation and endoplasmic-reticulum-generated oxidative stress. EMBO Rep 7(3):271–275. https://doi.org/10.1038/sj.embor.7400645
238. Jessop CE, Bulleid NJ (2004) Glutathione directly reduces an oxidoreductase in the endoplasmic reticulum of mammalian cells. J Biol Chem 279(53):55341–55347. https://doi.org/10.1074/jbc.M411409200
239. Jocelyn PC, Kamminga A (1974) The non-protein thiol of rat liver mitochondria. Biochim Biophys Acta 343(2):356–362
240. Schnellmann RG (1991) Renal mitochondrial glutathione transport. Life Sci 49(5):393–398
241. Holmgren A, Björnstedt M (1995) Thioredoxin and thioredoxin reductase. Methods Enzymol 252:199–208
242. Lee S, Kim SM, Lee RT (2013) Thioredoxin and thioredoxin target proteins: from molecular mechanisms to functional significance. Antioxid Redox Signal 18(10):1165–1207. https://doi.org/10.1089/ars.2011.4322
243. Perkins A, Nelson KJ, Parsonage D, Poole LB, Karplus PA (2015) Peroxiredoxins: guardians against oxidative stress and modulators of peroxide signaling. Trends Biochem Sci 40(8):435–445. https://doi.org/10.1016/j.tibs.2015.05.001
244. Kim K, Kim IH, Lee KY, Rhee SG, Stadtman ER (1988) The isolation and purification of a specific "protector" protein which inhibits enzyme inactivation by a thiol/Fe(III)/O2 mixed-function oxidation system. J Biol Chem 263(10):4704–4711
245. Knoops B, Argyropoulou V, Becker S, Ferté L, Kuznetsova O (2016) Multiple roles of peroxiredoxins in inflammation. Mol Cells 39(1):60–64. https://doi.org/10.14348/molcells.2016.2341
246. Woo HA, Yim SH, Shin DH, Kang D, Yu DY, Rhee SG (2010) Inactivation of peroxiredoxin I by phosphorylation allows localized H_2O_2 accumulation for cell signaling. Cell 140(4):517–528. https://doi.org/10.1016/j.cell.2010.01.009
247. Rhee SG, Woo HA (2011) Multiple functions of peroxiredoxins: peroxidases, sensors and regulators of the intracellular messenger H_2O_2, and protein chaperones. Antioxid Redox Signal 15(3):781–794. https://doi.org/10.1089/ars.2010.3393
248. Edgar RS, Green EW, Zhao Y, van Ooijen G, Olmedo M, Qin X, Xu Y, Pan M, Valekunja UK, Feeney KA, Maywood ES, Hastings MH, Baliga NS, Merrow M, Millar AJ, Johnson CH, Kyriacou CP, O'Neill JS, Reddy AB (2012) Peroxiredoxins are conserved markers of circadian rhythms. Nature 485(7399):459–464. https://doi.org/10.1038/nature11088
249. Hanschmann EM, Godoy JR, Berndt C, Hudemann C, Lillig CH (2013) Thioredoxins, glutaredoxins, and peroxiredoxins-molecular mechanisms and health significance: from

cofactors to antioxidants to redox signaling. Antioxid Redox Signal 19(13):1539–1605. https://doi.org/10.1089/ars.2012.4599
250. Knoops B, Clippe A, Bogard C, Arsalane K, Wattiez R, Hermans C, Duconseille E, Falmagne P, Bernard A (1999) Cloning and characterization of AOEB166, a novel mammalian antioxidant enzyme of the peroxiredoxin family. J Biol Chem 274(43):30451–30458
251. Song JJ, Rhee JG, Suntharalingam M, Walsh SA, Spitz DR, Lee YJ (2002) Role of glutaredoxin in metabolic oxidative stress. Glutaredoxin as a sensor of oxidative stress mediated by H_2O_2. J Biol Chem 277(48):46566–46575. https://doi.org/10.1074/jbc.M206826200
252. Lu J, Holmgren A (2012) Thioredoxin system in cell death progression. Antioxid Redox Signal 17(12):1738–1747. https://doi.org/10.1089/ars.2012.4650
253. Lu J, Holmgren A (2014) The thioredoxin antioxidant system. Free Radic Biol Med 66:75–87. https://doi.org/10.1016/j.freeradbiomed.2013.07.036
254. Nightingale H, Kemp K, Gray E, Hares K, Mallam E, Scolding N, Wilkins A (2012) Changes in expression of the antioxidant enzyme SOD3 occur upon differentiation of human bone marrow-derived mesenchymal stem cells in vitro. Stem Cells Dev 21(11):2026–2035. https://doi.org/10.1089/scd.2011.0516
255. Ye ZW, Zhang J, Townsend DM, Tew KD (2015) Oxidative stress, redox regulation and diseases of cellular differentiation. Biochim Biophys Acta 1850(8):1607–1621. https://doi.org/10.1016/j.bbagen.2014.11.010
256. Putnam CD, Arvai AS, Bourne Y, Tainer JA (2000) Active and inhibited human catalase structures: ligand and NADPH binding and catalytic mechanism. J Mol Biol 296(1):295–309. https://doi.org/10.1006/jmbi.1999.3458
257. Nicholls P (2012) Classical catalase: ancient and modern. Arch Biochem Biophys 525(2):95–101. https://doi.org/10.1016/j.abb.2012.01.015
258. Liao AC, Craver BM, Tseng BP, Tran KK, Parihar VK, Acharya MM, Limoli CL (2013) Mitochondrial-targeted human catalase affords neuroprotection from proton irradiation. Radiat Res 180(1):1–6. https://doi.org/10.1667/rr3339.1
259. Dai DF, Santana LF, Vermulst M, Tomazela DM, Emond MJ, MacCoss MJ, Gollahon K, Martin GM, Loeb LA, Ladiges WC, Rabinovitch PS (2009) Overexpression of catalase targeted to mitochondria attenuates murine cardiac aging. Circulation 119(21):2789–2797. https://doi.org/10.1161/circulationaha.108.822403
260. Nogueira NP, Saraiva FM, Sultano PE, Cunha PR, Laranja GA, Justo GA, Sabino KC, Coelho MG, Rossini A, Atella GC, Paes MC (2015) Proliferation and differentiation of *Trypanosoma cruzi* inside its vector have a new trigger: redox status. PLoS One 10(2):e0116712. https://doi.org/10.1371/journal.pone.0116712
261. Sterkel M, Oliveira JHM, Bottino-Rojas V, Paiva-Silva GO, Oliveira PL (2017) The dose makes the poison: nutritional overload determines the life traits of blood-feeding arthropods. Trends Parasitol 33(8):633–644. https://doi.org/10.1016/j.pt.2017.04.008
262. Carneiro PP, Conceição J, Macedo M, Magalhães V, Carvalho EM, Bacellar O (2016) The role of nitric oxide and reactive oxygen species in the killing of *Leishmania braziliensis* by monocytes from patients with Cutaneous Leishmaniasis. PLoS One 11(2):e0148084. https://doi.org/10.1371/journal.pone.0148084
263. Kierszenbaum F, Knecht E, Budzko DB, Pizzimenti MC (1974) Phagocytosis: a defense mechanism against infection with *Trypanosoma cruzi*. J Immunol 112(5):1839–1844
264. Alvarez MN, Peluffo G, Piacenza L, Radi R (2011) Intraphagosomal peroxynitrite as a macrophage-derived cytotoxin against internalized *Trypanosoma cruzi*: consequences for oxidative killing and role of microbial peroxiredoxins in infectivity. J Biol Chem 286(8):6627–6640. https://doi.org/10.1074/jbc.M110.167247
265. Guiñazú N, Carrera-Silva EA, Becerra MC, Pellegrini A, Albesa I, Gea S (2010) Induction of NADPH oxidase activity and reactive oxygen species production by a single *Trypanosoma cruzi* antigen. Int J Parasitol 40(13):1531–1538. https://doi.org/10.1016/j.ijpara.2010.05.012
266. Alvarez MN, Piacenza L, Irigoín F, Peluffo G, Radi R (2004) Macrophage-derived peroxynitrite diffusion and toxicity to *Trypanosoma cruzi*. Arch Biochem Biophys 432(2):222–232. https://doi.org/10.1016/j.abb.2004.09.015

267. Fridovich I (1997) Superoxide anion radical (O2-.), superoxide dismutases, and related matters. J Biol Chem 272(30):18515–18517
268. Babior BM (1984) The respiratory burst of phagocytes. J Clin Invest 73(3):599–601. https://doi.org/10.1172/jci111249
269. Bedard K, Krause KH (2007) The NOX family of ROS-generating NADPH oxidases: physiology and pathophysiology. Physiol Rev 87(1):245–313. https://doi.org/10.1152/physrev.00044.2005
270. de Carvalho TU, de Souza W (1987) Cytochemical localization of NADH and NADPH oxidases during interaction of *Trypanosoma cruzi* with activated macrophages. Parasitol Res 73(3):213–217
271. Brandes RP, Weissmann N, Schröder K (2014) Nox family NADPH oxidases: molecular mechanisms of activation. Free Radic Biol Med 76:208–226. https://doi.org/10.1016/j.freeradbiomed.2014.07.046
272. Van Assche T, Deschacht M, da Luz RA, Maes L, Cos P (2011) Leishmania-macrophage interactions: insights into the redox biology. Free Radic Biol Med 51(2):337–351. https://doi.org/10.1016/j.freeradbiomed.2011.05.011
273. Peluffo G, Piacenza L, Irigoín F, Alvarez MN, Radi R (2004) L-arginine metabolism during interaction of *Trypanosoma cruzi* with host cells. Trends Parasitol 20(8):363–369. https://doi.org/10.1016/j.pt.2004.05.010
274. Radi R (2013) Peroxynitrite, a stealthy biological oxidant. J Biol Chem 288(37):26464–26472. https://doi.org/10.1074/jbc.R113.472936
275. Cassina A, Radi R (1996) Differential inhibitory action of nitric oxide and peroxynitrite on mitochondrial electron transport. Arch Biochem Biophys 328(2):309–316. https://doi.org/10.1006/abbi.1996.0178
276. Ferrer-Sueta G, Radi R (2009) Chemical biology of peroxynitrite: kinetics, diffusion, and radicals. ACS Chem Biol 4(3):161–177. https://doi.org/10.1021/cb800279q
277. Piacenza L, Peluffo G, Alvarez MN, Martínez A, Radi R (2013) *Trypanosoma cruzi* antioxidant enzymes as virulence factors in Chagas disease. Antioxid Redox Signal 19(7):723–734. https://doi.org/10.1089/ars.2012.4618
278. Piacenza L, Alvarez MN, Peluffo G, Radi R (2009) Fighting the oxidative assault: the *Trypanosoma cruzi* journey to infection. Curr Opin Microbiol 12(4):415–421. https://doi.org/10.1016/j.mib.2009.06.011
279. Mattner J, Wandersee-Steinhäuser A, Pahl A, Röllinghoff M, Majeau GR, Hochman PS, Bogdan C (2004) Protection against progressive leishmaniasis by IFN-beta. J Immunol 172(12):7574–7582
280. Costa VM, Torres KC, Mendonça RZ, Gresser I, Gollob KJ, Abrahamsohn IA (2006) Type I IFNs stimulate nitric oxide production and resistance to *Trypanosoma cruzi* infection. J Immunol 177(5):3193–3200
281. Kinkade JM, Pember SO, Barnes KC, Shapira R, Spitznagel JK, Martin LE (1983) Differential distribution of distinct forms of myeloperoxidase in different azurophilic granule subpopulations from human neutrophils. Biochem Biophys Res Commun 114(1):296–303
282. Anand U, Anand CV (2012) Myeloperoxidase: a new twist to an old tale. Indian J Clin Biochem 27(2):107–109. https://doi.org/10.1007/s12291-012-0220-0
283. Lazarević-Pasti T, Leskovac A, Vasić V (2015) Myeloperoxidase Inhibitors as Potential Drugs. Curr Drug Metab 16(3):168–190
284. Segal AW (2005) How neutrophils kill microbes. Annu Rev Immunol 23:197–223. https://doi.org/10.1146/annurev.immunol.23.021704.115653
285. Piacenza L, Peluffo G, Alvarez MN, Kelly JM, Wilkinson SR, Radi R (2008) Peroxiredoxins play a major role in protecting *Trypanosoma cruzi* against macrophage- and endogenously-derived peroxynitrite. Biochem J 410(2):359–368. https://doi.org/10.1042/bj20071138
286. Chandrasekar B, Melby PC, Troyer DA, Colston JT, Freeman GL (1998) Temporal expression of pro-inflammatory cytokines and inducible nitric oxide synthase in experimental acute Chagasic cardiomyopathy. Am J Pathol 152(4):925–934

287. Machado FS, Martins GA, Aliberti JC, Mestriner FL, Cunha FQ, Silva JS (2000) *Trypanosoma cruzi*-infected cardiomyocytes produce chemokines and cytokines that trigger potent nitric oxide-dependent trypanocidal activity. Circulation 102(24):3003–3008
288. Luquetti AO, Miles MA, Rassi A, de Rezende JM, de Souza AA, Póvoa MM, Rodrigues I (1986) *Trypanosoma cruzi*: zymodemes associated with acute and chronic Chagas' disease in central Brazil. Trans R Soc Trop Med Hyg 80(3):462–470
289. Piacenza L, Zago MP, Peluffo G, Alvarez MN, Basombrio MA, Radi R (2009) Enzymes of the antioxidant network as novel determiners of *Trypanosoma cruzi* virulence. Int J Parasitol 39(13):1455–1464. https://doi.org/10.1016/j.ijpara.2009.05.010
290. McNeely TB, Turco SJ (1990) Requirement of lipophosphoglycan for intracellular survival of *Leishmania donovani* within human monocytes. J Immunol 144(7):2745–2750
291. Lodge R, Diallo TO, Descoteaux A (2006) *Leishmania donovani* lipophosphoglycan blocks NADPH oxidase assembly at the phagosome membrane. Cell Microbiol 8(12):1922–1931. https://doi.org/10.1111/j.1462-5822.2006.00758.x
292. Proudfoot L, Nikolaev AV, Feng GJ, Wei WQ, Ferguson MA, Brimacombe JS, Liew FY (1996) Regulation of the expression of nitric oxide synthase and leishmanicidal activity by glycoconjugates of Leishmania lipophosphoglycan in murine macrophages. Proc Natl Acad Sci U S A 93(20):10984–10989
293. Kumar R, Pai K, Sundar S (2001) Reactive oxygen intermediates, nitrite and IFN-gamma in Indian visceral leishmaniasis. Clin Exp Immunol 124(2):262–265
294. Kumar P, Pai K, Pandey HP, Sundar S (2002) NADH-oxidase, NADPH-oxidase and myeloperoxidase activity of visceral leishmaniasis patients. J Med Microbiol 51(10):832–836. https://doi.org/10.1099/0022-1317-51-10-832
295. DeLeo FR, Burritt JB, Yu L, Jesaitis AJ, Dinauer MC, Nauseef WM (2000) Processing and maturation of flavocytochrome b558 include incorporation of heme as a prerequisite for heterodimer assembly. J Biol Chem 275(18):13986–13993
296. Pham NK, Mouriz J, Kima PE (2005) *Leishmania pifanoi* amastigotes avoid macrophage production of superoxide by inducing heme degradation. Infect Immun 73(12):8322–8333. https://doi.org/10.1128/iai.73.12.8322-8333.2005
297. McNeely TB, Rosen G, Londner MV, Turco SJ (1989) Inhibitory effects on protein kinase C activity by lipophosphoglycan fragments and glycosylphosphatidylinositol antigens of the protozoan parasite Leishmania. Biochem J 259(2):601–604
298. Watson F, Robinson J, Edwards SW (1991) Protein kinase C-dependent and -independent activation of the NADPH oxidase of human neutrophils. J Biol Chem 266(12):7432–7439
299. da Silva AL, Moretti NS, Ramos TC, de Jesus TC, Zhang M, Castilho BA, Schenkman S (2015) A membrane-bound eIF2 alpha kinase located in endosomes is regulated by heme and controls differentiation and ROS levels in *Trypanosoma cruzi*. PLoS Pathog 11(2):e1004618. https://doi.org/10.1371/journal.ppat.1004618
300. Mittra B, Cortez M, Haydock A, Ramasamy G, Myler PJ, Andrews NW (2013) Iron uptake controls the generation of Leishmania infective forms through regulation of ROS levels. J Exp Med 210(2):401–416. https://doi.org/10.1084/jem.20121368
301. Graça-Souza AV, Maya-Monteiro C, Paiva-Silva GO, Braz GR, Paes MC, Sorgine MH, Oliveira MF, Oliveira PL (2006) Adaptations against heme toxicity in blood-feeding arthropods. Insect Biochem Mol Biol 36(4):322–335. https://doi.org/10.1016/j.ibmb.2006.01.009
302. Rodrigues JC, Bernardes CF, Visbal G, Urbina JA, Vercesi AE, de Souza W (2007) Sterol methenyl transferase inhibitors alter the ultrastructure and function of the *Leishmania amazonensis* mitochondrion leading to potent growth inhibition. Protist 158(4):447–456. https://doi.org/10.1016/j.protis.2007.05.004
303. Rodrigues JC, de Souza W (2008) Ultrastructural alterations in organelles of parasitic protozoa induced by different classes of metabolic inhibitors. Curr Pharm Des 14(9):925–938
304. Sen N, Majumder HK (2008) Mitochondrion of protozoan parasite emerges as potent therapeutic target: exciting drugs are on the horizon. Curr Pharm Des 14(9):839–846

305. Cavalli A, Bolognesi ML (2009) Neglected tropical diseases: multi-target-directed ligands in the search for novel lead candidates against Trypanosoma and Leishmania. J Med Chem 52(23):7339–7359. https://doi.org/10.1021/jm9004835
306. Menna-Barreto RF, Corrêa JR, Cascabulho CM, Fernandes MC, Pinto AV, Soares MJ, De Castro SL (2009) Naphthoimidazoles promote different death phenotypes in *Trypanosoma cruzi*. Parasitology 136(5):499–510. https://doi.org/10.1017/s0031182009005745
307. de Souza W, Attias M, Rodrigues JC (2009) Particularities of mitochondrial structure in parasitic protists (Apicomplexa and Kinetoplastida). Int J Biochem Cell Biol 41(10):2069–2080. https://doi.org/10.1016/j.biocel.2009.04.007
308. Vannier-Santos MA, De Castro SL (2009) Electron microscopy in antiparasitic chemotherapy: a (close) view to a kill. Curr Drug Targets 10(3):246–260
309. Fidalgo LM, Gille L (2011) Mitochondria and trypanosomatids: targets and drugs. Pharm Res 28(11):2758–2770. https://doi.org/10.1007/s11095-011-0586-3
310. Buckner FS, Urbina JA (2012) Recent developments in sterol 14-demethylase Inhibitors for Chagas disease. Int J Parasitol Drugs Drug Resist 2:236–242. https://doi.org/10.1016/j.ijpddr.2011.12.002
311. Sangenito LS, Menna-Barreto RFS, Oliveira ACS, d'Avila-Levy CM, Branquinha MH, ALS S (2018) Primary evidences of the mechanisms of action of HIV aspartyl peptidase inhibitors on *Trypanosoma cruzi* trypomastigote forms. Int J Antimicrob Agents. https://doi.org/10.1016/j.ijantimicag.2018.03.021
312. Menna-Barreto RF, de Castro SL (2014) The double-edged sword in pathogenic trypanosomatids: the pivotal role of mitochondria in oxidative stress and bioenergetics. Biomed Res Int 2014:614014. https://doi.org/10.1155/2014/614014
313. Docampo R, Gadelha FR, Moreno SN, Benaim G, Hoffmann ME, Vercesi AE (1993) Disruption of Ca2+ homeostasis in *Trypanosoma cruzi* by crystal violet. J Eukaryot Microbiol 40(3):311–316
314. Affranchino JL, De Tarlovsky MN, Stoppani AO (1985) Respiratory control in mitochondria from *Trypanosoma cruzi*. Mol Biochem Parasitol 16(3):289–298
315. Peloso EF, Vitor SC, Ribeiro LH, Piñeyro MD, Robello C, Gadelha FR (2011) Role of *Trypanosoma cruzi* peroxiredoxins in mitochondrial bioenergetics. J Bioenerg Biomembr 43(4):419–424. https://doi.org/10.1007/s10863-011-9365-4
316. Luque-Ortega JR, Reuther P, Rivas L, Dardonville C (2010) New benzophenone-derived bisphosphonium salts as leishmanicidal leads targeting mitochondria through inhibition of respiratory complex II. J Med Chem 53(4):1788–1798. https://doi.org/10.1021/jm901677h
317. Carvalho L, Luque-Ortega JR, López-Martín C, Castanys S, Rivas L, Gamarro F (2011) The 8-aminoquinoline analogue sitamaquine causes oxidative stress in *Leishmania donovani* promastigotes by targeting succinate dehydrogenase. Antimicrob Agents Chemother 55(9):4204–4210. https://doi.org/10.1128/aac.00520-11
318. Luque-Ortega JR, Rivas L (2007) Miltefosine (hexadecylphosphocholine) inhibits cytochrome c oxidase in *Leishmania donovani* promastigotes. Antimicrob Agents Chemother 51(4):1327–1332. https://doi.org/10.1128/aac.01415-06
319. Carvalho L, Luque-Ortega JR, Manzano JI, Castanys S, Rivas L, Gamarro F (2010) Tafenoquine, an antiplasmodial 8-aminoquinoline, targets leishmania respiratory complex III and induces apoptosis. Antimicrob Agents Chemother 54(12):5344–5351. https://doi.org/10.1128/aac.00790-10
320. Njogu RM, Whittaker CJ, Hill GC (1980) Evidence for a branched electron transport chain in *Trypanosoma brucei*. Mol Biochem Parasitol 1(1):13–29
321. Menna-Barreto RF, Goncalves RL, Costa EM, Silva RS, Pinto AV, Oliveira MF, de Castro SL (2009) The effects on *Trypanosoma cruzi* of novel synthetic naphthoquinones are mediated by mitochondrial dysfunction. Free Radic Biol Med 47(5):644–653. https://doi.org/10.1016/j.freeradbiomed.2009.06.004
322. Moore AL, Shiba T, Young L, Harada S, Kita K, Ito K (2013) Unraveling the heater: new insights into the structure of the alternative oxidase. Annu Rev Plant Biol 64:637–663. https://doi.org/10.1146/annurev-arplant-042811-105432

323. Nihei C, Fukai Y, Kita K (2002) Trypanosome alternative oxidase as a target of chemotherapy. Biochim Biophys Acta 1587(2–3):234–239
324. Minagawa N, Yabu Y, Kita K, Nagai K, Ohta N, Meguro K, Sakajo S, Yoshimoto A (1996) An antibiotic, ascofuranone, specifically inhibits respiration and in vitro growth of long slender bloodstream forms of *Trypanosoma brucei brucei*. Mol Biochem Parasitol 81(2):127–136
325. Tsuda A, Witola WH, Ohashi K, Onuma M (2005) Expression of alternative oxidase inhibits programmed cell death-like phenomenon in bloodstream form of *Trypanosoma brucei rhodesiense*. Parasitol Int 54(4):243–251. https://doi.org/10.1016/j.parint.2005.06.007
326. Yabu Y, Suzuki T, Nihei C, Minagawa N, Hosokawa T, Nagai K, Kita K, Ohta N (2006) Chemotherapeutic efficacy of ascofuranone in *Trypanosoma vivax*-infected mice without glycerol. Parasitol Int 55(1):39–43. https://doi.org/10.1016/j.parint.2005.09.003
327. Villa-Pulgarín JA, Gajate C, Botet J, Jimenez A, Justies N, Varela-M RE, Cuesta-Marbán Á, Müller I, Modolell M, Revuelta JL, Mollinedo F (2017) Mitochondria and lipid raft-located FOF1-ATP synthase as major therapeutic targets in the antileishmanial and anticancer activities of ether lipid edelfosine. PLoS Negl Trop Dis 11(8):e0005805. https://doi.org/10.1371/journal.pntd.0005805
328. Powis G (1987) Metabolism and reactions of quinoid anticancer agents. Pharmacol Ther 35(1–2):57–162
329. O'Brien PJ (1991) Molecular mechanisms of quinone cytotoxicity. Chem Biol Interact 80(1):1–41
330. Monks TJ, Jones DC (2002) The metabolism and toxicity of quinones, quinonimines, quinone methides, and quinone-thioethers. Curr Drug Metab 3(4):425–438
331. Lima NM, Correia CS, Leon LL, Machado GM, Madeira MeF, Santana AE, Goulart MO (2004) Antileishmanial activity of lapachol analogues. Mem Inst Oswaldo Cruz 99(7):757–761. doi:/S0074-02762004000700017
332. da Silva Júnior EN, de Melo IM, Diogo EB, Costa VA, de Souza Filho JD, Valença WO, Camara CA, de Oliveira RN, de Araujo AS, Emery FS, dos Santos MR, de Simone CA, Menna-Barreto RF, de Castro SL (2012) On the search for potential anti-*Trypanosoma cruzi* drugs: synthesis and biological evaluation of 2-hydroxy-3-methylamino and 1,2,3-triazolic naphthoquinoidal compounds obtained by click chemistry reactions. Eur J Med Chem 52:304–312. https://doi.org/10.1016/j.ejmech.2012.03.039
333. da Silva Júnior EN, Jardim GAM, Menna-Barreto RFS, de Castro SL (2014) Anti-*Trypanosoma cruzi* compounds: our contribution for the evaluation and insights on the mode of action of naphthoquinones and derivatives. J Braz Chem Soc 25(10):1780–1798. https://doi.org/10.5935/0103-5053.20140180
334. Ramírez-Macías I, Marín C, Es-Samti H, Fernández A, Guardia JJ, Zentar H, Agil A, Chahboun R, Alvarez-Manzaneda E, Sánchez-Moreno M (2012) Taiwaniaquinoid and abietane quinone derivatives with trypanocidal activity against *T. cruzi* and *Leishmania* spp. Parasitol Int 61(3):405–413. https://doi.org/10.1016/j.parint.2012.02.001
335. Diogo EB, Dias GG, Rodrigues BL, Guimarães TT, Valença WO, Camara CA, de Oliveira RN, da Silva MG, Ferreira VF, de Paiva YG, Goulart MO, Menna-Barreto RF, de Castro SL, da Silva Júnior EN (2013) Synthesis and anti-*Trypanosoma cruzi* activity of naphthoquinone-containing triazoles: electrochemical studies on the effects of the quinoidal moiety. Bioorg Med Chem 21(21):6337–6348. https://doi.org/10.1016/j.bmc.2013.08.055
336. Lam CF, Pearce AN, Tan SH, Kaiser M, Copp BR (2013) Discovery and evaluation of thiazinoquinones as anti-protozoal agents. Mar Drugs 11(9):3472–3499. https://doi.org/10.3390/md11093472
337. Pieretti S, Haanstra JR, Mazet M, Perozzo R, Bergamini C, Prati F, Fato R, Lenaz G, Capranico G, Brun R, Bakker BM, Michels PA, Scapozza L, Bolognesi ML, Cavalli A (2013) Naphthoquinone derivatives exert their antitrypanosomal activity via a multi-target mechanism. PLoS Negl Trop Dis 7(1):e2012. https://doi.org/10.1371/journal.pntd.0002012

338. Hata Y, Ebrahimi SN, De Mieri M, Zimmermann S, Mokoka T, Naidoo D, Fouche G, Maharaj V, Kaiser M, Brun R, Potterat O, Hamburger M (2014) Antitrypanosomal isoflavan quinones from *Abrus precatorius*. Fitoterapia 93:81–87. https://doi.org/10.1016/j.fitote.2013.12.015
339. Ellendorff T, Brun R, Kaiser M, Sendker J, Schmidt TJ (2015) PLS-prediction and confirmation of hydrojuglone glucoside as the antitrypanosomal constituent of *Juglans* Spp. Molecules 20(6):10082–10094. https://doi.org/10.3390/molecules200610082
340. Boveris A, Docampo R, Turrens JF, Stoppani AO (1977) Effect of beta and alpha-lapachone on the production of H_2O_2 and on the growth of *Trypanosoma cruzi*. Rev Asoc Argent Microbiol 9(2):54–61
341. Cruz FS, Docampo R, de Souza W (1978) Effect of beta-lapachone on hydrogen peroxide production in *Trypanosoma cruzi*. Acta Trop 35(1):35–40
342. Molina Portela MP, Fernandez Villamil SH, Perissinotti LJ, Stoppani AO (1996) Redox cycling of o-naphthoquinones in trypanosomatids. Superoxide and hydrogen peroxide production. Biochem Pharmacol 52(12):1875–1882
343. Salomão K, De Santana NA, Molina MT, De Castro SL, Menna-Barreto RF (2013) *Trypanosoma cruzi* mitochondrial swelling and membrane potential collapse as primary evidence of the mode of action of naphthoquinone analogues. BMC Microbiol 13:196. https://doi.org/10.1186/1471-2180-13-196
344. Lara LS, Moreira CS, Calvet CM, Lechuga GC, Souza RS, Bourguignon SC, Ferreira VF, Rocha D, Pereira MCS (2018) Efficacy of 2-hydroxy-3-phenylsulfanylmethyl-[1,4]-naphthoquinone derivatives against different *Trypanosoma cruzi* discrete type units: Identification of a promising hit compound. Eur J Med Chem 144:572–581. https://doi.org/10.1016/j.ejmech.2017.12.052
345. Fonseca-Silva F, Inacio JD, Canto-Cavalheiro MM, Almeida-Amaral EE (2011) Reactive oxygen species production and mitochondrial dysfunction contribute to quercetin induced death in *Leishmania amazonensis*. PLoS One 6(2):e14666. https://doi.org/10.1371/journal.pone.0014666
346. Fonseca-Silva F, Canto-Cavalheiro MM, Menna-Barreto RF, Almeida-Amaral EE (2015) Effect of Apigenin on *Leishmania amazonensis* is associated with reactive oxygen species production followed by mitochondrial dysfunction. J Nat Prod 78(4):880–884. https://doi.org/10.1021/acs.jnatprod.5b00011
347. Fonseca-Silva F, Inacio JD, Canto-Cavalheiro MM, Menna-Barreto RF, Almeida-Amaral EE (2016) Oral efficacy of apigenin against cutaneous leishmaniasis: involvement of reactive oxygen species and autophagy as a mechanism of action. PLoS Negl Trop Dis 10(2):e0004442. https://doi.org/10.1371/journal.pntd.0004442
348. Inacio JD, Canto-Cavalheiro MM, Almeida-Amaral EE (2013) In vitro and in vivo effects of (-)-epigallocatechin 3-O-gallate on *Leishmania amazonensis*. J Nat Prod 76(10):1993–1996. https://doi.org/10.1021/np400624d
349. Inacio JD, Gervazoni L, Canto-Cavalheiro MM, Almeida-Amaral EE (2014) The effect of (-)-epigallocatechin 3-O--gallate in vitro and in vivo in *Leishmania braziliensis*: involvement of reactive oxygen species as a mechanism of action. PLoS Negl Trop Dis 8(8):e3093. https://doi.org/10.1371/journal.pntd.0003093
350. Martins SC, Lazarin-Bidóia D, Desoti VC, Falzirolli H, da Silva CC, Ueda-Nakamura T, Silva SO, Nakamura CV (2016) 1,3,4-Thiadiazole derivatives of R-(+)-limonene benzaldehyde-thiosemicarbazones cause death in *Trypanosoma cruzi* through oxidative stress. Microbes Infect 18(12):787–797. https://doi.org/10.1016/j.micinf.2016.07.007
351. Soares RO, Echevarria A, Bellieny MS, Pinho RT, de Leo RM, Seguins WS, Machado GM, Canto-Cavalheiro MM, Leon LL (2011) Evaluation of thiosemicarbazones and semicarbazones as potential agents anti-*Trypanosoma cruzi*. Exp Parasitol 129(4):381–387. https://doi.org/10.1016/j.exppara.2011.08.019

352. Britta EA, Scariot DB, Falzirolli H, Ueda-Nakamura T, Silva CC, Filho BP, Borsali R, Nakamura CV (2014) Cell death and ultrastructural alterations in *Leishmania amazonensis* caused by new compound 4-Nitrobenzaldehyde thiosemicarbazone derived from S-limonene. BMC Microbiol 14:236. https://doi.org/10.1186/s12866-014-0236-0
353. Lapier M, Zuniga-Lopez MC, Aguilera-Venegas B, Adam R, Abarca B, Ballesteros R, Lopez-Munoz R, Maya JD, Olea-Azar C (2017) Evaluation of the Novel Antichagasic Activity of [1,2,3]Triazolo[1,5-a]pyridine Derivatives. Curr Top Med Chem 17(4):399–411
354. Ribeiro GA, Cunha-Júnior EF, Pinheiro RO, da Silva SA, Canto-Cavalheiro MM, da Silva AJ, Costa PR, Netto CD, Melo RC, Almeida-Amaral EE, Torres-Santos EC (2013) LQB-118, an orally active pterocarpanquinone, induces selective oxidative stress and apoptosis in *Leishmania amazonensis*. J Antimicrob Chemother 68(4):789–799. https://doi.org/10.1093/jac/dks498
355. Mendonça DVC, Lage DP, Calixto SL, Ottoni FM, Tavares GSV, Ludolf F, Chávez-Fumagalli MA, Schneider MS, Duarte MC, Tavares CAP, Alves RJ, Coimbra ES, Coelho EAF (2018) Antileishmanial activity of a naphthoquinone derivate against promastigote and amastigote stages of *Leishmania infantum* and *Leishmania amazonensis* and its mechanism of action against *L. amazonensis* species. Parasitol Res 117(2):391–403. https://doi.org/10.1007/s00436-017-5713-6
356. Tavares GSV, Mendonça DVC, Lage DP, Granato JDT, Ottoni FM, Ludolf F, Chávez-Fumagalli MA, Duarte MC, Tavares CAP, Alves RJ, Coimbra ES, Coelho EAF (2018) Antileishmanial activity, cytotoxicity and mechanism of action of clioquinol against *Leishmania infantum* and *Leishmania amazonensis* species. Basic Clin Pharmacol Toxicol. https://doi.org/10.1111/bcpt.12990
357. Cunha-Júnior EF, Andrade-Neto VV, Lima ML, da Costa-Silva TA, Galisteo Junior AJ, Abengózar MA, Barbas C, Rivas L, Almeida-Amaral EE, Tempone AG, Torres-Santos EC (2017) Cyclobenzaprine Raises ROS levels in *Leishmania infantum* and reduces parasite burden in infected mice. PLoS Negl Trop Dis 11(1):e0005281. https://doi.org/10.1371/journal.pntd.0005281
358. Polak A, Richle R (1978) Mode of action of the 2-nitroimidazole derivative benznidazole. Ann Trop Med Parasitol 72(1):45–54
359. Docampo R, Moreno SN (1984) Free radical metabolites in the mode of action of chemotherapeutic agents and phagocytic cells on *Trypanosoma cruzi*. Rev Infect Dis 6(2):223–238
360. Moreno SN, Docampo R, Mason RP, Leon W, Stoppani AO (1982) Different behaviors of benznidazole as free radical generator with mammalian and *Trypanosoma cruzi* microsomal preparations. Arch Biochem Biophys 218(2):585–591
361. Díaz de Toranzo EG, Castro JA, Franke de Cazzulo BM, Cazzulo JJ (1988) Interaction of benznidazole reactive metabolites with nuclear and kinetoplastic DNA, proteins and lipids from *Trypanosoma cruzi*. Experientia 44(10):880–881
362. Docampo R, Stoppani AO (1979) Generation of superoxide anion and hydrogen peroxide induced by nifurtimox in *Trypanosoma cruzi*. Arch Biochem Biophys 197(1):317–321
363. Moreno SN, Mason RP, Docampo R (1984) Reduction of nifurtimox and nitrofurantoin to free radical metabolites by rat liver mitochondria. Evidence of an outer membrane-located nitroreductase. J Biol Chem 259(10):6298–6305
364. Boiani M, Piacenza L, Hernández P, Boiani L, Cerecetto H, González M, Denicola A (2010) Mode of action of nifurtimox and N-oxide-containing heterocycles against *Trypanosoma cruzi*: is oxidative stress involved? Biochem Pharmacol 79(12):1736–1745. https://doi.org/10.1016/j.bcp.2010.02.009
365. Wilkinson SR, Taylor MC, Horn D, Kelly JM, Cheeseman I (2008) A mechanism for cross-resistance to nifurtimox and benznidazole in trypanosomes. Proc Natl Acad Sci U S A 105(13):5022–5027. https://doi.org/10.1073/pnas.0711014105
366. Wilkinson SR, Kelly JM (2009) Trypanocidal drugs: mechanisms, resistance and new targets. Expert Rev Mol Med 11:e31. https://doi.org/10.1017/s1462399409001252

367. Hall BS, Wilkinson SR (2012) Activation of benznidazole by trypanosomal type I nitroreductases results in glyoxal formation. Antimicrob Agents Chemother 56(1):115–123. https://doi.org/10.1128/aac.05135-11
368. Hall BS, Bot C, Wilkinson SR (2011) Nifurtimox activation by trypanosomal type I nitroreductases generates cytotoxic nitrile metabolites. J Biol Chem 286(15):13088–13095. https://doi.org/10.1074/jbc.M111.230847
369. Channon JY, Roberts MB, Blackwell JM (1984) A study of the differential respiratory burst activity elicited by promastigotes and amastigotes of *Leishmania donovani* in murine resident peritoneal macrophages. Immunology 53(2):345–355
370. Cardoni RL, Antunez MI, Morales C, Nantes IR (1997) Release of reactive oxygen species by phagocytic cells in response to live parasites in mice infected with *Trypanosoma cruzi*. Am J Trop Med Hyg 56(3):329–334
371. Horta MF, Mendes BP, Roma EH, Noronha FS, Macêdo JP, Oliveira LS, Duarte MM, Vieira LQ (2012) Reactive oxygen species and nitric oxide in cutaneous leishmaniasis. J Parasitol Res 2012:203818. https://doi.org/10.1155/2012/203818
372. Charmoy M, Megnekou R, Allenbach C, Zweifel C, Perez C, Monnat K, Breton M, Ronet C, Launois P, Tacchini-Cottier F (2007) *Leishmania major* induces distinct neutrophil phenotypes in mice that are resistant or susceptible to infection. J Leukoc Biol 82(2):288–299. https://doi.org/10.1189/jlb.0706440
373. Turrens JF (2004) Oxidative stress and antioxidant defenses: a target for the treatment of diseases caused by parasitic protozoa. Mol Aspects Med 25(1-2):211–220. https://doi.org/10.1016/j.mam.2004.02.021
374. Machado-Silva A, Cerqueira PG, Grazielle-Silva V, Gadelha FR, Peloso EF, Teixeira SM, Machado CR (2016) How *Trypanosoma cruzi* deals with oxidative stress: Antioxidant defence and DNA repair pathways. Mutat Res Rev Mutat Res 767:8–22. https://doi.org/10.1016/j.mrrev.2015.12.003
375. Brookes PS, Levonen AL, Shiva S, Sarti P, Darley-Usmar VM (2002) Mitochondria: regulators of signal transduction by reactive oxygen and nitrogen species. Free Radic Biol Med 33(6):755–764

Oxidative Stress: A Boon or Bane for Trypanosomatids Diseases?

9

Dandugudumula Ramu and Shailja Singh

Abstract

Infectious diseases are menace to the mankind, having a major contribution to the human morbidity and mortality. Trypanosomatids have a pervasive effect in the world, causing devastating but neglected diseases such as leishmaniasis, Chagas disease, and African sleeping sickness affecting 27 million people worldwide with 150,000 deaths annually. Trypanosomatids developing drug resistance is the current bottleneck in providing promising chemotherapeutics for these diseases which forces the continuous quest for new drugs and drug candidates. Balancing redox homeostasis is crucial for cell survival which has various implications in the biology of these parasites. Reactive oxygen species (ROS) act as signaling molecules, involving in various pathways and crucial for survival. Conversely, various chemotherapeutic drugs against trypanosomatids-caused ROS induction result in oxidative stress, eventually leading to apoptotic manifestations. Oxidative stress is one of the host defense mechanisms to control the infection, while detoxification is one of the crucial counteracts at the parasite front for successful host-parasite interaction. Therefore, oxidative stress is a good tool for better understanding of parasite biology, pathogenesis, and host-pathogen interactions. It is noteworthy that trypanosomatids have divergence from all other prokaryotes and eukaryotes at their redox system, majorly trypanothione-trypanothione reductase (TR)-based redox metabolism. The absence of this system in mammalians and structural/functional differences from host enzymes make it a lucrative target for studying its role in oxidative stress control and also to develop effective chemotherapeutics. One of the causes for

D. Ramu
Department of Life Sciences, School of Natural Sciences, Shiv Nadar University, Greater Noida, Uttar Pradesh, India

S. Singh (✉)
Special Centre for Molecular Medicine, Jawaharlal Nehru University, New Delhi, India
e-mail: shailjasingh@mail.jnu.ac.in

S. Chakraborti et al. (eds.), *Oxidative Stress in Microbial Diseases*,
https://doi.org/10.1007/978-981-13-8763-0_9

drug resistance of trypanosomatids is due to their action of inducing oxidative stress which in turn activates repair mechanisms resulting in the development of drug resistance. Hence, studying oxidative stress mechanism of trypanosomatids gives insights into drug resistance, which is an impendence in attaining efficacious chemotherapy. In this chapter, we have tried to give an outline of the significance of redox stress and its role in different cellular metabolisms of trypanosomatids, with a special focus on trypanothione-trypanothione reductase (TR)-based redox system as a peculiar system to study trypanosomatids oxidative stress mechanism, also for drug designing.

Keywords
Trypanosomatids · Oxidative stress · Trypanothione-trypanothione reductase · Reactive oxygen species (ROS) · Drug resistance · Host-pathogen interaction

9.1 Introduction

Protozoan parasites trypanosomatids group comprising *Trypanosoma brucei*, *T. cruzi*, and *Leishmania* spp. are etiological agents which cause several neglected diseases such as African trypanosomiasis (HAT), Chagas disease and leishmaniasis, respectively. These diseases have high morbidity and mortality affecting around 27 million people globally [1, 2]. In the absence of a licensed vaccine, development of drug therapies gave poor translational outcomes which are being a major concern to control these NTDs [3].

All of these parasites have complex life cycles, with significant differences in biochemistry, morphology, and cell biology between life cycle stages and in some cases between species. Interestingly, trypanosomatids show some peculiar characteristics with respect to their biological features, such as kinetoplast DNA, different antioxidant and energetic enzymes, different morphological forms in host and vector etc. In this regard, exploring pathways would be a strategic tool to gain insights into their implications in parasite biology and pathogenesis. This strategy is robust for drug discovery and design. Several pathways of trypanosomatids are being explored for a better understanding of their differences from the mammalian counterpart; these pathways include glycolytic pathway, DNA repairing pathway, sterol biosynthesis pathway, folate synthesis pathway, and polyamine synthesis pathway [4–8].

An imbalance between the antioxidants and oxidizing agents results in oxidative stress. Redox metabolism plays a major role in cell survival through reducing oxidative stress caused by reactive oxygen species (ROS), called as oxidative defense system. ROS causes DNA damage leading to apoptosis or death of cells and also has a role in pathogenesis of parasites [9]. For instance, protozoan parasite *Leishmania* is able to evade the immune system to perpetuate infection; thus interplay of *Leishmania* and host oxidative stress has a major role in infection. It has been evidenced that *Leishmania* inhibits neutrophil oxidative metabolism during the establishment of infection. However, after the disease establishment, there is an induction of oxidative stress by parasite [10]. Trypanosomatids have a unique system to maintain the oxidative stress, which is based on trypanothione-trypanothione reductase (TR) system. This system is specific to trypanosomatids and differs from all other organisms. Trypanosomatids redox system involved four major thiols: trypanothione, glutathionylspermidine, glutathione, and ovothiol. However, all these thiol systems are involved in maintaining the parasite redox homeostasis, but among all, trypanothione is the most powerful reducing agent as compared to other thiols. Trypanothione is kept in its reduced state by TR [11–13]. Thus, T (SH)2-trypanothione reductase-based thiol redox system draws more attention as a tool to understand the role of redox metabolism in trypanosomatids biology and pathogenesis, also as a candidate for drug/vaccine designing [14, 15]. In this chapter, we attempted to give insights into the significance of oxidative stress in trypanosomatids biology with special focus on trypanothione-trypanothione reductase-mediated redox homeostasis.

9.2 Oxidative and Antioxidative Metabolisms of Trypanosomatids and Their Differences from Mammalian System

Oxidative stress (OS) occurs due to the shift in balance between oxidizing agents and antioxidant defense. During OS excessive reactive oxygen species (ROS) are formed in cells, which could exceed the normal antioxidant capacity. The level of ROS is not controlled by internal antioxidants' defense mechanisms that leads to oxidative stress. This oxidative damage affects proteins, lipids, and DNA which could lead to cytotoxicity and genotoxicity [16]. In trypanosomatids interplay of parasite and host oxidative stress plays a major role during infection and pathogenesis. Trypanosomatids have aerobic respiration and survive only in the presence of oxygen. In aerobic respiration, energy is produced through molecular oxygen as it acts as electron acceptor, which further forms two molecules of water. However, partial reduction of molecular oxygen by 1, 2, or 3 electrons results in the formation of superoxide anion and hydroxyl radical, respectively. During normal conditions, antioxidant mechanism maintains all these reactive oxygen species levels at constant steady-state concentrations; a shift in the balance in these levels results in the oxidative stress [17–19].

Trypanosomatids electron transport chain has major differences as compared to the mammalian counterpart. The respiratory chain of African as well as in all stages of American trypanosomes has divergence from the mammalian counterpart. Previous investigations through inhibitors study demonstrated that Complex III is not involved in passing the electrons from NADH to cytochrome c. Additionally, trypanosomatids either lack Complex I or have substantial functional variation [20–23]. Moreover, even the Krebs cycle of trypanosomatids works in a different fashion with respect to mammalian Krebs cycle. Few Krebs cycle enzymes such as α-ketoglutarate dehydrogenase which is involved in NADH production in mammalian system were found missing, while other dehydrogenases work in different biochemical ways, like isocitrate dehydrogenase reduction of NADP instead of NAD. Interestingly, NADH-fumarate reductase capable of Krebs cycle involves in reduction of fumarate to succinate is unique to trypanosomatids and absent in mammalian cells [24, 25]. All these variations of oxidative stress machinery from mammalian system make it unique to trypanosomatids. Hence, exploring oxidative stress mechanism of trypanosomatids is a good tool to gain insights about its role in parasite biology and pathogenesis. Enzymes of oxidative metabolism of trypanosomatids could be drug agents, but the main disadvantage of this approach is that these enzymes are specific to trypanosomatids, but their substrates are key metabolites of the mammalian system, and there is a scope of interference of these inhibitors to intermediate metabolism of mammalian cells [26, 27].

9.3 Trypanothione-Mediated Metabolism in Trypanosomatids

Trypanothione is the form of *bis*-glutathione conjugated by spermidine that exclusively occurs in kinetoplastids, such as trypanosomes and *Leishmania* [28]. Trypanothione biosynthesis occurs in three steps; after the synthesis of glutathione and spermidine individually, two glutathione molecules are conjugated by spermidine to form *bis*(glutathionyl)spermidine trypanothione [29]. Trypanothione disulfide is reduced by a flavoenzyme trypanothione reductase (TR) at the cost of NADPH; this resembles the glutathione/glutathione reductase couple of mammalian system [30, 31]. As GR is absent in trypanosomatids, the only connection between the NADPH-thiol-based redox systems is TR in these parasites. Trypanothione is a reducing agent of thioredoxin tryparedoxin, and dithiol proteins, which delivers reducing elements for the synthesis of deoxyribonucleotides, and also serves as an electron source for RR, an essential enzyme for DNA precursor biosynthesis. UMSBP involves in the replication of kDNA minicircles. TXN system reactivates oxidized UMSBP, which further binds to the minicircle origin sequence suggesting that in vivo kDNA replication undergoes redox regulation via TXN [32, 33]. T(SH)2 is the key molecule for hydroperoxides, metals, and drug detoxification and thus maintains the redox homeostasis which is important for survival [34, 35]. Ascorbic acid is a strong antioxidant and radical scavenger. Trypanothione is an effective reducing agent of dehydroascorbate that maintains the ascorbic acid level. Many findings suggested that increased levels of trypanothione are required for metal and drug extrusion [36, 37]. High level of trypanothione was found to have a role in the amplification of PGPA gene which is known to play a role in *Leishmania* drug resistance to trivalent arsenite and antimonite [38, 39]. In *Leishmania* it is reported that T(SH)2 reduction of NO and Fe into nontoxic stable dinitrosyl iron complex is 600 times more efficient than mammalian GSH reductase system [32,

33]. In conclusion, aside from oxidative stress, trypanothione-mediated oxidative metabolism has various implications in trypanosomatids biology.

9.4 Trypanothione-Trypanothione Reductase System

The *Leishmania* parasite survives in the hazardous environment of macrophage in a mammalian host. Reactive nitrogen and oxygen species act as immune effectors as a part of host defense against parasite, but parasite fugitive strategy from these reactive oxygen and nitrogen species is yet to be elucidated. Recent investigations revealed that *Leishmania* has unique defense mechanism known as trypanothione/trypanothione reductase (TR) system to reduce the respiratory burst of macrophages. Three antioxidant enzyme cascades of trypanothione metabolism are required to counteract the mammalian antioxidant mechanism [40, 41]. Trypanothione/trypanothione reductase (TR) system of trypanosomatids is the counterpart to the human glutathione/glutathione reductase system [42]. TR has a key role in maintaining the redox balance in trypanosomatid parasites by reducing trypanothione which connects the thiol-NADPH-based redox systems [43–46]. TR disulfide substrate binding site is much wider and more negatively charged with respect to GR that accommodates the positively charged trypanothione disulfide and thus gives substrate specificity to the enzyme.

9.5 Trypanothione Reductase: A Key Regulator of Trypanosomatids Oxidative Stress and Bona Fide Antitrypanosomal Drug Agent

Trypanothione reductase is well-characterized enzyme from protein family of FAD-dependent NADPH oxidoreductases. TR is a homodimer of two 52 kDa monomers with both FAD and NADPH binding sites that regulate the whole thiol pool of trypanosomatids, thus crucial for the maintenance of redox homeostasis [47]. The three-dimensional structure of TR was unraveled in both free form and complex forms with substrates and competitive inhibitor mepacrine. The major difference between the host and parasite enzymes is fully specific for their respective disulfide substrate binding site. TR disulfide substrate binding site is largely and more negatively charged with respect to GR which is specific only to glutathionylspermidine conjugates, while glutathione is specific for GR with a net charge of −2. This substrate specificity is due to the substitution of Ala34, Arg37, Ile113, Asn117, and Arg347 of human GR with Glu18, Trp21, Ser109, Met113, and Ala343 in TR, making active site of the parasite enzyme more negatively charged and hydrophobic. Additionally, TR disulfide substrate binding site is wider as compared to GR with $2.2 \times 2.0 \times 2.8\ nm^3$ width. Cofactor binding site is another structural difference between TR and host GR as Tyr114 and Tyr197 are highly conserved sites in the cofactor binding site of human GR. Though these two residues are conserved in TR at position 110 and 198, Leu337 and Phe203 are substituted by an ionizable side chain Met333 and Met202, respectively. Position 287 has an additional basic side chain in TR. In consideration of all these structural differences of TR from GR, designing of structure-based drug with focus on substrate/cofactor binding site without affecting GR activity would be a rational approach in drug development of trypanosomatids [48].

Along with these differences, TR's absence in mammalian system and its key role in thiol-based redox homeostasis make trypanosomatids' oxidative metabolism a peculiar aspect from other eukaryotes, also making TR as a lucrative agent for redox biology study and drug design [49]. Interestingly, overexpression of TR in transfected *Leishmania donovani* promastigotes was found to be insignificant to the sensitivity of H_2O_2, inferring that the formation of T(SH)2 from its disulfide upon oxidative stress is not the rate-determining aspect of the defense system. In *T. cruzi*, antisense modulation of TR led to the gene arrangements due to the overexpression of TR. In *L. donovani*, the first allele of TR was successfully deleted, but the second allele could not be deleted, due to the third copy generation, thus failing to produce null mutants. Additionally, 15% downregulation of TR did not seem sufficient to affect the parasite survival, so it is suggested to develop inhibitors to attain >85% inhibition to affect parasite survival [50]. However, 15% inhibition was found to be adequate to get parasite sensitive to oxidative burst of macrophages, still keeping TR as bona fide target to destruct oxidative metabolism of trypanosomatids.

One disadvantage of targeting TR to study the oxidative metabolism of trypanosomatids is its concentration variation from trypanosomes to *Leishmania* spp. which is ranging from 0.3 mM to 3.0 mM. TR inhibition leads to the accumulation of T(S)2. It is necessary to know the level of T(S)2/T(SH)2 parasites can withstand. In this case, competitive/irreversible inhibition of TR with Ki values in nM range would be a better strategy for studying TR inhibition-mediated oxidative machinery of trypanosomatids [51].

9.6 Concluding Remarks

This chapter is a comprehensive review of oxidative stress mechanism and its role in trypanosomatids biology. We tried to give insights into oxidative and antioxidative metabolisms of trypanosomatids and their differences from the mammalian system. We focused on trypanothione role in various cellular metabolisms of trypanosomatids with special emphasis on trypanothione-trypanothione reductase-based redox system, a major redox system which is a peculiar feature of trypanosomatids. Trypanothione reductase is a key enzyme involved in redox metabolism belonging to trypanosomatids. We pictured TR as a lucrative target not only to study to oxidative mechanism of trypanosomatids but also as drug agent to develop chemotherapy against trypanosomatid diseases.

References

1. McCall L-I, McKerrow JH (2014) Determinants of disease phenotype in trypanosomatid parasites. Trends Parasitol 30(7):342–349
2. WHO (2015) Investing to overcome the global impact of neglected tropical diseases: third WHO report on neglected diseases. World Health Organisation, Geneva
3. Field MC, Horn D, Fairlamb AH, Ferguson MAJ, Gray DW, Read KD, De Rycker M et al (2017) Anti-trypanosomatid drug discovery: an ongoing challenge and a continuing need. Nat Rev Microbiol 15(4):217
4. Lazarin-Bidóia D, Desoti VC, Ueda-Nakamura T, Dias Filho BP, Nakamura CV, & Silva SO (2013) Further evidence of the trypanocidal action of eupomatenoid-5: confirmation of involvement of reactive oxygen species and mitochondria owing to a reduction in trypanothione reductase activity. Free Radic Biol Med 60, 17–28
5. Haanstra JR, González-Marcano EB, Gualdrón-López M, Michels PAM (2016) Biogenesis, maintenance and dynamics of glycosomes in trypanosomatid parasites. Biochimica et Biophysica Acta (BBA)-Mol Cell Res 1863(5):1038–1048
6. McConville MJ, Naderer T (2011) Metabolic pathways required for the intracellular survival of Leishmania. Annu Rev Microbiol 6:543–561
7. Singh N, Kumar M, Singh RK (2012) Leishmaniasis: current status of available drugs and new potential drug targets. Asian Pac J Trop Med 5(6):485–497
8. Chawla B, Madhubala R (2010) Drug targets in Leishmania. J Parasit Dis 34(1):1–13
9. Pohanka M (2013) Role of oxidative stress in infectious diseases. A review. Folia Microbiol 58(6):503–513
10. Almeida BFM, Narciso LG, Melo LM, Preve PP, Bosco AM, de Lima VMF, Ciarlini PC (2013) Leishmaniasis causes oxidative stress and alteration of oxidative metabolism and viability of neutrophils in dogs. Vet J 198(3):599–605
11. Krauth-Siegel RL, Comini MA (2008) Redox control in trypanosomatids, parasitic protozoa with trypanothione-based thiol metabolism. Biochimica et Biophysica Acta (BBA)-Gen Subj 1780(11):1236–1248
12. Comini MA, Flohé L (2013) Trypanothione-based redox metabolism of Trypanosomatids. Trypanosomatid Dis Mol Routes Drug Discov:167–199
13. Irigoín F, Cibils L, Comini MA, Wilkinson SR, Flohé L, Radi R (2008) Insights into the redox biology of Trypanosoma cruzi: Trypanothione metabolism and oxidant detoxification. Free Radic Biol Med 45(6):733–742
14. Krauth-Siegel RL, Comini MA (2008) Redox control in trypanosomatids, parasitic protozoa with trypanothione-based thiol metabolism. Biochimica et Biophysica Acta (BBA)-Gen Subj 1780(11):1236–1248
15. Sarwar HS, Akhtar S, Sohail MF, Naveed Z, Rafay M, Nadhman A, Yasinzai M, Shahnaz G (2017) Redox biology of Leishmania and macrophage targeted nanoparticles for therapy. Nanomedicine 12(14):1713–1725
16. da Silva MS, Segatto M, Pavani RS, Gutierrez-Rodrigues F, Bispo V d S, de Medeiros MHG, Calado RT, Elias MC, Cano MIN (2017) Consequences of acute oxidative stress in Leishmania amazonensis: from telomere shortening to the selection of the fittest parasites. Biochimica et Biophysica Acta (BBA)-Mol Cell Res 1864(1):138–150
17. Turrens JF (2004) Oxidative stress and antioxidant defenses: a target for the treatment of diseases caused by parasitic protozoa. Mol Asp Med 25(1–2):211–220
18. Chance B, Sies H, Boveris A (1979) Hydroperoxide metabolism in mammalian organs. Physiol Rev 59(3):527–605
19. Turrens JF (2003) Mitochondrial formation of reactive oxygen species. J Physiol 552(2):335–344
20. Turrens JF (1989) The role of succinate in the respiratory chain of Trypanosoma brucei procyclic trypomastigotes. Biochem J 259(2):363–368

21. Denicola-Seoane A, Rubbo H, Prodanov E, Turrens JF (1992) Succinate-dependent metabolism in Trypanosoma cruzi epimastigotes. Mol Biochem Parasitol 54(1):43–50
22. Santhamma KR, Bhaduri A (1995) Characterization of the respiratory chain of Leishmania donovani promastigotes. Mol Biochem Parasitol 75(1):43–53
23. Fairlamb AH, Cerami A (1992) Metabolism and functions of trypanothione in the Kinetoplastida. Ann Rev Microbiol 46(1):695–729
24. Boveris A, Hertig CM, Turrens JF (1986) Fumarate reductase and other mitochondrial activities in Trypanosoma cruzi. Mol Biochem Parasitol 19(2):163–169
25. Fry M, Beesley JE (1991) Mitochondria of mammalian Plasmodium spp. Parasitology 102(1):17–26
26. Turrens JF (1987) Possible role of the NADH-fumarate reductase in superoxide anion and hydrogen peroxide production in Trypanosoma brucei. Mol Biochem Parasitol 25(1):55–60
27. Fang J, Beattie DS (2002) Rotenone-insensitive NADH dehydrogenase is a potential source of superoxide in procyclic Trypanosoma brucei mitochondria. Mol Biochem Parasitol 123(2):135–142
28. Fairlamb AH, Blackburn P, Ulrich P, Chait BT, Cerami A (1985) Trypanothione: a novel bis (glutathionyl) spermidine cofactor for glutathione reductase in trypanosomatids. Science 227(4693):1485–1487
29. Grondin K, Haimeur A, Mukhopadhyay R, Rosen BP, Ouellette M (1997) Co-amplification of the γ-glutamylcysteine synthetase gene gsh1 and of the ABC transporter gene pgpA in arsenite-resistant Leishmania tarentolae. EMBO J 16(11):3057–3065
30. Ariyanayagam MR, Fairlamb AH (2002) Characterization of recombinant glutathionylspermidine synthetase/amidase from Crithidia fasciculata. Biochem J 364(3):679–686
31. Oza SL, Tetaud E, Ariyanayagam MR, Warnon SS, Fairlamb AH (2002) A single enzyme catalyses formation of trypanothione from glutathione and spermidine in Trypanosoma cruzi. J Biol Chem 277(39):35853–35861
32. Milman N, Motyka SA, Englund PT, Robinson D, Shlomai J (2007) Mitochondrial origin-binding protein UMSBP mediates DNA replication and segregation in trypanosomes. Proc Natl Acad Sci 104(49):19250–19255
33. Onn I, Milman-Shtepel N, Shlomai J (2004) Redox potential regulates binding of universal minicircle sequence binding protein at the kinetoplast DNA replication origin. Eukaryot Cell 3(2):277–287
34. Légaré D, Papadopoulou B, Roy G, Mukhopadhyay R, Haimeur A, Dey S, Grondin K, Brochu C, Rosen BP, Ouellette M (1997) Efflux systems and increased Trypanothione levels in Arsenite-resistant Leishmania. Exp Parasitol 87(3):275–282
35. Krauth-Siegel RL, Meiering SK, Schmidt H (2003) The parasite-specific trypanothione metabolism of Trypanosoma and Leishmania. Biol Chem 384(4):539–549
36. Halliwell B (2001) Vitamin C and genomic stability. Mutat Res/Fundament Mol Mech Mutagen 475(1):29–35
37. Clark D, Albrecht M, Arévalo J (1994) Ascorbate variations and dehydroascorbate reductase activity in Trypanosoma cruzi epimastigotes and trypomastigotes. Mol Biochem Parasitol 66(1):143–145
38. Ouellette M, Hettema E, Wüst D, Fase-Fowler F, Borst P (1991) Direct and inverted DNA repeats associated with P-glycoprotein gene amplification in drug resistant Leishmania. EMBO J 10(4):1009–1016
39. Mukhopadhyay R, Dey S, Xu N, Gage D, Lightbody J, Ouellette M, Rosen BP (1996) Trypanothione overproduction and resistance to antimonials and arsenicals in Leishmania. Proc Natl Acad Sci 93(19):10383–10387
40. Atwood JA, Weatherly DB, Minning TA, Bundy B, Cavola C, Opperdoes FR, Orlando R, Tarleton RL (2005) The Trypanosoma cruzi proteome. Science 309(5733):473–476
41. Piacenza L, Peluffo G, Alvarez MN, Kelly JM, Wilkinson SR, Radi R (2008) Peroxiredoxins play a major role in protecting Trypanosoma cruzi against macrophage-and endogenously-derived peroxynitrite. Biochem J 410(2):359–368

42. Singh BK, Sarkar N, Jagannadham MV, Dubey VK (2008) Modeled structure of trypanothione reductase of Leishmania infantum. BMB Rep 41(6):444–447
43. Comini MA, Guerrero SA, Haile S, Menge U, Lünsdorf H, Flohé L (2004) Validation of Trypanosoma brucei trypanothione synthetase as drug target. Free Radic Biol Med 36(10):1289–1302
44. Fairlamb AH, Blackburn P, Ulrich P, Chait BT, Cerami A (1985) Trypanothione: a novel bis (glutathionyl) spermidine cofactor for glutathione reductase in trypanosomatids. Science 227(4693):1485–1487
45. Williams CH Jr (1992) Lipoamide dehydrogenase, glutathione reductase, thioredoxin reductase, and mercuric ion reductase—a family of flavoenzyme transhydrogenases. Chem Biochem Flavoenzym 3:121–211
46. Singh BK, Sarkar N, Jagannadham MV, Dubey VK (2008) Modeled structure of trypanothione reductase of Leishmania infantum. BMB Rep 41(6):444–447
47. Ghisla S, Massey V (1989) Mechanisms of flavoprotein-catalyzed reactions. In: EJB reviews. Springer, Berlin/Heidelberg, pp 29–45
48. Schmidt A, Krauth-Siegel R (2002) Enzymes of the trypanothione metabolism as targets for antitrypanosomal drug development. Curr Topic Med Chem 2(11):1239–1259. Schneider, Erasmus, Yaw-Huei Hsiang, and Leroy
49. Singh N, Kumar M, Singh RK (2012) Leishmaniasis: current status of available drugs and new potential drug targets. Asian Pac J Trop Med 5(6):485–497
50. Tovar J, Cunningham ML, Smith AC, Croft SL, Fairlamb AH (1998) Down-regulation of Leishmania donovani trypanothione reductase by heterologous expression of a trans-dominant mutant homologue: effect on parasite intracellular survival. Proc Natl Acad Sci 95(9):5311–5316
51. Fairlamb AH (2003) Chemotherapy of human African trypanosomiasis: current and future prospects. Trends Parasitol 19(11):488–494

Redox Regulatory Circuits as Targets for Therapeutic Intervention of Bancroftian Filariasis: Biochemical, Molecular, and Pharmacological Perspectives

10

Suprabhat Mukherjee, Nikhilesh Joardar, and Santi P. Sinha Babu

Abstract

Bancroftian filariasis is one of the oldest human diseases that are still posing a myriad threat to mankind with an increasing number of new infections, morbidities, disabilities, and socioeconomic loss in each year. *Wuchereria bancrofti* is the causative parasite that infects human through vector-mediated transmission and parasitizes the human lymphatics leading to incompetence, lymphostasis, lymphatic dysfunction (hydrocele and lymphedema), and interstitial fibrosis and promotes vulnerability to opportunistic microbial infections resulting in elephantiasis. The worm usually resides for years in the human body and this is due to their unique ability to escape from the host defense system. In this connection, an efficient and highly versatile redox regulatory system in the filarial parasite is considered as the most crucial factor behind their survivability inside the host. The redox regulatory circuit is typically composed of enzymatic and nonenzymatic antioxidants which ameliorate the lethal effect of the prooxidants (reactive oxygen and nitrogen species) generated by the host immune cells. Therefore, modern antifilarial research aims to break the shield of the filarial redox regula-

Suprabhat Mukherjee and Nikhilesh Joardar have been contributed equally with all other contributors.

S. Mukherjee
Parasitology Laboratory, Department of Zoology, Visva-Bharati University, Santiniketan, West Bengal, India

Department of Animal Science, Kazi Nazrul University, Asansol, West Bengal, India

N. Joardar · S. P. Sinha Babu (✉)
Parasitology Laboratory, Department of Zoology, Visva-Bharati University, Santiniketan, West Bengal, India
e-mail: spsinhababu@gmail.com

S. Chakraborti et al. (eds.), *Oxidative Stress in Microbial Diseases*,
https://doi.org/10.1007/978-981-13-8763-0_10

tory system through targeted chemo- and immunotherapeutic approaches. Due to the emergence of the drug resistance against available antifilarials and alterations in the pathogenesis, therapeutic intervention of bancroftian filariasis through triggering the redox-regulatory circuits usher to be a new hope for the mankind. In this chapter, we have discussed the recent advancements and potential of redox regulation targeted intervention of lymphatic filariasis by majorly emphasizing the biochemical, molecular, and pharmacological perspectives.

Keywords
Bancroftian filariasis · Redox regulatory circuit · Antioxidants · Reactive oxygen species · Therapeutic intervention

Abbreviations

AP1	Activator Protein 1
ASK	Apoptosis signaling kinase
BNF	Bancroftian filariasis
CAMKII	Calmodulin Kinase II
CAT	Catalase
CED	Cell death abnormal
Cys	Cysteine
DC	Dendritic cell
DEC	Diethylcarbamazine citrate
EGL	Egg laying normal
GPELF	Global Program to Eliminate Lymphatic Filariasis
GPx	Glutathione peroxidase
GR	Glutathione reductase
Grx	Glutaredoxin
GSH	Glutathione
GST	Glutathione S-transferase
IL-10	Interleukin 10
iNOS	Inducible nitric oxide synthase
IVM	Ivermectin
LF	Lymphatic filariasis
MDA	Mass Drug Administration
Mf	Microfilaria
Mϕ	Macrophage
NF-κB	Nuclear factor kappa B
NTD	Neglected tropical disease
OSH	Ovothiol
PAMP	Pathogen-associated molecular pattern
PBMC	Peripheral blood mononuclear cell
PDI	Protein disulfide isomerase

Plrx	Plasmoredoxin
PRR	Pattern recognition receptor
Prx	Peroxiredoxin
PTM	Posttranslational modification
RNS	Reactive nitrogen species
ROS	Reactive oxygen species
SOD	Superoxide dismutase
Trx	Thioredoxin
TrxR	Thioredoxin reductase
TSH	Trypanothione
TXN	Tryparedoxin
WHO	World Health Organization

10.1 Introduction

Lymphatic filariasis (LF) is a deforming and disabling human disease caused by the mosquito-transmitted nematode parasites, *Wuchereria bancrofti*, *Brugia malayi* and *B. timori*. *W.* bancrofti has been the major causative parasite contributing to 90% of the total LF cases, and this form of LF is known as bancroftian filariasis (BNF) [1]. Currently, BNF is enlisted in the list of 'neglected tropical diseases (NTDs) as it is mainly prevalent amongst the underprivileged population [1]. Like other NTDs, BNF excessively affects the vulnerable populations and the intensely disfiguring perceptible outcome of the disease occurs afterward in life and can lead to everlasting disability [2, 3]. Bancroftian filarial patients are not only physically disabled but also suffer from mental, social, and financial crises [3–5].

At present, the prevalence map constitutes 947 million people in 54 countries, out of which an estimated male population of 25 million display hydrocele and over 15 million people with lymphedema [1]. In addition to these, at least 36 million individuals remain with all of the chronic manifestations of BNF [1]. According to the report of the National Vector Borne Disease Control Programme (Directorate General of Health Services, Ministry of Health & Family Welfare), in India, presently 17 states are affected with LF among which Uttar Pradesh, Bihar, Odhisa, and Tamil Nadu face the worst scenario [6]. Interestingly, 99.4% of the LF cases in India are caused by *W.* bancrofti and rests are caused by *B. malayi*. Since 2004, the Health Ministry has been carrying out mass drug administration as part of the Hathipaon Mukt Bharat (Filaria Free India) program for preventive medication [6].

In its life cycle, *W.* bancrofti utilizes human as a prime host and mosquito as intermediate. The infected mosquito releases the infective larvae (L3) of *W.* bancrofti which invades through the skin, mature into the adults (L5) and adults release L1 larvae that enter into circulation [4]. The nematode parasitizes the human lymphatics leading to incompetence, lymphostasis, lymphatic dysfunction (hydrocele and lymphedema), and interstitial fibrosis and promotes vulnerability to opportunistic microbial infections resulting in elephantiasis [7]. The present treatment regimen for treating BNF includes administration of broad-spectrum anthelmintics like albendazole and anti-*Wolbachia* antibiotic, i.e., doxycycline through mass drug administration (MDA) program by WHO and morbidity management [3]. The Global Program to Eliminate Lymphatic Filariasis (GPELF) also includes other drugs like diethylcarbamazine (DEC) and ivermectin (IVM) [8]. Major drawbacks in using these drugs have been their limited efficacies on the adult filarids, systemic and inflammatory adverse reactions; multiple rounds of dosing and prolonged treatment duration [9–12]. Alongside these, the emergence of resistance against the available anthelmintics and alteration in the mode of pathogenesis are also considered as added complications [7, 13]. Considering the alarming conditions, new approaches are urgently needed to combat the tide.

The most unique feature of the filarial parasite is its ability to cope up with the oxidative challenges posed by the host, and this is central behind the prolonged survival of the filarial worm inside the human body [14, 15]. Being a blood-borne parasite, *W.* bancrofti has evolved several unique mechanisms of escaping from the host defense system that potentially comprises a fascinating and highly versatile antioxidant system [15]. Therefore, targeted manipulation of the filarial redox machinery is emerging as an effective target to develop antifilarial chemotherapeutics or vaccines. Interestingly, the redox system is not a single-component system rather it contains functionally diverse biomolecules [16]. Therefore, triggering such components with chemo- and/or immunotherapeutics appears to be effective in controlling the drug-resistant strains. In this chapter, major emphasis is given to present a comprehensive view on the promises of the redox regulatory circuits of filarial parasites as a therapeutic target for the intervention of bancroftian filariasis based on the existing and upcoming approaches up to cellular and molecular level.

10.2 Redox Biology of Filariasis

A unique and versatile redox regulatory circuit is considered to be one of the principal reasons behind the long-term survivability of the filarial parasites inside the host [17]. Filarial redox system is mainly comprised of enzymatic and nonenzymatic components. A list of major redox regulators of filarial parasites has been depicted in this section.

10.2.1 Nonenzymatic Antioxidants

10.2.1.1 Low Molecular Weight Thiol Compounds

Cysteine (Cys) with its sulfhydryl group (-SH) is immensely important in filarial physiology due to its crucial role in protein folding. Upon donating the hydrogens, Cys in coordination with protein disulfide isomerases (PDI) forms disulfide bonds in proteins [16, 18]. However, in metazoan parasites, like filarial, covalent post-translational modifications (PTMs) of the cysteine residues are the major regulators of redox homeostasis and signaling by linking the redox state of the antioxidant proteins with their function inside the cell [19]. Interestingly, cysteine modifications play vital roles in various cellular functioning in case of human parasites that potentially includes functioning and trafficking of the redox regulatory proteins [19]. On the other hand, selenocysteine, a rare amino acid is the manufacturer of an elite group of proteins, selenoproteins, is also sharing equal importance in enzyme-mediated redox reactions [20]. Biochemically, the presence of selenium atom makes significant differences in both functioning and structure of the selenoproteins [16]. Intriguingly, free-living nematodes (*Caenorhabditis* elegans) have been shown to contain a single selenoprotein, namely, thioredoxin reductase [21, 22]. However, different parasitic nematode species including filarial parasites encode selenoprotein with an unusual selenocysteine insertion sequence (SECIS) that possesses a G in spite of the canonical A in the conserved region [23].

Glutathione (GSH) is considered being the most studied and most efficacious antioxidants of the filarial parasite. GSH is a tripeptide consisting of cysteine, glutamic acid, and glycine, having a gamma-glutamyl peptide bond between the glutamate and cysteine residues, which protects it from peptidase activities [24]. Glutathione system plays an important role in cellular antioxidant defense and detoxification of reactive oxygen species (ROS), reactive nitrogen species (RNS), electrophiles, as well as their toxic metabolites [25]. Interestingly, the entire glutathione system enables the detoxification of xenobiotic compounds and plays a vital candidature in drug resistance and pathology [25]. GSH maintain the redox balance of the cellular compartments though removal of ROS [26, 27], regeneration of the reduced state of ascorbic acid [26, 27], transport of the amino acids through the γ-glutamyl cycle [28], the formation of deoxyribonucleotides [29] etc., by acting as a reductant [29]. GSH also helps in the removal of the xenobiotic compounds by increasing the solubility of the xenobiotic or by covalent conjugation of the xenobiotic with that of the glutathione S-transferase [30, 31]. GSH via glutathionylation of the protein participates in the cellular signaling process [32]. Interestingly, the ratio of the oxidized (GSSG) and the reduced glutathione determines the overall cellular redox potential [30]. However, inside the cellular compartments like endoplasmic reticulum, vacuoles, and mitochondria, the independent pool of GSH maintain different GSH/GSSG ratio which reveals a difference in the intracellular redox environment [27]. Moreover, levels of GSH are crucial for the finest functioning of the immune system, T-cell activation, differentiation, and a vital regulator of the cell cycle [25]. Reduction in the GSH causes damage by oxidative and nitrosative stress; hypernitrosylation; increased levels of proinflammatory mediators and inflammatory potential; dysfunctions of intracellular signaling pathways (p53, NFκB, and JNK); decrement of cell proliferation and DNA synthesis and controls several other cellular events [25]. Intriguingly, it has been reported earlier by Morris et al., [25] that redundancy in the level of GSH results in significant alterations in the homeostatic control of the immune system, oxidative and nitrosative stress pathways.

Apart from GSH, ovothiol (OSH) and trypanothione (TSH) are also considered important in redox biology [33, 34]. OSHl, a mercapto-histidine, comprised of OSH-A, OSH-B, and OSH-C variants where a methyl group is absent in the

OSH-A, and one and two methyl groups are present in the rest two variants respectively [16]. This peptide is mainly involved in the H_2O_2 reduction with a consequence of the generation of the disulfide form of the compound [33]. On the other side, TSH is a dithiol containing two GSH molecules covalently joined through amide bonds to a spermidine molecule [35]. TSH with the help of peroxiredoxins is the main electron donor to tryparedoxin for the removal of hydroperoxides [36, 37].

10.2.2 Thiol-Containing Proteins

Thioredoxin is a low molecular weight protein (~12 kDa) and is present in the cytosol, mitochondria, and nucleus and is even present in the extracellular environment of the parasite [38]. Thioredoxin system has a potent role in controlling cellular redox equilibria and consequently the cellular fate. The multiple functions of the thioredoxin include synthesis of deoxyribonucleotides [39], detoxification of H_2O_2 [40], regulation of activities of the transcription factor (AP2, NFκB) [41], regeneration of methionine sulfoxide [42], function as a cytokine in the oxidative stress condition [43]. Most interestingly, the active CXXC redox motif present in the Trx acts as a "redox rheostat" [44]. Alike others, filarial Trx has the ability to maintain the reducing equivalents required for the antioxidant enzymes [45]. Unlike kinetoplastida apicomplexan parasites, the presence of tryparedoxin (a form of Trx), and plasmoredoxin in filarial parasite has not been reported.

Glutaredoxins (Grxs) are another group of low molecular weight redox-active proteins that potentially drive disulfide reduction and thiol-disulfide exchange reactions [16] in filarial nematodes. The proteins act in coordination with GSH, glutathione reductase (GR) and glutathione peroxidase (GPx). Alike another nematode-specific GPX, filarial GPX (as reported in *B. pahangi*) is glycosylated and it is specifically expressed after infection of the mammalian host and represents the major surface protein of adult lymphatic filarial parasites [46]. It has been reported that the inactive forms such as the dimeric and tetrameric structures of Grx are dependent on the interaction of the iron-sulfur redox centers (2Fe-2S) [47]. As a member of redoxin family, it has the ability to reduce the protein disulfides in a GSH dependent and/or deglutathionylation process (a process extremely important under oxidative stress condition in which a number of proteins become glutathionylated) and maintain the redox homeostasis inside the host body [48]. Thiol (disulfide)-based antioxidants are important mediators of redox-regulation in nematode parasites [16]. However, information about thiol-dependent antioxidant systems in filarids is scarce. In this context, most of the information is based on the studies conducted on *Onchocerca volvulus* and *Brugia malayi* [25]. So far, thioredoxins and glutathione are known as the major thiols that perform antioxidative functions.

10.2.3 Antioxidant Enzymes in Filarial Parasites

The enzymatic components constitute mainly the antioxidant enzymes such as glutathione S-transferases (GST), glutathione peroxidases (GPx), superoxide dismutase (SOD), catalase, and thioredoxin reductase (TrxR). GST (E.C. 2.5.1.18), the phase II detoxification enzyme, is a key member of the antioxidant family that controls or directly involved in the survivability of the filarial parasites. It protects the parasites from external assaults such as chemotherapeutics, immune effectors, and several free radicals generated by the different metabolic pathways of the host. Moreover, GST is also capable of combating the endogenous cyto- and xenotoxins (e.g., malondialdehyde, the end product of lipid peroxidation) [49]. This enzyme acts by catalyzing the nucleophilic addition of the tripeptide glutathione to the target moieties [49]. Apart from the aforementioned function, GST also plays a profound role in the development of resistance to antibiotics, anthelmintics, and other antiparasitic drugs [50, 51]. In parasitic nematodes (including filarids), Nu-class GST is known to act as heme-binding ligandins [52]. In particular, GST has been isolated and characterized from several filarial parasites that include a recent report [53] on *Setaria cervi*.

Antioxidant enzymes are of immense importance for the survival of long-lived parasites, especially the filarial parasites that resides inside the host body where ROS-mediated killing of the invading pathogen is of prime choice by the host [54–56]. These enzymes are important in terms of prolonged survival of the nematodes in host resulting in chronic infections. A study conducted by Saxena et al. [57] revealed that filarial parasites are unable to completely oxidize substrate to CO_2 and H_2O however, intriguingly pyruvate is decarboxylated and gives rise to respiratory CO_2 [58]. Microfilariae (Mf) of *Litomosoides carinii*, *Brugia pahangi*, *Setaria cervi* and *Acanthocheilonema viteae* are reliant to a complimentary oxygen-dependent energy-generating system within their mitochondria [57, 59]. In vitro studies have established that ROS are injurious for the parasites because they are known to induce apoptotic pathway [7, 60–66]. ROS-mediated induction of apoptosis is a much-studied aspect in the field of antifilarial drug development [7, 60–68]. Filarial parasites are equipped with both aerobic and anaerobic mode of metabolism [69], and because of their O_2 consumption, they are vulnerable to both self- and host-generated ROS [67, 68]. Studies have shown that considerable amounts of glutathione S-transferase, glutathione reductase activities are present in crude *S. cervi* homogenate and SOD, catalase and glutathione peroxidase, glutathione reductase and glucose-6-phosphate dehydrogenase are also found to be present in *S. cervi*, *L. carinii*, and *A. viteae* as well which glorify their potentials in oxidative stress defense [60–68]. *L. carinii* contained catalase, glucose-6-phosphate dehydrogenase, GPx and SOD [70]. In dog heartworm, *Dirofilaria immitis* Cu/Zn SOD is present in both the Mf and adult stages [71]. In filarial parasite particulate type, gamma-glutamyltranspeptidase (γ-GT) is present in the intestinal tract and body wall of the parasite with comparatively higher activity in cuticle-hypodermis-muscle suggesting that body wall has an important role in antioxidant defense and transport of amino acids [57]. Cytochrome P450 is lacking in helminth parasites and thus GSTs are the major detoxification enzymes of this group of parasites [72–75]. Information about the antioxidant enzymes in nematodes is very scarce [16]. Interestingly, catalase activity was found to be present in the crude extract of *B. malayi* [76] although, no sequence for CAT from any human filarial parasite was found in the gene data bank [76]. On the other hand, a secretory Cys-dependent GPx has been found with preferential activity toward fatty acid and phospholipidhydroperoxides in *B. malayi* and *D. immitis* [77]. The existence of Prx1a and Prx1b in both the larval as well as the adult stages of *O. volvulus* and *B. malayi* were studied [78, 79]. Trxr and GR are the major disulfide reductases found in the nematodes. Although the nematode GR is very much similar to that of the mammalian GR in structure and function, GR from *C.* elegans [80], *S.* digitata [81], *A. suum* [82], and *O. volvulus* [83] has been elucidated so far, and it has been concluded that differences in response pattern to arsenicals are the key feature behind the druggability of filarial GR [16, 81]. Presence of TrxR in free-living and parasitic species [84–87] has been reported. Selenocysteine-dependent and selenocysteine-independent isoforms in these nematodes were found to be present and localized in the cytosol and mitochondrial compartments [88–91].

10.2.4 Redox System-Host Immunity Crosstalk: Scope in Drug Development

Filarial thiol-dependent redox pathways are mainly operated by thioredoxin (TrX) and glutathione (GSH) systems [85, 92]. Particularly, studies in *B. malayi* (another human parasite) have shown constitutive expression of TRX throughout the life cycle of the parasite and such an abundance was postulated to play a key role in the host immune evasion strategy of the parasite [45, 93]. A unique feature of filarial TRXs is their immune regulatory attributes as they promote secretion of cytokines from host immune cells and also regulate the DNA-binding activity of the immunologically active transcription factors like NF-Kb [45]. In addition, filarial nematodes possess another thiol-specific antioxidant protein cluster namely thioredoxin peroxidase/thiol-specific antioxidant (TPX/TSA) [92]. In accordance with these proteins, TPX can also play a protective role scavenging the free radicals generated by the immune cells [79]. In addition, a 25-kDa TPX from *B. malayi* found to exist in all stages of the parasite [79]. In a previous report by Mukhopadhyay et al. [94] showed that poorer NF-κB signaling results in survival of filarids inside the host; therefore, it is evident that oxidative pathways also cross talk with the NF-κB axis. However, redox signaling pathways, their cross talks and relevance in the host-parasite interaction in the human model has not been elucidated in detail to date.

10.3 Filarial Redox Regulatory Pathways as Potential Therapeutic Targets

Being the eukaryotes, human and *W. bancrofti* possess similar aerobic energy metabolism pathways wherein oxygen is the final electron acceptor. ROS is generated as a byproduct and importantly, a number of important cellular processes. However, dysregulation of the generation rather overproduction results in oxidative stress. The highly versatile antioxidant machinery is therefore immensely important in filarial survival and therefore considered as an efficacious target for chemo- and/or immunotherapeutic intervention of BNF.

10.3.1 Redox Signaling Pathways for Antifilarial Chemotherapy

Filarial parasites are heme auxotrophs and they require uptake and transport of exogenous heme for incorporation in hemoproteins [52], and therefore, GST appears to be a meaningful target for antifilarial drugs. The advantage of using filarial GST for drug development is the high degree of dissimilarity in the amino acid sequence of GST between human and filarial parasites. For example, GST of *S. digitata* shares only 43% homology with human [95]. The mechanistic insights of thiol-based redox regulatory system in the filarial parasites have been demonstrated in Figs. 10.1 and 10.2.

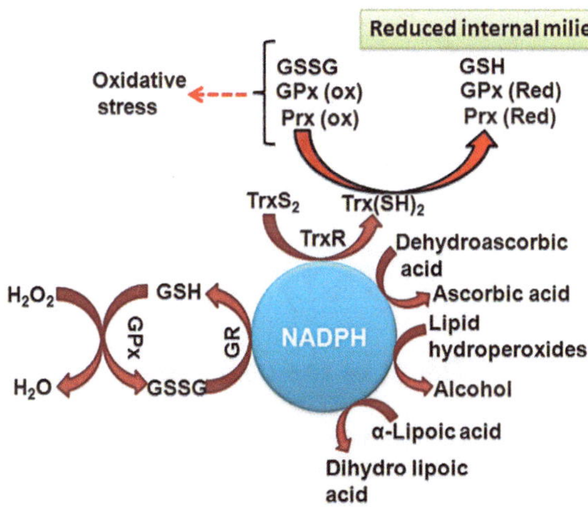

Fig. 10.1 Thiol-based redox regulatory circuit and its functional significance in the filarial nematode. Filarial antioxidant defense system is majorly operated by the thiol proteins principally comprised of glutathione (GSH) and thioredoxin (TRX). These proteins in reduced form contain -SH group that maintain the redox homeostasis through providing the H atom to the radicals. After their actions, these thiol proteins converted to their reduced state by the action of respective reductase enzymes, viz., glutathione reductase (GR) and thioredoxin reductase (TrxR)

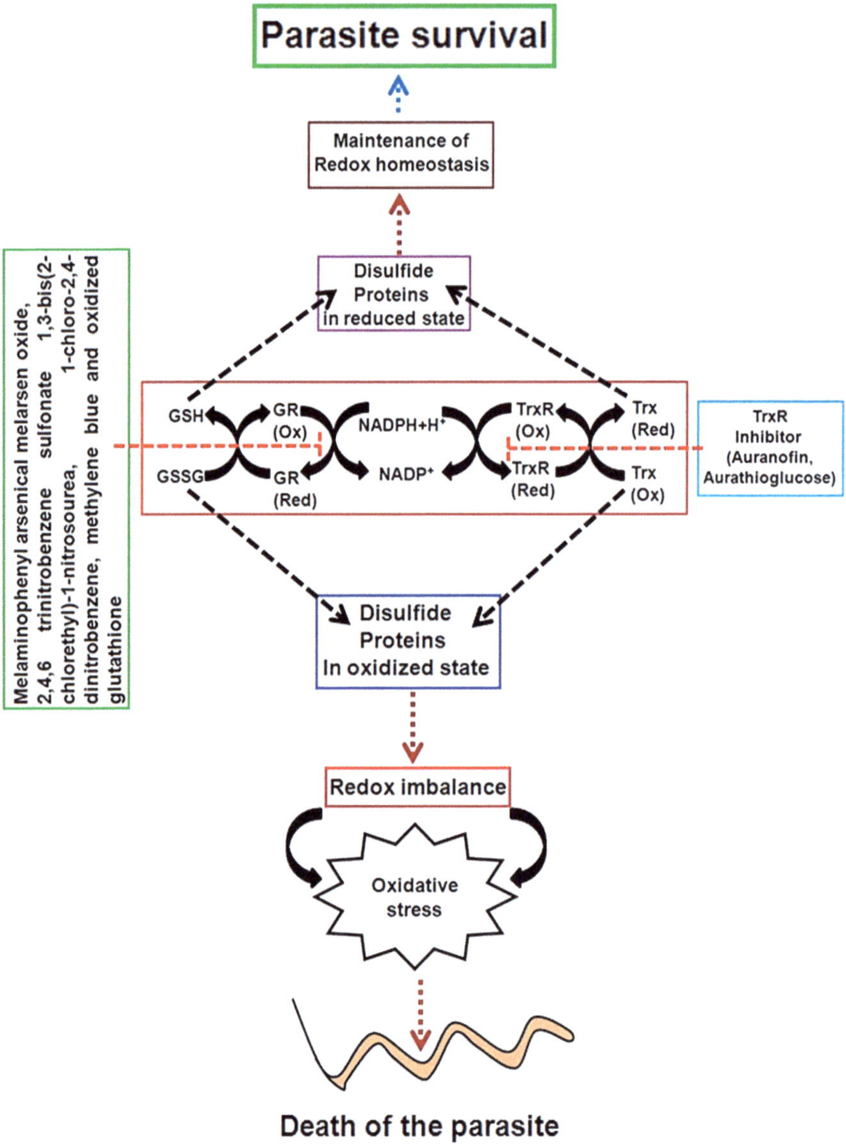

Fig. 10.2 Targeted perturbation of redox-homeostasis as a drug target to trigger filarial infection. Efficiency of the redox regulators in the removal of noxious radicals is crucial in determining the fate of the infectious filarial parasite inside the host body. Taking the advantage of this, the redox regulation circuits have been chosen as efficacious target for the antifilarial chemotherapy. Major proportions of currently available antifilarials result impairment in the redox cycling of thiol proteins (TRX and GSH) and/or direct inhibition of their activities

10.4 Targeted Induction of Oxidative Stress as a Drug Designing Approach

Recent research on the development of potential antifilarials from natural sources (curcumin, ferulic acid, ursolic acid, acaciaside isomers) and/or synthetic compounds having a core structure of natural compounds (resveratrol analogs, quinolone-fused sulfonamide, carbamoyl thiolates, cinnamoyl derivatives) have been found to induce oxidative stress. The interaction between the phytochemical(s)/synthetic moiety and cell membrane of the parasite is considered as the key phenomenon behind eliciting an antifilarial response [7, 54, 60, 65, 96]. Such an interaction results in peroxidation of the phospholipids present in the cell membrane and generates highly reactive lipid peroxide radicals [7, 54, 60, 65, 67, 68, 96]. Oxidative damage of the membrane not only disrupts the membrane integrity but also initiates free radical chain reaction through generating lipid peroxide radicals [97]. Cellular insights of the induction of lipid peroxidation and subsequent initiation of free radical chain reaction in filarial parasite after treatment with a diverse group of phytochemicals and synthetic compounds have been presented in Fig. 10.3.

Increased generation of lipid peroxides after treatment of several filarial parasites including *W.* bancrofti is usually inferred from a high level of malondialdehyde (MDA), the end product of lipid peroxidation [7, 54, 60, 65, 67, 68, 96, 98]. MDA is a neutral molecule generated at the termination reaction of the free radical chain reaction but the short-lived and highly reactive lipid peroxide radicals utilize electron from several important molecules (hydroxyl ion; nitric oxide) from the cells and the level of cellular free radicals dramatically enhanced [99]. Intriguingly, these free radicals are chemically diverse and can damage all sorts of biomolecules carbohydrates, proteins, lipids, nucleic acids as well as other important molecules like vitamins). Being the major functional biomolecule of a living cell, oxidative damages in protein structure is considered as the major cause of cellular dysfunctions and death [54, 68]. In this connection, oxidative damages to the cellular antioxidant proteins (glutathione, thioredoxins) are the critical parameters. In antifilarial drug development research, these proteins have been documented as major cellular targets [60]. However, a dearth in our knowledge on the biochemistry of such parasitic proteins in filarial nematodes, majorly the thiol-based proteins, is considered as a major setback. Recently, presence and kinetic characteristics of low molecular weight thioredoxin reductase (TrxR) have been reported in a model filarid, *Setaria cervi* [85]. The unique feature of the filarial redox system is the presence of both glutathione (GSH)- and thioredoxin (Trx)-based antioxidant systems. In most of the available literature, the GSH-based antioxidant system has been studied for their role in the context of antifilarial drug-induced induction of oxidative stress. Our previous studies deciphering the effect of natural antifilarials and natural-product inspired synthetic compound demonstrated depletion of GSH [54–56, 61–64, 96]. Due to such depletion in GSH, the parasite is unable to cope up with the oxidative milieu and as a result, free radicals start damaging the biomolecules [60, 98]. In most of the cases, after exposure to antifilarial compound, levels of the potent antioxidant enzymes (GST, GPX, SOD, and catalase) get upregulated but the level of

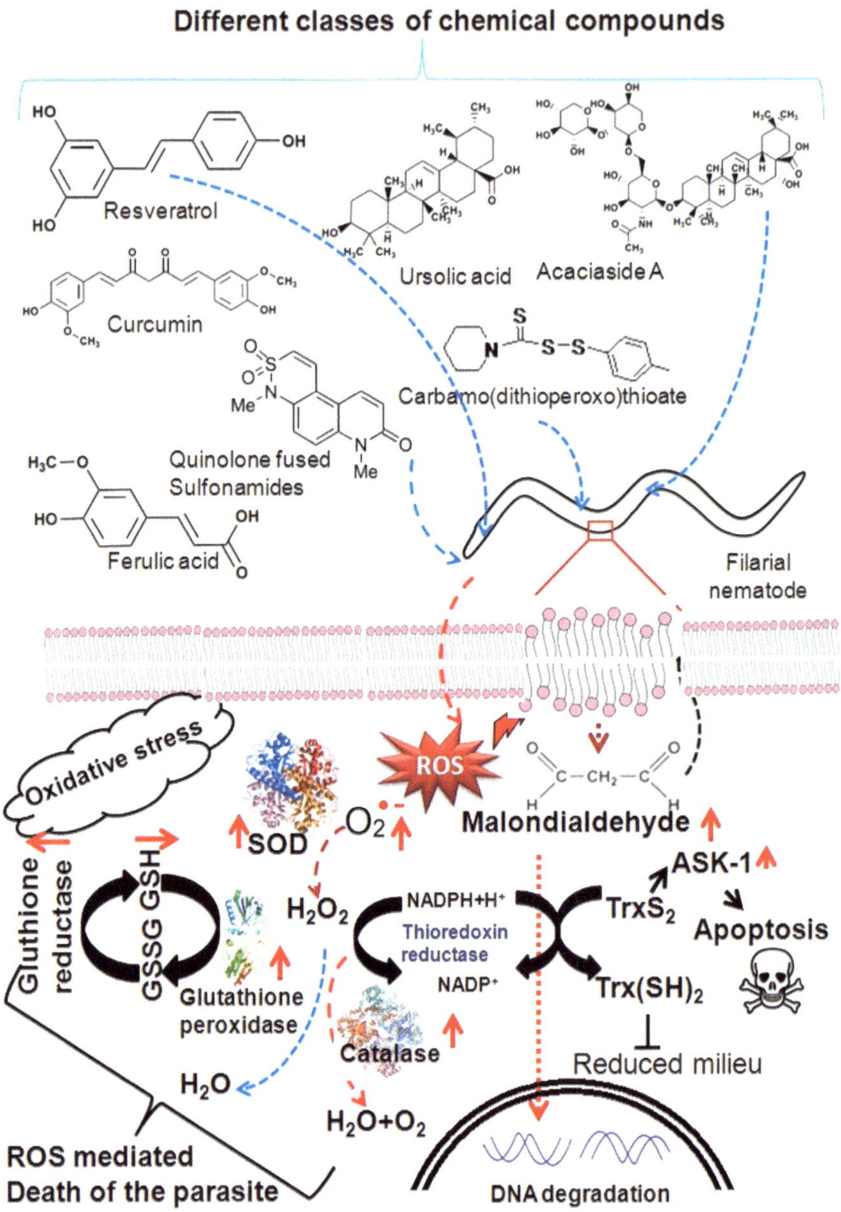

Fig. 10.3 Manipulation of redox regulatory pathways in filarial nematodes for the development of antifilarial agents. Antifilarial compounds from natural sources and/or synthetic compounds interact with the parasite membrane and result in induction of lipid peroxidation. Lipid peroxidation not only results damage of the membrane architecture but also causes initiation of the free radical chain reactions which damages all sorts of biomolecules. The high prooxidative milieu inside the parasite further governs activation of the cell death pathways, i.e., apoptosis and autophagy

oxidative stress is beyond the ameliorative property of the filarial antioxidant system. A pictorial representation on the principal mode of actions of the compounds on the antioxidant enzymes have been furnished in Fig. 10.3.

However, the molecular synergies with other potential antioxidants like TRX and/or cross talk between GSH and TRX system are not yet studied. Considering the impact of ROS in inducing worm mortality, chemotherapeutics have now been designed to trigger oxidative stress inside the filarial worm [60, 67, 68].

Mechanistic exploration of the induction of parasite death due to ROS revealed a direct signaling cross talk between ROS generation and induction of cell death [54–56, 60–64, 96]. The first and most studied signaling pathway is the programmed cell death, i.e., apoptosis. Two principal routes of cell death have been demonstrated to be induced in response to ROS. Firstly, ROS can signal increment in the expressions of the filarial proapoptotic proteins such as egg-laying defective (EGL-1) and cell death abnormal (CED-3 and CED-4) proteins while downregulation of the expression of CED-9 [54–56, 60, 96]. The induction of egl-1 is considered as a key protein behind the decision of a parasitic cell to undergo apoptotic death [100]. Under normal circumstances, CED-9 blocks apoptosis by inhibiting the activation of CED-4 [100]. However, during oxidative stress, egl-1 subdues the activity of CED-9 which allows CED-4 to activate CED-3 [100]. CED-3 is the effector caspase in nematode and executes apoptosis by promoting DNase activation and proteasomal degradation of important cellular proteins [100]. Secondly, ROS can signal the activation of the extrinsic pathway of apoptosis [54]. ROS can directly induce DNA damage, often visible as DNA fragmentation, which activates p53 expression [60]. This p53 activation promotes the release of cytochrome-c from mitochondria that further undergoes through the caspase-8/9 pathway to activate caspase-3 [54, 60]. In filarial nematodes, the concomitant existence of CED-pathway and caspase-3 [54, 60].

Interestingly, recent observation on ROS-induced activation of death signaling in filarial parasites revealed the occurrence of autophagy. Although the exact connection between ROS and autophagy is unknown, it has been demonstrated that high oxidative condition facilitates activation of mTOR which in turn activates beclin1 following PI3KC3 [101]. Influence of ROS in activation of cell death pathways and major regulatory pathways in filarial redox system as targets of antifilarial chemotherapy have been depicted in Fig. 10.4.

10.5 Redox Signaling Pathways for the Immunotherapeutic Intervention of Filariasis

A study by Veerapathram et al. [102] demonstrated the experimental evidence on the vaccine potential of GSTs from human lymphatic filarial parasites in jird model. GST purified from the filarid *S.* digitata has been reported to result in a mixed type of Th1/Th2 response upon stimulation of human peripheral blood mononuclear cells (PBMCs) [95]. Due to the high degree of dissimilarity between filarial and human GST, an antibody raised against filarial GST can selectively neutralize the enzymatic activity of filarial GST to induce cytotoxicity in Mf [95]. Interestingly,

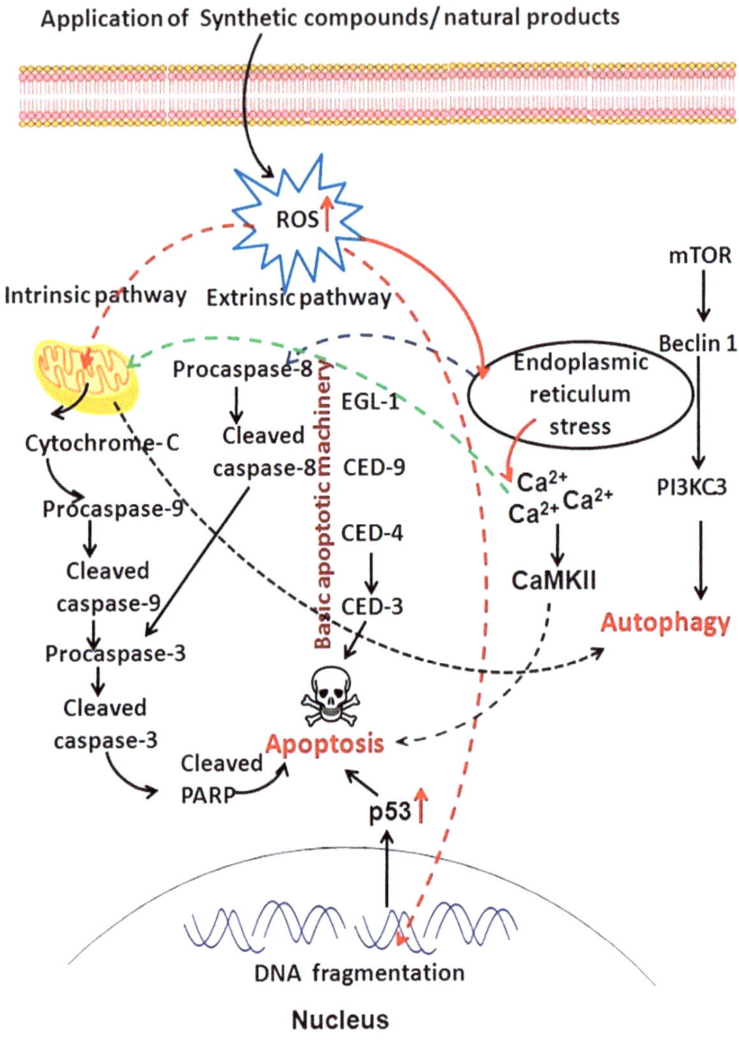

Fig. 10.4 Signaling cross talks between redox signaling and cell death pathways in filarial parasites. Oxidative stress induced by the antifilarial compounds result in activation of diverse signaling pathways that cross talk amongst each other in an interconnected manner. Chemotherapeutics induced reactive oxygen species (ROS) generated inside the filarial parasite signals activation major cell death pathway, i.e., EGL-1/CED-9/CED-4/CED-3 pathway. In addition, current experimental evidence also demonstrated activation of extrinsic pathway of apoptosis, i.e., cytochrome-c/caspase-9/caspase-3/PARP axis in the filarial pathway. Moreover, ROS damages genomic DNA that signals p53 activation to result induction of cell death. ROS also possesses a putative impact in inducing autophagy through mTOR/beclin1 pathway

immunization with *S.* digitata GST revealed clearance of *S.* digitata Mf from the circulation in *Mastomys* coucha [95]. Thioredoxin system, especially TRX is also slowly emerging a potential target for immunotherapeutic intervention of BNF. In a study by Prince et al. [92], recombinant *W.* bancrofti thioredoxin (WbTRX) and thioredoxin peroxidase (WbTPX) have been found to confer protection against *B. malayi* L3 larvae in *M.* coucha model. Antioxidant enzymes belonging to thiol antioxidant system are crucial to coping up with the tremendous oxidative stress generated by the host [103].

10.6 Current Trends of Targeting Oxidative Stress for Treating Filariasis

10.6.1 Redox System As a Tool for Immunopharmacological Manipulation of Antifilarial Host Immunity

A number of filarial redox regulatory proteins have found potent as a vaccine candidate. *B. malayi* thioredoxin (WbTRX), is a member of the 16-kDa nematode-TRX family proteins [45]. Filarial TRXs lack secretory signal sequences, although, these proteins are usually detected in the excretory-secretory components [45]. Owing to its smaller molecular size and secretory property, this protein was attempted for the use as single- as well as multiple-mode vaccinations [92]. Experimental evidence in *Dirofilaria immitis* (a filarid of dog) suggested TPX as a protective antigen [104]. Furthermore, studies in *M. coucha* model, both TRX and TPX, were found to display their prophylactic potential in combination against *B. malayi*-infective L3 larvae. *W.* bancrofti shares a high degree of homology with *B. malayi* antigens and therefore such strategy is likely to be effective in human as well. Studies with *W.* bancrofti TPX revealed a strong humoral response in *M.* coucha [92]. Therefore, thiol proteins could be considered for developing antifilarial vaccine. The cross talk between the host antigen presenting cells (Mϕ and DC) and the filarial parasites with concomitant generation of ROS and/or RNS with subsequent generation of oxidative stress inside parasite causing substantial damages in them has been depicted in the Fig. 10.5.

10.6.2 Bioactive Nanoparticles for Targeting Filarial Redox System

Currently, noble-metal (silver and gold) nanoparticles have been designed and found effective in targeting redox pathways in filarial parasites [98, 105, 106]. In this context, polymer-capped and surface modified nanoparticles have been found to induce ROS generation which in turn resulted in activation of apoptotic pathways and autophagy in filarial parasites (99, 106, 107). Apart from direct therapeutic applications, metal nanoparticles have now been used for delivery of drugs in several infectious diseases [107]. Such a strategy could be useful in targeting *W.* bancrofti-infected individuals.

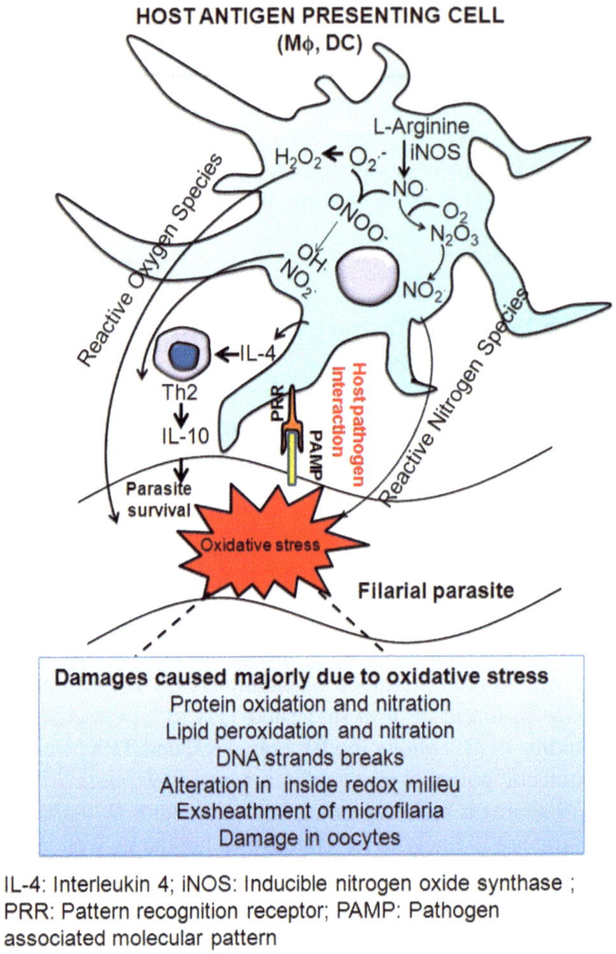

Fig. 10.5 Immunological cross talks between the antigen-presenting cell and a filarial parasite with concomitant generation of oxidative stress inside filarial parasite. After recognition of the pathogen-associated molecular pattern (PAMP) of the invading filarial parasite through pattern recognition receptor (PRR), the host immune cells sensitize itself for the production of reactive oxygen species (ROS) and/or reactive nitrogen species (RNS) by a series of complex biochemical process that readily affect the parasite. The oxidative stress generated through ROS and/or RNS by the host immune cells thus provides the first line of defense to the body

10.6.3 Targeting Filarial Posttranslational Modification System for ROS-Dependent Intervention Approaches

Most of the blood-borne parasites of human undergo through several physiological, morphological, and metabolic changes throughout their life cycles to make themselves compatible in the changing environments inside the host [108]. During such changes, posttranslational modification (PTMs) of redox regulatory proteins of parasite plays a key role. PTM of cysteine residues in proteins, specifically, S-glutathionylation and S-nitrosylation) regulate redox regulation activity of several proteins. Alongside these, PTMs through palmitoylation and prenylation influence trafficking and subcellular localization of the parasitic proteins. Importantly, the mechanisms are different from those of human. Thus, understanding of these PTMs in filarial redox signaling can provide a new dimension in antifilarial drug development.

10.6.4 Therapeutic Promises of the Mimics of the Cellular Antioxidant Enzymes

Exploration of the role of cellular thioredoxin based redox system in the regulation of the enzyme ribonucleotide reductase (the enzyme involved in the biosynthesis of deoxyribonucleotides). In this connection, chemical moieties mimicking cellular antioxidants/antioxidant enzymes are proving to be an effective choice. Ebselen, a synthetic organoselenium mimic of GPx, has emerged as a future drug to combat infectious bacteria by disrupting their redox regulatory mechanisms [109]. Recently, a study by Zou et al. [110] demonstrated the synergistic effect of ebselen against multidrug resistant bacteria infection by disrupting the redox homeostasis especially regulated at the level of DNA synthesis. However, use of antioxidant-mimics for treating parasitic infections has not been reported to date. However, considering the effectiveness of the ebselen on bacteria, the approach seems to be efficacious on the filarial nematodes too as because the survivability, fertility, pathogenesis alongside other physiological processes are dependent on the endosymbiotic bacteria, i.e., *Wolbachia*. The redox-based intervention strategy is therefore expected to provide a meaningful outcome in filarial research.

10.7 Prospects and Challenges

Considering the landmark discoveries in thiol-based antioxidants research, most of the studies emphasizing targeted manipulation of the filarial redox regulatory pathways were achieved either in vitro or in a rodent model [54–56, 60–64, 111, 112]. In recent years, GSH- and Trx-dependent systems have been considered for treating gram-negative and gram-positive bacteria separately, especially in respect of multidrug-resistant strains [113]. Therefore, considering the importance of thiol-based antioxidant systems, GSH and TRX are presently considered as potential targets for developing antifilarial chemotherapeutics in future and the molecular reason is the high degree of sequence dissimilarities to their mammalian counterparts. Apart from the prospects made till date on antifilarial research, there remain certain challenges which are still posing hurdles in the progress. The major hurdle in antifilarial research is that W. bancrofti cannot be maintained in laboratory condition, as the parasite is a typical blood auxotroph. Moreover, the definite animal model animal is also lacking for *W.* bancrofti; therefore, maximum reports are presented with only in vitro data. Animal models for *Brugia* species and *Acanthocheilonema viteae*, *Litomosoides sigmodontis* are available which cannot simulate the immunopathology of BNF and in vivo validation of laboratory experiments is also poor [114]. Previously *Presbytis entellus* (Indian leaf monkey) used to consider as an excellent model for filarial research [115], but banning of using the monkey in India as a model animal posed a serious concern in the study. Due to the unavailability of adult parasites, studies of *W.* bancrofti are only limited to Mf, the L1 larval stage, and therefore, government agencies should come forward to provide access to the clinical specimens from the hospital. Availability of efficacious therapeutic lead is another problem in redox-targeted antifilarial therapy. Although a number of chemical compounds and vaccines with high antifilarial activity have been proposed, unfortunately, most of the agents have failed at the early phase of the clinical trial due to toxic side effects as well as immunological reactions.

10.8 Concluding Remarks

Despite the tremendous effort of WHO, bancroftian filariasis still remains as a major cause of human health setback, especially in developing countries. Development of new therapeutic agents/approaches and exploration of new drug targets are currently considered as the major highlighted areas in filarial research. In this context, the filarial redox system has come out as a new efficacious target for developing effective antifilarials. Alike other eukaryotes, reactive oxygen species (ROS) are produced majorly as byproducts in filarial parasites, and ROS do have important roles in them. Intriguingly, the host immune system selectively triggers generation of ROS as a key "weapon" against the parasite. The parasite has also developed a versatile antioxidant defense to counteract the noxious effect of host-posed oxidative stress. Efficiency and adequacy of the antioxidant are one of the deciding factors behind the survivability of the parasite inside the host. In view of the essentiality of the redox system in parasite physiology, targeted chemo- and/or immunotherapy triggering parasite's redox system is considered as an effective option. Molecular dissection of the mechanism of action of several chemotherapeutics revealed that chemotherapeutics-induced ROS potentially activates the cell death pathways, majorly the programmed cell death (apoptosis) and also autophagy. Widespread use of the redox-targeted therapeutic intervention strategy has not been popularized due to the dearth in understanding the parasite redox-biology. Moreover, the majority of the chemical compound and/or vaccine candidates have failed in the earlier phases of clinical trials due to side effects. Therefore, future research should be emphasized more on exploring the detailed architecture of the redox-regulatory circuits in *W.* bancrofti, which will facilitate drug discovery by presenting new targets. At the same time, the discovery of novel and benign therapeutic molecules is required to hit the new targets. Taken *en masse*, "bench" to "bed" translation of the research is expected to bring a smile in the face of the unfortunate patients.

References

1. World Health Organization (WHO). Lymphatic filariasis. www.who.int/mediacentre/factsheets/fs102/en/ Accessed 6 Dec 2017
2. Mukherjee S, Joardar N, Sinha Babu SP (2018) Current trends in targeted chemo- and immunotherapy against bancroftian filariasis: biochemical, molecular and pharmacological perspectives. Panima Publishing Corporation, New Delhi, pp 397–414
3. Mukherjee S, Sinha Babu SP (2015) Neurofilariasis. Elsevier, Asia Pacific, pp 279–290
4. Simonsen PE, Fischer PU, Hoerauf A, Weil GJ (2014) The Filariases. Elsevier, London, pp 737–765
5. Ramaiah KD, Das PK, Michael E, Guyatt H (2000) The economic burden of lymphatic filariasis in India. Parasitol Today 6:251–253
6. National Vector Borne Disease Control Programme (Directorate General of Health Services, Ministry of Health & Family Welfare). http://www.nvbdcp.gov.in/filariasis-new.html. Accessed 13 May 2018
7. Gucchait A, Joardar N, Parida PK, Roy P, Mukherjee N, Yesuvadian R, Dutta A, Sinha Babu SP, Jana K, Misra AK (2018) Development of novel anti-filarial agents using carbamo(dithioperoxo) thioate derivatives. Eur J Med Chem 143:598–610
8. Ottesen EA, Hooper PJ, Bradley M, Biswas G (2008) The global programme to eliminate lymphatic filariasis: health impact after 8 years. PLoS Negl Trop Dis 2:e317
9. Shenoy RK (2008) Clinical and pathological aspects of filarial lymphedema and its management. Korean J Parasitol 46:119–125
10. Mukherjee S, Mukherjee Sa, Bhattacharya S, Sinha Babu SP (2017) Surface proteins of *Setaria cervi* induce inflammation in macrophage through Toll-like receptor 4 (TLR4)-mediated signalling pathway. Parasite Immunol 39:e12389. https://doi.org/10.1111/pim.12389
11. Shakya S, Bajpai P, Sharma S, Misra-Bhattacharya S (2008) Prior killing of intracellular bacteria *Wolbachia* reduces inflammatory reactions and improves antifilarial efficacy of diethylcarbamazine in rodent model of *Brugia malayi*. Parasitol Res 102:963–972
12. Mackenzie CD, Geary TG, Gerlach JA (2003) Possible pathogenic pathways in the adverse clinical events seen following ivermectin administration to onchocerciasis patients. Filaria J 2(Suppl 1):S5
13. McCarthy J (2005) Is anthelmintic resistance a threat to the program to eliminate lymphatic filariasis? Am J Trop Med Hyg 73:232–233
14. Kwarteng A, Ahuno ST, Akoto FO (2016) Killing filarial nematode parasites: role of treatment options and host immune response. Infect Dis Poverty 5:86
15. Babu S, Nutman TB (2012) Immunopathogenesis of lymphatic filarial disease. Semin Immunopathol 34:847–861
16. Guevara-Flores A, Martínez-González JJ, Rendón JL, Arenal IP (2017) The architecture of thiol antioxidant systems among invertebrate parasites. Molecules 22:259–295
17. Perry RN, Wharton DA (2011) Molecular and physiological basis of nematode survival. CAB International, Cambridge, MA
18. Gruber CW, Masa C, Heras B, Martin JL, Craik DJ (2006) Protein disulfide isomerase: The structure of oxidative folding. Trends Biochem Sci 31:455–464
19. Jortzik E, Wang L, Becker K (2012) Thiol-based posttranslational modifications in parasites. Antioxid Redox Signal 17:657–673
20. Novoselov SV, Kryukov GV, Xu XM, Carlson BA, Hatfield DL, Gladyshev VN (2007) Selenoprotein H is a nucleolar thioredoxin-like protein with a unique expression pattern. J Biol Chem 282:11960–11968
21. Gladyshev VN, Krause M, Xu XM, Korotkov KV, Kryukov GV, Sun QA, Lee BJ, Wootton JC, Hatfield DL (1999) Selenocysteine-containing thioredoxin reductase in *C. elegans*. Biochem Biophys Res Commun 259:244–249

22. Taskov K, Chapple C, Kryukov GV, Castellano S, Lobanov AV, Korotkov KV, Guigó R, Gladyshev VN (2005) Nematode selenoproteome: the use of the selenocysteine insertion system to decode one codon in an animal genome? Nucleic Acids Res 33:2227–2238
23. Salinas G, Lobanov AV, Gladyshev VN (2006) Selenoproteins in parasites. Springer Science+Business Media LLC, Boston, pp 355–366
24. Griffith OW (1999) Biologic and pharmacologic regulation of mammalian glutathione synthesis. Free Radic Biol Med 27:922–935
25. Morris G, Anderson G, Dean O, Berk M, Galecki P, Martin-Subero M, Maes M (2014) The glutathione system: a new drug target in neuroimmune disorders. Mol Neurobiol 50:1059–1084
26. Shen D, Dalton TP, Nebert DW, Shertzer HG (2005) Glutathione redox state regulate mitochondrial reactive oxygen production. J Biol Chem 27:25305–25312
27. Blokhina O, Virolainen E, Fagerstedt KV (2003) Antioxidants, oxidative damage and oxygen deprivation stress: A review. Ann Bot 91:179–194
28. Hanigan MH (2014) Gamma-glutamyl transpeptidase: redox regulation and drug resistance. Adv Cancer Res 122:103–141
29. Gon S, Faulkner MJ, Beckwith J (2006) *In vivo* requirement of glutaredoxins and thioredoxins in the reduction of the ribonucleotide reductases of *Escherichia coli*. Antioxid Redox Signal 8:735–742
30. Rendón JL, Juárez O (2008) Glutathione reductase: structural, catalytic and functional aspects. Transworld Research Network, Kerala, India, pp 317–349
31. Torres-Rivera A, Landa A (2008) Glutathione transferases from parasites: A biochemical view. Acta Trop 105:99–112
32. Hill BG, Ramana KV, Cai J, Bhatnagar A, Srivastava SK (2010) Measurement and identificationof S-glutathiolated proteins. Methods Enzymol 473:179–197
33. Turner E, Kelvit RE, Hager LJ, Shapiro BM (1987) Ovothiols, a family of redox-active mercaptohistidines compounds from marine invertebrate eggs. Biochemist 26:40208–44036
34. Van Laer K, Hamilton CJ, Messens J (2013) Low-molecular-weight thiols in thiol-disulfide exchange. Antioxid Redox Signal 8:1642–1653
35. Fairlamb AH, Cerami A (1985) Identification of a novel, thiol-containing co-factor essential for glutathione reductase enzyme activity in trypanosomatids. Mol Biochem Parasitol 14:187–198
36. Nogoceke E, Gommel DU, Kiess M, Kalisz HM, Flohé LA (1997) Unique cascade of oxidoreductases catalyses trypanothione-mediated peroxide metabolism in *Crithidia fasciculate*. Biol Chem 378:827–836
37. Hillebrand H, Schmidt A, Krauth-Siegel RL (2003) A second class of peroxidases linked to the trypanothione metabolism. J Biol Chem 278:6809–6815
38. Go YM, Jones DP (2008) Redox compartmentalization in eukaryotic cells. Biochim Biophys Acta 1780:1273–1290
39. Laurent TC, Moore EC, Reichard P (1964) Enzymatic synthesis of deoxyribonucleotides. IV. Isolation and characterization of thioredoxin, the hydrogen donor from *Escherichia coli*. J Biol Chem 239:3436–3444
40. Line K, Isupov MN, Garcia-Rodriguez E, Maggioli G, Parra F, Littlechild JA (2008) The *Fasciola hepatica* thioredoxin: High resolution structure reveals two oxidation states. Mol Biochem Parasitol 161:44–48
41. Sen R, Baltimore D (1986) Inducibility of kappa immunoglobulin enhancer-binding protein Nf-kappa B by a posttranslational mechanism. Cell 26:921–928
42. Arias DG, Cabeza MS, Erben ED, Carranza PG, Lujan HD, Téllez-Iñón MT, Iglesias AA, Guerrero SA (2011) Functional characterization of methionine sulfoxide reductase from *Trypanosoma* spp. Free Radic Biol Med 50:37–46
43. Pekkari K, Holmgren A (2004) Truncated thioredoxin: physiological functions and mechanism. Antioxid Redox Signal 6:53–61
44. Quan S, Schneider I, Pan J, von Hacht A, Bardwell JCA (2007) The CXXC motif is more than a redox rheostat. J Biol Chem 282:28823–28833

45. Kunchithapautham K, Padmavathi B, Narayanan RB, Kaliraj P, Scott AL (2003) Thioredoxin from *Brugia malayi:* defining a 16-Kilodalton class of thioredoxins from nematodes. Infect Immun 71:4119–4126
46. Selkirk ME, Smith VP, Thomas GR, Gounaris K (1998) Resistance of filarial nematode parasites to oxidativestress. Int J Parasitol 28:1315–1332
47. Berndt C, Hudemann C, Hanschmann EM, Axelsson R, Holmgren A (2007) How does iron-sulfur cluster coordination regulate the activity of human glutaredoxin 2? Antioxid Redox Signal 9:151–157
48. Kehr S, Jortzik E, Delahunty C, Yates JR, Rahlfs S, Becker K (2011) Protein S-glutathionylation in malaria parasites. Antioxid Redox Signal 15:2855–2865
49. van Rossum AJ, Brophy PM, Tait A, Barrett J, Jefferies JR (2001) Proteomic identification of glutathione S-transferases from the model nematode *Caenorhabditis elegans*. Proteomics 1:1463–1468
50. Liebau E, Eckelt VH, Wildenburg G, Teesdale-Spittle P, Brophy PM, Walter RD, Henkle-Dührsen K (1997) Structural and functional analysis of a glutathione S-transferase from *Ascaris suum*. Biochem J 324(Pt 2):659–666
51. Liebau E, Eschbach ML, Tawe W, Sommer A, Fischer P, Walter RD, Henkle-Dührsen K (2000) Identification of a stress-responsive *Onchocerca volvulus* glutathione S-transferase (Ov-GST-3) by RT-PCR differential display. Mol Biochem Parasitol 109:101–110
52. Perally S, Lacourse EJ, Campbell AM, Brophy PM (2008) Heme transport and detoxification in nematodes: subproteomics evidence of differential role of glutathione transferases. J Proteome Res 7:4557–4565
53. Ahmad R, Srivastava AK (2008) Inhibition of Filarial Glutathione-S-transferase by various classes of compounds and their evaluation as novel antifilarial agents. Helminthologia 45:114
54. Mukherjee N, Parida PK, Santra A, Ghosh T, Dutta A, Jana K, Misra AK, Sinha Babu SP (2016) Oxidative stress plays major role in mediating apoptosis in filarial nematode *Setaria cervi* in the presence of trans-stilbene derivatives. Free Radic Biol Med 93:130–144
55. Saini P, Gayen P, Nayak A, Kumar D, Mukherjee N, Pal BC, Sinha Babu SP (2012) Effect of ferulic acid from *Hibiscus mutabilis* on filarial parasite *Setaria cervi*: molecular and biochemical approaches. Parasitol Int 61:520–531
56. Mukherjee N, Joardar N, Sinha Babu SP (2018) Antifilarial activity of azadirachtin fuelled through reactive oxygen species induced apoptosis: a thorough molecular study on *Setaria cervi*. J Helminthol 93:519–528
57. Saxena J, Pandya U, Gupta D, Shukla OP (1996) Antioxidant system in adults and microfilariae of the bovine filarial parasite *Setaria cervi*. J Parasit Dis 20:159–162
58. Barrett J (1983) Biochemistry of filarial worms. Helminth Abstr Ser An anim hum helminthol 52:1–18
59. Cheah KS (1976) In biochemistry of parasites and host parasite relationships. Elsevier, North Holland Biomedical Press, Amsterdam, pp 133–143
60. Mukherjee S, Joardar N, Mondal S, Schiefer A, Hoerauf A, Pfarr K, Sinha Babu SP (2018) Quinolone fused sulfonamides as a novel benign antifilarial agent. Sci Rep 8:12073. https://doi.org/10.1038/s41598-018-30610-7
61. Nayak A, Gayen P, Saini P, Mukherjee N, Sinha Babu SP (2012) Molecular evidence of curcumin induced apoptosis in the filarial worm *Setaria cervi*. Parasitol Res 111:1173–1186
62. Saini P, Gayen P, Kumar D, Nayak A, Mukherjee N, Mukherjee S, Pal BC, SinhaBabu SP (2014) Antifilarial effect of ursolic acid from *Nyctanthes arbortristis*: molecular and biochemical evidences. Parasitol Int 63:717–728
63. Mukherjee N, Saini P, Mukherjee S, Roy P, Gayen P, Sinha Babu SP (2014) Ethanolic extract of *Azadirachta indica* (A. Juss.) causing apoptosis by ROS upregulation in *Dirofilariaimmitis* microfilaria. Res Vet Sci 97:309–317
64. Mukherjee N, Saini P, Mukherjee S, Roy P, Sinha Babu SP (2014) *In vitro* antifilarial activity of *Azadirachta indica* aqueous extract through reactive oxygen species enhancement. Asian Pac J Trop Med 7:841–848

65. Roy P, Dhara D, Parida PK, Kar RK, Bhunia A, Jana K, Sinha Babu SP, Misra AK (2016) C-cinnamoyl glycosides as a new class of anti-filarial agents. Eur J Med Chem 114:308–317
66. Giri BR, Roy B (2016) Praziquantel induced oxidative stress and apoptosis-like cell death in *Raillietina echinobothrida*. Acta Trop 159:50–57
67. Mukherjee N, Mukherjee S, Saini P, Roy P, Sinha Babu SP (2016) Phenolics and terpenoids; the promising new search for anthelmintics: A critical review. Mini Rev Med Chem 16:1415–1441
68. Mukherjee S, Mukherjee N, Gayen P, Roy P, Sinha Babu SP (2016) Metabolic inhibitors as antiparasitic drugs: pharmacological, biochemical and molecular perspectives. Curr Drug Metab 17:937–970
69. Bueding E (1949) Studies on the metabolism of the filarial worm, *Litomosoides carinii*. J Exp Med 89:107–130
70. Batra S, Singh SP, Srivastava VML, Chatterjee RK (1989) Xanthine oxidase, superoxide dismut ase, catalase and lipi d peroxidation in *Mastomys natalensis:* effect of *Dipetalonema viteae* infection. Ind J Exp Biol 27:1067–1070
71. Callahan HL, Crouch RK, James ER (1991) Hydrogen peroxide is the toxic product of *Onchocerca* microfilaria. J Cell Biol 107:3532–3538
72. Precions WY, Barrett J (1989) The possible absence of cytochrome P-450 linked xenobiotic metabolism in helminth. Biochem Biophy Acta 992:215–222
73. Brophy PM, Barrett L (1990) Glutathione transferase in helminth. Parasitology 100:345–349
74. Yawetz A, Agosin M (1981) Purification of glutathione-stransferase of *Trypanosoma cruzi*. Comp Biochem Physiol 68:237–243
75. Jaffe LJ, Lambert RA (1986) Glutathine-s-transferase in adult *Dirofilaria immitis* and *Brugia phangai*. Mol Biochem Parasitol 20:199–206
76. Ou X, Thomas GR, Chacón MR, Tang L, Selkirk ME (1995) *Brugia malayi*: differential susceptibility to and metabolism of hydrogen peroxide in adults and microfilariae. Exp Parasitol 80:530–540
77. Selkirk ME, Smith VP, Thomas GR, Gounaris K (1998) Resistance of filarial nematode parasites to oxidative stress. Int J Parasitol 28:1315–1332
78. Lu W, Egerton GL, Bianco AE, Williams SA (1998) Thioredoxin peroxidase from *Onchocerca volvulus*: A major hydrogen peroxide detoxifying enzyme in filarial parasites. Mol Biochem Parasitol 91:221–235
79. Ghosh I, Eisinger SW, Raghavan N, Scott AL (1998) Thioredoxin peroxidases from *Brugia malayi*. Mol Biochem Parasitol 91:207–220
80. Lüersen K, Stegehake D, Daniel J, Drescher M, Ajonina I, Hertel P, Woltersdorf C, Liebau E (2013) The glutathione reductase GSR-1 determines stress tolerance and longevity in *Caenorhabditis elegans*. PLoS One 8:e60731
81. Müller S, Walter RD, Fairlamb AH (1995) Differential susceptibility of filarial and human erythrocyte glutathione reductase to inhibition by the trivalent organic arsenial melarsen oxide. Mol Biochem Parasitol 71:211–219
82. Komuniecki R, Bruchhaus I, Ilg T, Wilson K, Zhang Y, Fairlamb AH (1992) Purification of glutathione reductase from muscle of the parasitic nematode *Ascaris suum*. Mol Biochem Parasitol 51:331–333
83. Müller S, Gilberger TW, Fairlamb AH, Walter RD (1997) Molecular characterization and expression of *Onchocerca volvulus* glutathione reductase. Biochem J 325:645–651
84. Li W, Bandyopadhyay J, Hwaang HS, Park BJ, Cho JH, Lee JI, Ahnn J, Lee SK (2012) Two thioredoxin reductases, trxr-1 and trxr-2, have differential physiological roles in *Caenorhabditis elegans*. Mol Cells 34:209–218
85. Joardar N, Mukherjee S, Sinha Babu SP (2018) Thioredoxin reductase from the bovine filarial parasite *Setaria cervi*: Studies on its localization and optimization of the extraction. Int J Biol Macromol 107:2375–2384
86. Yadav SS, Srikanth E, Singh N, Rathaur S (2013) Identification of GR and TrxR systems in *Setaria cervi*: Purification and characterization of glutathione reductase. Parasitol Int 62:193–198

87. Hudson AL, Sotirchos IM, Davey MW (2010) Substrate specificity of the mitochondrial thioredoxin reductase of the parasitic nematode *Haemonchus contortus*. Parasitol Res 107:487–493
88. Bondareva AA, Capecchi MR, Iverson SV, Li Y, Lopez NI, Lucas O, Merril GF, Prigge JR, Siders AM, Wakamiya M et al (2007) Effects of thioredoxin reductase-1 deletion on embryogenesis and transcriptome. Free Radic Biol Med 43:911–923
89. Stenvall J, Fierro-Gonzalez JC, Swoboda P, Saamarthy K, Cheng Q, Cacho-Valadez B, Arner ES, Persson OP, Miranda-Vizuete A, Tuck S (2011) Selenoprotein TRXR-1 and GSR-1 are essential for removal of old cuticle during molting in *Caenorhabditis elegans*. Proc Natl Acad Sci USA 108:1064–1069
90. Cacho-Valadez B, Munoz-Lobato F, Pedrajas JR, Cabello J, Fierro-Gonzalez JC, Navas P, Swoboda P, Link CD, Miranda-Vizuete A (2012) The characterization of the *Caenorhabditis elegans* mitochondrial thioredoxin system uncovers an unexpected protective role of thioredoxin reductase 2 in beta-amyloid peptide toxicity. Antioxid Redox Signal 16:1384–1400
91. Mora-Lorca JA, Sáenz-Narciso B, Gaffney CJ, Naranjo-Galindo FJ, Pedrajas JR, Guerrero-Gómez D, Dobrzynska A, Askjaer P, Szewczyk NJ, Cabello J et al (2016) Glutathione reductase gsr-1 is an essential gene required for *Caenorhabditis elegans* early embryonic development. Free Radic Biol Med 96:446–461
92. Prince PR, Madhumathi J, Anugraha G, Jeyaprita PJ, Reddy MVR, Kaliraj P (2013) Tandem antioxidant enzymes confer synergistic protective responses in experimental filariasis. J Helminthol 88:1–9
93. Madhumathi J, Prince PR, Nageswara Rao D, Kaliraj P (2010) Dominant T-cell epitopes of filarial BmALT-2 and their cytokine profile in BALB/c mice. Parasite Immunol 32:760–763
94. Mukhopadhyay S, Srivastava VML, Murthy PK, Hasnain SE (2004) Poorer NF-κB signaling by microfilariae in macrophages from BALB/c mice affects their ability to produce cytotoxic levels of nitric oxide to kill microfilariae. FEBS Lett 567:275–280
95. Bal M, Mandal N, Achary KG, Das MK, Kar SK (2011) Immunoprophylactic potential of filarial glutathione-s-transferase in lymphatic filariaisis. Asian Pac J Trop Med 4:185–191
96. Mukherjee N, Mukherjee S, Saini P, Roy P, Sinha Babu SP (2014) Antifilarial effects of polyphenol rich ethanolic extract from the leaves of *Azadirachta indica* through molecular and biochemical approaches describing reactive oxygen species (ROS) mediated apoptosis of *Setaria cervi*. Exp Parasitol 136:41–58
97. Nandi B, Roy S, Bhattacharya S, Sinha Babu SP (2004) Free radicals mediated membrane damage by the saponins acaciaside A and acaciaside B. Phytother Res 18:191–194
98. Chowdhury P, Roy B, Mukherjee N, Mukherjee S, Joardar N, Mondal MK, Roy DP, Sinha Babu SP (2018) Chitosan biopolymer functionalized gold nanoparticles with controlled cytotoxicity and improved antifilarial efficacy. Adv Compos Hybrid Mater. https://doi.org/10.1007/s42114-018-0040-7
99. Mukherjee S, Mukherjee N, Saini P, Roy P, Sinha Babu SP (2015) Ginger extract ameliorates phosphamidon induced hepatotoxicity. Indian J Exp Biol 53:574–584
100. Lettre G, Hengartner MO (2006) Developmental apoptosis in *C. elegans*: a complex CEDnario. Nat Rev Mol Cell Biol 7:97–108
101. Kang R, Zeh HJ, Lotze MT, Tang D (2011) The beclin 1 network regulates autophagy and apoptosis. Cell Death Differ 18(4):571–580
102. Veerapathran A, Dakshinamoorthy G, Gnanasekar M, Reddy MVR, Kalyanasundaram R (2009) Evaluation of *Wuchereria bancrofti* GST as a vaccine candidate for lymphatic filariasis. PLoS Negl Trop Dis 3:e457. https://doi.org/10.1371/journal.pntd.0000457
103. Chiumiento L, Bruschi F (2009) Enzymatic antioxidant systems in helminth parasites. Parasitol Res 105:593–603
104. Klimowski L, Chandrashekar R, Tripp CA (1998) Molecular cloning, expression and enzymatic activity of a thioredoxin peroxidase from *Dirofilaria immitis*. Mol Biochem Parasitol 90:297–306

105. Roy P, Saha SK, Gayen P, Chowdhury P, Sinha Babu SP (2018) Exploration of antifilarial activity of gold nanoparticle against human and bovine filarial parasites: A nanomedicinal mechanistic approach. Colloids Surf B Biointerfaces 161:236–243
106. Roy B, Mukherjee S, Mukherjee N, Chowdhury P, Sinha Babu SP (2014) Design and green synthesis of polymer inspired nanoparticles for the evaluation of their antimicrobial and antifilarial efficiency. RSC Adv 4:34487
107. Sundar S, Prajapati VK (2012) Drug targeting to infectious diseases by nanoparticles surface functionalized with special biomolecules. Curr Med Chem 19:3196–3202
108. Ginger ML (2006) Niche metabolism in parasitic protozoa. Philos Trans R Soc Lond Ser B Biol Sci 361:101–118
109. Gustafsson TN, Osman H, Werngren J, Hoffner S, Engman L, Holmgren A (2016) Ebselen and analogs as inhibitors of *Bacillus anthracis* thioredoxin reductase and bactericidal antibacterials targeting *Bacillus* species, *Staphylococcus aureus* and *Mycobacterium tuberculosis*. Biochim Biophys Acta 1860:1265–1271
110. Zou L, Jun W, Gao Y, Ren X, Rottenberg ME, Lu J, Holmgren A (2018) Synergistic antibacterial activity of silver with antibiotics correlating with the upregulation of the ROS production. Sci Rep 8:11131. https://doi.org/10.1038/s41598-018-29313-w
111. Misra S, Verma M, Mishra SK, Srivastava S, Lakshmi V, Misra-Bhattacharya S (2011) Gedunin and photogedunin of *Xylocarpus granatum* possess antifilarial activity against human lymphatic filarial parasite *Brugia malayi* in experimental rodent host. Parasitol Res 109:1351
112. Misra N, Sharma M, Raj K, Dangi A, Srivastava S, Misra-Bhattacharya S (2007) Chemical constituents and antifilarial activity of *Lantana camara* against human lymphatic filariid *Brugia malayi* and rodent filariid *Acanthocheilonema viteae* maintained in rodent hosts. Parasitol Res 100:434–489
113. Ren X, Zou L, Holmgren A (2018) Selenocysteine in mammalian thioredoxin reductase and application of ebselen as a therapeutic. Free Radic Biol Med 127:238–247. https://doi.org/10.1016/j.freeradbiomed.2018.05.081
114. Allen JE, Adjei O, Bain O, Hoerauf A, Hoffmann WH, Makepeace BL et al (2008) Of mice, cattle, and humans: the immunology and treatment of river blindness. PLoS Negl Trop Dis 2:e217. https://doi.org/10.1371/journal.pntd.0000217
115. Murthy PK, Srivastava K, Murthy PSR (2004) Responses of *Brugia malayi* – Indian leaf monkey (*Presbytis entellus*), a non-human primate model of filariasis, to diethylcarbamazine, ivermectin and CDRI compound 82-437. Curr Sci 86:432–439

11 Oxidative Stress and Antioxidant Defense Mechanism in the Human Enteric Protozoan Parasite *Entamoeba histolytica*

Ghulam Jeelani and Tomoyoshi Nozaki

Abstract

Entamoeba histolytica is an intestinal protozoan parasite that causes amebic dysentery and liver abscess in millions of inhabitants of endemic areas. *E. histolytica* trophozoites are exposed to highly toxic reactive oxygen and nitrogen species during tissue invasion. The capacity of *E. histolytica* trophozoites to survive reactive oxygen and nitrogen species is integral to its pathogenic potential and disease outcome. *E. histolytica* lacks most of the components of canonical eukaryotic anti-oxidative defense systems including catalase, glutathione, and its metabolic enzymes, and L-cysteine is the major intracellular low molecular mass thiol. However, this parasite possesses a functional thioredoxin system composed of thioredoxin and thioredoxin reductase, which is critical for maintaining cellular redox balance and antioxidant function. Major enzymes involved in the redox balancing and the antioxidative pathways have been proven to be essential for the pathogen and, therefore, fulfill the prerequisite for a potential drug target. In this chapter, we summarized the currently available knowledge on the oxidative/nitrosative stress during *E. histolytica* infections, redox regulation and detoxification mechanisms for oxidative and nitrosative species in *E. histolytica*, and its potential use as a target for drug discovery against amebiasis.

Keywords

Entamoeba histolytica · Oxidative stress · Reactive oxygen species · Reactive nitrogen species · Antioxidant · Redox · Thiol, Cysteine, Drug target

G. Jeelani · T. Nozaki (✉)
Department of Biomedical Chemistry, Graduate School of Medicine, The University of Tokyo, Tokyo, Japan
e-mail: nozaki@m.u-tokyo.ac.jp

S. Chakraborti et al. (eds.), *Oxidative Stress in Microbial Diseases*,
https://doi.org/10.1007/978-981-13-8763-0_11

11.1 Introduction

Amebiasis is a major cause of morbidity and mortality around the world, caused by a unicellular parasitic protozoan *E. histolytica*. Infections by this parasite cause amebic colitis (dysentery or diarrhea) and amebic liver abscess [1]. It is estimated that annually 50 million people worldwide suffer from invasive amebic infection, causing 40,000–100,000 deaths per year [2]. Amebic dysentery is considered as the third most common cause of parasitic death worldwide after malaria and schistosomiasis [3]. The life cycle of *Entamoeba* species consists of two main stages: the motile dividing trophozoite stage responsible for disease manifestations and the dormant transmissible cyst stage [4]. Infection in humans occurs following ingestion of mature cysts in water, food, or hands contaminated with faeces. In the human body, the cysts differentiate into trophozoites in the small intestine, which then colonize the large intestine and proliferate. The trophozoites then invade the gut mucin and epithelial barrier [5].

E. histolytica trophozoites were shown to be sensitive to oxygen, but are able to tolerate low levels of oxygen tension and therefore anaerobic/microaerophilic. During colonization, tissue invasion, and extra intestinal propagation, the trophozoites are exposed to various reactive oxygen species (ROS) and reactive nitrogen species (RNS) [6]. Also during its life cycle, the parasites encounter a wide variety of environmental stresses, such as changes in pH, pO2, and nutrient (e.g., glucose and amino acids) concentrations, elevated temperature (heat shock), and host immune system including the released of oxidative and nitrosative species from neutrophils and macrophages (Fig. 11.1) [7]. To be a successful parasite, *E. histolytica* must adapt to the ever-changing host environments, as represented by oxidative stress. The response toward oxidative stress by *E. histolytica* has been unveiled at the molecular levels, yet not fully understood. In this chapter, we summarize the currently available knowledge on the sources of oxidative stress during parasite proliferation and mammalian infections and also the antioxidative defense systems this parasite utilizes in order to cope with the oxidative stress.

11.2 Source of Oxidative and Nitrosative Stress in *Entamoeba histolytica*

E. histolytica trophozoites primarily reside in the colon, and during tissue invasion, the parasite must maintain intracellular hypoxia within the oxygenated human tissues and also defend against the host immune attack [8]. *E. histolytica* trophozoites invade the host colonic mucosal barrier, leading to significant tissue destruction and inflammation, with release of pro-infammatory cytokines, ROS, and RNS from the host's activated immune cells [9]. These reactive species are the major cytotoxic effectors for killing *E. histolytica* trophozoites. ROS and RNS inactivate key metabolic enzymes by oxidation and nitrosylation, inhibit energy generation via glycolysis, trigger stress responses by fragmenting the endoplasmic reticulum (ER), and reduce the activity of certain virulence factors [10].

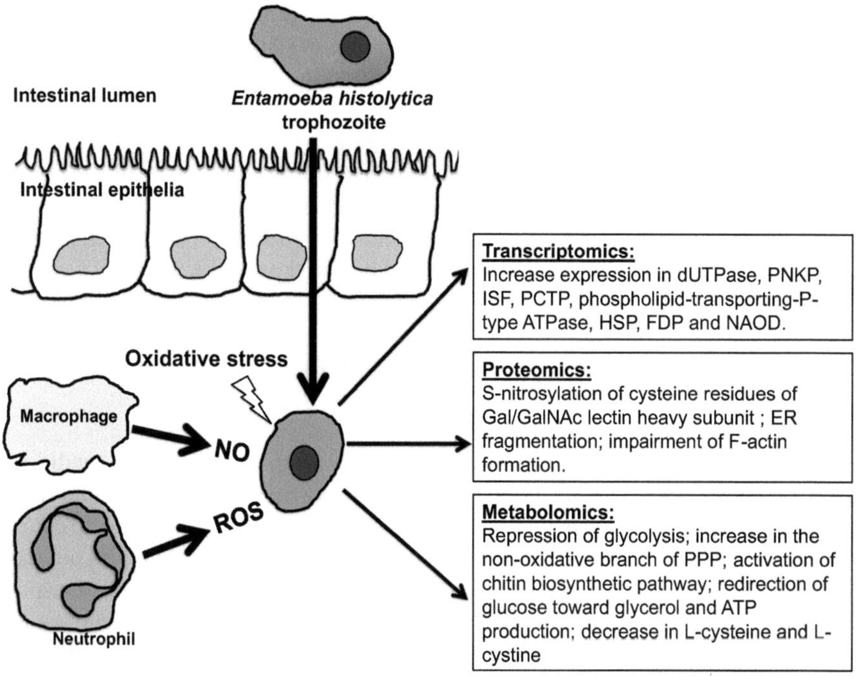

Fig. 11.1 Schematic illustration of *E. histolytica* invasion and summary of major findings by previous omics studies in response of *E. histolytica* trophozoites to oxidative stress. *E. histolytica* trophozoites invade the intestinal epithelial layer of the host, where they are exposed to ROS and RNS released by immune cells. Major experimental findings by transcriptomics, proteomics, and metabolomics analyses on the response by *E. histolytica* trophozoites triggered by oxidative stress are summarized in the boxes. *NO* nitric oxide, *ROS* reactive oxygen species, *dUTPase* deoxyuridine triphosphate nucletotidehydrolase, *PNKP* polynucleotide kinase-3-phosphatase, *ISF* iron-sulfur flavoprotein, *PCTP* phosphatidylcholine transport-like protein, *Hsp* heat shock protein, *FDP* flavodiiron protein, *NAOD* N-acetyl ornithine deacetylase, *ER* endoplasmic reticulum, *Gal/GalNAc* galactose/N-acetyl D-galactosamine, *PPP* pentose phosphate pathway

11.3 Correlation Between Pathogenesis and Oxidative/ Nitrosative Stress

Pathogenicity of *E. histolytica* trophozoites has been generally attributed to adherence-dependent killing of target mammalian cells/tissues as well as destruction of components of the host's extra-cellular matrix [11]. A direct link of oxidative and nitrosative stress management of the parasite with its pathogenesis has been suggested [12] by the following observations: (i) virulent *E. histolytica* strains showed higher oxygen resistance compared to avirulent, less virulent, or attenuated strains [14]; (ii) in virulent strains, genes that are known to be involved in oxidative stress defense were more expressed compared to avirulent/less virulent strains [13]; (iii) oxidative and nitrosative stress modulated a large set of *E. histolytica* genes including virulence-associated genes, as revealed by transcriptomic analyses [10,

15, 16]; (iv) the higher numbers of ROS and RNS responsive genes in the virulent strain compared to the avirulent strain [16]; and (v) high protein expression of superoxide dismutase (SOD) and peroxiredoxin in the virulent strain compared to an avirulent strain, suggest a possible role of these proteins in virulence, as revealed by proteomic analysis [17]. In the above-mentioned study [16], the most affected changes were observed in signaling/regulation pathways and repair of misfolded proteins, nucleic acids, and lipids [16]. The list of stress-regulated genes also includes several genes involved in virulence-related cellular processes such as ability to adhere to ingest and destroy human and bacterial cells [12, 16].

11.4 Mechanisms of the Parasite to Deal with Oxidative and Nitrosative Stress

E. histolytica trophozoites are able to tolerate up to 5% oxygen in culture conditions [18]. Oxygen is toxic to *E. histolytica* trophozoites when it is transformed into ROS, which then interact with proteins, nucleic acids, and lipids [19]. Therefore, to detoxify ROS *E. histolytica* possesses an enzymatic and non-enzymatic complex antioxidant system. As this parasite lacks a conventional respiratory electron transport chain that reduces oxygen to water [20], its mitochondrion-related organelle known as mitosome [21] does not produce ROS. However, *E. histolytica* lack most of the components involved in antioxidant defense mechanisms, including catalase, peroxidase, glutathione, and the glutathione-recycling enzymes such as glutathione

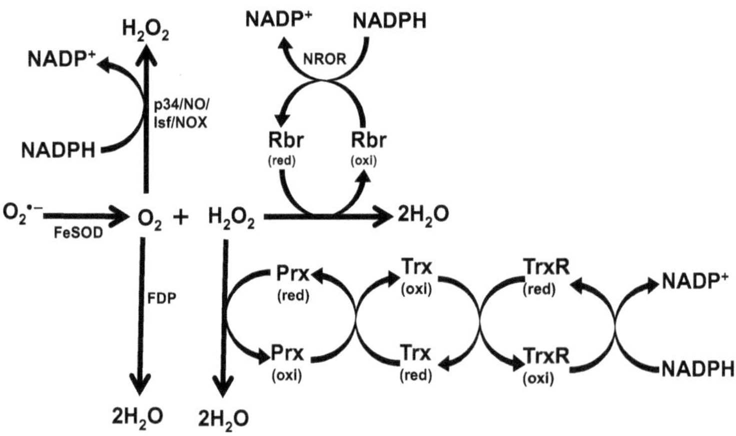

Fig. 11.2 Characterized antioxidative defense system in *E. histolytica. FeSOD* iron-containing superoxide dismutase, *p34* 34-kDa NADPH: flavin oxidoreductase, *NO* NADPH-dependent oxidoreductase, *Isf* iron-sulfur flavoprotein, *NOX* NADH oxidase, *NROR* NAD(P)H-dependent rubredoxin reductase, *FDP* flavodiiron protein, *Rbr (ox) and Rbr (red)* oxidized and reduced rubrerythrin, *Prx (ox) and Prx (red)* oxidized and reduced peroxiredoxin, *Trx (ox) and Trx (red)* oxidized and reduced thioredoxin, *TrxR (ox) and TrxR (red)* oxidized and reduced thioredoxin reductase, respectively

peroxidase and glutathione reductase or glutaredoxin [22, 23]. Instead, *E. histolytica* genome encodes enzymes that are involved in the detoxification of ROS, such as thioredoxin (Trx) [24], thioredoxin reductase (TrxR) [25], peroxidedoxin (Prx) [26], rubrerythrin (Rr) [27], flavodiiron proteins (FDP) [28], NADPH: flavin oxidoreductase (p34) [29], NADPH-dependent oxidoreductase (NO) [30], hybrid-cluster protein [31], and Fe-SOD [32] (Fig. 11.2). *E. histolytica* trophozoites are capable of reducing O_2 to H_2O_2 catalyzed by p34, NO, iron sulfur flavoprotein (Isf) [33] and NADH oxidase (NOX) [34] (Fig. 11.2). The H_2O_2 generated can be further converted to H_2O by the action of Prx [26], Rr [27], and/or TrxR enzymes (Fig. 11.2) [25]. Superoxide radical anion ($2O_2^{\bullet-}$) is also reduced to O_2 and H_2O_2 by SOD [32].

E. histolytica possesses several oxygen sensitive metabolic enzymes such as pyruvate: ferredoxin oxidoreductase (PFOR), which replaces the multienzyme complex pyruvate dehydrogenase present in aerobic organisms [35]. *Entamoeba* PFOR contains iron-sulfur clusters in their catalytic center, which are easily damaged by oxygen and several ROS-derived superoxide radical anion ($O_2^{\bullet-}$) or hydrogen peroxide [36, 37]. PFOR is essential for energy generation in this organism, and involved in the oxidative decarboxylation of pyruvate to acetyl-coenzyme A, which is further converted to ethanol and acetate, the end products of glucose metabolism in this parasite [35]. Since glycolysis is the main pathway for ATP generation in *E. histolytica*, PFOR inhibition by O_2 has deleterious effects on energy metabolism in this parasite.

It has been suggested that *E. histolytica* ROS resistance can be improved by the ROS-scavengers generated from ingested bacteria. Recently, Varet et al. [38] analyzed the interplay between *E. histolytica* trophozoites and enteric bacteria (commensals, pathogens, and probiotics). They found that live *Escherichia coli* O55 protected *E. histolytica* trophozoites against oxidative stress and this bacterium exerts strong influence on the transcriptome of oxidatively stressed *E. histolytica* trophozites [38]. In addition to oxidative stress, *E. histolytica* also encounters nitrosative stress generated by neutrophils and macrophages to kill the parasites [39]. NO reacts with $O_2^{\bullet-}$ to form peroxynitrite ($ONOO^-$), a RNS which causes a various harmful effects such as deamination of nucleotides and irreversible inhibition of metalloenzymes [40]. It has been demonstrated that prolonged exposure of *E. histolytica* trophozoites to sodium nitroprusside (SNP), sodium nitrite ($NaNO_2$), and sodium nitrate ($NaNO_3$), induce apoptosis in the parasite [41]. A short exposure of the trophozoites to these substances exerted cytotoxicity, although essential pathogenesis-associated functions such as erythrophagocytosis and proteolytic activity remained unaffected [41], indicating that the endomembrane and cytoskeletal systems involved in endocytosis and exocytosis were unaffected under NO stress. It was shown that NO inhibits PFOR activity and also induces fragmentation of the endoplasmic reticulum [10]. A pathway for NO degradation has not yet been identified in *E. histolytica*; however, it was demonstrated that *E. histolytica* peroxiredoxins could reduce peroxynitrite to nitrite and water [24]. It has also been shown that *E. histolytica* DNA methyltransferase (Ehmeth) plays a role in resistance to nitrosative stress [42]. It was proposed that enolase, a glycolytic protein that binds to Ehmeth and inhibits its activity, was reduced in *S*-nitrosoglutathione

(GSNO)-treated *E. histolytica*, thereby leading to higher tRNA^{Asp} methylation. Furthermore, it was shown that NO resistance in *E. histolytica* trophozoites could be accomplished by the methylation of tRNA^{Asp} [42].

11.5 Omics Approaches to Unravel the Stress–Response Mechanisms in *E. histolytica*

Oxidative stress response of *E. histolytica* is not yet fully understood at the molecular level. However, in order to address stress response mechanisms in *E. histolytica*, various "omic" studies such as transcriptomics (transcriptome), proteomics (proteome), and metabolomics (metabolome) have been carried out.

11.5.1 Transcriptome Analysis

A study using a whole-genome DNA microarray analysis was used to determine the transcriptional response of *E. histolytica* trophozoites to NO or H_2O_2 [28]. Numerous genes have been shown to be differentially expressed in *E. histolytica* exposed to oxidative or nitrosative stress [28]. However, a majority of the genes regulated by oxidative stress were annotated as hypothetical proteins [28]. Many genes encoding ubiquitin-conjugating enzymes, heat shock proteins, protein kinases, and small GTPases showed similar patterns of up- and downregulation with NO or H_2O_2 exposure. Two genes (EHI_116940 and EHI_034560) coding for deoxyuridine triphosphate nucletotidehydrolase (dUTPase), which are implicated for DNA integrity, and a putative homologue of polynucleotide kinase-3-phosphatase (EHI_148430), involved in the repair of damaged DNA strand gaps and nicks, were upregulated in the parasites exposed to OS. Three genes encoding iron-sulfur flavoproteins (EHI_138480, EHI_025710, and EHI_192040) that were involved in combating oxidative stress in other anaerobes by reducing O_2 and H_2O_2 to water [28] were upregulated by H_2O_2. The *E. histolytica* genome encodes 7 independent genes for ISFs [4]. In general, anaerobic prokaryotes possess ISFs, which participate in combating the oxidative stress by removing molecular oxygen and hydrogen peroxide. *E. histolytica* and *T. vaginalis* are the only two known eukaryotes that possess ISF genes among eukaryotes for which the genome information is available [43]. Pearson et al. examined the promoter regions of 57 amoebic genes that had increased expression in response to hydrogen peroxide exposure [44]. They found that a transcription factor that binds to a specific motif (AAACCTCAATGAAGA) in the promoter regions of various hydrogen peroxide-responsive genes plays a role in coordinately regulating gene expression in response to hydrogen peroxide exposure [44]. Genes induced by nitrosative stress are associated with signaling/regulatory processes and nucleic acids, proteins and lipids repairs [28]. Genes involved in lipid metabolism such as phosphatidylcholine transport-like protein (EHI_025400) and a phospholipid-transporting P-type ATPase (EHI_188210) homologous to aminophospholipid translocases, and genes encoding chaperone like heat shock proteins

(HSP70, EHI_052860; HSP101, EHI_022620; HSP20, EHI_055680) were upregulated by nitrosative stress [28]. Furthermore, two homologous genes (EHI_189480 and EHI_164520) encoding iron-sulfur flavoproteins and one of the four FDPs (EHI_129890), which have bacterial-type oxygen or nitric oxide reductases activity, were upregulated by nitrosative stress [28]. *E. histolytica* trophozoites can develop resistance to cytotoxic concentrations of NO in vitro and it has been suggested that the resistance phenotype could support the establishment of the parasite within the large intestine, where there is a constant attack by macrophages and NO [45]. A transcriptomic analysis of NO-adapted trophozoites revealed the molecular basis of anti-NO response of *E. histolytica* [46]. It was shown that among 332 genes that were differentially expressed, some genes showed upregulation including acetylornithine deacetylase (EHI_114340), genes associated with signal transduction, such as RAP/RAN GTPase-activating protein (EHI_079910) and serine/threonine kinase 3 (EHI_096560) as well as genes related to cytoskeleton proteins such as actin (EHI_126190) and filopodin [46]. Furthermore, it was shown that N-acetyl ornithine deacetylase (NAOD), which catalyzes the deacetylation of N-acetyl-L-ornithine to yield ornithine and acetate, is involved in the adaptation of *E. histolytica* trophozoites to nitrosative stress [46].

11.5.2 Proteomic Analysis

Using the advantages of proteomics, a large-scale identification of *S*-nitrosylated (SNO) proteins by SNO-RAC (resin-assisted capture) analysis was carried out for *E. histolytica* trophozoites that were treated with the NO donor [47]. SNO-RAC analysis revealed proteins that are involved in glycolysis, gluconeogenesis, translation, protein transport, and adherence to target cells (the heavy subunit of Gal/GalNAc lectin) are among the *S*-nitrosylated proteins that were enriched by SNO-RAC. It was shown that the cysteine residues of the carbohydrate recognition domain in Gal/GalNAc lectin were S-nitrosylation, which hindered the attachment of *E. histolytica* trophozoites to mammalian cells [47]. Proteomics studies using NO-adapted trophozoites using SNO-RAC identified 242 putative SNO proteins. Among them, proteins involved in calcium binding, enzyme modulation, redox homeostasis, and actin cytoskeleton were identified [48]. In NO-adapted trophozoites, the formation of actin filaments (F-actin) is diminished, which apparently led to an impairment in erythrocyte ingestion [48].

11.5.3 Metabolomic Analysis

Metabolomic analysis of *E. histolytica* trophozoites treated with hydrogen peroxide (H_2O_2) or paraquat (PQ) revealed the dynamics of metabolic response in coping with oxidative stress [37]. The metabolome analysis showed that response to hydrogen peroxide and PQ treatment exhibited similar patterns in terms of the species of metabolites. Oxidative stress caused changes in the metabolites involved in glycolysis, chitin biosynthesis, nucleotide, and amino acid metabolism [37]. It was shown that oxidative stress inactivates enzymatic activity of some important metabolic enzymes (PFOR, phosphoglycerate mutase, and NAD^+-dependent alcohol dehydrogenase) ceasing glycolysis and rerouting of metabolic flux toward glycerol synthesis, chitin biosynthesis, and the non-oxidative branch of the pentose phosphate pathway [37]. The first two enzymes of the oxidative pentose phosphate pathway, glucose 6-phosphate dehydrogenase and transaldolases, are missing in *E. histolytica* based on the genome information [20], suggesting that a hexose monophosphate shunt pathway unlikely exists. The metabolomic study showed that most of the intermediates of the non-oxidative branch of the pentose phosphate pathway, i.e., erythrose 4-phosphate, ribulose 5-phosphate, and ribose 5- phosphate, and sedoheptulose 7-phosphate increased under oxidative stress [37]. It was proposed that ribose 5-phosphate, which may be produced from sedoheptulose 7-phosphate, is a precursor of NAD^+, which is used to synthesize NADP by NAD kinase [49]. Inhibition of glycolysis and fermentation by oxidative stress caused a decrement in the levels of the nucleoside triphosphates (ATP, GTP, UTP, and CTP). However, the levels of nucleoside monophosphates (AMP, CMP, GMP, and IMP) were increased. Both H_2O_2- and PQ-mediated oxidative stress caused a decrease in L-cysteine and L-cystine levels, whereas the level of cysteine S-sulfinate was slightly increased. These observations indicate that L-cysteine is most likely the major effective scavenger of oxygen free radicals in *E. histolytica* [37].

Furthermore, this metabolomics study also revealed the existence of functional glycerol biosynthetic pathway in *E. histolytica*, by demonstrating that glycerol 3-phosphate is one of the most highly upregulated metabolites under oxidative stress [37]. The intracellular level of glycerol also dramatically increased upon oxidative stress. The drastic accumulation of glycerol 3-phosphate, together with the increase in other upstream glycolytic intermediates, upon oxidative stress undoubtedly suggests the existence of functional glycerol 3-phosphate dehydrogenase in this parasite. Further work is needed to determine the fate and functional implication of the increased formation of glycerol 3-phosphate and glycerol upon oxidative stress.

11.6 Antioxidative Defense Mechanisms in *E. histolytica* Trophozoites

In aerobic organisms, reactive oxygen and nitrogen species are constantly produced as byproducts of aerobic metabolism. These reactive toxic substances can disturb the cellular redox homeostasis of the organisms. Low molecular weight thiols and redox active protein thiols play an important role as redox buffers that respond to alterations in the intracellular redox potential [50]. A typical antioxidant defense system comprising non-enzymatic and enzymatic constituents such as catalase, peroxidase, glutathione, and glutathione recycling enzymes, all of which are almost ubiquitously found in a wide range of organisms [51], is missing in *E. histolytica* [20]. Instead, as described above, this parasite possesses a functional thioredoxin redox system consisting of TrxR, thioredoxin (Trx), and Prx. Trx and L-cysteine are believed to be the important redox regulator of *E. histolytica* [22].

11.6.1 Thioredoxin Redox System

The thioredoxin system in *E. histolytica* comprises thioredoxin, NADPH, thioredoxin reductase, coupled with peroxiredoxin [24, 25] (Fig. 11.2). This system participates in various redox reactions within the cell and is considered a common redox messenger interacting with different proteins. Thioredoxins are small ubiquitous redox proteins that are found in nearly all organisms including unicellular parasites and play a key role in redox signaling and oxidative stress responses [52]. Thioredoxins contain a conserved CXXC motif containing two cysteine residues in its active site [53]. The thioredoxin system eventually functions as an electron donor for the reduction in a diverse array of cellular redox reactions. In *E. histolytica* trophozoites, the participation of TrxR and Trx is essential for maintaining a redox balance in the parasite cytosol [24, 25]. The *E. histolytica* genome (http://amoedb.org/amoeba/) reveals 22 Trx genes. Among these 22 Trx genes, 14 genes apparently encode proteins of less than 200 amino acids. Four of the 14 proteins (EHI_096200, EHI_053840, EHI_004490, and EHI_042900) contain the conventional Trx consensus sequence (CGPC) and are thus assumed to be authentic Trx, while 8 genes encode larger proteins of 300–700 amino acids [54]. Only two Trx genes (EHI_004490, EHI_053840), which possess conventional Trx consensus sequence (WCGPC), were fully characterized [24, 55]. Trxs are kept in the reduced state by TrxR, which catalyzes the reduction of oxidized Trx by transferring reducing equivalents from NADPH to its bound FAD and its redox active disulfide [25] (Fig. 11.2). Trxs are not only involved in detoxification of reactive oxygen species but may also be involved in regulating enzymatic activities through posttranslational modifications [56]. It has been reported that reducing agents such as L-cysteine and Trxs completely restored the activity of UDP-glucose pyrophosphorylase, which is involved in the synthesis of UDP-glucose, inactivated by sodium nitroprusside and hydrogen peroxide [56].

The *E. histolytica* genome encodes a single *TrxR* (EHI_155440) gene. It has been reported that *Entamoeba* TrxR contains an active dithiol/disulfide center (CXXC) as well as the coenzyme-binding motif (GGGGDAA) [25]. In the presence of O_2, TrxR also exhibited a H_2O_2-generating NAD(P)H oxidase activity, following a mechanism found to be different from the disulfide reduction [25, 56]. TrxR can use either NADPH or NADH as reducing electron donors for its disulfide reductase activity. However, *E. histolytica* TrxR showed approximately tenfold higher affinity toward NADPH compared to NADH, while the activity with NADH should also be noted. Trxs from other organisms exhibit preference toward NADPH over NADH by two to three orders of magnitude [25, 56–58].

As L-cysteine is the major thiol in *E. histolytica* trophozoites, it is presumed that the formation of S-nitrosothiols would depend on intracellular L-cysteine concentrations [22]. In the parasite, the cysteinyl radical species would react with exogenous NO and generate S-nitrosothiols (CySNO). This metabolite is important for S-nitrosylation or S-thiolation of cellular proteins. It has been shown that amebic TrxR can catalyze direct NADPH-dependent reduction of CySNO in vitro, and thus is considered to play an important role in regulating the intracellular levels of CySNO under oxidative or nitrosative stress conditions [25].

11.6.2 Peroxiredoxins (Prx)

Peroxiredoxins (Prxs) are a class of antioxidant enzymes that play important roles in regulating peroxide levels in the cell. Prxs catalyze the reduction of H_2O_2, alkylhydroperoxides, and peroxynitrite to H_2O using electron donors provided by thiols like Trxs (Fig. 11.2) [26, 55]. Prxs contain a conserved cysteine residue at the active site that undergoes peroxide-dependent oxidation and thiol-dependent reduction during catalysis. Prx was characterized in the commensal non-invasive ameba, *E. dispar* [59] and potentially free living *E. moshkovskii* [60], as well as *E. histolytica*. The surface localized Prx from *E. histolytica* exhibited peroxidase activity and was suggested to play a role in the survival of the parasite in oxidative environments during tissue invasion [26, 60, 61].

11.6.3 Rubreryrthrin (Rr)

Rubrerythrins (Rrs) are non-heme di-iron proteins involved in detoxification of H_2O_2. Rrs contain the ferritin-like diiron cluster and rubredoxin-like iron sulfur (FeS4) center [62]. *E. histolytica* Rr (EHI_134810) exhibits peroxidase activity and possesses all residues implicated in chelating the iron atoms. *E. histolytica* Rr was shown to be localized to mitosomes; it was not demonstrated, however, by proteome of the purified mitosomes [63, 64], and thus, its localization needs to be confirmed. A potential physiological protein receptor of Rr has not been identified in the *E. histolytica* genome [20, 28]. However, a protein (EHI_153000) has been characterized to possess Rr reductase activity, which catalyzes the NAD(P)H-dependent reduction of homologous and heterologous Rrs as well as *E. histolytica* flavodiiron protein (EhFDP1) [65]. This enzyme can use either NADPH or NADH as a cofactor for the reductase or oxidase activity, while the Rr reductase activity was higher with NADH than with NADPH [65].

11.6.4 Flavodiiron Proteins (FDPs)

Flavodiiron proteins (FDPs) constitute a large superfamily of enzymes with an important role in the response to oxidative and nitrosative stress in a wide variety of organisms, by transferring electrons to oxygen or nitric oxide [66]. Substrate preference of FDPs is not yet fully understood; some FDPs have a preference for oxygen, whereas others prefer NO [67, 68]. The *E. histolytica* genome possesses four genes encoding FDPs (EhFdp1, EHI_096710; EhFdp2, EHI_064530; EhFdp3, EHI_129890; EhFdp1, EHI_159860), which are homologous to bacterial-type oxygen or nitric oxide reductases. EhFdp1 has been fully characterized to possess poor NO reductase activity, but efficient oxygen reductase activity without releasing ROS [68].

11.7 Low Molecular Mass Thiols in *E. histolytica*

11.7.1 Cysteine/S-Methyl Cysteine

Low molecular weight thiols such as glutathione and L-cysteine serve as a redox buffer to protect the cells against a variety of stress conditions [43, 69]. As described above, *E. histolytica* is completely devoid of glutathione metabolism and L-cysteine is the primary low-molecular-weight thiol that is essential for the parasite growth, survival, attachment, elongation, motility, gene regulation, and antioxidative stress defense [4, 22, 23]. In vitro cultivation of *E. histolytica* trophozoites requires higher concentration of L-cysteine, which can be substituted with L-cystine, D-cysteine, or L-ascorbic acid, suggesting that the extracellular L-cysteine and/or L-cystine plays an important role in parasite growth [70]. L-Cysteine can be synthesized from L-homocysteine using the reverse trans-sulfuration pathway or from serine via the de novo L-cysteine biosynthesis pathway [71]. *E. histolytica* lacks a part of both the forward and reverse trans-sulfuration pathways [72]. In *E. histolytica*, L-cysteine is produced via a sulfur assimilatory de novo L-cysteine biosynthetic pathway, which is generally present in bacteria and plants [73, 74]. This pathway consists of two steps mediating a conversion of L-serine to L-cysteine via O-acetyl-L-serine (OAS) catalyzed by serine acetyltransferase (SAT, EC 2.3.1.30) and cysteine synthase [CS, OAS (thiol) lyase, (EC 4.2.99.8)]. The *E. histolytica* genome encodes three isoforms each for SAT and CS [75]. It has been reported that the three SAT isoforms are inhibited by L-cysteine through a feedback mechanism, which helps to maintain appropriate L-cysteine concentrations in this parasite [76]. A metabolome study using L-cysteine-lacking medium revealed drastic accumulation of OAS, and production of S-methyl cysteine (SMC) instead of L-cysteine [77]. SMC is a sulfur-containing amino acid that is commonly found in relatively large amounts in several legumes, considered as sulfur storage [78]. It was reported that oxidative stress such as H_2O_2 and paraquat caused increase in the SMC level, signifying the participation of this metabolite in response to oxidative stress [37]. Recently, it has been reported that SMC is involved in the antioxidative stress mechanism in *E. histolytica* trophozoites [75]. It is speculated that SMC is non-enzymatically converted to SMC sulfoxide when exposed to H_2O_2 or O_2 and SMC sulfoxide is further enzymatically catabolized into pyruvate, ammonia, and alkylthiosulnates [79]. It has been shown that L-cysteine deprivation causes three- to fourfold increase in the ROS level in *E. histolytica* trophozoites, indicating that L-cysteine is an important antioxidant [77]. As mentioned above, DNA microarray analysis was carried out to understand the effect of L-cysteine deprivation on gene expression in *E. histolytica* [33].

L-Cysteine deprivation affected the expression of a number of genes, but the expression of genes involved in L-cysteine metabolism and oxidative stress defense were not significantly changed, suggesting that *E. histolytica* may rely on posttranscriptional or posttranslational mechanisms to regulate these pathways to combat the oxidative stress [33].

11.7.2 Thioproline (Thiazolidine-4-Carboxylic Acid)

As explained above, *E. histolytica* trophozoites require high levels of extracellular L-cysteine for its growth and survival. However, intracellular L-cysteine must be maintained at relatively low levels due to the toxicity of L-cysteine attributable to the highly reactive thiol group of L-cysteine, which easily reacts with electrophilic compounds [80]. Therefore, the intracellular concentrations of L-cysteine need to be maintained in an adequate range. The previous metabolomic study using stable isotope labeled L-cysteine revealed that extracellular L-cysteine in *E. histolytica* was metabolized into three unknown metabolites that were neither cystine nor alanine [81]. These metabolites were identified as thiazolidine-4-carboxylic acid (T4C), 2-methyl thiazolidine-4-carboxylic acid, and 2-ethyl-thiazolidine-4-carboxylic acid, the condensation products of L-cysteine with aldehydes [81]. It has also been reported that a significant amount of L-cysteine was oxidized to L-cystine and L-cystine was incorporated into the cell [81], which was then converted back to L-cysteine intracellularly by NADPH-dependent oxidoreductase (EhNO2) [30]. The synthesis of 2-(*R*)-thiazolidine-4-carboxylic acid derivatives in *E. histolytica* allows regulation of the intracellular L-cysteine levels and also functions as a mechanism for the detoxification of metabolically produced aldehydes in the cell. T4C is also called as thioproline because its molecular structure is similar to proline [82]. T4C scavenges free radicals and thus plays a vital role in protecting membranes and other oxidation-prone structures in the cell from oxidative damage [83]. In *E. histolytica*, it has been reported that these thiazolidine derivatives, T4C in particular, can decrease the intracellular ROS levels and thereby protect the parasite against oxidative stress [81].

11.8 Redox Pathways as Targets for Drug Development Against Amebiasis

Metronidazole, a 5-nitroimidazole drug, has been used over several decades in the treatment of infections caused by microaerophilic protozoan parasites, including *E. histolytica* [84]. However, metronidazole has low efficiency against asymptomatic cyst carriers and may also cause some unwanted effects, such as nausea, vomiting, dizziness, headache, insomnia, drowsiness, and hypersensitivity reactions [85]. In addition, *E. histolytica* trophozoites can tolerate sub-therapeutic levels of metronidazole and gain resistance in vitro [86, 87]. It has been demonstrated that resistance to metronidazole is associated with increased expression of antioxidative enzymes such as SOD and Prx and decreased expression of ferredoxin (Fdx) and p34 [86]. It was also reported that Prx, SOD, and Fdx are involved in metronidazole resistance in *E. histolytica* [86].

Redox pathways in parasites are, in general, thought to be good drug targets because, in most cases, enzymes involved in redox pathways significantly differ between the parasites and their hosts. It has been established that reduced (activated) metronidazole covalently binds to TrxR and Trx, thereby inhibiting the disulfide-reducing activity of the TrxR/Trx system, although TrxR nitroreductase activity, responsible for metronidazole activation, is not modulated by covalent adduct formation [88]. It has been proposed that the TrxR/Trx system is a promising target for antiparasitic drug development [89]. TrxR plays an important role in redox regulation of peroxiredoxin via thioredoxin (Fig. 11.2). A high-throughput screening of a panel of FDA-approved compounds for the amebicidal activity in vitro led to an identification of auranofin (an FDA-approved drug for rheumatoid arthritis). Auranofin is currently investigated in a phase I trial [90, 91]. Auranofin targets Trx reductase, thus preventing the reduction of Trx and enhancing sensitivity of trophozoites to ROS-mediated killing [90].

As partially stated above, L-cysteine biosynthetic pathway plays a crucial role in growth, survival, and pathogenicity of various pathogens including *E. histolytica*, *Trichomonas vaginalis*, *Leishmania donovani,* and *Salmonella typhimurium* [75, 92, 93]. The absence of this pathway in mammals rationalizes its potential exploitation for therapeutic and prophylactic drug targets against these pathogens. As described above, the de novo cysteine biosynthetic pathway in *E. histolytica* involves two key enzymes: SAT and CS, both promising targets against *E. histolytica* [75]. Both enzymes have been proven to be essential for growth and antioxidative defense of the parasite [75, 94]. Simultaneous gene silencing of the three CS genes inhibited growth of the *E. histolytica* trophozoites [75]. However, the role of individual CS isotypes (i.e., CS1/2 and CS3) is not understood. Similarly, the role of SAT isotypes also remains elusive because gene silencing of single SAT isotypes was successful only for SAT3, where its gene silencing showed relatively severe growth phenotype compared to SAT1/2 double gene silencing [75]. It is predicted but remains to be shown if simultaneous repression of all three SAT isotypes causes abolished growth. The crystal structures of representative SAT1 and CS1 isotypes have been previously determined [95, 96].

Several studies have been conducted to identify inhibitors targeting the cysteine biosynthetic pathway. In silico screening that was carried out using *Escherichia coli* SAT against the National Cancer Institute Chemical Database (https://cactus.nci.nih.gov/) yielded 11 compounds, out of which three available compounds were examined for biological activity. These three compounds (NCI-128884; NCI-29607; NCI-653543) were shown to partially block the enzymatic activity of *E. coli* SAT and also inhibit the growth of *E. histolytica* trophozoites, without affecting that of mammalian cells [97]. In silico screening of CS inhibitors from natural compounds in ZINC database (http://zinc.docking.org/) based on the available CS1 crystal structure identified 15 compounds. Furthermore, in vitro screening of commercially available compounds for CS inhibitors identified one compound (ZINC08931589) that inhibited CS activity by 73% at 100 μM [98]. Microbe-derived natural compounds from Kitasato Natural Products Library were recently screened against the recombinant *E. histolytica* CS1 and CS3 [97]. Nine compounds inhibited either one or both enzymes with the IC_{50} values of 0.31–490 μM. Among them, seven compounds share a naphthoquinone moiety, suggesting the importance of this moiety for binding to the active site of *E. histolytica* CS1 and EhCS3. More than 9000 fungal and actinomycete culture broth extracts were also screened against recombinant *E. histolytica* CS1 and CS3, and 2 compounds, xanthofulvin (which inhibited CS1 and CS3 with the IC_{50} values of 7.9–110 μM) and exophilic from fungal broths (which inhibited CS1 with the IC_{50} values of 24 μM but showed no inhibition against CS3 even at 2.5 mM), were identified [99]. Altogether, screening for SAT and CS inhibitors can potentially lead to a discovery of new anti-amebic compounds with a novel mode of action.

11.9 Concluding Remarks

Antioxidative stress management coupled with the redox system plays an important role in a wide range of biochemical processes, and a panel of many enzymes is indispensable in redox regulation in *E. histolytica*. Since many of oxidative stress-induced genes are annotated as hypothetical genes with no predictable functions [16], it is needed to develop robust genetic, molecular biological, or biochemical methods that allow us elucidate what roles these modulated hypothetical proteins play in response to oxidative stress. Thus, despite its importance, the redox-coupled antioxidative defense system has only started to be unveiled. For instance, although Trx/Trx reductase/Prx and FDP play important roles in the redox regulation, only one or two Trxs and FDPs each of multiple isotypes (e.g., 22 Trxs and 4 FDPs) have been studied. Other uncharacterized Trx plays an important role in oxidative stress defense or regulation of the activities of other redox-regulated proteins. Similarly, although *E. histolytica* genome encodes 7 isotypes of ISFs, detailed biochemical and functional analyses need to be done for most of them. Lastly, it is also important to identify and characterize novel antioxidative defense mechanisms unique to *E. histolytica* which are absent in humans in order to exploit as new drug targets.

References

1. Haque R, Huston CD, Hughes M, Houpt E, Petri W (2003) Amebiasis. N Engl J Med 348:1565–1573
2. Lozano R, Naghavi M, Foreman K, Lim S, Shibuya K, Aboyans V et al (2012) Global and regional mortality from 235 causes of death for 20 age groups in 1990 and 2010: a systematic analysis for the global burden of disease study 2010. Lancet 380:2095–2128
3. Ghasemi E, Rahdar M, Rostami M (2015) Prevalence of Entamoeba histolytica/dispar in drinking water in the city of shush, Khuzestan Province in 2011. Int J Curr Microbiol App Sci 4:582–588

4. Jeelani G, Sato D, Nozaki T (2015) Metabolomic analysis of Entamoeba biology. In: Nozaki T, Bhattacharya A (eds) Amebiasis: biology and pathogenesis of Entamoeba. Springer, Tokyo, pp 331–349
5. Ralston KS, Petri WA Jr (2011) Tissue destruction and invasion by *Entamoeba histolytica*. Trends Parasitol 27:254–263
6. Stanley SL Jr (2003) Amoebiasis Lancet 361:1025–1034
7. Nagaraja S, Ankri S (2018) Utilization of different omic approaches to unravel stress response mechanisms in the parasite *Entamoeba histolytica*. Front Cell Infect Microbiol 8:19
8. Pineda E, Perdomo D (2017) *Entamoeba histolytica* under oxidative stress: what countermeasure mechanisms are in place? Cell 6:44
9. Begum S, Quach J, Chadee K (2015) Immune evasion mechanisms of *Entamoeba histolytica*: progression to disease. Front Microbiol 6:1394
10. Santi-Rocca J, Smith S, Weber C, Pineda E, Hon CC, Saavedra E, Olivos-Garcia A, Rousseau S, Dillies MA, Coppee JY, Guillen N (2012) Endoplasmic reticulum stress-sensing mechanism is activated in *Entamoeba histolytica* upon treatment with nitric oxide. PLoS One 7:e31777
11. Nakada-Tsukui K, Nozaki T (2016) Immune response of amebiasis and immune evasion by *Entamoeba histolytica*. Front Immunol 7:175
12. Rastew E, Vicente JB, Singh U (2012) Oxidative stress resistance genes contribute to the pathogenic potential of the anaerobic protozoan parasite, *Entamoeba histolytica*. Int J Parasitol 42:1007–1015
13. MacFarlane RC, Singh U (2006) Identification of differentially expressed genes in virulent and nonvirulent Entamoeba species: potential implications for amebic pathogenesis. Infect Immun 74:340–351
14. Ramos-Martinez E, Olivos-Garcia A, Saavedra E, Nequiz M, Sanchez EC, Tello E, El-Hafidi M, Saralegui A, Pineda E, Delgado J, Montfort I, Perez-Tamayo R (2009) *Entamoeba histolytica*: oxygen resistance and virulcncc. Int J Parasitol 39:693–702
15. Akbar MA, Chatterjee NS, Sen P, Debnath A, Pal A, Bera T, Das P (2004) Genes induced by a high-oxygen environment in *Entamoeba histolytica*. Mol Biochem Parasitol 133:187–196
16. Vicente JB, Ehrenkaufer GM, Saraiva LM, Teixeira M, Singh U (2009) *Entamoeba histolytica* modulates a complex repertoire of novel genes in response to oxidative and nitrosative stresses: implications for amebic pathogenesis. Cell Microbiol 11:51–69
17. Biller L, Schmidt H, Krause E, Gelhaus C, Matthiesen J, Handal G, Lotter H, Janssen O, Tannich E, Bruchhaus I (2009) Comparison of two genetically related *Entamoeba histolytica* cell lines derived from the same isolate with different pathogenic properties. Proteomics 9:4107–4120
18. Band RN, Cirrito H (1979) Growth response of axenic *Entamoeba histolytica* to hydrogen, carbon dioxide, and oxygen. J Protozool 26:282–286
19. Avery SV (2011) Molecular targets of oxidative stress. Biochem J 434:201–210
20. Loftus B, Anderson I, Davies R, Alsmark UC, Samuelson J et al (2005) The genome of the protist parasite *Entamoeba histolytica*. Nature 433:865–868
21. Makiuchi T, Mi-ichi F, Nozaki T (2015) Mitosomes in *Entamoeba histolytica*. In: Nozaki T, Bhattacharya A (eds) Amebiasis: biology and pathogenesis of Entamoeba. Springer, Tokyo, pp 305–327
22. Jeelani G, Nozaki T (2016) Entamoeba thiol-based redox metabolism: a potential target for drug development. Mol Biochem Parasitol 206:39–45
23. Fahey RC, Newton GL, Arrick B, Overdank-Bogart T, Aley SB (1984) *Entamoeba histolytica*: a eukaryote without glutathione metabolism. Science 224:70–72
24. Arias DG, Gutierrez CE, Iglesias AA, Guerrero SA (2007) Thioredoxin-linked metabolism in *Entamoeba histolytica*. Free Radic Biol Med 42:1496–1505
25. Arias DG, Regner EL, Iglesias AA, Guerrero SA (2012) *Entamoeba histolytica* thioredoxin reductase: molecular and functional characterization of its atypical properties. Biochim Biophys Acta 1820:1859–1866
26. Bruchhaus I, Richter S, Tannich E (1997) Removal of hydrogen peroxide by the 29 kDa protein of *Entamoeba histolytica*. Biochem J 326:785–789

27. Maralikova B, Ali V, Nakada-Tsukui K, Nozaki T, van der Giezen M, Henze K, Tovar J (2010) Bacterial-type oxygen detoxification and iron–sulfur cluster assembly in amoebal relict mitochondria. Cell Microbiol 12:331–342
28. Vicente JB, Tran V, Pinto L, Teixeira M, Singh U (2012) A detoxifying oxygen reductase in the anaerobic protozoan *Entamoeba histolytica*. Eukaryot Cell 11:1112–1118
29. Bruchhaus I, Richter S, Tannich E (1998) Recombinant expression and biochemical characterization of an NADPH:flavin oxidoreductase from *Entamoeba histolytica*. Biochem J 330:1217–1221
30. Jeelani G, Husain A, Sato D, Ali V, Suematsu M, Soga T, Nozaki T (2010) Two atypical L-cysteine-regulated NADPH-dependent oxidoreductases involved in redox maintenance, L-cystine and iron reduction, and metronidazole activation in the enteric protozoan Entamoeba histolytica. J Biol Chem 285:26889–26899
31. Andersson JO, Hirt RP, Foster PG, Roger AJ (2006) Evolution of four gene families with patchy phylogenetic distributions: influx of genes into protist genomes. BMC Evol Biol 6:27
32. Bruchhaus I, Brattig NW, Tannich E (1992) Recombinant expression, purification and biochemical characterization of a superoxide dismutase from Entamoeba histolytica. Arch Med Res 23:27–29
33. Husain A, Jeelani G, Sato D, Nozaki T (2011) Global analysis of gene expression in response to L-cysteine deprivation in the anaerobic protozoan parasite Entamoeba histolytica. BMC Genomics 12:275
34. Sim S, Yong TS, Park SJ, Im KI, Kong Y, Ryu JS, Min DY, Shin MH (2005) NADPH oxidase-derived reactive oxygen species-mediated activation of ERK1/2 is required for apoptosis of human neutrophils induced by Entamoeba histolytica. J Immunol 174:4279–4288
35. Reeves RE (1984) Metabolism of *Entamoeba histolytica* Schaudinn, 1903. Adv Parasitol 23:105–142
36. Imlay JA (2006) Iron-Sulphur clusters and the problem with oxygen. Mol Microbiol 59:1073–1082
37. Husain A, Sato D, Jeelani G, Soga T, Nozaki T (2012) Dramatic increase in glycerol biosynthesis upon oxidative stress in the anaerobic protozoan parasite *Entamoeba histolytica*. PLoS Negl Trop Dis 6:e1831
38. Varet H, Shaulov Y, Sismeiro O, Trebicz-Geffen M, Legendre R, Coppée JY, Ankri S, Guillen N (2018) Enteric bacteria boost defences against oxidative stress in Entamoeba histolytica. Sci Rep 8:9042
39. Begum S, Quach J, Chadee K (2015) Immune evasion mechanisms of *Entamoeba histolytica*: progression to disease. Front Microbiol 6:1394
40. Radi R (2013) Peroxynitrite, a stealthy biological oxidant. J Biol Chem 288:26464–26472
41. Ramos E, Olivos-Garcia A, Nequiz M, Saavedra E, Tello E, Saralegui A, Montfort I, Perez Tamayo R (2007) *Entamoeba histolytica*: apoptosis induced in vitro by nitric oxide species. Exp Parasitol 116:257–265
42. Hertz R, Tovy A, Kirschenbaum M, Geffen M, Nozaki T, Adir N et al (2014) The Entamoeba histolytica Dnmt2 homolog (Ehmeth) confers resistance to nitrosative stress. Eukaryot Cell 13:494–503
43. Krauth-Siegel RL, Leroux AE (2012) Low molecular mass antioxidants in parasites. Antioxid Redox Signal 17:583–607
44. Pearson RJ, Morf L, Singh U (2013) Regulation of H2O2 stress-responsive genes through a novel transcription factor in the protozoan pathogen *Entamoeba histolytica*. J Biol Chem 288:4462–4474
45. Lin JY, Chadee K (1992) Macrophage cytotoxicity against Entamoeba histolytica trophozoites is mediated by nitric oxide from L-arginine. J Immunol 148:3999–4005
46. Shahi P, Trebicz-Geffen M, Nagaraja S, Hertz R, Baumel-Alterzon S, Methling K, Lalk M, Mazumder M, Samudrala G, Ankri S (2017) Corrigendum: N-acetyl ornithine deacetylase is a moonlighting protein and is involved in the adaptation of *Entamoeba histolytica* to nitrosative stress. Sci Rep 7:45802

47. Hertz R, Ben Lulu S, Shahi P, Trebicz-Geffen M, Benhar M, Ankri S (2014) Proteomic identification of S-nitrosylated proteins in the parasite Entamoeba histolytica by resin-assisted capture: insights into the regulation of the Gal/GalNAc lectin by nitric oxide. PLoS One 9:e91518
48. Trebicz-Geffen M, Shahi P, Nagaraja S, Vanunu S, Manor S, Avrahami A, Ankri S (2017) Identification of S-Nitrosylated (SNO) proteins in *Entamoeba histolytica* adapted to nitrosative stress: insights into the role of SNO actin and *in vitro* virulence. Front Cell Infect Microbiol 7:192
49. Jeelani G, Husain A, Sato D, Soga T, Suematsu M, Nozaki T (2013) Biochemical and functional characterization of novel NADH kinase in the enteric protozoan parasite Entamoeba histolytica. Biochimie 95:309–319
50. Lillig CH, Holmgren A (2007) Thioredoxin and related molecules from biology to health and disease. Antioxid Redox Signal 9:25–47
51. Mehlotra RK (1996) Antioxidant defense mechanisms in parasitic protozoa. Crit Rev Microbiol 22:295–314
52. Michelet L, Zaffagnini M, Lemaire SD (2009) Thioredoxins and related proteins. In: Stern DB (ed) The chlamydomonas sourcebook: organellar and metabolic processes, 2nd edn. Academic Press, Kidlington, pp 401–443
53. Holmgren A, Soderberg BO, Eklund H, Branden CI (1975) Three dimensional structure of Escherichia coli thioredoxin-S2 to 2.8 a resolution. Proc Natl Acad Sci U S A 72:2305–2309
54. Duchene M (2015) Metronidazole and the redox biochemistry of Entamoeba histolytica. In: Nozaki T, Bhattacharya A (eds) Amebiasis: biology and pathogenesis of Entamoeba. Springer, Tokyo, pp 523–541
55. Arias DG, Carranza PG, Lujan HD, Iglesias AA, Guerrero SA (2008) Immunolocalization and enzymatic functional characterization of the thioredoxin system in Entamoeba histolytica. Free Radic Biol Med 45:32–39
56. Martínez LI, Piattoni CV, Garay SA, Rodrígues DE, Guerrero SA, Iglesias AA (2011) Redox regulation of UDP-glucose pyrophosphorylase from Entamoeba histolytica. Biochimie 93:260–268
57. Cheng Z, Arscott LD, Ballou DP, Williams CH Jr (2007) The relationship of the redox potentials of thioredoxin and thioredoxin reductase from Drosophila melanogaster to the enzymatic mechanism: reduced thioredoxin is the reductant of glutathione in Drosophila. Biochemistry 46:7875–7885
58. Alkhalfioui F, Renard M, Montrichard F (2007) Unique properties of NADP-thioredoxin reductase C in legumes. J Exp Bot 58:969–978
59. Tachibana H, Cheng XJ (2000) Entamoeba dispar: cloning and characterization of peroxiredoxin genes. Exp Parasitol 94:51–55
60. Cheng XJ, Yoshihara E, Takeuchi T, Tachibana H (2004) Molecular characterization of peroxiredoxin from Entamoeba moshkovskii and a comparison with Entamoeba histolytica. Mol Biochem Parasitol 138:195–203
61. Choi MH, Sajed D, Poole L, Hirata K, Herdman S, Torian BE, Reed SL (2005) An unusual surface peroxiredoxin protects invasive Entamoeba histolytica from oxidant attack. Mol Biochem Parasitol 143:80–89
62. deMare F, Kurtz DM, Nordlund P (1996) The structure of *Desulfovibrio vulgaris* rubrerythrin reveals a unique combination of rubredoxin-like FeS4 and ferritin-like diiron domains. Nat Struct Biol 3:539–546
63. Maralikova B, Ali V, Nakada-Tsukui K, Nozaki T, van der Giezen M, Henze K, Tovar J (2010) Bacterial-type oxygen detoxification and iron-sulfur cluster assembly in amoebal relict mitochondria. Cell Microbiol 12:331–342
64. Mi-ichi F, Abu Yousuf M, Nakada-Tsukui K, Nozaki T (2009) Mitosomes in Entamoeba histolytica contain a sulfate activation pathway. Proc Natl Acad Sci U S A 106:21731–21736
65. Cabeza MS, Guerrero SA, Iglesias AA, Arias DG (2015) New enzymatic pathways for the reduction of reactive oxygen species in Entamoeba histolytica. Biochim Biophys Acta 1850:1233–1244

66. Saraiva LM, Vicente JB, Teixeira M (2004) The role of the flavodiiron proteins in microbial nitric oxide detoxification. Adv Microb Physiol 49:77–129
67. Gonçalves VL, Vicente JB, Pinto L, Romão CV, Frazão C, Sarti P, Giuffrè A, Teixeira M (2014) Flavodiiron oxygen reductase from Entamoeba histolytica: modulation of substrate preference by tyrosine 271 and lysine 53. J Biol Chem 289:28260–28270
68. Vicente JB, Tran V, Pinto L, Teixeira M, Singh U (2012) A detoxifying oxygen reductase in the anaerobic protozoan Entamoeba histolytica. Eukaryot Cell 11:1112–1118
69. Kurutas EB (2016) The importance of antioxidants which play the role in cellular response against oxidative/nitrosative stress: current state. Nutr J 15:71
70. Gillin FD, Diamond LS (1981) *Entamoeba histolytica* and *Giardia lamblia*: growth responses to reducing agents. Exp Parasitol 51:382–391
71. Nozaki T, Ali V, Tokoro M (2005) Sulfur-containing amino acid metabolism in parasitic protozoa. Adv Parasitol 60:1–99
72. Jeelani G, Nozaki T (2014) Metabolomic analysis of Entamoeba: applications and implications. Curr Opin Microbiol 20:118–124
73. Byrne CR, Monroe RS, Ward KA, Kredich NM (1988) DNA sequences of the cysK regions of Salmonella typhimurium and Escherichia coli and linkage of the cysK regions to ptsH. J Bacteriol 170:3150–3157
74. Saito K, Miura N, Yamazaki M, Hirano H, Murakoshi I (1992) Molecular cloning and bacterial expression of cDNA encoding a plant cysteine synthase. Proc Natl Acad Sci U S A 89:8078–8082
75. Jeelani G, Sato D, Soga T, Nozaki T (2017) Genetic, metabolomic and transcriptomic analyses of the de novo L-cysteine biosynthetic pathway in the enteric protozoan parasite Entamoeba histolytica. Sci Rep 7:15649
76. Hussain S, Ali V, Jeelani G, Nozaki T (2009) Isoform-dependent feedback regulation of serine O-acetyltransferase isoenzymes involved in L- cysteine biosynthesis of Entamoeba histolytica. Mol Biochem Parasitol 163:39–47
77. Husain A, Sato D, Jeelani G, Mi-ichi F, Ali V, Suematsu M, Soga T, Nozaki T (2010) Metabolome analysis revealed increase in S-methylcysteine and phosphatidylisopropanolamine synthesis upon L-cysteine deprivation in the anaerobic protozoan parasite Entamoeba histolytica. J Biol Chem 285:39160–39170
78. Rebeille F, Jabrin S, Bligny R, Loizeau K, Gambonnet B, Van Wilder V, Douce R, Ravanel S (2006) Methionine catabolism in Arabidopsis cells is initiated by a gamma-cleavage process and leads to S-methylcysteine and isoleucine syntheses. Proc Natl Acad Sci U S A 103:15687–15692
79. Mae T, Ohira K, Fujiwara A (1971) Metabolism of S-methylcysteine and its sulfoxide in chinese cabbage, *Brassica pekinensis* Rupr. Plant Cell Physiol 12:881–887
80. Park S, Imlay JA (2003) High levels of intracellular cysteine promote oxidative DNA damage by driving the Fenton reaction. J Bacteriol 185:942–950
81. Jeelani G, Sato D, Soga T, Watanabe H, Nozaki T (2014) Mass spectrometric analysis of L-cysteine metabolism: physiological role and fate of L-cysteine in the enteric protozoan parasite *Entamoeba histolytica*. MBio 5:e01995–e01914
82. Elthon TE, Stewart CR (1984) Effects of the proline analogue L-thiazolidine-4-carboxylic acid on proline metabolism. Plant Physiol 74:213–218
83. Weber HU, Fleming JF, Miquel J (1982) Thiazolidine-4-carboxylic acid, a physiologic sulfhydryl antioxidant with potential value in geriatric medicine. Arch Gerontol Geriatr 1:299–310
84. Freeman CD, Klutman NE, Lamp KC (1997) Metronidazole. A therapeutic review and update. Drugs 54:679–708
85. Cudmore SL, Delgaty KL, Hayward-McClelland SF, Petrin DP, Garber GE (2004) Treatment of infections caused by metronidazole-resistant *Trichomonas vaginalis*. Clin Microbiol Rev 17:783–793
86. Wassmann C, Hellberg A, Tannich E, Bruchhaus I (1999) Metronidazole resistance in the protozoan parasite Entamoeba histolytica is associated with increased expression of iron-containing

superoxide dismutase and peroxiredoxin and decreased expression of ferredoxin 1 and flavin reductase. J Biol Chem 274:26051–26056
87. Penuliar GM, Nakada-Tsukui K, Nozaki T (2015) Phenotypic and transcriptional profiling in Entamoeba histolytica reveal costs to fitness and adaptive responses associated with metronidazole resistance. Front Microbiol 6:354
88. Leitsch D, Kolarich D, Wilson IB, Altmann F, Duchêne M (2007) Nitroimidazole action in Entamoeba histolytica: a central role for thioredoxin reductase. PLoS Biol 5:e211
89. Muller S, Liebau E, Walter RD, Krauth-Siegel RL (2003) Thiol-based redox metabolism of protozoan parasites. Trends Parasitol 19:320–328
90. Debnath A, Parsonage D, Andrade RM, He C, Cobo ER, Hirata K, Chen S, García-Rivera G, Orozco E, Martínez MB, Gunatilleke SS, Barrios AM, Arkin MR, Poole LB, McKerrow JH, Reed SL (2012) A high through put drug screen for Entamoeba histolytica identifies a new lead and target. Nat Med 18:956–960
91. Capparelli EV, Bricker-Ford R, Rogers MJ, McKerrowJH RSL (2016) Phase I clinical trail results of auranofin, a novel antiparasitic agent. Antimicrob Agents Chemother 61(1):e01947
92. Spyrakis F, Singh R, Cozzini P, Campanini B, Salsi E, Felici P, Raboni S, Benedetti P, Cruciani G, Kellogg GE, Cook PF, Mozzarelli A (2013) Isozyme-specific ligands for O-acetylserine sulfhydrylase, a novel antibiotic target. PLoS One 8:e77558
93. Williams RA, Westrop GD, Coombs GH (2009) Two pathways for cysteine biosynthesis in *Leishmania major*. Biochem J 420:451–462
94. Nozaki T, Asai T, Sanchez LB, Kobayashi S, Nakazawa M, Takeuchi T (1999) Characterization of the gene encoding serine acetyltransferase, a regulated enzyme of cysteine biosynthesis from the protist parasites Entamoeba histolytica and Entamoeba dispar. Regulation and possible function of the cysteine biosynthetic pathway in Entamoeba. J Biol Chem 274:32445–32452
95. Kumar S, Raj I, Nagpal I, Subbarao N, Gourinath S (2011) Structural and biochemical studies of serine acetyltransferase reveal why the parasite Entamoeba histolytica cannot form a cysteine synthase complex. J Biol Chem 286:12533–12541
96. Chinthalapudi K, Kumar M, Kumar S, Jain S, Alam N, Gourinath S (2008) Crystal structure of native O-acetyl-serine sulfhydrylase from Entamoeba histolytica and its complex with cysteine: structural evidence for cysteine binding and lack of interactions with serine acetyl transferase. Proteins 72:1222–1232
97. Agarwal SM, Jain R, Bhattacharya A, Azam A (2008) Inhibitors of Escherichia coli serine acetyltransferase block proliferation of Entamoeba histolytica trophozoites. Int J Parasitol 38:137–141
98. Nagpal I, Raj I, Subbarao N, Gourinath S (2012) Virtual screening, identification and in vitro testing of novel inhibitors of O-acetyl-L-serine sulfhydrylase of Entamoeba histolytica. PLoS One 7:e30305
99. Mori M, Jeelani G, Masuda Y, Sakai K, Tsukui K, Waluyo D, Tarwadi WY, Nonaka K, Matsumoto A, Ōmura S, Nozaki T, Shiomi K (2015) Identification of natural inhibitors of Entamoeba histolytica cysteine synthase from microbial secondary metabolites. Front Microbiol 6:962

Oxidative Stress in Protozoan Parasites: A Close Surveillance of Proteases and Endogenous Protease Inhibitors in Host-Parasite Interaction

12

Dibyapriya Roy Chowdhury, Pijush Kanti Pramanik, and Tapati Chakraborti

Abstract

Oxidative stress is one of the physiological conditions where the reactive oxygen species (ROS) production overcomes the capacity of antioxidant defences and causes damage to key biomolecules. The major global parasites bare some key factors involved in the pathogenesis of parasitic diseases regarding oxidative stress. In this context, parasitic proteases play central roles in host-parasite interactions. Both parasitic proteases and host proteases are deeply implicated in the clinical manifestation of parasite survival and in the host immune modulation. Additionally, the involvement of endogenous parasitic protease inhibitors in the understanding of progression of infection under oxidative stress of host cannot be ignored as well. Therefore, the present review provides some insights into some important parasitic proteases with the contribution of the versatile roles of parasitic endogenous protease inhibitors in regulation of host proteases in oxidative stress in the context of host-parasite interaction.

Keywords

Endogenous parasite protease · Endogenous parasite protease inhibitor · Host protease · Immune modulation · Oxidative stress · Reactive oxygen species

Dibyapriya Roy Chowdhury and Pijush Kanti Pramanik contributed equally with all other contributors.

D. Roy Chowdhury · P. K. Pramanik · T. Chakraborti (✉)
Department of Biochemistry and Biophysics, University of Kalyani, Kalyani, West Bengal, India
e-mail: tcbiochem@gmail.com

S. Chakraborti et al. (eds.), *Oxidative Stress in Microbial Diseases*,
https://doi.org/10.1007/978-981-13-8763-0_12

Abbreviations

CD	cluster of differentiation
COX	cyclooxygenase
ICP	inhibitor of cysteine protease
IFN	interferon
IL	interleukin
ISP	inhibitor of serine protease
MAP	mitogen-activated protein
NLR	NOD-like receptor
NO	nitric oxide
PTP	protein tyrosine phosphatase
ROS	reactive oxygen species
SOCS	suppressor of cytokine signalling
STAT	signal transducer and activator of transcription
TGF	transforming growth factor
TLR	toll-like receptor
TNF	tumour necrosis factor

12.1 Introduction

Oxidative stress is an imperative physiological manifestation with imbalance between oxidants and antioxidants, which is in favour of oxidants over antioxidants. Since long time, oxidative stress has been considered as one of the most burning topics in biological research worldwide [1]. The mere concept of oxidative stress was first formulated in 1985 in the introductory chapter of the book entitled *Oxidative Stress* [2]. Subsequently, in the next year, the review *Biochemistry of Oxidative Stress* makes the knowledge of redox biology more apparent by presenting the enormous information on pro-oxidants and antioxidants, their endogenous and exogenous sources and finally metabolic sinks between infection and host health status [3]. In these events, excessive generation of ROS plays crucial roles in the pathogenesis of various diseases followed by maintenance of cellular homeostasis. Additionally, oxidative stress due to ROS generation promotes cellular injury and tissue damage. Consequently, the cellular damages lead to immune response alteration to microbes and ultimately altered susceptibility to bacterial, viral and parasitic infections [4]. Generally, in most of the parasitic invasions, these ROS productions are of great concern in the propagation of the infection. The excess ROS in the parasitised cells depends on the nutritional status of the host and the degree of parasitic infestations. Thus, the redox system plays fundamental and indispensable roles for parasite survival within their host either via ROS generation or by inhibition of cellular antioxidant systems which will cause redox imbalance and results in parasite's death.

The relationship between oxidative stress and immune function of the parasitised host is well established. The generation of oxidative stress is believed to be an important part of immune activation. The potent immune cells capture and ingest intracellular microbes into phagolysosomes, which eventually kill intracellular pathogens by compartmentalised oxidative stress [5]. In general, parasites themselves can also be directly responsible for oxidant release through degradation products of their own metabolism.

In protozoan parasites, proteases play the crucial role in a broad range of biological functions including parasite survival in host cells, immune modulation in the mammalian host, migration through host tissues and also the promotion of infection. Identification and characterisation of protease-mediated processes in protozoan parasites are very promising [6]. In this consequence, we have dedicated ourselves to review on the parasite as well as host-derived proteases in the involvement of infectivity through oxidative stress. It is also focused on selective endogenous protease inhibitors of the key proteases of pathogenic parasites and their involvement in oxidative stress in the issue of infectivity. The role of parasitic proteases, host proteases and endogenous protease inhibitors from parasites are also addressed herein in the participation of pathogenesis as well as in the redox inequality in the parasitised host. These intracellular protease inhibitors are mostly proteinaceous in nature, present in various protozoan species as a weapon to survive within host body by inhibiting host cell proteases. Therefore, attention has also been given to identify the gaps in our current knowledge and future prediction in parasitological research.

12.2 Host Protease and Oxidative Damage

Both host and protozoan contain proteases for their own purposes. Host proteases aid in killing protozoan cells, and in opposition, protozoan parasites use their proteases to counteract host defence system. Proteases are mainly categorised into four types: (1) cysteine protease, (2) serine protease, (3) aspartyl protease and (4) metalloprotease. During this parasitic attack, host cysteine and serine proteases have been found to participate in defence system against them [7].

Generally, after protozoan infection in the body, host cysteine proteases mount hostility against the parasites by degrading protozoan antigen protein causing apoptosis of host cell to halt further parasite spread to neighbouring cells [8]. Among several cysteine proteases, only certain cathepsin enzymes like cathepsin B, cathepsin K and cathepsin L are found having connection in oxidative stress mainly induction of ROS generation, which is a special mode of action of proteases against protozoan like *Leishmania*, *Plasmodium* and *Toxoplasma* [8–10]. The cathepsins which have been found to kill protozoa are stored within lysosome and also released from there [11]. A special activity of cathepsin B is ROS-mediated activation of a cellular inflammasome named NLRP3, a multiprotein complex which detects a protozoa by its specific "pathogen-associated molecular patterns" (PAMPs) and activate pro-inflammatory cytokines. Hence, during infection of the parasites like

Leishmania, *Entamoeba* and *Plasmodium,* NLRP3 activates pro-IL-1β and pro-IL-18 inflammatory cytokines to form active IL-1β and IL-18 which induce ROS generation in host cells [12, 13].

The host serine proteases which aid in immunity against protozoa are mostly granzymes (GzM). Indeed, in host cell, a "triad" of enzymes namely perforin, granulysin and granzymes work together. During infection by *Trypanosoma cruzi*, *Leishmania major* and *Toxoplasma gondii*, at first perforin transports the other two enzymes in infected cells, and consequently granulysin carries granzyme into intracellular protozoa by disruption of parasite cell membrane. Among different GzMs, only GzMA and GzMB have been found to act on intracellular parasites where they proteolytically help in ROS generation for killing the parasites. Here, the proteolytic degradation of dehydrogenase and reductase enzymes of protozoa occurs, but no such proteolytic modification is found in thiol reductase in these events of oxidative damage [14].

In addition to granzyme, neutrophil elastase, a serine protease is secreted by neutrophils and macrophages during inflammation. The parasite uptake process is facilitated via the interaction of the complement type 3 receptor CR3 and toll-like receptor TLR4 at the surface of macrophages and initially TLR4 interacts with CD11b, a subunit of CR3. Furthermore, activation of TLR4 in macrophages occurs by neutrophil elastase which also leads to the production of ROS and consequently provokes partial elimination of intracellular parasites. Thus, neutrophil elastase is found to be involved in the intracellular killing of *Leishmania* parasites and plays a protective role in host responses to *Leishmania* infection [15, 16]. Besides, neutrophil elastase also appears to be involved in the production of NO and TNF-α by macrophages and consequently increased the trypanocidal activity [17]. However, among the remaining two other types of host proteases, aspartyl protease (e.g. cathepsin D) and metalloproteases have not been found to contribute to oxidative damage during protozoan attack in host body.

12.3 Protozoan Proteases and Oxidative Damage

Different types of cysteine, serine and metalloproteases expressed by protozoan parasites play vital roles of parasite survival and infectivity. Protozoans, which are susceptible to ROS injury, employ their proteases to devastate the ROS-generating machinery of host to survive. Alternatively, protozoans, which prefer ROS for their survival, utilise the proteases involved in ROS increment in host cells. Herein, we have also focused on the contribution of some major parasitic proteases in oxidative stress.

12.3.1 Trypanosoma

Among the different trypanosomal species from the order *Kinetoplastida*, only cysteine protease plays a fundamental role in oxidative stress of parasitised host cells.

Cruzipain2 and cruzain, the *Trypanosoma cruzi* cysteine proteases, are found to be involved in the production of ROS [18]. However, Rhodesain and TbCatB are the secreted cysteine proteases in *Trypanosoma brucei* which are not established yet in the matter of oxidative stress-related mechanisms [19]. Paradoxically, in case of *Trypanosoma*, ROS act as a protecting and promoting agent for infection [20]. Hence to stay alive, intracellular *Trypanosoma cruzi* induce macrophages to produce ROS. This ROS oxidise antioxidant enzyme peroxiredoxin 2, which relocate the oxidative equivalents from oxidised peroxiredoxin2 to STAT3 protein, and as a result, different STAT3 are joined by disulfide bond to form an oligomeric structure. Ultimately this oligomerisation prevent the activity of STAT to work as transcription factor, hence in host cell, different pathways are altered. ROS also activates CaM-kinase II like pathway, which is helpful for epimastigotes of *Trypanosoma cruzi* to stay in the host cell. Another important point is that ROS oxidise guanine base of DNA and produce 8-oxo-dGTP. It has been seen that parasites which have 8-oxo-dGTP pyrophosphohydrolase show higher rate of survival in host macrophages than other intracellular parasites. This is because 8-oxo-dGTP hydrolysis prevents any lesion formation in DNA, so that replication of DNA can be maintained at higher rate [21, 22] and thus ROS production appears as one of the supportive options for survival of *Trypanosoma cruzi*. In host Raw macrophage 264.7 cell line, cruzipain increases ROS by stimulating the host phagocytic cell by assembling different subunits of NADPH oxidase enzyme i.e. membrane bound subunits $p22^{phox}$, NOX2 and cytosolic subunits $p40^{phox}$, $p47^{phox}$, $p67^{phox}$ into phagolysosomes. This enzyme transfers electron from NADPH to oxygen, thus forming superoxide anion $O2^{\bullet-}$ [23]. As well, cruzipain also increases the ROS level via increased secretion of pro-inflammatory cytokines IL-6 and IL-1β from host cell [23].

12.3.2 Plasmodium

In *Plasmodium falciparum,* falcipain 1, falcipain 2, falcipain 3, dipeptidyl peptidases are the major cysteine proteases. Another species *Plasmodium vivax* contains three types of cysteine proteases vivapain 2, vivapain 3 and vivapain 4. Among them, only falcipain 2 has yet been proven to be involved in oxidative stress mechanisms [24–27]. In case of malaria infection, hydroxyl radical (OH^-) is induced in liver tissues and consequently provokes the apoptosis like pathway via oxidative stress. In fact, an immense source of ROS during malaria infection is heme group from haemoglobin of host body. Plasmodial falcipain 2 is involved in haemoglobin degradation resulting in release of heme, which in turn is involved in increased production of ROS [28, 29].

Plasmodium falciparum also contains a variety of serine proteases viz. SERA 1, SERA2, SERA3, SERA4, SERA5, SUB 1, SUB2, PfDegP. Among them only PfDegP saves protozoa during ROS attack, while parasites stay within RBC. This enzyme is not retained in *Plasmodium berghei* or *Plasmodium yoelii*. However, in *Plasmodium falciparum,* PfDegP favours parasite growth by lowering ROS level in host. The fundamental mechanism of this ROS lowering is not clear, but a remarkable point has been noted that PfDegP remains associated with enolase and superoxide dismutase (SOD) in parasite cells. Hence, it may be possible that PfDegP executes its supportive role in parasite survival, somehow in association with those enzymes [27, 30–32].

12.3.3 Leishmania

In *Leishmania* species, GP63 is a major surface metalloprotease is abundantly expressed [33–36]. *Leishmania* GP63 is a very powerful protease that acts as a crucial virulence factor which greatly influences macrophage physiology and host cell signalling pathways [35, 37].

The exact mechanism that *Leishmania* use to resist the toxic effects of ROS is yet not clear. But, some parasite surface molecules have been found to be imperative in protecting parasites by scavenging toxic effects of ROS or inhibit macrophage responses. Among those surface molecules, metalloprotease GP63 is the most important one. Evidence suggests that *Leishmania* GP63 acts as a vital prerequisite for the inhibition of macrophage IL-1β production within host macrophages. This impairment of cytokine secretion occurs through downregulation of ROS and inhibition of NLRP3 inflammasome function [38]. GP63 accompanied ROS inhibition is eventually linked with an impaired PKC-mediated protein phosphorylation. Impairment of the classical PKC signalling within the host macrophages is one of the major controlling strategies through which the parasite exerts its immunosuppressive effects. Evidence suggests that the activity of PKC was attenuated by the proteolytic effect of GP63 and leads to a diminished release of ROS, thus facilitating survival of *Leishmania* [37, 39].

Leishmania species contains cysteine proteases also play massive role in oxidative stress in host cell. The most common cysteine proteases in *Leishmania* species are CP and Ldccys2 [40]. CP is further categorised as CPA, CPB and CPC, but, among them, roles of only CPB have been established to be linked with oxidative stress. *Leishmania tropica* and *Leishmania mexicana* contain cysteine protease CPB2.8, a subtype of CPB. For survival of protozoan cells, CPB 2.8 increase IL-4 level in host cell and also induce Th2 response, which in turn reduce ROS generation from host macrophages thus favouring *Leishmania* survival in host cells [41, 42]. In case of *Leishmania mexicana*, among different cysteine proteases, CPB is the most studied enzyme related to lowering oxidative damage in host [43]. *Leishmania mexicana*-derived CPB is responsible for lowering of IFN-γ production and hence reduce ROS attack for their survival. Generally, CPB cleaves pro-TGF-β and activates it. This activation of TGF-β eventually prevents secretion of IFN-γ via suppressing ROS production and facilitates parasite survival in macrophages [44]. Further, CPB intensifies secretion of various leishmanial enzymes such as trypanothione reductase, tryparedoxin and tryparedoxin peroxidase. Those enzymes nullify H2O2 produced in host macrophages and in turn help in survival of parasite in host macrophages. Thus, CPB can lower down ROS also by enhancing those enzymes [45–47]. Prevention of ROS has another effect of saving the cysteine residue of active site of PTP enzyme, which in turn prevents the apoptotic pathway mediated by MAP kinase, so that parasites retain in host macrophages [48].

The intracellular serine protease of *Leishmania donovani* (LdSP) is responsible for lowering ROS production in host macrophage cells. Inhibition of LdSP with anti LdSP antibody and classical serine protease inhibitor aprotinin, the ROS level were increased in host macrophages along with increment of inflammatory cytokines [49]. LdSP has been shown to induce expression of COX-2, which in turn elevate prostaglandin E2. Being an immunosuppressive agent, prostaglandin E2 lowers down inflammatory cytokine production as well as ROS generation from host macrophages resulting parasite survival [50].

12.3.4 Toxoplasma

Toxoplasma gondii, a ubiquitous apicomplexan parasite has cysteine protease which is similar to mammalian cathepsin L, cathepsin B and cathepsin C. Those proteases are termed as TgCPL, TgCPB, TgCPC1, TgCPC2, TgCPC3 [51]. Though these proteins are homolog to cathepsin, but their contribution in oxidative stress in parasite biology has not yet been clearly defined.

Toxoplasma contains serine protease named TgDegp which is associated with Iron superoxide dismutase and enolase. During oxidative damage, this complex is induced and confers protection, but the specific mechanisms are still under debate [52].

12.4 Protozoan Endogenous Protease Inhibitor in Oxidative Stress

A special weapon which has been evolved by protozoan parasites to counteract host protease action on them is endogenous protease inhibitor. Nevertheless, versatility exists in endogenous protease inhibitors of protozoa and their target proteases. *Entamoeba histolytica* contain protease inhibitor EhICP against protozoan cysteine protease [53]. Some types of protozoa contain endogenous protease inhibitor against both host and parasite protease, e.g. *Plasmodium* [24]. Lastly, some other types possess protease inhibitors which exclusively have activity only against host cell protease, e.g. *Leishmania* [10]. In synopsis, several protozoan species contain proteases as well as endogenous protease inhibitors which are important for their infectivity in host cells.

Protozoan cells have endogenous cysteine protease inhibitors against cathepsins. Endogenous inhibitors of cysteine proteases, abridged as ICP, are distributed in protozoan cells [54, 55]. Endogenous inhibitors, which are specific for parasitic protease, ultimately regulate protease activity for the survival of parasites themselves in host cell [56, 57]. Hitherto, 18 protozoan species have been documented to contain protease inhibitors [58]. Most of the protozoan cysteine protease inhibitors bear homology to structure and sequence to trypanosomal endogenous protease inhibitor chagasin, which confers a similarity between modes of action of the protozoan cysteine protease inhibitors [59].

Likewise, endogenous serine protease inhibitors are generally designated as ISP. Protozoan like *Toxoplasma gondii* [60] and *Leishmania* contain endogenous serine protease inhibitor, e.g. TgPI and LdISP, respectively [61, 62].

Involvement of protozoan endogenous protease inhibitor in oxidative stress is not well explored and also contains less number of reports in parasites regarding the aforesaid topics. Still, to make an insight on endogenous protease inhibitors concerning the oxidative stress, a few reports on some protozoan species, viz. *Trypanosoma*, *Plasmodium*, *Leishmania* and *Toxoplasma gondii*, have been discussed in this section. Distribution of protease inhibitors in different species and their mode of action are illustrated below:

12.4.1 Trypanosoma

Trypanosoma cruzi contains 110 amino acid long endogenous cysteine protease inhibitor viz. chagasin [63] which is the first member discovered within family of cysteine protease inhibitors. Chagasin can bind to host protease cathepsin B, cathepsin L and protozoan protease cruzipain, but the binding affinity with cathepsin L and cathepsin B is different [64]. As mentioned previously, cathepsin is responsible for ROS-dependent NLRP3 inflammasome formation; thus prevention of cathepsin by chagasin lowers down NLRP3 formation, and ultimately, there is a check in the production of ROS by IL-1β and IL-18 in host cells. Indeed, the binding of endogenous protease inhibitor chagasin and target cysteine proteases is dependent on the complementary binding between them [8, 59, 63, 64]. In trypomastigotes, chagasin resides in flagellar pocket as well as vesicle of cytoplasm, and in amastigote, it exists in cell surface. Cruzipain is also located similarly at cell surface of amastigotes. As a result, binding between cruzipain and chagasin occurs easily in case of amastigotes but slighter amount in trypomastigotes [65]. Consequently, in amastigotes cruzipain mediated oxidative stress is controlled by chagasin. Hence, the earlier description of ROS as a defence mechanism in *Trypanosoma cruzi* might be a resultant effect of this interaction of chagasin with cruzipain in this event [21]. Therefore, it may arise a question that chagasin may exert a negative role for survival of parasites. On the other hand, inflammasome NLRP3, which is induced by cathepsin, is responsible for killing of *Trypanosoma cruzi* in macrophages by production of NO rather than ROS [66]. NO combines with ROS to produce peroxynitrile which kills *Trypanosoma cruzi* [67]. Thus, blocking of cathepsin by chagasin assists *Trypanosoma cruzi* for their survival. *Trypanosoma brucei*, on other hand has only ICP which is mainly important during growth and differentiation [68].

12.4.2 Plasmodium

The plasmodial species like *Plasmodium falciparum*, *Plasmodium berghei*, *Plasmodium yoelii* contain endogenous cysteine protease inhibitors ICPs [24, 25, 69]. Plasmodial ICPs bear a structural similarity to chagasin i.e. endogenous protease inhibitor in those plasmodial species can inhibit both host cysteine protease as well as protozoan cysteine protease like chagasin [24, 69]. ICPs, distributed in different species are designated as falstatin or PfICP in *Plasmodium falciparum*, PbICP in *Plasmodium berghei* [9, 24], PyICP in *Plasmodium yoelii* [69].

All the plasmodial ICPs inhibit target host proteases in almost same pattern, for example, PbICP uses its C-terminal region to block host cysteine protease by complementary binding, and in other cases, cathepsin L is restrained by falstatin as similar to action of chagasin [25].

PfICP or falstatin also helps to establish infectivity of protozoa in host body. Falstatin, present in merozoites of *Plasmodium falciparum* [24], blocks host proteases named cathepsin K and cathepsin L. In addition to this, protozoan cysteine protease namely falcipain 2 and falcipain 3 are also inhibited by falstatin [70]. As mentioned earlier, falcipain 2 is associated with oxidative damage by heme, released by proteolysis of haemoglobin [27, 28]. Hence, inhibition of falcipain 2 by falstatin helps in protozoan infectivity [27, 70]. On the other hand, PbICP has no such important role in oxidative stress but it interferes with the host cell apoptosis machinery through the blockage of host cell cysteine proteases [24].

12.4.3 Leishmania

Among various *Leishmania* species, *Leishmania mexicana* contains cysteine protease inhibitor [71], named as LmICP [9]. It acts on proteases in a fashion similar to chagasin [54]. Besteiro et al. proved that two different protozoan cysteine protease viz. CPA and CPB, and host cysteine protease cathepsin L, are target of LmICP of *Leishmania mexicana*. LmICP are also able to block cathepsin B, though this binding affinity is lesser than cathepsin L [10]. But a paradox exists in case of LmICP which helps to increase ROS and thus [71] it cannot abet any benefit to survival of the parasites [68].

In *Leishmania donovani*, a well-known serine protease inhibitor named as LdISP [61] was isolated from Indian strain of *Leishmania donovani* (MHOM/IN/1983/AG83) in our laboratory and found to inhibit the protease named trypsin and neutrophil elastase in host. This serine protease inhibitor, named LdISP belongs to the type of protease inhibitor "ecotin". This ecotin type of endogenous serine protease inhibitor is also present at *Leishmania major*, *Leishmania braziliensis* and *Leishmania infantum.* In *Leishmania donovani* LdISP is located in near pocket of flagella in promastigotes [Fig. 12.1]. LdISP acts as a vital agent in progression of infection in infected macrophages. LdISP causes the diminished level of both ROS and NO in *Leishmania donovani* infected macrophages [Fig. 12.2], and the treatment of infected macrophages with LdISP helps in the promotion of infection with marginal

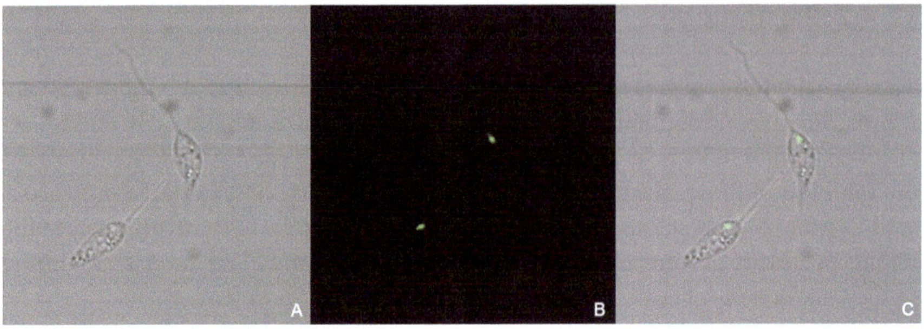

Fig. 12.1 Localisation of LdISP in *Leishmania donovani* promastigotes using FITC-conjugated anti-LdISP Ab. FITC signalling indicates that LdISP is localised near flagellar pocket. (**a**) cell without any stain; (**b**) FITC channel; (**c**) Merged

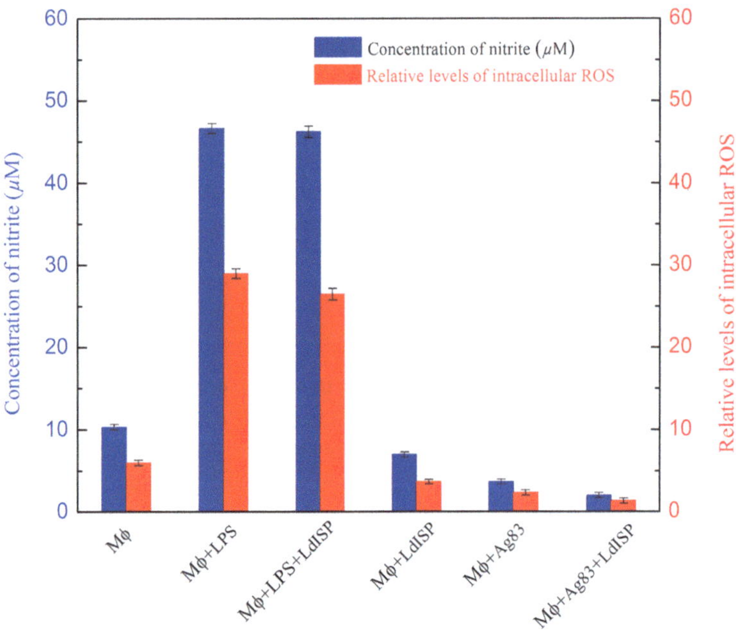

Fig. 12.2 Nitric oxide/ROS level was measured by Griess reagent and H_2DCFDA, respectively, in infected as well as LdISP (60 ng/ml) treated RAW 264.7 cells. LPS (10 ng/ml) was used as positive control. Data represented are significant compared with infected control

increase in parasite load inside macrophages by increasing in granularity and size of the host cells [Fig. 12.3]. For the validation of the role of LdISP in the promotion of infection by lowering ROS within the host macrophages, the known *Leishmania major* endogenous serine protease inhibitor ECOT2_LEIIN was compared with LdISP by docking with human neutrophil elastase (HNE). As LdISP have significant sequence resemblance with ECOT2_LEIIN, the docking of ECOT2_LEIIN

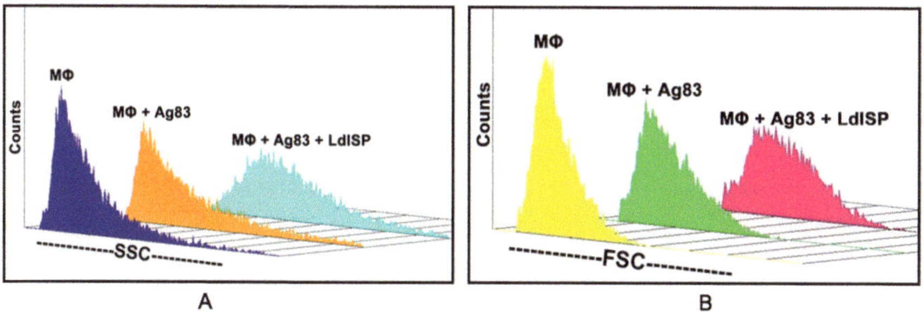

Fig. 12.3 Cytogram depicting increase in parasite load. (**a**) Unstained infected macrophages. (**b**) PE-stained (anti-CD11b Ab) infected macrophages

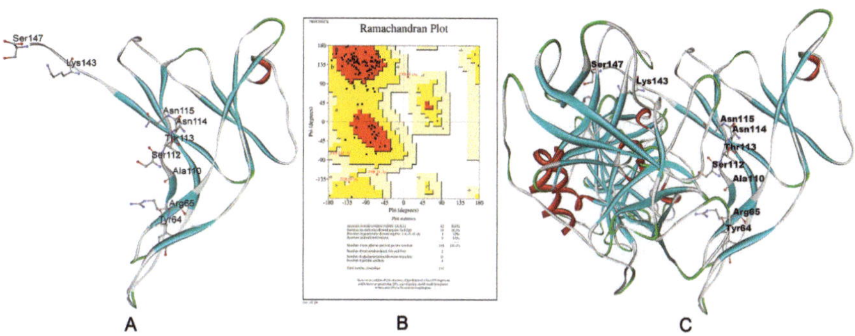

Fig. 12.4 In silico prediction of amino acids present in the inhibitory cleft. (**a**) Optimised modelled structure of ECOT2_LEIIN using MODELLER. (**b**) Validation result of the modelled structure using Ramachandran plot as obtained from PROCHECK analysis. The generated model showed that 96.5% residues are in the allowed region with no amino acid in the disallowed region, which indicates a good stereo chemical fitness of the model. (**c**) Docking of neutrophil elastase with ECOT2_LEIIN

with HNE depicted that the amino acid residues (Tyr64, Arg65, Ala110, Ser112, Thr113, Asn114, Asn115, Lys143, Ser147) of ECOT2_LEIIN significantly involved in the inhibitory interaction with HNE which is shown in Fig. 12.4. The amino acid residues from the derived comparative inhibitory cleft structure are amusingly matched with the fragmented sequences of LdISP. In the Inhibitory interaction of LdISP with HNE, the comparison of the amino acid residues resulting from the protein mass fingerprinting (red) and docking results (green) has been illustrated in Fig. 12.5. Thus, LdISP seems a very potent endogenous inhibitor which curbs the host's antimicrobial efficacy by lowering oxidative stress during intracellular survival of *Leishmania donovani* [61, 72].

In *Leishmania major*, they have endogenous serine protease inhibitor named ISP1 and ISP2. Only ISP2 blocks neutrophil elastase activity by negatively regulating the TLR4-neutrophil elastase signalling pathway, thus ultimately curtailing ROS production in host macrophage [73].

1 10	11 20	21 30	31 40	41 50	51 60
MPAGMSGAAG	KTLANFKAPY	PEPTSQQRRY	VIFLDPKGYR	KELNDYKVEL	IPGRVEKVDG

61 70	71 80	81 90	91 100	101 110	111 120
ANVYRMGGSI	EERTIDGWGY	TYYVVTLTTM	SGTLMMPLGD	AALKRPRFVA	MSTNNLYRYN
Tyr64, Arg65				Ala110	Ser112, Thr113, Asn114, Asn115

121 130	131 140	141 150	151 160
SQLPIVVYTP	KDGELRYRIW	TVKSTGSGTA	KSINAREM
		Lys143, Ser147	

Fig. 12.5 Sequence of ECOT2_LEIIN (black), super imposed with fragments of LdISP sequence (red). The amino acid residues present in the inhibitory cleft (green) derived from in silico docking with HNE (Human Neutrophil Elastase)

12.4.4 Toxoplasma

Cysteine protease inhibitors, present in *Toxoplasma* are Toxostatin 1 and Toxostatin 2. These inhibitors are responsible for binding host cell cathepsin enzyme. Toxostatin also have been reported to bind protozoan protease viz. cathepsin B like TgCPB and cathepsin L like TgCPL. Similar to plasmodial ICP, binding to own cysteine protease is liable for protozoan growth regulation. But, Toxostatin is different from other cysteine protease inhibitors in lacking any homology to chagasin [58].

In *Toxoplasma gondii*, the presence of different classes of endogenous serine protease inhibitors such as kazal, serpin etc. binds to host protease. Among those inhibitors, a kazal type inhibitor TgPI2 and TgPI1 has been found to inhibit host serine protease trypsin [70]. Though, this has illustrated the significant inhibitory roles of TgPI2 and TgPI1 in oxidative stress regulation, but the detailed mechanisms in this scenario are still under debate and will need more in depth information.

12.5 Conclusion and Future Perspective

Oxidative stress is the collection of damages caused by ROS which kills protozoa viz. *Leishmania*, *Plasmodium*, *Toxoplasma*. In contrast, ROS helps in parasite persistence of *Trypanosoma* within the host. For the survival, protozoa which are susceptible to ROS tend to counteract oxidative damages, thus they have also employed mechanisms to halt inflammatory cytokine secretion and prompt anti-inflammatory cytokine secretion. At the same time to counterbalance protease action on them; they induce an intracellular protease inhibitor. Thus, host cell protease and protozoan protease inhibitor set up equilibrium in between them. On the whole, excessive formation of ROS in state of oxidative stress plays an important role in the pathogenesis of the parasitic infection and parasitic proteases and endogenous protease inhibitors are genuinely involved in the issue of oxidative stress. Therefore, the present article might bring a new track in future for better understanding of the biology of host–parasite interactions.

Acknowledgements The authors thank the Department of Biotechnology, Government of West Bengal [Grant No: 75 (sanc) BTP/Budget/RD 38/2017], and the Indian Council of Medical Research, Government of India [Grant number: Fellowship/21/2018-ECD-II].

Conflict of Interest The authors declare no conflict of interest.

References

1. Breitenbach M, Eckl P (2015) Introduction to oxidative stress in biomedical and biological research. Biomol Ther 5(2):1169–1177
2. Sies H (1985) Introductory remarks. In: Sies H (ed) Oxidative stress. Academic Press, London, pp 1–8. https://doi.org/10.1016/B978-0-12-642760-8.50005-3
3. Sies H (1986) Biochemistry of oxidative stress. Angew Chem Int Ed Eng 25:1058–1071
4. Rahal A, Kumar A, Singh V et al (2014) Oxidative stress, prooxidants, and antioxidants: the interplay. Biomed Res Int 2014:761264
5. van de Crommenacker J, Richardson DS, Koltz AM et al (2012) Parasitic infection and oxidative status are associated and vary with breeding activity in the Seychelles warbler. Proc Biol Sci 279(1733):1466–1476
6. Schröder J, Noack S, Marhöfer RJ et al (2013) Identification of semicarbazones, thiosemicarbazones and triazine nitriles as inhibitors of Leishmania mexicana cysteine protease CPB. PLoS One 8(10):e77460
7. McKerrow JH, Rosenthal PJ, Swenerton R et al (2008) Development of protease inhibitors for protozoan infections. Curr Opin Infect Dis 21(6):668–672
8. Redzynia I, Ljunggren A, Abrahamson M et al (2008) Displacement of the occluding loop by the parasite protein, chagasin, results in efficient inhibition of human cathepsin B. J Biol Chem 283(33):22815–22825
9. Hansen G, Heitmann A, Witt T et al (2011) Structural basis for the regulation of cysteine-protease activity by a new class of protease inhibitors in Plasmodium. Structure 19(7):919–929
10. Besteiro S, Coombs GH, Mottram JC (2004) A potential role for ICP, a leishmanial inhibitor of cysteine peptidases, in the interaction between host and parasite. Mol Microbiol 54(5):1224–1236
11. Droga-Mazovec G, Bojic L, Petelin A et al (2008) Cysteine cathepsins trigger caspase-dependent cell death through cleavage of bid and antiapoptotic Bcl-2 homologues. J Biol Chem 283(27):19140–19150
12. Bai H, Yang B, Wu Y et al (2018) Cathepsin B links oxidative stress to the activation of NLRP3 inflammasome. Exp Cell Res 362(1):180–187
13. Gurung P, Kanneganti TD (2016) Immune responses against protozoan parasites: a focus on the emerging role of nod-like receptors. Cell Mol Life Sci 73(16):3035–3051
14. Dotiwala F, Mulik S, Polidoro RB et al (2016) Killer lymphocytes use granulysin, perforin and granzymes to kill intracellular parasites. Nat Med 22(2):210–216
15. Ribeiro-Gomes FL, Moniz-de-Souza MC, Alexandre-Moreira MS et al (2007) Neutrophils activate macrophages for intracellular killing of Leishmania major through recruitment of TLR4 by neutrophil elastase. J Immunol 179(6):3988–3994

16. Faria MS, Reis FC, Lima AP (2012) Toll-like receptors in leishmania infections: guardians or promoters? J Parasitol Res 2012:930257
17. Luna-Gomes T, Filardy AA, Rocha JD et al (2014) Neutrophils increase or reduce parasite burden in Trypanosoma cruzi-infected macrophages, depending on host strain: role of neutrophil elastase. PLoS One 9(3):e90582
18. Lima AP, dos Reis FC, Serveau C et al (2001) Cysteine protease isoforms from Trypanosoma cruzi, cruzipain 2 and cruzain, present different substrate preference and susceptibility to inhibitors. Mol Biochem Parasitol 114(1):41–52
19. Mott BT, Ferreira RS, Simeonov A et al (2010) Identification and optimization of inhibitors of Trypanosomal cysteine proteases: cruzain, rhodesain, and TbCatB. J Med Chem 53(1):52–60
20. Paiva CN, Feijó DF, Dutra FF et al (2012) Oxidative stress fuels Trypanosoma cruzi infection in mice. J Clin Invest 122(7):2531–2542
21. Goes GR, Rocha PS, Diniz AR et al (2016) Trypanosoma cruzi needs a signal provided by reactive oxygen species to infect macrophages. PLoS Negl Trop Dis 10(4):e0004555
22. Nogueira NP, Saraiva FMS, Oliveira MP et al (2017) Heme modulates Trypanosoma cruzi bioenergetics inducing mitochondrial ROS production. Free Radic Biol Med 108:183–191
23. Guiñazú N, Carrera-Silva EA, Becerra MC et al (2010) Induction of NADPH oxidase activity and reactive oxygen species production by a single Trypanosoma cruzi antigen. Int J Parasitol 40(13):1531–1538
24. Rennenberg A, Lehmann C, Heitmann A et al (2010) Exoerythrocytic Plasmodium parasites secrete a cysteine protease inhibitor involved in sporozoite invasion and capable of blocking cell death of host hepatocytes. PLoS Pathog 6(3):e1000825
25. Sundararaj S, Saxena AK, Sharma R et al (2014) Cross-talk between malarial cysteine proteases and falstatin: the BC loop as a hot-spot target. PLoS One 9(4):e93008
26. Lehmann C, Heitmann A, Mishra S et al (2014) A cysteine protease inhibitor of plasmodium berghei is essential for exo-erythrocytic development. PLoS Pathog 10(8):e1004336
27. McCoubrie JE, Miller SK, Sargeant T et al (2007) Evidence for a common role for the serine-type Plasmodium falciparum serine repeat antigen proteases: implications for vaccine and drug design. Infect Immun 75(12):5565–5574
28. Antoine T, Fisher N, Amewu R et al (2014) Rapid kill of malaria parasites by artemisinin and semi-synthetic endoperoxides involves ROS-dependent depolarization of the membrane potential. J Antimicrob Chemother 69(4):1005–1016
29. Percário S, Moreira DR, Gomes BA et al (2012) Oxidative stress in malaria. Int J Mol Sci 13(12):16346–16372
30. Sajid M, Withers-Martinez C, Blackman MJ (2000) Maturation and specificity of *Plasmodium falciparum* subtilisin-like protease-1, a malaria merozoite subtilisin-like serine protease. J Biol Chem 275(1):631–641
31. Barale JC, Blisnick T, Fujioka H et al (1999) Plasmodium falciparum subtilisin-like protease 2, a merozoite candidate for the merozoite surface protein 1-42 maturase. Proc Natl Acad Sci U S A 96(11):6445–6450
32. Sharma S, Jadli M, Singh A et al (2014) A secretory multifunctional serine protease, DegP of Plasmodium falciparum, plays an important role in thermo-oxidative stress, parasite growth and development. FEBS J 281(6):1679–1699
33. Fong D, Chang KP (1982) Surface antigenic change during differentiation of a parasitic protozoan, Leishmania mexicana: identification by monoclonal antibodies. Proc Natl Acad Sci U S A 79(23):7366–7370
34. Bouvier J, Etges RJ, Bordier C (1985) Identification and purification of membrane and soluble forms of the major surface protein of Leishmania promastigotes. J Biol Chem 260(29):15504–15509
35. Isnard A, Shio MT, Olivier M (2012) Impact of Leishmania metalloprotease GP63 on macrophage signaling. Front Cell Infect Microbiol 16(2):72
36. Chang CS, Inserra TJ, Kink JA et al (1986) Expression and size heterogeneity of a 63 kilodalton membrane glycoprotein during growth and transformation of Leishmania mexicana amazonensis. Mol Biochem Parasitol 18(2):197–210

37. Olivier M, Atayde VD, Isnard A et al (2012) Leishmania virulence factors: focus on the metalloprotease GP63. Microbes Infect 14(15):1377–1389
38. Shio MT, Christian JG, Jung JY et al (2015) PKC/ROS-mediated NLRP3 inflammasome activation is attenuated by Leishmania zinc-metalloprotease during infection. PLoS Negl Trop Dis 9(6):e0003868
39. Chawla M, Vishwakarma RA (2003) Alkylacylglycerolipid domain of GPI molecules of Leishmania is responsible for inhibition of PKC-mediated c-fos expression. J Lipid Res 44(3):594–600
40. Mundodi V, Kucknoor AS, Gedamu L (2005) Role of Leishmania (Leishmania) chagasi amastigote cysteine protease in intracellular parasite survival: studies by gene disruption and antisense mRNA inhibition. BMC Mol Biol 6:3
41. Pollock KG, McNeil KS, Mottram JC et al (2003) The Leishmania mexicana cysteine protease, CPB2.8, induces potent Th2 responses. J Immunol 170(4):1746–1753
42. Buxbaum LU, Denise H, Coombs GH et al (2003) Cysteine protease B of Leishmania mexicana inhibits host Th1 responses and protective immunity. J Immunol 171(7):3711–3717
43. Denise H, McNeil K, Brooks DR et al (2003) Expression of multiple CPB genes encoding cysteine proteases is required for Leishmania mexicana virulence in vivo. Infect Immun 71(6):3190–3195
44. Somanna A, Mundodi V, Gedamu L (2002) Functional analysis of cathepsin B-like cysteine proteases from Leishmania donovani complex. Evidence for the activation of latent transforming growth factor beta. J Biol Chem 277(28):25305–25312
45. Iyer JP, Kaprakkaden A, Choudhary ML et al (2008) Crucial role of cytosolic tryparedoxin peroxidase in Leishmania donovani survival, drug response and virulence. Mol Microbiol 68(2):372–391
46. Dumas C, Ouellette M, Tovar J et al (1997) Disruption of the trypanothione reductase gene of Leishmania decreases its ability to survive oxidative stress in macrophages. EMBO J 16(10):2590–2598
47. Gerbaba TK, Gedamu L (2013) Cathepsin B gene disruption induced Leishmania donovani proteome remodeling implies cathepsin B role in secretome regulation. PLoS One 8(11):e79951
48. Srivastav S, Basu Ball W, Gupta P et al (2014) Leishmania donovani prevents oxidative burst-mediated apoptosis of host macrophages through selective induction of suppressors of cytokine signaling (SOCS) proteins. J Biol Chem 289(2):1092–1105
49. Choudhury R, Das P, Bhaumik SK et al (2010) In situ immunolocalization and stage-dependent expression of a secretory serine protease in Leishmania donovani and its role as a vaccine candidate. Clin Vaccine Immunol 17(4):660–667
50. Das P, De T, Chakraborti T (2014) Leishmania donovani secretory serine protease alters macrophage inflammatory response via COX-2 mediated PGE-2 production. Indian J Biochem Biophys 51(6):542–551
51. Dou Z, Carruthers VB (2011) Cathepsin proteases in toxoplasma gondii. Adv Exp Med Biol 712:49–61
52. Lentini G, El Hajj H, Papoin J et al (2017) Characterization of toxoplasma DegP, a rhoptry serine protease crucial for lethal infection in mice. PLoS One 12(12):e0189556
53. Sato D, Nakada-Tsukui K, Okada M et al (2006) Two cysteine protease inhibitors, EhICP1 and 2, localized in distinct compartments, negatively regulate secretion in Entamoeba histolytica. FEBS Lett 580(22):5306–5312
54. Smith BO, Picken NC, Westrop GD et al (2006) The structure of Leishmania mexicana ICP provides evidence for convergent evolution of cysteine peptidase inhibitors. J Biol Chem 281(9):5821–5828
55. Sanderson SJ, Westrop GD, Scharfstein J et al (2003) Functional conservation of a natural cysteine peptidase inhibitor in protozoan and bacterial pathogens. FEBS Lett 542(1–3):12–16
56. Sarić M, Vahrmann A, Bruchhaus I et al (2006) The second cysteine protease inhibitor, EhICP2, has a different localization in trophozoites of Entamoeba histolytica than EhICP1. Parasitol Res 100(1):171–174

57. Santos CC, Scharfstein J, Lima AP (2006) Role of chagasin-like inhibitors as endogenous regulators of cysteine proteases in parasitic protozoa. Parasitol Res 99(4):323–324
58. Costa TF, Lima AP (2016) Natural cysteine protease inhibitors in protozoa: fifteen years of the chagasin family. Biochimie 122:197–207
59. Wang SX, Pandey KC, Scharfstein J et al (2007) The structure of chagasin in complex with a cysteine protease clarifies the binding mode and evolution of an inhibitor family. Structure 15(5):535–543
60. Pszenny V, Davis PH, Zhou XW et al (2012) Targeted disruption of toxoplasma gondii serine protease inhibitor 1 increases bradyzoite cyst formation in vitro and parasite tissue burden in mice. Infect Immun 80(3):1156–1165
61. Alam MN, Das P, De T et al (2016) Identification and characterization of a Leishmania donovani serine protease inhibitor: possible role in regulation of host serine proteases. Life Sci 1(144):218–225
62. Martinvalet D (2015) ROS signaling during granzyme B-mediated apoptosis. Mol Cell Oncol 2(3):e992639
63. Ljunggren A, Redzynia I, Alvarez-Fernandez M et al (2007) Crystal structure of the parasite protease inhibitor chagasin in complex with a host target cysteine protease. J Mol Biol 371(1):137–153
64. dos Reis FC, Smith BO, Santos CC et al (2008) The role of conserved residues of chagasin in the inhibition of cysteine peptidases. FEBS Lett 582(4):485–490
65. Monteiro AC, Abrahamson M, Lima AP et al (2001) Identification, characterization and localization of chagasin, a tight-binding cysteine protease inhibitor in Trypanosoma cruzi. J Cell Sci 114(Pt 21):3933–3942
66. Gonçalves VM, Matteucci KC, Buzzo CL et al (2013) NLRP3 controls Trypanosoma cruzi infection through a Caspase-1-dependent IL-1R-independent NO production. PLoS Negl Trop Dis 7(10):e2469
67. Paiva CN, Medei E, Bozza MT (2018) ROS and Trypanosoma cruzi: fuel to infection, poison to the heart. PLoS Pathog 14(4):e1006928
68. Jecna L, Svarovska A, Besteiro S et al (2009) Inhibitor of cysteine peptidase does not influence the development of Leishmania mexicana in Lutzomyia longipalpis. J Med Entomol 46(3):605–609
69. Pei Y, Miller JL, Lindner SE et al (2013) Plasmodium yoelii inhibitor of cysteine proteases is exported to exomembrane structures and interacts with yoelipain-2 during asexual blood-stage development. Cell Microbiol 15(9):1508–1526
70. Pandey KC, Singh N, Arastu-Kapur S et al (2006) Falstatin, a cysteine protease inhibitor of Plasmodium falciparum, facilitates erythrocyte invasion. PLoS Pathog 2(11):e117
71. Bryson K, Besteiro S, McGachy HA et al (2009) Overexpression of the natural inhibitor of cysteine peptidases in Leishmania mexicana leads to reduced virulence and a Th1 response. Infect Immun 77(7):2971–2978
72. Alam MN, Chakraborti S, Paik D et al (2018) Functional attribution of LdISP, an endogenous serine protease inhibitor from Leishmania donovani in promoting infection. Biochimie 147:105–113
73. Alves CR, Souza RS, Charret KDS et al (2018) Understanding serine proteases implications on Leishmania spp lifecycle. Exp Parasitol 184:67–81

Oxidative Stress and Antioxidants in Host Defense in Leishmaniasis

13

Carlos Kusano Bucalen Ferrari

Abstract

Leishmaniasis, a tropical neglected disease of the poor and underserved populations, is characterized by skin lesions and ulcers or a clinical pattern of visceral/internal complications which can be lethal. Since innate response is important in hanseniasis, this work reviewed and updated the role of reactive oxygen/nitrogen species in different clinical forms of leishmaniasis (cutaneous, mucocutaneous, and visceral), as well as their role in phagocyte free radicals generators (NADPH-oxidase, iNOS, myeloperoxidase, mitochondrial pathways, and extracellular traps). Knowledge of multiple leishmanial antioxidant responses that can rescue parasites from death and promote resistance to treatment is essential to develop genetic and pharmacologic leishmanicidal strategies.

Keywords

Cutaneous leishmaniasis · Visceral leishmaniasis · *L. major* · *L. amazonensis* · *L. braziliensis* · Reactive oxygen species · Antioxidant

13.1 Introduction and Epidemiology

Leishmaniasis, a sandlfy-transmitted disease caused by many intracelular protozoa from *Leishmania* spp., comprises an important neglected tropical disease [1]. The two major parasite forms are the promastigote, which lives in the sandfly gut, and the amastigote that survives into the cytoplasm of monocytes and macrophages of the human and mammalian hosts [2].

C. K. B. Ferrari (✉)
Instituto de Ciências Biológicas e da Saúde (ICBS), Campus Universitário do Araguaia, Universidade Federal de Mato Grosso (UFMT), Barra do Garças, MT, Brazil
e-mail: drcarlosferrari.ufmt@gmail.com

S. Chakraborti et al. (eds.), *Oxidative Stress in Microbial Diseases*,
https://doi.org/10.1007/978-981-13-8763-0_13

Three major clinical patterns of Leishmaniasis are distinguished [3]:

Cutaneous or tegumentary leishmaniasis (CL): it comprises the majority of cases and it is characterized by skin lesions which commonly evolute to ulcerations and permanent scars on the exposed body parts bitten by the mosquitoes. Americas, Middle East, Central Asia, and Mediterranean basin comprise 95% of the cases with emphasis to Brazil, Afghanistan, Algeria, Colombia, Iran, Nicaragua, Peru, and Syria. 600,000 to 1.2 million of new cases are estimated each year. In Brazil, the major agents comprise *Leishmania (Leishmania) amazonensis*, *L. (Viannia) guyanensis*, *L. (Viannia) braziliensis*, and *L. (Viannia) naiffii*, whereas in other parts of the world, the major ethiologic agents are *L. major*, *L. tropica*, *L. aethiopica*, *L. mexicana*, *L. panamensis*, and *L. guyanensis*, among others [4–6].

Mucocutaneous leishmaniasis (MCL): It is characterized by partial or total destruction of the mucous membranes of the nose, throat, and mouth, including lesions in pharynx, and larynx. The same *Leishmania* species involved in CL are also found in MCL patients.

Visceral leishmaniasis (VL) or Kala-Azar: this form could be lethal if untreated. Its incidence is rising in Brazil [8]. In the Americas, 96% of cases have been reported in Brazil. In the world, 90% of VL cases have been reported from India, Bangladesh, Sudan, South Sudan, Ethiopia, and Brazil. 50,000 to 90,000 new cases have been estimated world-wide. *Leishmania (Leishmania) chagasi* and *L. (Infantum) chagasi* are the major causative agents of VL [7, 8]. *L. donovani* causes both cutaneous and visceral leishmaniasis in the old world [9].

In the Americas, two-thirds of patients are male and about 15% of children are affected [10]. Raising incidences of total leishmaniasis were found in Colombia (31.3%), Peru (33%), and Nicaragua (181.8%) [10].

In Brazil, although the total incidence of leishmaniasis has been declined by 34%, the incidence of VL has been increased [8, 10]. In many inner counties of Brazil, CL affects mainly men, in productive age, working in the agricultural sector or mining, and living in rural or semi-rural areas [11].

Higher co-infection rates due to leishmaniasis and HIV-infection have been reported in Brazil, Ethiopia, and Bihar State from India [3].

Since reactive oxygen and nitrogen species display important roles in innate response to pathogens, the objective of the present work was to review the role of oxidative/nitrosative stress in leishmaniasis as well as the role of antioxidant systems in parasite escape against phagocyte engulfment and free radical killing.

We performed a non-systematic review, from 1998 to 2018, using the following databases: Pubmed/Medline (https://www.ncbi.nlm.nih.gov/pubmed/), Scielo (www.scielo.org), and Google scholar (https://scholar.google.com.br). The combined keywords used were leishmaniasis and oxidative stress, oxidation and immune and leishmaniasis, leishmaniasis and nitrosative stress, myeloperoxidase and leishmaniasis, NADPH-oxidase and leishmaniasis, antioxidants and leishmaniasis, and antioxidant capacity and leishmaniasis.

13.2 How Free Radicals and Reactive Species Are Formed?

Adding an electron to the molecular oxygen (O_2) results in production of the superoxide anion (O_2^-), which is released by the mitochondrial respiratory chain and through the activity of the membrane NADPH-oxidase in activated phagocytes (monocytes/macrophages and neutrophils) [12]. Superoxide anion reaction with molecular oxygen in the presence of the superoxide dismutase (SOD) enzyme yields hydrogen peroxide (H_2O_2) which, combined with chloride (Cl^-), in the presence of MPO, synthezises hypochlorous acid (HOCl), an oxidant produced by inflammatory neutrophils at the site of infection [13]. Therefore, in phagocytes, NADPH-oxidase is characterized as a phagocyte-oxidase with an iron-containing protein (HEME protein) and it represents an important defense mechanism in microbial and parasite killing by professional phagocytes [14, 15].

When HOCl reacts with hydrogen peroxide, singlet oxygen ($^1O_2$), a very toxic tissue-damaging reactive species, is formed [16]. Reaction of L-arginine with molecular oxygen, NADPH, and nitric oxide synthase (NOS) produces nitric oxide (NO), which in turn, in combination with superoxide anion, yields peroxynitrite ($OONO^-$), a very toxic free radical [17].

13.3 Oxidative Relationships Between Neutrophils and Leishmaniasis

Initially, promastigotes that were introduced into the bloodstream due to sandfly bite are phagocytosed by neutrophils. In many circumstances, but not all, neutrophils are able to activate their respiratory burst, releasing superoxide anion, oxygen free radicals, nitric oxide, and peroxynitrite which kill promastigotes from various *Leishmania* species [18, 19]. Neutrophil depletion increases parasite multiplication in mice [18]. Neutrophils can kill *Leishmania* parasites by both oxygen free radical-dependent and oxygen free radical-independent pathways [19, 20].

13.3.1 Neutrophil Extracellular Traps (NETS) and Leishmania Killing

Triggered by H_2O_2 and other oxidative/nitrosative compounds, extracellular traps are fragments of nuclear and mitochondrial DNA triggered, composed by chromatin, histones, and antimicrobial proteins, that kill some microrganisms and activate the innate immunity [15, 21–23]. It is important to know that a Mexican study with *M. tuberculosis* showed ET were not capable of destroying bacilli which contributed to pathogen persistence [24]. In *Leishmania amazonensis* infection models, NETS can efficiently kill parasites as noted by some studies, but a few organisms are resistant to NETS and can survive and replicate [25–27]. These NETS-induced killing pathways of *Leishmania amazonensis* promastigotes are dependent and independent of ROS generation by NADPH-oxidase [28]. It should be noted that NETS contained expressive ammounts of MPO [29].

13.3.2 Other Relationships Between Neutrophils and Oxidation in Leishmaniasis

When amastigotes are phagocytosed by neutrophils, they induce a massive oxidative and nitrosative stress which leads to neutrophil damage and death [30].

Comparing the role of amastigotes of *L. braziliensis* and *L. amazonensis*, the first parasite species was able to boost neutrophil activation, respiratory burst, synthesis of IL-10 and IL-22 interleukins, and microbicidal activity, whereas *L. amazonensis* did not trigger neutrophil effectiveness [31]. Another study corroborated the concept that *L. braziliensis*, but not *L. major*, trigger neutrophil activation, respiratory burst, degranulation, and immune activation [32].

Heme proteins, but not free iron, induced neutrophil activation with raised activity of myeloperoxidase and neutrophil elastase production as well release of FR which compromised cell viability and caused human cell death in *L. infantum* infection [33].

Notwithstanding, a further study suggested that neutrophils are not so microbicidal but display an important pro-inflammatory role in human neutrophils exposed to *L. braziliensis* [34]. In fact, in some circumstances and depending on different *Leishmania* species, amastigotes phagocytosed by some neutrophils can induce massive oxidative and nitrosative stresses which lead to neutrophil damage and death [30].

13.4 Oxidative Relationships Between Macrophages and Leishmaniasis

As occurs with neutrophils, *Leishmania* spp. can either succumb or survive and escape to the killing effects of free radicals released by the respiratory burst of macrophages.

Oxygen, nitrogen, and chlorine free radicals derived from the macrophage respiratory burst products are toxic to *Leishmania*, but the parasite has developed mechanisms to escape from oxidative and nitrosative stress such as the lipophosphoglycan from membranes which disrupts superoxide anion release from the NADPH-oxidase [35]. In lipophosphoglycan null *L. infantum* strain, there is a strong activation of the nuclear factor kappa-beta (NFkB)-induced production of nitric oxide [36].

In fact, a previous work showed that intracellular infection of monocytes by *L. major*, *L. mexicana*, and *L. donovani* decreased production of superoxide anion and hydrogen peroxide [37]. In patients with active form of VL, suppressed activity of the macrophages enzymes NADH-oxidase, NADPH-oxidase, and MPO is found [38].

13.5 Dual Role of Nitric Oxide in Leishmania: Killing or Survival?

Nitric oxide was proved to be an effective parasite-killing molecule in human infection by *Leishmania major* in a previous study [39].

Although activated macrophages are important in chronic immune response to *L. braziliensis*, their relevance has been associated with releasing of cytokines and T-cell chemotaxis but not nitric oxide or superoxide anion discharge [40].

In fact, while inhibition of nitric oxide and ROS enhanced parasite survival, delivery of ROS did not improve tissue infection, and nitric oxide was associated with worsening tissue damage in CL patients [41]. It seems that this paradox is associated with different genetic strains of *Leishmania braziliensis*. There are nitric oxide-susceptible and nitric oxide-resistant *L. braziliensis* strains, and those resistant strains are also resistant to glucantime therapy [42].

Another study pointed out that ROS but not NO were important in *L. braziliensis* death in monocytes activated by IFN-y, which are typically found in chronic immune responses [43]. On the other hand, *L. amazonensis* infection in mice caused massive liver oxidative stress, increasing SOD levels, advanced glycation end products, inflammatory cytokines, and thiol oxidation and nitrotyrosine oxidation [44].

In mice infected with *L. major*, nitric oxide decreased replication of amastigotes inside macrophages as well as diminished lesion size and parasite dissemination to liver, spleen, and adjacent lymph nodes [45].

Thus, the effectiveness of NO in *Leishmania* killing or its failure depends on many factors, such as the kind of *Leishmania* species, the kind of clinical leishmaniasis (CL, MCL, and VL), the adequate or suppressed phagocytes' function, the absence or presence of defects on parasite killing effector pathways, and development of *Leishmania* spp. escape mechanisms against effector ROS/RNS pathways.

Concomitant deficiency of the phagocyte-oxidase (NADPH-oxidase complex) and nitric oxide pathway strongly decreased the effectiveness of phagocytosis and this was associated with infection dissemination and worsen prognosis [46].

Another mechanism of escape by *Leishmania* spp. against NO killing is represented by expression of hepatic arginase, which results in consumption of L-arginine, the substrate of nitric oxide synthase. Then, overexpression of arginase decreases production of nitric oxide which is associated with augment of visceral leishmania infection [47]. Inhibition of arginase lowers replication of *Leishmania* spp. into the cytoplasm of macrophages [48].

13.6 Different *Leishmania* Spp. and Clinical Forms: Susceptibility or Resistance?

As discussed above, the effectiveness of phagocytes in *Leishmania* killing or its survival depends on the clinical form of disease and the Leishmania species.

Cutaneous Leishmaniasis, Oxidative Stress, and Phagocyte Effectiveness.

Studying cutaneous leishmaniasis (CL) patients, raised levels of lipid peroxidation (measured by MDA content) and nitric oxide and diminished levels of SOD and GPX were observed [49]. Another work which studied possible changes on antioxidant enzyme catalase (CAT), uric acid, and MDA levels in CL patients reported enhanced levels of both uric acid and MDA [50]. In the same report, CAT levels tend to be decreased among CL patients, but there was no statistical significance ($p = 0.115$). Intermediate markers of lipid peroxidation, the lipoperoxides, were found to be increased in CL patients as well MDA and DNA damage [51].

During the course of leishmania infection (cutaneous or visceral), the parasite genes also increase the expression of antioxidant proteins and enzymes in order to protect parasites against phagocytes ROS/RNS [35]. In this regard, it is important to suppress the antioxidant response system of the parasites to improve the efficacy of therapeutic agents. Then, genetic inhibition of the antioxidant response genes glutathione synthetase, glutathione-S-transferase 1, and ATP-binding cassette B in human macrophages improved the efficacy of glucantime in *Leishmania braziliensis* killing [52].

On the other hand, it is important to know that parasite can also trigger massive oxidative stress, worsening tissue damage and disease. In this respect, *L. amazonensis* infection in mice caused massive liver oxidative stress, increasing SOD levels, advanced glycation end products, inflammatory cytokines, and thiol oxidation and nitrotyrosine oxidation [44].

13.6.1 Visceral Leishmaniasis, Oxidative Stress, and Phagocyte Effectiveness

Oxidative stress occurs in VL, as noted by increased levels of erythrocyte and urinary malondialdehyde (MDA), a lipid peroxidation product [53–55]. As a compensatory antioxidant defense, GSH levels increased in VL [53–55].

At least part of *L. donovani* strains is susceptible to ROS/RNS effector pathways and glucantime treatment can enhance the release of oxidative and nitrosative compounds increasing parasites' killing. Before treatment regimen, TNF-α and NO are effective in parasite clearance, whereas after treatment, TNF-α, NO, IFN-γ, and IL-10 contributed to parasite elimination [56]. Glucantime, one of the therapeutic drugs used in the treatment of VL, is able to induce the production of H_2O_2 and other reactive oxygen species via respiratory burst in macrophages infected by *Leishmania donovani* [57].

However, the worsening of VL is associated with defects on phagocyte function as well as parasite overexpression of antioxidant defensive proteins, enzymes, and factors. According to the abovementioned information, VL is associated with inhibition of the two phagocyte enzyme systems (NADH-oxidase/NADPH-oxidase; myeloperoxidase), which are associated with impaired effectiveness of phagocytosis [38].

In this regard, when promastigotes of *L. donovani* were exposed to oxidative and nitrosative stress, the parasites develop an antioxidant defensive response with increased expression of trypanothione reductase, trypanothione synthetase, tryparedoxin peroxidase, glutathione peroxidase, peroxidoxin, endoplasmic reticulum oxidoreductin, small heat schock protein 20, and ascorbate peroxidase [58].

In *Leishmania donovani* and other *Leishmania* spp., glucose-6-phosphate dehydrogenase transfers electrons to NADP+ to form NADPH. This reaction is necessary to maintain trypanoredoxin and reduced thiol content. Thus, parasite triggers an antioxidant microenvironment in the cytoplasm of host cell macrophages decreasing oxidative and nitrosative stress as well as cell death. Consequently, there is massive parasite survival and replication in the cytoplasm of infected macrophages [59, 60].

Resistance of parasite to drug treatment is associated with raised activation of parasite antioxidant systems, and consequently resistance to oxidative stress.

A study reported that two isolated strains of *Leishmania donovani*, which were resistant to sodium stilbogluconate, comprising both promastigotes (sandfly parasite) and amastigotes (human parasite) forms, had also greater resistance to oxidative stress and highest infective capability [61].

Another report showed that *L. donovani* resistance to oxidative stress in macrophages is linked to overexpression of a base-excision DNA repair enzyme—the uracil DNA glycosylase [62].

Excessive thiol levels were present in antimony-resistant VL caused by *L. infantum* [63].

13.7 Parasite Antioxidant Defenses: Escape Mechanisms Against ROS/RNS

Trypanosoma brucei, *T. cruzi*, and *L. donovani* present the alternative oxidase, di-iron protein that uses ubiquinol to perform a four-electron reduction of oxygen to water, deactivating reactive species and free radicals [64].

An iron-SOD is essential in eliminating the macrophage superoxide and survival of *Leishmania* spp. This parasite also has a glutathione peroxidase-like enzyme (without selenium) [65]. This iron-SOD confered resistance to oxidative stress and to antimony treatment to *L. V. braziliensis* and *L. L. infantum* [66].

Trypanosomatids, such as *Leishmania* spp. and *Trypanosoma* spp. present peroxisome-like organelles, the glycosomes which control oxidation of carbohydrates, proteins, and lipids [64].

Beyond this protective compounds, *Leishmania* also present the trypanothione, a dithiol compound comprised of two glutathione molecules linked to one spermidine, and its reductase form, a mitochondrial isoform [67]. Furthermore, *Leishmania* spp. also have peroxiredoxins, especially the tryparedoxin peroxidase, essential to metabolize H_2O_2 into non-toxic compounds [67, 68]. It has been sugested that overexpression of tryparedoxin protein is associated with resistance to amphotericin B in a *L. donovani* strain [69].

In *Leishmania donovani,* the mitochondrial tryparedoxin peroxidase (mtp) and cytosolic tryparedoxin peroxidase (ctp) have different roles for parasite survival.

Even though mtp protects the parasite against endogenous respiratory oxidative stress, ctp guarantees parasite metabolism of exogenous ROS/RNS with improved replication rates into infected macrophages [70]. In the same study, resistance to both glutacantime and miltefosine was associated with overexpression of mtp and ctp in *L. donovani*. Those results corroborate previous findings of Iver et al., which observed that ctp protected promastigotes of *L. donovani* against oxidative stress and confered resistance to glucantime therapy [71]. Another study reported that raised expression of ctp protected *L. braziliensis* but not *L. infantum* against glucantime therapy. In the same study, ctp expression provided tolerance to H_2O_2-induced toxicity among *L. braziliensis* and moderate resistance among *L. infantum* [72].

Resistance of *L. donovani* clinical strains to miltefosine is linked to diminished production of free radicals and overexpression of six genes: tryparedoxin peroxidase, trypanothione synthetase, cytochrome b5 reductase, phosphoglucomutase, multidrug resistance-like protein and lipase precursor-like protein [73].

As exposed above, there is efficient leishmanicidal free radical and oxidative/nitrosative mechanisms, whereas some *Leishmania* species and strains express diverse antioxidant genes producing different antioxidant proteins that rescue parasite against phagocyte-induced death. Those mechanisms are summarized in Fig. 13.1.

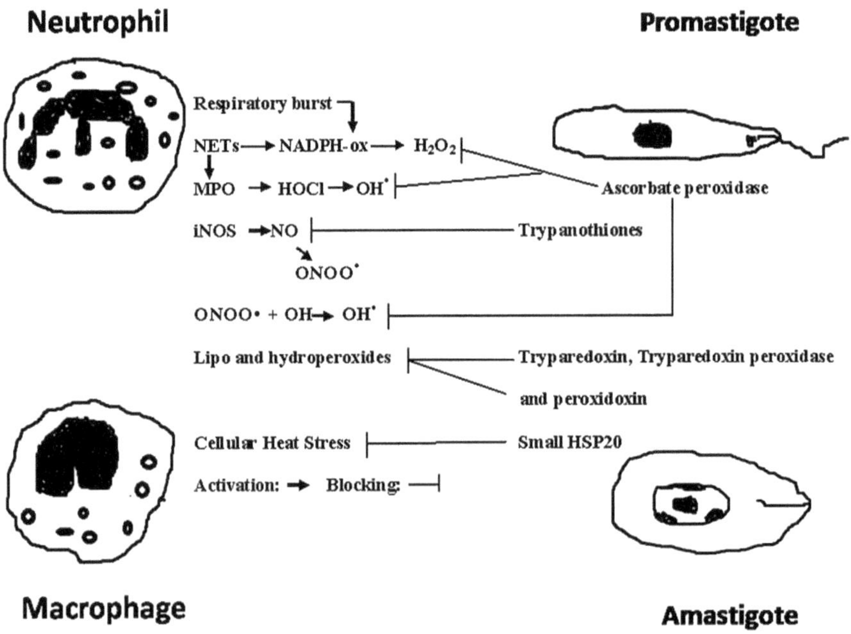

Fig. 13.1 Phagocyte leishmanicidal mechanisms *versus* antioxidant defense mechanisms of *Leishmania spp*

13.8 Conclusion

Although reactive oxygen, nitrogen, and chlorine species are effetive against *L. major* and few strains of *L. braziliensis*, it seems that *L. donovani* and *L. amazonensis*, and some strains of *L. braziliensis* are resistant to respiratory burst-induced killing. That resistance to reactive species and free radicals seems to be related to expression of diverse antioxidant response genes.

References

1. Alvar J, Vélez ID, Bern C et al (2012) Leishmaniasis worldwide and global estimates of its incidence. PLoS One 7(5):e35671. https://doi.org/10.1371/journal.pone.0035671
2. Gossage SM, Rogers ME, Bates PA (2003) Two separate growth phases during the development of Leishmania in sand flies: implications for understanding the life cycle. Int J Parasitol 33:1027–1034
3. WHO (2018) Leishmaniasis. WHO fact sheet. Available at: http://www.who.int/news-room/fact-sheets/detail/leishmaniasis [07/03/2018]
4. Reithinger R, Dujardin J-C, Louzir H et al (2007) Cutaneous leishmaniasis. Lancet Infect Dis 7:581–596
5. Coelho LIC, Paes M, Guerra JA et al (2011) Characterization of *Leishmania spp.* causing cutaneous leishmaniasis in Manaus, Amazonas, Brazil. Parasitol Res 108(3):671–677
6. Kone AK, Niare DS, Thera MA et al (2016) Epidemiology of the outbreak, vectors and reservoirs of cutaneous leishmaniasis in Mali: a systematic review and meta-analysis. Asian Pacific J Trop Med 9(10):985–990
7. Harhay MO, Olliaro PL, Costa DL, Costa CHN (2011) Urban parasitology: visceral leishmaniasis in Brazil. Trends Parasitol 27(9):403–409
8. Reis LL, Balieiro AAS, Fonseca FR, Gonçalves MJF (2017) Changes in the epidemiology of visceral leishmaniasis in Brazil from 2001 to 2014. Rev Soc Bras Med Trop 50(5):638–645
9. Kariyawasam K, Selvapandiyan A, Siriwardana HVYD et al (2018) Dermotropic *Leishmania donovani* in Sri Lanka: visceralizing potential in clinical and preclinical studies. Parasitology 145(4):443–452
10. Organização Panamericana da Saúde (OPAS) (2018) Leishmanioses. Informe epidemiológico das Américas. Informe de Leishmanioses, n.6. Washington, PAHO, 7p
11. França EL, Mandadori MN, França JL, Botelho ACF, Ferrari CKB, Honorio-França AC (2009) Epidemiological aspects of American cutaneous Leishmaniasis in the city of Juína, Mato Grosso state, Brazil. Scientia Médica 19(3):103–107
12. Bagaitkar J, Huang J, Zeng MY et al (2018) NADPH oxidase activation regulates apoptotic neutrophil clearance by murine macrophages. Blood 2018:blood-2017-09-809004. https://doi.org/10.1182/blood-2017-09-809004
13. Halliwell B (1994) Free radicals, antioxidants and human disease: curiosity, cause or consequence? Lancet 344(8924):721–724
14. Minakami R, Sumimoto H (2006) Phagocytosis-coupled activation of the superoxide-producing phagocyte oxidase, a member of the nadph oxidase (nox). Int J Hematol 84:193–198
15. Ferrari CKB, Souto PCS, França EL, Honorio-França AC (2011) Oxidative and nitrosative stress on phagocytes' function: from effective defense to immunity evasion mechanisms. Arch Immunol Ther Exp 59(6):441–448
16. Phaniendra A, Jestadi DB, Periyasami L (2015) Free radicals: properties, sources, targets, and their implication in various diseases. Indian J Clin Biochem 30(1):11–26
17. Ferrari CK, França EL, Honorio-França AC (2009) Nitric oxide, health and disease. J Appl Biomed 7:163–173

18. Novais FO, Santiago RC, Báfica A et al (2009) Neutrophils and macrophages cooperate in host defense against Leishmania braziliensis infection. J Immunol 183(12):8088–8098
19. Carmo EVS, Katz S, Barbiéri CL (2010) Neutrophils reduce the parasite burden in Leishmania (Leishmania) amazonensis-infected macrophages. PLoS One 5(11):e13815. https://doi.org/10.1371/journal.pone.0013815
20. Oliveira CI, Brodskyn CI (2012) The immunobiology of Leishmania braziliensis infection. Front Immunol 3:article 145. https://doi.org/10.3389/fimmu.2012.00145
21. Brinkmann V, Reichard U, Goosmann C et al (2004) Neutrophil extracellular traps kill bacteria. Science 2004 303(5663):1532–1535
22. Papayannopoulos V, Zychlinsky A (2009) NETs: a new strategy for using old weapons. Trend Immunol 30(11):513–521
23. von Köckritz-Blickwede M, Nizet V (2009) Innate immunity turned inside-out: antimicrobial defense by phagocyte extracellular traps. J Mol Med 87:775–783
24. Ramos-Kichik V, Mondragón-Flores R, Mondragón-Castelán M et al (2009) Neutrophil extracellular traps are induced by *Mycobacterium tuberculosis*. Tuberculosis 89:29–37
25. Guimarães-Costa AB, Nascimento MTC, Froment GS et al (2009) Leishmania amazonensis promastigotes induce and are killed by neutrophil extracellular traps. Proc Natl Acad Sci U S A 106(16):6748–6753
26. Horta MF, Mendes BP, Roma EH et al (2012) Reactive oxygen species and nitric oxide in cutaneous Leishmaniasis. J Parasitol Res Article ID 203818.:11 pages. https://doi.org/10.1155/2012/203818
27. Silva MS, Segatto M, Pavani RS et al (2017) Consequences of acute oxidative stress in *Leishmania amazonensis*: from telomere shortening to the selection of the fittest parasites. Biochem Biophys Acta 1864:138–150
28. Rochael NC, Guimarães-Costa AB, Nascimento MTC et al (2015) Classical ROS-dependent and early/rapid ROS-independent release of neutrophil extracellular traps triggered by *Leishmania* parasites. Sci Rep 5:18302. https://doi.org/10.1038/srep18302
29. Parker H, Winterbourn CC (2013) Reactive oxidants and myeloperoxidase and their involvement in neutrophil extracellular traps. Front Immunol 3:424
30. Carlsen ED, Hay C, Henard CA, Popov V, Garg NJ, Soong L (2013) Leishmania amazonensis amastigotes trigger neutrophil activation but resist neutrophil microbicidal mechanisms. Infect Immun 81(11):3966–3974
31. Carlsen ED, Jie Z, Liang Y et al (2015) Interactions between neutrophils and Leishmania braziliensis amastigotes facilitate cell activation and parasite clearance. J Innate Immun 7(4):354–363
32. Falcão SA, Weinkopff T, Hurrell BP et al (2015) Exposure to Leishmania braziliensis triggers neutrophil activation and apoptosis. PLoS Negl Trop Dis 9(3):e0003601. https://doi.org/10.1371/journal.pntd.0003601
33. Quintela-Carvalho G, Luz NF, Celes FS et al (2017) Heme drives oxidative stress-induced cell death in human neutrophils infected with *Leishmania infantum*. Front Immunol 8:1620. https://doi.org/10.3389/fimmu.2017.01620
34. Conceição J, Davis R, Carneiro PP et al (2016) Characterization of neutrophil function in human cutaneous leishmaniasis caused by Leishmania braziliensis. PLoS Negl Trop Dis 10(5):e0004715. https://doi.org/10.1371/journal.pntd.0004715
35. Van Assche T, Deschacht M, da Luz RA, Maes L, Cos P (2011) Leishmania-macrophage interactions: insights into the redox biology. Free Rad Biol Med 51(2):337–351
36. Lázaro-Souza M, Matte C, Lima JB et al (2018) Leishmania infantum lipophosphoglycan-deficient mutants: a tool to study host cell-parasite interplay. Front Microbiol 9:626. https://doi.org/10.3389/fmicb.2018.00626
37. Passwell JH, Shor R, Smolen J, Jaffe CL (1994) Infection of human monocytes by Leishmania results in a defective oxidative burst. Int J Exp Pathol 75(4):277–284
38. Kumar P, Pai K, Pandey HP, Sundar S (2002) NADH-oxidase, NADPH-oxidase and myeloperoxidase activity of visceral leishmaniasis patients. J Med Microbiol 51:832–836

39. Vouldoukis I, Riveros-Moreno V, Dugas B et al (1995) The killing of Leishmania major by human macrophages is mediated by nitric oxide induced after ligation of the Fc epsilon RII/CD23 surface antigen. Proc Natl Acad Sci U S A 92(17):7804–7808
40. Giudice A, Vendrame C, Bezerra C et al (2012) Macrophages participate in host protection and the disease pathology associated with Leishmania braziliensis infection. BMC Infect Dis 12:75. https://doi.org/10.1186/1471-2334-12-75
41. Carneiro PP, Conceição J, Macedo M, Magalhães V, Carvalho EM, Bacellar O (2016) The role of nitric oxide and reactive oxygen species in the killing of Leishmania braziliensis by monocytes from patients with cutaneous leishmaniasis. PLoS One 11(2):e0148084. https://doi.org/10.1371/journal.pone.0148084
42. Souza AS, Giudice A, Pereira JM et al (2010) Resistance of Leishmania (Viannia) braziliensis to nitric oxide: correlation with antimony therapy and TNF-alpha production. BMC Infect Dis 10:209. https://doi.org/10.1186/1471-2334-10-209
43. Novais FO, Nguyen BT, Beiting DP et al (2014) Human classical monocytes control the intracellular stage of Leishmania braziliensis by reactive oxygen species. J Infect Dis 209(8):1288–1296
44. Gasparotto J, Kunzler A, Senger MR et al (2017) N-acetyl-cysteine inhibits liver oxidative stress markers in BALB/c mice infected with *Leishmania amazonensis*. Mem Inst Oswaldo Cruz 112(2):146–154
45. Nahrevanian H, Jalalian M, Farahmand M, Assmar M, Rastaghi ARE, Sayyah M (2012) Inhibition of murine systemic leishmaniasis by acetyl salicylic acid via nitric oxide immunomodulation. Iran J Parasitol 7(2):21–28
46. Murray HW, Xiang Z, Ma X (2006) Responses to *Leishmania donovani* in mice deficient in both phagocyte oxidase and inducible nitric oxide synthase. Am J Trop Med 74:1013–1015
47. Yizengaw E, Getahun M, Tajebe F et al (2016) Visceral leishmaniasis patients display altered composition and maturity of neutrophils as well as impaired neutrophil effector functions. Front Immunol 7:517
48. Iniesta V, Gómez-Nieto C, Corraliza I (2001) The inhibition of arginase by Nw-hydroxy-L-arginine controls the growth of *Leishmania* inside macrophages. J Exp Med 193(6):777–783
49. Serarslan G, Yilmaz HR, Söğüt S (2005) Serum antioxidant activities, malondialdehyde and nitric oxide in human cutaneous leishmaniasis. Clin Exp Dermatol 30(3):267–271
50. Asmaa Q, Al-Shamerii S, Al-Tag M, Al-Shameni A, Li Y, Osman BH (2017) Parasitological and biochemical studies on cutaneous leishmaniasis in Shara'b district, Taiz, Yemen. Ann Clin Microbiol Antimicrob 16:47. https://doi.org/10.1186/s12941-017-0224-y
51. Kocyigit A, Keles H, Selek S, Guzel S, Celik H, Erel O (2005) Increased DNA damage and oxidative stress in patients with cutaneous leishmaniasis. Mutat Res 585(1–2):71–78
52. Téllez J, Romero I, Soares MJ, Steindel M, Romanha AJ (2017) Knockdown of host antioxidant defense genes enhances the effect of glucantime on intracellular Leishmania braziliensis in human macrophages. Antimicrob Agent Chemother 61(7):e02099–e02016. https://doi.org/10.1128/AAC.02099-16
53. Neupane DP, Majhi S, Chandra L, Rijal S, Baral N (2008) Erythrocyte glutathione status in human visceral leishmaniasis. Indian J Biochem 23(1):95–97
54. Jasim BT, Al-Azzauy AAM (2009) Oxidative stress and antioxidant status in human infected with Kala-azar. AJPS 6(1):116–122
55. Oliveira MJC, Silva Junior GB, Sampaio AM et al (2014) Preliminary study of tubuloglomerular dysfunction and evidence of renal inflammation in patients with visceral Leishmaniasis. Am J Trop Med Hyg 91(5):908–911
56. Gatto M, Abreu MM, Tasca KI et al (2015) The involvment of TLR2 and TLR4 in cytokine and nitric oxide production in visceral leishmaniasis patients before and after treatment with anti-leishmanial drugs. PLoS One 10(2):e0117977. https://doi.org/10.1371/journal.pone.0117977
57. Basu JM, Mookerjee A, Sen P et al (2006) Sodium antimony gluconate induces generation of reactive oxygen species and nitric oxide via phosphoinositide 3-kinase and mitogen-activated protein kinase activation in *Leishmania donovani*-infected macrophages. Antimicrob Agent Chemother 50(5):1788–1797

58. Sardar AH, Kumar S, Kumar A et al (2013) Proteome changes associated with Leishmania donovani promastigote adaptation to oxidative and nitrosative stresses. J Proteome 81:185–199
59. Ghosh AK, Sardar AH, Mandal A et al (2015) Metabolic reconfiguration of the central glucose metabolism: a crucial strategy of Leishmania donovani for its survival during oxidative stress. FASEB J 29(5):2081–2098
60. Ghosh AK, Saini S, Das S et al (2017) Glucose-6-phosphate dehydrogenase and trypanothione reductase interaction protects Leishmania donovani from metalloid mediated oxidative stress. Free Rad Biol Med 106:10–23
61. Vanaerschot M, Maes I, Ouakad M et al (2010) Linking *in vitro* and *in vitro* survival of clinical *Leishmania donovani* strains. PLoS One 5(8):e12211. https://doi.org/10.1371/journal.pone.0012211
62. Mishra A, Khan MI, Jha PK et al (2018) Oxidative stress-mediated overexpression of uracil DNA glycosylase in Leishmania donovani confers tolerance against antileishmanial drugs. Oxidat Med Cel Longevit 2018:article ID 4074357. https://doi.org/10.1155/2018/4074357
63. Magalhães LS, Bomfim LGS, Mota SG et al (2018) Increased thiol levels in antimony-resistant Leishmania infantum isolated from treatment-refractory visceral leishmaniasis in Brazil. Mem Inst Oswaldo Cruz 113(2):119–125
64. Menna-Barreto RFS, Castro SL (2014) The double-edged sword in pathogenic trypanosomatids: the pivotal role of mitochondria in oxidative stress and bioenergetics. Biomed Res Int 2014:Article ID 614014. https://doi.org/10.1155/2014/614014
65. Ghosh S, Goswami S, Adhya S (2003) Role of superoxide dismutase in survival of Leishmania within the macrophage. Biochem J 369.(pt.3:447–452
66. Tessarollo NG, Andrade JM, Moreira DS, Murta SM (2015) Functional analysis of iron superoxide dismutase-A in wild-type and antimony-resistant *Leishmania braziliensis* and *Leishmania infantum* lines. Parasitol Int 64(2):125–129
67. Krauth-Siegel LR, Comini MA, Schlecker T (2007) The trypanothione system. Subcell Biochem 44:231–251
68. Beig M, Oellien F, Garoff L, Noack S, Krauth-Siegel RL, Selzer PM (2015) Trypanothione reductase: a target protein for a combined in vitro and in silico screening approach. PLoS Negl Trop Dis 9(6):e0003773. https://doi.org/10.1371/journal.pntd.0003773
69. Suman SS, Equbal A, Zaidi A et al (2016) Up-regulation of cytosolic tryparedoxin in Amp B resistant isolates of Leishmania donovani and its interaction with cytosolic tryparedoxin peroxidase. Biochimie 121:312–325
70. Das S, Giri S, Sundar S, Shaha C (2018) Functional involvement of *Leishmania donovani* tryparedoxin peroxidases during infection and drug treatment. Antimicrob Agent Chemother 62(1):e00806–e00817
71. Iver JP, Kaprakkaden A, Choudhary ML, Shaha C (2008) Crucial role of cytosolic tryparedoxin peroxidase in *Leishmania donovani* survival, drug response and virulence. Mol Microbiol 68(2):372–391. https://doi.org/10.1111/j.1365-2958.2008.06154.x
72. Andrade JM, Murta SM (2014) Functional analysis of cytosolic tryparedoxin peroxidase in antimony-resistant and –susceptible *Leishmania braziliensis* and *Leishmania infantum* lines. Parasit Vectors 7:406. https://doi.org/10.1186/1756-3305-7-406
73. Deep DK, Singh R, Bhandari V et al (2017) Increased miltefosine tolerance in clinical isolates of *Leishmania donovani* is associated with reduced drug accumulation, increased infectivity and resistance to oxidative stress. PLoS Negl Trop Dis 11(6):e0005641. https://doi.org/10.1371/journal.pntd.000564

Oxidative Stress in *Entamoeba histolytica* 14

Somasri Dam, Pinaki Biswas, and Raktim Ghosh

Abstract

Entamoeba histolytica is a human pathogen, responsible for invasive amoebiasis and dysentery. This chapter aims to describe the effect of various stresses especially oxidative and nitrosative stress on this organism. This parasite is subjected to several types of stress throughout its life cycle and also during the invasion of human tissues as a result of host's response to the infection. For successful infection, it must produce an adaptive response against host defense mechanisms. *E. histolytica* is microaerophilic, but during tissue invasion, it is exposed to high oxygen content in well-perfused tissues. This parasite has its own antioxidant strategy to protect itself against reactive oxygen and nitrogen species generated by both host and parasite. *E. histolytica* doesn't have most of the antioxidant defense mechanisms such as glutathione peroxidase, glutathione reductase, and catalase. Instead, it manages the antioxidant components from engulfed bacteria and red blood cells. L-cysteine takes a major role to protect the trophozoites of *E. histolytica* from oxidative stress and nitrosative stress as the parasite lacks glutathione, a major thiol in eukaryotes. During oxidative stress caused by H_2O_2, 286 genes have been found to be upregulated in *E. histolytica* HM1: IMSS. In response to environmental stresses like glucose starvation, serum starvation, iron starvation, heat shock, and UV irradiation, several responsible genes are either downregulated or upregulated in this pathogen.

Keywords

Entamoeba histolytica · Oxidative stress · Nitrosative stress · Starvation · L-Cysteine

S. Dam (✉) · P. Biswas · R. Ghosh
Department of Microbiology, University of Burdwan, Burdwan, West Bengal, India
e-mail: dam_somasri@rediffmail.com

S. Chakraborti et al. (eds.), *Oxidative Stress in Microbial Diseases*,
https://doi.org/10.1007/978-981-13-8763-0_14

Abbreviations

ALDO	Aldolase
Eh	*Entamoeba histolytica*
GAPDH	Glyceraldehyde 3-P dehydrogenase
HK	Hexokinase
HPI	Hexose phosphate isomerase
ORP	Oxygen reduction pathway
UDP-GPP	UDP-glucose pyrophosphorylase
PFK	Phosphofructokinase
TPI	Triose-phosphate isomerase
PGK	Phosphoglycerate kinase
PGAM	Phosphoglycerate mutase
ENO	Enolase
PPDK	Pyruvate phosphate dikinase
PK	Pyruvate kinase
ADHE	Bifunctional aldehyde/alcohol dehydrogenase
AcCoAS	ADP-forming acetyl-CoA synthetase
GK	Glycerol kinase
GPP	Glycerol 3-phosphate phosphatase
PFOR	Pyruvate:ferredoxin oxidoreductase
Fd(ox) and Fd(red)	Oxidized and reduced ferredoxin
NO	NADPH-dependent oxidoreductase
FDP	Flavodiiron protein
p34	34-kDa NADPH:flavin oxidoreductase
FeSOD	Iron-containing dismutase
TrxR(ox) and TrxR(red)	Oxidized and reduced thioredoxin reductase
Trx	Thioredoxin
Rbr(ox) and Rb(red)	Oxidized and reduced rubrerythrin
NROR	NAD(P)H-dependent rubredoxin reductase
Prx(ox) and Prx(red)	Oxidized and reduced peroxiredoxin
ISF	Iron-sulfur flavoprotein
HCP	hybrid cluster protein
Eh34/p34	Flavin oxidoreductase
Trans-PMET	Transplasma membrane electron transport
TPQ-7	Thermoplasmaquinone-7
PNT	Pyridine nucleotide transhydrogenase
CoA	Coenzyme A
ATP	Adenosine triphosphate
$NAD(P)^+$	Oxidized nicotinamide adenine dinucleotide (phosphate)
NAD(P)H	Reduced nicotinamide adenine dinucleotide (phosphate)

14.1 Introduction

Entamoeba histolytica causes amoebic colitis and liver abscess worldwide. This tissue-lysing amoeba is the causative agent for amoebiasis which persists as a global health problem leading to 50 million clinical cases and 40,000–100,000 deaths annually [1]. Amoebiasis occurs worldwide and it is a major problem where poor sanitation causes contamination of drinking water and food with feces [1–4]. It is the second most lethal disease in the world caused by protozoan parasite after malaria [5]. Amoebic colitis was known to the ancients. The disease was first documented in a Sanskrit description as bloody mucoid diarrhea during 3000 BCE [6]. It took more than 2000 years to be invented. Amoeba was first identified as a reason of dysentery in 1875, when the St Petersburg physician Fedor Aleksandrovich Lösch identified the trophozoites of amoeba in the stool of a farmer with a fatal case of dysentery [3].

E. histolytica has high potential for invading and destroying the tissue in human colon causing hemorrhagic colitis. During residing in human gut, the parasite lives in the environment of reduced oxygen pressure. The production of reactive oxygen species (ROS) is an important component of the innate immune defense against microbial infections, including amoebiasis [7]. In this case, the parasite experiences high amount of exogenous ROS during tissue invasion of the host, which may lead to metabolic malfunctions. Several defense mechanisms including enzymatic and non-enzymatic components help to protect the pathogen against the oxidative stress.

14.2 Stress-Related Studies in *Entamoeba histolytica*

E. histolytica undergoes different types of environmental stresses during its life cycle. These can be classified mainly into seven categories, such as oxidative and nitrosative stress, glucose starvation, cysteine starvation, iron starvation, heat shock, and UV irradiation [8]. The parasite is challenged in host environment mostly due to the major fluctuations in oxygen level and glucose concentration level. The activation of innate immune response against *E. histolytica* directs to the production of ROS and NO by macrophages. This helps *E. histolytica* to become capable of adapting the oxidative stress [9]. Most of the eukaryotic organisms respond to environmental stress by reducing or switching off the protein synthesis process. The exception is synthesis of heat shock proteins and some transcription factors [10]. *E. histolytica* is not an exception. Table 14.1 summarizes the role of proteins and enteric bacteria during various stresses in this parasite. *E. histolytica* shows a specific heat shock response with overall reduction in gene transcription including the genes responsible for its virulence, such as galactose/N-acetylgalactosamine (Gal/GalNAc) lectin, certain cysteine proteinases, and a 20-kDa antigen [11]. Ultraviolet light produces cyclobutane pyrimidine dimers, oxidized bases, single-strand breaks, and also DNA double-strand breaks (DSB) [12, 13]. DSB is a critical injury to DNA, and it activates a complex network of proteins that may arrest cell cycle to enhance DNA repair mechanisms [14]. A cDNA microarray study on global

Table 14.1 *E. histolytica* proteins during various stresses

Type of stress	Protein involved	Summary of study	References
Heat shock Oxidative stress	Ehssp1	Stress condition upregulates the expression of polymorphic copies of Ehssp1, antigenic in invasive amoebiasis Extent of polymorphism differs between pathogenic and non-pathogenic strain	[130]
Oxidative stress.	Eh29 HSP70A2 EhSOD EhCP5 G protein Peptidylprolyl isomerase	In response to oxidative stress, several genes are upregulated in a time-dependent manner to protect *E. histolytica* during invasion	[43]
Oxidative stress.	Eh29	29-kDa surface antigen, exhibits protective antioxidant activities Survival and pathogenesis of *E. histolytica* through invasion increase with increased expression of Eh29	[58]
Oxidative stress Nitrosative stress	Heat-shock proteins Ubiquitin-conjugating enzymes Protein kinases and small GTPases	During oxidative and nitrosative stress, a large set of genes are either upregulated or downregulated and their expression may differ in pathogenic and non-pathogenic strains of *E. histolytica*	[44]
Oxidative stress	EhPFOR EhADH2	Trophozoites under oxidative stress modify their energy metabolism by regulating the steps of glycolysis	[48]
Oxidative stress Nitrosative stress	EhFdp1	*E. histolytica* genome contains four genes encoding flavodiiron proteins which are likely to be acquired by horizontal gene transfer from prokaryotes EhFdp1 is cytoplasmic, having specific and high oxygen reductase activity but poor nitric oxide reductase activity	[131]
Oxidative stress Nitrosative stress	EhTRXR.	EhTRXR catalyzes the NAD(P) H-dependent reduction of thioredoxins and S-nitrosothiols	[132]
Oxidative stress Nitrosative stress	EhSIAF EhPTPA	In oxidative and nitrosative stressed condition, *E. histolytica* shows higher transcription level of EhSIAF and EhPTPA EhSIAF and EhPTPA are responsible for increasing adherence to the host cells and reduction in motility	[77]
Oxidative stress	H_2O_2-regulatory motif (HRM) binding protein	Overexpression of the transcription factor, HRM-binding protein increases the virulence potential in *E. histolytica* This is a special transcription factor controlling the transcriptional regulatory network in response to oxidative stress induced by hydrogen peroxide	[133]

(continued)

Table 14.1 (continued)

Type of stress	Protein involved	Summary of study	References
Oxidative stress	EhPFOR EhPrx EhFDP EhHSP70	Virulent strain of *Entamoeba* differs from non-virulent strains in gene expression level of several proteins related to oxidative stress HSP70 protects *E. histolytica* from oxidative damage in O_2 exposure and may contribute to its virulence	[134]
Oxidative stress	Arginase Gal/GalNAc	A total, 154 number of oxidized proteins were detected which are functionally associated with antioxidant activity Oxidation of Gal/GalNAc has inhibitory effect on adherence of *E. histolytica* trophozoites to host cells Novel function of arginase in protection of parasite against oxidative stress has been identified	[17]
Oxidative stress	EhTrx EhFDP1 EhRbr EhPrx EhTrxR	Proteins of thiol-based redox system counteract with oxidative and nitrosative stress and maintain intracellular redox potential during invasion, colonization, and disease progression Proteins of thiol-based redox system can be used as a suitable drug target for disruption of redox balance in parasite	[61]
Oxidative stress	Serine-rich *Entamoeba histolytica* protein (SREHP) Eh29 Gal/GalNAc lectin	In silico analysis of protein related to stress response in *E. histolytica* can serve future prospect in developing of diagnostic markers, drug design, and vaccine research	[135]
Oxidative stress ER stress Thermal stress	EhPDI	Stressful condition inhibits EhPDI activity resulting in unfolding, aggregation, and inability to assist the refolding of proteins	[136]
Oxidative stress	Thioredoxin Peroxiredoxin Nitroreductase Oxidoreductase	Enteric bacteria except probiotics protect the parasite from oxidative stress	[83]

transcriptional response to the DNA damage of *E. histolytica* showed that most gene regulation in response to UV-C irradiation is operating at post-translational level. Importantly, several genes encoding Fe–S clusters containing proteins involved in stress adaptation were found to be upregulated and some genes encoding cytoskeleton proteins were downregulated upon UV exposure. This is probably due to the fact that *E. histolytica* actin dynamics is arrested to permit proper repairing of DNA damage caused by UV-C irradiation [15].

14.2.1 Oxidative Stress

Among the parasitic protozoa, *Entamoeba histolytica* is unique because of its in vitro nutritional requirements and in vivo pathogenicity mechanisms. Until now, very little is known regarding molecular factors and physicochemical conditions that regulate the growth, establishment, and successful invasion of this parasite in the host. It can be grown in vitro under the condition of low oxygen concentration and it is susceptible to metabolically produced toxic oxygen derivatives and ROS. The parasite may also consume low amount of oxygen and therefore produce some toxic reactive oxygen derivatives. *E. histolytica* lacks catalase, glutathione reductase, and glutathione peroxidase, the major defense mechanisms in other aerotolerant cells. For this reason, *E. histolytica* has evolved itself with some alternative approaches. It phagocytoses bacteria during the commensal phase, and during invasive phase, it consumes red blood cells (RBCs). These engulfed bacteria or RBCs, through their intact form or antioxidant component form, detoxify the ROS produced in their intracellular environment during invasion phase [16]. In a study, 154 number of proteins have been found to be oxidized in oxidatively stressed trophozoites of *E. histolytica* in resin-assisted capture coupled to mass spectrometric analysis [17]. Unlike many organisms, *E. histolytica* lacks glutathione which plays as major thiol in most eukaryotes. In this parasite, L-cysteine works as a principal low molecular weight thiol [18]. Growth medium lacking L-cysteine elevates intercellular ROS level by more than threefold, implicating the significance of this amino acid in the antioxidant defense against the oxidative stress [19]. Apart from this, L-cysteine protects the trophozoites of *E. histolytica* from oxidative shock generated by treatment with the drug metronidazole [20]. L-cysteine can be obtained either through a de novo synthesis pathway or via uptake from extracellular environment. The de novo synthesis is carried out in two steps catalyzed by serine acetyltransferase and cysteine synthetase [21, 22]. Experimental data showed that internal concentration of L-cysteine is almost undetectable when the trophozoites are cultured in a medium devoid of cysteine. This clearly implies that the de novo synthesis pathway itself is not sufficient to maintain intracellular concentration up to the mark and it mostly depends on external uptake. During nitrosative stress, the trophozoites of *E. histolytica* overexpress cysteine synthetase to satisfy the increased amino acid demand. Fig. 14.1 describes the various antioxidant pathways operating in *E. histolytica.*

14.2.2 Glucose Starvation

E. histolytica faces glucose starvation in human colon because the available glucose is very low due to high absorptive capacity of the glucose transporters in small intestine [23–25]. In response to glucose starvation, several genes related to glycolysis are downregulated, and the genes involved in degradation of stored carbohydrates are upregulated. From the proteomics and transcriptomics study, it has been found that mainly three genes/proteins get upregulated, namely, Gal/GalNAc lectin,

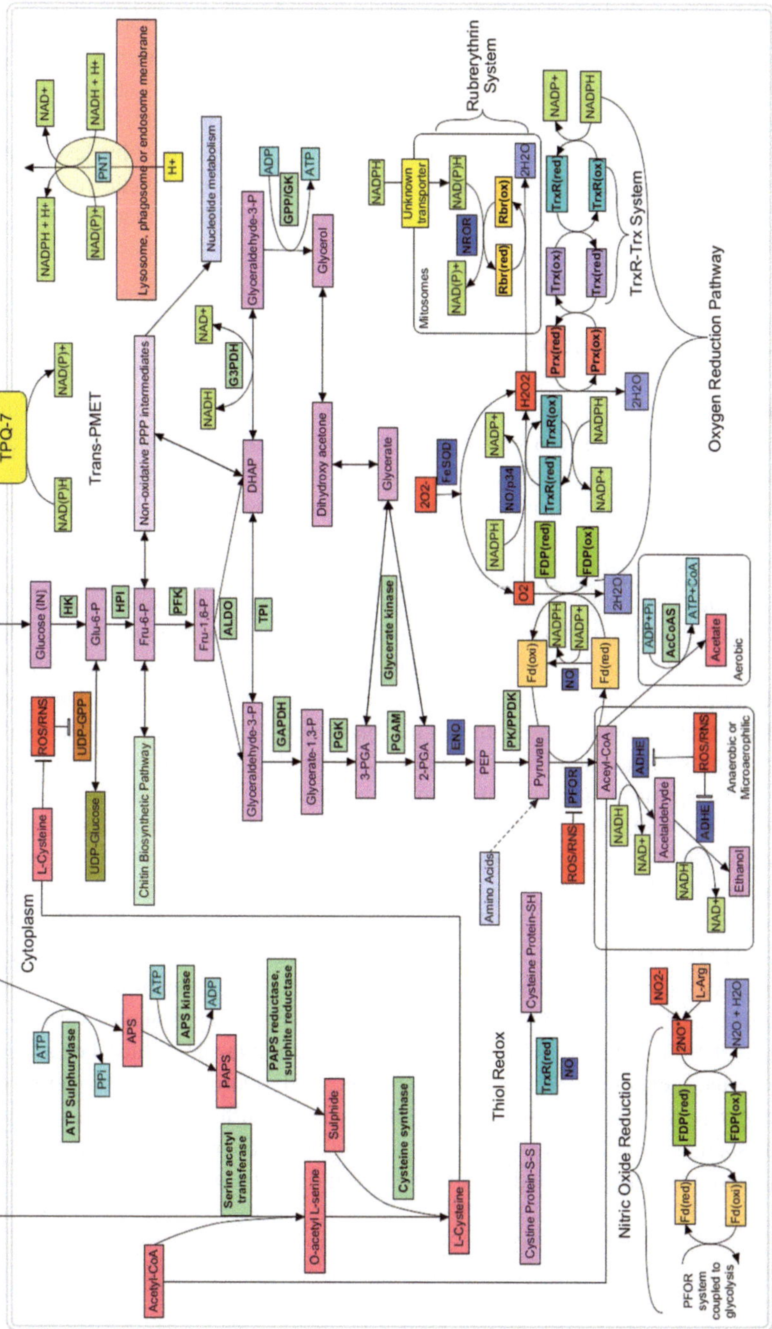

Fig. 14.1 Schematic representation of antioxidant pathways in *Entamoeba histolytica*

dihydropyrimidine dehydrogenase (DPD), and MGL-1 [26, 27]. DPD is involved in the degradation of pyrimidine. So, its increased expression in *E. histolytica* during glucose starvation may be due to the energy production by degradation of pyrimidines [28, 29]. Another carbon source to catch up the needs for energy is to break down the colonic mucin by β-amylase [30]. Glucose starvation enhances virulence property by upregulation of virulence factors such as Gal/GalNAc lectins and a cysteine protease, EhCP-A4, that helps the parasite to invade host by destroying the intracellular matrix [26]. Glucose starvation stimulates epigenetic regulation by building up the shuttling of glycolytic enzyme enolase in the nucleus and inhibition of Ehmeth [27].

14.2.3 Iron Starvation

Iron is one of the essential elements used for growth of *E. histolytica*. Ferric ammonium citrate is used as iron source in medium preparation for maintaining the axenic culture of *E. histolytica* in laboratory condition [31]. When present in a host, iron demand is satisfied by scavenging iron from host's normal gut flora and also from the iron-containing proteins of host such as hemoglobin and ferratin [32]. NifS, NifU, and rubrerythrin form the iron-sulfur cluster in this parasite [33, 34]. These are essential for proper enzymatic activity of superoxide dismutase, alcohol dehydrogenase 2, and ferredoxin [35]. In vitro experiments showed that low iron concentration significantly decreases the adherence property and cytopathic activity in the parasite [36, 37]. Thus, iron is exclusively and directly linked to its pathogenicity and no other cationic salts are known for this effect [37]. It is evident from transcriptome analysis that when grown in absence of iron, the trophozoites show an increase in transcriptional activity of some proteins, such as cysteine proteinases (CP-EHI_01850, CP-A5, and CP-A7), translation elongation factors, and ribosomal proteins. Stress signal from iron deprivation upregulates the expression of androgen-inducible Gene 1 (AIG1), acyl-CoA synthetase, ComEC protein, and NADPH-dependent oxidoreductase (EhNO2) [38]. Three genes of AIG1 family, EHI_195260, EHI_115160, and EHI_022500, are highly expressed during liver abscess in cell line study signifying its connection with virulence [39]. Reduction from cystine to cysteine is done by EhNO2 to meet the cellular level of cysteine for various functions. EhNO2 reduces the amoebiasis drug, metronidazole, which becomes activated and generates toxic reactive species [40]. During iron starvation of *E. histolytica*, there is an increased gene expression of different transport family proteins such as ABC family of transport proteins, P-glycoprotein-5, and major family transporters to increase the influx of iron from different sources [38, 41].

14.3 Genes Upregulated and Downregulated During Oxidative and Nitrosative Stress

E. histolytica should have different defense strategies to cope with the major stresses caused by cytotoxic reactive species. ROS and RNS are produced during the tissue invasion process by the parasite as it moves from anaerobic colonic lumen to oxygen-rich colonic tissues [3]. ROS and RNS target several cellular proteins, nucleic acids, and lipids [42].

Upon exposure to high oxidative and nitrosative stresses and during the transition from anaerobic to aerobic metabolism, *E. histolytica* upregulates several genes [43–45]. Some of the genes for detoxification of ROS and RNS have been acquired probably from prokaryotes by horizontal gene transfer mechanism [46, 47]. Whole genome microarray revealed that a considerable number of genes are upregulated during oxidative stress in both pathogenic strain of HM1: IMSS (ATCC 30459) and non-pathogenic strain Rahman (ATCC 30886) [44]. However, non-pathogenic strain *E. histolytica* Rahman showed relatively lower number of transcriptional changes. In *E. histolytica* HM1: IMSS, 286 genes were upregulated upon exposure to H_2O_2, whereas for nitrosative stress, the number was 1036 genes. 164 genes overlapped in both stress conditions. 102 genes in total were found to be downregulated due to oxidative stress. Although most of these genes were hypothetical proteins, Rab family GTPase (XM_645246), cyclin (XM_647175), and mitotic inducer phosphatase (XM_644512) were among the known proteins which were downregulated by oxidative stress [44].

14.4 Role of Cellular Proteins

14.4.1 Pyruvate: Ferredoxin Oxidoreductase (EhPFOR)

Controlling steps of energy metabolism in *E. histolytica* are affected by oxidative stress [48]. The organism has neither tricarboxylic acid cycle nor oxidative phosphorylation steps. ATPs are mainly generated by glycolysis with pyruvate as end product. Pyruvate is decarboxylated oxidatively by pyruvate: ferredoxin oxidoreductase (EhPFOR; E.C 1.2.7.1) to acetyl-CoA. In microaerophilic condition, acetyl-CoA is further reduced to acetaldehyde and ethanol, catalyzed by a bifunctional NADH-dependent aldehyde-alcohol dehydrogenase (EhADH2), whereas in aerobic condition, it produces ethanol and acetate by EhADH2 and ADP-forming acetyl-CoA synthetase (AcCoAS, E.C 6.2.13) [49–52]. A characteristic feature of anaerobic protozoan parasites like *Entamoeba* [51], *Giardia* [53], and *Trichomonas* [54] is the absence of pyruvate dehydrogenase complex which converts pyruvate to acetyl-CoA. PFOR connects glycolysis with carbohydrate fermentation. EhPFOR can be detected in plasma membrane and cytoplasmic structures of the trophozoites [48]. Interestingly, it has high preference for pyruvate over other oxoacids [48]. Enzymatic activities of EhPFOR and EhADH2 are susceptible to reactive oxygen species.

14.4.2 Thiol-Dependent Peroxidase/Peroxiredoxin (Eh29/EhPrx)

E. histolytica lacks catalase and glutathione reductase machinery to deal with oxidative stress, but it has a cysteine-rich thiol-dependent peroxiredoxin, EhPrx, or Eh29, a surface antigen located in its outer membrane [55]. This 29-kDa thiol-dependent peroxiredoxin of *E. histolytica* is homologous to alkyl hydroperoxide C-22 protein, AhpC of *Salmonella enterica* serovar typhimurium, and a thiol-specific antioxidant protein of *Saccharomyces cerevisiae* [56]. Under oxidative stress, this protein helps to detoxify the system by reducing peroxides and peroxynitrites [57]. As a membrane-bound peroxyredoxin, Eh29 removes H_2O_2 generated in metabolic processes [58]. A conserved cysteine residue in this 29-kDa protein participates in a cyclic process of peroxide-dependent oxidation and thiol-dependent reduction to protect from H_2O_2 generated by oxidative stress both from internal and external sources [57, 59]. During tissue adherence and invasion, it binds to cytosolic domain of galactose/N-acetylgalactosamine lectin present in surface membrane which carries the signal to counteract the parasite from oxidative attack by the activated host phagocytic cells and epithelial cells, promoting a successful invasion [58, 60, 61].

14.4.3 Flavodiiron Protein (EhFDP1)

Flavodiiron proteins (FDPs) are fundamental constituents of a wide variety of detoxifying enzyme families that work as oxygen and/or nitric oxide reductases [62–64]. Substrate specificity of FDPs is still not clear. Some are selective towards oxygen, a few proteins are selective to nitric oxide, and others have equal selectivity for both. Most prokaryotic anaerobes have FDP-encoding genes and a few anaerobic protozoa contain single or multiple homologs of FDP in their genome. In *E. histolytica* genome, four genes have been identified encoding FDPs [65], whereas *Giardia intestinalis* has one and *Trichomonas vaginalis* has four homologs of FDP. In protozoa, FDPs mainly act as oxygen reductases [47, 66–68]. These genes are probably acquired from prokaryotes via lateral gene transfer [46, 47].

14.4.4 G Protein

G proteins along with other oxidative stress-related proteins is overexpressed during exposure to highly oxygenated environment compared to normally grown *E. histolytica* in axenic culture. A series of signal transduction cascades function during the oxidative stress of this parasite that control the activity of related transcription factors. G protein regulates expression of several genes which are necessary to maintain normal cellular functions to overcome oxidative stress [43].

14.4.5 DNA Methyl Transferase (Ehmeth)

Nitric oxide (NO), released by natural killer cells, activated macrophages, and other phagocytic cells, has a crucial role in host's defense mechanism against pathogens by inhibiting protein synthesis machinery. The responsible factors for this inhibitory activity include NO-mediated cleavage of 18S and 28S rRNA [69] and NO-induced phosphorylation of eukaryotic initiation factor eIF-2α [70]. To overcome the deleterious effect of NO, *E. histolytica* increases the expression of Ehmeth, the cytosine-5-methyltransferase from Dnmt2 family. Experimental data suggests that Ehmeth has a strong connection with Ehmeth-mediated tRNA methylation and process of protein synthesis in this parasite under nitrosative stress [71]. Methylation at 5′-cytosine probably increases tRNA stability and upregulation of 40S and 60S ribosomal proteins maintains the level of protein synthesis. Nitrosative stress-mediated epigenetic regulation of gene expression is also possible [72]. S-Nitrosylation in cysteine residues 228–229 of Ehmeth negatively regulates the activity of Dnmt2 domain and prevents the formation of Ehmeth-enolase complex [73]. In complex form, enolase suppresses the activity of tRNA methylation of Ehmeth. tRNA stability by methylation does not occur in normal or unstressed condition and, therefore, Ehmeth largely remains in Ehmeth-enolase complex in this parasite. During nitrosative stress, S-nitrosylation in cysteine residues inhibits the complex formation and invokes the stress response for survival [71]. *E. histolytica* takes typical strategy like downregulation of protein synthesis to cease energy wasting and building up the intracellular toxicity by production of damaged or misfolded proteins [11, 74].

14.4.6 N-Acetyl Ornithine Deacetylase (EhNAOD)

During acute nitrosative stress, the expression of various redox proteins is increased in the trophozoites of *E. histolytica* [45]. The transcriptome study (RNA-seq) of *E. histolytica* treated with NO donor drug S-nitrosoglutathione (GSNO) shows upregulation of 208 genes, including N-acetyl ornithine deacetylase (EhNAOD, XP_649738.2, Pathema Id: EHI_114340). The trophozoites overexpressing glyceraldehyde 3-phosphate dehydrogenase are more sensitive to nitrosative stress than the control, and EhNAOD helps the parasite to neutralize the detrimental effect of glyceraldehyde 3-phosphate dehydrogenase during nitrosative stress [75]. *E. histolytica* depends on its dynamic actin-made cytoskeleton to move within different compartments of human body [76]. The genes which are associated with actin family cytoskeletal proteins are found to be upregulated during the adaptation of *E. histolytica* to nitrosative stress (Shahi et al. 2016b). The cysteine residues of actin are susceptible to oxidation.

14.4.7 Stress-Induced Adhesion Factor (EhSIAF), Phospholipid Transporting P-Type ATPase/Flippase (EhPTPA), and Arginase

A stress-induced adhesion factor (EhSIAF, XP_649092.1) and a phospholipid transporting P-type ATPase/flippase (EhPTPA, XP_653689.1) of *E. histolytica* are among the responsive genes against oxidative and nitrosative stress [44]. Overexpression of these two proteins enhances the survival of parasite under oxidative stress [77]. From transwell motility assay, it has been found that motility of trophozoites is reduced significantly due to the overexpression of EhSIAF and EhPTPA suggesting a boost in adherence property of the parasite [77]. Conversion of L-arginine to L-ornithine by enzymatic activity is an important source of L-ornithine for this parasite. Ornithine is required in the synthesis of polyamines, and these polyamines have protective role against nitrosative and oxidative stress. Arginase activity is very much essential for *E. histolytica*'s resistance to nitrosative stress [78] and oxidative stress [17]. Importantly, arginase activity is highly inhibited during oxidative stress of *E. histolytica* trophozoites, and arginase-overexpressing *E. histolytica* trophozoites are more resistant to the oxidative stress than control.

14.5 Effect of Enteric Bacteria on the Oxidative Stress Response

The human parasite *E. histolytica* lives in large intestine with many other microorganisms. *E. histolytica* can feed on bacteria and its pathogenicity may also depend on the microbiota present in the intestine [79]. The bacteria with appropriate recognition molecules are selectively ingested by the parasite [16, 80]. Presence of some bacteria may help the growth and pathogenesis of *E. histolytica*, whereas the presences of some filamentous bacteria are harmful for this parasite [79, 81, 82]. In amoebiasis, acute inflammation occurs during which ROS and RNS are released as a result of host immune response. This anaerobic parasite must be able to fight this oxidative stress to establish its pathogenicity. The study of *E. histolytica* transcriptome induced by oxidative stress and live *Enterobacteriaceae* reveals that *E. coli*, *Salmonella enterica* (an enteropathogen often found as co-infections with *E. histolytica*), and *Enterococcus faecalis* (present in human microbiota) protect *E. histolytica* against oxidative stress but not *Lactobacillus acidophilus* (a popular probiotic) [83]. Pre-incubation of *E. histolytica* with this probiotic is associated with the non-protective response involving some signalling molecules like oxidoreductases, kinases, and regulators of small GTPases [83]. In response to oxidative stress, 1402 genes including trafficking factors and small Rab GTPases are strongly downregulated and 1169 genes including ribosomal proteins, translation factors, hydrolases, and peptidases are strongly upregulated. However, no change of expression for these genes was observed upon pre-incubation with live bacteria (*E. coli* O55). LRR proteins of *E. histolytica* play an important role in response to the oxidative stress and live bacteria [83].

14.6 Effects of Stress on the Cellular Events of *E. histolytica*

14.6.1 Effect of Oxidative and Nitrosative Stress on Adherence and Motility

Adherence capacity of oxidatively stressed trophozoites of *E. histolytica* to the HeLa cell monolayer is remarkably less than that of the untreated trophozoites [75]. The Gal/GalNac lectin consists of a heavy subunit, Hgl (170 kDa), and a light subunit, Lgl (35/31 kDa). Hgl mediates *E. histolytica* adherence. Under oxidative stress of trophozoites, Hgl can be oxidized and this may be the reason for reduced adherence and low motility of this parasite [17]. Motility is necessary for survival [84, 85], expressing pathogenicity, and intracellular trafficking of virulence factors [86]. Motility of parasite depends on dynamic actin cytoskeleton and other associated proteins like ARP2/complex whose function is affected by oxidative stress and nitrosative stress [87, 88].

14.6.2 Effect of Oxidative and Nitrosative Stress on Metabolic Activity

Cellular dysfunction of organisms is caused by exposure to various ROS and RNS as they damage the structural and functional units of cells such as proteins, lipids, and DNA [89–91]. Oxidative stress may promote apoptosis in higher eukaryotes, and this happens in *E. histolytica* also [91–94]. Antioxidant enzymes and free radical scavengers build up the powerful mechanisms to defend the cells [95]. From microbial world to higher eukaryotic system, cells avoid the destructive consequences of oxidative stress by a global modification of antioxidants and other metabolic enzymes. Metabolic alterations play a very important role to counteract the adverse effects of oxidative stress [96]. *E. histolytica* trophozoites face challenges by ROS and RNS during the steps of tissue invasion, colonization, and extraintestinal propagation to establish infection [3, 97]. The common antioxidant detoxification systems are absent in *E. histolytica* [18, 65, 98], but the genome of *E. histolytica* is equipped with genes encoding rubrerythrin, peroxiredoxin, hybrid-cluster protein, superoxide dismutase, and flavodiiron proteins for detoxification of ROS and RNS [58, 64, 65]. In *E. histolytica*, pyridine nucleotide transhydrogenase synthesizes NADPH from NADH by utilizing the driving force of electrochemical proton gradient across the membrane [99, 100]. A detailed study on oxidative stress-mediated metabolomic changes in *E. histolytica* reveals that various glycolytic intermediates, such as glucose 6-phosphate, fructose 6-phosphate, and pyruvate, are accumulated in the parasite due to free radical-mediated inactivation of glycolysis-related enzymes [101–103]. *E. histolytica* may accumulate G 6-P, F 6-P, and DHAP on experimentally created nitrosative stress and this can mediate apoptosis in this parasite [102, 103]. In *E. histolytica*, pentose phosphate pathway (PPP) is absent due to the lack of glucose 6-phosphate dehydrogenase (G6PD) and transaldolases [65], and most intermediates of the non-oxidative branch of PPP – erythrose 4-phosphate (E4-P), ribose 5-phosphate (ribose 5-P), ribulose 5-phosphate (ribulose 5-P), and sedoheptulose 7-phosphate (S7-P) – are therefore increased upon oxidative stress [101]. On the other hand, *E. histolytica* has an alternative non-oxidative hexose-pentose interconversion pathway which does not require transaldolases, but depends mainly on three enzymes: phosphofructokinase, transketolase, and aldolase [104]. Ribose 5-phosphate produced in this pathway can serve as the precursor of NAD, which is further used by NAD kinase to generate NADP [105]. Pyridine nucleotide trans-hydrogenase catalyzes the conversion of NADP to NADPH using electrochemical proton gradient [99].

Life cycle of *E. histolytica* can be observed in two distinct morphological forms: trophozoite form which is proliferative, non-infective, but pathogenic and cyst form which is dormant, non-pathogenic, but infective. Cyst surface is shielded with chitin-made cell wall [106, 107]. Oxidative stress activates chitin biosynthetic pathway by two influential intermediates N-acetylglucosamine 6-phosphate (GlcNAc 6-P) and N-acetylglucosamine 1-phosphate (GlcNAc 1-P). Oxidative stress serves as the environmental stimulus to trigger the production of glucosamine 6-phosphate isomerase, an enzyme that converts GlcNAc 6-P and GlcNAc 1-P, and thus cyst-like structures are formed in *E. histolytica* via encystation process [108]. Genome sequence of *E. histolytica* suggests that ATP can be generated from amino acid catabolism during energy crisis and glucose starvation [109].

Sulfur-containing amino acids and their derivatives are synthesized and degraded by unique pathways in *E. histolytica*. Oxidatively stressed trophozoites show increased production of S-methylcysteine from O-acetylserine and methanethiol in a time-dependent manner [22, 110]. Functional trans-sulfuration pathways are missing, but the parasite possesses methionine γ-lyase for the degradation of sulfur-containing amino acids [111–114]. Unique de novo methanethiol and sulfur (sulfide) assimilatory pathways for synthesis of S-methylcysteine and L-cysteine are present in *E. histolytica* [110–112]. NADPH-dependent oxidoreductase acts as cystine reductase to restore L-cysteine [40]. O-phosphoserine, a precursor of L-serine and L-cysteine, increases significantly during oxidative stress [111, 112]. Glycolytic enzymes are highly susceptible to inhibition by ROS [115, 116]. In *E. histolytica*, activities of PFOR, phosphoglycerate mutase, and NAD^+-dependent alcohol dehydrogenase (ADH) are decreased greatly by oxidative stress, whereas there is a comparative marginal decrease in activities of GAPDH, TPI, PGK, ENO, and PPDK. *E. histolytica* can make glycerol from glucose, similar to other protozoa like *T. vaginalis*, *Trypanosoma brucei*, and *Plasmodium falciparum* [117–119]. Glycerol functions as an efficient free radical scavenger and it may protect against oxidative stress. One of the alternative mechanisms is the conversion of glycerol to dihydroxyacetone by glycerol dehydrogenase or glyceraldehyde by glyceraldehyde reductases [101].

14.6.3 Stress Response and Non-coding RNA

Short noncoding regulatory RNAs are present in *E. histolytica* [120–123]. These have an influential role in various cellular processes [124, 125]. EhslncRNA, long non-coding RNA, helps *E. histolytica* cells to overcome stressful conditions [126]. Various stresses like serum starvation, oxygen, and heat stress increase the expression of EhslncRNA indicating its role as a general stress regulator [127]. A 3′-5′ exoribonuclease, EhRrp6, in *E. histolytica* is lost from nucleus in response to stress including heat, oxidative stress, and serum starvation [128].

14.7 Conclusion

E. histolytica has developed multiple mechanisms to adapt oxidative stress, nitrosative stress, glucose starvation, serum starvation, UV exposure, and iron starvation. To survive in harsh environment, it increases or decreases the expression of specific genes. Some of these are common in different stresses, whereas the regulation of some genes is stress specific. A high number of genes are upregulated or downregulated as instant response of the parasite to a particular stress, such as after treatment with H_2O_2 or dipropylenetriamine-NONOate [44]. Moreover, it has the ability to adapt to diverse stresses, and various proteins are involved in this adaptation (Table 14.1). Dihydropyrimidine dehydrogenase helps to adapt in low glucose environment [26] and N-acetyl ornithine deacetylase is involved to overcome nitrosative stress of *E. histolytica* [129]. The activity of PFOR and ADH2 is highly inhibited when *E. histolytica* is subjected to high oxidative stress [102]. During nitrosative stress, Ehmeth-mediated $tRNA^{Asp}$ methylation is crucial to maintain active protein synthesis [71]. In absence of glutathione system and catalase, human pathogen *E. histolytica* overcomes the oxidative stress during invasion into the host cells with the help of thioredoxin system, L-cysteine, Eh29, and other stress-responsive proteins (Table 14.1).

Acknowledgments The financial assistance from BRNS, DAE (Project No. 37 (1)/20/15/2014-BRNS), and CSIR (Project No. 27(0322)/17/EMR-II) is thankfully acknowledged.

Legend to Figure 14.1: *E. histolytica* insulates itself from O_2 exposure, ROS, and RNS primarily by different enzymatic and redox reactions. In glycolytic pathway, glucose is converted into pyruvate via a series of biochemical reactions. Eventually, pyruvate is converted to acetyl-CoA catalyzed by PFOR, which is susceptible for ROS and RNS. In anaerobic or microaerophilic condition, ADHE converts acetyl-CoA into ethanol, but during high oxygen exposure, it produces acetate, indicating that activity of ADHE is inhibited by ROS/RNS. ROS/RNS inhibits the enzyme UDP-GPP. L-cysteine protects the parasite against harmful consequences of ROS/RNS attack by shielding the enzymes to maintain their normal metabolic pathways. L-cysteine can be synthesized in a de novo pathway. A dramatic increase in glycerol biosynthesis occurs due to inactivation of PFOR by ROS/RNS during extensive oxidative stress. Thus, conversion from pyruvate into acetyl-CoA is halted and sends a feedback to stop pyruvate production and activates the machinery towards glycerol biosynthesis. ROS/RNS (O_2, H_2O_2, O_2^-, $NO^{\cdot}$) are neutralized in coordination of several proteins like EhPFOR, Eh29/EhPrx, EhFDP, G protein, Ehmeth, EhNAOD, EhSIAF, EhPTPA, EhTrxR, EhSOD, EhADHE, EhHSP70, EhRbr, EhCP5, and peptidylprolyl isomerase to maintain the intracellular redox state. In glycolysis, a pool of NADH and electrons are generated from reduced Fd, which can be subjected to transhydrogenation by NADP-dependent oxidoreductase or transhydrogenase to produce NADPH and may serve as the reductive equivalent donor to reduce O_2 and ROS by the antioxidant machinery. Divalent O_2 is converted to H_2O_2 with the help of Eh34, EhTrxR, and EhNO. Later on, H_2O_2 produces H_2O by EhPrx which is reduced by Eh34 or Trx. Fd can act as an alternative source of electron to facilitate the reduction of O_2 to H_2O without ROS generation via FDP system. FDP enzyme system can also detoxify the NO·generated by the parasite or from host immune response. Reduced Trx is converted to its oxidized form by TrxR through NADPH oxidation. EhSOD works to clean the harmful intracellular O_2^- by ORP. EhRbr with association of NROR protect mitosomes by converting H_2O_2 into H_2O. ISF and HCP have potential antioxidant capacity. *E. histolytica* can also decrease the redox potential of extracellular environment through the trans-PMET.

References

1. Ackers JCG, Diamond L, Duchene M, Cantellano M et al (1997) WHO/PAHO/UNESCO report. A consultation with experts on amoebiasis. Mexico City, Mexico 28-29 January, 1997. Epidemiol Bull 18(1):13–14
2. Haque R, Huston CD, Hughes M, Houpt E, Petri WA Jr (2003) Amebiasis. N Engl J Med 348(16):1565–1573. https://doi.org/10.1056/NEJMra022710
3. Stanley SL Jr (2003) Amoebiasis. Lancet 361(9362):1025–1034. https://doi.org/10.1016/S0140-6736(03)12830-9
4. Marie C, Petri WA Jr (2013) Amoebic dysentery. BMJ clinical evidence 2013, pii: 0918:1–16
5. Marie C, Petri WA Jr (2014) Regulation of virulence of *Entamoeba histolytica*. Annu Rev Microbiol 68:493–520. https://doi.org/10.1146/annurev-micro-091313-103550
6. Jarnagin WR (2012) Blumgart's surgery of the liver. Pancreas, Biliary Tract and
7. Paiva CN, Bozza MT (2014) Are reactive oxygen species always detrimental to pathogens? Antioxid Redox Signal 20(6):1000–1037. https://doi.org/10.1089/ars.2013.5447
8. Nagaraja S, Ankri S (2018) Utilization of different omic approaches to unravel stress response mechanisms in the parasite *Entamoeba histolytica*. Front Cell Infect Microbiol 8:19. https://doi.org/10.3389/fcimb.2018.00019
9. Mortimer L, Chadee K (2010) The immunopathogenesis of *Entamoeba histolytica*, vol 126. Exp Parasitol, vol 3, 2010/03/23 edn. doi:https://doi.org/10.1016/j.exppara.2010.03.005
10. Anderson P, Kedersha N (2002) Stressful initiations. J Cell Sci 115(Pt 16):3227–3234
11. Weber C, Guigon G, Bouchier C, Frangeul L, Moreira S, Sismeiro O, Gouyette C, Mirelman D, Coppee JY, Guillen N (2006) Stress by heat shock induces massive down regulation of genes and allows differential allelic expression of the Gal/GalNAc lectin in *Entamoeba histolytica*. Eukaryot Cell 5(5):871–875. https://doi.org/10.1128/EC.5.5.871-875.2006
12. Lisby M, Rothstein R (2004) DNA damage checkpoint and repair centers. Curr Opin Cell Biol 16(3):328–334. https://doi.org/10.1016/j.ceb.2004.03.011
13. Lisby M, Rothstein R (2004) DNA repair: keeping it together. Curr Biol 14(23):R994–R996. https://doi.org/10.1016/j.cub.2004.11.020
14. Lettier G, Feng Q, de Mayolo AA, Erdeniz N, Reid RJ, Lisby M, Mortensen UH, Rothstein R (2006) The role of DNA double-strand breaks in spontaneous homologous recombination in *S. cerevisiae*. PLoS Genet 2(11):e194. https://doi.org/10.1371/journal.pgen.0020194
15. Weber C, Marchat LA, Guillen N, Lopez-Camarillo C (2009) Effects of DNA damage induced by UV irradiation on gene expression in the protozoan parasite *Entamoeba histolytica*. Mol Biochem Parasitol 164(2):165–169. https://doi.org/10.1016/j.molbiopara.2008.12.005
16. Bracha R, Mirelman D (1984) Virulence of *Entamoeba histolytica* trophozoites. Effects of bacteria, microaerobic conditions, and metronidazole. J Exp Med 160(2):353–368. https://doi.org/10.1084/jem.160.2.353

17. Shahi P, Trebicz-Geffen M, Nagaraja S, Alterzon-Baumel S, Hertz R, Methling K, Lalk M, Ankri S (2016) Proteomic identification of oxidized proteins in *Entamoeba histolytica* by resin-assisted capture: insights into the role of arginase in resistance to oxidative stress. PLoS Negl Trop Dis 10(1):e0004340. https://doi.org/10.1371/journal.pntd.0004340
18. Fahey RC, Newton GL, Arrick B, Overdank-Bogart T, Aley SB (1984) *Entamoeba histolytica*: a eukaryote without glutathione metabolism. Science 224(4644):70–72. https://doi.org/10.1126/science.6322306
19. Pineda E, Perdomo D (2017) *Entamoeba histolytica* under oxidative stress: what countermeasure mechanisms are in place? Cell 6(4). https://doi.org/10.3390/cells6040044
20. Leitsch D, Kolarich D, Wilson IB, Altmann F, Duchene M (2007) Nitroimidazole action in *Entamoeba histolytica*: a central role for thioredoxin reductase. PLoS Biol 5(8):e211. https://doi.org/10.1371/journal.pbio.0050211
21. Pye VE, Tingey AP, Robson RL, Moody PC (2004) The structure and mechanism of serine acetyltransferase from *Escherichia coli*. J Biol Chem 279(39):40729–40736. https://doi.org/10.1074/jbc.M403751200
22. Hussain S, Ali V, Jeelani G, Nozaki T (2009) Isoform-dependent feedback regulation of serine O-acetyltransferase isoenzymes involved in L-cysteine biosynthesis of *Entamoeba histolytica*. Mol Biochem Parasitol 163(1):39–47. https://doi.org/10.1016/j.molbiopara.2008.09.006
23. Hirayama A, Kami K, Sugimoto M, Sugawara M, Toki N, Onozuka H, Kinoshita T, Saito N, Ochiai A, Tomita M, Esumi H, Soga T (2009) Quantitative metabolome profiling of colon and stomach cancer microenvironment by capillary electrophoresis time-of-flight mass spectrometry. Cancer Res 69(11):4918–4925. https://doi.org/10.1158/0008-5472.CAN-08-4806
24. Kellett GL, Brot-Laroche E, Mace OJ, Leturque A (2008) Sugar absorption in the intestine: the role of GLUT2. Annu Rev Nutr 28:35–54. https://doi.org/10.1146/annurev.nutr.28.061807.155518
25. Cummings JH, Macfarlane GT (1997) Role of intestinal bacteria in nutrient metabolism. JPEN J Parenter Enteral Nutr 21(6):357–365. https://doi.org/10.1177/0148607197021006357
26. Baumel-Alterzon S, Weber C, Guillen N, Ankri S (2013) Identification of dihydropyrimidine dehydrogenase as a virulence factor essential for the survival of *Entamoeba histolytica* in glucose-poor environments. Cell Microbiol 15(1):130–144. https://doi.org/10.1111/cmi.12036
27. Tovy A, Hertz R, Siman-Tov R, Syan S, Faust D, Guillen N, Ankri S (2011) Glucose starvation boosts *Entamoeba histolytica* virulence. PLoS Negl Trop Dis 5(8):e1247. https://doi.org/10.1371/journal.pntd.0001247
28. Beaulande M, Tarbouriech N, Hartlein M (1998) Human cytosolic asparaginyl-tRNA synthetase: cDNA sequence, functional expression in *Escherichia coli* and characterization as human autoantigen. Nucleic Acids Res 26(2):521–524. https://doi.org/10.1093/nar/26.2.521
29. Girjes AA, Hobson K, Chen P, Lavin MF (1995) Cloning and characterization of cDNA encoding a human arginyl-tRNA synthetase. Gene 164(2):347–350. https://doi.org/10.1016/0378-1119(95)00502-W
30. Thibeaux R, Weber C, Hon CC, Dillies MA, Ave P, Coppee JY, Labruyere E, Guillen N (2013) Identification of the virulence landscape essential for *Entamoeba histolytica* invasion of the human colon. PLoS Pathog 9(12):e1003824. https://doi.org/10.1371/journal.ppat.1003824
31. Park SJ, Lee SM, Lee J, Yong TS (2001) Differential gene expression by iron-limitation in *Entamoeba histolytica*. Mol Biochem Parasitol 114(2):257–260. https://doi.org/10.1016/S0166-6851(01)00264-X
32. Wandersman C, Delepelaire P (2004) Bacterial iron sources: from siderophores to hemophores. Annu Rev Microbiol 58:611–647. https://doi.org/10.1146/annurev.micro.58.030603.123811
33. Maralikova B, Ali V, Nakada-Tsukui K, Nozaki T, van der Giezen M, Henze K, Tovar J (2010) Bacterial-type oxygen detoxification and iron-sulfur cluster assembly in amoebal relict mitochondria. Cell Microbiol 12(3):331–342. https://doi.org/10.1111/j.1462-5822.2009.01397.x

34. Ali V, Shigeta Y, Tokumoto U, Takahashi Y, Nozaki T (2004) An intestinal parasitic protist, *Entamoeba histolytica*, possesses a non-redundant nitrogen fixation-like system for iron-sulfur cluster assembly under anaerobic conditions. J Biol Chem 279(16):16863–16874. https://doi.org/10.1074/jbc.M313314200
35. Tannich E, Bruchhaus I, Walter RD, Horstmann RD (1991) Pathogenic and nonpathogenic *Entamoeba histolytica*: identification and molecular cloning of an iron-containing superoxide dismutase. Mol Biochem Parasitol 49(1):61–71. https://doi.org/10.1016/0166-6851(91)90130-X
36. Espinosa A, Perdrizet G, Paz YMCG, Lanfranchi R, Phay M (2009) Effects of iron depletion on *Entamoeba histolytica* alcohol dehydrogenase 2 (EhADH2) and trophozoite growth: implications for antiamoebic therapy. J Antimicrob Chemother 63(4):675–678. https://doi.org/10.1093/jac/dkp015
37. Lee J, Park SJ, Yong TS (2008) Effect of iron on adherence and cytotoxicity of *Entamoeba histolytica* to CHO cell monolayers. Korean J Parasitol 46(1):37–40. https://doi.org/10.3347/kjp.2008.46.1.37
38. Hernandez-Cuevas NA, Weber C, Hon CC, Guillen N (2014) Gene expression profiling in *Entamoeba histolytica* identifies key components in iron uptake and metabolism. PLoS One 9(9):e107102. https://doi.org/10.1371/journal.pone.0107102
39. Biller L, Davis PH, Tillack M, Matthiesen J, Lotter H, Stanley SL Jr, Tannich E, Bruchhaus I (2010) Differences in the transcriptome signatures of two genetically related *Entamoeba histolytica* cell lines derived from the same isolate with different pathogenic properties. BMC Genomics 11:63. https://doi.org/10.1186/1471-2164-11-63
40. Jeelani G, Husain A, Sato D, Ali V, Suematsu M, Soga T, Nozaki T (2010) Two atypical L-cysteine-regulated NADPH-dependent oxidoreductases involved in redox maintenance, L-cystine and iron reduction, and metronidazole activation in the enteric protozoan *Entamoeba histolytica*. J Biol Chem 285(35):26889–26899. https://doi.org/10.1074/jbc.M110.106310
41. Perez-Victoria JM, Parodi-Talice A, Torres C, Gamarro F, Castanys S (2001) ABC transporters in the protozoan parasite *Leishmania*. Int microbiol 4(3):159–166. https://doi.org/10.1007/s10123-001-0031-2
42. Halliwell B, Gutteridge JMC (2015) Free radicals in biology and medicine, 5th edn. Oxford University Press, Oxford. https://doi.org/10.1093/acprof:oso/9780198717478.001.0001
43. Akbar MA, Chatterjee NS, Sen P, Debnath A, Pal A, Bera T, Das P (2004) Genes induced by a high-oxygen environment in *Entamoeba histolytica*. Mol Biochem Parasitol 133(2):187–196. https://doi.org/10.1016/j.molbiopara.2003.10.006
44. Vicente JB, Ehrenkaufer GM, Saraiva LM, Teixeira M, Singh U (2009) *Entamoeba histolytica* modulates a complex repertoire of novel genes in response to oxidative and nitrosative stresses: implications for amebic pathogenesis. Cell Microbiol 11(1):51–69. https://doi.org/10.1111/j.1462-5822.2008.01236.x
45. Santi-Rocca J, Smith S, Weber C, Pineda E, Hon CC, Saavedra E, Olivos-Garcia A, Rousseau S, Dillies MA, Coppee JY, Guillen N (2012) Endoplasmic reticulum stress-sensing mechanism is activated in *Entamoeba histolytica* upon treatment with nitric oxide. PLoS One 7(2):e31777. https://doi.org/10.1371/journal.pone.0031777
46. Andersson JO, Hirt RP, Foster PG, Roger AJ (2006) Evolution of four gene families with patchy phylogenetic distributions: influx of genes into protist genomes. BMC Evol Biol 6:27. https://doi.org/10.1186/1471-2148-6-27
47. Andersson JO, Sjogren AM, Davis LA, Embley TM, Roger AJ (2003) Phylogenetic analyses of diplomonad genes reveal frequent lateral gene transfers affecting eukaryotes. Curr Biol 13(2):94–104. https://doi.org/10.1016/S0960-9822(03)00003-4
48. Pineda E, Encalada R, Rodriguez-Zavala JS, Olivos-Garcia A, Moreno-Sanchez R, Saavedra E (2010) Pyruvate:ferredoxin oxidoreductase and bifunctional aldehyde-alcohol dehydrogenase are essential for energy metabolism under oxidative stress in *Entamoeba histolytica*. FEBS J 277(16):3382–3395. https://doi.org/10.1111/j.1742-4658.2010.07743.x

49. Reeves RE (1984) Metabolism of *Entamoeba histolytica* Schaudinn, 1903. Adv Parasitol 23:105–142
50. Montalvo FE, Reeves RE, Warren LG (1971) Aerobic and anaerobic metabolism in *Entamoeba histolytica*. Exp Parasitol 30(2):249–256. https://doi.org/10.1016/0014-4894(71)90089-0
51. Reeves RE, Warren LG, Susskind B, Lo HS (1977) An energy-conserving pyruvate-to-acetate pathway in *Entamoeba histolytica*. Pyruvate synthase and a new acetate thiokinase. J Biol Chem 252(2):726–731
52. Lo HS, Reeves RE (1978) Pyruvate-to-ethanol pathway in *Entamoeba histolytica*. Biochem J 171(1):225–230. https://doi.org/10.1042/bj1710225
53. Townson SM, Upcroft JA, Upcroft P (1996) Characterisation and purification of pyruvate:ferredoxin oxidoreductase from *Giardia duodenalis*. Mol Biochem Parasitol 79(2):183–193. https://doi.org/10.1016/0166-6851(96)02661-8
54. Williams K, Lowe PN, Leadlay PF (1987) Purification and characterization of pyruvate: ferredoxin oxidoreductase from the anaerobic protozoon *Trichomonas vaginalis*. Biochem J 246(2):529–536. https://doi.org/10.1042/bj2460529
55. Bruchhaus I, Tannich E (1993) Analysis of the genomic sequence encoding the 29-kDa cysteine-rich protein of *Entamoeba histolytica*. Trop Med Parasitol 44(2):116–118
56. Flores BM, Batzer MA, Stein MA, Petersen C, Diedrich DL, Torian BE (1993) Structural analysis and demonstration of the 29 kDa antigen of pathogenic *Entamoeba histolytica* as the major accessible free thiol-containing surface protein. Mol Microbiol 7(5):755–763. https://doi.org/10.1111/j.1365-2958.1993.tb01166.x
57. Bruchhaus I, Richter S, Tannich E (1997) Removal of hydrogen peroxide by the 29 kDa protein of *Entamoeba histolytica*. Biochem J 326(Pt 3):785–789. https://doi.org/10.1042/bj3260785
58. Sen A, Chatterjee NS, Akbar MA, Nandi N, Das P (2007) The 29-kilodalton thiol-dependent peroxidase of *Entamoeba histolytica* is a factor involved in pathogenesis and survival of the parasite during oxidative stress. Eukaryot Cell 6(4):664–673. https://doi.org/10.1128/EC.00308-06
59. Cheng XJ, Yoshihara E, Takeuchi T, Tachibana H (2004) Molecular characterization of peroxiredoxin from *Entamoeba moshkovskii* and a comparison with *Entamoeba histolytica*. Mol Biochem Parasitol 138(2):195–203. https://doi.org/10.1016/j.molbiopara.2004.08.009
60. Hughes MA, Lee CW, Holm CF, Ghosh S, Mills A, Lockhart LA, Reed SL, Mann BJ (2003) Identification of *Entamoeba histolytica* thiol-specific antioxidant as a GalNAc lectin-associated protein. Mol Biochem Parasitol 127(2):113–120. https://doi.org/10.1016/S0166-6851(02)00326-2
61. Jeelani G, Nozaki T (2016) *Entamoeba* thiol-based redox metabolism: A potential target for drug development. Mol Biochem Parasitol 206(1–2):39–45. https://doi.org/10.1016/j.molbiopara.2016.01.004
62. Kurtz DM (2007) Flavo-diiron enzymes: nitric oxide or dioxygen reductases? Dalton T 37:4115–4121. https://doi.org/10.1039/b710047g
63. Vicente JB, Justino MC, Goncalves VL, Saraiva LM, Teixeira M (2008) Biochemical, spectroscopic, and thermodynamic properties of flavodiiron proteins. Methods Enzymol 437:21–45. https://doi.org/10.1016/S0076-6879(07)37002-X
64. Saraiva LM, Vicente JB, Teixeira M (2004) The role of the flavodiiron proteins in microbial nitric oxide detoxification. Adv Microb Physiol 49:77–129. https://doi.org/10.1016/S0065-2911(04)49002-X
65. Loftus B, Anderson I, Davies R, Alsmark UC, Samuelson J, Amedeo P, Roncaglia P, Berriman M, Hirt RP, Mann BJ, Nozaki T, Suh B, Pop M, Duchene M, Ackers J, Tannich E, Leippe M, Hofer M, Bruchhaus I, Willhoeft U, Bhattacharya A, Chillingworth T, Churcher C, Hance Z, Harris B, Harris D, Jagels K, Moule S, Mungall K, Ormond D, Squares R, Whitehead S, Quail MA, Rabbinowitsch E, Norbertczak H, Price C, Wang Z, Guillen N, Gilchrist C, Stroup SE, Bhattacharya S, Lohia A, Foster PG, Sicheritz-Ponten T, Weber C, Singh U, Mukherjee C, El-Sayed NM, Petri WA Jr, Clark CG, Embley TM, Barrell B, Fraser CM, Hall N (2005)

The genome of the protist parasite *Entamoeba histolytica*. Nature 433(7028):865–868. https://doi.org/10.1038/nature03291
66. Di Matteo A, Scandurra FM, Testa F, Forte E, Sarti P, Brunori M, Giuffre A (2008) The O2-scavenging flavodiiron protein in the human parasite *Giardia intestinalis*. J Biol Chem 283(7):4061–4068. https://doi.org/10.1074/jbc.M705605200
67. Smutna T, Goncalves VL, Saraiva LM, Tachezy J, Teixeira M, Hrdy I (2009) Flavodiiron protein from *Trichomonas vaginalis* hydrogenosomes: the terminal oxygen reductase. Eukaryot Cell 8(1):47–55. https://doi.org/10.1128/EC.00276-08
68. Mastronicola D, Giuffre A, Testa F, Mura A, Forte E, Bordi E, Pucillo LP, Fiori PL, Sarti P (2011) *Giardia intestinalis* escapes oxidative stress by colonizing the small intestine: A molecular hypothesis. IUBMB Life 63(1):21–25. https://doi.org/10.1002/iub.409
69. Cai CQ, Guo H, Schroeder RA, Punzalan C, Kuo PC (2000) Nitric oxide-dependent ribosomal RNA cleavage is associated with inhibition of ribosomal peptidyl transferase activity in ANA-1 murine macrophages. J Immunol 165(7):3978–3984. https://doi.org/10.4049/jimmunol.165.7.3978
70. Kim YM, Son K, Hong SJ, Green A, Chen JJ, Tzeng E, Hierholzer C, Billiar TR (1998) Inhibition of protein synthesis by nitric oxide correlates with cytostatic activity: nitric oxide induces phosphorylation of initiation factor eIF-2 alpha. Mol Med 4(3):179–190
71. Hertz R, Tovy A, Kirschenbaum M, Geffen M, Nozaki T, Adir N, Ankri S (2014) The *Entamoeba histolytica* Dnmt2 homolog (Ehmeth) confers resistance to nitrosative stress. Eukaryot Cell 13(4):494–503. https://doi.org/10.1128/EC.00031-14
72. Illi B, Colussi C, Grasselli A, Farsetti A, Capogrossi MC, Gaetano C (2009) NO sparks off chromatin: tales of a multifaceted epigenetic regulator. Pharmacol Ther 123(3):344–352. https://doi.org/10.1016/j.pharmthera.2009.05.003
73. Tovy A, Siman Tov R, Gaentzsch R, Helm M, Ankri S (2010) A new nuclear function of the *Entamoeba histolytica* glycolytic enzyme enolase: the metabolic regulation of cytosine-5 methyltransferase 2 (Dnmt2) activity. PLoS Pathog 6(2):e1000775. https://doi.org/10.1371/journal.ppat.1000775
74. Vogel C, Silva GM, Marcotte EM (2011) Protein expression regulation under oxidative stress. Mol Cell Proteomics: MCP 10(12):M111009217. https://doi.org/10.1074/mcp.M111.009217
75. Shahi P, Trebicz-Geffen M, Nagaraja S, Hertz R, Alterzon-Baumel S, Methling K, Lalk M, Mazumder M, Samudrala G, Ankri S (2016) N-acetyl ornithine deacetylase is a moonlighting protein and is involved in the adaptation of Entamoeba histolytica to nitrosative stress. Sci Rep 6:36323. https://doi.org/10.1038/srep36323
76. Aguilar-Rojas A, Olivo-Marin JC, Guillen N (2016) The motility of *Entamoeba histolytica*: finding ways to understand intestinal amoebiasis. Curr Opin Microbiol 34:24–30. https://doi.org/10.1016/j.mib.2016.07.016
77. Rastew E, Vicente JB, Singh U (2012) Oxidative stress resistance genes contribute to the pathogenic potential of the anaerobic protozoan parasite, *Entamoeba histolytica*. Int J Parasitol 42(11):1007–1015. https://doi.org/10.1016/j.ijpara.2012.08.006
78. Elnekave K, Siman-Tov R, Ankri S (2003) Consumption of L-arginine mediated by *Entamoeba histolytica* L-arginase (EhArg) inhibits amoebicidal activity and nitric oxide production by activated macrophages. Parasite Immunol 25(11–12):597–608. https://doi.org/10.1111/j.0141-9838.2004.00669.x
79. Burgess SL, Petri WA Jr (2016) The intestinal bacterial microbiome and *E. histolytica* infection. Curr Trop Med Rep 3:71–74. https://doi.org/10.1007/s40475-016-0083-1
80. Mirelman D, Feingold C, Wexler A, Bracha R (1983) Interactions between *Entamoeba histolytica*, bacteria and intestinal cells. Ciba Found Symp 99:2–30
81. Paniagua GL, Monroy E, Garcia-Gonzalez O, Alonso J, Negrete E, Vaca S (2007) Two or more enteropathogens are associated with diarrhoea in Mexican children. Ann Clin Microbiol Antimicrob 6:17. https://doi.org/10.1186/1476-0711-6-17
82. Gilchrist CA, Petri SE, Schneider BN, Reichman DJ, Jiang N, Begum S, Watanabe K, Jansen CS, Elliott KP, Burgess SL, Ma JZ, Alam M, Kabir M, Haque R, Petri WA Jr (2016) Role

of the gut microbiota of children in diarrhea due to the protozoan *Parasite Entamoeba histolytica*. J Infect Dis 213(10):1579–1585. https://doi.org/10.1093/infdis/jiv772

83. Varet H, Shaulov Y, Sismeiro O, Trebicz-Geffen M, Legendre R, Coppee JY, Ankri S, Guillen N (2018) Enteric bacteria boost defences against oxidative stress in *Entamoeba histolytica*. Sci Rep 8(1):9042. https://doi.org/10.1038/s41598-018-27086-w
84. Arhets P, Gounon P, Sansonetti P, Guillen N (1995) Myosin II is involved in capping and uroid formation in the human pathogen *Entamoeba histolytica*. Infect Immun 63(11):4358–4367
85. Coudrier E, Amblard F, Zimmer C, Roux P, Olivo-Marin JC, Rigothier MC, Guillen N (2005) Myosin II and the Gal-GalNAc lectin play a crucial role in tissue invasion by *Entamoeba histolytica*. Cell Microbiol 7(1):19–27. https://doi.org/10.1111/j.1462-5822.2004.00426.x
86. Perdomo D, Manich M, Syan S, Olivo-Marin JC, Dufour AC, Guillen N (2016) Intracellular traffic of the lysine and glutamic acid rich protein KERP1 reveals features of endomembrane organization in *Entamoeba histolytica*. Cell Microbiol 18(8):1134–1152. https://doi.org/10.1111/cmi.12576
87. Kumar N, Somlata MM, Dutta P, Maiti S, Gourinath S (2014) EhCoactosin stabilizes actin filaments in the protist parasite *Entamoeba histolytica*. PLoS Pathog 10(9):e1004362. https://doi.org/10.1371/journal.ppat.1004362
88. Hertz R, Ben Lulu S, Shahi P, Trebicz-Geffen M, Benhar M, Ankri S (2014) Proteomic identification of S-nitrosylated proteins in the parasite *Entamoeba histolytica* by resin-assisted capture: insights into the regulation of the Gal/GalNAc lectin by nitric oxide. PLoS One 9(3):e91518. https://doi.org/10.1371/journal.pone.0091518
89. Imlay JA (2003) Pathways of oxidative damage. Annu Rev Microbiol 57:395–418. https://doi.org/10.1146/annurev.micro.57.030502.090938
90. Apel K, Hirt H (2004) Reactive oxygen species: metabolism, oxidative stress, and signal transduction. Annu Rev Plant Biol 55:373–399. https://doi.org/10.1146/annurev.arplant.55.031903.141701
91. Perrone GG, Tan SX, Dawes IW (2008) Reactive oxygen species and yeast apoptosis. Biochim Biophys Acta 1783(7):1354–1368. https://doi.org/10.1016/j.bbamcr.2008.01.023
92. Clement MV, Pervaiz S (1999) Reactive oxygen intermediates regulate cellular response to apoptotic stimuli: an hypothesis. Free Radic Res 30(4):247–252. https://doi.org/10.1080/10715769900300271
93. Nandi N, Sen A, Banerjee R, Kumar S, Kumar V, Ghosh AN, Das P (2010) Hydrogen peroxide induces apoptosis-like death in *Entamoeba histolytica* trophozoites. Microbiology 156(Pt 7):1926–1941. https://doi.org/10.1099/mic.0.034066-0
94. Ghosh AS, Dutta S, Raha S (2010) Hydrogen peroxide-induced apoptosis-like cell death in *Entamoeba histolytica*. Parasitol Int 59(2):166–172. https://doi.org/10.1016/j.parint.2010.01.001
95. Temple MD, Perrone GG, Dawes IW (2005) Complex cellular responses to reactive oxygen species. Trends Cell Biol 15(6):319–326. https://doi.org/10.1016/j.tcb.2005.04.003
96. Ralser M, Wamelink MM, Kowald A, Gerisch B, Heeren G, Struys EA, Klipp E, Jakobs C, Breitenbach M, Lehrach H, Krobitsch S (2007) Dynamic rerouting of the carbohydrate flux is key to counteracting oxidative stress. J Biol 6(4):10. https://doi.org/10.1186/jbiol61
97. Bogdan C, Rollinghoff M, Diefenbach A (2000) Reactive oxygen and reactive nitrogen intermediates in innate and specific immunity. Curr Opin Immunol 12(1):64–76. https://doi.org/10.1016/S0952-7915(99)00052-7
98. Mehlotra RK (1996) Antioxidant defense mechanisms in parasitic protozoa. Crit Rev Microbiol 22(4):295–314. https://doi.org/10.3109/10408419609105484
99. Jackson JB (1991) The proton-translocating nicotinamide adenine dinucleotide transhydrogenase. J Bioenerg Biomembr 23(5):715–741. https://doi.org/10.1007/BF00785998
100. Yousuf MA, Mi-ichi F, Nakada-Tsukui K, Nozaki T (2010) Localization and targeting of an unusual pyridine nucleotide transhydrogenase in *Entamoeba histolytica*. Eukaryot Cell 9(6):926–933. https://doi.org/10.1128/EC.00011-10

101. Husain A, Sato D, Jeelani G, Soga T, Nozaki T (2012) Dramatic increase in glycerol biosynthesis upon oxidative stress in the anaerobic protozoan parasite *Entamoeba histolytica*. PLoS Negl Trop Dis 6(9):e1831. https://doi.org/10.1371/journal.pntd.0001831
102. Ramos-Martinez E, Olivos-Garcia A, Saavedra E, Nequiz M, Sanchez EC, Tello E, El-Hafidi M, Saralegui A, Pineda E, Delgado J, Montfort I, Perez-Tamayo R (2009) *Entamoeba histolytica*: oxygen resistance and virulence. Int J Parasitol 39(6):693–702. https://doi.org/10.1016/j.ijpara.2008.11.004
103. Ramos E, Olivos-Garcia A, Nequiz M, Saavedra E, Tello E, Saralegui A, Montfort I, Perez Tamayo R (2007) *Entamoeba histolytica*: apoptosis induced in vitro by nitric oxide species. Exp Parasitol 116(3):257–265. https://doi.org/10.1016/j.exppara.2007.01.011
104. Susskind BM, Warren LG, Reeves RE (1982) A pathway for the interconversion of hexose and pentose in the parasitic amoeba *Entamoeba histolytica*. Biochem J 204(1):191–196
105. Pollak N, Dolle C, Ziegler M (2007) The power to reduce: pyridine nucleotides--small molecules with a multitude of functions. Biochem J 402(2):205–218. https://doi.org/10.1042/BJ20061638
106. Chavez-Munguia B, Martinez-Palomo A (2011) High-resolution electron microscopical study of cyst walls of *Entamoeba* spp. J Eukaryot Microbiol 58(6):480–486. https://doi.org/10.1111/j.1550-7408.2011.00576.x
107. Arroyo-Begovich A, Carabez-Trejo A, Ruiz-Herrera J (1978) Composition of the cell wall of *Entamoeba invadens* cysts. Arch Invest Med (Mex), 9(Suppl 1). 1978/01/01 edn. https://doi.org/10.2307/3280662
108. Aguilar-Diaz H, Diaz-Gallardo M, Laclette JP, Carrero JC (2010) In vitro induction of *Entamoeba histolytica* cyst-like structures from trophozoites. PLoS Negl Trop Dis 4(2):e607. https://doi.org/10.1371/journal.pntd.0000607
109. Anderson IJ, Loftus BJ (2005) *Entamoeba histolytica*: observations on metabolism based on the genome sequence. Exp Parasitol 110(3):173–177. https://doi.org/10.1016/j.exppara.2005.03.010
110. Husain A, Sato D, Jeelani G, Mi-ichi F, Ali V, Suematsu M, Soga T, Nozaki T (2010) Metabolome analysis revealed increase in S-methylcysteine and phosphatidylisopropanolamine synthesis upon L-cysteine deprivation in the anaerobic protozoan parasite *Entamoeba histolytica*. J Biol Chem 285(50):39160–39170. https://doi.org/10.1074/jbc.M110.167304
111. Nozaki T, Ali V, Tokoro M (2005) Sulfur-containing amino acid metabolism in parasitic protozoa. Adv Parasitol 60:1–99. https://doi.org/10.1016/S0065-308X(05)60001-2
112. Ali V, Nozaki T (2007) Current therapeutics, their problems, and sulfur-containing-amino-acid metabolism as a novel target against infections by "amitochondriate" protozoan parasites. Clin Microbiol Rev 20(1):164–187. https://doi.org/10.1128/CMR.00019-06
113. Tokoro M, Asai T, Kobayashi S, Takeuchi T, Nozaki T (2003) Identification and characterization of two isoenzymes of methionine gamma-lyase from *Entamoeba histolytica*: a key enzyme of sulfur-amino acid degradation in an anaerobic parasitic protist that lacks forward and reverse trans-sulfuration pathways. J Biol Chem 278(43):42717–42727. https://doi.org/10.1074/jbc.M212414200
114. Husain A, Jeelani G, Sato D, Nozaki T (2011) Global analysis of gene expression in response to L-Cysteine deprivation in the anaerobic protozoan parasite *Entamoeba histolytica*. BMC Genomics 12:275. https://doi.org/10.1186/1471-2164-12-275
115. Colussi C, Albertini MC, Coppola S, Rovidati S, Galli F, Ghibelli L (2000) H2O2-induced block of glycolysis as an active ADP-ribosylation reaction protecting cells from apoptosis. FASEB J 14(14):2266–2276. https://doi.org/10.1096/fj.00-0074com
116. Shenton D, Grant CM (2003) Protein S-thiolation targets glycolysis and protein synthesis in response to oxidative stress in the yeast *Saccharomyces cerevisiae*. Biochem J 374(Pt 2):513–519. https://doi.org/10.1042/BJ20030414
117. Lian LY, Al-Helal M, Roslaini AM, Fisher N, Bray PG, Ward SA, Biagini GA (2009) Glycerol: an unexpected major metabolite of energy metabolism by the human malaria parasite. Malar J 8:38. https://doi.org/10.1186/1475-2875-8-38

118. Hammond DJ, Bowman IB (1980) Studies on glycerol kinase and its role in ATP synthesis in *Trypanosoma brucei*. Mol Biochem Parasitol 2(2):77–91. https://doi.org/10.1016/0166-6851(80)90033-X
119. Chapman A, Linstead DJ, Lloyd D, Williams J (1985) 13C-NMR reveals glycerol as an unexpected major metabolite of the protozoan parasite *Trichomonas vaginalis*. FEBS Lett 191(2):287–292. https://doi.org/10.1016/0014-5793(85)80026-0
120. Mar-Aguilar F, Trevino V, Salinas-Hernandez JE, Tamez-Guerrero MM, Barron-Gonzalez MP, Morales-Rubio E, Trevino-Neavez J, Verduzco-Martinez JA, Morales-Vallarta MR, Resendez-Perez D (2013) Identification and characterization of microRNAs from *Entamoeba histolytica* HM1-IMSS. PLoS One 8(7):e68202. https://doi.org/10.1371/journal.pone.0068202
121. Morf L, Pearson RJ, Wang AS, Singh U (2013) Robust gene silencing mediated by antisense small RNAs in the pathogenic protist *Entamoeba histolytica*. Nucleic Acids Res 41(20):9424–9437. https://doi.org/10.1093/nar/gkt717
122. Zhang H, Alramini H, Tran V, Singh U (2011) Nucleus-localized antisense small RNAs with 5′-polyphosphate termini regulate long term transcriptional gene silencing in *Entamoeba histolytica* G3 strain. J Biol Chem 286(52):44467–44479. https://doi.org/10.1074/jbc.M111.278184
123. Zhang H, Ehrenkaufer GM, Pompey JM, Hackney JA, Singh U (2008) Small RNAs with 5′-polyphosphate termini associate with a Piwi-related protein and regulate gene expression in the single-celled eukaryote *Entamoeba histolytica*. PLoS Pathog 4(11):e1000219. https://doi.org/10.1371/journal.ppat.1000219
124. Geisler S, Coller J (2013) RNA in unexpected places: long non-coding RNA functions in diverse cellular contexts. Nat Rev Mol Cell Biol 14(11):699–712. https://doi.org/10.1038/nrm3679
125. Rinn JL, Chang HY (2012) Genome regulation by long noncoding RNAs. Annu Rev Biochem 81:145–166. https://doi.org/10.1146/annurev-biochem-051410-092902
126. Shrimal S, Bhattacharya S, Bhattacharya A (2010) Serum-dependent selective expression of EhTMKB1-9, a member of *Entamoeba histolytica* B1 family of transmembrane kinases. PLoS Pathog 6(6):e1000929. https://doi.org/10.1371/journal.ppat.1000929
127. Saha A, Bhattacharya S, Bhattacharya A (2016) Serum stress responsive gene EhslncRNA of *Entamoeba histolytica* is a novel long noncoding RNA. Sci Rep 6:27476. https://doi.org/10.1038/srep27476
128. Ray AK, Naiyer S, Singh SS, Bhattacharya A, Bhattacharya S (2018) Application of SHAPE reveals in vivo RNA folding under normal and growth-stressed conditions in the human parasite *Entamoeba histolytica*. Mol Biochem Parasitol 219:42–51. https://doi.org/10.1016/j.molbiopara.2017.11.003
129. Shahi P, Trebicz-Geffen M, Nagaraja S, Hertz R, Baumel-Alterzon S, Methling K, Lalk M, Mazumder M, Samudrala G, Ankri S (2016) N-acetyl ornithine deacetylase is a moonlighting protein and is involved in the adaptation of *Entamoeba histolytica* to nitrosative stress. Sci Rep 6:36323. https://doi.org/10.1038/srep36323
130. Satish S, Bakre AA, Bhattacharya S, Bhattacharya A (2003) Stress-dependent expression of a polymorphic, charged antigen in the protozoan parasite *Entamoeba histolytica*. Infect Immun 71(8):4472–4486. https://doi.org/10.1128/IAI.71.8.4472-4486.2003
131. Vicente JB, Tran V, Pinto L, Teixeira M, Singh U (2012) A detoxifying oxygen reductase in the anaerobic protozoan *Entamoeba histolytica*. Eukaryot Cell 11(9):1112–1118. https://doi.org/10.1128/EC.00149-12
132. Arias DG, Regner EL, Iglesias AA, Guerrero SA (2012) *Entamoeba histolytica* thioredoxin reductase: molecular and functional characterization of its atypical properties. Biochim Biophys Acta 1820(12):1859–1866. https://doi.org/10.1016/j.bbagen.2012.08.020
133. Pearson RJ, Morf L, Singh U (2013) Regulation of H2O2 stress-responsive genes through a novel transcription factor in the protozoan pathogen *Entamoeba histolytica*. J Biol Chem 288(6):4462–4474. https://doi.org/10.1074/jbc.M112.423467

134. Santos F, Nequiz M, Hernandez-Cuevas NA, Hernandez K, Pineda E, Encalada R, Guillen N, Luis-Garcia E, Saralegui A, Saavedra E, Perez-Tamayo R, Olivos-Garcia A (2015) Maintenance of intracellular hypoxia and adequate heat shock response are essential requirements for pathogenicity and virulence of *Entamoeba histolytica*. Cell Microbiol 17(7):1037–1051. https://doi.org/10.1111/cmi.12419
135. Manochitra K, Parija SC (2017) In-silico prediction and modeling of the *Entamoeba histolytica* proteins: Serine-rich *Entamoeba histolytica* protein and 29 kDa Cysteine-rich protease. PeerJ 5:e3160. https://doi.org/10.7717/peerj.3160
136. Mares RE, Ramos MA (2018) An amebic protein disulfide isomerase (PDI) complements the yeast PDI1 mutation but is unable to support cell viability under ER or thermal stress. FEBS open bio 8(1):49–55. https://doi.org/10.1002/2211-5463.12350

Oxidative Stress Regulation in *Giardia lamblia*

15

Dibyendu Raj and Sandipan Ganguly

Abstract

Giardia lamblia is a common gut parasite that infects the human gastrointestinal tract and causes 280 million cases of diarrhea every year. There are several reports on oxidative stress management in *G. lamblia* and how *Giardia* establishes its pathogenesis against the high oxygen tension in the gut. Here we have discussed about four oxidative stress-generating conditions which are actually mimicking the environment where *Giardia* resides. This chapter explored the fact that pyruvate, a crucial intermediary metabolite, acts as an antioxidant. This chapter has also elucidated the conserved and stress-specific responses of the trophozoite under different stressed conditions. The chapter will unveil the information generated by proteomic study. The chapter will also reveal how oxidative stress regulation works in *Giardia* trophozoites.

Keywords

Giardia lamblia · Oxidative stress · Cysteine-ascorbate · Antioxidant · Pyruvate · Reactive oxygen species

D. Raj
Vivekananda College, Kolkata, India

Division of Parasitology, ICMR-National Institute of Cholera and Enteric Diseases, Kolkata, India

S. Ganguly (✉)
Division of Parasitology, ICMR-National Institute of Cholera and Enteric Diseases, Kolkata, India
e-mail: sandipanganguly@gmail.com

S. Chakraborti et al. (eds.), *Oxidative Stress in Microbial Diseases*,
https://doi.org/10.1007/978-981-13-8763-0_15

15.1 Introduction

In 1681, *Giardia* was first depicted by Antonie Van Leeuwenhoek during the study of his own diarrheal stool under compound microscope [27]. Nowadays, *Giardia* is familiar as a very important re-emerging protozoan parasite of the upper intestinal tract, causing acute illness to humans worldwide [86]. The only species *Giardia lamblia* (Synonym *G. intestinalis* and *G. duodenalis*) generally infects humans with an estimated 280 million cases every year [44]. In developing countries like Asia, Africa, and Latin America, 500,000 new cases are reported per year [26]. In children of below 5 years, *Giardia* infections have significant effects on growth, nutritional development, and comprehensive abilities [79]. The World Health Organization (WHO) incorporated *Giardia* in the "Neglected Diseases Initiative" in 2004, because giardiasis adds to the microbial disease burden especially in the developing world, impeding development and socioeconomic advancement in these countries [74]. The enteric pathogen *G. duodenalis* is also very common in livestock, wild life, dogs, and cats [81]. Eight genotypes (assemblages) have been identified for *G. duodenalis* designated from A to H [32]. Humans are infected only by assemblages A and B, but *Giardia* are also capable of infecting most mammals [87]; thus, *G. duodenalis* is a parasite with zoonotic potential. Zoonotic risk factors include the exposure to recreational waters, drinking untreated water polluted with fecal material from humans or animals, and close contact with infected animals [93]. Cysts are the most infective form of the parasite *Giardia*. They do not require maturation, and they are immediately infectious upon their release from infected hosts into the environment [32]. They have the low infective dose required to cause giardiasis (i.e., 10–100 cysts) [70], zoonosis has raised public health concerns, and strategies have been recommended to prevent zoonotic transmission [94]. In their life cycle, *Giardia* has two alternate stages: the trophozoite and cyst. The cyst predominates in the environment due to the hard cell wall (CW) that protects it from harsh environment. The trophozoite grows and thrives inside the small intestines; however, because of its fragile nature, it breaks quickly after leaving the host [87]. Humans are infected by the intake of *Giardia* cysts via contaminated meal and unhygienic water or direct or indirect fecal oral contact with infected hosts [82]. The cysts produce in the duodenum and jejunum where trophozoites emerge, multiply, and attach to the microvillus brush border of epithelial cells [57]. Trophozoite attachment and consequent interaction with enterocytes activates pathogenic mechanisms such as the stimulation of functional and structural changes in the intestinal epithelia, the modulation of host metabolism in favor of parasite growth and differentiation, and triggering of host immune defense in response to the parasite or certain parasite's molecules [63]. Collectively, the interplay between these factors determines the clinical outcomes of *Giardia* infection, which ranges from asymptomatic carriage to very sensitive acute and chronic diarrhea [3]. To date, the accurate mechanism of infection by which it causes disease is completely unrevealed, as the parasite does not invade cells or produce any proteins with direct cellular toxicity [3]. It is also unclear why symptoms of giardiasis vary among individuals [18]. Nevertheless, it has been suggested that this might be ascribed to both host (i.e., immune status, age, and nutrition) and parasite factors (i.e., differences in the virulence of *Giardia* isolates) [38].

The *Giardia* infection can be of two types: asymptomatic which means it has no such manifestation of exact symptoms and symptomatic, with much clinical appearance like diarrhea, epigastric pain, nausea, bloating, flatulence, and steatorrhea [60]. Some patients with *Giardia* infection may develop a chronic infection with constant diarrhea, malabsorption, malnutrition, weight loss in infants, and reduced growth rate in children [92]. It lacks conventional mechanisms of oxidative stress regulation, including catalase, superoxide dismutase, peroxidase, and glutathione, which are functional in nearly all eukaryotes. *Giardia duodenalis*, an aerotolerant protozoon, generates energy by fermentative metabolism. It is devoid of true mitochondria, peroxisomes, and the Golgi body [30]. The anaerobic *G. lamblia* rely on fermentative metabolism for ATP production as an energy source [14]. Carbohydrates (glucose) are converted to various organic compounds that include alanine, ethanol, acetate, and CO_2 [8]. Pyruvate formed by glycolysis is a key intermediate of energy production and is the precursor of various metabolic end products which are dependent on the ambient environment for oxygen pressure [64]. Like *Entamoeba histolytica*, *Trichomonas vaginalis*, and *Dasytricha ruminantium*, in *Giardia*, the conversion of pyruvate to acetate, an essential metabolic sequence, occurs in the cytosol, the site of all steps of energy metabolism. The energy of the thioester bond of acetyl-CoA is always conserved by substrate-level phosphorylation. Thus, acetate formation plays a crucial role in *Giardia*'s energy metabolism. The rates of overall metabolic pathway are alike under aerobic and anaerobic conditions in *Giardia* and the metabolism remains fermentative [46]. As a result, some minor shifts occur in the overall carbon metabolism, and for that reason the proportion of more oxidized end products increased (Muller 1988) and it has also been hypothesized that change in the redox state of NAD(P)H pools affects the relative rates of end product formation [54]. Cells have a capability to restructure their transcriptomes in order to acclimatize to the environmental circumstances by sensing the endogenous level of various metabolites. In the environment of high oxygen concentration, *Giardia lamblia* consume oxygen up to a threshold level depending on the various species, above which consumption is detained due to the generation of ROS [6, 47].

The detailed mechanism by which the parasite could relieve the detoxification of reactive oxygen species (ROS) formed under oxidative stress is unknown until today. In this context, three approaches have been preferred to generate oxidative stress in the trophozoite in vitro. First, hydrogen peroxide (H_2O_2), which is a well-known chemical reagent, produces free oxygen radicals effortlessly. Second one, a modified medium that lacks cysteine and ascorbic acid. One of the most important requirements for axenic growth of *G. duodenalis* trophozoites is the relatively high concentration of cysteine (16 mM). Cysteine also provides a partial protection from the toxicity of oxygen that is not seen with other reducing agents, including cystine, it helps to attach on the surface of the glass tube also and therefore appears to be a specific effect of cysteine [35, 36]. Cysteine is not synthesized by the de novo pathway and is not synthesized from cystine. It seems to be imported into the cell by passive diffusion, although active transport may account for some of the acquirement of cysteine [49]. The third one is metronidazole, a frequently used drug against parasitic infections in the most developed countries, which has been administered to generate oxidative stress. It kills parasite by generating the free nitro radical within the organism [14]. Pyruvate:ferredoxin oxidoreductase (PFOR) of *Giardia* has been shown to reduce metronidazole to its functionally active radical state via ferredoxin-I in an in vitro system using purified components [89]. Metronidazole is known to apply its toxic effect through unstable intermediate products like nitro and nitroso-free radicals and superoxide anions. The reactive intermediates also interact with DNA, causing DNA degradation, strand breakages, and cross-linking with the nucleic acid and the release of nucleotide phosphates and also protein damages [73].

ROS generation is such a massive effected within a cell; it can regulate not only the metabolic gene expressions but also the expressions of genes for transcription, translation, cell division, and even the mode of cell death [50]. Programmed cell death (PCD) is a well-regulated cellular phenomenon that has been widely characterized in multicellular organisms. Programmed cell death has also been observed in an increasing number of unicellular eukaryotes including the trypanosomatids, *Dictyostelium*, *Plasmodium*, *T. vaginalis*, *E. histolytica*, and *Blastocystis* [15, 21, 83]. This process can be initiated by various factors both external and internal, and these are coordinated by a complex network of modulators. There are several switches that act as trigger like cellular stress, growth factor (serum) deprivation, chemotherapeutic agents, receptor ligand binding, and deregulation of cell division and development or differentiation [25]. The most common types of programmed cell death known are apoptosis and autophagy. Apoptosis (type I PCD) involves an organized cascade of biochemical proceedings leading to characteristic changes in cell morphology which includes proteolytic cleavage by different caspases, cell shrinkage, DNA internucleosomal degradation, phosphatidylserine exposure, blebbing of the plasma membrane, development of apoptotic bodies, and loss of mitochondrial membrane potential with cytochrome c release to the cytosolic environment [56]. All the changes ultimately lead to cell death. Apoptosis has been differentiated into two forms like caspase-dependent and caspase-independent apoptotic pathway. Firstly, the caspase-independent pathway involves the induction of mitochondrial membrane permeabilization (MMP) and the release of apoptosis-inducing factor (AIF), which is under control of the Bcl-2 family of proteins. Secondly, in the caspase-dependent pathway, after MMP induction, cytochrome-c reallocates from mitochondria in the cytosol to activate caspase-9, in association with ATP and the cytosolic factor Apaf-1 [48]. The other one is autophagy (type II PCD) that engages in the autophagosomal-lysosomal system. Autophagosomes are double-membrane vesicles and are responsible for the engulfment of cytoplasmic components, during turnover of organelles. After autophagosome-lysosome fusion, an autophagolysosome is produced in which cellular constituents are digested. Autophagy is vital to maintain the metabolic equilibrium and the maintenance of cellular structures during cell growth and development. The deregulation of such equilibrium can instigate cell death as a result of prolonged starvation under oxidative stress. It has been reported that limited self-digestion of cellular materials, including different cellular components, can help individual cells to supply energy for its survival for several days. However, if the situations do not improve, self-digestion continues and ultimately results in autophagic cell death [15]. Clearly these processes vary from that of necrosis which is characteristically connected with severe cell injury causing remarkable alterations in mitochondrial function, cytoplasmic vacuolization, and, eventually, the collapse of the plasma membrane. The salvage process of dead cells in necrosis by comparison is slow and less synchronized, due to the absence of specific cell signals, leading to a significant inflammatory response [56]. It is important to understand how programmed cell death evolved in unicellular organism with respect to the pathways as we have many evidences of PCD in single-celled organisms including bacteria [43]. Although there are a considerable number of reports in this area on organisms such as *Dictyostelium*, *Trypanosoma*, *Leishmania*, and *Plasmodium*, our perceptive of the pathways involved in many organisms is variable. *Giardia* is an aerotolerant protozoan parasite that resides in the gut of humans and animals. It lacks mitochondria but contains mitosomes (thought to be relic mitochondria). As the organism possesses many typical characteristics such as a distinct nucleus and nuclear membrane, cytoskeleton, and endomembrane system, it belongs to the mesokaryotes; however, other aspects of the cell such as SSU rRNA and some key metabolic enzymes are prokaryotic-like (Svard et al. 2003). Although controversial, the mainstream scientists believe that *Giardia* as a eukaryote that has diverged at or just after mitochondrial acquirement and this has made *Giardia* an important organism for research into the understanding of evolution in the eukaryotes [87]. It has been reported that organisms use diverse pathways of programmed cell death to activate self-destruction. Several studies specify that *Giardia* undergoes programmed cell death in response to different stressed conditions and that there are two distinct non-necrotic forms, apoptosis-like (Type I) PCD and autophagy-like (Type II) PCD. Our report sustains studies by other investigators [21, 34, 66, 78] who have established through the use of staining that a type of programmed cell death happens in *Giardia*. However, in our previous study, we have hypothesized that apoptotic-like cell death occurs via a mechanism(s) independent of caspases. We were unable to identify caspase-9 activity using a range of approaches including, Western blot, and the use of specific fluorescent substrates and inhibitor assays. It was verified and confirmed that *Giardia* possesses two forms of cell death—apoptosis and autophagy [23]. Amusingly, a group proposed the presence of caspases and a caspase-dependent mechanism by using CaspaTag and fluorescence microscopy. This method, although frequently used, has been revealed by several authors to have debatable specificity [24]. There are several studies that have been documented on the lack of caspase genes in the genome of unicellular eukaryotes and the control of an apoptotic pathway independent of caspases. Nowadays there have been various reports highlighting the connection of metacaspases in an apoptotic-like programmed cell death in unicellular organisms. Metacaspases have shown a well-established efficient activity in yeast [53], but work in *Trypanosoma brucei* [39] and *Plasmodium berghei* [45] has been unconvincing [5]. As *Giardia* does not have mitochondria, caspase-independent apoptosis may occur. However, *Giardia* has a remnant structure termed the mitosome, but this organelle is unable to generate energy, and no clear metabolic or signaling role has been linked with it [71]. We were unable to recognize any components like apoptotic-related genes including caspases and metacaspases after genome analysis of *Giardia*. In addition to apoptotic-like PCD, starvation-induced death confirmed evidence of autophagy (type II PCD) in *Giardia*. This observation was supported by morphological studies and bioinformatics analysis. The recent data from genome analysis suggested that the pathway of autophagy shows some conservation and also the pathway in *Giardia* is not complete. This is in clear disparity to our hypothesis on "apoptotic"-like mechanism(s). However, these data must be looked at carefully because the divergence in *Giardia* proteins may make these types of analyses biased towards those genes that have significant amount of homology. The exact mechanistic pathway of programmed cell death (PCD) in amitochondriate *Giardia* is still unknown. The recent studies found some of the end results which are somehow similar to known pathways in other unicellular organisms. Our study strongly supports the claim that programmed cell death instigated prior to multicellularity and divergence of the unicellular eukaryotes. These processes would have appeared to be evolutionarily conserved; however, the mechanisms that strengthen these pathways are not common to all eukaryotes. Finally, it is understandable that to explain such pathways in the protozoans will necessitate a global analysis of transcriptomics, proteomics, and metabolomics. Elucidation of these pathways has a great importance in *Giardia* research not only in terms of evolution but also in terms of recognizing novel approaches for the treatment of *giardia*sis and for the control of many harmful protozoan parasites. Effective targeting at the molecular level should direct to disease control and produce few adverse side effects in the host, predominantly as the mechanisms involved in both the mammalian and *Giardia* systems would seem to be significantly different. It has been recognized some metabolic genes which actively participate in the oxidative stress management of *Giardia lamblia*. This study highlighted on the effect of changes in the level of intermediary metabolites of *Giardia lamblia* in causing stress relieved and protected by transducing signals to genes to maintain redox homeostasis. Such studies are a crucial part in the assimilation of transcriptomics, proteomics, and metabolomics of *Giardia*.

15.2 *Giardia* Survival Under Hydrogen Peroxide Stress

Extracellular pyruvate defends *Giardia* trophozoites against the toxicity induced by intracellular reactive oxygen species (ROS) produced during the treatments of hydrogen peroxide (H_2O_2) [68]. In vitro and in vivo studies confirm the antioxidant effect of different α-ketoacids on several cell types. Pyruvate can pass through across the biological membrane, and therefore it has been assumed that the level of intracellular pyruvate increased due to the administration of pyruvate [9]. Hydrogen peroxide is uncharged, and therefore it can penetrate freely through the plasma membrane and it was shown to increase the rate of intracellular production of reactive oxygen species [68]. Similarly, the rates of reactive oxygen intermediates are increased due to the addition of menadione. It has been reported that reactive oxygen species induced by menadione in the trophozoites were reduced by the administration of pyruvate [68]. It was also shown that pyruvate plays a direct role as an antioxidant in the trophozoites as the pyruvate reduces the generation of total fluorescence generated by oxidation of the H_2DCFDA detected by spectrofluorometer. It has been shown that pyruvate and related α-ketoacids recover the trophozoites after H_2O_2 exposure. Several observations suggest that the protective effect of pyruvate results from its ability to interact with hydrogen peroxide and produce acetate, H_2O, and CO_2, rather than from the enrichment of energy metabolism of *Giardia*.

The protective consequence of pyruvate was emulated by several α-ketoacids, which showed the ability of pyruvate to react with hydrogen peroxide the ability to react with hydrogen peroxide; these compounds include mannitol, which is not an energy substrate. The defensive result of the different α-ketoacids due to oxidative stress was closely interrelated with their capability to suppress ROS [68]. It was reported that the consequences of elevated ROS generation was correlated with a decrease in the generation of ATP as well as protein synthesis and increase in DNA breakage and lipid peroxidation [40, 91].

The decline of ROS level may be attributed either to their reduced generation or to the amplification in antioxidant levels. We have reported previously based on linoleic acid autoxidation in micelle system [22] that pyruvate cannot inhibit lipid peroxidation but also no induction period was observed [68]. Indeed, in a micelle system, the autoxidation process engrosses lipid radicals such as peroxyl radical (ROO·) and alkoxyl radical (RO·). The levels of conjugated dienes amplified rapidly to reach a maximum level and then remained constant. The addition of pyruvate at several concentrations was without any effect on the formation of conjugated dienes during the indicated time period. Thus, it can be presumed that pyruvate did not scavenge lipid radicals.

Lipid peroxidation was used as an indicator for quantification of the damage that occurs in cell membranes as a result of lipid damage by free radicals [73]. In this system, pyruvate was capable to restrain lipid peroxidation. This strangeness between a cellular antilipoperoxidant consequence of pyruvate and its ineffectiveness to prevent diene production in micelles could be elucidated by the property of pyruvate to scavenge radicals such as $H_2O_2^{-\cdot}$, $O_2^{-\cdot}$, and $^{\cdot}OH$ but not lipid radicals [33]. In *Giardia* trophozoites, hydroxyl and superoxide radicals are formed during different oxidative stresses. These radicals then commence membrane lipid peroxidation. Thus, pyruvate, by scavenging hydroxyl and superoxide radicals, prohibits lipid peroxidation [68]. In brief, pyruvate plays a role at the initiation step of lipid peroxidation but not at the proliferation step, which produces lipid radicals. Such results are not unexpected because pyruvate is not liposoluble. Vitamin C, a well-known antioxidant intermediary metabolite, reacts rapidly with $O_2^{-\cdot}$ and $^{\cdot}OH$ but not with peroxyl radical [37]. In contrast with the known liposoluble metabolite, vitamin E is an effective scavenger of lipid radicals [19]. It was very surprising to us that acetate exhibited inhibition of linoleic acid autoxidation, i.e., with increased concentration of acetate, diene formation reduces.

Intracellular pyruvate concentration was augmented with time and, after several hours, it was observed to be decreasing [68]. The elevated pyruvate concentrations upon oxidative stress may be because of the inactivation of pyruvate:ferredoxin oxidoreductase, an enzyme which is susceptible to oxidative stress [88]. The study [68] provided new insights into the role of pyruvate in *Giardia* trophozoites during oxidative stress. Oxidative stress inhibits such regulation, whereas exogenously administered pyruvate reinstates it. As we have already established, the effect of pyruvate against H_2O_2 toxicity may be attributable to its ability to degrade H_2O_2 during a nonenzymatic oxidative decarboxylation leading to the production of CO_2, H_2O, and acetate [17, 41]. This reaction may occur in the intracellular medium, leading to the degradation of equal amounts of H_2O_2 and pyruvate. On the contrary, due to its high membrane permeability, H_2O_2 may drop down energy metabolism through degradation of intracellular pyruvate (and other α-ketoacids).

Stimulation of DNA fragmentation by oxidative stress is a familiar phenomenon mostly caused by reactive oxygen species. A working antioxidant system is necessary to diminish such damage. In *Giardia lamblia*, in spite of the deficiency of an effect of glutathione, pyruvate has advantageous property on DNA damages under oxidative stress. Predominantly, the augmented uptake of pyruvate might tolerate *Giardia* trophozoites to preserve their antioxidative abilities in a way independent of glutathione cycle during oxidative stress. DNA fragmentation decreases quicker in stressed trophozoites of *Giardia* previously incubated with pyruvate, signifying stimulation of DNA repair. Some studies established that pyruvate has antioxidant mechanism such as the commencement of different anti-apoptotic pathways.

Fundamentally, destruction of the regulation of the intracellular metabolite and the enhanced rate of pyruvate uptake stimulated during oxygen exposure demonstrated that modification of pyruvate metabolism has taken place. The increased uptake might be due to an increased necessity to preserve antioxidant potential in *Giardia lamblia*. However, it might be also the consequences of a metabolic requirement. The activities of pyruvate dikinase, pyruvate kinase, and also malate dehydrogenase were increased in *Giardia* due to pyruvate production. This indicates enhancement of glycolysis activity in the presence of pyruvate. Improved glycolysis might therefore continue the ATP generation during oxidative stress. Such an enhancement of the intracellular pyruvate in response to high O_2 concentration has been reported previously to have protective effects against oxidative stress [40].

This study has revealed the dynamics of the transcriptional and metabolic regulatory networks during oxidative stress for the first time. Management of oxidative stress was thought to be restricted by NADH oxidase, flavodiiron protein, etc. [51]. However, after the transcriptomic data analysis, it has been concluded that the oxidative stress regulation is not only restricted within stress-scavenging genes but also some metabolic genes and diverse types of other proteins take a significant role in ROS detoxification [67].

We have observed that the pyruvate metabolism pathway is regulated by the different experimental conditions differently. However, the precise mechanism underlying the pyruvate effects on oxidative stress regulation in *Giardia lamblia* has yet to be further examined.

As *Giardia* does not contain the components of respiratory chain (e.g., cytochromes), by controlling pyruvate level, it remains the detrimental consequences of ROS at bay. Modification of the fate of pyruvate in one way or the other can be significant for homeostatic response of *Giardia* under oxidative stress. This could adjust the execution of the antioxidant system and have protective effects against DNA damage induced by oxidative stresses. Modifications of pyruvate metabolism are observed in *Giardia* owing to high oxygen environment. This could be beneficial for *Giardia* trophozoites in such harsh conditions.

15.3 *Giardia* Survival Under Cysteine-Ascorbate Deprivation

The study was demonstrated that pyruvate addition protects *Giardia* trophozoites against the toxicity stimulated by cysteine-ascorbate deprivation. It was previously confirmed that cysteine-ascorbate-deprived medium produces reactive oxygen species (ROS) in *Giardia* [67]. Cysteine-ascorbate-deprived medium generates ROS intracellularly in *Giardia*, and the reactive oxygen species was also augmented by the addition of synthetic quinine menadione [69]. The study proved that intracellular generation of ROS decreases with the administration of pyruvate and also attenuates the menadione-induced ROS production. The defensive effect of pyruvate was reproduced by several α-ketoacids, which distribute with pyruvate the ability to react with intracellular reactive oxygen species; these compounds include mannitol, which is not an energy molecule. From the experimental data, we have observed that pyruvate showed a higher competence to scavenge ROS [68]. Hydroxyl and superoxide radicals are generated in *Giardia* trophozoites during cysteine-ascorbate stress. These radicals then initiate membrane lipid peroxidation.

15.4 Trophozoites of *Giardia* Under Metronidazole Treatment

Metronidazole has been used for the treatment of infections for giardiasis. Anaerobic parasitic infections caused by different protozoan parasite respond favorably to metronidazole therapy. The trophozoites fight against oxidative stress generated by metronidazole. Metronidazole reduction was driven by pyruvate, but progressive damage to the radical-generating system was observed. There are several enzymes involved in response to metronidazole treatment in *Giardia* such as pyruvate ferredoxin oxidoreductase, NADH oxidase, and peroredoxin. Exogenous addition of physiologically relevant concentration of pyruvate was shown to induce the rate of ROS production in *Giardia* suspension treated with metronidazole. Our results provide the proof of evidence that exogenously administered pyruvate also induced lipid peroxidation of *Giardia* under stress. Pyruvate can reduce metronidazole and form different types of nitroso radical derivatives which can damage DNA. We have shown the expression levels of different metabolic genes which are significantly upregulated or downregulated during metronidazole treatment. This proposed that these genes are involved in combating against metronidazole.

15.5 Effects of Albendazole on *Giardia* Trophozoites

Parasitic diseases caused by protozoa and helminths correspond to a severe health trouble worldwide. The management of this parasitic infection is carried out mainly by drug therapy. The drug albendazole is a broad-spectrum benzimidazole with an anti-parasitic effect and it is of very low cost which makes this drug an appropriate candidate for mass drug administration programs to deworm children in endemic countries. However, up to 400 mg doses for five times are required to clear infections by *G. lamblia*, restraining the efficacy of these campaigns to reduce loads of giardiasis. The use of suboptimal doses may cause and spread albendazole-resistant trophozoites of *Giardia* [92]. Similarly, many parasites including *Giardia* show cross-resistance to various drugs. Therefore, it is very crucial to know the pathway of mechanisms involved in the cytotoxicity of the drugs used; this concept is needed to recommend high-efficiency drug administration and new and potential drugs against resistant strains [92]. The effects of albendazole on *Giardia* has been reported, but, albendazole toxicity is also reported in some patients having a problem in liver, as well as other side effects [59].

15.6 Proteomics Study of *Giardia* Trophozoites Under Oxidative Stress

Giardia lamblia, an anaerobic, unicellular protozoan and a causative agent of diarrheal disease, inhabits in the lumen of small intestinal in close collocation to epithelial cells. Since the concrete pathway of disease mechanisms of giardiasis is inadequately understood, exposing the parasite under different oxidative stress conditions to mimic the environment like the intestinal tract of the host provides clues to understand the pathogenesis. Here we have examined the hypothesis that the exposure of *G. lamblia* trophozoites to different oxidative stresses might lead to excretion of specific proteins. In a proteomics study, *Giardia lamblia* at its untreated conditions were compared with the trophozoites treated with hydrogen peroxide, cysteine-ascorbate-deprived medium, and metronidazole by using two-dimensional gel electrophoresis. The increased protein spots in the extracts of treated trophozoites compared to untreated trophozoites were identified by MALDI-TOF mass spectrometry. Expression patterns of five of the identified proteins were determined due to exposure to hydrogen peroxide, cysteine-ascorbate deprived medium, and metronidazole by real-time PCR. The released *Giardia lamblia* proteins were localized to the cytoplasm and embedded in the plasma membrane of trophozoites.

15.7 Conclusion

The genes identified from the transcriptomic study playing a role in oxidative stress regulation can be used for new drug targets. Arginine deiminase, released during oxidative stress, interferes directly in the mucosal immune system. As arginine deiminase is absent in human, it could be a good target for drug designing. Pyruvate can alter the effective lifetime of reactive oxygen species by scavenging them. Pyruvate attenuates ROS production in the *Giardia* trophozoites. Lipid peroxidation was inhibited by exogenously added pyruvate in stressed *Giardia* trophozoites. It had a high level of necessity for pyruvate during oxidative stress. Pyruvate reduced the number of DNA breaks. Pyruvate plays a role in DNA stability and repair in *Giardia* trophozoites during oxidative stress. Oxidative stress-induced calcium-dependent arachidonic acid-mediated apoptotic-like death occurs in *Giardia lamblia*.

Giardia, a microaerophilic, unicellular, early branching eukaryote, the most common enteric parasite in case of human and other animals, and the causative agent of giardiasis, has to establish mechanisms to prevent themselves from oxygen pressure in the human gut where it resides.

It does not contain any conventional oxidative stress management pathway, including different enzymes to act on different radicals to diminish them and glutathione cycling, which are present in most eukaryotes. NADH oxidase, a major component of electron transport chain of *Giardia*, protects oxygen-labile proteins such as pyruvate:ferredoxin oxidoreductase against oxidative stress by a reduced intracellular environment. Most importantly, it has the arginine dihydrolase pathway, which is also present in a number of anaerobic prokaryotes which provides substrate-level phosphorylation and effectively active to make a major contribution for ATP production. The study has been focused on the genes that are differentially expressed during several experimental conditions and regulate the stress management pathway differently to achieve redox homeostasis. Identification of some new unique genes in oxidative stress regulation may help in new drug designing for this common enteric parasite prone to drug resistance. Moreover, these data proposed the major role of this early divergent ancient eukaryote in anaerobic to aerobic organism evolution.

Giardia lamblia, aerotolerant, produces energy by fermentative metabolism. It has to survive against detrimental oxidative stress, and hence it has to build up defensive strategies to fight with elevated oxygen pressure. It was reported previously that exogenously added pyruvate was able to prevent lipid peroxidation of *Giardia* under stress and effects of pyruvate were concentration-dependent, but it has no effect on lipid peroxidation observed in micelle model. We have verified trophozoites have the capacity to control intracellular level of pyruvate during oxidative stress. Pyruvate recovers trophozoites of *Giardia* from oxidative stress by diminishing the number of DNA damage and might favor DNA repair [70].

Recently, the role of pyruvate as a physiological antioxidant on oxidative stress in *Giardia* trophozoites by cysteine-ascorbate-deprived medium was investigated [69]. In this study, we have examined the effects of pyruvate administration, after

DNA damage as a result of stress from cysteine-ascorbate-deprived medium, in *Giardia* trophozoites. The pyruvate concentrations at individual time points were determined during cysteine-ascorbate deprivation in *Giardia*. The exogenous administration of a physiologically appropriate concentration of pyruvate to trophozoite suspensions was shown to sequester the rate of ROS production. We have hypothesized that *Giardia* protects itself from harmful consequences of ROS by maintaining the intracellular pyruvate concentration. Trophozoites were recovered from the consequences of oxidative stress due to pyruvate accumulation by reducing the number of DNA breaks that might favor DNA repair.

Albendazole stimulates the production of intracellular ROS in *Giardia* trophozoites under oxidative stress condition where the major affected biomolecule is DNA. This involves hydroxyl radical-resulted double-strand DNA damage. This harm in turn leads to cell cycle abnormalities and finally apoptotic-like cell death. These explanations permit us to increase our understanding on the cytotoxic mechanism of albendazole in *Giardia* and opens up a new chance to design a rational drug for giardiasis and allowing for antioxidant responses as possible mechanisms of multidrug resistance. In this circumstance, generating ROS or inhibiting endogenous antioxidant enzymes would be a coherent approach to develop new anti-giardial drugs as formerly anticipated for other parasites [65].

References

1. Adam RD (2001) The biology of *Giardia spp*. Clin Microbiol Rev 55:706–732
2. Andrae U, Singh J, Ziegler-Skylakakis K (1985) Pyruvate and related α-ketoacids protect mammalian cells in culture against hydrogen peroxide-induced cytotoxicity. Toxicol Lett 28:93–98
3. Ankarklev J, Jerlström-Hultqvist J, Ringqvist E, Troell K, Svärd SG (2010) Behind the smile: cell biology and disease mechanisms of Giardia species. Nat Rev Microbiol 8:413–422
4. Atkinson HJ (1980) Respiration. In: Zuckerman BH (ed) Nematodes as biological models. Academic, London, pp 101–138
5. Atkinson HJ, Babbitt PC, Sajid M (2009) The global cysteine peptidase landscape in parasites. Trends Parasitol 25:573–581
6. Bar-Or D, Rael LT, Lau EP, Rao NK, Thomas GW, Winkler JV, Yukl RL, Kingston RG, Curtis CG (2001) An analog of the human albumin N-terminus (Asp-AlaHis-Lys) prevents formation of copper-induced reactive oxygen species. Biochem Biophys Res Commun 284:856–862
7. Biagini GA, Suller MTE, Finlay BJ, Lloyd D (1997) Oxygen uptake and antioxidant responses of the free-living diplomonad Hexamita sp. J Eukaryot Microbiol 44:447–453
8. Biagini GA, Mclntyre PS, Finlay BJ, Lloyd D (1998) Carbohydrate and amino acid fermentation in the free-living primitive protozoon *Hexamita sp*. Appl Environ Microbiol 64:203–207
9. Biagini GA, Park JH, Lloyd D, Edwards MR (2001) The antioxidant potential of pyruvate in the amitochondriate diplomonads *Giardia intestinalis* and *Hexamita inflata*. Microbiology 147:3359–3365
10. Brown PO, Botstein D (1999) Exploring the new world of the genome with DNA microarrays. Nat Genet 21:33–37. [Review]
11. Brown DM, Upcroft JA, Upcroft P (1993) Cysteine is the major low molecular weight thiol in *Giardia duodenalis*. Mol Biochem Parasitol 61:155–158
12. Brown DM, Upcroft JA, Upcroft P (1995) Free radical detoxification in *Giardia duodenalis*. Mol Biochem Parasitol 72(1–2):47–56

13. Brown DM, Upcroft JA, Upcroft P (1996) A thioredoxin reductase-class of disulphide reductase in the protozoan parasite Giardia duodenalis. Mol Biochem Parasitol 83(2):211–220
14. Brown DM, Upcroft JA, Edwards MR, Upcroft P (1998) Anaerobic bacterial metabolism in the ancient eukaryote *Giardia duodenalis*. Int J Parasitol 28(1):149–164
15. Bruchhaus I, Roeder T, Rennenberg A, Heussler VT (2007) Protozoan parasites: programmed cell death as a mechanism of parasitism. Trends Parasitol 23(8):376–383
16. Bruderer T, Wehrli C, Kohler P (1996) Cloning and characterization of the gene encoding pyruvate phosphate dikinase from *Giardia duodenalis*. Mol Biochem Parasitol 77:225–233
17. Bunton CA (1949) Oxidation of α-diketones and α-keto acids by hydrogen peroxide. Nature 144–163:444
18. Buret AG (2007) Mechanisms of epithelial dysfunction in giardiasis. Gut 56:316–317
19. Burton GW, Ingold KU (1989) Vitamin E as an in vitro and in vivo antioxidant. Ann N Y Acad Sci 570:7–22
20. Cacciò SM, Ryan U (2008) Molecular epidemiology of giardiasis. Mol Biochem Parasitol 160(2):75–80
21. Chose O, Sarde CO, Gerbod D, Viscogliosi E, Roseto A (2003) Programmed cell death in parasitic protozoans that lack mitochondria. Trends Parasitol 19:559–564
22. Cillard J, Cillard P, Cormier M, Girre L (1980) Alpha-tochopherol prooxidant effect in aqueous media: increased autoxidation rate of linoleic acid. J Am Oil Chem Soc 57:252–255
23. Corrêa Mde P, Ferreira AP, Gollner AM, Rodrigues MF, Guerra MC (2009) Markers expression of cell proliferation and apoptosis in basal cell carcinoma. An Bras Dermatol 84(6):606–614
24. Darzynkiewicz Z, Pozarowski P (2007) All that glitters is not gold: all that FLICA binds is not caspase. A caution in data interpretation--and new opportunities. Cytometry A 71(8):536–537
25. Debrabant A, Lee N, Bertholet S, Duncan R, Nakhasi HL (2003) Programmed cell death in trypanosomatids and other unicellular organisms. Int J Parasitol 33:257–267
26. Dib HH, Lu SQ, Wen SF (2008) Prevalence of *Giardia lamblia* with or without diarrhea in South East, South East Asia and the Far East. Parasitol Res 103:239–251
27. Dobell C (1920) The discovery of the intestinal protozoa of man. Proc R Soc Med 1920:13
28. Eckmann L (2003) Mucosal defences against Giardia. Parasite Immunol 25(5):259–270
29. Edwards MR, Schofield PJ, O'Sullivan WJ, Costello M (1992) Arginine metabolism during culture of *Giardia intestinalis*. Mol Biochem Parasitol 53(1–2):97–103
30. Elias EV, Quiroga R, Gottig N, Nakanishi H, Nash TE, Neiman A, Lujan HD (2008) Characterization of SNAREs determines the absence of a typical Golgi apparatus in the ancient eukaryote *Giardia lamblia*. J Biol Chem 283(51):35996–36010
31. Evans PJ, Gallesi D, Mathieu C, Hernandez MJ, Felipe M, Halliwell B, Puppo A (1999) Oxidative stress occurs during soybean nodule senescence. Planta 208:73–79
32. Feng Y, Xiao L (2011) Zoonotic potential and molecular epidemiology of Giardia species and giardiasis. Clin Microbiol Rev 24:110–140
33. Floriano-Sánchez E, Villanueva C, Medina-Campos ON, Rocha D, Sánchez-González DJ, Cárdenas-Rodríguez N, Pedraza-Chaverrí J (2006) Nordihydroguaiaretic acid is a potent in vitro scavenger of peroxynitrite, singlet oxygen, hydroxyl radical, superoxide anion and hypochlorous acid and prevents in vivo ozone-induced tyrosine nitration in lungs. Free Radic Res 40(5):523–533
34. Ghosh E, Ghosh A, Ghosh AN, Nozaki T, Ganguly S (2009) Oxidative stress-induced cell cycle blockage and a protease-independent programmed cell death in microaerophilic Giardia lamblia. Drug Des Devel Ther 3:103–110
35. Gillin FD, Reiner DS (1982) Attachment of the flagellate Giardia lamblia: role of reducing agents, serum, temperature, and ionic composition. Mol Cell Biol 2(4):369–377
36. Gillin FD, Reiner DS, Levy RB, Henkart PA (1984) Thiol groups on the surface of anaerobic parasitic protozoa. Mol Biol Biochem Parasitol 13(1):1–12
37. Groussard C, Morell I, Chevanne M, Monnier M, Cillard J, Delamarche A (2000) Free radical scavenging and antioxidant effects of lactate ion: an in vitro study. J Appl Physiol 89:169–175
38. Haque R, Roy S, Kabir M, Stroup SE, Mondal D, Houpt ER (2005) Giardia assemblage a infection and diarrhea in Bangladesh. J Infect Dis 192:2171–2173

39. Helms MJ, Ambit A, Appleton P, Tetley L, Coombs GH, Mottram JC (2006) Bloodstream form Trypanosoma brucei depend upon multiple metacaspases associated with RAB11-positive endosomes. J Cell Sci 119(6):1105–1117
40. Herbener GH (1976) A morphometric study of age-dependent changes in mitochondrial population of mouse liver and heart. J Gerontol 31:8–12
41. Holleman AF (1904) Notice sur l'action de l'eau oxygénée sur les acides αcétoniques et sur les dicétones 1.2. Recl Trav Chim Pays Bas 23:169–171
42. Jaroll EL, Paget TA (1995) Carbohydrate and amino acid metabolism in *Giardia*: a review. Fol Parasitol 42:81–89
43. Koonin EV, Aravind L (2002) Origin and evolution of eukaryotic apoptosis: the bacterial connection. Review Cell Death Differentiation 9:394–404
44. Lane S. And Lloyd D. (2002) Current trends in research into the waterborne parasite *Giardia*; Crit Rev Microbiol; 28: 123–147
45. Le Chat L, Sinden RE, Dessens JT (2007) The role of metacaspase 1 in Plasmodium berghei development and apoptosis. Mol Biochem Parasitol 153(1):41–47
46. Lindmark DG (1980) Energy metabolism of the anaerobic protozoan *Giardia lamblia*. Mol Biochem Parasitol 1:1–12
47. Lloyd D, Harris JC, Maroulis S, Biagini GA, Wadley RB, Turner MP, Edwards MR (2000) The microaerophilic flagellate Giardia intestinalis: oxygen and its reaction products collapse membrane potential and cause cytotoxicity. Microbiology 146(Pt 12):3109–3118
48. Lorenzo HK, Susin SA (2004) Mitochondrial effectors in caspase-independent cell death. FEBS Lett 557:14–20
49. Lujan HD, Byrd LG, Mowatt MR, Nash TE (1994) Serum Cohn fraction IV-1 supports the growth of Giardia lamblia in vitro. Infect Immun 62(10):4664–4666
50. Ma Q (2010) Transcriptional responses to oxidative stress: pathological and toxicological implications. Pharmacol Ther 125:376–393
51. Mastronicola D, Giuffrè A, Testa F, Mura A, Forte E, Bordi E, Pucillo LP, Fiori PL, Sarti P (2011) *Giardia intestinalis* escapes oxidative stress by colonizing the small intestine: a molecular hypothesis. IUBMB Life 63(1):21–25
52. Mastronicola D, Falabella M, Testa F, Pucillo LP, Teixeira M, Sarti P, Saraiva LM, Giuffrè A (2014) Functional characterization of peroxiredoxins from the human protozoan parasite *Giardia* intestinalis. PLoS Negl Trop Dis 8(1):e2631
53. Mazzoni C, Falcone C (2008) Caspase-dependent apoptosis in yeast. Biochim Biophys Acta 1783(7):1320–1327
54. Mead JM (1976) Free radical mechanisms of lipid damage and consequences for cellular membranes. Academic, London, pp 51–68
55. Mendis A, Thompson R, Reynoldson J, Armson A, Meloni B, Gunsberg S (1992) The uptake and conversion of 1-[$U^{14}C$-] aspartate and L-[$U^{14}C$-] to $^{14}CO_2$ by intact trophozoites of *Giardia duodenalis*. Comp Biochem Physiol 102:235–239
56. Menna-Barreto RF, Goncalves RL, Costa EM, Silva RS, Pinto AV, Oliveira MF, de Castro SL (2009) The effects onTrypanosoma cruzi of novel synthetic naphthoquinones are mediated by mitochondrial dysfunction. Free Radic Biol Med 47:644–653
57. Müller M (1988) Energy metabolism of protozoa without mitochondria. Annu Rev Microbiol 42:465–488
58. Müller N, von Allmen N (2005) Recent insights into the mucosal reactions associated with Giardia lamblia infections. Int J Parasitol 35(13):1339–1347
59. Nandi M, Sarkar S (2013) Albendazole-induced recurrent hepatitis. Indian Pediatr:1064–1050. https://doi.org/10.1007/s13312-013-0285-8
60. Nash TE (2013) Unraveling how Giardia infections cause disease. J Clin Invest 123(6):2346–2347
61. Nash TE, Aggarwal A (1986) Cytotoxicity of monoclonal antibodies to a subset of Giardia isolates. J Immunol 136(7):2628–2632
62. Nash TE, Banks SM, Alling DW, Merritt JW Jr, Conrad JT (1990) Frequency of variant antigens in Giardia lamblia. Exp Parasitol 1(4):415–421

63. Ortega-Pierres G, Smith HV, Cacciò SM, Thompson RC (2009) New tools provide further insights into Giardia and Cryptosporidium biology. Trends Parasitol 25(9):410–416
64. Paget, T. A., Kelly, M. L., Jarrol, E. L., Lindmark, D. G. And Lloyd D. (1993): The effects of oxygen on fermentation in *Giardia lamblia*. Mol Biochem Parasitol, 57: 65–72
65. Pal C, Bandyopadhyay U (2012) Redox-active antiparasitic drugs. Antioxid Redox Signal 17:555–582. https://doi.org/10.1089/ars.2011.4436
66. Pérez-Arriaga L, Mendoza-Magaña ML, Cortés-Zárate R, Corona-Rivera A, Bobadilla-Morales L, Troyo-Sanromán R, Ramírez-Herrera MA (2006) Cytotoxic effect of curcumin on Giardia lamblia trophozoites. Acta Trop 98:152–161
67. Raj D, Ghosh E, Mukherjee AK, Nozaki T, Ganguly S (2014) Differential gene expression in Giardia lamblia under oxidative stress: significance in eukaryotic evolution. Gene 535(2):131–139
68. Raj D, Saini P, Nozaki T, Ganguly S (2015) Involvement of pyruvate on oxidative stress management in the microaerophilic protozoan parasite *Giardia lamblia*. Int J Adv Res 3:1148–1166
69. Raj D, Chowdhury P, Sarkar R, Nakano YS, Okamoto K, Dutta S, Nozaki T, Ganguly S (2017) Pyruvate protects *Giardia* trophozoites from cysteine-ascorbate deprived medium induced cytotoxicity. Korean J Parasitol 1:1–9
70. Rendtorff RC (1954) The experimental transmission of human intestinal protozoan parasites. II. Giardia lamblia cysts given in capsules. Am J Hyg 59(2):209–220
71. Rosa Ide A, Einicker-Lamas M, Bernardo RR, Benchimol M (2008) Cardiolipin, a lipid found in mitochondria, hydrogenosomes and bacteria was not detected in Giardia lamblia. Exp Parasitol 120(3):215–220
72. Sandhu H, Mahajan RC, Ganguly NK (2004) Flow cytometric assessment of the effect of drugs on *Giardia lamblia* trophozoites in vitro. Mol Cell Biochem 265:151–160
73. Sandhu SK, Kaur G (2003) Mitochondrial electron transport chain complexes in aging rat brain and lymphocytes. Biogerontology 4:19–29
74. Savioli L, Smith H, Thompson A (2006) Giardia and cryptosporidium join the 'neglected diseases initiative'. Trends Parasitol 22:203–208
75. Schofield PJ, Costello M, Edwards MR, O'Sullivan WJ (1990) The arginine dihydrolase pathway is present in *Giardia intestinalis*. Int J Parasitol 20:697–699
76. Schuessel K, Frey C, Jourdan C, Keil V, Weber CC, Muller F, Muller WE, Echest A (2006) Aging sensitizes toward ROS formation and lipid peroxidation in PSIM146L transgenic mice. Free Radic Biol Med 40:850–862
77. Scott KG, Meddings JB, Kirk DR, Lees-Miller SP, Buret AG (2002) Intestinal infection with *Giardia* spp. reduces epithelial barrier function in a myosin light chain kinase-dependent fashion. Gastroenterology 123:1179–1190
78. Shemarova IV (2010) Signalling mechanisms of apoptosis like programmed cell death in unicellular eukaryotes. Comp Biochem Physiol B Biochem Mol Biol 155:341–353
79. Silva RR, da Silva CA, de Jesus Pereira CA, de Carvalho Nicolato RL, Negrao-Correa D, Lamounier JA, Carneiro M (2009) Association between nutritional status, environmental and socio-economic factors and Giardia lamblia infections among children aged 6–71 months in Brazil. Trans R Soc Trop Med Hyg 103:512–519
80. Svärd SG, Hagblom P, Palm JE (2003) *Giardia lamblia* – a model organism for eukaryotic cell differentiation. FEMS Microbiol Lett 218:3–7
81. Sweeny JP, Ryan UM, Robertson ID, Jacobson C (2011) Cryptosporidium and Giardia associated with reduced lamb carcase productivity. Vet Parasitol 182:127–139
82. Takizawa MG, Falavigna DL, Gomes ML (2009) Enteroparasitosis and their ethnographic relationship to food handlers in a tourist and economic center in Parana, Southern Brazil. Rev Inst Med Trop Sao Paulo 51:31–35
83. Tan KSW, Nasirudeen AMA (2005) Protozoan programmed cell death – insights from Blastocystis deathstyles. Trends Parasitol 21:547–550
84. Tekwani BL, Mehlotra RK (1999) Molecular basis of defence against oxidative stress in *Entamoeba histolytica* and *Giardia lamblia*. Microbes Infect 1:385–394. [review]

85. Tellez A, Winiecka-Krusnell J, Paniagua M, Linder E (2003) Antibodies in mother's milk protect children against giardiasis. Scand J Infect Dis 35(5):322–325
86. Thompson RC (2000) Giardiasis as a re-emerging infectious disease and its zoonotic potential. Int J Parasitol 30(12–13):1259–1267
87. Thompson EB, Webb MS, Miller AL, Fofanov Y, Johnson BH (2004) Identification of genes leading to glucocorticoid-induced leukemic cell death. Lipids 39(8):821–825
88. Townson SM, Upcroft JA, Upcroft P (1996) Characterisation and purification of pyruvate:ferredoxin oxidoreductase from Giardia duodenalis. Mol Biochem Parasitol 79(2):183–193
89. Upcroft J, Upcroft P (1998) My favorite cell: Giardia. Bio Essays 20:256–263, John Wiley & Sons, Inc.
90. Von Allmen N, Bienz M, Hemphill A, Müller N (2004) Experimental infections of neonatal mice with cysts of Giardia lamblia clone GS/M-83-H7 are associated with an antigenic reset of the parasite. Infect Immun 72(8):4763–4771
91. Vorbeck M, Martin A, Park J, Townsend J (1982) Aging related decrease in hepatic cytochrome oxidase of the Fischer 344 rat. Arch Biochem Biophys 214:67–79
92. Watkins RR, Eckmann L (2014) Treatment of giardiasis: current status and future directions. Curr Infect Dis Rep 16:396. https://doi.org/10.1007/s11908-0140396-y
93. Xiao L, Fayer R (2008) Molecular characterisation of species and genotypes of Cryptosporidium and Giardia. Int J Parasitol 38:1239–1255
94. Yoder JS, Gargano JW, Wallace RM, Beach MJ (2012) Giardiasis surveillance-United States, 2009–2010. Centers for disease control and prevention (CDC). MMWR Surveill Summ 61:13–23

Role of Reactive Oxygen Species in Infection by the Intracellular *Leishmania* Parasites

16

Moumita Basu and Pijush K. Das

Abstract

While encountering pathogens, generation of reactive oxygen species (ROS) is one of the earliest antimicrobial defence mechanisms put forward by phagocytic cells. The elevated level of ROS is expected to dampen the persistence of the microbes within the host cell. As a direct approach, ROS damages pathogen's DNA through distortion of bases. In an indirect way, it activates pro-inflammatory cytokines, thus creating an environment not suitable for the microorganisms. ROS has also been found to cause apoptosis of the host cell, thereby ensuring complete elimination of the pathogen. However, intra-macrophage protozoan parasite, *Leishmania* sp., efficiently subverts this antimicrobial defence mechanism of macrophages for their successful survival and propagation of the fatal disease, Leishmaniasis. They either activate negative regulatory molecules of macrophages to impair the activation of pro-oxidant enzymes or upregulate antioxidant molecules to neutralise the oxidative burst. This chapter mainly highlights the important role played by ROS in defence against intracellular infection and how *Leishmania* overcomes this antimicrobial defence arsenal and successfully survives within macrophages.

Keywords

Reactive oxygen species · *Leishmania sp* · Host defence

M. Basu
Department of Biochemistry, University of Calcutta, Kolkata, India

P. K. Das (✉)
Indian Institute of Chemical Biology, Kolkata, India
e-mail: pijush52@gmail.com

S. Chakraborti et al. (eds.), *Oxidative Stress in Microbial Diseases*,
https://doi.org/10.1007/978-981-13-8763-0_16

16.1 Introduction

16.1.1 Leishmaniasis: The Disease Caused by *Leishmania sp.*

The disease Leishmaniasis is a tropical disease mainly transmitted by bite of *Leishmania* infected female plebotomine sand flies. *Leishmania sp.* is a digenetic protozoan parasite, as it requires two hosts with two distinct developmental stages. Within the gut of its intermediate host sand fly, *Leishmania* resides as flagellated promastigote form whereas the aflagellated infective form amastigote survives in the phagolysosomes of mammalian macrophages [1]. Depending upon the parasite species and host strains, the disease has three clinical forms, namely, cutaneous, mucocutaneous and visceral leishmaniasis. Cutaneous leishmaniasis (CL) is the most common form of leishmaniasis caused by *Leishmania major* and *Leishmania mexicana* complex (*Leishmania amazonensis* and *Leishmania pifanoi*). The features of the disease are skin lesions, mainly ulcers, on exposed parts of the body. Mucocutaneous leishmaniasis is caused by *Leishmania brazillensis* leading to partial or total destruction of mucous membranes of the nose, mouth and throat. Amongst these three clinical forms, visceral leishmaniasis (VL) or Kala-azar is the most severe form and if not properly treated, becomes fatal. In East Africa and Indian subcontinent, it is mainly caused by *Leishmania donovani* and in Europe and North America by *Leishmania infantum chagasi*. According to WHO, VL is the second largest parasitic disease after malaria [2, 3]. In this form of leishmaniasis, the internal organs like spleen, liver and bone marrow get affected. Apart from the disease itself, Post-kala-azar dermal leishmaniasis (PKDL) is also a complication of visceral leishmaniasis caused by *Leishmania donovani* [4]. The clinical symptoms include skin rashes consisting of macules, papules or nodules in an individual recovered from VL.

16.1.2 Reactive Oxygen Species (ROS): Antimicrobial Defence Arsenal

When a phagocyte or cells like lung epithelial cells encounter a foreign pathogen, the immediate response of the cell is to generate high amount of ROS which acquaint an antimicrobial defence against a broad range of infectious organisms. Upon engulfment of the microbe, large quantities of ROS are delivered within phagosomes thereby preventing the pathogen to escape to the cytosol. ROS produced can eliminate pathogens directly by oxidative means or indirectly by stimulation of non-oxidative mechanisms like pattern recognition receptor signalling, autophagy, neutrophil extracellular trap (NET) formation and T-lymphocyte responses like activation, anergy and apoptosis [5]. Excess production of ROS damages DNA, protein, and lipids, and thereby causes cell death. One of the most important cellular targets of ROS is DNA [6]. Base oxidation, particularly guanine, may be mutagenic in nature [7], and blocking lesions or strand breaks may be lethal unless they are repaired [8, 9]. Iron-sulphur cluster-containing proteins are also vulnerable to ROS damage [10] and may substantially restrict metabolic pathways even if the damage is not microbicidal. However, a complete picture of exact mechanism in which ROS defends the host cell from the invading microbe is not completely clear till date.

ROS has been found to effectively combat certain microbes. Patients of chronic granulomatous disease (CGD) have deficiency of ROS generation and are found to be highly susceptible to infection by various microbes like *Salmonella enterica*, *Staphylococcus aureus*, *Serratia marcescens* and *Aspergillus* spp. [11, 12]. Opportunistic pathogens such *as Salmonella enteric serovar typhimurium (Salmonella typhimurium), Staphylococcus aureus, L. monocytogenes, Franscisella tularensis* and *S. marcescens* successfully infect when there is a lack of functional gp91phox protein, the key component of major cellular ROS-producing enzyme, NADPH oxidase (NOX2) [11, 13]. *S. aureus*is is the most common bacterial pathogen leading to CGD [13]. A study using *S. enteric* demonstrated that its base excision repair system was required to resist killing by ROS. *L. monocytogenes* has been found to show unrestricted growth in NOX2 (gp91phox)$^{-/-}$ mice and developed high bacterial burden in spleen and liver as well as increased micro abscesses [14]. The facultative intracellular pathogen, *F. tularensis* was also found to cause oxidative stress [15]. Similar to the case of *L. monocytogenes* infection, mice with NOX2 (gp91phox) $^{-/-}$ gene were found to be susceptible to *F. tularensis* infection, with significantly higher lung and spleen bacterial count, and were found to die 1 day sooner than wild-type mice [16].

Not only bacteria, viral infections are also successfully combated by the level of ROS in infected cell. Autophagosome induction in infected fibroblasts with Chikungunya virus (CHIKV), an arbovirus, was found to be diminished in infected cells treated with N-acetyl cysteine (NAC), a chemical scavenger of ROS, emphasising the prominent role of ROS in preventing infection [17]. The herpes simplex virus 1 (HSV1) growth in corneal cells was also found to be hampered by H_2O_2-releasing granulocytes [18]. Respiratory epithelial cells upon influenza virus infection demonstrated induced mitochondrial ROS. This ROS induction leads to type III interferon generation and attainment of an antiviral state-limiting replication of the virus within infected cell [19]. ROS also play a role in host defence against *Mycobacteria* most likely via ROS-dependent signalling pathways, such as pro-inflammatory cytokine production, induction of autophagy and granuloma formation [20]. Apart from bacterial and viral infections, certain fungal infections are also found to flourish upon impaired NOX2 function. CGD patients have susceptibility to *Aspergillus* and *Candida* [21] and myeloperoxidase (MPO) deficient individuals are found to have increased susceptibility to *Candida sp*. [22, 23].

16.2 ROS Production by Host During *Leishmania* Infection

During blood feeding, the sand fly inoculates the metacyclic promastigotes in to the body of secondary hosts, i.e. mammals where they are phagocytosed by neutrophils, dendritic cells and mainly macrophages [24–26]. Neutrophils are of very small life span and infected neutrophils mainly serve to silently transfer the parasites to macrophages avoiding cell activation [27]. Macrophages and neutrophils both play a very important role in the host immune defence. Neutrophils and macrophages generate ROS upon phagocytosis of a foreign pathogen ensuring elimination of the pathogen. Thus, persistence of a parasite within these host cells well equipped with anti-microbial defence arsenal is apparently almost impossible. However, *Leishmania* not only survives within macrophages, but it also successfully establishes the infection. Therefore, detailed study regarding how *Leishmania* escapes oxidative burst-mediated host defence is extremely important for designing effective therapeutic strategies against intracellular pathogens.

Thus, this chapter mainly focuses on how *Leishmania* avoids the detrimental effects of ROS, the very first line of defence within host cells, i.e. neutrophils and macrophages. Interestingly, not all species of *Leishmania* are equally efficient in the subversion process and the chapter also tries to highlight the differences found for species variation. Moreover, stage-specific differential behaviour has also been depicted for same species, and this review tries to provide a comprehensive picture on how *Leishmania* encounters oxidative bust-mediated host response.

16.2.1 Neutrophils

16.2.1.1 Suppression of ROS Generation

Neutrophils are the most abundant leukocytes. Upon an infection, neutrophils are rapidly recruited to the sites of infection. They are equipped with an array of antimicrobial arsenals that are sufficient to kill the pathogen following infection. Upon *Leishmania* infection by sandfly bite or inoculation of the parasite by needle (experimental leishmaniasis), neutrophils are found to be recruited to that site and majority of the parasites are rapidly engulfed by them. However, several studies clearly suggested that *Leishmania* sp. evolved an array of mechanisms to slip out of neutrophil-mediated death and utilise these cells as a medium for better invasion of the host.

Exposure of C57BL/6 mice to *L. major* promastigotes by infected sand fly bites or inoculation by needle led to a rapid infiltration by neutrophils into the skin as observed by dynamic two photon intravital microscopy and flow cytometry [28]. The invading neutrophils were found to phagocytose the promastigotes but the phagocytosed *L. major* promastigotes remained viable and could contribute to further infection and disease progression [28]. In this study, it has been shown that there is no oxidative stress in neutrophils during the early hours of infection [28].

According to the study, *L. major* utilises the neutrophils for subsequent efficient unidentifiable entry inside macrophage cells [28]. Another study, also by ultrastructure approaches, showed that promastigotes of *L. major* upon engulfment interact with the neutrophils in a manner that these neutrophils did not evoke ROS generation [29]. The antimicrobial action of polymorphonuclear neutrophils is determined by the cytoplasmic granules. During *L. major* infection, the azurophilic granules (primary or peroxidase-positive granules) were found to fuse with the parasite-containing phagosomes in a manner such that it did not lead to parasite elimination [29]. They found that the tertiary and specific granules, which are responsible for vacuole acidification and superoxide anion generation, did not fuse with *Leishmania*-containing phagosomes. *L. major* upon phagocytosis by the neutrophils did not elicit superoxide generation or mobilisation of tertiary and specific granules [29]. Thus, *L. major* promastigotes upon uptake by the neutrophils regulated the fusion of cytoplasmic granules with the parasite containing phagosomes which favoured the persistence of the parasites within the neutrophils [29]. Contrary to these observations, mid-log phase and metacyclic promastigotes of *L. major* have been found to induce significant respiratory burst in rabbit peritoneal neutrophils upon infection [30]. Similarly, promastigotes of *L. donovani* and *L. mexicana* also evoked large respiratory responses by rabbit peritoneal neutrophils as evident by luminometry [30]. Co-culture of inflammatory neutrophils and bone-marrow neutrophils with both stationary phase and metacyclic *L. braziliensis* promastigotes led to increased superoxide production in comparison to incubation with dead promastigotes [31]. Moreover, *L. braziliensis* infection induced apoptosis of these inflammatory neutrophils but co-culture with *L. major* did not induce apoptosis [31]. In another study, it was found that neutrophils internalised *L. braziliensis* amastigotes more competently in comparison to *L. amazonensis* amastigotes and lead to oxidative burst which was absent in case of *L. amazonensis* amastigote infection [32].

16.2.1.2 Neutrophil Extracellular Trap (NET) Release

NADPH oxidase (NOX2) is a multi-subunit enzyme complex. The enzyme complex remains dormant in resting phagocytic cells. During phagocytosis of a foreign particle, the enzyme becomes activated and produces superoxide which may lead to generation of further harmful ROS, thus playing a key role in pathogen elimination. In addition to release of superoxide anions, NADPH oxidase of neutrophils is also found to be critical for classical ROS-dependent NET formation. These NETs extruded from neutrophils upon activation is also an important antimicrobial arsenal as these are composed of a DNA backbone associated with microbicidal proteins. NETs entrap the pathogens preventing their spread and also may lead to death of the entrapped pathogen [33]. Infection of human neutrophils with *L. amazonensis* evoked production of both ROS- dependent and ROS-independent NET formation leading to the elimination of the parasites [34]. *L. donovani* promastigotes, on the other hand, were found to be able to combat the antimicrobial activity of neutrophil-released NETs [35]. Thus, it may be suggested that ROS production by infected neutrophils depends upon the species and the stage of *Leishmania* infecting the host and ROS generation by neutrophils is inhibited by the cunning parasite upon infection.

16.2.2 Macrophages

The first host defence put forward by macrophage cells upon infection is huge oxidative burst as there is an enormous production of ROS upon phagocytosis of any foreign particle. Phagocytosis of *Leishmania* species are not exception to this universal rule [36]. ROS generation is induced by host cell upon phagocytosis or recognition of ligands of pattern recognition receptors (PRRs) and can ultimately result in apoptosis of the cell ensuring complete parasite clearance along with the infected cell [37]. However, when ROS production in *Leishmania* infected macrophage cells was assessed during early hours of infection, it was found to vary depending upon the parasite species and their various stages. While, *L. major* promastigotes trigger superoxide generation by macrophages [38–40], purified metacyclic *L. major* promastigotes on the other hand generate nominal superoxide. Similarly, although both *L. mexicana and L. chgasi* parasites in their promastigote forms trigger macrophage superoxide production [40, 41], *L. donovani* promastigotes have shown to generate limited superoxide [42–44]. Insufficient ROS production thus leads to inappropriate defence mechanism and the parasite gets a suitable niche for its persistence.

16.2.2.1 NADPH Oxidase

Once any pathogen is phagocytosed, it must be able to withstand the oxidative burst, subvert the sudden respiratory burst or escape the phagosome to survive. In order to reduce ROS level of a cell, there must be upregulation of an antioxidant enzyme or downregulation of a pro-oxidant enzyme within the infected macrophage cell. The major pro-oxidant enzyme of phagocytes is NADPH oxidase. In order to maintain a low level of ROS within infected cells, the parasite must downregulate or impair the activation of this enzyme complex. As discussed earlier, NADPH oxidase is inactive in quiescent cells. The only membrane-bound component of NADPH oxidase, cytochrome b_{558}, is composed of gp91phox and p22phox subunits and is stored in intracellular granules, whereas cytosolic components, p67phox, p47phox and p40phox, form a ternary cytosolic complex [45]. Immediately upon a phagocytic stimulation, another cytosolic member of NADPH oxidase, Rac and the ternary complex bind to the phagosome independently. Activated Rac then binds to gp91phox. p47phox, a regulatory subunit of the enzyme complex, undergoes phosphorylation during phagocytic stimulation and leads to ternary complex and membrane complex interaction. Activated Rac then finally binds to p67phox leading to complete NADPH oxidase assembly on the phagosome which is essential for the release of optimal amounts of superoxide. Several studies have shown that *Leishmania*-infected macrophages fail to assemble the NADPH oxidase enzyme complex leading to failure in production of ROS [46–49]. A study by Pham et al. has suggested that this failure of ROS production is the result of defective maturation of gp91phox [46]. This subunit is actually synthesised as 58-kDa polypeptide which after glycosylation becomes a 65-kDa molecule in the endoplasmic reticulum [47]. The 65-kDa molecule then gets trafficked through the trans-Golgi network getting additionally glycosylated and acquires heme groups. This leads to emergence of a 91 kDa molecule [47–49]. The complete maturation of gp91phox increases the affinity for the other membrane resident subunit p22phox. Thus, the maturation of the functional subunit gp91phox is mainly dependent upon its acquisition of heme which is responsible for complete maturation of gp91phox and its association with p22phox. Since the acquisition of heme is the deciding factor, availability of heme in the infected cell is very crucial. The study by Pham et.al showed that *L. pifanoi* amastigotes upon infection did not induce superoxide production [46]. *L. pifanoi* upon infection induced the expression of heme-oxygenase-1 (HO-1), the rate-limiting enzyme of heme degradation, which decreased the cellular heme content resulting in blockage of the complete maturation of gp91phox, and ultimately preventing the NADPH oxidase enzyme complex assembly [47].

The NADPH oxidase enzyme activation is also dependent upon the regulatory cytosolic components p67phox, p47phox, p40phox and Rac 1 or 2 depending upon cell type. Lipophosphoglycan (LPG), the surface glycolipid of *L. donovani,* was also found to hamper the assembly and function of the NADPH oxidase complex [50]. Phagocytosis of both wild-type and LPG-defective *L. donovani* promastigotes resulted in the release of similar levels of superoxide. However, only the wild-type promastigotes but not the LPG-defective mutants, inhibited superoxide generation at the phagosome [50]. Confocal microscopy imaging revealed that the cytosolic components p47phox and p67phox of NADPH oxidase were excluded from phagosomes upon phagocytosis of wild-type promastigotes in LPG-dependent manner. However, phosphorylation of p47phox and the formation of complexes containing p47phox and p67phox were similar for both wild-type and LPG-defective promastigotes infection [50]. Amastigotes of *L. donovani* also induced defective phosphorylation of the NADPH oxidase cytosolic component p47phox, leading to a defective phagosomal recruitment of p67phox and p47phox and ultimately hampering the activity of NADPH oxidase [51].

Studies with *L. donovani* promastigotes have shown an immediate oxidative burst after internalisation within macrophages starting from 15-min postinfection, which was followed by steadily decreased ROS level and within 1 h of postinfection the ROS generation was at very low level [52, 53]. Moreover, protein kinase C was found to be rapidly and transiently translocated to bone marrow-derived macrophage membrane following parasite attachment, but these translocations were absent in parasite internalised infected cells suggesting active suppression of protein kinase C-mediated downstream oxidative events upon *L. donovani* infection [52]. Infection with *L. chagasi* or treatment with LPG were also found to trigger HO-1 production by murine macrophages [54]. Upon *L. chagasi* infection, macrophages from HO-1 knockout mice presented significantly lower parasite load when compared with those from wild-type mice [54]. Upregulation of HO-1 by cobalt protoporphyrin IX on the other hand diminished the production of TNF-α and ROS by infected murine macrophages along with increased Cu/Zn superoxide dismutase expression [54].

In a study by Blos et al., the ability of macrophages from wild-type and gp91phox−/− mice to combat *L. major* infection was assessed [55]. Macrophages from both type of mice were found to equally affect *L. major* survival upon cytokine activation and although the requirement of NADPH oxidase was essential for elimination of *L. major* in spleen but it was not the case in containment of parasites in draining lymph nodes [55]. In a study involving infection of peritoneal macrophages from C57BL/6 mice with *L. amazonensis*, although infection was found to induce ROS generation, the amount of ROS generated had no effect on parasite killing [56].

16.2.2.2 Mitochondrial ROS

Apart from NADPH oxidase, another important source of ROS is mitochondria. Mitochondria are major producers of ROS and hence contributes greatly to the cellular oxidative burst. The mitochondrial ROS generation is monitored by uncoupling proteins (UCP2 and UCP3) and adenine nucleotide translocase (ANT) [57], found within the inner mitochondrial membrane. To persist within the host cells, apart from phagosomal ROS, *Leishmania* successfully neutralises mitochondrial ROS production. *L. donovani* infection leads to a significant induction of uncoupling protein 2 (UCP2) which negatively regulates mitochondrial ROS [58]. It was observed that in UCP2 knocked-down cells, there was p38 and ERK1/2 activation resulting from ROS-mediated inhibition of protein tyrosine phosphatases (PTPs) [58]. Silencing of UCP2 in *L. donovani* infected mice also led to decrease of Src homology 2 domain-containing tyrosine phosphatase 1 and PTP-1B activity as well as pro-inflammatory cytokine response leading to decreased parasite burden. UCP2 silencing in macrophages even led to significant IL-10 and TGF-β downregulation and enhanced TNF-α and IL-12 expression post *L. donovani* infection [58]. Other uncouplers such as ANT1 and UCP3 were also evaluated, but the expression level of these proteins did not change during *L. donovani* infection [58]. Recently, it has been documented that *L. donovani* is capable of restricting the activation of the well characterised and most critical inflammasome, NLRP3 by upregulating the deubiquitinating enzyme A20 and UCP2 of inner mitochondrial membrane [59]. A20 inhibits the inflammasome priming step by inhibiting the transcription factor NF-κB, whereas UCP2 negatively regulates ROS generation and impairs the formation of inflammasome complex during infection [59].

16.3 Effect of ROS on Intracellular Survival of *Leishmania*

Leishmania sp. upon infection manipulates the host's capacity to generate ROS. ROS either directly affects the cellular organisation of the parasite or may even induce apoptosis of the infected host cell to ensure complete elimination of the unwanted pathogen along with infected cell. There are various factors apart from phagocytosis of pathogen which can induce ROS generation from a cell. Stimulation of cells with

ROS inducers like PMA, H_2O_2 and iron has been found to affect the intracellular persistence of these parasites. In a study, treatment of neutrophils with phorbol myristate acetate (PMA), a ROS inducer, induced elimination of the *L. braziliensis* [32]. However, the same study has shown that *L. amazonensis* amastigotes were found to be highly resistant to PMA-induced microbicidal activities [32]. Treatment of *L. donovani* infected RAW 264.7 macrophage cells with 400 μM of H_2O_2, another well-known inducer of ROS, showed significantly upregulated level of ROS within infected cells [53]. However, this increased amount of ROS did not hamper the parasite persistence within infected cells as *L. donovani* inhibited the subsequent H_2O_2-mediated apoptosis of these cells [53]. In this regard, our study revealed that *Leishmania* parasite employs suppressor of cytokine signalling (SOCS) group of proteins to subvert macrophage apoptotic machinery through participation of thioredoxin and protein tyrosine phosphatase (PTP) [53]. Subsequently, a host protein myeloid cell leukaemia 1 (MCL-1) has been identified, which plays an important role in inhibiting mitochondria-dependent macrophage apoptosis [60]. It has been shown in this study that infection by *Leishmania* induced translocation of MCL-1 to mitochondria, where it interacts with the major pro-apoptotic protein BAK and prevents BAK-BAK homo-oligomer formation thereby preventing cytochrome *c* release-mediated mitochondrial dysfunction [60]. Another important factor that can regulate the amount of ROS within a host cell is the presence of free iron within the cell. In BALB/c mice, the persistence of *L. major* was found to be inhibited by iron-induced oxidative burst by neutrophils [61, 62].

16.4 ROS-Generating Compounds: Role for Antileishmanial Therapy

Leishmaniasis is one of the deadliest diseases and is severely fatal if left untreated. Various therapeutic molecules were used as antileishmanial therapy but had little success, due to various reasons like toxicity, chemo-resistance and their high cost. Pentavalent antimony complexes like sodium stibogluconate and sodium antimony gluconate (SAG) were used for more than 60 years. These compounds resulted in the production of ROS during early hours of treatment leading to killing of the parasites [63]. Currently, there are two antileishmanial front-line therapies miltefosine and amphotericin B. Generation of ROS is the first line of defence put forward by the host cells against any foreign pathogen. Thus, if a medication can boost this antimicrobial property; it may turn out to be a very effective drug not only for Leishmaniasis but for other infectious diseases also. Several natural compounds such as plant extracts deploy potent immunomodulatory potential, which could be a convenient adjunct to conventional chemotherapy for treatment of Leishmaniasis. Recent studies have demonstrated significant immunomodulatory effects and antileishmanial activity of polyphenols in vitro [64–66]. Quercetin is a plant polyphenol found in many fruits, vegetables, leaves and grains. Quercetin was found to have potential antileishmanial activity against the intracellular amastigote form of

L. amazonensis [67]. Treatment with the compound in *L. amazonensis*-infected macrophages led to reduced infection index along with increased ROS generation in a dose-dependent manner. Quercetin thus acted as a potential antileishmanial compound because of its pro-oxidant properties. Quercetin has been found to inhibit the growth of both promastigotes and amastigotes of *L. donovani* by the blockade of DNA synthesis [67–69]. Ghosh et al. have shown the effective role of curdlan, a naturally occurring herbal immunomodulator that could completely cure experimental VL through significantly enhanced production of NO [70]. Along with induction of disease-resolving Th1 cytokines, curdlan stimulates the production of IL-23, which helps in the stabilisation and differentiation of Th17 cytokine, IL-17 [70]. The leaves of *Nectandra cuspidate* were found to have high antioxidant capacity and presented activity against *L. amazonensis* along with low cytotoxicity in mammalian host cells [71]. Mahanine, a carbazole alkaloid isolated from an Indian edible medicinal plant, is an anticancer agent with minimal effect on normal cells. Oral administration of mahanine in *L. donovani* infected mice exhibited almost complete reduction of parasite burden, upregulation of NO/iNOS/ROS/IL-12 and T cell proliferation [72]. Mahanine, therefore, can act as a potent antileishmanial molecule capable of inducing ROS and together with modulation of host's immune response could be developed as an inexpensive and nontoxic therapeutic.

16.5 Conclusion

Leishmania is an obligate intracellular parasite, which efficiently subverts the major antimicrobial arsenals of the host neutrophils and macrophages. ROS is the first line of defence of the host cell which directly or indirectly ensures complete elimination of the intracellular foreign pathogen. The level of ROS within a phagocyte is very crucial as high amount of ROS within a cell can hamper its own survival. Whenever the host cell engulfs a pathogen, there is an immediate oxidative burst which either directly hampers the pathogen or indirectly leads to a pro-inflammatory environment or induces apoptosis to ensure complete elimination of the pathogen. *Leishmania* upon infection modulates the signalling pathways of the host cell in its own favour resulting in persistence of the parasite within the host cell and disease progression (Fig. 16.1). *Leishmania* upon infection efficiently reduces the ROS level in the infected cell. The parasite either directly hampers the complete association of the main ROS-generating enzyme of phagocyte, i.e. NADPH oxidase or indirectly increases the expression level of anti-oxidant enzymes heme oxygenase-1 (HO-1) and superoxide dismutase. *Leishmania* sp. upon infection even affects other ROS-generating pathways like mitochondrial ROS to ensure its survival within macrophage cells. Thus, this chapter showcases a compact picture of the status of the infected phagocyte upon infection by *Leishmania sp.* and the information might have the potential towards developing therapeutics not only for leishmaniasis but in general for other macrophage-associated diseases.

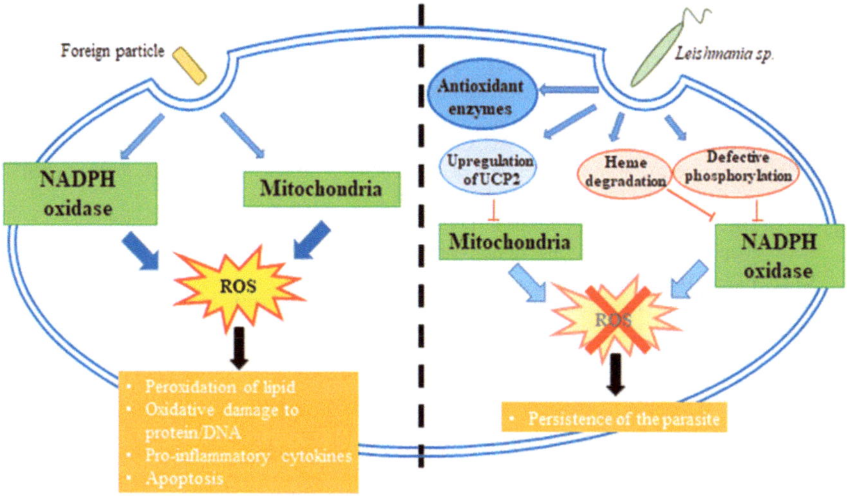

Fig. 16.1 Schematic representation of the difference of ROS production ability by host cell during entry of any foreign particle or infection with *Leishmania sp.* A host cell recognises a foreign particle and activates the reactive oxygen species (ROS)-generating enzymes: NAD(P)H oxidase and the mitochondrial enzymes. These activations lead to generation of ample amount of ROS sufficient to destroy the foreign particle. The amount of ROS generated may even lead to apoptosis of the host cell bearing the foreign particle. However, when the host cell gets infected with *Leishmania sp.*, the ROS generation by host cell gets inhibited by several pathways. Upregulation of host antioxidant enzymes occurs, leading to a general neutralisation of the ROS. Moreover, *Leishmania sp.* specifically degrades the cellular heme and hampers phosphorylation of a subunit of NAD(P)H oxidase, which are necessary for the proper function of the enzyme. *Leishmania* infection also causes specific upregulation of Uncoupling protein 2 (UCP2) leading to downregulation of mitochondrial ROS generation. Thus, reduced ROS level in *Leishmania* infected cell allows persistence of the parasite within cell and in disease progression

Acknowledgements We apologise to all the research groups whose work might not have been discussed or cited in the chapter. This work was supported by the Department of Science and Technology Grant No. EMR/2014/000287, Department of Biotechnology Grant No. BT/PR10289/BRB/10/1257/2013, National Academy of Sciences, India (NASI) Senior Scientist Platinum Jubilee Fellowship and Council of Scientific and Industrial Research (CSIR), Govt. of India.

References

1. Kaye P, Scott P (2011) Leishmaniasis: complexity at the host-pathogen interface. Nat Rev Microbiol 9:604–615
2. Jha PK, Vankalakunti M, Siddini V, Babu K, Ballal SH (2012) Postrenal transplant laryngeal and visceral leishmaniasis - a case report and review of the literature. Indian J Nephrol 22:301–303
3. Bin Mohanna MA (2015) Leishmaniasis, malaria, and schistosomiasis concurrently in an 8-year-old boy. Saudi Med J 36:494–496
4. Zijlstra EE (2016) The immunology of post-kala-azar dermal leishmaniasis (PKDL). Parasit Vectors 9:464

5. Belikov AV, Schraven B, Simeoni L (2015) T cells and reactive oxygen species. J Biomed Sci 22:85
6. Imlay JA, Linn S (1988) DNA damage and oxygen radical toxicity. Science 240:1302–1309
7. Imlay JA (2008) Cellular defenses against superoxide and hydrogen peroxide. Annu Rev Biochem 77:755–776
8. Galhardo RS, Almeida CE, Leitao AC, Cabral-Neto JB (2000) Repair of DNA lesions induced by hydrogen peroxide in the presence of iron chelators in *Escherichia coli*: participation of endonuclease IV and Fpg. J Bacteriol 182:1964–1968
9. Lloyd DR, Carmichael PL, Phillips DH (1998) Comparison of the formation of 8-hydroxy-2′-deoxyguanosine and single- and double-strand breaks in DNA mediated by Fenton reactions. Chem Res Toxicol 11:420–427
10. Imlay JA (2006) Iron-Sulphur clusters and the problem with oxygen. Mol Microbiol 59:1073–1082
11. Ben-Ari J, Wolach O, Gavrieli R, Wolach B (2012) Infections associated with chronic granulomatous disease: linking genetics to phenotypic expression. Expert Rev Anti-Infect Ther 10:881–894
12. Fang FC (2011) Antimicrobial actions of reactive oxygen species. MBio 2:e00141–e00111
13. Soler-Palacin P, Margareto C, Llobet P et al (2007) Chronic granulomatous disease in pediatric patients: 25 years of experience. Allergol Immunopathol (Madr) 35:83–89
14. Dinauer MC, Deck MB, Unanue ER (1997) Mice lacking reduced nicotinamide adenine dinucleotide phosphate oxidase activity show increased susceptibility to early infection with *Listeria monocytogenes*. J Immunol 158:5581–5583
15. Pohanka M, Pavlis O, Ruttkay-Nedecky B et al (2012) Tularemia progression accompanied with oxidative stress and antioxidant alteration in spleen and liver of BALB/c mice. J Microbiol 50:401–408
16. KuoLee R, Harris G, Conlan JW, Chen W (2011) Role of neutrophils and NADPH phagocyte oxidase in host defense against respiratory infection with virulent Francisella tularensis in mice. Microbes Infect 13:447–456
17. Joubert PE, Werneke S, de la Calle C et al (2012) Chikungunya-induced cell death is limited by ER and oxidative stress-induced autophagy. Autophagy 8:1261–1263
18. Hayashi K, Hooper LC, Okuno T, Takada Y, Hooks JJ (2012) Inhibition of HSV-1 by chemoattracted neutrophils: supernatants of corneal epithelial cells (HCE) and macrophages (THP-1) treated with virus components chemoattract neutrophils (PMN), and supernatants of PMN treated with these conditioned media inhibit viral growth. Arch Virol 157:1377–1381
19. Kim HJ, Kim CH, Ryu JH et al (2013) Reactive oxygen species induce antiviral innate immune response through IFN-lambda regulation in human nasal epithelial cells. Am J Respir Cell Mol Biol 49:855–865
20. Deffert C, Cachat J, Krause KH (2014) Phagocyte NADPH oxidase, chronic granulomatous disease and mycobacterial infections. Cell Microbiol 16:1168–1178
21. Paiva CN, Bozza MT (2014) Are reactive oxygen species always detrimental to pathogens? Antioxid Redox Signal 20:1000–1037
22. Lanza F (1998) Clinical manifestation of myeloperoxidase deficiency. J Mol Med (Berl) 76:676–681
23. Lehrer RI, Cline MJ (1969) Leukocyte myeloperoxidase deficiency and disseminated candidiasis: the role of myeloperoxidase in resistance to Candida infection. J Clin Invest 48:1478–1488
24. Filardy AA, Pires DR, DosReis GA (2011) Macrophages and neutrophils cooperate in immune responses to *Leishmania* infection. Cell Mol Life Sci 68:1863–1870
25. Peters NC, Sacks DL (2009) The impact of vector-mediated neutrophil recruitment on cutaneous leishmaniasis. Cell Microbiol 11:1290–1296
26. Sacks D, Noben-Trauth N (2002) The immunology of susceptibility and resistance to *Leishmania major* in mice. Nat Rev Immunol 2:845–858
27. van Zandbergen G, Klinger M, Mueller A et al (2004) Cutting edge: neutrophil granulocyte serves as a vector for *Leishmania* entry into macrophages. J Immunol 173:6521–6525

28. Peters NC, Egen JG, Secundino N et al (2008) In vivo imaging reveals an essential role for neutrophils in leishmaniasis transmitted by sand flies. Science 321:970–974
29. Mollinedo F, Janssen H, de la Iglesia-Vicente J, Villa-Pulgarin JA, Calafat J (2010) Selective fusion of azurophilic granules with *Leishmania*-containing phagosomes in human neutrophils. J Biol Chem 285:34528–34536
30. Mallinson DJ, Lackie JM, Coombs GH (1989) The oxidative response of rabbit peritoneal neutrophils to leishmanias and other trypanosomatids. Int J Parasitol 19:639–645
31. Falcao SA, Weinkopff T, Hurrell BP et al (2015) Exposure to *Leishmania braziliensis* triggers neutrophil activation and apoptosis. PLoS Negl Trop Dis 9:e0003601
32. Carlsen ED, Jie Z, Liang Y et al (2015) Interactions between neutrophils and *Leishmania braziliensis* amastigotes facilitate cell activation and parasite clearance. J Innate Immun 7:354–363
33. Brinkmann V, Reichard U, Goosmann C et al (2004) Neutrophil extracellular traps kill bacteria. Science 303:1532–1535
34. Rochael NC, Guimaraes-Costa AB, Nascimento MT et al (2015) Classical ROS-dependent and early/rapid ROS-independent release of neutrophil extracellular traps triggered by *Leishmania* parasites. Sci Rep 5:18302
35. Gabriel C, McMaster WR, Girard D, Descoteaux A (2010) *Leishmania donovani* promastigotes evade the antimicrobial activity of neutrophil extracellular traps. J Immunol 185:4319–4327
36. Horta MF, Mendes BP, Roma EH et al (2012) Reactive oxygen species and nitric oxide in cutaneous leishmaniasis. J Parasitol Res 2012:203818
37. Lind L, Ljunghall S (1992) Serum urate and renal function in different forms of hypercalcemia. Exp Clin Endocrinol 99:87–90
38. Da Silva RP, Hall BF, Joiner KA, Sacks DL (1989) CR1, the C3b receptor, mediates binding of infective *Leishmania major* metacyclic promastigotes to human macrophages. J Immunol 143:617–622
39. Kantengwa S, Muller I, Louis J, Polla BS (1995) Infection of human and murine macrophages with *Leishmania major* is associated with early parasite heat shock protein synthesis but fails to induce a host cell stress response. Immunol Cell Biol 73:73–80
40. Mallinson DJ, Coombs GH (1989) Interaction of leishmania metacyclics with macrophages. Int J Parasitol 19:647–656
41. Gantt KR, Goldman TL, McCormick ML et al (2001) Oxidative responses of human and murine macrophages during phagocytosis of *Leishmania chagasi*. J Immunol 167:893–901
42. Brandonisio O, Panaro MA, Marzio R, Marangi A, Faliero SM, Jirillo E (1994) Impairment of the human phagocyte oxidative responses caused by *Leishmania* lipophosphoglycan (LPG): in vitro studies. FEMS Immunol Med Microbiol 8:57–62
43. Channon JY, Roberts MB, Blackwell JM (1984) A study of the differential respiratory burst activity elicited by promastigotes and amastigotes of *Leishmania donovani* in murine resident peritoneal macrophages. Immunology 53:345–355
44. Descoteaux A, Turco SJ (1993) The lipophosphoglycan of *Leishmania* and macrophage protein kinase C. Parasitol Today 9:468–471
45. Leto TL, Morand S, Hurt D, Ueyama T (2009) Targeting and regulation of reactive oxygen species generation by Nox family NADPH oxidases. Antioxid Redox Signal 11:2607–2619
46. Pham NK, Mouriz J, Kima PE (2005) *Leishmania pifanoi* amastigotes avoid macrophage production of superoxide by inducing heme degradation. Infect Immun 73:8322–8333
47. Yu L, DeLeo FR, Biberstine-Kinkade KJ, Renee J, Nauseef WM, Dinauer MC (1999) Biosynthesis of flavocytochrome b558. gp91(phox) is synthesized as a 65-kDa precursor (p65) in the endoplasmic reticulum. J Biol Chem 274:4364–4369
48. DeLeo FR, Burritt JB, Yu L, Jesaitis AJ, Dinauer MC, Nauseef WM (2000) Processing and maturation of flavocytochrome b558 include incorporation of heme as a prerequisite for heterodimer assembly. J Biol Chem 275:13986–13993
49. Nauseef WM (2004) Assembly of the phagocyte NADPH oxidase. Histochem Cell Biol 122:277–291

50. Lodge R, Diallo TO, Descoteaux A (2006) *Leishmania donovani* lipophosphoglycan blocks NADPH oxidase assembly at the phagosome membrane. Cell Microbiol 8:1922–1931
51. Lodge R, Descoteaux A (2006) Phagocytosis of *Leishmania donovani* amastigotes is Rac1 dependent and occurs in the absence of NADPH oxidase activation. Eur J Immunol 36:2735–2744
52. Bhunia AK, Sarkar D, Das PK (1996) *Leishmania donovani* attachment stimulates PKC-mediated oxidative events in bone marrow-derived macrophages. J Eukaryot Microbiol 43:373–379
53. Srivastav S, Basu Ball W, Gupta P, Giri J, Ukil A, Das PK (2014) *Leishmania donovani* prevents oxidative burst-mediated apoptosis of host macrophages through selective induction of suppressors of cytokine signaling (SOCS) proteins. J Biol Chem 289:1092–1105
54. Luz NF, Andrade BB, Feijo DF et al (2012) Heme oxygenase-1 promotes the persistence of *Leishmania chagasi* infection. J Immunol 188:4460–4467
55. Blos M, Schleicher U, Soares Rocha FJ, Meissner U, Rollinghoff M, Bogdan C (2003) Organ-specific and stage-dependent control of *Leishmania major* infection by inducible nitric oxide synthase and phagocyte NADPH oxidase. Eur J Immunol 33:1224–1234
56. Roma EH, Macedo JP, Goes GR et al (2016) Impact of reactive oxygen species (ROS) on the control of parasite loads and inflammation in *Leishmania amazonensis* infection. Parasit Vectors 9:193
57. Echtay KS, Esteves TC, Pakay JL et al (2003) A signalling role for 4-hydroxy-2-nonenal in regulation of mitochondrial uncoupling. EMBO J 22:4103–4110
58. Basu Ball W, Kar S, Mukherjee M, Chande AG, Mukhopadhyaya R, Das PK (2011) Uncoupling protein 2 negatively regulates mitochondrial reactive oxygen species generation and induces phosphatase-mediated anti-inflammatory response in experimental visceral leishmaniasis. J Immunol 187:1322–1332
59. Gupta AK, Ghosh K, Palit S, Barua J, Das PK, Ukil A (2017) *Leishmania donovani* inhibits inflammasome-dependent macrophage activation by exploiting the negative regulatory proteins A20 and UCP2. FASEB J 31:5087–5101
60. Giri J, Srivastav S, Basu M, Palit S, Gupta P, Ukil A (2016) *Leishmania donovani* exploits myeloid cell leukemia 1 (MCL-1) protein to prevent mitochondria-dependent host cell apoptosis. J Biol Chem 291:3496–3507
61. Bisti S, Konidou G, Boelaert J, Lebastard M, Soteriadou K (2006) The prevention of the growth of *Leishmania major* progeny in BALB/c iron-loaded mice: a process coupled to increased oxidative burst, the amplitude and duration of which depend on initial parasite developmental stage and dose. Microbes Infect 8:1464–1472
62. Bisti S, Soteriadou K (2006) Is the reactive oxygen species-dependent-NF-kappaB activation observed in iron-loaded BALB/c mice a key process preventing growth of *Leishmania major* progeny and tissue-damage? Microbes Infect 8:1473–1482
63. Mookerjee Basu J, Mookerjee A, Sen P et al (2006) Sodium antimony gluconate induces generation of reactive oxygen species and nitric oxide via phosphoinositide 3-kinase and mitogen-activated protein kinase activation in *Leishmania donovani*-infected macrophages. Antimicrob Agents Chemother 50:1788–1797
64. Passero LF, Laurenti MD, Santos-Gomes G, Soares Campos BL, Sartorelli P, Lago JH (2014) Plants used in traditional medicine: extracts and secondary metabolites exhibiting antileishmanial activity. Curr Clin Pharmacol 9:187–204
65. Radtke OA, Foo LY, Lu Y, Kiderlen AF, Kolodziej H (2003) Evaluation of sage phenolics for their antileishmanial activity and modulatory effects on interleukin-6, interferon and tumour necrosis factor-alpha-release in RAW 264.7 cells. Z Naturforsch C 58:395–400
66. Miranda MM, Panis C, da Silva SS et al (2015) Kaurenoic acid possesses Leishmanicidal activity by triggering a NLRP12/IL-1beta/cNOS/NO pathway. Mediat Inflamm 2015:392918
67. Fonseca-Silva F, Inacio JD, Canto-Cavalheiro MM, Almeida-Amaral EE (2011) Reactive oxygen species production and mitochondrial dysfunction contribute to quercetin induced death in *Leishmania amazonensis*. PLoS One 6:e14666

68. Wang G, Wang JJ, Chen XL et al (2013) The JAK2/STAT3 and mitochondrial pathways are essential for quercetin nanoliposome-induced C6 glioma cell death. Cell Death Dis 4:e746
69. Passero LF, Bonfim-Melo A, Corbett CE et al (2011) Anti-leishmanial effects of purified compounds from aerial parts of *Baccharis uncinella* C. DC. (Asteraceae). Parasitol Res 108:529–536
70. Ghosh K, Sharma G, Saha A, Kar S, Das PK, Ukil A (2013) Successful therapy of visceral leishmaniasis with curdlan involves T-helper 17 cytokines. J Infect Dis 207:1016–1025
71. Ferreira L, Oliveira MM, Faleiro FL et al (2017) Antileishmanial and antioxidant potential of fractions and isolated compounds from *Nectandra cuspidata*. Nat Prod Res:1–4
72. Roy S, Dutta D, Satyavarapu EM et al (2017) Mahanine exerts in vitro and in vivo antileishmanial activity by modulation of redox homeostasis. Sci Rep 7:4141

Part III

Bacterial Diseases

Oxidative Stress and Brucellosis

17

Hasan Karsen

Abstract

Brucellosis is a zoonotic infection caused by brucella bacteria, which are facultative and intracellular pathogens and can be seen worldwide. According to the World Health Organization (WHO), there are more than 500,000 new cases globally each year. It is estimated that in reality there are four times more cases worldwide that are undiagnosed or unreported. It is one of the most important socioeconomic problems in several developing countries. Several previous studies have reported that in addition to economic losses, infections in the host cause several pathologies in brucellosis. The changes in host tissue and cells occur with an increase in oxidative stress due to brucellosis.

While brucellosis reduces antioxidant enzymes and molecules (TAS, magnesium, manganese, arylesterase, paraoxonase, native thiol, total thiol, disulfide, glutathione, PDE_4, zinc), it increases oxidative enzymes and molecules (TOS, Cu, MDA, ceruloplasmin, LOOH, OSI, NO). Thus, oxidative stress increases causing lipid peroxidation, protein denaturation, and DNA breakage. It can also lead to metabolic changes and atherosclerosis in the host.

To prevent all these negative effects of brucellosis on the host, an early start to primary brucellosis treatment should be made and the use of supplemental therapies.

Keywords

Oxidative stress · Brucellosis

H. Karsen (✉)
Department of Infectious Diseases and Clinical Microbiology, School of Medicine, Harran University, Sanliurfa, Turkey
e-mail: hasankarsen@hotmail.com

S. Chakraborti et al. (eds.), *Oxidative Stress in Microbial Diseases*,
https://doi.org/10.1007/978-981-13-8763-0_17

17.1 Introduction

Brucellosis is a zoonotic infection caused by brucella bacteria, which are facultative and intracellular pathogens [1]. This disease can be seen in males and females of all age groups and continues to be a major public health problem worldwide [2]. It is one of the most important socioeconomic problems in several developing countries and is widespread particularly in the Mediterranean Basin, Africa, the Middle East, the Arabian Peninsula, the Indian subcontinent, parts of South America and some regions of Central Asia [3]. According to the World Health Organization (WHO), there are more than 500,000 new cases globally each year. It is estimated that in reality, there are four times more cases worldwide that are undiagnosed or unreported [4–7]. The main factors of infection and transmission of brucellosis are dairy products, environmental hygiene, and socioeconomic conditions. The most common mode of transmission is human contact with infected animals [1]. Several previous studies have reported that in addition to economic losses, infections in the host cause several pathologies in brucellosis [8].

The changes in host tissue and cells occurring with an increase in oxidative stress due to brucellosis were evaluated in this paper. To better understand the subject, it is useful to first look at the oxidative and antioxidative systems.

17.2 Free Radicals and Antioxidants

17.2.1 Free Radicals

Free radicals are highly reactive chemical products containing an electron not matched to the trajectory of an atom or molecule [9]. Each organelle of a cell can produce free radicals at different amounts. Reactive oxygen species (ROS) from free radicals are mostly produced by phagocytic cells in the organism. Phagocytic cells produce ROS by reducing oxygen when various metabolic inflammations are stimulated. Although free radical reactions are necessary for the defense mechanisms of cells such as neutrophils and macrophages in immune system cells, excessive production of free radicals results in tissue damage and cell death [10]. By affecting the lipid components in the cell membrane, free radicals cause lipid peroxidation [11]. In addition, by entering into a reaction with cell components, which inhibits the main functions of the cell, there is a reduction in protein synthesis, and this is known to cause the formation of organic peroxides that block DNA [12]. Free radicals also cause impairments in the function of mitochondria [13].

17.2.2 Reactive Oxygen Species (ROS)

Molecular oxygen (O_2) is defined as diradical. Superoxide (O_2–) radical is formed with the electron reduction of molecular oxygen as an indeterminate structure. Although superoxide anion is a free radical, it is not greatly harmful. Its importance is a source of hydrogen peroxide (H_2O_2) and reducer of temporary metal ions. Superoxide has effects such as the regulation of the bactericidal activity of neutrophils, apoptosis, inflammation, and vascular functions. Decreased superoxide levels can lead to an increased susceptibility to bacterial infections in the macrophages [14]. H_2O_2 radical is formed with a direct reduction of the oxygen molecule or with the addition of an electron to the superoxide anion (superoxide dismutation). Hydrogen peroxide is another component of the leukocyte defense against bacteria. H_2O_2 in the environment is produced and expressed by cells undergoing phagocytosis. The concentration of H_2O_2 has a very important place in the antibacterial mechanism [15]. Hydroxyl radical (OH-) is formed from hydrogen peroxide [16]. As it can easily pass the membrane barriers in the cell nucleus and mutagenically affects DNA, hydroxyl is the most effective radical [17].

17.2.3 Hypochlorous Acid (HOCl)

Hypochlorous acid (HOCl), which is an effective antibacterial agent, is formed by myeloperoxidase enzyme catalyzing chloride oxidation in the presence of hydrogen peroxide. HOCl is expressed by phagocytic cells and has an important role in the cytotoxic effect on bacteria. Activated neutrophils, monocytes, and macrophages produce superoxide radical, and the radical production of phagocytic cells plays an important role in the mechanisms of killing bacteria [18].

17.2.4 Lipid Peroxidation

Lipids are the most sensitive biomolecules against the effects of free radicals. Fatty acids and bonds of unsaturated cholesterol can react with free radicals and cause peroxidation in the cell membrane. Lipid peroxidation continuing spontaneously progresses in the form of a chain reaction and is extremely harmful. OH- radical is accepted as a trigger factor starting the chain reaction in lipid peroxidation [19]. By breaking a hydrogen atom (H+) from the methylene molecule of a fatty acid, OH- radical forms a lipid radical (L-), and the lipid radical reacts with oxygen to form a lipid peroxide radical (LOO-), which then reacts with other unsaturated fatty acids. Thus, the chain reaction starts [20]. The formation of lipid and lipid peroxide radicals in cell membranes is accepted as an important characteristic of the cell damage caused by ROS. Malondialdehyde (MDA) is formed as a result of the peroxidation of fatty acids. Various studies have shown that MDA and other oxygen radicals have a mutagenic effect on bacteria by reacting with DNA. In some studies, it has been shown that there is a slight increase in lipid peroxidation in brucellosis disease [19].

17.2.5 Nitric Oxide (NO)

While other free radicals are harmful to cells, NO at low concentrations has a role in important physiological functions [21]. In macrophages activated by bacterial lipopolysaccharides, NO synthesized from L-arginine by nitric oxide synthase (NOS) plays a role in the very important mechanism against intracellular bacteria [22]. NO at a high concentration has a cytotoxic effect with the de-animation of cell DNA in bacteria [23]. When there is inflammation or bacterial infection, NO is induced by cytokines or endotoxins and produced at high amounts in the long term.

The NO produced has an antimicrobial effect in macrophages and is therefore an important part of the nonspecific host defense system [24]. In human brucella infections, it is known that the expression of bactericidally effective NO is not sufficiently stimulated [25]. Therefore, pathogens show resistance to NO-origin oxidative mechanisms and gain a significant advantage by protecting themselves from antimicrobial attacks [26]. As a result of the interaction of nitric oxide and superoxide radicals, peroxynitrite (ONOO−) is formed, which is an extremely strong oxidative molecule. Thus, the physiological effect of nitric oxide is lost and an oxidative effect emerges. ONOO- is mainly responsible for NO toxicity. This free radical causes DNA breakage and oxidation in lipids [27].

17.2.6 Xanthine Oxidase (XO) Enzyme

Free radicals emerge during the catalytic cycle of this enzyme. When the phosphorylation of ADP to ATP is decreased associated with ischemia, ADP is destroyed, and the purine base transforms xanthine oxidase to hypoxanthine as there is an oxidase effect. When the activity of xanthine oxidase is seen as oxidase, molecular oxygen is used when transforming hypoxanthine to xanthine and xanthine to uric acid and molecular oxygen reduces to O_2 – anion. Superoxide anion is important in the synthesis of H_2O_2 vs OH- radicals [28].

17.2.7 Antioxidant System

Antioxidants are substances which prevent or can delay the oxidation of molecules such as proteins, lipids, carbohydrates, and DNA in the cells, and this is known as the antioxidant system [29]. Superoxide dismutase (SOD), catalase (CAT), and glutathione peroxidase (GSH-Px) are important enzymatic antioxidants of the intracellular defense system [30]. Vitamins (vitamin E, vitamin C, carotenoids), glutathione (GSH), natural flavonoids, melatonin, and some other molecules are among the nonenzymatic intracellular antioxidants [31].

17.2.7.1 SOD

SOD is an antioxidant enzyme that catalyzes the transformation of superoxide radical to hydrogen peroxide [32]. SOD keeps the O_2- levels in the cellular compartments under control and plays a role in the intracellular death of phagocytosed bacteria [33].

17.2.7.2 CAT

CAT is eliminated for the prevention of OH- formation that forms H_2O_2 in the cell. H_2O_2 transforms back to water and oxygen [32]. SOD and CAT are important in brucellosis as they remove the nitrogen radicals and free oxygen expressed by brucella agents with an antibacterial effect.

17.2.7.3 Glutathione Peroxidase (GSH-Px)

Glutathione peroxidase is an enzyme responsible for the reduction of intracellular hydroperoxides [34]. By converting H_2O_2 to water when reduced glutathione increases, the formation of methemoglobin is prevented, and by protecting membrane lipids against peroxide anion, the integrity of the cell membrane and hemoglobin is protected from oxidative stress [35]. Glutathione is an important intracellular antioxidant, and it is found within the cell in reduced form (GSH). They are reduced as oxidants against endogenously produced peroxides. Glutathione peroxidase catalyzes this reaction. For glutathione to be able to effectively protect the cell, the greater part must be kept in reduced form. This reaction catalyzes glutathione reductase [36]. In the oxidized form, it functions in free radical inhibition, the stabilization of reduced sulfydryl groups and in the regeneration of ascorbatin with tocopherol [37].

17.2.7.4 Myeloperoxidase (MPO)

Myeloperoxidase is an enzyme from one of the antibacterial defense mechanisms containing Fe-Hem in the primary lysosomal granules of neutrophils and monocytes. MPO has an important role in the death of phagocytosed bacteria. MPO shows an antibacterial effect related to oxygen together with H_2O_2. With the effect of various stimulants, phagocytes empty granules containing MPO into the phagocytic vacuole in intracellular spaces. In the presence of H_2O_2, MPO catalyzes the oxidation of chloride, iodide, and bromide, thus forming hypochlorous acid (HOCl), hypoiodic acid (HOI), and hypobromic acid (HOBr). These components and salts are strong antioxidants and, by reacting with biologically important molecules, produce toxic agents that affect the microorganism [38]. Catalase, peroxidase, superoxide dismutase, glutathione, and thioredoxin form the components of the antioxidant metabolic strategy in almost all locations [39].

17.3 Interaction Between Brucellosis and Host Tissue

Brucella bacteria have the capacity for long life and replication in the macrophages of the host that has been invaded. This leads to chronic infection and consequently various pathologies in the host [40]. Macrophages and polymorphonuclear leukocytes have a primary role in the elimination of brucellosis in the host cell [1]. Macrophages mobilized for brucellosis elimination activate cytokines, chemokines, and free radicals in the cell and cause the production of extremely toxic and destructive reactive nitrogen species (RNS) and reactive oxygen species (ROS). As a result of this, there is an increase in free radical products and a decrease in antioxidants. The increase created in oxidative stress has a role in the pathogenesis of brucellosis [41, 42]. Thus, cellular apoptosis is induced and various pathological changes occur in the plasma membrane.

Most of the studies that have examined the interaction between brucellosis and the oxidative system and the outcomes are evaluated below in chronological order.

The study entitled "Serum copper and zinc concentrations in patients with brucellosis" by Cesur S et al. in 2005 stated that while serum copper (Cu) was increased in brucellosis patients, the zinc (Zn) level was decreased. As serum Zn and Cu could have a role in brucellosis recovery, the serum concentrations of patients should be examined before and after treatment [43].

In a 2006 experimental rat study by Melek IM et al., entitled "Evaluation of oxidative stress and inflammation in long-term *Brucella melitensis* infection," it was reported that brucellosis caused lipid peroxidation in the host cell membrane and induced NO production without changing antioxidant enzyme activity [44]. In the presence of microbial infection and inflammation in some cells, the defense mechanisms, immunomodulation, and autotoxicity of the host are related to NO overproduction. MDA concentration, as a marker of nitric oxide and lipid peroxidation, was reported to be significantly increased following brucella innoculation [45].

In the 2007 study entitled "Investigation of serum nitric oxide and malondialdehyde levels in cattle infected with *Brucella abortus*," Nisbet C et al. stated that brucellosis increases lipopolysaccharide levels by stimulating NO synthesis from macrophages and causes excessive radical production by increasing the MDA level, thereby affecting membrane lipids. It was concluded that MDA could be used as a marker of tissue damage [46].

In another rat study by Erdoğan et al. in 2008, entitled "The effects of increased cAMP content on inflammation, oxidative stress and PDE_4 transcripts during *Brucella melitensis* infection," it was reported that brucellosis suppressed phosphodiesterase (PDE), reduced antioxidant enzymes, and increased lipid peroxidation, and when the cAMP level was increased, these negative effects were reversed. The severe oxidative effect of brucellosis was demonstrated in these studies [47].

In the 2009 study by Amal H Ali, entitled "The Effect of Brucellosis on Lipid Profile and Oxidant-Antioxidants Status," it was reported that brucellosis caused atherosclerosis at a high rate by increasing the plasma total cholesterol, LDL cholesterol, and the LDL/HDL ratio. The study concluded that brucellosis patients are at high risk of metabolic and cardiovascular diseases [48].

Apostolou F et al. published a study in 2009, entitled "Persistence of an atherogenic lipid profile after treatment of acute infection with brucella," in which it was shown that by decreasing the antioxidative PON1 and arylesterase, brucellosis causes atherosclerosis with changes made in lipids, lipoproteins, and the related enzymes. Despite successful brucellosis treatment, this was reported not to have fully recovered at even 4 months later [49].

In 2009, Serefhanoglu K et al. conducted a study entitled "Evaluation of oxidative status in patients with brucellosis." It was reported that brucellosis increased the oxidative stress index (OSI = Total oxidative status/Total antioxidative capacity), MDA and total peroxide, and the total antioxidative capacity (TAC) level was decreased [50].

The study entitled "Serum zinc and copper concentrations in brucellosis patient" conducted by Mobaien A et al. was published in 2010. Zn and Cu are known to have a regulatory effect on the immune system and the serum concentrations of these micronutrients have been shown to be altered in patients with certain diseases. In this study, it was demonstrated that the Cu level was elevated and the Zn level was decreased in brucellosis patients, and this was reported to be a precursor of oxidative stress [51].

A previous study was conducted in our clinic in 2011, entitled "Serum paraoxonase and arylesterase activities and oxidant status in patients with brucellosis." The results showed that serum paraoxonase and arylesterase activity were significantly reduced in brucellosis patients compared to a control group (Fig. 17.1), and the oxidative status was increased. This change was determined to have a role in the pathogenesis of atherosclerosis that could develop in brucellosis patients [52].

Another study in our clinic in 2011 by Leman, Karaağaç et al. was entitled "Decreasing oxidative stress in response to treatment in patients with brucellosis: could it be used to monitor treatment?" It was concluded in this study that the increase in oxidative stress in brucellosis as a result of the decrease in TAS and increase in TOS caused severe damage to the body, and even if this damage recovered with 6 weeks of treatment, it would not reduce to the level of a normal healthy person. This is illustrated in Fig. 17.2 as OSI = TOS/TAS). It was also reported that

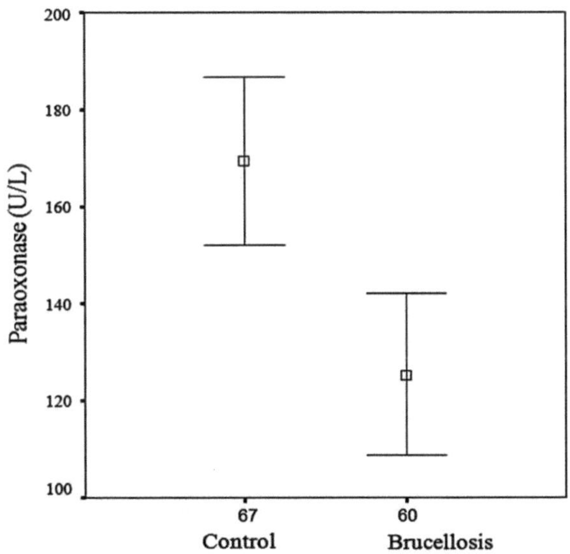

Fig. 17.1 The paraoxonase activity of the patient and control groups (Karsen et al. [52])

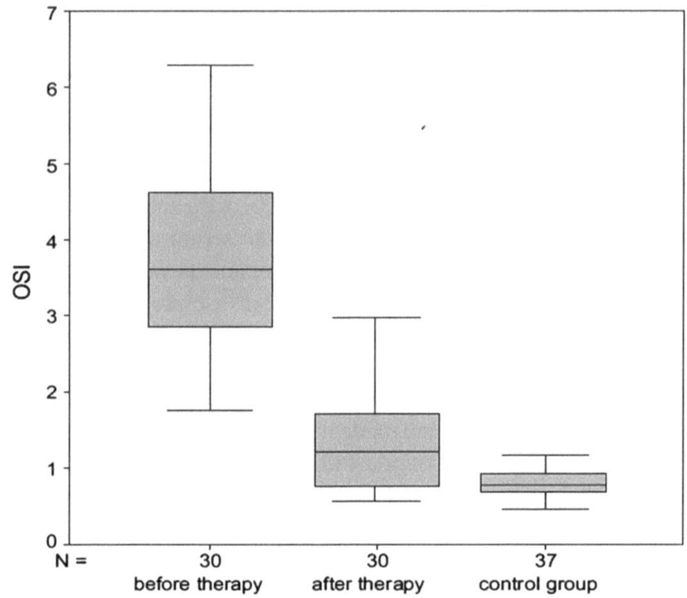

Fig. 17.2 OSI levels before and after treatment in the patients and in the control group (Karaagac et al. [53])

it might be possible to use these oxidant and antioxidant parameters to monitor treatment [53].

A study entitled "Oxidant and antioxidant parameters in patients with Brucella canis" by Murat Usta et al. was published in 2012. The findings showed that compared with smooth *Brucella species* infection, *B. canis* generate a low TOS and OSI index in humans, and it was concluded that this difference could contribute to the understanding of the pathogenesis of *B. canis* infection in man [54].

In the 2012 study by Gholami M et al. entitled, "What is the total antioxidant status in the plasma of Brucellosis patients?", decreased TAC levels in patients with brucellosis demonstrated exposure to potent oxidative stress. It was reported that TAC could be a useful early marker of oxidative stress in the monitoring and optimization of antioxidant therapy in the management of patients with brucellosis [55].

The study entitled "Serum myeloperoxidase activity and oxidative stress in patients with acute brucellosis" by Karahocagil MK et al. was published in 2012. As MPO has a central role in inflammation and oxidative stress, it has a function in the initiation and progression of atherosclerosis. Associated with the increase in oxidative stress in brucellosis patients, serum MPO activity increases and catalase activity decreases. This causes the development of atherosclerosis in acute brucellosis patients. It was therefore concluded that in patients with acute brucellosis, serum MPO activity should be measured as a biomarker of the pathogenesis of atherosclerosis [56].

In the 2014 study by Özlem Demirpence et al., entitled "Serum paraoxonase, TAS, TOS and ceruloplasmin in brucellosis," it was reported that in brucella infections, while paraoxonase and TAS decreased, TOS and ceruloplasmin (CP) increased, and this caused an impairment in the oxidant and antioxidant balance. Imbalanced oxidant and antioxidant status has a role in the development of atherosclerosis, cancer, and several chronic degenerative diseases. The study concluded that oxidant-antioxidant system markers could be used to determine the role of brucella infection in such diseases and to monitor response to treatment [8].

In a 2015 study of brucellosis patients by Ramazan Esen et al. entitled "Paraoxonase activity, total thiol levels, and oxidative status," it was concluded that while the total thiol level and PON1 activity decreased in acute brucellosis, oxidative stress was increased. From these results, decreased total thiol levels and PON1 activity were determined to be associated with an imbalance in oxidant-antioxidant status in brucellosis patients [57].

Servet Kolgelier et al. conducted a study in 2017 entitled "Impaired Thiol-Disulfide Balance in Acute Brucellosis" and the thiol-disulfide balance was seen to have a significant role in various processes, such as antioxidant defense, immune response, modulation of enzyme activity, and apoptosis [58]. Cytokine production is inhibited by thiol antioxidants such as glutathione and N-acetylcysteine [59]. Furthermore, greater sensitivity to infection, inflammation, or impaired immunity results from depletion of low GSH levels [60].

In acute brucellosis, there is a strong negative ratio between ceruloplasmin, which is an acute phase reactant, and native thiol and total thiol. The disulfide/total thiol ratio is also imbalanced, and this shows the imbalance between the oxidant and antioxidant systems. The low native and total thiol serum levels demonstrate a reduced effect against oxidative stress and inflammation. In brucellosis patients, the increase in disulfide/native thiol and disulfide/total thiol ratios and the decrease in the native/total thiol ratio show increased oxidative stress. Correction of the thiol-disulfide balance in brucellosis may create a response to the potential anti-inflammatory process. Therefore, the administration of supportive treatments containing thiol for the restoration of thiol in brucellosis patients is controversial. New methods may emerge in the treatment of brucellosis [61].

The study entitled "Brucellosis Causes Alteration in Trace Elements and Oxidative Stress Factors" was conducted by Naser Zanganeh et al. in 2018.

Micronutrients, as trace elements, are vital for living organisms to survive and are known to have a significant role in many biochemical reactions in the body. Therefore, when levels are disrupted, many biological functions are also disrupted. Micronutrients include elements such as zinc, iron, manganese, copper, magnesium, cobalt, and chromium [62]. Zinc (Zn) is an essential element that has an important role in the functions of several enzymes, including DNA polymerase and RNA polymerase. As an antioxidant, it also has an important function in protection against free radicals [63]. Copper (Cu) is another trace element with pro-oxidant and antioxidant properties. As a part of the Cu-Zn SOD enzyme, it can accelerate the formation of free radicals through the Haber-Weiss and Fenton reaction [14]. Cu is also a co-factor in more than 300 enzymatic reactions, and has a role in glucose

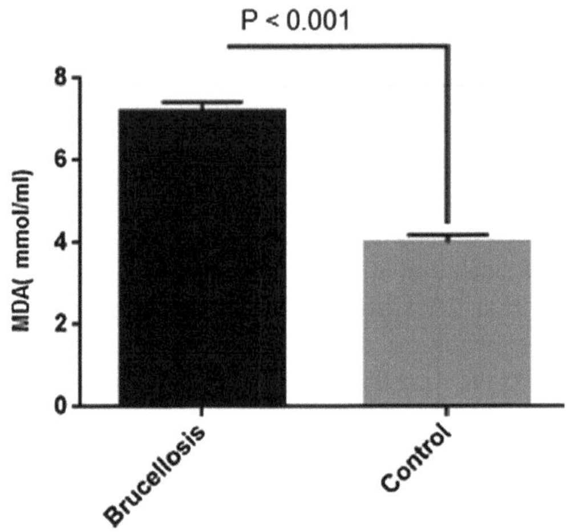

Fig. 17.3 MDA level in the case and control groups (Zanganeh et al. [66])

metabolism, nucleic acid synthesis, ATP metabolism, muscle contraction, and the membrane transport system [64]. Manganese (Mn) is the major part of many enzymes and the activator of several others. Mitochondrial superoxide dismutase is an enzyme that contains Mn and contributes to antioxidant defense. Furthermore, Mn is involved in several body functions, including the metabolism of carbohydrates, amino acids, and cholesterol [65].

Brucellosis creates damage with a toxic effect on some serum trace elements, and this has been reported to increase serum Cu and MDA and reduce magnesium, manganese, and zinc. Consequently, oxidative stress is formed [66] (Fig. 17.3).

As seen from the findings of all the abovementioned studies, oxidative stress is increased and the antioxidative capacity is decreased in the host during brucellosis. However, despite the increase in oxidative stress, this may not be sufficient to eradicate the brucella bacteria from the host cell, and the damage is caused to the host tissue and cells.

17.4 Conclusion

While brucellosis reduces antioxidant enzymes and molecules (TAS, magnesium, manganese, arylesterase, paraoxonase, native thiol, total thiol, disulfide, glutathione, PDE_4, zinc), it increases oxidative enzymes and molecules (TOS, Cu, MDA, ceruloplasmin, LOOH, OSI, NO). Thus, oxidative stress increases causing lipid peroxidation, protein denaturation, and DNA breakage. The breakup of DNA is a process that can be a precursor of malignancy. It can also lead to metabolic changes and atherosclerosis in the host. To prevent all these negative effects of brucellosis on the host, an early start to primary brucellosis treatment should be made and the use of supplemental therapies such as Vitamin C, Vitamin E, tocopherol and glutathione, which are antioxidant, is controversial.

References

1. Young EJ (2005) Brucella species In: Mandell GL, Bennett JE, Dolin R (eds) Principles and practice of infectious diseases, 6th edn. Churchill Livingstone. Philadelphia, pp 2669–2672
2. Mirnejad R, Mohammadi M, Majdi A et al (2013) Molecular typing of Brucella melitensis and *B. abortus* from human blood samples using PCR-RFLP method, Jundishapur. J Microbiol 6:1e5
3. Rahman M, Faruk M, Her M et al (2011) Prevalence of brucellosis in ruminants in Bangladesh. Vet Med 56:379e385
4. Nimri LF (2003) Diagnosis of recent and relapsed cases of human brucellosis by PCR assay. BMC Infect Dis 3:1e7
5. Mirnejad R, Jazi FM, Mostafaei S et al (2017) Molecular investigation of virulence factors of Brucella melitensis and Brucella abortus strains isolated from clinical and non-clinical samples. Microb Pathog 109:8e14
6. Purwar S (2007) Human brucellosis: a burden of half-million cases per year. South Med J 100:1074e1075
7. Samaha H, Al-Rowaily M, Khoudair RM et al (2008) Multicenter study of brucellosis in Egypt. Emerg Infect Dis 14(1916e):1918
8. Demirpençe Ö, Sevim B, Yıldırım M et al (2014) Serum paraoxonase, TAS, TOS and ceruloplasmin in brucellosis. Int J Clin Exp Med 7:1592–1597
9. Woods JR, Plessinger MA, Miller RK et al (2001) Vitamins C and E: missing links in preventing premature rupture of membranes. Am J Obstet Gynecol 185:5–10
10. Halliwell B, Gutteridge JM, Cross CE et al (1992) Free radicals, antioxidants and human disease: where are we now? J Lab Clin Med 119(6):598–620
11. Mansour MA, Mostafa AM, Nagi MN et al (2002) Protective effect of aminoguanidine against nephrotoxicity induced by cisplatin in normal rats. Comp Biochem Physiol C 132:123–128
12. Santovito G, Irato P, Piccinni E et al (2000) Relationship between metallothionein and metal contents in red–blooded and white-blooded antarctic teleosts. Polar Biol 23:383–391
13. Leibbrandt MEI, Grushenka HIW, Metz AL et al (1995) Critical subcellular targets of cisplatin and related platinum analogs in rat proximal tubule cells. Kidney Int 48:761–770
14. Cherubini A, Ruggiero C, Polidori MC et al (2005) Potential markers of oxidative stress in stroke. Free Radic Biol Med 39:841–852
15. Desideri A, Falconi M (2003) Prokaryotic Cu, Zn superoxidies dismutases. Biochem Soc Trans 31:1322–1325
16. Reddy R, Yao JK (1999) Schizophrenia-role of oxidative stress and essential fatty acids. In: Kitap: Basu TK, Temple NJ, Garg ML (eds) Antioxidants in human health and disease. CABI Publishing, Walingford
17. Parke DV (1999) Nutritional antioxidants and disease prevention; mechanism of action. In Antioxidants in human health and disease. CABI, Walingford, pp 215–226
18. Carr AC, McCall MR, Frei B et al (2000) Oxidation of LDL by myeloperoxidase and reactive nitrogen species reaction pathways and antioxidant protection. Arterioscl Thromb Vasc Biol 20:1716–1723
19. Marnett LJ (2002) Oxy radicals, lipid peroxidation and DNA damage. Toxicology 181:219–222
20. Young IS, Woodside JV (2001) Antioxidants in health and disease. J Clin Pathol 54:176–186
21. Lowenstein CJ, Dinerman JL, Snyder SH et al (1994) Nitric oxide: a physiologic messenger. Ann Intern Med 120:227–237
22. Ghafourifar P, Cadenas E (2005) Mitochondrial nitric oxide synthase. Trends Pharmacol Sci 26:190–195
23. Lin JY, Chadee K (1992) Macrophage cytotoxicity against Entamoeba histolytica trophozoites is mediated by nitric oxide from L-arginine. J Immunol 148:3999–4005
24. Pratic'o D (2005) Antioxidants and endothelium protection. Atherosclerosis 181:215–224

25. Caron E, Peyrard TU, Köhler S et al (1994) Cabane S, Liautard JP and Dornand J. live Brucella spp. fail to induce tumor necrosis factor alpha excretion upon infection of U937-derived phagocytes. Inf Immun 62:5267–5274
26. Orozco G, Lopez-Nevot MAS, Caballero A et al (2003) Inducible nitric oxide synthase promoter polymorphism sn human brucellosis. Microbes Infect 5:1165–1169
27. Yigit M, Sogut O, Tataroglu Ö et al (2018) Oxidative/antioxidative status, lymphocyte DNA damage, and urotensin-2 receptor level in patients with migraine attacks. Neuropsychiatr Dis Treat 24:367–374
28. Valko M, Izakovic M, Mazur M (2004) Role of oxygen radicals in DNA damage and cancer incidence. Mol Cell Biochem 266:37–56
29. Powell SR (2000) The antioxidant properties of zinc. J Nutr 130:1447–1454
30. Mates JM, Perez-Gomez C, De INC et al (1999) Antioxidant enzymes and human diseases. Clin Biochem 32:595–603
31. McCall MR, Frei B (1999) Can antioxidant vitamins materially reduce oxidative damage in humans? Free Rad Biol Med 26:1034–1105
32. McIntyre M, Bohr DF, Dominiczak AF et al (1999) Endothelial function in hypertension: the role of superoxide anion. Hypertension 34:539–545
33. Landis GN, Tower J (2005) Superoxide dismutase evolution and life span regulation. Ech Ageing Dev 126:365–379
34. Armstrong D (1998) Methods in molecular biology, vol 108. Humana Press, Toronto
35. Akkus I, Kalak S, Vural H et al (1996) Leukocyte lipid peroxidation, superoxide dismutase, glutathione peroxidase and serum and leukocyte vitamin c levels of patients with type II diabetes mellitus. Clin Chim Acta 244:221–227
36. Cherubini A, Ruggiero C, Polidori MC et al (2005) Potential markers of oxidative stress in stroke. Free Radic Biol Med 39:841–852
37. Hampton MB, Kettle AJ, Winterbourn CC et al (1998) Inside the neutrophil phaposome: oxidants, myeloperoxidase and bacterial killing. Blood 92:3007–3017
38. Malle E, Waeg G, Schreiber R et al (2000) Immunohistochemical evidence for the myeloperoxidase/H_2O_2/halide system in human atherosclerotic lesions: colocalization of myeloperoxidase and hypochlorite-modified proteins. Eur J Biochem 267:4495–4503
39. Kim JA, Sha Z, Mayfıeld JE (2000) Regulation of Brucella abortus catalase. Infect Immun 68:3681–3866
40. Franco MP, Mulder M, Gilman RH et al (2007) Human brucellosis. Lancet Infect Dis 7:775–786
41. Kataria N, Kataria AK, Maan R et al (2010) Evaluation of oxidative stress in Brucella infected cows. J Stress Physiol Biochem 6:19–25
42. Halliwell B, Gutteridge JM (1999) Free radicals in biology and medicine, 3rd edn. Oxford Science Publications, London
43. Cesur S, Kocaturk PA, Kavas GO et al (2005) Serum copper and zinc concentrations in patients with brucellosis. J Infect 50:31–33
44. Melek IM, Erdogan S, Celik S et al (2006) Evaluation of oxidative stress and inflammation in long term Brucella melitensis infection. Mol Cell Biochem 293:203–209
45. Nathan C, Xie QW (1994) Regulation of biosynthesis of nitric oxide. J Biol Chem 269:13725–13728
46. Nisbet C, Yarım GF, Çiftçi A (2007) Investigation of serum nitric oxide and malondialdehyde levels in cattle infected with Brucella abortus. Ankara Üniv Vet Fak Derg 54:159–163
47. Erdogan S, Aslantas O, Celik S (2008) The effects of increased cAMP content on inflammation, oxidative stress and PDE4 transcripts during Brucella melitensis infection. Veterin Sci 84:18–25
48. Ali AH (2009) The effect of Brucellosis on lipid profile and oxidant-antioxidants status. Iraqi J Pharm Sci 18(Suppl)
49. Apostolou F, Gazi IF, Kostoula A (2009) Persistence of an atherogenic lipid profile after treatment of acute infection with Brucella. J Lipid Res 50:2532–2539

50. Serefhanoglu K, Taskin A, Turan H (2009) Evaluation of oxidative status in patients with brucellosis. Braz J Infect Dis 13:249–251
51. Mobaien A, Hajiabdolbaghi M, Jafari S et al (2010) Serum zinc and copper concentrations in brucellosis patient. Iran J Clin Infect Dis 5:96–100
52. Karsen H, Karahocagil MK, Akdeniz H et al (2011) Serum paraoxonase and arylesterase activities and oxidant status in patients with brucellosis. Afr J Microbiol Res 5:1701–1706
53. Karaagac L, Koruk ST, Koruk I et al (2011) Decreasing oxidative stress in response to treatment in patients with brucellosis: could it be used to monitor treatment? Int J Infect Dis 15:346–349
54. Usta M, Aras Z, Tas A (2012) Oxidant and antioxidant parameters in patients with Brucella canis. Clin Biochem 45:366–367
55. Gholami M, Hasanjani Roushan M, Mahjoub S et al (2012) How is total antioxidant status in plasma of patients with brucellosis? Caspian J Intern Med 3:363–367
56. Karahocagil MK, Aslan M, Ceylan MR et al (2012) Serum myeloperoxidase activity and oxidative stress in patients with acute brucellosis. Clin Biochem 45:733–736
57. Esen R, Aslan M, Kucukoglu ME et al (2015) Serum paraoxonase activity, total thiols levels, and oxidative status in patients with acute brucellosis. Wien Klin Wochenschr 127:427–433
58. Valko M, Leibfritz D, Moncol J et al (2007) Free radicals and antioxidants in normal physiological functions and human disease. Int J Biochem Cell Biol 39:44–84
59. Hasegawa S, Ichiyama T, Sonaka I et al (2012) Cysteine, histidine and glycine exhibit anti-inflammatory effects in human coronary arterial endothelial cells. Clin Exp Immunol 167:269–274
60. Ghezzi P (2011) Role of glutathione in immunity and inflammation in the lung. Int J Gen Med 4:105–113
61. Kolgelier S, Ergin M, Demir LS, Inkaya AC, Nazlim AD, Alisik M, Erel O et al (2017) Impaired thiol-disulfide balance in acute Brucellosis. Jpn J Infect Dis 70:258–262
62. Ghasemi H, Karimi J, Goodarzi MT et al (2016) Seminal plasma zinc and magnesium levels and their relation to spermatozoa parameters in semen of diabetic men. Int J Diab Dev Ctries 36:34–39
63. Song Y, Leonard SW, Traber MG et al (2009) Zinc deficiency affects DNA damage, oxidative stress, antioxidant defenses, and DNA repair in rats. J Nutr 139:1626–1631
64. Swaminathan R (2003) Magnesium metabolism and its disorders. Clin Biochem Rev 24:47
65. Lawrence GD, Sawyer DT (1987) The chemistry of biological manganese. Coord Chem Rev 27:173–193
66. Zanganeh N, Siahpoushi E, Kheiripour N et al (2018) Brucellosis causes alteration in trace elements and oxidative stress factors. Biol Trace Elem Res 182:204–208

Oxidative Stress and Antioxidant Supplementation on Immunity in Hansen's Disease (Leprosy)

18

Carlos Kusano Bucalen Ferrari

Abstract

Hanseniasis comprises one of the main causes of physical disabilities due to its potential to strike neural lesions, foot deformities, amputations, and mutilations. Since innate response is important in hanseniasis, this work reviewed and updated the role of reactive oxygen/nitrogen species in different clinical forms of hanseniasis, as well as their role in different phagocyte free radicals' generator systems (NADPH-oxidase, mitochondrial pathways, myeloperoxidase, extracellular traps, iNOS). Since hanseniasis can induce antioxidant depletion, possible benefits of antioxidant nutritional supplementation are also discussed.

Keywords

Leprosy · Macrophages · Neutrophils · Nitric oxide · Free radicals · Multibacillary

18.1 Introduction

Hanseniasis, a chronic infectious disease, caused by *Mycobacterium leprae* is considered one of the main causes of physical disabilities due to its potential to trigger neural lesions, foot deformities, amputations, and mutilations [1–3].

According to the World Health Organization (WHO) report from 2016, 143 countries reported 214,783 new cases of the disease, which represents a detection rate of 2.9 cases per 100 thousand inhabitants [4]. In Brazil, 25,218 new cases were reported, making a detection rate of 12.2/100 thousand inhabitants which is higher than that found in India (10.23/100 thousand inhabitants) [4, 5]. Brazil has the

C. K. B. Ferrari (✉)
Instituto de Ciências Biológicas e da Saúde (ICBS), Campus Universitário do Araguaia, Universidade Federal de Mato Grosso (UFMT), Barra do Garças, MT, Brazil
e-mail: drcarlosferrari.ufmt@gmail.com

S. Chakraborti et al. (eds.), *Oxidative Stress in Microbial Diseases*,
https://doi.org/10.1007/978-981-13-8763-0_18

Table 18.1 Ten nations with more new cases of hanseniasis (2016)

Country	Region	Number of new cases
India	Southeast Asia	135,485
Brazil	South America	25,218
Indonesia	Southeast Asia	16,826
Democratic Republic of Congo	Africa	3742
Ethiopia	Africa	3692
Nepal	Southeast Asia	3054
Bangladesh	Southeast Asia	3000
Myanmar	Southeast Asia	2609
Madagascar	Africa	1780
Mozambique	Africa	1289

Source: [4]

second highest new number of hanseniasis cases, but the highest incidence of the disease in the world [4].

In Brazil, Mato Grosso state has the highest incidence (150/100 thousand inhabitants) [6] which is 14.7-fold bigger than in India. In Mato Grosso, cities like Barra do Garças has higher prevalence and incidence of hanseniasis which is commonly found since the early age with a cure rate of only 77% [7, 8].

The ten nations with more new cases of hanseniasis are presented in Table 18.1.

Since reactive oxygen and nitrogen species display important roles in innate response to pathogens, the objective of the present work was to review the role of oxidative/nitrosative stress in hanseniasis and the possible benefits of antioxidant supplementation in this disease.

We performed a non-systematic review, from 1998 to 2018, using the following databases: PubMed/Medline (https://www.ncbi.nlm.nih.gov/pubmed/), SciELO (www.scielo.org), and Google Scholar (https://scholar.google.com.br). The keywords used were hanseniasis, leprosy, oxidative stress, nitrosative stress, myeloperoxidase and leprosy, NADPH-oxidase and leprosy, and antioxidant capacity and leprosy.

18.2 Oxygen, Nitrogen, and Other Reactive Species

Cell survival, adaptation, degeneration, and death depend on genetic, epigenetic, and metabolic factors and processes. For guarantee of cell viability and life span, there is a fine-tune regulation of the oxidation-related molecules and the cellular and nutritional antioxidants. This means that a proper balance of the production of free radicals and other reactive oxygen (ROS), nitrogen (RNS) and chlorine species (RCS), mainly by mitochondria, but also released by endoplasmic reticulum and other organelles), should be scavenged or removed by the enzymatic and non-enzymatic (mainly nutritional antioxidants) cellular antioxidant defense systems. The rupture of this balance due to the excessive release of reactive species and/or due to decreased amounts of antioxidants results in oxidative and nitrosative stress in cells and tissues, both associated with cell degeneration and death, and a large number of pathophysiological processes and diseases [9–12].

Any molecule or atom that has one or more incomplete orbitals is considered a free radical (FR). In this sense, free radicals can gain electrons oxidizing another atom or molecule, or they can release them resulting in reduction of an element or molecule.

Some ROS are FR [superoxide anion (O_2^-), hydroxyl radical ($^{\bullet}OH$), nitric oxide ($NO^{\bullet}$), peroxynitrite ($ONOO^-$), while many other reactive oxygen species are not free radicals [hydrogen peroxide (H_2O_2), singlet oxygen ($^1O_2$)]. Anyway, ROS, RNS, and FR can induce oxidative changes in proteins and enzymes (protein peroxidation), lipids (lipid peroxidation), nucleic acids (DNA/RNA peroxidation), and biomembranes causing reversible or irreversible changes in cells leading to pathological states and diseases [13–15].

Lipid peroxidation releases peroxides as primary biomarkers and metabolites and malondialdehyde (MDA), 4-hydroxynonenal, and other aldehydes as secondary biomarkers, whereas protein peroxidation yields protein carbonyls as markers of oxidative damage [16–18].

Since ROS/RNS display important roles in innate response to pathogens, the objective of the present work was to review the role of oxidative/nitrosative stress in hanseniasis and the possible benefits of antioxidant supplementation in this disease.

18.3 How Are Free Radicals and Reactive Species Synthesized?

The reaction of adding an electron to the molecular oxygen (O_2) leads to the formation of the superoxide anion (O_2^-), a relatively weak reactive specie, produced especially in the mitochondrial respiratory chain and in other electron transport systems in biomembranes of cellular organelles and, secondly, by the reaction of the O_2 with proteins with thiol groups, reducing agents, in the presence of free iron, or by the autoxidation of oxyhemoglobin in methemoglobin [13, 14]. In hanseniasis, it is very common to find a higher oxidative stress which oxidizes hemoglobin to produce methemoglobin and Heinz bodies [19, 20].

In inflammatory processes, activated phagocytes, fibroblasts, and lymphocytes also produce O_2^- via respiratory burst [21, 22]. The superoxide anion can also react with nitric oxide (NO) to form the peroxynitrite, a potent oxidant of SH groups, which can undergo decomposition and give the hydroxyl radical (OH), another high-power oxidant [14, 23].

The reaction between H^+ and + O_2 reaction can give the hydroperoxyl (HO_2) radical, which is also an extremely oxidizing agent [24].

Finally, in mitochondria, microsomes, and peroxisomes, the spontaneous reaction of O_2 with two electrons and two hydrogen atoms in the presence of the superoxide enzyme (SOD) produces the hydrogen peroxide (H_2O_2) [25].

Two molecules of H_2O_2 are sequentially converted to two water molecules plus one molecule of O_2 by the catalase enzyme [26, 27].

18.4 The Five Basic Mechanisms of Free Radical Killing of Microbial Pathogens and Tissue Damage

The five basic immunopathogenic mechanisms for microbial killing have been recognized, and they comprise [28]:

1. Nitric oxide-peroxynitrite pathway.
2. Myeloperoxidase.
3. NADPH-oxidase.
4. Mitochondrial H_2O_2 pathway.
5. Extracellular traps: nuclear and mitochondrial DNA-based structures triggered by H_2O_2 and other factors that kill some microorganisms and activate the innate immunity [29]. It is important to know that a Mexican study with *M. tuberculosis* showed ET were not capable of destroying bacilli which contributed to pathogen persistence [30].

18.4.1 The Nitric Oxide-Peroxynitrite Pathway in Hanseniasis

Since the discovery of the conversion of the arginine (Arg) into nitric oxide (NO) by the action of the nitric oxide synthases (NOS) [31], those enzymes were found to be largely found in different human tissues. Macrophage (mNOS), endothelial (eNOS), neuronal (nNOS), inducible (iNOS), and mitochondrial nitric oxide synthase (mtNOS) isoforms [15] work to convert arginine in the NO radical ($NO^•$) that quickly reacts with superoxide releasing peroxynitrite ($ONOO^•$), one of the most deleterious free radical [15, 32, 33]. It should be emphasized that iNOS isoform is activated in infectious and inflammatory processes.

On contrary to *Mycobacterium tuberculosis*, *M. leprae* is susceptible to nitric oxide killing [34]. However, into the course of disease, there is a decrement on nitric oxide levels, but treatment can recover them to normal levels [35].

Studying deficiency of nitric oxide in hanseniasis infection, one research pointed out that this deficiency is partially offset by increased expression of macrophage protein-1 and interleukins 4 and 10 [36]. However, the other investigation reported that nitric oxide knockout mice presented granulomas with an area of 10 times higher than normal mice [37].

However, more studies are necessary to elucidate the possible roles of nitric oxide and peroxynitrite in Hanseniasis pathogenesis and prognostics.

18.4.2 Myeloperoxidase in Hanseniasis

Myeloperoxidase (MPO), found in macrophages and neutrophils, is a heme-binding-enzyme, that yields hypochlorous acid (HOCl), a toxic substance for fungi and bacteria, and also combines with superoxide anion releasing chlorine anion and another deleterious free radical: the hydroxyl radical (·OH) [38, 39]. Furthermore, when HOCl chemically reacts with Fe^{2+}, large amounts of hydroxyl radicals are released; and Fe^{2+} also reacts with O_2^- yielding H_2O_2, and both compounds are very cytotoxic to pathogens trapped in the phagosome [40, 41]. The toxic phagosome cocktail also includes ·OH, $^1O_2$, and ozone (O_3) which contribute to destroy bacteria and fungi [42, 43]. MPO-null mice are susceptible to *E. coli* [44]. In the same manner, macrophages of MPO-deficient mice presented a higher number of viable *M. leprae* into the cytoplasm [45]. Administration of exogenous peroxidase triggered MPO activity by phagocytes, reduced leukocytosis, and had no adverse liver effects [46].

18.4.3 NADPH-Oxidase, Mitochondrial H_2O_2, and Hanseniasis

Macrophages, neutrophils, and other phagocytes present in plasmatic membranes the NADPH-oxidase enzyme (an HEME protein) that converts four electrons and $4O_2$ into $4O_2^-$ radicals, a mechanism called "oxidative burst" [22, 43]. Half of the O_2^- chemically reacts with atomic hydrogen producing two molecules of hydrogen peroxide which can destroy microbial intracellular pathogens [47–49]. NADPH-oxidase deficiency to produce reactive oxygen species has been associated with chronic granulomatous diseases and abscesses caused by bacteria and fungi [50, 51].

In murine leprosy, H_2O_2 is essential for phagocytic killing. Deficiency of this molecule is associated with *M. leprae* survival and worse outcome of experimental murine infection [52].

Mitochondrial releasing of superoxide and subsequently formation of hydrogen peroxide is essential to kill pathogens but also to trigger proliferation, differentiation, growth, and cell survival of immune cells [53, 54]. Then, ROS are not necessarily harmful in the course of activation of the immune system against infections.

A microarray study reported that type 1 reaction was linked to the expression of 275 genes, while type 2 was associated with 517 different genes, and the SOD_2 gene represented a genetic biomarker of both type 1 and type 2 hanseniasis reactions [55]. Following the same approach, it was suggested that people carrying the SOD2 gene are at increased risk of developing hanseniasis, since this defective gene produces lesser amounts of ROS/RNS, which inhibits apoptosis of *M. leprae* and permits bacterial survival and infection success [56, 57].

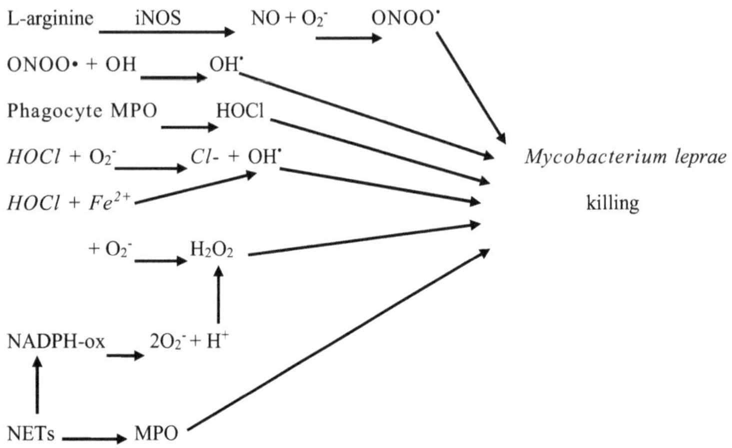

Fig. 18.1 Effector pathways of oxygen, nitrogen, and chlorine species in *Mycobacterium leprae* killing. Legend: *iNOS* inducible isoform of nitric oxide synthase, *HOCl* hypochlorous acid, *MPO* myeloperoxidase, *NADPH-ox* NADPH-oxidase, *NETs* neutrophil extracellular traps, *NO* nitric oxide, ONOO• peroxynitrite, O_2^-: superoxide anion, OH• hydroxyl radical, $^1O_2$ singlet oxygen

Then, the role of both nitrogen, oxygen, and chlorine reactive species and free radicals in killing *M. leprae* is summarized in Fig. 18.1.

Beyond defects on phagocytes' enzymatic systems, *M. leprae* express a phenolic glycolipid that is involved in microbial pathogenesis. That molecule suppresses the activity of NADPH-oxidase, xanthine-oxidase, and myeloperoxidase in phagocytes, decreasing the effectivity of phagocytosis and mycobacteria killing [58, 59]. Notwithstanding, macrophages react against phenolic glycolipid and produces excessive amounts of nitric oxide, which, in turn, causes nerve damaging in the host [60, 61, 62]. Furthermore, *M. leprae* triggers cytosolic accumulation of cholesterol and other lipids in macrophages and Schwann cells, transforming them into foamy cells. Those lipid droplets help mycobacteria to escape from microbicidal activity of macrophages and can potentially induce atherogenesis and strongly decrease the levels of nitric oxide [63].

Considering the importance of pathogenic and escaping pathways of *M. leprae* within the host, those mechanisms are represented in Fig. 18.2.

18.5 Clinical Studies of Hanseniasis and Oxidative and Nitrosative Stress

Hanseniasis could be clinically classified into the following groups [58]:

1. Indeterminate: presenting non-specific hypochromic spots.
2. Tuberculoid: with annular erythematous plaques.
3. Borderline or dimorphous: erythematous plaques and polymorphic lesions.

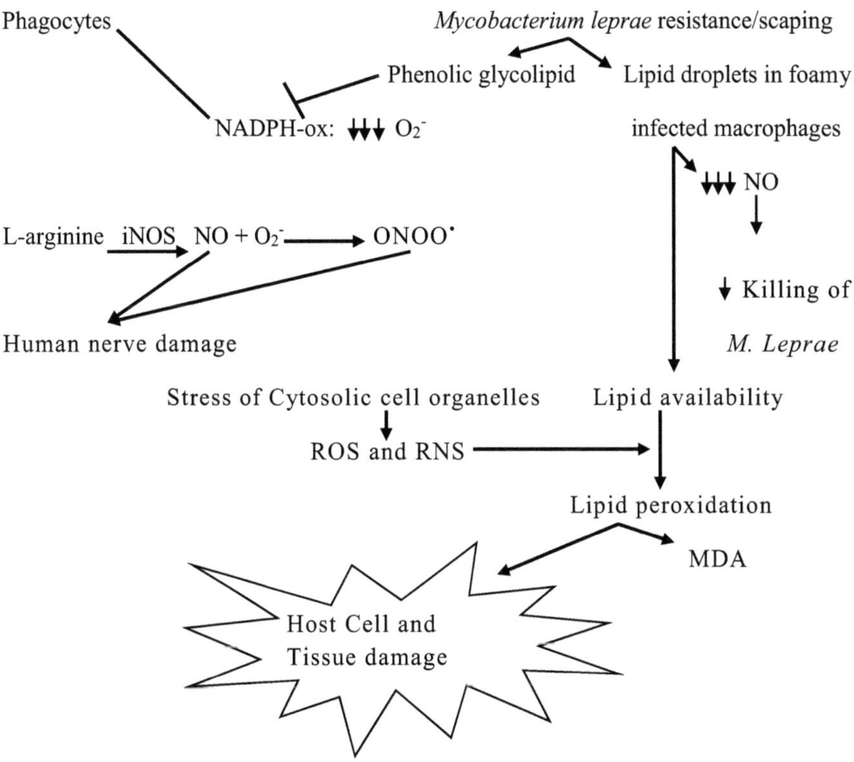

Fig. 18.2 Pathophysiological roles of ROS/RNS and bacterial escaping mechanisms against reactive species in hanseniasis. Legend: Activation —— or ⟶, Blocking ⊣, *MDA* malondialdehyde, *NADPH-ox* NADPH-oxidase, *NO* nitric oxide, ONOO• peroxynitrite, O_2^- superoxide anion

4. Lepromatous: the most disseminated clinical form and more serious than the others. It presents a great number of macrophages infected by a large number of living *M. leprae* bacilli. It could form madarosis, dry and barely hypochromic spots and ichthyosiform lesions.

A simple classification based on the number of skin lesions was proposed by WHO in 1994 and comprises only two groups [64]:

1. Paucibacillary hanseniasis (PbH): less than 5 skin lesions.
2. Multibacillary hanseniasis (MbH): more than 5 skin lesions.

Studying oxidative stress in patients with paucibacillary or multibacillary clinical forms of hanseniasis, an increase on malondialdehyde levels in both clinical forms and a decrease on SOD levels in multibacillary patients only were reported [65].

It has been suggested that in tuberculoid hanseniasis the oxidative stress is lower since SOD levels are preserved, whereas in lepromatous hanseniasis oxidative stress is higher and SOD levels are depleted [66].

Other study, covering both Pb and Mb patients also reported raised levels of lactate dehydrogenase (LDH) and malondialdehyde as well as decreased levels of SOD, CAT, GSH, and total antioxidant capacity [67]. In the same report, clinical improvement of treated patients was associated with recovering of antioxidant balance.

Multidrug treatment (MDT) of hanseniasis (clofazimine, dapsone, and rifampicin) did no beneficial effect on increased oxidative stress, as measured by protein carbonyls and lipid peroxidation, but patients who received vitamin E supplementation had improved antioxidant balance [68]. However, a Brazilian study observed no protective effect of vitamin E against MDT-induced oxidative stress in multibacillary hanseniasis patients [19].

The same positive effect of MDT treatment was noted in type 1 hanseniasis patients (reverse reaction type), a typical cellular immunity reaction frequently seen in borderline patients [69]. This work reported that MDT decreased oxidative stress and improved both GSH and total antioxidant capacity levels [69].

Another study which evaluated the effects of MDT reported that there was a decrease on both CAT and MDA levels, whereas MDT augmented GSH content [70]. It is important to emphasize that in this study, MDT did not improve SOD and the total antioxidant capacity as measured by the trolox-equivalent antioxidant assay (TEAC).

Another study supported the concept that LH is associated with increased oxidative stress and decreased antioxidant defense as measured by blood GSH, and erythrocyte GPX and GR activities when compared to other clinical forms of hanseniasis [71]. An inverse association between bacteriological index and CAT levels was detected in another study [72].

The above-mentioned studies showed an antioxidant imbalance, especially in blood and serum. An Egyptian study demonstrated that beyond the blood, the infected tissues of hanseniasis patients presented also increased levels of MDA and diminished SOD content [73].

In patients of Pb hanseniasis, a mild form of disease, vitamin C content is higher compared to Mb hanseniasis [74].

In any way, the pathogenesis of hanseniasis, especially the multibacillary form, produces a might oxidative stress which can deplete vitamins C, A, and E [74–77]. In this respect, lepromatous form had the lowest vitamin A levels when compared to indeterminate, borderline, and tuberculoid disease forms, whereas both clinical presentations were linked to increased lipid peroxidation [75].

Other more recent studies have supported the concept that Mb hanseniasis has sustained greater oxidative stress than the Pb form of pathology [78, 79].

18.6 Hanseniasis, Uric Acid, and Antioxidant Imbalance

In many physiological circumstances, uric acid is the major antioxidant defense in the blood [80–83]. Produced in purine catabolism, through xanthine-oxidoreductase, uric acid has important antioxidant activities [84, 85].

When oxidative, nitrosative, and chlorine stress are increased, there is a consumption of the circulating uric acid levels [83].

It is important to note that in some biological systems, uric acid activates the pro-oxidant enzyme NADPH-oxidase, which triggers massive production of oxygen and nitrogen free radicals and reactive species [86, 87].

A study with hanseniasis patients with type 1, type 2 (erythema nodosum leprosum), and no immune reactions reported reduced levels of uric acid in type 2 patients compared to type 1 and control subjects [88]. This result is different from a previous study which reported raised serum uric acid levels among Nigerian hanseniasis patients submitted to MDT treatment [76].

18.7 Hanseniasis as a Cause of Antioxidant Depletion

Evaluating 23 newly diagnosed hanseniasis patients which were compared with 20 healthy subjects (control), serum levels of nitric oxide, methemoglobin, MDA, GSH, SOD, CAT, and trolox-equivalent antioxidant capacity (TEAC) were dosed [89]. Authors found enhanced levels of both nitric oxide and TEAC and no other significant difference between groups [89].

Recently, an Indian investigation reported that hanseniasis was positively associated with SOD, MDA, NO and total oxidant status levels, and inversely associated with total antioxidant capacity [90].

Evaluating the possible roles of nitric oxide and its metabolites in hanseniasis prognostics, an Egyptian study revealed that untreated MbH and type 1 patients presented higher nitric oxide values in comparison to Pb untreated subjects and control group. Treatment decreased NO_2 levels in both hanseniasis subjects compared to the control group [91].

The severity of hanseniasis was correlated with increased levels of lipid peroxidation, as measured by MDA levels, and reduced levels of SOD in skin biopsies of 41 untreated Egyptian patients [92].

At baseline, tuberculoid hanseniasis (TH) and lepromatous hanseniasis (LH) patients presented elevated levels of MDA compared to control subjects [88]. Furthermore, although TH patients presented a slight decrease on zinc levels, LH subjects had a very poor zinc status. Zinc supplementation by 4 months completely restored zinc levels of TH patients and partially restored zinc levels in LH [93].

A Brazilian study compared PbH and MbH with control subjects and reported that MDA was higher and vitamin E was lower among both hanseniasis groups, whereas GSH was higher in control subjects compared to MbH patients with no differences between hanseniasis groups [94]. The same study showed no significant differences in copper, selenium, and zinc, whereas magnesium content was deficient in all studied groups [94].

A recent Indonesian research with 19 PbH and 11 MbH patients demonstrated a lower Selenium level among MbH patients compared to PbH subjects [95].

Since some hanseniasis patients suffer from specific nutritional deficiencies, it is important to summarize the role of a few of nutrients.

Vitamin C is a cofactor for hydroxylation of amino acids for collagen synthesis and an antioxidant which scavenges free radicals through its powerful reducing action [96]. Vitamin E is a lipid membrane protector which also scavenges reactive species and free radicals against lipid peroxidative damage [96]. It is usually depleted in hanseniasis patients. Vitamin A is also an inhibitor of lipid peroxidation in lipid membranes and organelles, which could be diminished in hanseniasis [96]. Selenium is a cofactor of the enzyme selenium-glutathione peroxidase and acts as an antioxidant and stimulator of both innate and acquired immunity, displaying antiviral, anti-tumoral, and cardioprotective effects [97–99]. Zinc is an immune cofactor; an antioxidant and anti-inflammatory agent, which could be linked to the structure of metallothioneins; a group of metal-chelating proteins; as well as SOD enzyme which catabolizes the superoxide anion to hydrogen peroxide [96, 100–102].

18.8 Should Hanseniasis Patients Receive Nutritional Supplementation?

In many studies, recovering of the antioxidant balance and reduction of the oxidative and nitrosative stress were associated with better clinical outcomes [67, 69, 75, 90, 91, 93, 94].

Nutritional supplementation should only be used in patients with clinically detected deficiency. But they can be benefited with supplementation with vitamins A, C, E, as well as selenium and zinc.

18.9 Conclusion

It is very clear that *M. leprae* infection causes oxidative stress due to its colonization and replication process but also due to the innate immune response against the bacteria. Then, it is important to note that therapies based on suppression of oxidative stress can in fact break down the innate immunity and also disrupt some effector reactions of the acquired immunity. Then, it is important to avoid overexpression of the ROS and RNS, but their complete suppression can be very harmful to the organism.

References

1. Slim FJ, Keukenkamp R, van Schie CH, Faber WR, Nollet F (2011) Foot Impairments and limitations in walking activities in people affected by leprosy. J Rehabil Med 43(1):32–38
2. Nsagha DS, Bissek AZ, Nsagha SM et al (2011) Social stigma as an epidemiological determinant for leprosy elimination in Cameroon. J Public Health Afr 2(1):e10. https://doi.org/10.4081/jphia.2011.e10
3. Faria L, Santos LAC (2015) A hanseníase e sua história no Brasil: a história de um "flagelo nacional". Hist Cienc Saude-Manguinhos 22(4):1491–1495
4. WHO (2016) Global leprosy update, 2016: accelerating reduction of disease burden. Weekly Epidemiol Rec 92(35):501–520
5. Worldometers (2018) India population. Available at: http://www.worldometers.info/world-population/india-population/ (06/03/2018)
6. da Saúde M (2018) Secretaria de Vigilância Sanitária. Caracterização da situação epidemiológica da hanseníase e diferenças por sexo, Brasil, 2012–2016. Bol Epidemiol 49(4):1–10
7. Santos MJS, Ferrari CKB, de Toledo OR, de Moraes EV, David FL (2012) Leprosy among children and adolescents under 15 years-old in a city of Legal Amazon, Brazil. Indian J Leprosy 84:265–269
8. Fonseca GAA, Silva TC, Ferrari GSL, Ferrari CKB (2013) Epidemiological aspects of leprosy in a city of Legal Amazon, Brazil. Int J Sci Nat 4(4):576–578
9. Wiedau-Pazos M, Goto JJ, Rabizadeh S et al (1996) Altered reactivity of superoxide dismutase in familial amyotrophic lateral sclerosis. Science 271(5248):515–518
10. Sandoval M, Zhang X-J, Liu X, Mannick EE, Clark DA, Miller MJS (1997) Peroxynitrite-induced apoptosis in T84 and RAW 264 cells: attenuation by L-ascorbic acid. Free Rad Biol Med 22(3):489–495
11. Ferrari CKB (2000) Free radicals, lipid peroxidation and antioxidants in apoptosis: implications in cancer, cardiovascular and neurological diseases. Biologia 55(6):581–590
12. da Silva WJM, Ferrari CKB (2011) Mitochondrial metabolism, free radicals and aging. Rev Bras Geriatr Gerontol 14(3):441–451
13. Duthie GG (1993) Lipid peroxidation. Eur J Clin Nutr 47(11):759–764
14. Halliwell B (1994) Free radicals, antioxidants and human disease: curiosity, cause or consequence? Lancet 344(8924):721–724
15. Ferrari CK, França EL, Honorio-França AC (2009) Nitric oxide, health and disease. J Appl Biomed 7:163–173
16. Ferrari CKB (1998) Lipid oxidation in food and biological systems: general mechanisms and nutritional and pathological implications. Rev Nutr 11(1):3–14
17. Ferrari CKB (2001) Oxidative stress pathophysiology: searching for an effective antioxidant protection. Int Med J 8(3):175–184
18. Suzuki YJ, Carini M, Butterfield DA (2010) Protein carbonylation. Antiox Redox Signal 12(3):323–325
19. Rimoli LF, Godoy MF (2011) Efetividade da vitamina E sobre o estresse oxidativo, em hansenianos da forma multibacilar sob tratamento. Hansen Int 36(1):17–21
20. Schalcher TR, Borges RS, Coleman MD et al (2014) Clinical oxidative stress during leprosy multidrug therapy: impact of dapsone oxidation. PLoS One 9(1):e85712. https://doi.org/10.1371/journal.pone.0085712
21. Babior BM, Cumutte JT, Kipnes RS (1975) Biological defense mechanisms. Evidence for the participation of superoxide in bacterial killing by xanthine oxidase. J Lab Clin Med 85(2):235–244
22. Babior BM (1984) The respiratory burst of phagocytes. J Clin Invest 73:599–601
23. Halliwell B, Chirico S (1993) Lipid peroxidation: its mechanism, measurement, and significance. Am J Clin Nutr 57(5):S715–S725
24. Kanner J (1994) Oxidative processes in meat and meat products: quality implications. Meat Sci 36(1/2):169–189

25. Esterbauer H (1993) Cytotoxicity and genotoxicity of lipid-oxidation products. Am J Clin Nutr 57(5):S779–S786
26. Beers RF Jr, Sizër IW (1952) A spectrophotometric method for measuring the breakdown of hydrogen peroxide by catalase. J Biol Chem 195:133–140
27. Lee JN, Dutta RK, Maharjan Y et al (2018) Catalase inhibition induces pexophagy through ROS accumulation. Biochem Biophys Res Commun 501(3):696–702
28. Ferrari CKB, Souto PCS, França EL, Honorio-França AC (2011) Oxidative and nitrosative stress on phagocytes' function: from effective defense to immunity evasion mechanisms. Arch Immunol Ther Exp 59(6):441–448
29. von Köckritz-Blickwede M, Nizet V (2009) Innate immunity turned inside-out: antimicrobial defense by phagocyte extracellular traps. J Mol Med 87:775–783
30. Ramos-Kichik V, Mondragón-Flores R, Mondragón-Castelán M (2009) Neutrophil extracellular traps are induced by *Mycobacterium tuberculosis*. Tuberculosis 89:29–37
31. Palmer RMJ, Ashton DS, Moncada S (1988) Vascular endothelial cells synthesize nitric oxide from arginine. Nature 333:664–666
32. Ahmad R, Rasheed Z, Ahsan H (2009) Biochemical and cellular toxicology of peroxynitrite: implications in cell death and autoimmune phenomenon. Immunopharmacol Immunotoxicol 31(3):388–396
33. Khan MA, Alam K, Zafaryab M, Rizvi MA (2017) Peroxynitrite-modified histone as a pathophysiological biomarker in autoimmune diseases. Biochimie 140:1–9
34. Visca P, Fabozzi G, Milani M, Bolognesi M, Ascenzi P (2002) Nitric oxide and *Mycobacterium leprae* pathogenicity. IUBMB Life 54(3):95–99
35. Boga P, Shety VP, Khan Y (2001) Nitric oxide metabolites in sera of patients across the spectrum of leprosy. Indian J Lepr 82(3):123–129
36. Adams LB, Scollard DM, Ray NA et al (2002) The study of Mycobacterium leprae infection in interferon-gamma gene-disrupted mice as a model to explore the immunopathologic spectrum of leprosy. J Infect Dis 185(Suppl.1):S1–S8
37. Adams LB, Job CK, Krahenbuhl JL (2000) Role of inducible nitric oxide synthase in resistance to *Mycobacterium leprae* in mice. Infect Immun 68(9):5462–5465
38. Vladimirov YA, Proskurnina EV (2009) Free radicals and cell chemiluminescence. Biochemistry 74:1545–1566
39. Nauseef WM (2014) Myeloperoxidase in human neutrophil host defense. Cell Microbiol 16(8):1146–1155
40. Nappi AJ, Vass E (2002) Interactions of iron with reactive intermediates of oxygen and nitrogen. Dev Neurosci 24:134–142
41. Schürmann N, Forrer P, Casse O et al (2017) Myeloperoxidase targets oxidative host attacks to *Salmonella* and prevents collateral tissue damage. Nature Microbiol 2:16268
42. Klebanoff SJ (2005) (2005) Myeloperoxidase: friend and foe. J Leukoc Biol 77:598–625
43. Segal AW (2005) How neutrophils kill microbes. Annu Rev Immunol 23:197–223
44. Brovkovych V, Gao X-P, Ong E et al (2008) Augmented inducible nitric oxide synthase expression and increased NO production reduce sepsis-induced lung injury and mortality in myeloperoxidase-null mice. Am J Physiol Lung Cell Mol Physiol 295:L96–L103
45. Maslov AK (2000) Phagocytic myeloperoxidase in leprosy pathogenesis. Int J Lepr Other Mycobact Dis 68(1):71–73
46. Maslov AK, Luzhnova AS (2000) Effects of peroxidase therapy on functional state of the liver and phagocytes and blood cell counts in mice with experimental leprosy. Bull Exp Biol Med 130(1):682–686
47. Escorza MAQ, Salinas JVC (2009) La capacidad antioxidante total. Bases y aplicaciones REB 28:89–101
48. França-Botelho AC, França EL, Honório-França AC et al (2006) (2006) Phagocytosis of *Giardia lamblia* trophozoites by human colostral leukocytes. Acta Paediatr 95:438–443
49. Hii CS (2007) Ferrante A (2007) Regulation of the NADPH oxidase activity and antimicrobial function of neutrophils by arachidonic acid. Arch Immunol Ther Exp 55:99–110

50. Gozalo AS, Hofmann VJ, Brinster LR et al (2010) Spontaneous Staphylococcus xylosus infection in mice deficient in NADPH oxidase and comparison with other laboratory mouse strains. J Am Assoc Lab Anim Sci 49:480–486
51. Holland SM (2010) (2010) Chronic granulomatous disease. Clin Rev Allerg Immunol 38:3–10
52. Rojas-Espinosa (2009) Chapter 4: Murine leprosy revisited. In: Tomioka H (ed) Current topics on the profiles of host immunological response to Mycobacterial infections. Kerala, India, Research SignPost, pp 97–140
53. Chieber M, Chandel NS (2014) ROS function in redox signaling and oxidative stress. Curr Biol 24(10):R453–R462
54. Yang Y, Bazhin AV, Werner J, Karakhanova S (2013) Reactive oxygen species in the immune system. Int Rev Immunol 32(3):249–270
55. Dupnik KM, Bair TB, Maia AO et al (2015) Transcriptional changes that characterize the immune reactions of leprosy. J Infect Dis 211:1658–1676
56. Guerreiro LT, Robottom-Ferreira AB, Ribeiro-Alves M et al (2013) Gene expression profiling specifies chemokine, mitochondrial and lipid metabolism signatures in leprosy. PLoS One 8(6):e64748. https://doi.org/10.1371/journal.pone.0064748
57. Ramos GB, Salomão H, Francio AS, Fava VM, Werneck RI, Mira MT (2016) Association analysis suggests *SOD2* as a newly Identified candidate gene associated with leprosy susceptibility. J Infect Dis 214(3):475–478
58. Lastória JC, de Abreu MAMM (2014) Leprosy: review of the epidemiological, clinical and etiopathogenetic aspects- part 1. An Bras Dermatol 89(2):205–218
59. Launois P, Blum L, Dieye A, Milan J, Sarthou JL, Bach MA (1989) Phenolic glycolipid-1 from *M. leprae* inhibits oxygen free radical production by human mononuclear cells. Res Immunol 140(9):847–855
60. Vachula M, Holzer TJ, Andersen BR (1989) Suppression of monocyte oxidative response by phenolic glycolipid I of Mycobacterium leprae. J Immunol 142(5):1696–1701
61. Cambier CJ, O'Leary SM, O'Sullivan MP, Keane J, Ramakrishnan L (2017) Phenolic glycolipid facilitates mycobacterial escape from microbicidal tissue-resident macrophages. Immunity 47:552–565
62. Madigan CA, Cambier CJ, Kelly-Scumpia KM, Sagasti A, Modin RL, Ramakrishnan L (2017) Phenolic glycolipid initiates nerve damage in leprosy. Cell 170:973–985
63. Mattos KA, Sarno EN, Pessolani MCV, Bozza PT (2012) Deciphering the contribution of lipid droplets in leprosy: multifunctional organelles with roles in *Mycobacterium leprae* pathogenesis. Mem Inst Oswaldo Cruz 107(suppl.1):156–166
64. WHO (1994) Chemotherapy of leprosy. Technical report series 847. World Health Organization, Geneva
65. Jyothi P, Riyaz N, Nandakumar G, Binitha MP (2008) A study of oxidative stress in paucibacillary and multibacillary leprosy. Indian J Dermatol Venereol Leprol 74(1):80
66. Bhadwat VR, Borade VB (2000) Increased lipid peroxidation in lepromatous leprosy. Indian J Dermatol Venereol Leprol 66(3):121–125
67. Reddy YN, Murthy SV, Krishna DR, Prabhakar MC (2003) Oxidative stress and anti-oxidant status in leprosy patients. Indian J Lepr 75(4):307–316
68. Vijayaraghavan R, Suribabu CS, Sekar B et al (2005) Protective role of vitamin E on the oxidative stress in Hansen's disease (leprosy) patients. Eur J Clin Nutr 59(10):1121–1128
69. Chhabra N, Bhattacharya SN, Singal A, Ahmed RS, Verma P (2015) Profile of oxidative stress in response to treatment for type 1 reaction. Lepr Ver 86:80–88
70. Schalcher TR, Borges RS, Coleman MD et al (2014) Clinical oxidative stress during leprosy multidrug therapy: impact of dapsone oxidation. PLoS One 9(1):e85712. https://doi.org/10.1371/journal.pone.0085712
71. Prasad CV, Kodiwadmath MV, Kodiwadmath GB (2007) Erythrocyte superoxide dismutase, catalase activities and hydrogen peroxide induced lipid peroxidation in leprosy. Lepr Rev 78(4):391–397

72. Prasad CV, Kodiwadmath MV, Kodiwadmath GB (2008) Erythrocyte glutathione peroxidase, glutathione reductase activities and blood glutathione content in leprosy. J Infect 56(6):469–473
73. Abdel-Hafez HZ, Mohamed E-EM, Abd-Elghany AA (2010) Tissue and blood superoxide dismutase activity and malondialdehyde level in leprosy. J Eur Acad Dermatol Venereol 24(6):704–708
74. Patni V, Baliga S, Sawal S (2015) Saliva as a diagnostic tool for measurement of total antioxidant capacity in children with leprosy and Born to leprosy parent. Indian J Lepr 87(1):17–21
75. Lima ES, Roland IA, Maroja MF, Marcon JL (2007) Vitamin A and lipid peroxidation in patients with different forms of leprosy. Rev Inst Med Trop S. Paulo 49(4):211–214
76. Osadolor HB, Ihongbe JC (2008) Effect of leprosy on non-enzymatic antioxidants (vitamin C, vitamin E and uric acid) in (Edo State) Nigerian leprosy patients. Cont J Biomed Sci 2:1–5
77. Asalkar A, Girish S, Naoley R (2011) Protein oxidation and antioxidant vitamins in leprosy. Int J Pharm Sci Res 2(11):2870–2873
78. Prabhakar MC, Santhikrupa D, Manasa N, Rao OU (2013) Status of free radicals and antioxidants in leprosy patients. Indian J Lepr 85(1):5–9
79. Swathi M, Tagore R (2015) Study of oxidative stress in different forms of leprosy. Indian J Dermatol 60(3):321
80. Benzie IFF, Strain JJ (1996) The ferric reducing ability of plasma (FRAP) as a measure of "Antioxidant Power": the FRAP assay. Anal Biochem 239:70–76
81. Re R, Pellegrini N, Proteggente A, Pannala A, Yang M, Rice-Evans C (1999) Antioxidant activity applying an improved ABTS radical cation decolorization assay. Free Rad Biol Med 26(9/10):1231–1237
82. Bowman GL, Shannon J, Frei B, Kaye JA, Quinn JF (2010) Uric acid as a CNS antioxidant. J Alzheim Dis 19(4):1331–1336
83. El Ridi R, Tallima H (2017) Physiological functions and pathogenic potential of uric acid: a review. J Adv Res 8:487–493
84. Waring WS, Webb DJ, Maxwell SRJ (2001) Systemic uric acid administration increases serum antioxidant capacity in healthy volunteers. J Cardiovasc Pharmacol 38:365–371
85. Barsoum R, El-Khatib M (2017) Uric acid and life on earth. J Adv Res 8:471–474
86. Sautin YY, Nakagawa T, Zharikov S, Johnson RJ (2007) Adverse effects of the classic antioxidant uric acid in adipocytes: NADPH oxidase-mediated oxidative/nitrosative stress. Am J Physiol Cell Physiol 293(2):C584–C596
87. Sautin YY, Johnson RJ (2008) Uric acid: the oxidant-antioxidant paradox. Nucleos Nucleot Nucleic Acid 27(6):608–619
88. Morato-Conceiçao YT, Alves-junior ER, Arruda TA, Lopes JC, Fontes CJF (2016) Serum uric acid levels during leprosy reaction episodes. Peer J 4:4e1799. https://doi.org/10.7717/peerj.1799
89. Schalcher TR, Vieira JLF, Salgado CG, Borges RS, Monteiro MC (2013) Antioxidant factors, nitric oxide levels, and cellular damage in leprosy patients. Rev Soc Bras Med Trop 46(5):645–649
90. Raka I, Rastogi MK, Gahalaut P, Kaur J, Mishra N (2018) Enzymatic oxidative stress indicators and oxidative stress índex in patients of leprosy. Nepal J Dermatol Venereol Leprol 16(1):35–40
91. Elesawy FM, Mikhael NW, Sabry JH (2015) Serum nitric oxide metabolites in leprosy patients as a parameter of prognostic value. J Egypt Women Dermatol Soc 12:44–48
92. Abd-Elmaged WM, Hassan MH, Mostafa MA, Ahmed NS, Samy ES (2017) Lesional levels of superoxide dismutase and malondialdehyde in paucibacillary and multibacillary leprosy patients. J Egypt Women Dermatol Soc 14:156–160
93. Pradhan T, Kumari S (2015) Evaluation of oxidative status and zinc level in leprosy patients after zinc supplementation. Int J Biol Med Res 6(2):4984–4987
94. Oliveira FM, Barbosa Junior F, Jordão Junior AA, Foss NT, Navarro AM, Frade MAC (2015) Oxidative stress and micronutrients in leprosy. Rev Nutr 28(4):349–357

95. Partogi D, Dalimunthe DA, Hazlianda CP (2018) A study of Selenium in leprosy. Macedonian J Med Sci 6(3):485–487
96. Vázquez CMP, Mendes Netto RS, Barbosa KBF et al (2014) Micronutrients influencing the immune response in leprosy. Nutr Hosp 29(1):26–36
97. Ferrari CKB (2005) Minerals. From basic aspects to newly discovered physiological and nutritional actions. Evid Based Integrat Med 2(3):123–131
98. Rayman MP (2012) Selenium and human health. Lancet 379(9822):1256–1268
99. Ferrari CKB (2014) Nutrição, saúde e longevidade baseadas em evidências científicas. Plêiade 8(15):26–36
100. Arias ARL, Santos VG (2008) Metalotioneína: processos celulares e moleculares. Cad Saúde Col 16(4):701–716
101. Jarosz M, Olbert M, Wyszogrodzka G, Mlyniec K, Librowski T (2017) Antioxidant and anti-inflammatory effects of zinc. Zinc-dependent NF-κB signaling. Inflammopharmacology 25(1):11–24
102. Wessels I, Maywald M, Rink L (2017) Zinc as a gatekeeper of immune function. Nutrients 9(12):1286. https://doi.org/10.3390/nu9121286

Neonatal Septicemic *Escherichia coli* Protease (SslE) Induces Macrophage Activation and Polarization Through Induction of ROS and NO

19

Rima Tapader and Amit Pal

Abstract

Oxidative stress, characterized by the undue production of reactive oxygen and nitrogen species, is a serious burden in patients having sepsis. SslE, the cell surface associated and secreted lipoprotein with a metalloprotease domain, was identified and found to be widely distributed among the *E. coli* isolates causing neonatal septicemia. Importantly, SslE was shown to trigger secretion of various proinflammatory cytokines, chemokines, and other proinflammatory hallmarks like reactive oxygen species (ROS) and reactive nitrogen species (RNS) such as NO in mouse macrophages. SslE-induced production of ROS and NO was found to take place via TLR2–/TLR1-dependent manner with the downstream activation of NFκB and MAP kinase signaling molecules. Furthermore, increased expression of MHC II and other co-stimulatory molecules (CD80, CD86) was found on macrophages on induction with SslE, indicating macrophage activation and polarization toward M1 type. Hence, SslE has been identified as one of the factors for inducing production of ROS and NO in macrophages which actually activates host's innate immune defense against *E. coli* causing neonatal sepsis, proposing the use of SslE as an immunotherapeutic target against *E. coli* sepsis. Moreover, identification of the receptor and the signaling pathways for SslE-induced ROS and NO formation would be beneficial to prevent any oxidative stress generated due to SslE in *E. coli* sepsis.

Keywords

SslE (YghJ) · Neonatal sepsis · Reactive oxygen species (ROS) · Reactive nitrogen species (RNS) · Toll-like receptor 2 (TLR2) · NFκB · MAP kinase

R. Tapader · A. Pal (✉)
Division of Pathophysiology, National Institute of Cholera and Enteric Diseases (NICED), Beliaghata, Kolkata, India
e-mail: apal7364@rediffmail.com

S. Chakraborti et al. (eds.), *Oxidative Stress in Microbial Diseases*,
https://doi.org/10.1007/978-981-13-8763-0_19

19.1 Introduction

Incidence of systemic signs and symptoms of infection with detection of bacteria in blood or cerebrospinal fluid is clinically featured as sepsis. If this occurs within first 4 weeks of life, i.e., in the neonatal period, it is ascribed as neonatal sepsis [1, 2]. Sepsis in newborns continues to be a global health menace. Among the wide array of pathogens concerned, *E. coli* has emerged as the most threatening agent specifically in developing countries [3–5]. Increasing antibiotic resistance among the *E. coli* isolates, specifically the emergence of the multidrug-resistant *E. coli* sequence type ST131 clone holds critical challenge in prevention and management of *E. coli* infections including neonatal sepsis [6, 7]. Hence, researches on development of an effective and inexpensive vaccine against ExPECs demand serious attention. Therefore, studies on virulence determinants and understanding their detailed molecular mechanism of pathogenesis should be encouraged.

SslE (surface associated and secreted lipoprotein of *E. coli*) has been identified as a potent virulence determinant to contribute in the virulence of various intestinal and extraintestinal pathogenic *E. coli* (ExPEC) [8–11]. Importantly, SslE has been recognized as the most promising vaccine candidate for ExPEC by a "substractive reverse vaccinology" approach and was shown to provide almost complete protection from bacteremia in a mouse model of sepsis [12, 13]. Therefore, it is crucial to understand the molecular mechanism of SslE-induced activation of host's innate and adaptive responses in great detail.

Neonates primarily rely on their innate immune defense to fight against infection [14, 15] and TLRs, the evolutionary conserved pattern recognition receptors (PRRs) are the sentinel of the innate immune system of host [16, 17]. TLRs, the type I transmembrane proteins of the IL-1R (interleukin-1 receptor) family, possess an N-terminal leucine-rich repeat (LRR) domain for ligand binding, a single transmembrane domain, and a C-terminal intracellular signaling TIR (Toll/IL-1 receptor) domain. Thirteen mammalian TLRs have been identified till now (among which 10 are present in human and all 13 are present in mice), expressing in various cell types specifically cells of the innate defense such as neutrophils, dendritic cells, and macrophages and having distinct substrate specificities [16, 17].

Macrophages, the fore frontiers of the innate response [18, 19], get activated as they specifically recognize certain PAMPs through TLRs leading to the activation of NFκB and different MAP kinase signaling pathways, which finally trigger secretion of various proinflammatory cytokines, chemokines, reactive oxygen species (ROS), and reactive nitrogen species (RNS) reflecting activation and M1 polarization of macrophages, crucial in framing antibacterial defense of hosts [20–23]. This chapter focuses on the association of SslE with neonatal septicemic *E. coli* which includes identification, purification, and characterization of SslE from NSEC as well as its distribution among NSEC isolates. In addition, this chapter also explores the detailed molecular mechanism of SslE-induced production of ROS and RNS and subsequent activation of host's innate immune defense.

19.2 Neonatal Sepsis, a Global Burden

According to WHO, in the year 2016, sepsis and meningitis were accounted for 7% of deaths in children under 5 years and 15% of deaths in neonates [24]. As per Global Health Observatory Data, in the year 2016, neonatal death accounts for almost 46% (2.6 million) of all child deaths below 5 years of age, and neonatal sepsis stands as the fourth cause of neonatal death globally [24, 25]. Importantly, studies across the world show that 50% of the sepsis cases are severe and associated with death and in case of septic shock it is 60%. In some African countries, the mortality rate is as high as 100% [24]. The incidence of sepsis and the mortality rate is optimum in preterm or the prematurely born neonates (born before 28 days) [26, 27] specifically in the low-birth weight neonates [28–30]. In the United States, 36% of preterm neonates suffer at least one incidence of sepsis during their birth [28]. The burden of sepsis in newborns exists not only in developed countries, it continues to persist also in developing countries [31, 32]. In India, sepsis and pneumonia together stands for the commonest cause of neonatal mortality, accounting for almost 30.4% of total neonatal deaths [33]. Difficulty in diagnosis, limited diagnostic accuracy, maternal risk factors, neonatal susceptibility, increasing multidrug resistance, resource poor setting, vast repertoire of causative agent, all make the disease creepy.

Neonatal sepsis is classified into two types on the basis of onset of symptoms, early-onset sepsis (EOS) and late-onset sepsis (LOS) [34]. EOS, being more severe than LOS, occurs within 72 h of life and is caused by vertical transmission of microorganisms from mother before and during delivery [29]. LOS takes place after 72 h of life and the organisms are transmitted horizontally (nosocomial or community-acquired) in LOS [29, 35, 36]. This classification of sepsis holds immense clinical significance in understanding the modes of transmission of the disease, severity of the disease, causative organisms, and expected outcomes of the disease.

19.3 *E. coli*, a Major Causative Agent in Neonatal Sepsis

A wide variety of pathogens (both Gram-positive and Gram-negative) are actually involved in sepsis of newborns. However, the spectrum of organisms differs between EOS and LOS and also between developed and developing countries. In developing countries, the incidence of Gram-negative sepsis is more common than Gram-positive ones [37–39]. Studies of recent times have procured that Gram-negative organisms account for 18%–78% of all neonatal sepsis and risk of death is much higher in case of Gram-negative organisms than Gram-positive ones [40–43]. Among the Gram-negative organisms, *E. coli* has appeared as the commonest cause of sepsis in preterm neonates and the second most common cause in term neonates [3]. *E. coli* has also been found to be frequently associated with the cases of sepsis-related mortality specifically among very low-birth weight infants (24.5%) [44]. Additionally, *E. coli*, being able to colonize mother's genital tract and infect neonates at the time of delivery, has emerged as the key pathogen in case of early-onset neonatal sepsis [3, 5]. Of note, incidence of *E. coli* is also frequent in cases of late-onset sepsis [4].

19.4 Generation of Reactive Oxygen and Nitrogen Species in Sepsis

Under normal conditions, there exists an equilibrium between the generation of reactive oxygen and nitrogen species (ROS and RNS) such as superoxide anion, hydroxyl radical, hydrogen peroxide, nitric oxide, and peroxynitrite and elimination of such species by endogenous protective antioxidant compounds such as catalase, reduced glutathione (GSH), superoxide dismutase (SOD), vitamins C and E. Oxidative stress develops when this homeostatic balance gets disrupted due to overwhelming production of the free radicals and/or by limited antioxidative defense. Both of these can happen in sepsis [45–48]. The majority of ROS forms in mitochondria during respiration [48, 49] when there is incomplete four-electron reduction of oxygen to water during the process of ATP production, generating reactive oxygen species such as superoxide anion radical which is converted to hydrogen peroxide by SOD [48, 50]. Hydrogen peroxide generally converted to water by peroxidases and catalases. However, excess hydrogen peroxide on reaction with superoxide anion can form a much more reactive free radical, the hydroxyl radical [51]. During an infection, production of ROS is stimulated in phagocytic cells involved in proinflammation such as neutrophils and macrophages for host defense. These reactive species may directly act against pathogens or may act as secondary signaling molecules for inflammation and immune response [52]. In addition, in sepsis, ROS is produced in other ways such as respiratory burst, i.e., due to recruitment of more cells of the innate immune system to the site of infection, leading to an increased oxygen uptake finally resulting in production of higher amount of ROS at the site of infection [53, 54]. Xanthine oxidase activation as a result of ischemia and reperfusion as well as arachidonic acid metabolism also serves as sources of ROS in sepsis [48]. ROS acts by oxidation of cellular DNA, RNA, and proteins along with lipid peroxidation resulting in altered membrane permeability and membrane damage [55, 56].

Reactive nitrogen species (RNS) are another critical participant in generating oxidative stress. During infection, cells of the innate immune system such as macrophages and neutrophils also produce the short-lived free radical nitric oxide (NO) in abundance which is essential in host defense [57–59]. NO is synthesized from L-arginine by the enzyme NO synthase (NOS). There are three types of NOS: eNOS, nNOS, and iNOS. eNOS and nNOS are constitutively expressed and were originally characterized in endothelial cells and neurons respectively, whereas iNOS only express in infected or activated cells, hence, named as inducible NOS [60–63]. Expression of iNOS is induced by various cytokines produced during inflammation such as TNF-α, IFN-γ, and IL-1 [64, 65]. Being antibacterial, NO inhibits replication of pathogens [46, 66], can also react with superoxide to generate peroxynitrite which on decomposition may form hydroxyl radical [48, 66]. There are increasing evidences of formation of ROS and RNS in patients with sepsis. Increased amount of lipid peroxides with elevated redox reactive iron in plasma was found in septicemic patients compared to the healthy control [67–69]. In another study, patients with sepsis were also found to possess elevated level of free radicals and xanthine oxidase activation [70]. Batra et al. have reported increased activation of xanthine oxidase, superoxide dismutase, and glutathione peroxidase in newborns with sepsis, indicating enhanced production of ROS in these patients [71]. Moreover, Cowley et al. have found decreased antioxidant potential in patients having sepsis and secondary organ dysfunction, associated with non-survival [72].

19.5 Macrophage Polarization and Host Defense

Macrophages, the paramount component of innate immune defense, are appointed at the site of infection to efficiently eliminate invading pathogens [18, 19]. In addition to innate defense, macrophages also shape the adaptive response by bridging the innate with adaptive immunity [73]. Macrophages in response to diverse bacterial products or other environmental stimuli get activated and differentiated into specific functional phenotypes, either classically activated M1 macrophages or alternatively activated M2 macrophages [22]. M1 or type 1 macrophages upon encountering pathogens are able to produce several proinflammatory cytokines, chemokines as well as ROS and RNS and hence these proinflammatory macrophages have strong microbicidal activity, whereas M2 macrophages are associated with the release of anti-inflammatory cytokines and chemokines. Both M1 and M2 have distinct repertoire of secreted cytokines and chemokines. M1 polarized macrophages release an array of proinflammatory cytokines such as IL-1, IL-6, IL-12, and TNF-α as well as chemokines such as CCL3/MIP-1α, CCL4/MIP-1β, CCL5/RANTES, CXCL9/MIG, and CXCL-10/IP-10. In contrary, M2 macrophages are equipped with a set of anti-inflammatory cytokine such as IL-10 and chemokines such as TARC/CCl17, MDC/CCL22, eotaxin-2/CCL24, etc. [22, 74–76]. The secreted proinflammatoy cytokines from M1 macrophages attract various immune cells such as leukocytes and T cells to the site of infection for robust elimination of pathogens. Moreover, M1 macrophages induce T cell differentiation to Th1 subtype [76].

19.6 SslE Is a Cell Associated and Secreted Lipoprotein Having M60 Metalloprotease Domain of Diverse *E. coli*

SslE (YghJ), a cell surface associated and secreted lipoprotein of *E. coli* having a putative M60 metalloprotease domain, is reported to be conserved among diverse *E. coli* pathotypes [8–11]. A study conducted by Moriel et al. has identified ECOK1–3385 as a novel vaccine candidate for extraintestinal pathogenic *E. coli* (ExPEC) by reverse vaccinology and found that it could confer almost complete protection in a murine model of sepsis making it a potent immunogenic vaccine candidate for ExPECs causing sepsis [12, 13]. This ECOK1–3385 protein was later described as SslE [77] and is reported to be secreted via type II secretion system (T2SS), an exporting apparatus of Gram-negative bacteria to deliver different proteins including various virulence determinants [78]. *E. coli* possess two T2SS operons: T2SSα and T2SSβ. T2SSβ operon is composed of three genes (*yghJ*, *pppA*, and *yghG*). *yghJ* encodes the SslE protein [79]. Hence, SslE was formerly named as YghJ [80, 81]. SslE has been found to be associated with both intestinal and extraintestinal pathogenic *E. coli* where it contributed in the virulence of the producer organisms. In EPEC, it is involved in biofilm formation [10]. In ETEC, it is engaged in degradation of intestinal mucins such as MUC2, MUC3, and bovine submaxillary mucin to smoothen *E. coli* penetration of mucus layer and facilitates access to apical epithelial cells [8, 9, 77]. SslE from NSEC was found to cause significant fluid accumulation and extensive tissue hemorrhage in a mice ileal loop [82]. In addition, SslE-immunized mice were found to resist UTI as well as intestinal infections rendering SslE as a broadly protective vaccine antigen against *E. coli* infections [77]. Importantly, in patients infected with ETEC, SslE was identified as an immunogenic antigen [83, 84].

19.7 SslE (YghJ), a Cell Associated and Secreted Lipoprotein Having Metalloprotease Domain was Identified and Purified from Neonatal Septicemic *E. coli*

19.7.1 SslE Was Partially Purified and Identified from a Neonatal Septicemic *E. coli* (NSEC) Isolate

In search for a novel protease, SslE was partially purified from culture supernatant of a NSEC isolate EB260. Protein in culture supernatant was concentrated in an Amicon concentrator using ultrafiltration membrane (30 kDa cutoff) and purified using successive anion exchange (DE-52) (Fig. 19.1a) and gel filtration (Sephadex G-200) (Fig. 19.1b) chromatography. The pool I fraction from G-200 column that was positive for protease activity (Fig. 19.1c) showed the presence of three major bands on SDS-PAGE (Fig. 19.1d). All the three bands were identified by matrix-assisted laser desorption ionization time of flight (MALDI-TOF) mass spectrometry and analyzed using Mascot software. The 167 kDa band on SDS-PAGE showed the presence of *E. coli* lipoprotein SslE/YghJ containing M60 metalloprotease domain (Fig. 19.1e) with 36 kDa *E. coli* outer membrane protein C (OmpC) and 35 kDa *E. coli* outer membrane protein A (OmpA) (Fig. 19.1d).

19.7.2 Distribution of SslE Was Studied Among *E. coli* Isolates from Septicemic and Healthy Neonates

Both the genotypic and phenotypic distribution of SslE was studied among NSEC and fecal *E. coli* isolates of healthy neonates. Genotypic distribution was studied using PCR with overnight grown cell lysates from both septicemic and fecal isolates as templates. Out of the 40 septicemic isolates tested, 31 (77.5%) were found to harbor *yghJ* gene. Furthermore, the distribution of *yghJ* among different phylogroups of septicemic *E. coli* showed that *yghJ* was mostly prevalent among the isolates of

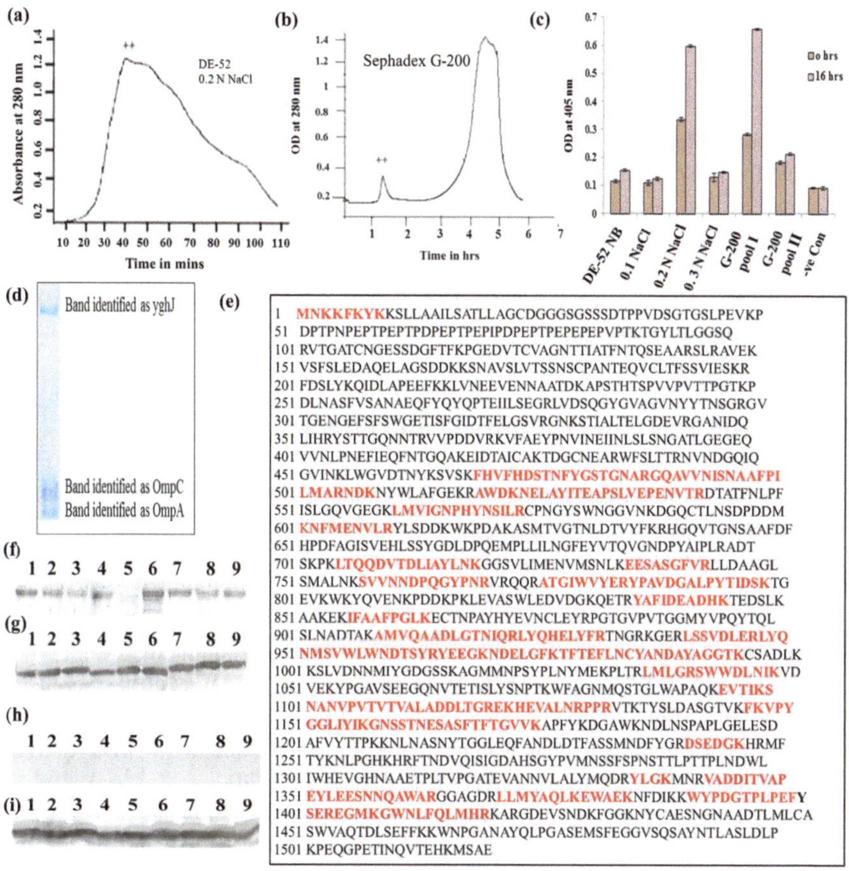

Fig. 19.1 Partial purification and identification of SslE from neonatal septicemic *E. coli* (NSEC) isolate and its distribution among septicemic and fecal *E. coli* isolates. Culture supernatant from NSEC isolate EB260 was concentrated and run on DE-52 anion exchange chromatography (**a**) and 0.2 N NaCl eluted fraction was positive for protease activity (indicated as '++' using AAPM-pNA oligopeptide substrate assay (**c**). Proteins in 0.2 N NaCl eluted fraction was further run on Sephadex G-200 column (**b**) and pool II showed protease activity (**c**) which on SDS-PAGE (7.5%) showed three bands identified as 167 kDa *E. coli* lipoprotein SslE having M60 metalloprotease domain, 36 kDa *E. coli* outer membrane protein C (OmpC) and 35 kDa *E. coli* outer membrane protein A (OmpA) (**d**). The peptides of the 167 kDa band that showed homology with SslE is indicated in bold red with the HExxH metalloprotease domain is underlined (**e**). Immunoblots were done to show distribution of SslE with TCA precipitated proteins of overnight grown culture supernatant of nine septicemic isolates (**f**) and nine fecal *E. coli* isolates (**h**) using anti-SslE antibody. OmpA was used as loading control for both septicemic (**g**) and fecal (**i**) isolates

phylogroup B2 (92%), followed by D (83%). A significant occurrence of *yghJ* was also found among the other phylogroups A (57%) and B1 (67%).

How far the fecal isolates were concerned, 15 isolates possessed *yghJ* gene out of 28 isolates (54%) tested with the highest occurrence in phylogroup B1 (57%). Hence, *yghJ* was found to be less prevalent among fecal isolates (54%) than septicemic isolates (77%). However, there exists a moderate occurrence of *yghJ* also among the fecal isolates which led us to further investigate the phenotypic distribution of YghJ. The expression and secretion of YghJ or SslE among both groups of isolates was determined by immunoblot analysis with overnight grown culture supernatants after TCA precipitation. Nine isolates from both the groups that were positive for *yghJ* were randomly selected for expression and secretion studies using immunoblot. For septicemic isolates, among 9 isolates tested, 8 isolates (89%) were positive for expression and secretion of SslE (Fig. 19.1f). Interestingly, in case of fecal isolates, 3 isolates out of 9 (33%) were positive for the secretion of SslE (Fig. 19.1h). OmpA was used as loading control for both septicemic (Fig. 19.1g) and fecal (Fig. 19.1i) isolates.

19.7.3 SslE Was Cloned, Purified, and Characterized

yghJ (~4.56-kb) was cloned into pBAD TOPO expression vector (Invitrogen) using TA cloning, a positive clone pTYghJ7 was selected and its entire 4.56-kb insert DNA was sequenced (GenBank accession number KX245009) which was highly homologous (98%) with the reported sequences of *yghJ* from other *E. coli* strains. pTYghJ7 was transformed into TOP10 *E. coli* cells and was induced optimally with 0.02% L-arabinose for 5 h. rSslE or rYghJ was purified by running the soluble fraction of whole cell lysate on Ni-NTA column (Fig. 19.2a) and a further Sephadex G-200 column (Fig. 19.2b). The first peak eluted from Sephadex G-200 was positive for protease activity and showed presence of a single band of 167 kDa rSslE on SDS-PAGE (Fig. 19.2c) and on NATIVE-PAGE (Fig. 19.2d).

SslE was proteolytically active against the oligopeptide substrate Ala-Ala-Pro-Met-p-nitroanilide with an optimum activity at pH 7.2 (Fig. 19.2e) and at temperatures 37 °C and 40 °C (Fig. 19.2f). The protease activity was completely abolished by EDTA (97%) and 1,10 phenanthroline (94%) and remained unchanged with PMSF, confirming that YghJ is a Zn++ containing metalloprotease.

19.8 SslE-Induced Proinflammation and Activation of Mouse Macrophages

19.8.1 Production of Different Proinflammatory Cytokines and Chemokines by SslE

RAW 264.7 cells (mouse macrophages) were incubated with SslE at 200 nM for various time points in presence or absence of Polymyxin B (PMB) at a dose of 50 μg/ml and the amount of cytokines expressed and secreted in the cell supernatant were then quantified by ELISA. SslE was found to stimulate the production of TNF-α at a much higher amount compared to the other cytokines studied. The level of TNF-α was found to decrease slightly at all the time points studied, even in the presence of PMB, indicating that LPS is not solely responsible for the production of

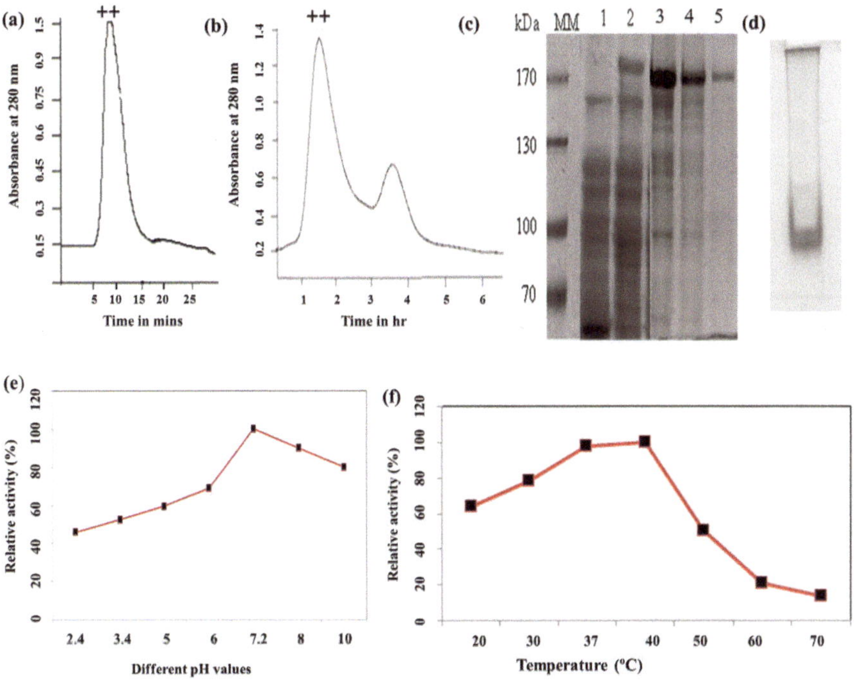

Fig. 19.2 Cloning, expression, purification, and characterization of SslE. *yghJ* (as the gene named for SslE) amplified from EB260 was cloned into pBAD TOPO vector using TA cloning and transformed into TOP10 *E. coli* cells. Proteins in the soluble fraction of recombinant clone after optimal induction with 0.02% L-arabinose was run on Ni-NTA column (**a**) and successive Sephadex G-200 column (**b**). "++" indicates fraction positive for protease activity. SDS-PAGE profile (**c**) shows uninduced lysate (Lane 1), induced lysate (Lane 2), Ni-NTA eluted fraction (Lanes 3 and 4) and pool I of Sephadex G-200 eluted fraction (Lane 5). Pool I eluted fraction of G-200 was also run on NATIVE-PAGE (**d**) showing presence of single band. Optimal pH (**e**) and temperature (**f**) of rSslE were determined using pNA oligopeptide substrate assay

TNF-α (Fig. 19.3a). On induction with SslE, TNF-α exhibited rapid kinetics of stimulation, being induced as early as 2 hrs of treatment and showed considerable increase as compared to control at 0 hrs of time point (244.4 ± 31 versus 60.5 ± 20.3 pg/ml; $p = 0.001$). The level was highest at 10 hrs of post-stimulation, being 14.6 ± 3.3 fold higher compared to control (883.52 ± 66 versus 60.5 ± 20.3 pg/ml; $p < 0.0001$) (Fig. 19.3a). In contrast to TNF-α, SslE induced relatively smaller amount of IL-1α and IL-1β in mouse macrophages (Fig. 19.3b, c). However, the production of these two cytokines was delayed by 1–2 hrs compared to TNF-α in response to SslE. Likewise TNF-α, the level of IL-1α was found to be optimally induced at 10 hrs of time point, being 22 ± 4 fold higher than the respective level of control at 0 hrs of time point (351 ± 30 vs 16 ± 7 pg/ml; $p < 0.0001$) and 17 ± 6 fold higher than the respective level of negative control (human recombinant elafin) at the same time point (351 ± 30 vs 21 ± 5 pg/ml; $p < 0.0001$) (Fig. 19.3b). Similarly, amount of secreted IL-1β was also optimally induced at 10 hrs of time point being 33 ± 13 fold and 15 ± 6 fold higher than the corresponding controls at 0 hrs of time

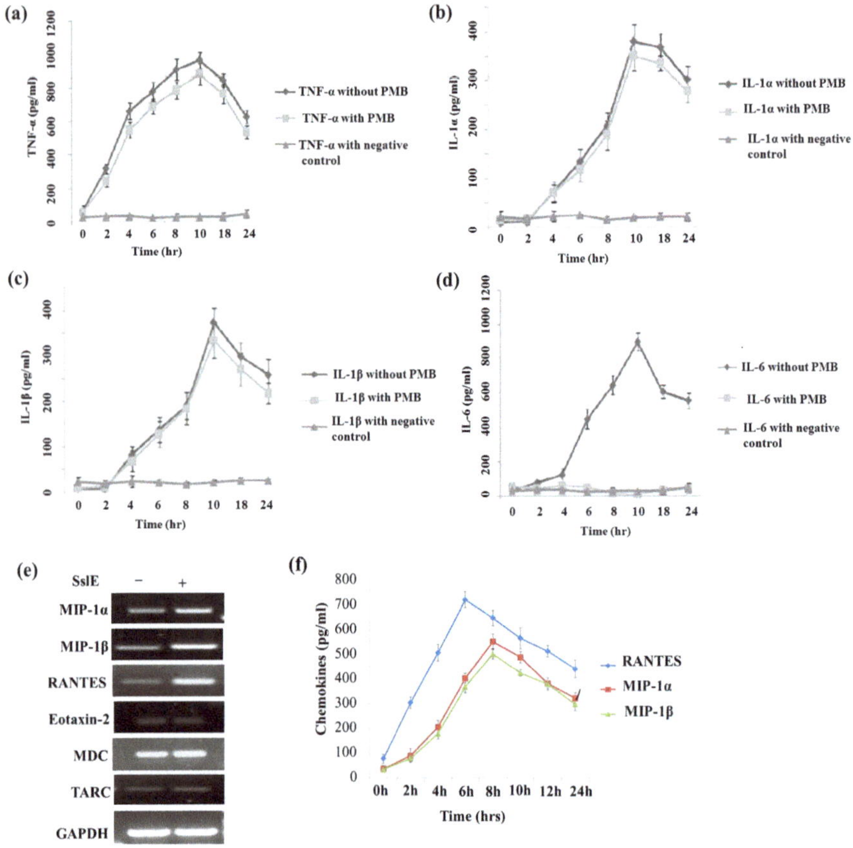

Fig. 19.3 SslE induces secretion of various proinflammatory cytokines and chemokines in RAW264.7 cells.RAW264.7 cells were treated with 200 nM of SslE with and without Polymyxin B (PMB) at a dose of 50 μg/ml and the level of secreted TNF-α (**a**), IL-1α (**b**), IL-1β (**c**) and IL-6 (**d**) were measured through ELISA. Recombinant human skin antimicrobial peptide elafin (Gene ID 5266) cloned, expressed, and purified similarly like SslE and showing no proinflammation was used as negative control. (**e**) RT-PCR was done using cDNA prepared from RNA of RAW264.7 cells treated or untreated with SslE for 6 h to check expression of RANTES, MIP-1α, MIP-1β, Eotaxin-2, MDC and TARC with GAPDH as control. (**f**) Kinetics of SslE-induced secretion of different proinflammatory chemokines by ELISA

point (333 ± 38 vs 10 ± 3 pg/ml; $p = 0.0001$) and negative control (recombinant elafin) at the similar time point (333 ± 38 vs 22 ± 6 pg/ml; $p < 0.0001$) respectively (Fig. 19.3c). On the other hand, SslE failed to stimulate the secretion of IL-6 (Fig. 19.3d) and IL-12p70 (not shown) from mouse macrophages. Furthermore, the effect of SslE on the expression of a major anti-inflammatory cytokine, IL-10 was elucidated. Interestingly, SslE did not stimulate secretion of IL-10 in mouse macrophages (not shown).

Proinflammatory chemokines are also associated with macrophage activation and polarization toward M1 type. SslE was found to stimulate overexpression of mRNAs for M1 chemokines such as RANTES, MIP-1α, and MIP-1β after 6 h of treatment on RAW264.7 cells compared to the untreated control (Fig. 19.3e). In contrary, no upregulation in the mRNA expression of M2 chemokines such as MDC, TARC, and eotaxin-2 could be found on treatment with SslE (Fig. 19.3e). In addition, the kinetics of the proinflammatory chemokines was also studied with culture supernatant from SslE-treated and SslE-untreated RAW cells. SslE induces rapid kinetics of stimulation for RANTES, being 3.9 ± 1.25 fold higher as early as 2 h of posttreatment than the control at 0 h of time point (304 ± 23.4 versus 77 ± 18.7 pg/ml; $p < 0.0001$) (Fig. 19.3f) and was optimum at 6 h, being 9 ± 2 fold higher than the control at 0 h (720 ± 33 versus 77 ± 18.7 pg/ml; $p = 0.0001$). The amount of secreted RANTES decreased only by 1.3-fold and 1.6-fold at 10 h and 24 h of post-induction, respectively, implicating a sustained release over the time period studied. Unlike RANTES, MIP-1α started secreting at 4 h of post-induction and reached a maximum level at 8 h, being 15.3 ± 2.2 fold higher than the control at 0 h (551 ± 30 versus 36 ± 13.6 pg/ml; $p < 0.0001$) (Fig. 19.3f). SslE induced similar kinetics for MIP-1β that also showed optimum secretion at 8 h of posttreatment, 14.7 ± 3.4 fold higher than control at 0 h (500 ± 21 versus 34 ± 6.2 pg/ml; $p < 0.0001$), and secretion of both these chemokines declined little and was steady till 24 h of detection (Fig. 19.3f).

19.8.2 Production of Reactive Oxygen and Nitrogen Species by SslE in Mouse Macrophages

The level of cellular ROS induced by SslE was measured by 2′,7′-dichlorofluorescein diacetate (DCFDA) staining. The nonpolar fluorogenic dye, DCFDA diffuses into cells and gets converted to a nonfluorescent derivative DCFH by cellular esterases. DCFH is oxidized by cellular ROS to a highly fluorescent compound 2′,7′-dichlorofluorescin (DCF). SslE (150 nM for 2 h) was found to increase cellular ROS level as detected by increased fluorescence of DCF compared to the untreated RAW cells viewed under a fluorescence microscope (Olympus–IX71) at a resolution of 40X (Fig. 19.4a).

SslE was also found to trigger the production of NO, as measured by the levels of accumulated nitrite, a stable oxidized end product of NO by Griess reaction. RAW 264.7 cells were incubated with different doses of SslE, and culture supernatant was collected 12 h of posttreatment, and the amount of nitrite was measured by Griess reaction (Fig. 19.4b). SslE exhibited a dose-dependent increase in the production of NO. The production of NO was optimum (55 ± 3 μM) with 500 nM of SslE and SslE at a dose of 62.5 nM was still able to trigger the production of 19 ± 2 μM of NO (Fig. 19.4b).

19.8.3 SslE Stimulated Overexpression of *iNOS*

In addition to ROS and NO, SslE was able to trigger the expression of the enzyme for NO synthesis in macrophages, the inducible NO synthase, *iNOS*. This further substantiates the ability of SslE to stimulate proinflammation. RAW 264.7 cells

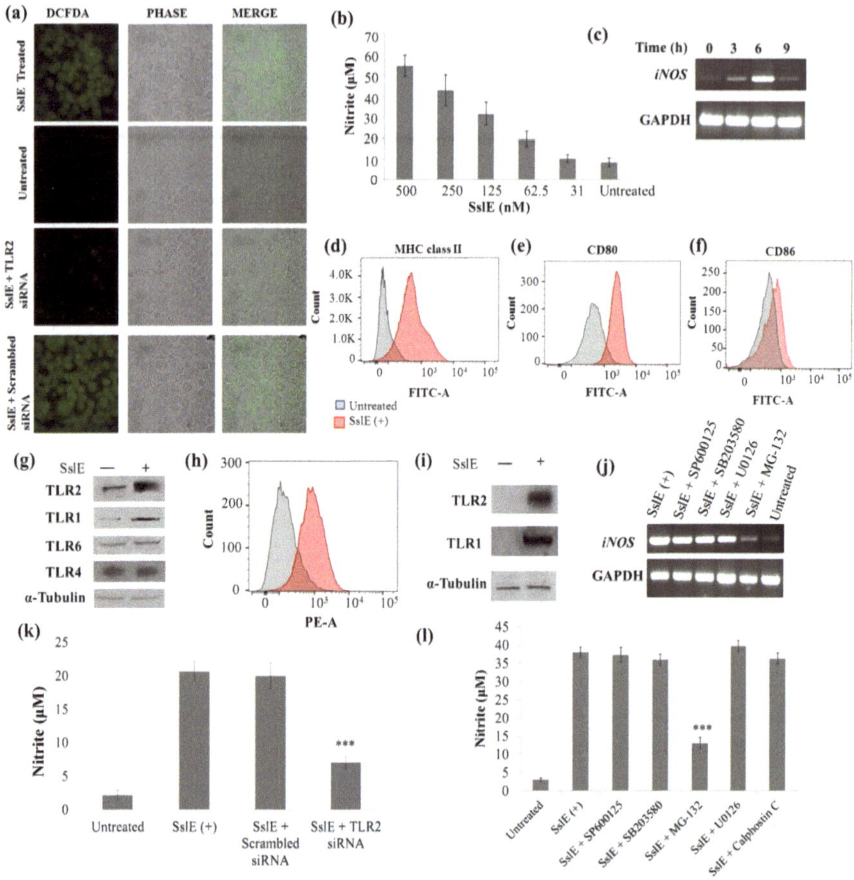

Fig. 19.4 SslE stimulates production of reactive oxygen and nitrogen species in a TLR2-dependent manner involving NFκB and overexpression of MHC II and other co-stimulatory molecules in RAW264.7 cells. To detect production of ROS, RAW264.7 cells were incubated with SslE (150 nM) for 2 h and stained with DCFDA (2′,7′-dichlorofluorescein acetate) for 20 min at 37 °C and viewed under a fluorescence microscope at 40 X resolution (**a**). Griess assay was done with culture supernatant of RAW264.7 cells treated with different concentrations of SslE for 8 h to quantify the amount of accumulated nitrite, the stable end product of NO (**b**). RT-PCR was done to determine the expression of *iNOS* on induction with SslE in RAW264.7 cells (**c**). To check the increased expression of MHC II, CD80, and CD86, SslE-treated (red) (for 6 h) or SslE-untreated (gray) RAW264.7 cells were stained with FITC-conjugated anti-MHC II, CD80, and CD86 antibodies for 20 min in dark at RT and analyzed by flow cytometry (**d**–**f**). Immunoblots were done to check overexpression of different TLRs with α-Tubulin as loading control in SslE-treated (for 6 h) RAW264.7 cells pre-incubated with PMB for 1 h (**g**). Flow cytometry was performed with cells stained with PE-conjugated anti-TLR2 antibody for 20 min at RT in dark (**h**). For immunoprecipitation, lysates of SslE-treated or SslE-untreated RAW264.7 were incubated with 2 μg of mouse anti-SslE antibody overnight at 4 °C. After immobilization of the immune complexes on Protein A/G plus Agarose beads, the proteins were detected by immunoblotting using anti-TLR2 and anti-TLR1 antibodies (**i**). To show involvement of signaling molecules in SslE-induced *iNOS* expression, RT-PCR was done with mRNA from RAW264.7 cells treated with JNK1/2 inhibitor SP600125 (10 μM), p38 inhibitor SB203580 (10 μM), ERK1/2 inhibitor U0126 (10 μM), and NFκB inhibitor MG-132 (2.5 μM) 1 h prior treatment with SslE (**j**). Griess assay was performed with supernatant from SslE-treated TLR2 siRNA transfected, scrambled siRNA-transfected and siRNA-untransfected cells (**k**) and RAW264.7 cells treated with SP600125, SB203580, U0126, and MG-132 (**l**). Statistical significance was calculated using the Student's t-test ($*P < 0.05$; $**P < 0.01$; $***P < 0.001$)

were treated with 150 nM of SslE for different time intervals, and RT-PCR was done to check the *iNOS* mRNA levels which were found to be upregulated in a time-dependent manner on treatment with SslE, being optimally expressed at 6 h of post-induction (Fig. 19.4c).

19.8.4 Overexpression of MHC II and Other Co-stimulatory Molecules on Macrophages by SslE

Overexpression of MHC II and other co-stimulatory molecules on macrophages signifies macrophage activation and M1 polarization. Flow cytometry revealed that SslE could induce 13-fold increased expression of MHC class II than untreated control (13 ± 2, $p < 0.0001$) as evidenced from relative upregulation in fluorescence intensity in SslE-treated cells (Fig. 19.4d). The other co-stimulatory molecules such as CD80 and CD86 exhibited a fivefold (5 ± 0.3, $p < 0.001$) (Fig. 19.4e) and twofold (2.4 ± 0.12, $p = 0.05$) (Fig. 19.4f) higher expression, respectively, on SslE-treated macrophages than untreated cells.

19.9 Signaling Pathway for SslE-Induced Proinflammation and Macrophage Activation

19.9.1 SslE Induces Overexpression of TLR2

It is a known fact that lipoproteins are potent agonist for TLR2. As SslE is a lipoprotein, the upregulation of TLR2 was determined on treatment with SslE on macrophages. Immunoblots indicated an increased expression of TLR2 on SslE-treated mouse macrophages for 6 h compared to the untreated control (Fig. 19.4g). Flow cytometry also showed a fivefold (5 ± 0.2, $p < 0.001$) greater expression of TLR2 on macrophages stimulated with SslE (Fig. 19.4h). It is known that TLR2 can't signal independently; it needs the co-expression of either TLR1 or TLR6. SslE was found to induce the higher expression of TLR1 and not TLR6 on RAW264.7 cells (Fig. 19.4g). Expression of TLR4 was also checked as TLR4 is a known ligand for LPS (Fig. 19.4g). Hence, no upregulation in TLR4 induction substantially proves that SslE is LPS free and the TLR2 upregulation is specifically due to SslE and not any contaminating LPS if present. The receptor specificity of SslE was further evaluated by a pull down assay in which immobilized anti-SslE antibody particularly pulled down TLR2 and TLR1 (Fig. 19.4i) but not TLR4 and TLR6 (not shown).

19.9.2 Involvement of TLR2 in SslE-Stimulated Production of ROS and NO

Culture supernatant after 12 h of posttreatment with SslE from TLR2 knocked down RAW264.7 cells showed significantly reduced level of nitrite accumulation as determined by Griess assay, whereas supernatant from SslE-treated scrambled siRNA-transfected cells showed similar level of nitrite production like in SslE-treated untransfected cells (Fig. 19.4k). This establishes that SslE-induced NO production in mouse macrophages is TLR2 mediated. Likewise, the production of ROS in RAW264.7 cells transfected with TLR2 siRNA was found to be remarkably decreased in the presence of SslE as indicated by less fluorescence of DCF in confocal microscopy as compared to the intense fluorescence in SslE-stimulated untransfected cells and cells transfected with scrambled siRNA, indicating that TLR2 is essential in ROS production in response to SslE (Fig. 19.4a).

19.9.3 SslE Induces NO Production in Mouse Macrophages Via NFκB

NFκB is not only the primary transcription factor involved in proinflammation but also is the major factor activated in TLR signaling. NFκB is also known to be involved in NO production. NFκB functions as a dimer formed by the interactions of two (p50 and p65) of the five Rel family proteins and is generally sequestered in cytosol by IκB in an unstimulated cell. In presence of stimulus, IκB gets phosphorylated, degraded, making NFκB dimer free to enter the nucleus. Different inhibitors against NFκB and each MAP kinase signaling molecule elucidated the signaling pathway involved in SslE-induced NO production. RAW264.7 cells on incubation with 10 μM U0126 (ERK1/2 inhibitor), 10 μM SP600125 (JNK1/2 inhibitor), 10 μM SB203580 (p38 inhibitor), and 2.5 μM MG-132 (inhibitor for NFκB) for 1 h prior to SslE treatment showed unaltered expression in *iNOS* mRNA in presence of U0126, SP600125, and SB203580 (Fig. 19.4j). In contrary, *iNOS* exhibited reduced expression in RT-PCR analysis of mRNAs isolated from cells incubated with MG-132 (Fig. 19.4j) which clearly implies the involvement of NFκB in SslE-mediated *iNOS* expression. Similarly, production of NO was unaltered in presence of p38, JNK1/2, and ERK1/2 inhibitors, while use of MG-132 caused 3.5 ± 0.75 fold (38 ± 1.5 vs 13 ± 2 μM, $p < 0.001$) reduced production of NO in mouse macrophages (Fig. 19.4l).

19.9.4 SslE-Induced Production of Proinflammatory Cytokines Is TLR2 Mediated Involving Both MAP Kinase and NFκB

The involvement of TLR2 in SslE-mediated cytokine secretion was further evident with the use of monoclonal antibody (mAb) against TLR2 and knocking down TLR2 with TLR2 siRNA in RAW264.7 cells (Fig. 19.5a). Secretion of all the cytokines was drastically abrogated in presence of mAb against TLR2. The secretion of IL-1α was reduced to 86%, IL-1β 82%, and TNF-α 87% (Fig. 19.5a). Additionally, secretion of the cytokines was also remarkably obstructed in TLR2 knocked down RAW264.7 cells with 66% inhibition in the secretion of IL-1α, 64.5% in case of IL-1β, and 63% inhibition in secretion of TNF-α (Fig. 19.5a). Furthermore, use of

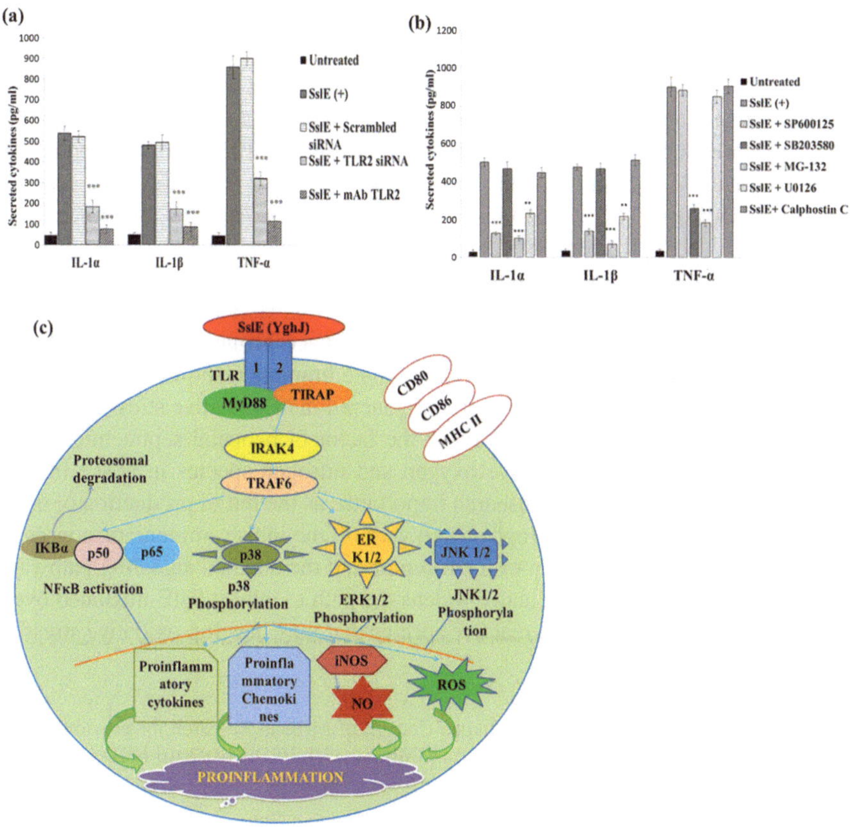

Fig. 19.5 SslE-induced cytokine secretion involves NFκB and MAP kinase signaling and is TLR2 dependent. (**a**) To confirm involvement of TLR2, RAW 264.7 cells were transfected with TLR2 siRNA, a control siRNA for 24 h prior to SslE treatment and also pre-incubated with anti-TLR2 mAb 1 h before treatment with SslE. After 8 h of treatment ELISA was done with the culture supernatants to measure secretion of TNF-α, IL-1α, and IL-1β. (**b**) To determine the signaling pathway, cells were treated with JNK1/2 inhibitor SP600125 (10 μM), p38 inhibitor SB203580 (10 μM), ERK1/2 inhibitor U0126 (10 μM), NFκB inhibitor MG-132 (2.5 μM), and PKC-α inhibitor Calphostin C (3 μM) for 1 h before treatment with SslE, and ELISA was done for various cytokines. (**c**) Schematic representation of SslE-induced proinflammation and macrophage activation shows SslE binds to TLR1/2 heterodimer, recruiting downstream adaptors MyD88, TIRAP, and TRAF6. This leads to the activation of p38, ERK1/2, and JNK1/2 as well as NFκB for the secretion of various proinflammatory cytokines such as TNF-α, IL-1α, and IL-1β; different M1 chemokines such as MIP-1α, MIP-1β, and RANTES; and other important proinflammatory hallmarks such as cellular ROS and NO. All these together with the increased expression of MHC II, CD80, and CD86 finally result in activation and M1 polarization of mouse macrophages

different inhibitors confirms the cytokine-specific involvement of NFκB and different MAP kinases such as ERK1/2, JNK1/2, and p38 (Fig. 19.5b). Hence, all these show that SslE-induced proinflammation is TLR2 mediated with activation of different MAP kinases and NFκB.

19.10 Conclusion

SslE, the cell associated and secreted lipoprotein having a M60 metalloprotease domain of neonatal septicemic *E. coli*, induces secretion of different proinflammatory hallmarks such as proinflammatory cytokines such as TNF-α, IL-1α, and IL-1β; proinflammatory chemokines like MIP-1α, MIP-1β, and RANTES; and reactive oxygen and nitrogen species in mouse macrophages accompanied with the increased expression of MHC II and other co-stimulatory molecules like CD80 and CD86 on macrophages. Furthermore, production of ROS and RNS was found to be mediated via TLR2 and through activation of NFκB and MAP kinases (Fig. 19.5c). All these actually suggest that SslE activates and polarizes macrophages toward the M1 type and hence plays a pivotal role in initiating innate immune response of host [85]. It is of note that there are multiple reports of generation of oxidative stress in septicemic patients. Therefore, identification of the factor of neonatal septicemic *E. coli* responsible for producing reactive oxygen and nitrogen species in host's macrophages should be of immense research importance as the initial production of these species actually activates macrophages to gear up the innate immune responses of host. In addition, identification of the receptor and the detailed signaling pathways of ROS and NO generation actually widens the path to inhibit SslE-mediated overwhelming production of these reactive species that may develop oxidative stress in host.

Acknowledgments RT acknowledges Indian Council of Medical Research for providing post-doctoral fellowship (ICMR PDF Fellowship) {ICMR No. [3/1/3/PDF(14)/2016-HRD]}.

References

1. Paolucci M, Landini MP, Sambri V (2012) How can the microbiologist help in diagnosing neonatal sepsis? Int J Pediatr:Article ID 120139
2. Edwards MS, Baker CJ (2004) Sepsis in the newborn. In: Gershon AA, Hotez PJ, Katz SL (eds) Krugman's infectious diseases of children, 11th edn. Mosby, Philadelphia, pp 545–561
3. Stoll BJ, Hansen NI, Sánchez PJ, Faix RG, Poindexter BB, Van Meurs KP et al (2011) Early onset neonatal sepsis: the burden of group B streptococcal and *E. coli* disease continues. Pediatrics 127(5):817–826
4. Viswanathan R, Singh AK, Basu S, Chatterjee S, Sardar S, Isaacs D (2012) Multi-drug resistant gram negative bacilli causing early neonatal sepsis in India. Arch Dis Child Fetal Neonatal Ed 97(3):F182–F187
5. Lin CY, Hsu CH, Huang FY, Chang JH, Hung HY, Kao HA et al (2011) The changing face of early-onset neonatal sepsis after the implementation of a maternal group B *Streptococcus* screening and intrapartum prophylaxis policy-a study in one medical center. Pediatr Neonatol 52(2):78–84
6. Olesen B, Hansen DS, Nilsson F, Frimodt-Møller J, Leihof RF, Carsten Struve C, Scheutz F, Johnston B, Krogfelt KA, Johnson JR (2013) Prevalence and characteristics of the epidemic multiresistant *Escherichia coli* ST131 clonal group among extended-Spectrum Beta-lactamase-producing *E. coli* isolates in Copenhagen, Denmark. J Clin Microbiol 51:1779–1785
7. Mathers AJ, Peirano G, Pitout JD (2015) *Escherichia coli* ST131: the quintessential example of an international multiresistant high-risk clone. Adv Appl Microbiol 90:109–154

8. Luo Q, Kumar P, Vickers TJ, Sheikh A, Lewis WG, Rasko DA, Sistrunk J, Fleckenstein JM (2014) Enterotoxigenic *Escherichia coli* secretes a highly conserved mucin-degrading metalloprotease to effectively engage intestinal epithelial cells. Infect Immun 82:509–521
9. Valri M, Rossi Paccani S, Kassendra M, Nesta B, Serino L, Pizza M, Soriani M (2015) Pathogenic *E. coli* exploits SslE mucinase activity to translocate through the mucosal barrier and get access to host cells. PloS One 10:e0117486. https://doi.org/10.1371/journal.pone.0117486
10. Baldi DL, Higginson EE, Hocking DM, Praszkier J, Cavaliere R, James CE, Bennett-Wood V, Azzopardi KI, Turnbull L, Lithgow T, Robins-Browne RM, Whitchurch CB, Tauscheka M (2012) The type II secretion system and its ubiquitous lipoprotein substrate, SslE, are required for biofilm formation and virulence of Enteropathogenic *Escherichia coli*. Infect Immun 80:2042e2052
11. Tapader R, Bose D, Basu P, Mondal M, Mondal A, Chatterjee NS, Dutta P, Basu S, Bhadra RK, Pal A (2016) Role in proinflammatory response of SslE, a secreted metalloprotease from neonatal septicemic *Escherichia coli*. Int J Med Microbiol 306:554–565
12. Moriel DG, Bertoldi I, Spagnuolo A, Marchi S, Rosini R, Nesta B, Pastorello I, Corea VA, Torricelli G, Cartocci E, Savino S, Scarselli M, Dobrindt U, Hacker J, Tettelin H, Tallon LJ, Sullivan S, Wieler LH, Ewers C, Pickard D, Dougan G, Fontana MR, Rappuoli R, Pizza M, Serino L (2010) Identification of protective and broadly conserved vaccine antigens from the genome of extraintestinal pathogenic *Escherichia coli*. Proc Natl Acad Sci 107:9072–9077
13. Moriel DG, Rosini R, Seib KL, Serino L, Pizza M, Rappuoli R (2012) *Escherichia coli*: great diversity around a common core. MBio 3:e00118–e00112. https://doi.org/10.1128/mBio.00118-12
14. Levy O (2007) Innate immunity of the newborn: basic mechanisms and clinical correlates. Nat Rev Immunol 7:379–390
15. Wynn JL, Levy O (2010) Role of innate host defenses in susceptibility to early-onset neonatal sepsis. Clin Perinatol 37:307–337
16. West AP, Koblansky AA, Ghosh S (2006) Recognition and signaling by toll-like receptors. Annu Rev Cell Dev Biol 22:409–437
17. Beutler BA (2009) TLRs and innate immunity. Blood 113:1399–1407
18. Kantari C, Pederzoli-Ribeil M, Witko-Sarsat V (2008) The role of neutrophils and monocytes in innate immunity. Trends Innate Immun 15:118–146
19. Twigg HL III (2004) Macrophages in innate and acquired immunity. Semin Respir Crit Care Med 25:21–31
20. Kopp E, Medzhitov R (2003) Recognition of microbial infection by toll-like receptors. Curr Opin Immunol 15:396–401
21. Symons A, Beinke S, Ley SC (2006) MAP kinase kinase kinases and innate immunity. Trends Immunol 27:40–48
22. Hor-Yue Tan, Wang N, Li S, Hong M, Wang X, Feng Y (2016) The reactive oxygen species in macrophage polarization: Reflecting its dual role in progression and treatment of human diseases. Oxid Med Cell Longev 2016:Article ID 2795090, 16 pages
23. Turner MD, Nedjai B, Hurst T, Pennington DJ (2014) Cytokines and chemokines: at the crossroads of cell signalling and inflammatory disease. Biochem Biophys Acta 1843:2563–2582
24. WHO Sepsis Technical Expert Meeting (2018) World Health Organization, Geneva
25. Child mortality and causes of death (2018) Global health observatory data. WHO, Geneva
26. Stoll BJ, Hansen NI, Bell EF, Shankaran S, Laptook AR, Walsh MC et al (2010) Neonatal outcomes of extremely preterm infants from the NICHD neonatal research network. Pediatrics 126(3):443–456
27. Cohen-Wolkowiez M, Moran C, Benjamin DK, Cotten CM, Clark RH, Benjamin DK Jr et al (2009) Early and late onset sepsis in late preterm infants. Pediatr Infect Dis J 28(12):1052–1056
28. Barton L, Hodgman JE, Pavlova Z (1999) Causes of death in the extremely low birth weight infant. Pediatrics 103(2):446–451

29. Hornik CP, Fort P, Clark RH, Watt K, Benjamin DK Jr, Smith PB, Manzoni P, Jacqz-Aigrain E, Kaguelidou F, Cohen-Wolkowiez M (2012) Early and late onset sepsis in very-low-birth-weight infants from a large group of neonatal intensive care units. Early Hum Dev 88:S69–S74
30. Liu L, Oza S, Hoga D, Peri J, Rudan I, Lawn JE, Cousens S, Mathers C, Black RE (2015) Global, regional, and national causes of child mortality in 2000–13, with projections to inform post-2015 priorities: an updated systematic analysis. Lancet 385:430–440
31. Black RE, Cousens S, Johnson HL (2010) Global, regional, and national causes of child mortality in 2008: a systematic analysis. Lancet 375(9730):1969–1987
32. Shehab El-Din EMR, Adel El-Sokkary MM, Bassiouny MR, Hassan R (2015) Epidemiology of neonatal sepsis and implicated pathogens: a study from Egypt. BioMed Res Int:Article ID 509484, 11 pages
33. Lahariya C, Paul VK (2010) Burden, differentials, and causes of child deaths in India. Indian J Pediatr 77(11):1312–1321
34. Bizzarro MJ, Raskind C, Baltimore RS, Gallagher PG (2005) Seventy-five years of neonatal sepsis at Yale:1928–2003. Pediatrics 116(3):595–602
35. Guilbert J, Levy C, Cohen R, Bacterial Meningitis Group, Delacourt C, Renolleau S, Flamant C (2010) Late and ultra late onset Streptococcus B meningitis: clinical and bacteriological data over 6 years in France. Acta Paediatr 99:47–51
36. Shah BA, Padbury JF (2014) Neonatal sepsis. Virulence 5(1):170–178
37. Shah SS, Ehrenkranz RA, Gallagher PG (1999) Increasing incidence of Gram-negative rod bacteremia in a newborn intensive care unit. Pediatr Infect Dis J 18(7):591–595
38. Bhat YR, Lewis LE, Vandana KE (2011) Bacterial isolates of early-onset neonatal sepsis and their antibiotic susceptibility pattern between 1998 and 2004: an audit from a center in India. Ital J Pediatr 37:32
39. Joshi SG, Ghole VS, Niphadkar KB (2000) Neonatal gram-negative bacteremia. Indian J Pediatr 67(1):27–32
40. Macharashvili N, Kourbatova E, Butsashvili M, Tsertsvadze T, McNutt LA, Leonard MK (2009) Etiology of neonatal bloodstream infections in Tbilisi, republic of Georgia. Int J Infect Dis 13(4):499–505
41. Multu M, Aslan Y, Saygin B, YilmazG BG, Koksal I (2011) Neonatal Sepsis caused by Gram-negative Bacteria in a neonatal intensive care unit: a six years analysis. HK J Paediatr 16:253–257
42. Roy I, Jain A, Kumar M, Agarwal SK (2002) Bacteriology of neonatal septicemia in a tertiary care Hospital of Northern India. Indian J Med Microbiol 20(3):156–159
43. Kaistha N, Mehta M, Singla N, Garg R, Chander J (2010) Neonatal septicemia isolates and resistance patterns in a tertiary care hospital of North India. J Infect Dev Ctries 4(1):55–57
44. Weston EJ, Pondo T, Lewis MM, Martell-Cleary P, Morin C, Jewell B et al (2011) The burden of invasive early-onset neonatal sepsis in the United States, 2005-2008. Pediatr Infect Dis J 30(11):937–941
45. Durackova Z (2009) Some current insights into oxidative stress. Physiol Res 59(4):459–470
46. Gutteridge JMC (1995) Lipid peroxidation and antioxidants as biomarkers of tissue damage. Clin Chem 41:1819–1828
47. Gutteridge JMC, Mitchell J (1999) Redox imbalance in the critically ill. Br Med Bull 55:49–75
48. Macdonald J, Galley HF, Webster NR (2003) Oxidative stress and gene expression in sepsis. Br J Anaesth 90(2):221–232
49. Poyton RO, Ball KA, Castello PR (2009) Mitochondrial generation of free radicals and hypoxic signaling. Trends Endocrinol Metab 20:332–340
50. Galley HF (2011) Oxidative stress and mitochondrial dysfunction in sepsis. Br J Anaesth 107(1):57–64
51. Webster NR, Nunn JF (1988) Molecular structure of free radicals and their importance in biological reactions. Br J Anaesth 60:98–108
52. Lau AT, Wang Y, Chiu JF (2008) Reactive oxygen species: current knowledge and applications in cancer research and therapeutic. J Cell Biochem 104(2):657–667
53. Hussain SP, Hofseth LJ, Harris CC (2003) Radical causes of cancer. Nat Rev Cancer 3:276–285

54. Coussens LM, Werb Z (2002) Inflammation and cancer. Nature 420:860–867
55. Reuter S, Gupta SC, Chaturvedi MM, Aggarwal BB (2010) Oxidative stress, inflammation, and cancer: how are they linked? Free Radic Biol Med 49(11):1603–1616
56. Zimmerman JJ (1995) Defining the role of oxyradicals in the pathogenesis of sepsis. Crit Care Med 23:616–617
57. Bogdan C, Rollinghoff M, Diefenbach A (2000) The role of nitric oxide in innate immunity. Immunol Rev 173:17–26
58. MacMicking J, Xie Q-W, Nathan C (1997) Nitric oxide and macrophage function. Annu Rev Immunol 15:323–350
59. Bogdan C (2000) The function of nitric oxide in the immune system. Handbook of experimental pharmacology. In: Mayer B (ed) Volume: nitric oxide. Springer, Heidelberg, pp 443–492
60. Forstermann U, Kleinert H (1995) Nitric oxide synthase: expression and expressional control of the three isoforms. Naunyn Schmiede bergs Arch Pharmacol 352:351–364
61. Marletta MA (1993) Nitric oxide synthase: structure and mechanism. J Biol Chem 268:12231–12234
62. Moncada S, Higgs A (1993) The L-arginine-nitric oxide pathway. N Engl J Med 329:2002–2012
63. Lowenstein CJ, Padalko E (2004) iNOS (NOS2) at a glance. J Cell Sci 117:2865–2867
64. Green SJ, Scheller LF, Marletta MA, Seguin MC, Klotz FW, Slayter M, Nelson BJ, Nacy CA (1994) Nitric oxide: cytokine regulation of nitric oxide in host resistance to intracellular pathogens. Immunol Lett 43:87–94
65. Liew FY, Wei XQ, Proudfoot L (1997) Cytokines and nitric oxide as effector molecules against parasitic infections. Philos Trans R Soc Lond B 352:1311–1315
66. Bogdan C, Rollinghoff M, Diefenbach A (2000) Reactive oxygen and reactive nitrogen intermediates in innate and specific immunity. Curr Opin Immunol 12:64–76
67. Galley HF, Webster NR (1996) Elevated serum bleomycin-detectable iron concentrations in patients with sepsis syndrome. Intensive Care Med 22:226–229
68. Galley HF, Davies MJ, Webster NR (1996) Ascorbyl radical formation in patients with sepsis: effect of ascorbate loading. Free Radic Biol Med 20:139–143
69. Galley HF, Howdle PD, Walker BE, Webster NR (1997) The effects of intravenous antioxidants in patients with septic shock. Free Radic Biol Med 23:768–774
70. Galley HF, Davies MJ, Webster NR (1996) Xanthine oxidase activity and free radical generation in patients with sepsis syndrome. Crit Care Med 24:1649–1653
71. Batra S, Kumar R, Seema, Kapoor AK, Ray G (2000) Alterations in antioxidant status during neonatal sepsis. Ann Trop Paediatr 20:27–33
72. Cowley HC, Bacon PJ, Goode HF, Webster NR, Jones JG, Menon DK (1996) Plasma antioxidant potential in severe sepsis: a comparison of survivors and non survivors. Crit Care Med 24:1179–1183
73. Hoebe K, Janssen E, Beutler B (2004) The interface between innate and adaptive immunity. Nat Immunol 5:971–974
74. Mantovani A, Sica A, Sozzani S, Allavena P, Vecchi A, Locati M (2004) The chemokine system in diverse forms of macrophage activation and polarization. Trends Immunol 25:677–686
75. Benoit M, Desnues B, Mege JL (2008) Macrophage polarization in bacterial infection. J Immunol 181:3733–3739
76. Zlotnik A, Yoshie O (2000) Chemokines: a new classification system and their role in immunity. Immunity 12:121–127
77. Nesta B, Valeri M, Spagnuolo A, Rosini R, Mora M, Donato P, Alteri CJ, Vecchio MD, Buccato S, Pezzicoli A, Bertoldi I, Buzzigoli L, Tuscano G, Falduto M, Rippa V, Ashhab Y, Bensi G, Fontana MR, Seib KL, HLT M, Pizza M, Soriani M, Serino L (2014) SslE elicits functional antibodies that impair *in vitro* mucinase activity and *in vivo* colonization by both intestinal and extraintestinal *Escherichia coli* strains. PLoS Pathog 10:e1004124. https://doi.org/10.1371/journal.ppat.1004124
78. Korotkov KV, Sandkvist M, Hol WG (2012) The type II secretion system: biogenesis, molecular, architrcture and mechanism. Nat Rev Microbiol 10:336–351

79. Strozen TG, Li G, Howard SP (2012) YghG (GspSbeta) is a novel pilot protein required for localization of the GspS beta type II secretion system secret in of enterotoxicogenic *Escherichia coli*. Infect Immun 80:2608–2622
80. Iguchi A, Thomson NR, Ogura Y, Saunders D, Ooka T, Henderson IR, Harris D, Asadulghani M, Kurokawa K, Dean P, Kenny B, Quail MA, Thurston S, Dougan G, Hayashi T, Parkhill J, Frankel G (2009) Complete genome sequence and comparative genome analysis of entero-pathogenic *Escherichia coli* O127:H6 strain E2348/69. J Bacteriol 191:347–354
81. Yang J, Baldi DL, Tauschek M, Stugnell RA, Robins-Browne RM (2007) Transcriptional regulation of the *yghJ-ppA-yghG-gspCDEFGHIJKLM* cluster, encoding the type II secretion pathway in enterotoxigenic *Escherichia coli*. J Bacteriol 189:142–150
82. Tapader R, Bose D, Pal A (2017) YghJ, the secreted metalloprotease of pathogenic *E. coli* induces hemorrhagic fluid accumulation in mice ileal loop. Microb Pathog 105:96–99
83. Roy K, Bartels S, Qadri F, Fleckenstein JM (2010) Enterotoxigenic *Escherichia coli* elicits immune responses to multiple surface proteins. Infect Immun 78:3027–3035
84. Luo Q, Quadri F, Kansal R, Rasko DA, Sheikh A, Fleckenstein JM (2015) Conservation and immunogenicity of novel antigens in diverse isolates of enterotoxigenic *Escherichia coli*. PloS Neg Trop Dis 9:e0003446. https://doi.org/10.1371/journal.pntd.0003446
85. Tapader R, Bose D, Dutta P, Das S, Pal A (2018) SslE (YghJ), a secreted lipoprotein of neonatal septicemic *E. coli* induces TLR2 mediated activation of macrophages with the involvement of NFκB and MAP kinase signaling. Infect Immun 86(9):e00399–e00318

Oxidative Stress, Pathophysiology, and Immunity in *Brucellosis*

20

Amit Kumar, Anu Rahal, and V. K. Gupta

Abstract

The *Brucella* being obligate intracellular parasite, its survival in host requires the ability to overcome host immune system as well as host oxidative stress mechanism. In general, *Brucellosis* induces rise in host oxidative stress with weakening of the host's antioxidant defense system. The survival of *Brucella* to host oxidative and antioxidant system depends on the presence of superoxide dismutase (SOD) and catalase in *Brucella*. Of these, SOD acts as a major antioxidant enzyme in *Brucella* pathophysiology. *Brucella* possesses two SODs, SodA and SodC, that directly detoxify superoxide radicals. SodA resides in the cytoplasmic compartment, while SodC in the periplasm of *Brucella* strains and exclusively detoxifies superoxide radical within the cellular compartments in which bacteria reside. The endogenously produced superoxide radicals of respiratory metabolism are detoxified by SodA, while exogenously generated during respiratory burst in host phagocytes are typically detoxified by SodC. The catalase neutralizes H_2O_2 produced during the process. The catalase is mainly constrained to the periplasm and provides protection against H_2O_2 produced during the immune response provoked against *Brucella* infection. Regulation of this enzyme along with detoxification of superoxide radicals with SOD is mainly responsible for the adaptation process of *Brucella* and allows its survival under hostile conditions.

A. Kumar (✉)
Department of Immunology & Defence Mechanism, College of Biotechnology, Sardar Vallabhbhai Patel University of Agriculture & Technology, Meerut, Uttar Pradesh, India
e-mail: balyan74@gmail.com

A. Rahal
Goat Health Division, ICAR-Central Institute for Research on Goats (CIRG), Mathura, Uttar Pradesh, India

V. K. Gupta
CADRAD, ICAR-Indian Veterinary Research Institute (IVRI), Bareilly, Uttar Pradesh, India

S. Chakraborti et al. (eds.), *Oxidative Stress in Microbial Diseases*,
https://doi.org/10.1007/978-981-13-8763-0_20

Keywords
Oxidative stress · Pathophysiology · Immunity · *Brucella* · *Brucellosis*

20.1 Introduction

Brucellosis, an important re-emerging infectious disease, is distributed worldwide. It is caused by bacterium *Brucella* and constitutes a very important zoosanitary and economic problem [1, 2]. Due to highly infectious nature and ability to transmit through aerosol, it has been a point of attraction as a potential biological warfare agent [3]. Genus *Brucella* initially included six species *B. abortus*, *B. melitensis*, *B. ovis*, *B. canis*, *B. suis*, and *B. neotomae* [4]. Later on, three new species isolated from marine mammals, namely, *B. microti* (*Microtus arvalis*), *B. pinnipedialis*, and *B. ceti*, were also included in genus *Brucella* [5] (Fig. 20.1). The species included in genus *Brucella* bear 94% similarity. These species show preferences for different host; however, interspecies transmission is also reported. The animal host acting as reservoir is largely responsible for the occurrence of human brucellosis [6]. The highest prevalence of human brucellosis coincides with the high rates of sheep and goats brucellosis, leading to the greatest incidence of infection in humans [7, 8]. *B. melitensis* is

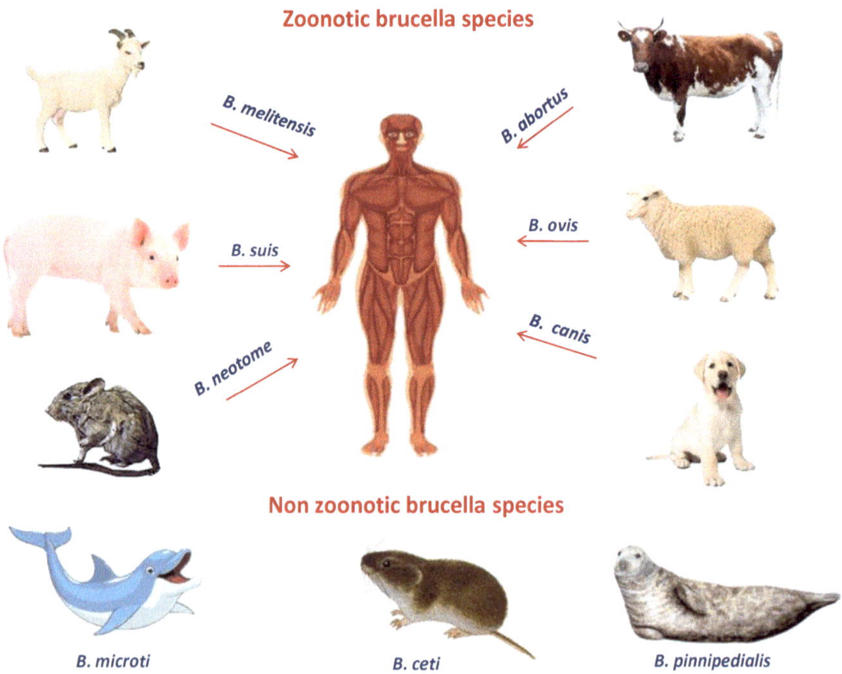

Fig. 20.1 Graphical presentation of the source of zoonotic as well as non-zoonotic *Brucella species*

most common cause of human brucellosis followed by *B. abortus* and *B. suis*. *B. canis* preferably infect dogs with a comparatively low zoonotic threat. *B. neotomae* and *B. ovis* are pathogens of desert rats and sheep and have not been found in human [9]. The main source of human infections is exposure to infected animals as *Brucellosis* is prevalent in many species of animals of economic importance, such as cattle, sheep, goats, and pigs [10]. Other animal species are supposed to be comparatively resistant to the infection, particularly of *B. melitensis*. However, other than sheep and goats, *B. melitensis* has also been reported in dogs [11], cattle [12, 13], camel [14], pigs [15], and wild animals like alpine ibex [16], chamois [17], and farm deer (*Odocoileus virginianus*) in New Zealand [18]. Due to high incidence of *Brucellosis* in developing and third world countries, economic impact, lack of prevention and control strategy, and difficulty in eradication, *Brucellosis* is considered as one of the seven most neglected zoonotic diseases that perpetuate poverty [19].

20.2 Antigenic Composition of *Brucella*

Similar to other bacterial pathogens, many antigenic components of bacterium *Brucella* have been duly characterized [20]. The presence of lipopolysaccharide (LPS) and surface antigens (designated as A, M in smooth strains and R in rough strain) dominates the antibody response. The unique structure of *Brucella* lipopolysaccharide (LPS) provides *Brucella* a very low endotoxicity that not only modulates immune response of host but also confers resistance to antimicrobial drugs. *Brucella* lipopolysaccharide (LPS) also acts as virulence factor and helps in its survival and intracellular replication [21, 22]. *B. abortus* and *B. suis* are predominated by antigen A, whereas *B. melitensis* predominates in antigen M.

The LPS of smooth phase strains (S-LPS) is comprised of lipid A containing two types of aminoglycose and distinctive fatty acids that exclude ß-hydroxymyristic acid. The core region of *Brucella* S-LPS contains glucose, quinovosamine, and mannose with an O chain that is comprised of a homopolymer of nearly 100 residues of 4-formamido-4,6-dideoxymannose (linked predominantly a-1,2 in A epitope-dominant strains with every fifth residue linked a-1,3 in M dominant strains) [23].

20.3 Oxidative Stress in *Brucella* Pathophysiology

Oxidative stress is presently viewed as a combination of physiological molecule as well as a cellular damage and stress-responsive signal. Oxidative stress stands for excessive production of reactive oxygen species (ROS) or inadequate ROS neutralization. Oxidative stress plays a critical role in the progress of clinical conditions and includes damage of DNA, peroxidation of lipid, deactivation of enzymes, and apoptosis and necrosis of cells [24, 25]. In *Brucellosis*, oxidative stress level gets increased with the weakening of body's antioxidant-based defense mechanism [26]. Prior to entry in macrophages, *Brucella* are opsonized inside the mammalian host with brucella-specific IgG, and the entry of *Brucella* into macrophages is mainly due to phagocytosis mediated by Fc receptor. Superoxide dismutase is the major antioxidant enzyme present in *Brucella*. Brucella possesses two superoxide dismutases (SODs), SodA [27] and SodC [28], which directly detoxify superoxide radicals. Depending upon their nature, SodA being a manganese SOD localizes in *Brucella* cytoplasmic compartment, while SodC, a copper/zinc SOD, localizes in the *Brucella* periplasm [29]. Being a charged molecule, superoxide does not easily pass through cellular membranes, and subsequently each SOD generally detoxifies superoxide radical formed intracellularly in the bacterial compartments where they reside [30, 31]. The superoxide molecules generated during respiratory metabolism as a by-product are specifically endogenously detoxified by SodA [32], while superoxide of exogenous origin aroused during the respiratory burst of phagocytes is typically detoxified by SodC [33, 34]. The SodA also decides susceptibility of *Brucella* to acidic pH during the early stages of their interactions with host macrophages when *Brucella* reside in phagosomal compartments [35]. The expression of *virB* genes that type IV secretion system is induced by the acidic pH developed in host macrophages [36]. The type IV secretion system secretes signaling molecules essentially required for proper BCVs trafficking in macrophages [37] and neutralizes pH in BCVs which prevents the intracellular *Brucella* from replicating [35]. Further, the virulence of *Brucella* strains is dependent on anaerobic and/or microaerobic respiration [38, 39]. The attenuation of *Brucella* in macrophages is also governed by the acid-sensitive nature of the SodA mutant. Thus, both lower O_2 availability and the reduced growth rate [40] may thwart endogenous superoxide levels attaining an optimum level at which a cytoplasmic SOD detoxify in *Brucella* strains after their intracellular establishment in the host. After brucellae succeed to infect a host, SodA becomes dispensable (Fig. 20.2).

The H_2O_2 produced during the process is neutralized by catalase. The site of catalase activity is mainly restricted to the periplasm of *Brucella* [41] and provides protection against hydrogen peroxide generated during the immune response provoked against infection. Regulation of catalase enzyme is required for the adaptation process of *Brucella* to survive and maintain under intimidating conditions. Due to their periplasmic location, catalase and Cu-Zn SOD are involved in protecting the *Brucella* from external oxidative compounds [41]. In an in vitro study conducted with neutrophil extracts to assess the worth of oxygen-dependent and oxygen-independent killing of *Brucella*, the former was found to be more effective [42]. Another study revealed protection of *Brucella* in cultured murine peritoneal macrophages by the addition of exogenous catalase [43] indirectly confirming the importance of catalase to *Brucella* survival in the host system. The survival of *Brucella* also depends on cellular density. This phenomenon is well proven and believed to be a consequence of higher rate of diffusion of H_2O_2 [44, 45]. In a report, H_2O_2 stress condition using 10 mM H_2O_2 at the A_{600} of 0.2–0.5 in 2-D gel electrophoresis resulted in more than 90% rate of survival as compared to 20.8% when H_2O_2 stress level was 10 mM H_2O_2 and the A_{600} was 0.01 [46]. Further, to minimize infectious oxidative stress, host antioxidant defense mechanism comes into play, in which purinergic enzymes play an important modulating role [47]. These purinergic enzymes are ATP metabolizing enzymes that modulate the inflammatory cascade

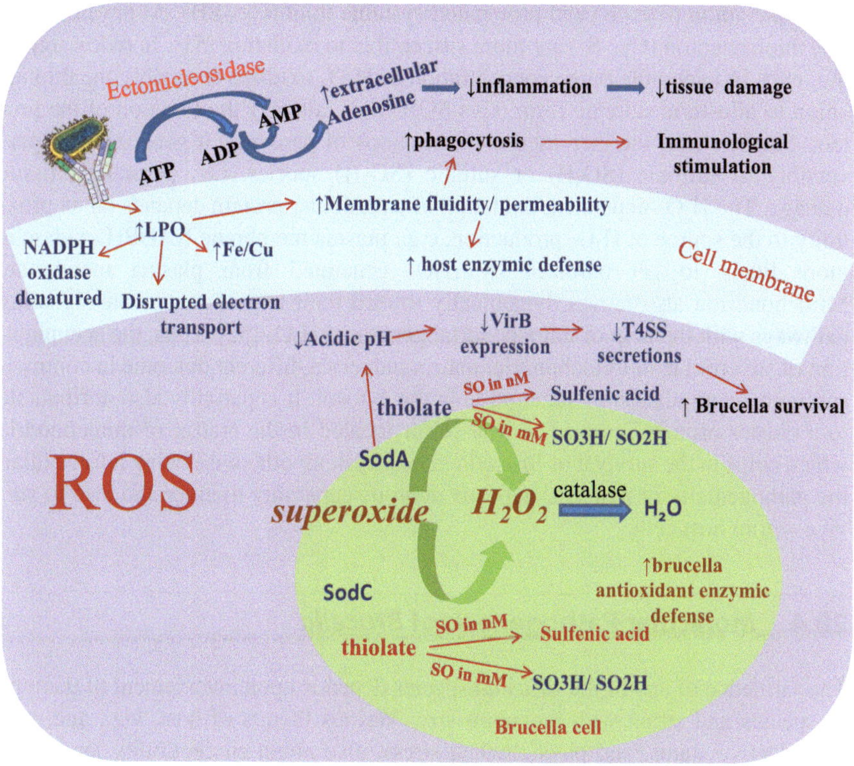

Fig. 20.2 Oxidative stress-based mechanism of *Brucella* survival in host cells

and thus, influence the tissue damage during oxidative stress. The ectonucleotidase activity also results in the breakdown of ATP and ADP to AMP and, finally, hydrolysis of AMP to adenosine, which is an endogenous activator of cellular innate antioxidant mechanism against injury. Moreover, the enzyme, adenosine deaminase, acts as an anti-inflammatory and immunosuppressive signaling molecule, i.e., in contradiction to ATP. Since $CD4^+$ lymphocytes and macrophages mainly reveal the increased activity of ADA, the ADA is also considered as marker of immune response.

The occurrence of oxidative stress or redox signaling-induced damage is determined by the presence of both the type of ROS present in the tissue and its level collectively. The superoxide accumulation in cells can inactivate and damage proteins carrying iron-sulfur clusters [48] suggesting superoxide accumulation is more associated with oxidative stress as compared to redox signaling. However, superoxide does not discriminate between proteins. In addition to this, a set of specific proteins are more sensitive to superoxide and activate signaling pathways and promote adaptation of elevated SODs or, alternatively, may initiate cellular death [49]. Moreover, redox signaling is responsible for hydrogen peroxidase-mediated oxidation of cysteine residues of proteins [50]. Based on pH, cysteine residues remain as a thiolate anion (Cys-S-) and protonated cysteine thiol (Cys-SH). At physiological pH, thiolate anion (Cys-S-) are more susceptible to oxidation [51]. In redox signaling, even in nanomole range concentration of H_2O_2 oxidizes reversibly the thiolate anion to allosteric sulfenic form (Cys-SOH) thus, altering the function of the protein. Further, with the increased concentrations of peroxide, it gets oxidized irreversibly to sulfonic (SO_3H) or sulfinic (SO_2H) species, i.e., permanent tissue damage. The H_2O_2-dependent oxidation of a particular protein depends on its proximity to the source of H_2O_2 production, e.g., plasma membrane NADPH oxidase is more likely to get oxidized by H_2O_2 generated from plasma membrane. Mitochondrion also moves dynamically toward their targets to activate signaling pathways with the help of mitochondrial generated H_2O_2 [52]. Thus, the accumulation of superoxide in mitochondrial matrix induces a different outcome in comparison with accumulation of superoxide in the cytosol. It is partially also attributable to a cluster protein enriched in iron-sulfur located in the matrix of mitochondria which helps in the survival of *Brucella* in host system; otherwise being intracellular, the pathogenicity of *Brucella* depends upon its capability to enter and also to survive within host cells.

20.4 Molecular Pathogenesis of *Brucella*

The virulence of any of the *Brucella* species depends upon involvement of the type of species and strain with inoculum size. Various factors of host, viz., age, sex, reproductive status, and physiological stress, also affect susceptibility leading to typical acute, intermediate, and complete resistance. *Brucella* lack in the common virulence attributes of bacteria like endotoxic lipopolysaccharide (LPS), capsules, exotoxins, cytolysins, plasmids, fimbria, lysogenic phages, and drug-resistant forms [53]. The ability to produce disease involves virulence mechanisms as prerequisite to invade host cell and governs its replication and intracellular survival [21, 54]. The invasion and survival of *Brucella* in host macrophages are the primary steps to establish infection. The main routes of entry for *Brucella* are upper respiratory tract, mainly through oropharynx mucous membranes, conjunctiva, and gastrointestinal mucosa. Other than these, it can also enter through the mucosa of the female or male genital tract. The digestive tract is the main entrance route, and it has been investigated in detail with possible virulence factors involved [55, 56]. *B. melitensis*, *B. abortus*, and *B. suis* carry genes encoded for urease to establish infection [56]. Urease leads to an increase of pH via ammonia production through urea hydrolysis. It is a multi-subunit enzyme responsible for nitrogen metabolism [57]. Two different non-adjacent operons are reported for urease, out of which *ure1* produces urease activity. The inactivation of urease causes strain attenuation when *Brucella* was inoculated through the digestive tract in mice [56]. Thus, urease is required to establish *Brucella* through mucous membranes and to reach macrophages to establish infection, though the role of urease in inhibiting phagosome acidification through ammonia is not established [58].

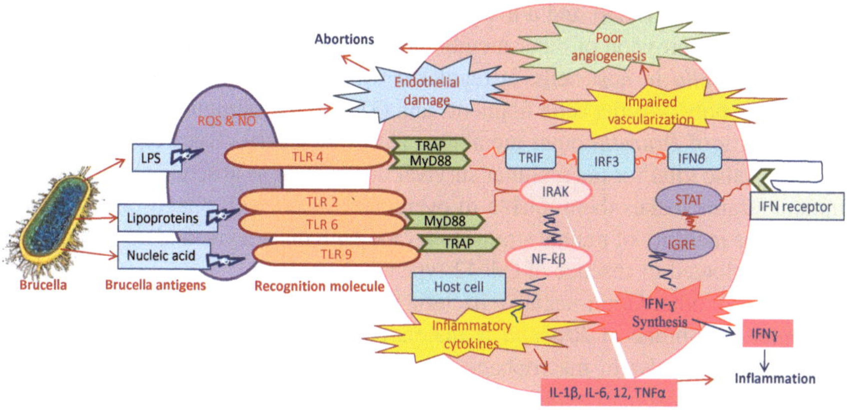

Fig. 20.3 Brucella and host cell interaction

The role of type IV secretion system (T4SS) and lipopolysaccharide (LPS) has been proven to produce gastrointestinal infection; however, reports suggest crossing of mutant strains similar to virulent strain, through M cells of intestinal mucosa [59]. The presence of bile in lower gastrointestinal tract inhibits the growth of non-enteric bacteria. Thus, presence of bile salt hydrolase enzymes is required to establish *Brucella* infection through oral route. The deletion of gene responsible for choloylglycine hydrolase in *B. abortus* is essential for the impairment of *Brucella* growth in the presence of bile salts and leads to attenuation in mice within 10 days post intragastric infection [60] (Fig. 20.3).

Brucella mainly infects reticuloendothelial system. So, after entry into host, it reaches to nearest lymph nodes through the process of phagocytosis mediated by lymphoepithelial cells located in gut-associated lymphoid tissues [61]. However, *Brucella* also elicits humoral immune response, but macrophages and T lymphocytes always play important role in bacterial evacuation. Initially when *Brucella* overcomes the body defenses, it produces bacteremia after 2–4 weeks of infection and can persist for up to 2 months of period leading to humoral immune response and increased number of granulocytes in circulation. Subsequently, *Brucella* reaches submucosa and gets ingested and phagocytosed by polymorphonuclear leukocytes and macrophages, respectively. The polymorphonuclear leukocytes generally fail to kill *Brucella* [54]. The organism is carried by the macrophages to different vital organs like spleen, lymph nodes, joints, kidneys, mammary glands and bone marrow and ultimately, localizes in these organs [62]. In macrophages, the organism inhibits fusion of phagosomes and lysosomes, and that allows it to replicate within compartments containing endoplasmic reticulum components. The whole process is mediated through type IV secretion system. During the process, the bacteria may destroy host cells to infect other cells when remain unchecked by macrophage microbicidal mechanisms. Interestingly, there are reports of replication of *Brucella* extracellularly in host tissues [63, 64]. During the process of internalization, *Brucella* relies on a two-protein-based regulatory system, which includes a regulator protein (BvrS) and a histidine-kinase activity performing sensor protein (BvrS). The expression of outer membrane proteins (Omps), required for the invasion of *Brucella* in host cells, is regulated by BvrS [65]. BvrR-based regulatory system not only governs actin filaments and GTPase recruitment but also maintains the integrity of *Brucella* outer membrane [65]. This BvrR-/BvrS-based regulatory system is also vital for intracellular survival of *Brucella* as absence of the system in mutant strains makes them unable to overcome fusion with phagosome and lysosome. This inability of mutant strains to invade both phagocytic and non-phagocytic cells is attributed to absence of Cdc42-type GTPases [65].

Similar to other Gram-negative bacteria the structural and functional integrity of the outer membrane of *Brucella* is maintained by LPS and contributes to initial survival of *Brucella* in macrophages [22]. A number of mechanisms have been reported in *Brucella* to avoid or suppress bactericidal reactions in macrophages [64]. The O-polysaccharide chain present in *Brucella* plays a critical role in its virulence so based on its presence and absence the phenotypes of *Brucella* spp. are characterized as smooth or rough type, respectively. The mutant or rough strains lacking in O-polysaccharide chain could not survive in macrophage cultures, as well as in mice [22]. Smooth type strains cause bacteria to escape host immune system leading to its intracellular survival by inhibiting the host cell apoptosis. The rough mutants (except *B. ovis* and *B. canis*) are reported to provoke necrosis in phagocytes particularly macrophage [66]. However, rough mutants originated from smooth virulent *Brucella* strains, more efficiently invade host cells. This might be because of natural virulence of some rough polysaccharide chain. Further, it is also imperative to consider the intracellular pathway of replication in smooth strains to understand the involvement of smooth type *Brucella* LPS in the initial process of development of *Brucella*-containing phagosome. The smooth LPS of *Brucella* carries the ability to block phagosome maturation through its interaction with lipid rafts and may contribute to the inhibition of phagolysosome fusion [34]. *Brucella* LPS overcomes host antimicrobial responses by impairing activity of complement, antibacterial peptide and by inhibiting the immune mediator synthesis [22, 67]. The evidences are lacking for the stimulation of tumor necrosis factor-α (TNF-α) or nitric oxide (NO) by *B. melitensis* LPS [68]. Whereas, the involvement of O-polysaccharides (present in LPS of smooth *Brucella*) in inhibition of host cell apoptosis is mediated by a TNF-α/independent mechanism. By inhibition of apoptosis, smooth LPS-containing *Brucella* easily escape host immune surveillance system. This further avoids the activation of immune system and triggering of antigen-presenting cells (APC) through various factors released from dead host cells. Thus, LPS-mediated properties constitute main virulence attributes governing intracellular replication and survival of virulent *Brucella* [64, 66]. The O-side chain of rough strains of *B. abortus* makes them more sensitive to neutralization through normal serum and complement-mediated attack.

Brucella LPS has also been reported to mediate protection against one of major bactericidal peptides mainly cationic in nature like lactoferrin, lysozyme, defensin NP-2, defensin-like antibiotic (Polymyxin B), cecropines, bactenecin-derived peptides and crude lysosomal extracts obtained from PMN cells [21]. Smooth LPS of

Brucella also contribute to resistance against lactoferrins and α-defensins like antimicrobial peptides and complement-mediated opsonization. It also provides resistance against important antibacterial mechanisms of phagocytes and PMN cells along with free radicals, nitric oxide, and lysozyme [22, 64]. The interaction of smooth *B. suis* with lipid rafts on macrophages surface leads to its entry into the cell, avoiding fusion with lysosomes while O-side chain lacking rough strains fuse hastily with lysosomes and thus do not enter via lipid rafts [22]. However, in contrast to rough mutants naturally occurring rough strains like *B. canis* and *B. ovis* that also lack LPS O-side chain in their LPS are pathogenic and cause chronic infections in natural hosts like dogs and rams, respectively and also colonize in spleen of laboratory animals [69, 70].

A number of *Brucella* genes involve in pathogenesis during *Brucellosis* and deletion or insertion of these genes results in attenuation of *Brucella*. To summarize, these include genes governing the expression of enzymes like mutases [phosphoglucomutase (*pgm*); phosphomannomutase (*pmm*)]; synthetases [perosamine synthetase (*rfbE*); cyclic-1,2 glucan synthetase (*cgs*); chorismate synthase (*aroC*); and de novo purine synthetase (*purE*)]; transferases [glycosyltransferase (*wboA*); cytochrome bd oxidase (*cydB*); Cu-Zn SOD (*sod*); protease (*lon*)]; and various proteins playing critical role in resistance mechanism like heat shock protein (*dnaK*); cytoplasmic membrane transport protein (*bacA*), host factor-1, RNA-binding protein (*hfq*); two-component regulatory protein (*bvrRS*); and type IV secretion system regulating proteins (*virB*). The expression of these genes depends upon multiple factors including host and environmental factors. The expression of these genes in host governs the pathogenesis of *Brucella* in host [65, 71].

20.5 Immune Response against *Brucella*

20.5.1 Innate Immunity

The acquired immune response is slow in onset. This period during early stage of infection is covered by nonspecific innate immunity, mediated by clonal selection of specific lymphocytes. During in vivo *Brucella* infection, innate immunity reduces the initial number of infective bacteria without memorization and provides favorable environment for generation of host Th1 immune response. To induce innate immunity, complement system plays an important role. Complement system is a cascade of plasma proteins that interact both with bacterial surfaces and with bound antibodies to directly kill or opsonize the pathogens. In Gram-negative bacteria like *Brucella*, a membrane attack complex mediates opsonization process [72]. LPS being a major surface virulent attribute mediates interaction of *Brucella* with complement components. The deposition of complement and complement-mediated killing of *Brucella* involves both classical and lectin pathways of the complement system [73, 74].

Similar to other bacterial infections, neutrophils interact with *Brucella* in human infection. Neutrophils, the first cells of immune system, lead to initial phagocytosis of opsonized attenuated and virulent *Brucella* strains [75, 76]. However, some of

Brucella may survive neutrophil-based phagocytosis in early infection [42]. The neutrophils being the first cell of host defense mechanism are also supposed to mediate the transportation of survived *Brucella* to lymphoid tissues. Similar to other bacterial pathogens, the survival of *Brucella* in host cells also depends upon its escape from ROI- and RNI-based bactericidal mechanism of macrophage. This ROI- and RNI-based system is activated by pro-inflammatory cytokines like interferon gamma (IFN-γ) and tumor necrosis factor-α (TNF-α). The respective increase and decrease of intracellular *B. abortus* organisms in methylene blue (an electron carrier)- and ROI inhibitor-treated murine macrophages also suggest the role of ROI- and RNI-based bactericidal mechanism in *Brucellosis* [43]. The inability of RNI inhibitor (NG-monomethyl-L-arginine) to stop anti-*Brucella* activities in macrophages suggests greater role of ROIs in vitro *Brucella* killing in comparison with RNIs.

Thus, the survival of *Brucella* overcoming this entire existing host antimicrobial mechanisms also suggests the presence of resistance against antimicrobial molecules like hypo halide, lactoferrin, lysozyme, phospholipase A2, cathelicidin, serprocidins, bacterial permeability increasing factor and defensins [77]. In *B. suis,* process of phagocytosis by mouse macrophages revealed superoxide production suggesting the role of SOD in *Brucella* phagocytosis [65, 78]. Thus, induction of oxidative stress during *Brucella* infection may support or inhibit *Brucella* colonization.

20.5.2 Adaptive Immunity

In *Brucellosis*, innate immunity is followed by adaptive immune response of host immune system. Anti-brucellosis adaptive immune response involves three different but interrelated host mechanisms: (a) IFN-γ-based activation of phagocytosis, (b) cytotoxicity for infected macrophages, and (c) Th1-type antibody isotype-based opsonization and phagocytosis. The $CD4^+$, $CD8^+$, and γδ T cells produce IFN-γ that activates the bactericidal function in macrophages leading to killing of intracellular *Brucella.* The second type of adaptive immunity is based on the cytotoxicity of γδ T and $CD8^+$ cells that kill the *Brucella*-infected macrophages. The third mechanism involves production of IgG2a and IgG3 (Th1-type antibody isotypes) that leads to opsonization of *Brucella* and ultimately facilitates phagocytosis. αβ $CD4^+$ and $CD8^+$ T cells mainly govern the adaptive immunity against *Brucella.* These cells play major role in secretion of IFN-γ, cytotoxic T-lymphocyte activity and IgG2a and IgG3 isotype switching [54, 61]. The role of $CD4^+$ and/or $CD8^+$ T cells in *Brucella* adaptive immunity has been contentious. An experiment conducted over immunized BALB/c mice revealed presence of all types $CD4^+$, $CD8^+$, and whole T-cell populations revealed lower splenic *Brucella* count (CFU) as compared to untreated mice after infection [79], suggesting the role of both T-cell populations in *Brucella* adaptive immunity. Moreover, $CD8^+$ T cells play critical role in controlling brucellosis as they produce Th-1-type cytokines. The Th-1-type cytokines like IFN-γ and IL-2 play important role in *Brucella* control in host cells [80].

Various experiments, conducted in murine models involving passive transfer of serum, suggested significant role of humoral immune response in brucellosis. The sera carrying antibodies against *Brucella* LPS provided passive protection to mice challenged with virulent culture of *B. abortus* [79, 81]. Similarly, passive immunization with OPS-specific monoclonal antibody (IgG2a) produced against *B. abortus* reduced splenic bacterial count and protected mice from challenge of virulent *B. abortus* [81]. Out of different antibody isotypes, IgG2a and IgG3 are predominantly detected in the infected mice and the natural host. This suggests presence/involvement of Th1-based immune response during *Brucella* infection [82]. Opsonization plays principal role in the antibody-mediated protection against *Brucella* infection. The increased intracellular killing of *Brucella* is coupled with Th1-based immune response despite the publication of frequent reports on the role of humoral response in resistance against brucellosis. It can be justified with the example of *B. abortus* RB51-based protection against brucellosis as strain lacks OPS. It suggests that OPS-specific antibodies are not required for protective immune response against brucellosis [21, 83].

References

1. Martins R, Da Costa IJM, Gamazo C (2012) Acellular vaccines for ovine brucellosis: a safer alternative against a worldwide disease. Expert Rev Vaccines 11(1):87–95
2. Singh BB, Dhand NK, Gill JP (2015) Economic losses occurring due to brucellosis in Indian livestock populations. Prev Vet Med 119(3–4):211–215
3. Doganay GD, Doganay M (2012) Brucella as a potential agent of bioterrorism. Recent Pat Antiinfect Drug Discov 8:27–33
4. Osterman B, Moriyon I (2006) International committee on systematics of prokaryotes subcommittee on the taxonomy of *Brucella*. Int J Syst Evol Microbiol 56:1173–1175
5. Foster G, Osterman BS, Godfroid J et al (2007) *Brucella ceti* sp. nov. and *Brucella pinnipedialis* sp. nov. for brucella strains with cetaceans and seals as their preferred hosts. Int J Syst Evol Microbiol 57:2688–2693
6. Blasco JM (1997) A review of the use of *B. melitensis* Rev 1 vaccine in adult sheep and goats. Prev Vet Med 31:275–283
7. England T, Kelly L, Jones R et al (2004) A simulation model of brucellosis spread in British cattle under several testing regimes. Prev Vet Med 63:63–73
8. CFSPH (2009) Undulant fever, Malta fever, Mediterranean fever, enzootic abortion, epizootic abortion, contagious abortion, bang's disease. The Centre for Food Security and Public Health, Brucellosis. www.cfsph.iastate.edu/Factsheets/pdfs/brucellosis
9. Godfroid J, Cloeckaert A, Liautard JP (2005) From the discovery of the Malta fever's agent to the discovery of a marine reservoir, brucellosis has continuous been a reemerging zoonosis. Vet Res 36:313–326
10. Cutler S, Whatmore A (2003) Progress in understanding brucellosis. Vet Rec 153:641–642
11. European Council Directive (ECD) (1991). On animal health conditions governing intracommunity trade in ovine and caprine animals 91/68/EEC L 46: 19–30
12. Meador VP, Hagemoser WA, Deyoe BL (1988) Histopathologic findings in *Brucella abortus*-infected, pregnant goats. Am J Vet Res 49:274–280
13. Alvarez J, Sáez JL, García N et al (2011) Management of an outbreak of brucellosis due to *B. melitensis* in dairy cattle in Spain. Res Vet Sci 90(2):208–211
14. Abbas B, Agab H (2002) A review of camel brucellosis. Prev Vet Med 55(1):47–56
15. Garin-Bastuji B, Hars J (2000) La brucellose porcine. Bull GTV 5:301–302

16. Ferroglio E, Tolari F, Bollo E et al (1998) Isolation of *Brucella melitensis* from alpine ibex. J Wildl Dis 34(2):400–402
17. Garin-Bastuji B, Oudar J, Richard Y et al (1990) Isolation of *Brucella melitensis* biovar 3 from a chamois (*Rupicapra rupicapra*) in the Southern French alps. J Wildl Dis 26:116–118
18. Ridler AL, West DM, Stafford KJ et al (2006) Persistence, serodiagnosis and effects on semen characteristics of artificial *Brucella ovis* infection in red deer stags. N Z Vet J 54:85–90
19. WHO (2006) Brucellosis in humans and animals. WHO Press, World Health Organization, 20 Avenue Appia, 1211 Geneva 27, Switzerland 16
20. Bundle DR, Cherwonogrodzky JW, Gidney MAJ et al (1989) Definition of *Brucella* A and M epitopes by monoclonal typing reagents and synthetic oligosaccharides. Infect Immun 57:2829–2836
21. Kumar A, Gupta VK, Verma AK et al (2016) Vaccines for caprine brucellosis: status and prospective. Int J Vaccines Vaccin 2(3):00030
22. Lapaque N, Moriyon I, Moreno E (2005) *Brucella* lipopolysaccharide acts as a virulence factor. Curr Opin Microbiol 8:60–66
23. Perry MB, Bundle DR (1990) Lipopolysaccharide antigens and carbohydrates of *Brucella*. In: Adams LG (ed) Advances in brucellosis research. Texas A & M University, Austin, pp 76–88
24. Birben E, Sahiner UM, Sackesen C et al (2012) Oxidative stress and antioxidant defense. World Allergy Organ J 5(1):9
25. Mohammadian A, Moradkhani S, Ataie S et al (2016) Antioxidative and hepatoprotective effects of hydroalcoholic extract of *Artemisia absinthium* L. in rat. J HerbMed Pharmacol 5(1):29–32
26. Serefhanoglu K, Taskin A, Turan H et al (2009) Evaluation of oxidative status in patients with brucellosis. Braz J Infect Dis 13(4):249–251
27. Sriranganathan N, Boyle SM, Schurig G et al (1991) Superoxide dismutases of virulent and avirulent strains of *Brucella abortus*. Vet Microbiol 26:359–366
28. Bricker BJ, Tabatabai LB, Judge BA et al (1990) Cloning, expression, and occurrence of the *Brucella* Cu-Zn superoxide dismutase. Infect Immun 58:2935–2939
29. Stabel TJ, Sha Z, Mayfield JE (1994) Periplasmic location of *Brucella abortus* Cu/Zn superoxide dismutase. Vet Microbiol 38:307–314
30. Sadosky AB, Wilson JW, Steinman HM et al (1994) The iron superoxide dismutase of *Legionella pneumophila* is essential for viability. J Bacteriol 176:3790–3799
31. St John G, Steinman HM (1996) Periplasmic copper-zinc superoxide dismutase of *Legionella pneumophila*: role in stationary phase survival. J Bacteriol 178:1578–1584
32. Imlay JA (2008) Cellular defenses against superoxide and hydrogen peroxide. Annu Rev Biochem 77:755–776
33. Piddington DL, Fang FC, Laessig T et al (2001) Cu,Zn superoxide dismutase of *Mycobacterium tuberculosis* contributes to survival in activated macrophages that are generating an oxidative burst. Infect Immun 69:4980–4987
34. Dunn KL, Farrant JL, Langford PR et al (2003) Bacterial [Cu,Zn]-cofactored superoxide dismutase protects opsonized, encapsulated Neisseria meningitidis from phagocytosis by human monocytes/ macrophages. Infect Immun 71:1604–1607
35. Porte F, Liautard JP, Kohler S (1999) Early acidification of phagosomes containing *Brucella suis* is essential for intracellular survival in murine macrophages. Infect Immun 67:4041–4047
36. Boschiroli ML, Ouahrani-Bettache S, Foulongne V et al (2002) The *Brucella suis* virB operon is induced intracellularly in macrophages. Proc Natl Acad Sci U S A 99:1544–1549
37. Celli J, de Chastellier C, Franchini DM et al (2003) *Brucella* evades macrophage killing via VirB-dependent sustained interactions with the endoplasmic reticulum. J Exp Med 198:545–556
38. Loisel-Meyer S, Jiménez de Bagüés MP, Bassères E et al (2006) Requirement of norD for *Brucella suis* virulence in a murine model of *in vitro* and *in vivo* infection. Infect Immun 74:1973–1976

39. Jimenez de Bagues MP, Loisel-Meyer S, Liautard JP et al (2007) Different roles of the two high-oxygen-affinity terminal oxidases of *Brucella suis*: cytochrome c oxidase, but not ubiquinol oxidase, is required for persistence in mice. Infect Immun 75:531–535
40. Roop RM, Gee JM, Robertson GT et al (2003) Brucella stationary-phase gene expression and virulence. Annu Rev Microbiol 57:57–76
41. Sha Z, Stabel TJ, Mayfield JE (1994) *Brucella abortus* catalase is a periplasmic protein lacking a standard signal sequence. J Bacteriol 176:7375–7377
42. Riley LK, Robertson DC (1984) Brucellacidal activity of human and bovine polymorphonuclear leukocyte granule extracts against smooth and rough strains of *Brucella abortus*. Infect Immun 46:231–236
43. Jiang X, Leonard B, Benson R et al (1993) Macrophage control of *Brucella abortus*: role of reactive oxygen intermediates and nitric oxide. Cell Immunol 151:309–319
44. Ma M, Eaton JW (1992) Multicellular oxidant defense in unicellular organisms. Proc Natl Acad Sci U S A 89:7924–7928
45. Buchmeier NA, Libby SJ, Xu Y et al (1995) DNA repair is more important than catalase for *Salmonella* virulence in mice. J Clin Investig 95:1047–1053
46. Kim JA, Sha Z, Mayfield JE (2000) Regulation of *Brucella abortus* catalase. Infect Immun 68(7):3861–3866
47. Perin G, Avero JFF, Diego RT et al (2017) Occurrence of oxidative stress in dairy cows seropositives for *Brucella abortus*. Microb Pathog 110:196e201
48. Fridovich I (1997) Superoxide anion radical (O2-), superoxide dismutases, and related matters. J Biol Chem 272:18515–18517
49. Chen Y, Azad MB, Gibson SB (2009) Superoxide is the major reactive oxygen species regulating autophagy. Cell Death Differ 16:1040–1052
50. Rhee S (2006) Cell signalling H_2O_2, a necessary evil for cell signalling. Science 312:1882–1883
51. Finkel T (2012) From sulfenylation to sulfhydration: what a thiolate needs to tolerate. Sci Signal 5:pe10
52. Al-Mehdi AB, Pastukh VM, Swiger BM et al (2012) Perinuclear mitochondrial clustering creates an oxidant-rich nuclear domain required for hypoxia-induced transcription. Sci Signal 2012(5):ra47
53. Sola-Landa A, Pizarro-Cerda J, Grilo MJ (1998) A two-componentregulatory system playing a critical role in plant pathogens and endosymbionts is present in *Brucella abortus* and controls cell invasion and virulence. Mol Microbiol 29:125–138
54. Grillo MJ, Blasco JM, Gorvel JP et al (2012) What have we learned from brucellosis in the mouse model? Vet Res 43:29
55. Delpino MV, Estein SM, Fossati CA et al (2007a) Partial protection against *Brucella* infection in mice by immunization with nonpathogenic alphaproteobacteria. Clin Vaccine Immunol 14(10):1296–1301
56. Bandara AB, Contreras A, Contreras-Rodriguez A (2007) *Brucella suis* urease encoded by ure1 but not ure2 is necessary for intestinal infection of BALB/c mice. BMC Microbiol 7:57
57. Collins CM, D' Orazio SE (1993) Bacterial ureases: structure, regulation of 42 expression and role in pathogenesis. Mol Microbiol 9:907–913
58. Sangari FJ, Seoane A, Rodríguez MC (2007) Characterization of the urease operon of *Brucella abortus* and assessment of its role in virulence of the bacterium. Infect Immun 75:774–7780
59. Paixao TA, Roux CM, den Hartigh AB (2009) Establishment of systemic *Brucella melitensis* infection through the digestive tract requires urease, the type IV secretion system, and lipopolysaccharide O-antigen. Infect Immun 77:4197–4208
60. Delpino MV, Marchesini MI, Estein SM (2007b) A bile salt hydrolase of *Brucella abortus* contributes to the establishment of a successful infection through the oral route in mice. Infect Immun 75:299–305
61. Grillo MJ, Marín CM, Barberán M et al (2009) Efficacy of bp26 and bp26/omp31 *B. melitensis* Rev.1 deletion mutants against *Brucella ovis* in rams. Vaccine 27:187–191
62. Elsbach P (1980) Degradation of microorganisms by phagocytic cells. Rev Infect Dis 2:106–128

63. Pizarro-Cerda J, Meresse S, Parton RG et al (1998) *Brucella abortus* transits through the autophagic pathway and replicates in the endoplasmic reticulum of nonprofessional phagocytes. Infect Immun 66:5711–5724
64. de Figueiredo P, Ficht TA, Rice-Ficht A et al (2015) Pathogenesis and immunobiology of brucellosis: review of *Brucella*-host interactions. Am J Pathol 185(6):1505–1517
65. Lopez-Goñi I, Guzmán-Verri C, Manterola L (2002) Regulation of *Brucella* virulence by the two-component system BvrR/BvrS. Vet Microbiol 90:329–339
66. Pei J, Turse JE, Wu Q et al (2006) *Brucella abortus* rough mutants induce macrophage oncosis that requires bacterial protein synthesis and direct interaction with the macrophage. Infect Immun 74:2667–2675
67. Forestier C, Deleuil F, Lapaque N et al (2000) *Brucella abortus* lipopolysaccharide in murine peritoneal macrophages acts as a down-regulator of T cell activation. J Immunol 165:5202–5210
68. Dornand J, Gross A, Lafont V et al (2002) The innate immune response against *Brucella* in humans. Vet Microbiol 90(1–4):383–394
69. Jimenez de Bagues MP, Marin CM, Barberan M et al (1993) Evaluation of vaccines and of antigen therapy in a mouse model for *Brucella ovis*. Vaccine 11:61–66
70. Caro-Hernandez P, Fernandez-Lago L, de Miguel MJ et al (2007) Role of the Omp25/Omp31 family in outer membrane properties and virulence of *Brucella ovis*. Infect Immun 75:4050–4061
71. Gupta VK, Nayakwadi S, Kumar A et al (2014) Markers for the molecular diagnosis of brucellosis in animals. Adv Anim Vet Sci 2(3S):31–39
72. Robertson M (1998) Innate immunity. Curr Biol 8:R595–R597
73. González D, Grilló MJ, De Miguel MJ et al (2008) Brucellosis vaccines: assessment of *Brucella melitensis* lipopolysaccharide rough mutants defective in core and O-polysaccharide synthesis and export. PLoS One 3(7):e2760
74. Rolán HG, Xavier MN, Santos RL et al (2009) Natural antibody contributes to host defense against an attenuated *Brucella abortus* virB mutant. Infect Immun 77:3004–3013
75. Young EJ, Borchert M, Kretzer FL et al (1985) Phagocytosis and killing of *Brucella* by human polymorphonuclear leukocytes. J Infect Dis 151:682–690
76. Kumar A, Gupta VK, Verma AK et al (2017) Assessment of hematological bio markers during vaccination and challenge of *Brucella melitensis* in goats. Int J Vaccines Vaccin 4(2):00078
77. Nathan C, Shiloh MU (2000) Reactive oxygen and nitrogen intermediates in the relationship between mammalian hosts and microbial pathogens. Proc Natl Acad Sci 97:8841–8848
78. Gay B, Sanchez-Teff S, Caravano R (1984) Ultrastructural localization of NADPH oxidase activity in murine peritoneal macrophages during phagocytosis of *Brucella* correlation with the production of superoxide anions. Virchows Arch B Cell Pathol Incl Mol Pathol 45:147–155
79. Araya LN, Elzer PH, Rowe GE et al (1989) Temporal development of protective cell-mediated and humoral immunity in BALB/c mice infected with *Brucella abortus*. J Immunol 143:3330–3337
80. Oliveira SC, Splitter A (1995) $CD8^+$ type 1 CD44hi CD45 RBlo T lymphocytes control intracellular *Brucella abortus* infection as demonstrated in major histocompatibility complex class I- and class II – deficient mice. Eur J Immunol 25:2551–2557
81. Montaraz JA, Winter AJ (1986) Comparison of living and nonliving vaccines for *Brucella abortus* in BALB/c mice. Infect Immun 53(2):245–251
82. Elzer PH, Jacobson RH, Nielsen KH et al (1994) BALB/c mice infected with *Brucella abortus* express protracted polyclonal responses of both IgG2a and IgG3 isotypes. Immunol Lett 42:145–150
83. Schurig GG, Roop RM Jr, Bagchi T et al (1991) Biological properties of RB51; a stable rough strain of *Brucella abortus*. Vet Microbiol 28:171–188

Oxidative Stress in *Candida albicans* Infection

21

Santanu Palchaudhuri and Dhrubajyoti Chattopadhyay

Abstract

Candidiasis, one of the most common hospital-acquired infections of humans, is predominantly caused by the fungi *Candida albicans*. In most healthy individuals, *C. albicans* can be found as part of the normal microbial flora in the skin, oral cavity, digestive tract, and female reproductive tract without causing any infection. However, in the case of immunocompromised patients or in situations where normal microbiota is disrupted, the fungi can cause mild to severe infection. After entering the host system, *C. albicans* cells are exposed to a wide variety of toxic chemicals, most important of which being the reactive oxygen species (ROS) produced by host phagocytes, such as neutrophils and macrophages. The pathogenic fungi, on its part, elicit a strong response against the ROS by activating/regulating multiple signaling pathways involving the stress-activated protein kinase Hog1, the DNA damage kinase Rad53, and the transcription factor Cap1 in order to survive under the oxidative stress condition. In this chapter, we have tried to depict a brief overview of this host–*C. albicans* interaction while focusing more on the oxidative stress generated by the host phagocytes and the response of the *C. albicans* cells against this deadly attack.

Keywords

Candida albicans · Oxidative stress · Cellular response · Cell signaling · Epigenetics

S. Palchaudhuri · D. Chattopadhyay (✉)
Amity Institute of Biotechnology, Amity University Kolkata, Kolkata, West Bengal, India
e-mail: spalchaudhuri@kol.amity.edu; dchattopadhyay@kol.amity.edu

S. Chakraborti et al. (eds.), *Oxidative Stress in Microbial Diseases*,
https://doi.org/10.1007/978-981-13-8763-0_21

21.1 Introduction

Candida albicans, an opportunistic fungal pathogen, exists as a part of the normal human microbiome and can be found on the skin, oral cavity, and gastrointestinal and urogenital tracts of most healthy individuals [1–3]. Although it is a harmless commensal organism in most humans, *Candida albicans* can cause infection, known as candidiasis under certain circumstances, such as in immunologically weak or immunocompromised individuals, in individuals whose microflora has been disturbed by antibiotic usage, or sometimes even in otherwise healthy individuals [4, 5]. Moreover, candida co-infection has also been found in pulmonary tuberculosis patients. Based on the severity of the disease, Candidiasis can be broadly categorized into two major classes: in less severe cases the superficial mucosal infections such as oral candidiasis (OC) and vulvovaginal candidiasis (thrush, VVC) can be observed as white spots in the infected regions. High percentage of OC can be observed in HIV-infected patients, in premature infants, and in aged people with defective immunity [4, 6–8]. Significantly, VVC has been found to be very common among women of childbearing age, some of whom experience at least one recurrent infection [4]. Superficial candida infection can also be found in skin and nails, resulting in the chronic mucocutaneaous candidiasis (CMC) [9]. Unlike superficial candidiasis, which is not life-threatening and easily treatable, the systemic candida infection exhibits a very high mortality rate in patients. Systematic candidiasis can be found in severely immunocompromised individuals such as in transplant patients, cancer patients receiving chemotherapeutic drugs, HIV patients, and also premature infants [10]. In these individuals, *Candida albicans* can survive in the bloodstream, which leads to the colonization of the fungus in various organs such as the kidney, liver, spleen, and brain [3, 11]. Although candidiasis can occur due to infection of other candida species such as *Candida glabrata*, *Candida krusei*, *Candida dubliniensis*, etc., *Candida albicans* has been found to be the principal and most common fungal infection causative agent of candidiasis [12].

21.2 HOST-*Candida albicans* Interaction

The mucosal epithelial layer is the first line of defense against *Candida* infection. It not only acts as a passive physical barrier, thereby preventing invasion of fungi, but also plays an active role by inducing production of chemokines and cytokines to recruit and/or activate various immune cells, thus triggering an immune response [13–16]. Once activated, the immune cells such as phagocytes engulf the invading *Candida* cells and process them through fusion with lysosome forming phagolysosome. Subsequently, the phagocytosed microorganisms are killed within the phagolysosome by hydrolytic enzymes, antimicrobial peptides, acidification, cationic influxes (K+) and most importantly by reactive oxygen species (ROS) and reactive nitrogen intermediates (RNI) produced by the host cell [17–21]. Neutrophils have also been found to kill invading *C. albicans* through the release of neutrophil extracellular traps (NETs) containing the antifungal peptide calprotectin [22].

Candida albicans, on its part, employ various methods to evade the host immune system [15]. For example, the fungi have been shown to downregulate epithelial toll-like receptors, such as TLR-4, required for fungi recognition [16], and to shield pathogen-associated molecular patterns (PAMPs) present on its surface from recognition by the phagocytes [23–25]. *C. albicans* have also been shown to switch to the filamentous hyphal state when comes in contact with epithelial cells [26, 27] to avoid phagocytosis by neutrophils and macrophages. Furthermore, switching from budding to filamentous form following engulfment by the phagocytes has also been observed in *C. albicans* cells as a part of their response toward oxidative stress [28, 29]. This helps the fungi not only in escaping from inside the phagocytes by breaking the phagosomal membrane but also in killing the phagocyte itself [30]. *C. albicans* have been shown to also inhibit/degrade complement system [31–36], inhibit phagolysosome formation, or modulate T-cell function [15, 37] in order to evade its host-mediated killing.

21.3 Increased Level of Reactive Oxygen and Nitrogen Species Produced by Activated Phagocytes Causes Oxidative/Nitrosative Stress in *C. albicans*

Reactive oxygen species or ROS are produced by the phagocytes such as macrophages and neutrophils in response to the invading fungi through a process called respiratory burst via activation of NADPH oxidase complex, which increases production of superoxide radicals in the extracellular space or in the phagolysosomal compartments [38]. The superoxide anion radical ($O2^{-\bullet}$) thus produced is then converted to hydrogen peroxide (H_2O_2) by superoxide dismutase or to hydroxyl anions (OH^-) and hydroxyl radicals ($OH^{\cdot}$) via the Haber–Weiss reaction. Along with the production of ROS, the phagocytes also produce nitric oxide (NO•) and nitrite through the action of inducible nitric oxide synthase (iNOS). The nitric oxide, in turn, reacts with superoxide radicals and generates more stable fungicide peroxynitrite (ONOO-) [20, 39–41]. Furthermore, hydrogen peroxide can react with chloride ions (Cl^-) to form hypochlorous acid (HOCl) in a myeloperoxidase-dependent manner. All these reactive compounds are known to attack biomolecules such as DNA, proteins, and lipids and subsequently exhibit microbicidal activity by inducing programmed cell death [42–44]. Interestingly, although ROS was previously thought of having only the microbicidal activity, current work from various labs using multiple pathogenic fungi, including *C. albicans*, has suggested other significant functions for these reactive species. For example, ROS has been found to recruit more phagocytes and other immune cells to the site of infection, thus enhancing phagocytosis and subsequent killing of invading microorganisms. Furthermore, ROS produced within the phagocytes have been found to induce autophagy, NETosis, and inflammasome activation [45–47]. ROS has also been found to play an immunomodulatory role to limit the damage of host tissues. Interestingly, a combination of cationic influx and oxidative burst as induced by activated phagocytes upon *C. albicans* engulfment has been found to be more potent in killing the fungi in healthy individuals when compared to individual effects [41]. This phenomenon has been termed as 'stress pathway interference' [48]. Collectively, ROS, RNI, and other toxic chemicals produced by the host phagocytes following engulfment of the *C. albicans* cells work together to kill the fungi while causing minimal damage to the host cells, thereby limiting the spread of infection.

Increased level of ROS generated by the activated phagocytes stimulates oxidative stress responses in the invading *C. albicans* through which the fungi try to nullify the microbicidal effects of ROS and survive. Specifically, *C. albicans* cells have been found to elicit a very strong transcriptional response under oxidative stress condition [49–51]. It is the battle between oxidative stress produced by the host and the response against it invoked by the fungi, which decides whether the fungi will survive or be killed. Significantly, increased level of ROS has also been found to activate the apoptosis process in the fungi leading to death of the invading microorganism [52].

Notably, as a part of the normal microbiome, *C. albicans* gets exposed to hydrogen peroxide producing microorganisms like *Lactobacillus* species [53, 54]. As a pathogen, it experiences a tremendous amount of oxidative stress caused by increased production of ROS/RNI inside the phagocytes. Even at the extracellular milieu, the fungi are exposed to high level of ROS secreted by activated phagocytes [55, 56]. Consequently, both the phagocytosed and non-phagocytosed *C. albicans* have been found to activate the oxidative stress response pathway in the presence of immune cells [56].

21.4 Oxidative Stress Response in *Candida albicans*

Exposure to ROS has been found to activate three major signaling pathways in *C. albicans* involving the stress-activated protein kinase (SAPK) Hog1, the DNA damage kinase Rad53, and the AP-1 family transcription factor Cap1.

21.4.1 The Hog1 SAPK Pathway

The high osmolarity glycerol (HOG) MAPK/SAPK pathway is one of the major signal transduction pathways in *C. albicans* which senses increased level of ROS and elicits a response that helps the fungi to adapt and survive under oxidative stress [57]. In the absence of oxidative stress, a *C. albicans* cell membrane-associated member of the two-component signaling pathway, Sln1p, autophosphorylates itself at the H-box histidine residue. Subsequently, the phosphate group at H-Box gets moved to an aspartate residue of its receiver domain. Eventually, the same phosphate group is then transferred from Sln1 to Ssk1 through a histidine-containing phosphor-transfer protein, Ypd1. The phosphorylated Ssk1, however, does not have the ability to activate Ssk2, as mentioned above. Consequently, activation of Hog1 does not happen in un-stressed cells in *C. albicans*. In the presence of oxidative stress, the ROS inhibits autophosphorylation of Sln1 leading to accumulation of unphosphorylated and active

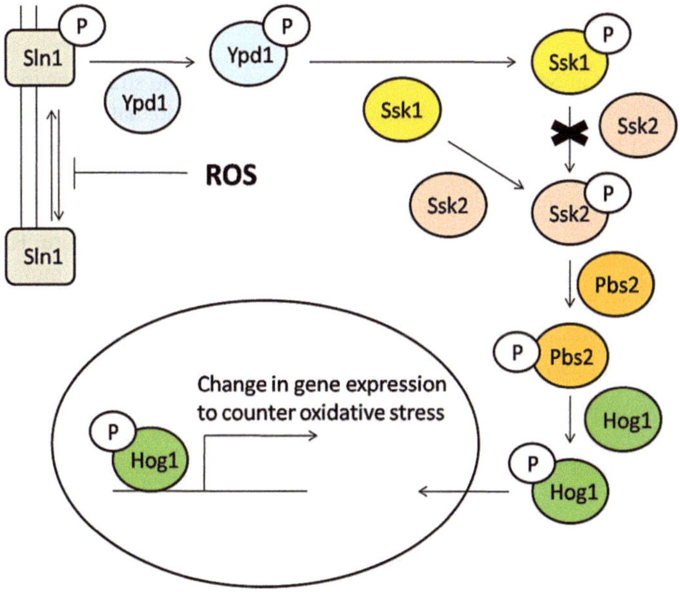

Fig. 21.1 Regulation of Hog1 both in presence and absence of oxidative stress

Ssk1. The unphosphorylated Ssk1 activates Ssk2 (MAPKK kinase), which in turn activates Pbs2p (MAPK kinase). Activated Pbs2 again in turn phosphorylates and activates the MAPK/SAPK Hog1 (Fig. 21.1). Interestingly, two more histidine kinases, Chk1 and Nik1, can be found in *C. albicans*, which have the potential to regulate Hog1p activity. Importantly, both Chk1 and Nik1 are required for virulence, morphogenesis, and cell wall biosynthesis [58–60]. The activity of Hog1 has also been found to be regulated by the redox-sensitive thioredoxin peroxidase enzyme, Tsa1, as well as thioredoxin enzyme, Trx1 [61]. It should also be noted that Hog1 has been shown to be activated and necessary only for high level of oxidative stress, but not low levels of oxidative stress [62, 63].

Following its activation under oxidative stress condition, the phosphorylated Hog1p translocates to the nucleus and activates transcription of genes required for adaptation under oxidative stress. In addition to the canonical Ssk2–Pbs2–Hog1 pathway, another sensor protein associated with cell membrane, Sho1p, can also activate Hog1p via its interaction with Pbs2p. Interestingly, even though increased phosphorylation and thereby activation of Hog1p and increased sensitivity of Δhog1 cells have been observed under oxidative stress condition in *C. albicans*, transcriptomic analysis could not identify any genes with known antioxidant activity that may be regulated by Hog1 [62–64]. It is possible that under oxidative stress condition, Hog1p may have a redundant function with other regulators. Alternatively, Hog1p may have yet unknown functions in the presence of oxidative stress. For example, a study conducted by Yin Z et al. [65] indicated that Hog1 might be required for prolonged expression of some proteins while recovering from stress elicited by hydrogen peroxide. Clearly, further work needs to be done to clearly understand the significance of Hog1 pathway in *C. albicans* under oxidative stress condition.

21.4.2 The Rad53 DNA Damage Pathway

As stated earlier in Sect. 21.2, *C. albicans* employ multiple strategies to adapt and survive inside the host. In one such mechanism, it was found that hydrogen peroxide exposure can lead to enhanced filamentous growth (hyperpolarized bud formation) in a RAD53 kinase-dependent manner in *C. albicans* to adapt and survive under oxidative stress condition [61, 66]. Filamentation of the fungi allows it to escape from and even kill the phagocytes. Upon hydrogen peroxide exposure, Rad53 was found to become activated via phosphorylation and Rad53 deletion mutant fails to form hyperpolarized bud upon hydrogen peroxide treatment [61, 67]. Interestingly, the thioredoxin protein Trx1 was found to inhibit RAD53 in the absence of oxidative stress, and oxidation and subsequent inactivation of Trx1 was found to be critical for the activation of RAD53 and hyperpolarized bud formation under oxidative stress [61] (Fig. 21.2). It is possible that oxidation of RAD53 in the presence of ROS is somehow activating the protein and subsequent filamentation of the fungi.

21.4.3 The Cap1 Transcriptional Regulator

Cap1 is a bZip transcription factor from AP-1 family and is ortholog of *S. cerevisiae* Yap1 and *S. pombe* Pap1. Significantly, Yap1, Pap1, and Cap1, all have been found to play critical role during oxidative stress response and multidrug resistance (MDR). In unstressed *C. albicans* cells, Cap1 remains in complex with a fungal-specific protein Ybp1 in the cytoplasm. Ybp1 functions to stabilize Cap1 in cytoplasm by preventing proteasome-mediated degradation [68]. Furthermore, the nuclear export signal (NES) present at the C-terminus of Cap1 promotes its interaction with the nuclear export factor, Crm1, which in turn expels Cap1 out of the nucleus to the cytoplasm and thereby prevents its nuclear accumulation. Within the phagocytes, exposure of the fungi to ROS promotes oxidation of specific Cysteine residues, leading to the disulfide bond formation between two cysteine-rich domains, namely, n-CRD and c-CRD in Cap1. This induces a conformational change in Cap1, resulting in the masking of NES. Following this oxidation, Cap1 can no longer interact with

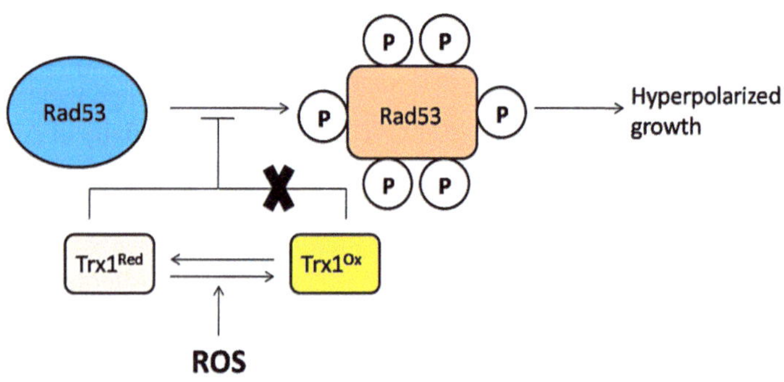

Fig. 21.2 Regulation of Rad53 under normal and oxidative stress condition

Crm1 and gets accumulated in the nucleus. It should be noted that Cap1 is not oxidized by the ROS directly. Instead, the oxidation of Cap1 is mediated by a glutathione peroxidase (Gpx)-like enzyme Gpx3, facilitated by Ybp1 [68]. Therefore, when exposed to ROS within phagocytes, the AP-1 transcription factor Cap1 gets oxidized mediated by Gpx3 and Ybp1 and accumulates in the nucleus. Once in the nucleus, Cap1 gets phosphorylated and induces genes required to survive and adapt under oxidative stress. Subsequently, this leads to filamentation within the phagosomes resulting in killing of the phagocytes. Interestingly, Cap1 transcription has also been found to be increased in the presence of human neutrophils and also when phagocytosed by human polymorphonuclear cells or PMNs [69, 70].

It should be noted that Sac6 fimbrin, an actin-binding protein (ABP), which mediates actin bundling and regulates actin dynamics and function also has a role in modulating Cap1 activity. In the absence of any oxidative stress, Sac6 facilitates formation of actin assemblies, which in turn binds Cap1, thereby blocking its transport to the nucleus. Consequently, oxidative stress-responsive genes do not get induced under such situation. However, increased level of ROS leads to disruption of the Sac6-actin-Cap1 interaction, resulting in the release of Cap1 [71]. The free Cap1 then enters the nucleus (upon oxidation-induced conformational change), becomes phosphorylated, and induces expression of antioxidative stress response genes. Figure 21.3 shows the regulation of Cap1 both in the presence and absence of oxidative stress.

Notably, a combination of ROS and cationic flux has been found to be more potent in killing microbial pathogens such as *C. albicans* by immune cells, when

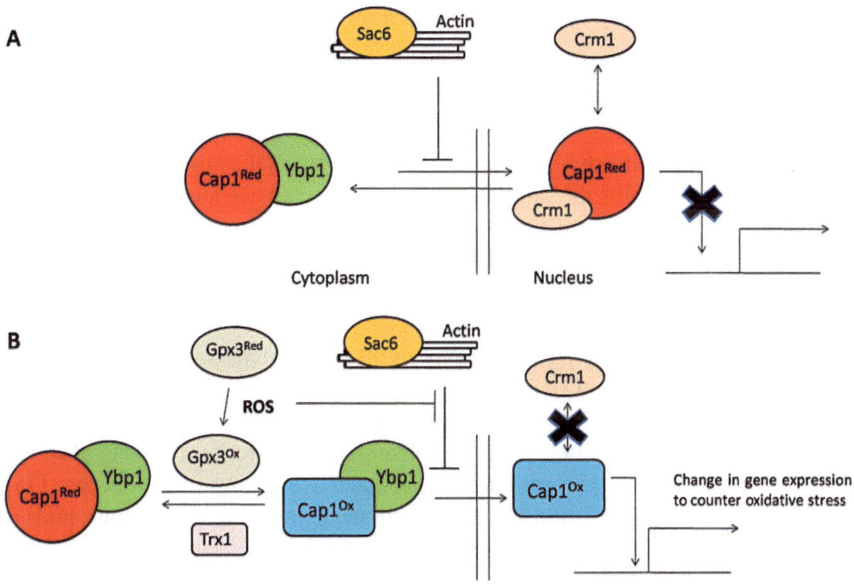

Fig. 21.3 Regulation and functioning of Cap1 transcription factor both in absence (A) and presence (B) of oxidative stress

compared with individual effects of either of them. It was further shown that when *C. albicans* cells were exposed to a combination of cation and ROS, Cap1 fails to accumulate in the nucleus. Consequently, no induction of the Cap1-regulated antioxidant gene expression was found under combinatorial stress involving cationic flux and ROS [48]. Additionally, combinatorial effect of ROS and cationic flux has also been found to influence Hog1 functionality. This at least partially explains why in healthy individuals, *C. albicans* fail to cause candidiasis, as phagocytes in these individuals tend to expose the engulfed fungi to a combination of different toxic chemicals including cations and ROS and thereby kill the pathogens very efficiently.

Cap1 has been found to be the most prominent transcriptional regulator utilized by *Candida albicans* to counteract the damaging effects of ROS. Wang et al. [72] have shown that hydrogen peroxide treatment leads to change in the expression of 89 genes in the fungi and 76 of these genes are expressed in a Cap1-dependent manner. The authors further showed that apart from activating the antioxidative response genes, Cap1 also regulated expression of genes involved in carbohydrate and energy metabolism, protein degradation, and drug resistance pathways when exposed to hydrogen peroxide [72]. Direct targets of Cap1 include key antioxidant genes such as catalase (CAT1), glutathione reductase (GLR1), glutathione peroxidase (GPX), glutamate cysteine ligase (GCS1), thioredoxin (TRX1), superoxide dismutase (SOD), as well as multidrug resistance genes such as MDR1 [73]. Interestingly, loss of Cap1 has been shown to effect virulence of *C. albicans* cells minimally in a disseminated mouse model, suggesting the presence of CAP1-independent mechanisms for survival of the fungi in vivo [74].

21.4.4 Effector Molecules Involved in Oxidative Stress Response in *C. albicans*

Various signaling pathways ultimately lead to activation of a large number of effector molecules that helps the fungi to survive under oxidative stress caused by ROS produced by the phagocytes. One of the key families of enzymes that are involved in countering ROS is superoxide dismutase (SOD), which catalyzes conversion of superoxide radical to hydrogen peroxide [75]. *C. albicans* has six SODs, SOD1–SOD6. Sod1–Sod3 are intracellular, and Sod4–Sod6 can be found on the cell surface, anchored by glycosylphosphatidylinositol (GPI). Interestingly, Sod2 and Sod3 are manganese dependent, whereas Sod1 and Sod4–Sod6 are Cu/Zn dependent. Among these SODs, Sod1, Sod4, and Sod5 have been found to be critical to counter oxidative stress. Expression of Sod1 has been found to increase following phagocytosis and is required to prevent macrophage-mediated killing of *C. albicans* [75]. Sod5 level has also been found to increase following phagocytosis by neutrophils as well as during yeast to hyphae phase transition [69, 76, 77].

In addition to SODs, other cell surface bound enzymes such as thiol-specific peroxidase, Tsa1 and the peroxide detoxifying enzyme catalase, Cat1 have also been found to play critical role to protect the fungi from oxidative stress produced by the host phagocytes [78, 79]. Furthermore, expression of Cat1, glutathione reductase, and glutaredoxins have also been found to be elevated following exposure of *C. albicans* to hydrogen peroxide indicating their involvement during oxidative stress response [80–83].

Another Important factor involved in oxidative stress response in *Candida albicans* is Thioredoxin (Trx1), a redox-sensitive antioxidant protein. Trx1 not only plays a critical role in detoxifying hydrogen peroxide but also acts as a sensor of oxidant level inside the cell. Furthermore, Trx1 has been shown to regulate oxidation status and thereby activity of all the three major proteins, namely, Hog1 MAPK, Rad53 Kinase, and Cap1 transcription factor, that are activated in *Candida albicans* cells following oxidative insult and are required to survive under oxidative stress condition produced by host phagocytes [61]. Trx1 has been shown to negatively regulate Cap1 and Rad53 in the absence of oxidative stress. However, in the presence of ROS, Trx1 becomes oxidized, thus facilitating activation of both Cap1 and Rad53. In contrast, both Trx1 and Tsa1 have been found to be required for activation of Hog1 under oxidative stress condition. Further work needs to be done to resolve this apparent contradiction and establish the mechanism of Trx1-mediated activation of Hog1 in the presence of ROS.

In addition to the three major signaling pathways, there are other key players that are involved in the oxidative stress response in *C. albicans* and are found to be required for the process. For example, the sterol 14α-demethylase (CYP51), encoded by the gene ERG11, has been shown to play a critical role during oxidative stress adaptation, hyphal elongation, and virulence in *C. albicans* [84]. It has been shown that ERG11-deleted *C. albicans* mutants are more sensitive toward hydrogen peroxide and were defective in clearing ROS. Additionally, these mutants were found to exhibit reduced filamentous and invasive growth and were killed my macrophages more efficiently.

The mRNA export factor Sus1 in *C. albicans* has been shown to play a significant role in oxidative stress tolerance [85]. Deletion of Sus1 in *C. albicans* resulted in hypersensitivity to oxidative stress, increased level of ROS accumulation, and attenuated virulence.

The spindle assembly checkpoint proteins Mps1 and Mad2 are also found to be important for oxidative stress response elicited by *C. albicans* [86, 87]. Mps1 has been shown to be essential for oxidative stress tolerance and Mad2 has been found to be essential for survival and pathogenicity of *Candida albicans*.

Further investigation is required to elucidate the mechanism of action of all these protein molecules under oxidative stress condition.

21.5 Epigenetic Regulation of Oxidative Stress Response in *C. albicans*

Although a number of studies were performed to decipher the oxidative stress response pathways involving the signaling molecules, the transcription factors, and the effector molecules, less focus was given to study the epigenetic modifications and their role during this process. In recent years, however, a number of reports are coming up, which are trying to establish the epigenetic regulation of gene expression including various histone modifications and the enzymes involved therein under oxidative stress condition. For example, Cap1 was found to regulate expression of its targets by modifying the chromatin structure at the gene promoters. Importantly, the Spt-Ada-Gcn5-acetyltransferase (SAGA) coactivator complex containing Ada2 has been shown to interact with Cap1 and is found to be recruited at some of the Cap1 target antioxidant genes. Upon recruitment, the SAGA complex having HAT activity acetylates at histone proteins at the target gene promoters [88–90]. Reduction in Ada2 level in *C. albicans* has been found to increase sensitivity toward oxidative stress and decrease histone acetylation on a number of genes. Furthermore, ChIP-on-chip analysis found Ada2 to be associated with 200 promoters of stress-responsive genes, including genes involved in oxidative stress response [89]. Importantly, the SAGA complex is also known to interact with TATA box binding protein, thus acting as an important coactivator for RNA polymerase enzyme.

Another protein complex with HAT activity, NuB4, has been found to have an opposite effect with negatively regulating expression of oxidative stress response genes in the absence of stress. NuB4 consists of histone acetyltransferase hat1, a chaperone hat2, and a histone chaperone Hif1 and is known to deposit histones into the damaged sites on DNA [91, 92]. Hat1 and hat2 form complex in the cytoplasm and acetylate H4K5 and K12. They also enter the nucleus, form complex with hif1, and thereby form the NuB4 complex [91]. Subsequently, the NuB4 complex interacts with other histone chaperones such as HIR and CAF-1 and function in histone deposition [93, 94]. Deletion of Hat1 has been found to increase expression of oxidative stress response genes, thereby increasing the resistance toward the stress [93]. The NuB4 complex is thus believed to repress expression of antioxidant genes by increasing the nucleosome occupancy at the promoter region of target genes.

A fungal-specific HAT, Rtt109, has also been shown to be required for oxidative stress resistance in *C. albicans*. Rtt109 mediates acetylation of H3K56 and is required for survival of *C. albicans* under DNA damaging condition. Deletion mutant of Rtt109 has been shown to exhibit increased susceptibility toward hydrogen peroxide-meditated killing of *C. albicans* cells by phagocytes and is found to be required for *C. albicans* pathogenesis [95]. Interestingly, increased expression of oxidative stress-responsive genes was observed in Rtt109 deletion mutants when exposed to hydrogen peroxide. This suggests that the ΔRtt109 cells can both sense oxidative stress and mount a response against it, but still could not survive phagocyte attack, possibly due to increased incidence of DNA damage.

21.6 Conclusion

In this chapter, we summarized all the processes that happen following *C. albicans* infection focusing more on oxidative stress responses elicited by the fungi. The oxidative stress caused by ROS produced by the phagocytes following fungal attack not only damages various biomolecules and changes them, but also have other function. This ensures killing of pathogenic fungi without causing much harm to the host organism. Multiple signaling pathways are activated in *C. albicans* following the ROS insult, resulting in expression of various oxidative stress-responsive genes, which would help the fungi to survive the stress. Although a lot is known about the response of the fungi to ROS, still there are gaps to be filled. For example, what is the role of Hog1 SAPK pathway in this? Hog1 is needed for sure, as the deletion mutant exhibits increased sensitivity toward oxidative stress. However, no antioxidant genes have been found to be regulated by this pathway. Further work needs to be done to get a clearer picture. Furthermore, some of the factors involved in the oxidative stress response are found to be indispensable for the virulence of the fungi, and some are not. A detailed study needs to be done to address this issue. Significantly, more focus could be given to study the epigenetic regulation of oxidative stress response. Especially, one can look at the role of Rtt109 in this process. Being a fungal-specific HAT, Rtt109 could be a potential therapeutic target for candidiasis treatment. Finally, efforts should also be made to get a holistic view of the events that unfold following exposure of *Candida albicans* to the ROS/RNI produced by the host phagocytes.

References

1. Odds FC (1988) *Candida* and *Candidiasis*. London, UK
2. Calderone RA, Fonzi WA (2001) Virulence factors of Candida albicans. Trends Microbiol 9:327–335
3. Calderone RA, Clancy CJ (2011) *Candida* and candidiasis. ASM Press, Washington, DC
4. Sobel JD (2007) Vulvovaginal candidosis. Lancet:369, 1961–1971
5. Revankar SG, Sobel JD (2012) Mucosal candidiasis. In: Calderone RA, Clancy CJ (eds) Candida and Candidiasis, 2nd edn. ASM Press, Washington, DC, pp 419–427
6. Anaissie EJ, McGinnis MR, Pfaller MA (2009) Clinical Mycology. Churchill Livingstone, London
7. Flevari A, Theodorakopoulou M, Velegraki A, Armaganidis A, Dimopoulos G (2013) Treatment of invasive candidiasis in the elderly: a review. Clin Interv Aging 8:1199–1208
8. Healy CM, Campbell JR, Zaccaria E, Baker CJ (2008) Fluconazole prophylaxis in extremely low birth weight neonates reduces invasive candidiasis mortality rates without emergence of fluconazole-resistant candida species. Pediatrics 121:703–710
9. Lilic D (2012) Unravelling fungal immunity through primary immune deficiencies. Curr Opin Microbiol 15:420–426
10. Koh AY, Kohler JR, Coggshall KT, van Rooijen N, Pier GB (2008) Mucosal damage and neutropenia are required for *Candida albicans* dissemination. PLoS Pathog 4:e35
11. Pfaller MA, Diekema DJ (2007) Epidemiology of invasive candidiasis: a persistent public health problem. Clin Microbiol Rev 20:133–163
12. Wingard JR (1995) Importance of *Candida* species other than *C. albicans* as pathogens in oncology patients. Clin Infect Dis 20:115–125
13. Abiko Y et al (2002) Upregulation of human beta-defensin 2 peptide expression in oral lichen planus, leukoplakia and candidiasis. An immunohistochemical study. Pathol Res Pract 198:537–542
14. Lilly EA, Leigh JE, Joseph SH, Fidel PL Jr (2006) Candida-induced oral epithelial cell responses. Mycopathologia 162:25–32

15. Shih-Chin Cheng, Joosten LAB, Kullberg B-J, Netea MG (2012) Infect Immun 80(4):1304. https://doi.org/10.1128/IAI.06146-11
16. Weindl G et al (2007) Human epithelial cells establish direct antifungal defense through TLR4-mediated signaling. J Clin Invest 117:3664–3672
17. Vazquez-Torres A, Jones-Carson J, Balish E (1996) Peroxynitrite contributes to the candidacidal activity of nitric oxide-producing macrophages. Infect Immun 64:3127–3133
18. Aratani Y et al (1999) Severe impairment in early host defense against *Candida albicans* in mice deficient in myeloperoxidase. Infect Immun 67:1828–1836
19. Reeves EP, Lu H, Jacobs HL, Messina CG, Bolsover S, Gabella G, Potma EO, Warley A, Roes J, Segal AW (2002) Killing activity of neutrophils is mediated through activation of proteases by K+ flux. Nature 416:291–297
20. Fang FC (2004) Antimicrobial reactive oxygen and nitrogen species: Concepts and controversies. Nat Rev Microbiol 2:820–832
21. Steinberg BE, Huynh KK, Grinstein S (2007) Phagosomal acidification: Measurement, manipulation and functional consequences. Biochem Soc Trans 35:1083–1087
22. Urban CF et al (2009) Neutrophil extracellular traps contain calprotectin, a cytosolic protein complex involved in host defense against Candida albicans. PLoS Pathog 5:e1000639
23. Gantner BN, Simmons RM, Underhill DM (2005) Dectin-1 mediates macrophage recognition of Candida albicans yeast but not filaments. EMBO J 24:1277–1286
24. Gow NA et al (2007) Immune recognition of Candida albicans betaglucan by dectin-1. J Infect Dis 196:1565–1571
25. McKenzie CG et al (2010) Contribution of *Candida albicans* cell wall components to recognition by and escape from murine macrophages. Infect Immun 78:1650–1658
26. Ray TL, Payne CD (1988) Scanning electron microscopy of epidermal adherence and cavitation in murine candidiasis: a role for *Candida* acid proteinase. Infect Immun 56:1942–1949
27. Wachtler B, Wilson D, Haedicke K, Dalle F, Hube B (2011) From attachment to damage: defined genes of Candida albicans mediate adhesion, invasion and damage during interaction with oral epithelial cells. PLoS One 6:e17046
28. Lorenz MC, Bender JA, Fink GR (2004) Transcriptional response of *Candida albicans* upon internalization by macrophages. Eukaryot Cell 3:1076–1087
29. Nasution O, Srinivasa K, Kim M, Kim YJ, Kim W, Jeong W, Choi W (2008) Hydrogen peroxide induces hyphal differentiation in *Candida albicans*. Eukaryot Cell 7:2008–2011
30. Lewis LE, Bain JM, Okai B, Gow NA, Erwig LP (2013) Live-cell video microscopy of fungal pathogen phagocytosis. J Vis Exp. https://doi.org/10.3791/50196
31. Gropp K et al (2009) The yeast Candida albicans evades human complement attack by secretion of aspartic proteases. Mol Immunol 47:465–475
32. Luo S et al (2011) The pH-regulated antigen 1 of Candida albicans binds the human complement inhibitor C4b-binding protein and mediates fungal complement evasion. J Biol Chem 286:8021–8029
33. Luo S, Hartmann A, Dahse HM, Skerka C, Zipfel PF (2010) Secreted pH-regulated antigen 1 of Candida albicans blocks activation and conversion of complement C3. J Immunol 185:2164–2173
34. Luo S, Poltermann S, Kunert A, Rupp S, Zipfel PF (2009) Immune evasion of the human pathogenic yeast Candida albicans: Pra1 is a Factor H, FHL-1 and plasminogen binding surface protein. Mol Immunol 47:541–550
35. Meri T et al (2004) The hyphal and yeast forms of *Candida albicans* bind the complement regulator C4b-binding protein. Infect Immun 72:6633–6641
36. Meri T et al (2002) The yeast *Candida albicans* binds complement regulators factor H and FHL-1. Infect Immun 70:5185–5192
37. Xiong J et al (2000) *Candida albicans* and *Candida krusei* differentially induce human blood mononuclear cell interleukin-12 and gamma interferon production. Infect Immun 68:2464–2469
38. Hogan D, Wheeler RT (2014 August) The complex roles of NADPH oxidases in fungal infection. Cell Microbiol 16(8):1156–1167

39. Brown AJ, Haynes K, Quinn J (2009) Nitrosative and oxidative stress responses in fungal pathogenicity. Curr Opin Microbiol 12:384–391
40. Ferrari CK, Souto PC, França EL, Honorio-França AC (2011) Oxidative and nitrosative stress on phagocytes' function: from effective defense to immunity evasion mechanisms. Arch Immunol Ther Exp 59:441–448
41. Kaloriti D, Tillmann A, Cook E, Jacobsen M, You T, Lenardon M, Ames L, Barahona M, Chandrasekaran K, Coghill G et al (2012) Combinatorial stresses kill pathogenic *Candida* species. Med Mycol 50:699–709
42. Cross CE, Halliwell B, Borish ET, Pryor WA, Ames BN, Saul RL, McCord JM, Harman D (1987 Oct) Oxygen radicals and human disease. Ann Intern Med 107(4):526–545
43. Stadtman ER, Levine RL (2003) Free radical-mediated oxidation of free amino acids and amino acid residues in proteins. Amino Acids 25(3–4):207–218. Epub 2003 July 29
44. Moradas-Ferreira P, Costa V (2000) Adaptive response of the yeast Saccharomyces cerevisiae to reactive oxygen species: defences, damage and death. Redox Rep 5(5):277–285
45. Steinberg BE, Huynh KK, Grinstein S (2007) Phagosomal acidification: Measurement, manipulation and functional consequences. Biochem Soc Trans 35:1083–1087
46. Dupre-Crochet S, Erard M, Nubetae O (2013) ROS production in phagocytes: why, when, and where? J Leukoc Biol 94:657–670
47. Nathan C, Cunningham-Bussel A (2013) Beyond oxidative stress: an immunologist's guide to reactive oxygen species. Nat Rev Immunol 13:349–361
48. Kaloriti D, Jacobsen M, Yin Z, Patterson M, Tillmann A, Smith DA, Cook E, You T, Grimm MJ, Bohovych I et al (2014) Mechanisms underlying the exquisite sensitivity of *Candida albicans* to combinatorial cationic and oxidative stress that enhances the potent fungicidal activity of phagocytes. MBio 5:e01334–e01314
49. Fradin C, Kretschmar M, Nichterlein T, Gaillardin C, d'Enfert C, Hube B (2003) Stage-specificngene expression of *Candida albicans* in human blood. Mol Microbiol 47:1523–1543
50. Lorenz MC, Bender JA, Fink GR (2004) Transcriptional response of *Candida albicans* upon internalization by macrophages. Eukaryot Cell 3:1076–1087
51. Enjalbert B, MacCallum DM, Odds FC, Brown AJ (2007) Niche-specific activation of the oxidative stress response by the pathogenic fungus Candida albicans. Infect Immun 75:2143–2151
52. Phillipe B et al (2003) Killing of *Aspergillus fumigatus* is mediated by reactive oxygen intermediates. Infect Immun 71:3034–3042
53. Collins EB, Hardt P (1980 May) Inhibition of *Candida albicans* by Lactobacillus acidophilus. J Dairy Sci 63(5):830–832
54. Fitzsimmons N, Berry DR (1994) Inhibition of *Candida albicans* by lactobacillus acidophilus: Evidence for the involvement of a peroxidase system. Microbios 80:125–133
55. Frohner IE, Bourgeois C, Yatsyk K, Majer O, Kuchler K (2009) Candida albicans cell surface superoxide dismutases degrade host-derived reactive oxygen species to escape innate immune surveillance. Mol Microbiol 71:240–252
56. Miramon P, Dunker C, Windecker H, Bohovych IM, Brown AJ, Kurzai O, Hube B (2012) Cellular responses of *Candida albicans* to phagocytosis and the extracellular activities of neutrophils are critical to counteract carbohydrate starvation, oxidative and nitrosative stress. PLoS One 7:e52850
57. Moye-Rowley S (2003) Regulation of the transcriptional response to oxidative stress in fungi: similarities and differences. Euk. Cell 2:381–389
58. Kruppa M et al (2003) The Chk1p of *Candida albicans* and its role in the regulation of cell wall synthesis. FEMS Yeast Res 3:289–299
59. Yamada-Okabe T et al (1999) Roles of three histidine kinase genes in hyphal development and virulence of the pathogenic fungus, Candida albicans. J Bacteriol 181:7243–7247
60. Calera JA, Calderone RA (1999) Histidine kinase, twocomponent signal transduction proteins of *Candida albicans* and the pathogenesis of candidosis. Mycoses 42:S49–S53
61. Da Silva Dantas A, Patterson MJ, Smith DA, Maccallum DM, Erwig LP, Morgan BA, Quinn J (2010) Thioredoxin regulates multiple hydrogen peroxide-induced signaling pathways in *Candida albicans*. Mol Cell Biol 30:4550–4563

62. Smith DA, Nicholls S, Morgan BA, Brown AJ, Quinn J (2004) A conserved stress-activated protein kinase regulates a core stress response in the human pathogen *Candida albicans*. Mol Biol Cell 15:4179–4190
63. Enjalbert B, Smith DA, Cornell MJ, Alam I, Nicholls S, Brown AJ, Quinn J (2006) Role of the Hog1 stress-activated protein kinase in the global transcriptional response to stress in the fungal pathogen *Candida albicans*. Mol Biol Cell 17:1018–1032
64. Alonso-Monge R et al (2003) The Hog1 MAP kinase is essential in the oxidative stress response and chlamydospore formation in *Candida albicans*. Euk Cell 2:351–361
65. Yin Z, Stead D, Walker J, Selway L, Smith DA, Brown AJ, Quinn J (2009 Oct) A proteomic analysis of the salt, cadmium and peroxide stress responses in Candida albicans and the role of the Hog1 stress-activated MAPK in regulating the stress-induced proteome. Proteomics 9(20):4686–4703
66. Shi QM, Wang YM, Zheng XD, Lee RT, Wang Y (2007) Critical role of DNA checkpoints in mediating genotoxic-stress-induced filamentous growth in *Candida albicans*. Mol Biol Cell 18:815–826
67. Loll-Krippleber R, d'Enfert C, Feri A, Diogo D, Perin A, Marcet-Houben M, Bougnoux ME, Legrand M (2014) A study of the DNA damage checkpoint in *Candida albicans*: Uncoupling of the functions of Rad53 in DNA repair, cell cycle regulation and genotoxic stress-induced polarized growth. Mol Microbiol 91:452–471
68. Patterson MJ, McKenzie CG, Smith DA, da Silva DA, Sherston S, Veal EA, Morgan BA, MacCallum DM, Erwig LP, Quinn J (2013) Ybp1 and Gpx3 signaling in Candida albicans govern hydrogen peroxide-induced oxidation of the Cap1 transcription factor and macrophage escape. Antioxid Redox Signal 19(18):2244–2260
69. Fradin C, De Groot P, MacCallum D, Schaller M, Klis F, Odds FC, Hube B (2005) Granulocytes govern the transcriptional response, morphology and proliferation of *Candida albicans* in human blood. Mol Microbiol 56:397–415
70. Alarco AM, Raymond M (1999) The bZip transcription factor Cap1p is involved in multidrug resistance and oxidative stress response in Candida albicans. J Bacteriol 181:700–708
71. Zhang B, Yu Q, Wang Y, Xiao C, Li J, Huo D, Zhang D, Jia C, Li M (1863) The Candida albicans fimbrin Sac6 regulates oxidative stress response (OSR) and morphogenesis at the transcriptional level. Biochim Biophys Acta 2016:2255–2266
72. Wang Y, Cao YY, Jia XM, Cao YB, Gao PH, Fu XP, Ying K, Chen WS, Jiang YY (2006 Apr 1) Cap1p is involved in multiple pathways of oxidative stress response in Candida albicans. Free Radic Biol Med 40(7):1201–1209
73. Znaidi S, Barker KS, Weber S, Alarco AM, Liu TT, Boucher G, Rogers PD, Raymond M (2009 Jun) Identification of the Candida albicans Cap1p regulon. Eukaryot Cell 8(6):806–820. https://doi.org/10.1128/EC.00002-09
74. Jain C, Pastor K, Gonzalez AY, Lorenz MC, Rao RP (2013) The role of *Candida albicans* AP-1 protein against host derived ROS in *in vivo* models of infection. Virulence 4:67–76
75. Hwang CS, Rhie GE, Oh JH, Huh WK, Yim HS, Kang SO (2002) Copper- and zinc-containing superoxide dismutase (Cu/ZnSOD) is required for the protection of Candida albicans against oxidative stresses and the expression of its full virulence. Microbiology 148:3705–3713
76. Martchenko M, Alarco AM, Harcus D, Whiteway M (2004) Superoxide dismutases in *Candida albicans*: transcriptional regulation and functional characterization of the hyphal-induced *SOD5* gene. Mol Biol Cell 15:456–467
77. Heilmann CJ, Sorgo AG, Siliakus AR, Dekker HL, Brul S, de Koster CG, de Koning LJ, Klis FM (2011) Hyphal induction in the human fungal pathogen *Candida albicans* reveals a characteristic wall protein profile. Microbiology 157:2297–2307
78. Urban C et al (2005) The moonlighting protein Tsa1p is implicated in oxidative stress response and in cell wall biogenesis in *Candida albicans*. Mol Microbiol 57:1318–1341
79. Crowe JD, Sievwright IK, Auld GC, Moore NR, Gow NA, Booth NA (2003) *Candida albicans* binds human plasminogen: Identification of eight plasminogen-binding proteins. Mol Microbiol 47:1637–1651

80. Enjalbert B, Nantel A, Whiteway M (2003) Stress-induced gene expression in *Candida albicans*: absence of a general stress response. Mol Biol Cell 14:1460–1467
81. González-Párraga P, Hernández JA, Argüelles JC (2003 Oct 30) Role of antioxidant enzymatic defences against oxidative stress H(2)O(2) and the acquisition of oxidative tolerance in Candida albicans. Yeast 20(14):1161–1169
82. Nakagawa Y, Kanbe T, Mizuguchi I (2003) Disruption of the human pathogenic yeast Candida albicans catalase gene decreases survival in mouse-model infection and elevates susceptibility to higher temperature and to detergents. Microbiol Immunol 47:395–403
83. Cue'llar-Cruz M, Lo'pez-Romero E, Ruiz-Baca E, Zazueta-Sandoval R (2014) Differential Response of Candida albicans and Candida glabrata to Oxidative and Nitrosative Stresses. Curr Microbiol 69:733–739
84. Wu Y, Wu M, Wang Y, Chen Y, Gao J, Ying C (2018 Nov) ERG11 couples oxidative stress adaptation, hyphal elongation and virulence in Candida albicans. FEMS Yeast Res 1:18(7)
85. Xiao C, Yu Q, Zhang B, Li J, Zhang D, Li M (2018) Role of the mRNA export factor Sus1 in oxidative stress tolerance in *Candida albicans*. Biochem Biophys Res Commun 496(2):253–259
86. Kamthan M, Nalla VK, Ruhela D, Kamthan A, Maiti P, Datta A (2014) Characterization of a putative spindle assembly checkpoint kinase Mps1, suggests its involvement in cell division, morphogenesis and oxidative stress tolerance in *Candida albicans*. PLoS One 9:e101517
87. Bai C, Ramanan N, Wang YM, Wang Y (2002) Spindle assembly checkpoint component CaMad2p is indispensable for *Candida albicans* survival and virulence in mice. Mol Microbiol 45:31–44
88. Ramírez-Zavala B, Mogavero S, Schöller E, Sasse C, Rogers PD, Morschhäuser J (2014) SAGA/ADA complex subunit Ada2 is required for Cap1- but not Mrr1-mediated upregulation of the Candida albicans multidrug efflux pump MDR1. Antimicrob Agents Chemother 58:5102–5110
89. Sellam A, Askew C, Epp E, Lavoie H, Whiteway M, Nantel A (2009) Genome-wide mapping of the coactivator Ada2p yields insight into the functional roles of SAGA/ADA complex in Candida albicans. Mol Biol Cell 20:2389–2400
90. Kim J, Park S, Lee J-S (2018) Epigenetic Control of Oxidative Stresses by Histone Acetyltransferases in Candida albicans. J Microbiol Biotechnol 28(2):181–189
91. Poveda A, Pamblanco M, Tafrov S, Tordera V, Sternglanz R, Sendra R (2004) Hif1 is a component of yeast histone acetyltransferase B, a complex mainly localized in the nucleus. J Biol Chem 279:16033–16043
92. Ge Z, Wang H, Parthun MR (2011) Nuclear Hat1p complex (NuB4) components participate in DNA repair-linked chromatin reassembly. J Biol Chem 286:16790–16799
93. Tscherner M, Zwolanek F, Jenull S, Sedlazeck FJ, Petryshyn A, Frohner IE et al (2015) The Candida albicans histone acetyltransferase Hat1 regulates stress resistance and virulence via distinct chromatin assembly pathways. PLoS Pathog 11:e1005218
94. Verzijlbergen KF, van Welsem T, Sie D, Lenstra TL, Turner DJ, Holstege FCP et al (2011) A barcode screen for epigenetic regulators reveals a role for the NuB4/HAT-B histone acetyltransferase complex in histone turnover. PLoS Genet 7:e1002284
95. Da Rosa L, Boyartchuk VL, Zhu LJ, Kaufman PD. 2010. Histone acetyltransferase Rtt109 is required for Candida albicans pathogenesis. Proc Natl Acad Sci USA 107:1594–1599

Part IV

Prevention and Therapeutics

Advances on Metabolism and Disposition of Benzimidazoles Anthelmintic in *Fasciola hepatica*: Its Contribution to the Phenomenon of Anthelmintic Resistance

22

H. Solana, S. Scarcella, and M. V. Solana

Abstract

The fascioliasis is an important zoonotic disease, particularly in underdeveloped countries. In fascioliasis, the anthelmintic control has been done mainly by the use of triclabendazole (TCBZ), which is metabolized into the anti-helmintic metabolite, sulphoxide in the host liver and targeted to the subcellular fractions of the parasite, *Fasciola hepatica* (Liver Fluke). The existence of genetically different populations of liver fluke could allow, against any selection pressure, natural or artificial (for use fasciolicides products and/or control measures), one or more populations of *F. hepatica* to be able to survive and create resistance or adaptability to such selective pressure. It is known that the uptake and effects of TCBZ and the sulfoxide metabolite is significantly greater in TCBZ-susceptible isolates in comparison to the TCBZ-resistant flukes. This result are analyzed in the present contribution.

Keywords

Fasciola hepatica · Benzimidazoles · Anthelmintic Resistance · Metabolism

An important zoonotic disease, particularly in underdeveloped countries, is the fascioliasis [1]. This parasitic infestation is considered by the World Health Organization as a neglected tropical disease affecting 2.4–17 million people in the earth [2]. The agent is *Fasciola hepatica*, a trematode digenean that is in addition a major pathogen

H. Solana (✉) · S. Scarcella · M.V. Solana
Lab. Biología Celular y Molecular, Dpto. de Cs. Biológicas, FCV-UNICEN-CIVETAN-CONICET, Tandil, Argentina
e-mail: hsolana@vet.unicen.edu.ar

S. Chakraborti et al. (eds.), *Oxidative Stress in Microbial Diseases*,
https://doi.org/10.1007/978-981-13-8763-0_22

for domestic and wild ruminants. The fascioliasis produce to losses estimated in U$S 2000–3000 million at year, associated to minor weight gains and fertility, reduced milk and wool manufacture or fatalities in the domestic animal production [3].

Due to a rise in reported human and livestock cases in numerous countries, a major emerging disease in the last few years has been considered to be the fascioliasis. This phenomenon is due to weather changes determining a different distribution of the intermediate host, the snail [4]. A wide range of anthelmintics with broad-spectrum and high efficacy is available commercially to control infestation by helminthes in ruminants. Anthelmintic compound benzimidazole (BZD) is widely used against nematode, cestode, and trematode parasites such as the fluke *Fasciola hepatica.*

Most of the flukicidal compounds have excellent activity against the liver fluke mature but are not effective against the immature stages. Triclabendazole (5-chloro-6(2,3-dichlorophenoxy)-2-(methylthio)- 1H-benzimidazole) (TCBZ) is a halogenated thiol derivative, which shows admirable efficacy against both adult stages and juvenile (immature) of *F. hepatica* [5, 6].

In *F. hepatica* infestation, the control is mostly with TCBZ [7]. Triclabendazole is metabolized at active metabolite sulfoxide (TCBZSO) in the liver of the host [8] but too by the parasite's subcellular fractions [9, 10]. For the control of adult stages of the liver fluke older than 12 weeks, albendazole (ABZ) a methyl-[(5-propylthio)-1H-benzimidazol-2-yl] carbamate is also an alternative. The induction of metabolizing enzymes could amplify the metabolic deactivation in the parasites facilitating the survival of resistant *F. hepatica* exposed to therapy [11].

The mechanisms of protection by the parasite include antioxidant and detoxifying enzymes that would suppress its oxidative killing. In *F. hepatica*, there is a lack of knowledge about the mechanisms of detoxification and anthelmintic resistance [12]. The helminthes parasites have diverse biochemical mechanisms for detoxification. The different parasites may elude drug effects by (1) alteration of receptors, (2) increase of the expression of efflux pumps, and/or (3) high expression of metabolic enzymatic systems [13].

The anthelmintic resistance including that *F. hepatica* to TCBZ is growing the entire world [14]. An increase in this parasitic disease has been reported in the last years due to weather changes and there are additional factors, like alterations of the environment as the utilization of artificial irrigation to improve the quality and quantity of fodder to animals, the floods that contribute to increase the prevalence determining a dissimilar distribution of the intermediary snail [15, 16].

The first description of resistance of liver fluke to TCBZ was reported in the 1980s [5]. Today, a great number of reports on *F. hepatica* resistance to TCBZ under field conditions in different countries such as Australia [17], Ireland [18], Scotland [19], Wales [20], the Netherlands [21, 22], Spain [23], Argentine [24], and Peru [25] have been reported.

The parasite resistance to BZDs has been associated to a particular amino acid replacement (phenylalanine to tyrosine) in the position 200 on the beta-tubulin molecule. On the other hand, sequencing of cDNAs of beta-tubulin from TCBZ -susceptible and -resistant flukes exposed no amino acid differences between the respective primary sequences [26]. The genome of liver fluke encodes at least five α- and six ß-tubulin isotypes that are expressed at the adult trematode. Of the sequences recognized, three α-isotypes and four ß-isotypes group with the other trematode tubulins in phylogenetic analysis, whereas the others are more different. The analysis of these sequences should allow their in vitro expression, which in turn may allow investigation as to which of these isotypes TCBZ binds. When isotype RT-PCR fragment sequences were compared between six individual liver flukes from the TCBZ-susceptible (*Cullompton* isolate) and seven individual flukes from the TCBZ-resistant isolates (*Sligo* and *Oberon*), these residues were conserved [27]. This confirmation of the lack of mutations described for other helminthes parasites on the target molecule (tubulin) leads to think that in the liver fluke the incident of resistance to TCBZ could be due to alterations in some of the detoxification pathways of the trematode.

The breakdown of drugs into living organisms through specialized enzyme systems is known as drug metabolism. The xenobiotic metabolism is the set of metabolic pathways that change the chemical composition of xenobiotics, (compounds foreign to an organism's normal biochemistry), such as any drug or poisonous. These pathways are a form of biotransformation present in the major groups of organisms, are considered to be of ancient source and act to detoxify toxic compounds. The presence of a semipermeable barrier determines that organisms are competent to develop detoxification systems using the common hydrophobicity of xenobiotics. These systems therefore resolve the specificity difficulty by possessing such broad substrate specificities that they metabolize almost any nonpolar compound. Useful metabolites are disqualified since they are polar and in general contain one or more charged groups.

The metabolic detoxification is divided into three phases:

(I) the modification, (II) the conjugation, and (III) the excretion. These reactions act in concert to expulse xenobiotics from the cells. The interactions with such enzymatic systems may drastically affect the disposition kinetics of different drugs and it clinical efficacy. The fluke, *Fasciola hepatica*, has a variety of detoxifying enzymes that can act on different xenobiotics. Within these xenobiotic metabolizing enzymes there are enzymes from the three known detoxification phases.

22.1 Activity of Xenobiotic Metabolizing Enzymes (XMEs)

Phase I 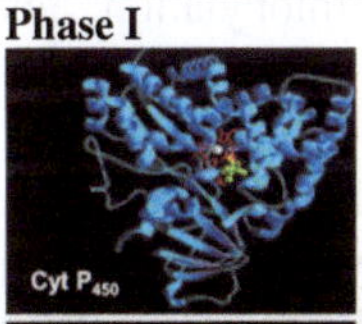 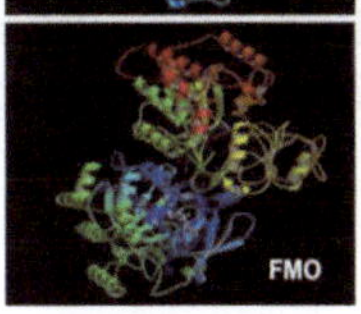	The polarity of the molecules is increased by introducing new functional groups. Is also termed no synthetic reactions may occur by removal of hydrogen or addition of oxygen, hydrolysis, reduction, oxidation, cyclization, and decyclization. Many products of Phase I are rapidly eliminated and undergo a reaction with an endogenous substrate and form a highly polar conjugate merged with the recently included functional group.

(continued)

<table>
<tr><td>Phase II
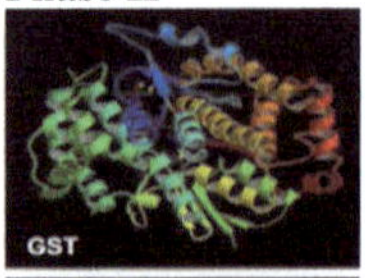

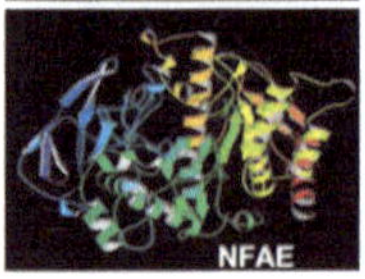
</td><td>Increases its solubility in water, inactivation of its biological activity. In phase II reactions, these activated xenobiotic metabolites are conjugated with charged species such as glucuronic acid, glutathione sulfate, or glycine.</td></tr>
<tr><td>Phase III
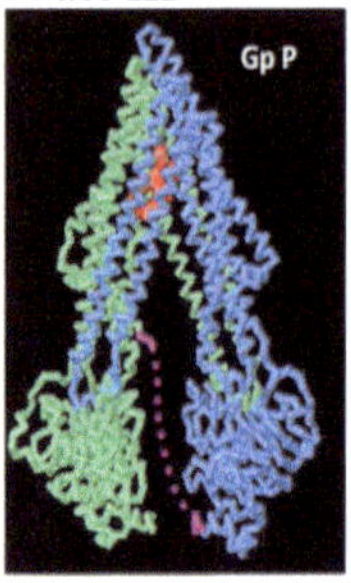
</td><td>Products from phase II and other xenobiotics are extracted into the extracellular medium. ATP-binding transmembrane transporters (example: P glycoprotein) catalyze transport of a broad variety of hydrophobic anions.</td></tr>
</table>

In **Phase I** the enzymes can oxidize a wide array of soft nucleophiles (heteroatom), such as phosphites, sulfides, and amines. This reaction requires oxygen, an NADPH, and an FAD prosthetic group. The enzymes of Phase I are typically flavin monooxygenase (FMO) and principally the Cyt P_{450}, a supergene family of enzymes. In this Phase both enzymes begin the detoxification procedure of the chemically lipophilic compound transforming into intermediate metabolites that are more hydrophilic.

In reactions of **Phase II** the metabolites produced in Phase I are transformed into hydrosolubles compounds that can be excreted throughout bile or urine. The reaction is catalyzed by a great group of transferases of broad-specificity, which in combined form can nearly metabolize any hydrophobic molecule with nucleophilic or electrophilic groups. This involves a number of types of nutrient-dependent reactions, including sulfation, glycination, glucuronidation, glutathionylation, and amino acid conjugation. The group of glutathione S-transferases (GSTs) is one of the most important enzymes families of this grouping.

The **Phase III** reactions are an additional process which is highly concentrated in the intestine. These reactions are known as antiporter activity. The best known and most studied antiporter proteins have been identified as P-glycoprotein (PGp). These efflux transporters work on specific substrates and can be induced increasing the transporters activity and be inhibited causing substrate levels to become higher. This efflux pump is also responsible for the transport of hydrophilic metabolites out of the cell after Phase II conjugation.

22.2 Advances on Metabolism and Disposition of BZDs in *Fasciola hepatica*

Disposition of ABZ and TCBZ	The trematodes have higher metabolic activity than cestodes and nematodes	[28]
	TCBZ alters the distribution of the microtubules in the testis tubules of *F. hepatica* of rats exposed to this drug. The disruption of the microtubules is one of its main mechanisms of action	[29, 30]
	The metabolism of some benzimidazoles anthelmintic is a quiral metabolism. There are differences in the generation of the chiral metabolite ABZ in rats and sheep (ABZSO + and ABZSO -), and similar circumstances occur with TCBZ (TCBZSO + and TCBZSO -)	[9, 39, 40]
	TCBZ is metabolized into its active metabolite sulfoxide by the liver of the host however also by the helminthes subcellular fractions. It has also been reported that liver flukes has significantly higher sulfoxidation than nematode and cestode parasites	[9, 30, 31, 37, 38]
	The fluke's cytosolic protein-bound TCBZ have 83% larger affinity than ABZ (44%). Albendazole bound to cytosolic proteins was displaced by its competence with an equivalent concentration of TCBZ ($P < 0.001$). The TCBZ displacement from its binding site(s) was not observed although concentrations of ABZ were threefold higher than the assayed TCBZ concentrations demonstrating that tubulin would not be the single target protein for TCBZ in trematodes	[29, 32]
	Tubulin would not be the unique site for trematodicidal action of TCBZ	[33]
	In the flukes resistant at TCBZ, there are a great diffusion and biotransformation of the parent drug and its oxidized metabolites	[9, 41]
	The liver fluke showed efficient oxidative biotransformation of the anthelmintic TCBZ into its sulfoxide derivative. This drug exhibits significantly higher sulfoxidative capacity as compared to *Taenia* spp. and *Ascaris* spp.	[34]
	In all organisms, xenobiotics metabolizing enzymes (XMEs) serve as an efficient defense against the potential negative action. The development of resistance is facilitated by the action of XMEs of Phase I and Phase II	[35]
	The mechanisms of parasite defense include antioxidant and detoxifying enzymes that suppress its oxidative killing. In *F. hepatica*, there is a lack of knowledge about the mechanisms of detoxification and anthelmintic resistance	[12]

Metabolism	**Phase I**	The zymogram of cytosolic fractions of liver fluke identified a protein 170 kDa containing the carboxylesterase activity. In *F. hepatica* exposed to TCBZ is observed a statistically significant increase in carboxylesterase activity at 24 and 48 hours post-treatment. The densitograms of the zymograms confirmed the enzyme induction under the experimental conditions of the assay	[36]
		In studies about TCBZ-R flukes, there is an enhanced ability for TCBZ oxidation at TCBZSO that confirmed a metabolic overexpression of **FMO (flavin monooxygenases)**	[41]
		Both TCBZ-resistant and susceptible *F. hepatica* metabolized ABZ into ABZSO. The oxidative action was 49% higher in resistant flukes ($P < 0.001$) with a predominant production of the (+) ABZSO enantiomer which is regularly created by the enzyme **FMO**. The results confirm a superior capacity for ABZ oxidation in TCBZ-resistant flukes. This oxidation is preferably carried out via **FMO**	[29, 32, 38]

Metabolism	**Phase II**	The increased activity of the **microsomal and cytosolic glutathione S-transferases** (cGST) (XEM Phase II) in the *Sligo* (TCBZ-R isolate) provides an understanding of the occurrence of resistance and adds information to the knowledge of the response that the parasites have exposure to diverse xenobiotics. The overexpression was detected in both GSTs (c**GST** and m**GST**) associated with previous data where the *Sligo* isolate overexpress **FMO** (XME Phase I) confirm that expression of the phenomenon of resistance at TCBZ in *F. hepatica* is a multi-enzymatic response relating more than one metabolic pathway	[13, 43, 44]
		Cytosolic GST The enzymatic activity of both TCBZ-R (*Oberon* and *Sligo* isolates) was 52% and 59% respectively higher ($P < 0.001$) than that in the *Cullompton* isolate (TCBZ susceptible)	[42]
		Isoenzymes of GST The activity of **GST mu** of the *Sligo* and *Oberon* isolate (both TCBZ-R) are respectively 26% and 17% higher than the activity of *Cullompton* isolate (TCBZ-S). **cGST pi** activity does not differ between the different strains tested. **cGST mu** not only is implicated in the metabolization of TCBZ however also presents an obvious contribution in the expression of the phenomenon of anthelmintic resistance in liver fluke	[42]

Metabolism	**Phase III**	The P-glycoproteins (PGp) of *F. hepatica* increased the cellular efflux of TCBZ	[13, 62]
		TCBZ-R isolates of *F. hepatica* have been shown to process TCBZ more rapidly. Under in vitro conditions, the resistant can be reversed, by the coadministration of inhibitors of P-glycoprotein	[62, 65]

Genetic	**Phase I**	For the identification and characterization of *fmo* and *Cyt* P_{450} the RT-PCR was performed with specific oligonucleotides designed for that purpose. Bands of 720 bp (gen *fmo*) and 930 bp (gen *Cyt* P_{450}) were obtained in the resistant (*Sligo*) and susceptible (*Cullompton*) isolates. When both isolates were compared, significant differences were found in *fmo* expression. In the *Sligo* isolate, the mRNA expression was 0.5 times superior to the *Cullompton* isolate. No significant differences were detected in the expression of the RNAm of the Cyt P_{450} enzyme	[41]
	Phase II	In the **GST mu gene** genetic analysis of the TCBZ susceptible isolate (*Cullompton*) and TCBZ-resistant isolate (*Sligo*), GST mu isozyme of the TCBZ-resistant isolate showed changes in three nucleotides and one amino acid at position 143	[45]
	Phase III	Recently it was detected the presence of numerous ABC transporters and multidrug-resistant genes. These finding open new ideas for investigating variants associated with the resistance of liver fluke to TCBZ	[46]
		A PGp mutation suggested that a single nucleotide polymorphism (SNPs) is associated with the resistant phenotype.	[66]
		The equal variation in Australian isolates of *F. hepatica* was not related with the resistance descript by European isolates	[67]
		The analysis of the presence/absence of the T687G SNP confirms the absence of this polymorphism in more than 40 specimens from 2 TCBZ-resistant isolates and 3 susceptible isolates across Latin America evaluated. The American samples showed other different SNPs than the previous ones and none of these SNPs detected showed a manifest connection with the phenomenon of resistance in the analyzed specimens. Evaluating the 42 kb of the FhPGP gene based on RNAseq data highlights that the variation has been underestimated, signifying that more detailed hard work is required in order to recognize the real markers of resistance	[68]

Variability	The analysis of *F. hepatica* isolates TCBZ-S and TCBZ-R to from dissimilar regions (America and Europe), using RAPDs-PCR technique to reveal genetic variability between the diverse isolates.	[61]
	The phenogram generated showed three main clusters, the most important contained European isolates (*Sligo* and *Cullompton*) showing a genetic space of 27.2 between them. Alternatively, the American isolates (*Cajamarca* and *Cedive*) showed a distance of 37.8 and 33.8, respectively. Formed each their individual group but visibly maintaining a closer genetic relationship among them than that to their counterparts of the Europe, with which this polymorphism would give this species enhanced adaptability against the host, as well as the environment. The existence of genetically different populations of *F. hepatica* allow, against any selection pressure, artificial or natural (the use of other control measures or other fasciolicide products), different populations of liver fluke to be able to survive and generate resistance or adaptability to such selective pressure.	
	The research over the isolates TCBZ-R of *F. hepatica* in different laboratories worldwide resulted in the identification of potential candidate genes and possible alternating biological pathway	[62, 63]
	The genes precise and loci involved in the phenomenon of TCBZ resistance are still to be defined. A genome-wide approach is currently underway to identify the principal genetic determinant of TCBZ resistance in liver fluke	[64]

22.3 Finding

Parasites have diverse biochemical mechanisms to eliminate toxic substances such as pharmaceuticals or other xenobiotics (detoxification system). Generally, helminthes may evade drug anthelmintic effects by (1) genetic changes of receptors (mutations), (2) overexpression of the membrane efflux pumps, and/or (3) overexpression of enzymatic systems of detoxification [13]. With reference to the detoxifying activity of certain metabolic enzymes of the parasite, the induction of this enzymes could enhance metabolic deactivation of the anthelmintic within parasites facilitating the survival of helminthes exposed to anthelmintic therapy (parasite resistance) [11]. The reaction that *F. hepatica* manifests to the different xenobiotics was already known but until relatively recently its implications in the development of anthelmintic resistance are unknown. The trematode *F. hepatica* uses its enzymatic battery to metabolize different xenobiotics but is also able to overexpress different detoxification pathways obtaining a greater metabolic capacity that allows it to overcome the pharmacological effect of the anthelmintics used for its control.

The metabolism of *Fasciola hepatica* transforms TCBZ at the active metabolite sulfoxide (TCBZ.SO) and subsequently to inactivate TCBZ sulfone (TCBZ.SO_2). This inactive metabolite is more in TCBZ-R isolates than TCBZ-S [13, 29]. It is known that the uptake of TCBZ by the liver fluke TCBZ-S is significantly superior than in TCBZ-R flukes, although the ABZ absorption is analogous in both isolates [13]. This reaction can be inverted by incubate parasite in presence of a P-glycoprotein substrate (ivermectin). The co-incubation with verapamil (another P-glycoprotein inhibitor) enhances severe tegumental lesions [49], confirming that the efflux bomb P-glycoprotein is one of the possible detoxifying mechanisms. The amplified tegument rupture also is seen when resistant flukes are incubated with methimazole (flavin monooxygenases inhibitor) [50]. The Cyt P_{450} inhibitor, the drug ketoconazole [51], has an analogous effect, suggesting that it may be connected with drug resistance. These metabolic ways may be upregulated in the TCBZ-resistant isolates. Furthermore, in worms treated with TCBZ, an increased activity of other detoxifying enzymes (glutathione S-transferase, carboxyl esterase, and carbonyl reductase) [44, 47] and when examined in comparison with the *Cullompton* isolate (TCBZ sensitive fluke) a superior GST reaction in *Sligo* isolate (TCBZ-R) [36] were observed. The proteomic study of liver flukes showed difference in the action of enzymes of detoxification, the energy metabolic and the structural proteins, confirm the pleiotropic character of the effects of TCBZ and reinforced proposal of several proteins expression [52]. The anthelmintic drug produces intricate effects on the trematode affecting morphology, physiology, and metabolism of liver fluke. To cope with these drugs, the TCBZ-resistant isolates of the liver fluke seem to utilize different detoxifying mechanisms [43, 53]. Recently into the studies of different drug resistance get ahead by using a genomics analysis advance [48], which has been to confirmed to be especially important in schistosomes [54–58]. Comparable hard work have been initiated in *F. hepatica* [59], starting from their well-characterized genome sequences of Europe isolates (TCBZ-S or TCBZ-R) [48, 60].

Polymorphism gives this species a better adaptability against the environment and against the host. Genetically different populations of *F. hepatica* allow that in front of any pressure of natural or artificial selection, one or more populations of the parasite survive by creating resistance or adaptability to this selective pressure. In recent transcriptomic study of liver fluke, isolates were competent to detect varied protein families and functions that demonstrate a different gene expression [48].

As a corollary of pharmacodynamic aspects, resistance in different *F. hepatica* isolates may depend on different mechanisms or targets. This highlights the need to study different isolates of *F. hepatica* in order to better recognize the effect of TCBZ and the different resistance mechanisms developed by this trematode helminthes. All the results obtained so far concerning the understanding of the phenomenon of resistance are an important step in that direction and in the near future will surely have an impact on parasite control.

References

1. Mas-Coma S (2004) Human fascioliasis. In: Cotruvo JA, Dufour A, Rees G, Bartram J, Carr R, Cliver DO, Craun R, Fayer R, Gannon VPJ (eds) Waterborne Zoonoses: identification, causes, and control. World Health Organisation/IWA Publishing, London, pp 305–322
2. Cwiklinski K, Dalton JP, Dufresne PJ et al (2015) The Fasciola hepatica genome: gene duplication and polymorphism reveals adaptation to the host environment and the capacity for rapid evolution. Genome Biol 16:71
3. Boray JC, McMichael DF (1961) The identity of the Australian Lymnaeid snail host of *Fasciola hepatica* L. and its response to environment. Mar Freshw Res [Internet] 12:150–163. Available from: http://www.publish.csiro.au/mf/MF9610150

4. Mas-Coma S, Valero MA (2009) BM. Climate change effects on trematodiases, with emphasis on zoonótica fascioliasis and schistosomiasis. Vet. Parasitology 163:264–366
5. Boray JC, De Bono D (1989) Drug resistance in Fasciola hepatica. In: Outteridge PM (ed) Advances in veterinary science. R. D. Richards, Artarmon, p 166
6. World Health Organization (2007) Report of the WHO informal meeting on use of triclabendazole in fascioliasis control. World Health Organization. http://www.who.int/neglected_diseases/preventive_chemotherapy/WHO_CDS_NTD_PCT_2007.1.pdf
7. Lacey E (1988) The role of the cytoskeletal protein, tubulin, in the mode of action and mechanism of drug resistance to benzimidazoles. Int J Parasitol 18:885–936
8. Virkel G, Lifschitz A, Sallovitz J, Pis A, Lanusse C (2006) Assessment of the main metabolism pathways for the flukicidal compound triclabendazole in sheep. J Vet Pharmacol Ther 29(3):213–223
9. Mottier L, Virkel G, Solana H, Alvarez L, Salles J, Lanusse C (2004) Triclabendazole biotransformation and comparative diffusion of the parent drug and its oxidized metabolites into Fasciola hepatica. Xenobiotica 34(11–12):1043–1057
10. Solana HD, Rodriquez JA, Lanusse CE (2001) Comparative metabolism of albendazole and albendazole sulphoxide by different helminth parasites. Parasitol Res 87:275–280
11. Brennan GP, Fairweather I, Trudgett A, Hoey E, McCoy, McConville M, Meaney M et al (2007) Understanding triclabendazole resistance. Exp Mol 82:104–109
12. Betts MJ, Russell RB (2003) Amino acid properties and consequences of substitutions. Chapter 14 Bioinforma. Genet:290e316
13. Alvarez L, Solana HD, Mottier ML, Virkel GL, Fairweather I, Lanusse CE (2005) Altered drug influx/efflux and enhanced metabolic activity in triclabendazole-resistant liver flukes. Parasitology 131:1–10. https://doi.org/10.1017/S0031182005007997
14. Mas-Coma S, Valero MA, Bargues MD (2009) Fasciola, lymnaeids and human fascioliasis, with a global overview on disease transmission, epidemiology, evolutionary genetics, 236 molecular epidemiology and control. Adv Parasitol 69:41–147
15. Bargues MD, Artigas P, Khoubbane M, Ortiz P, Naquira C, Mas-Coma S (2012) Molecular characterization of *Galba truncatula*, *Lymnaea neotropica* and *L. schirazensis* from Cajamarca, Peru and their potential role in transmission of human and animal fascioliasis. Parasites & Vectors 5:174
16. Mas-Coma S, Valero MA, Bargues MD (2009) Climate change effects on trematodiases, with emphasis on zoonótica fascioliasis and schistosomiasis. Vet Parasitol 163:264–280
17. Overend DJ, Bowen FL (1995) Resistance of Fasciola hepatica to triclabendazole. Austral Vet J 72:275–276. https://doi.org/10.1111/j.1751–0813.1995.tb03546.x
18. O'Brien DJ 1998) Fasciolosis: a threat to livestock. Irish Vet J 51:539–541. Overend DJ, Bowen FL (1995) Resistance of Fasciola hepatica to triclabendazole. Austral Vet J 72:275–276. https://doi.org/10.1111/j.1751–0813.1995.tb03546.x
19. Mitchell GB, Maris L, Bonniwell MA (1998) Triclabendazole resistant liver fluke in Scottish sheep. Vet Rec 143:399
20. Thomas I, Coles GC, Duffus K (2000) Triclabendazole-resistant Fasciola hepatica in south-West Wales. Vet Rec 146:200
21. Moll L, Gaasenbeek CP, Vellema P, Borgsteede FH, (2000) Resistance of Fasciola hepatica against triclabendazole in cattle and sheep in The Netherlands. Vet Parasitol 91:153–158. https://doi.org/10.1016/S0304-4017(00)00267-3
22. Gaasenbeek CP, Moll L, Cornelissen JB, Vellema P, Borgsteede FH (2001) An experimental study on triclabendazole resistance of Fasciola hepatica in sheep. Vet Parasitol 95:37–43. https://doi.org/10.1016/S0304–4017(00)00413-1
23. Alvarez-Sánchez MA, Mainar-Jaime RC, Pérez-García J, Rojo-Vázquez FA (2006) Resistance of Fasciola hepatica to triclabendazole and albendazole in sheep in Spain. Vet Rec 159:424–425
24. Olaechea F, Lovera V, Larroza M, Raffo F, Cabrera R (2011) Resistance of Fasciola hepatica against triclabendazole in cattle in Patagonia (Argentina). Vet Parasitol 178:364–366

25. Ortiz P, Scarcella S, Cerna C, Rosales C, Cabrera M, Guzmán M, Lamenza P, Solana H (2013) Resistance of *Fasciola hepatica* against Triclabendazole in cattle in Cajamarca (Peru): a clinical trial and an *in vivo* efficacy test in sheep. Vet Parasitol 195(1–2):118–121
26. Robinson MW, Trudgett A, Hoey EM, Fairweather I (2002) Triclabendazole-resistant Fasciola hepatica: ß-tubulin and response to in vitro treatment with triclabendazole. Parasitology 124:325–338. https://doi.org/10.1017/S003118200100124X
27. Ryan LA, Hoey E, Trudgett A, Fairweather I, Fuchs M, Robinson MW, Chambers E, Timson DJ, Ryan E, Feltwell T, Ivens A, Bentley G, Johnston D (2008) Fasciola hepatica expresses multiple a- and ß- tubulin isotypes. Mol Biochem Parasitol 159(1):73–78
28. Solana HD, Rodriguez JA, Lanusse CE (2001) Comparative metabolism of albendazole and albendazole sulphoxide by different helminth parasites. Parasitol Res 87:275–280
29. Solana H, Scarcella S, Virkel G, Ceriani C, Rodríguez J, Lanusse C (2009) Albendazole enantiomeric metabolism and binding to cytosolic proteins in the liver fluke Fasciola hepatica. Vet Res Commun 33:163–173
30. Scarcella S, Fiel C, Guzman M, Alzola R, Felipe A, Hanna R, Fairweather I, Mcconnell S, Solana H (2011) Reproductive disruption in Fasciola hepatica associated with incomplete efficacy of a new experimental formulation of Triclabendazole. Veterinary parasitology, vol 176. Elsevier Science BV, Amsterdam, pp 157–164
31. Virkel G, Lifschitz A, Sallovitz J, Pis A, Lanusse C (2006) Assessment of the main metabolism pathways for the flukicidal compound triclabendazole in sheep. J Vet Pharmacol Ther 29(3):213–223. https://doi.org/10.1111/j.1365–2885.2006.00735.x
32. Solana HD, Sallovitz JM, Lanusse CE, Rodríguez JA (2002) Enantioselective binding of albendazole sulphoxide to cytosolic proteins from helminth parasites. Methods Find Exp Clin Pharmacol 24(1):7–13. https://doi.org/10.1358/mf.2002.24.1.677121
33. Solana H, Rodríguez J, Lanusse C (2002) Triclabendazole interaction with Fasciola hepatica cytosolic protein. Biocell 26(3):402
34. Mottier L, Alvarez L, Fairweather I, Lanusse C (2006) Resistance-induced changes in triclabendazole transport in Fasciola hepatica: ivermectin reversal effect. J Parasitol 92(6):1355–1360
35. Cvilink V, Lamka J, Skálová L (2009) Xenobiotic metabolizing enzymes and metabolism of anthelmintics in helminths. Drug Metabol Rev 41:8–26. https://doi.org/10.1080/03602530802602880
36. Scarcella S, Lamenza P, Virkel G, Solana H (2012) Expression differential of microsomal and cytosolic glutathione-S-transferases in Fasciola hepatica resistant at triclabendazole. Mol Biochem Parasitol 181(1):37–39
37. Solana H, Najle R, Lanusse C, Rodríguez J (2000) Mode of action of Benzimidazole Anthelmintics: characterization of tubulin from different helminth parasites. Biocell 24:144–144
38. Solana H, Virkel G, Rodriguez J, Lanusse C (2003) Triclabendazole oxidation by the liver fluke Fasciola hepatica: identification of the metabolic pathways. Biocell 27(3):389
39. Delatour P, Benoit E, Caude I, Tambute A (1990) Species differences in the generation of the chiral sulphoxide metabolite of albendazole in sheep and rats. Chirality 2:156–160. https://doi.org/10.1002/chir.530020306
40. Solana HD, Sallovitz JM, Najle R, Rodriguez J, Lanusse C (2000) Liver sulphoxidative metabolism of albendazole in rat: enantioselectivity and effect of methimazole. Methods Find Exp Clin Pharmacol 22(2):83–88
41. Lamenza P, Ortiz Oblitas P, Ceriani C, Solana H (2013) Identification and characterization of phase I detoxification enzymes in isolates of *Fasciola hepatica* susceptible and resistant to triclabendazole. WAAVP 2013, Australia
42. Fernández V, Ortiz P, Solana MV, Solana H (2014) Differential activities of glutathione S-transferase isoenzymes in strains of Fasciola Hepatica susceptible and resistant to Triclabendazole. Am J Anim Vet Sci 9:177–181
43. Fernández V, Acevedo M, Solana H (2015) A multienzyme response is involved in the phenomenon of resistance to triclabendazole on Fasciola. J Bacteriol Parasitol 6:4

44. Scarcella S, Solana MV, Fernandez V, Lamenza P, Ceballos L, Solana H (2013) Increase of glutathione S-transferase, carboxyl esterase and carbonyl reductase in Fasciola hepatica recovered from triclabendazole treated sheep. Mol Biochem Parasitol 191:63–65
45. Fernández V, Estein S, Ortiz P, Luchessi P, Solana MV, Solana H (2015) A single amino acid substitution in isozyme GST mu in Triclabendazole resistant Fasciola hepatica (*Sligo* strain) can substantially influence the manifestation of anthelmintic resistance. Exp Parasitol 159:274–279
46. Solana M. Fernández V, Solana H (2017) Genetic-metabolic advances in the study of the anthelmintic resistance in *Fasciola hepatica*. J Bacteriol Parasitol 8(Suppl):4
47. Scarcella S, Miranda-Miranda E, Cossío-Bayúgar R, Ceballos L, Fernandez V, Solana H (2012) Increase of carboxylesterase activity in Fasciola hepatica recovered from Triclabendazole treated sheep. Mol Biochem Parasitol 181:37–39
48. Radio S, Fontenla S, Solana V, Matos Salim AC, Araújo FMG, Ortiz P, Hoban C, Miranda E, Gayo V, Pais FS, Solana H, Oliveira G, Smircich P, Tort JF (2018) Pleiotropic alterations in gene expression in Latin American Fasciola hepatica isolates with different susceptibility to drugs. Parasit Vectors 11(1):56. https://doi.org/10.1186/s13071-017-2553-2
49. Savage J, Meaney M, Brennan GP, Hoey E, Trudgett A, Fairweather I (2013) Effect of the P-glycoprotein inhibitor, R(+)-verapamil on the drug susceptibility of a triclabendazole-resistant isolate of Fasciola hepatica. Vet Parasitol 195:72–86
50. Devine C, Brennan G, Lanusse C, Alvarez L, Trudgett A, Hoey E et al (2009) Effect of the metabolic inhibitor, methimazole on the drug susceptibility of a triclabendazole-resistant isolate of Fasciola hepatica. Parasitology 136:183–192
51. Devine C, Brennan GP, Lanusse CE, Alvarez LI, Trudgett A, Hoey E et al (2011) Enhancement of triclabendazole action in vivo against a triclabendazole resistant isolate of Fasciola hepatica by co-treatment with ketoconazole. Vet Parasitol 177:305–315
52. Chemale G, Perally S, LaCourse E, Prescott M, Jones L, Ward D et al (2010) Comparative proteomic analysis of triclabendazole response in the liver fluke Fasciola hepatica. J Proteome Res 9:4940–4951
53. Matoušková P, Vokrál I, Lamka J, Skálová L (2016) The role of xenobiotic metabolizing enzymes in anthelmintic deactivation and resistance in helminths. Trends Parasitol 32:481–491
54. Valentim CLL, Cioli D, Chevalier FD, Cao X, Taylor AB, Holloway SP et al (2013) Genetic and molecular basis of drug resistance and species-specific drug action in schistosome parasites. Science 342:1385–1389
55. Chevalier FD, Valentim CL, LoVerde PT, Anderson TJ (2014) Efficient linkagemapping using exome capture and extreme QTL in schistosome parasites. BMC Genomics 15:617–629
56. You H, McManus DP, Hu W, Smout MJ, Brindley PJ, Gobert GN (2013) Transcriptional responses of in vivo praziquantel exposure in schistosomes identifies a functional role for calcium signaling pathway member CamKII. PLoS Pathog 9:e1003254
57. Kasinathan RS, Morgan WM, Greenberg RM (2010) Schistosoma mansoni express higher levels of multidrug resistance-associated protein 1 (SmMRP1) in juvenile worms and in response to praziquantel. Mol Biochem Parasitol 173:25–31
58. Hines-Kay J, Cupit PM, Sanchez MC, Rosenberg GH, Hanelt B, Cunningham C (2012) Transcriptional analysis of Schistosoma mansoni treated with praziquantel in vitro. Mol Biochem Parasitol 186:87–94
59. Hodgkinson J, Cwiklinski K, Beesley N, Paterson S, Williams D (2013) Identification of putative markers of triclabendazole resistance by a genome-wide analysis of genetically recombinant Fasciola hepatica. Parasitology 140:1523–1533
60. McNulty S, Tort J, Rinaldi G, Fischer K, Rosa B, Smircich P et al (2017) Genomes of Fasciola hepatica from the Americas reveal colonization with Neorickettsia endobacteria related to the agents of potomac horse and human sennetsu fevers. PLoS Genet 13:e1006537
61. Scarcella S, Miranda-Miranda E, Solana MV, Solana H (2015) Approach to molecular characterization of different strains of Fasciola hepatica using random amplified polymorphic DNA polymerase chain reaction. Parasitol Res 114(4):1341–1345

62. Brennan GP, Fairweather I, Trudgett A, Hoey E, McCoy M, McConville M, Meaney M, Robinson N, McFerran L, Ryan C, Lanusse C, Mottier L, Alvarez L, Solana H, Virkel G, Brophy PM (2007) Understanding triclabendazole resistance. Exp Mol Pathol 82:104–109. https://doi.org/10.1016/j.yexmp.2007.01.009
63. Kelley JM, Elliott TP, Beddoe T, Anderson G, Skuce P, Spithill TW (2016) Current threat of triclabendazole resistance in Fasciola hepatica. Trends Parasitol 32:458–469. https://doi.org/10.1016/j.pt.2016.03.002
64. Hodgkinson J, Cwiklinski K, Beesley NJ, Paterson S, Williams DJL (2013) Identification of putative markers of triclabendazole resistance by a genome-wide analysis of genetically recombinant Fasciola hepatica. Parasitology 140:1523–1533. https://doi.org/10.1017/S0031182013000528
65. Meaney M, Savage J, Brennan GP, Hoey E, Trudgett A, Fairweather I (2013) Increased susceptibility of a triclabendazole (TCBZ)-resistant isolate of to Fasciola hepatica TCBZ following co-incubation in vitro with the P-glycoprotein inhibitor, R (+)-verapamil. Parasitology 140:1287–1303. https://doi.org/10.1017/S0031182013000759
66. Wilkinson R., Law CJ, Hoey EM, Fairweather I, Brennan GP, Trudgett A (2012) An amino acid substitution in P-glycoprotein from triclabendazole-Fasciola hepatica resistant and triclabendazole-susceptible populations. Mol Biochem Parasitol 186:69–72. https://doi.org/10.1016/j.molbiopara.2012.08.008
67. Elliott TP, Spithill TW (2014) The T687G SNP in a P-glycoprotein gene of Fasciola hepatica is not associated with resistance to triclabendazole in two resistant Australian populations. Mol Biochem Parasitol 198:45–47. https://doi.org/10.1016/j.molbiopara.2014.11.006
68. Solana MV, Domínguez MF, Scarcella S, Radio S, Smircich P, Fernández S, Solana H, Tort JF (2018) Different SNPs in Fasciola hepatica P-glycoprotein from diverse Latin American populations are not associated with Triclabendazole resistance. Mol Biochem Parasitol 224:57–60

Oxidative Stress in Malarial Diseases: *Plasmodium*-Human Host Interactions and Therapeutic Interventions

23

Greanious Alfred Mavondo, Blessing Nkazimulo Mkhwanazi, Catharine Mhlanga, Rachael Dangarembizi, Nyasha Mukonowenzou, Mayibongwe Loius Mzingwane, and Joy Mavondo

Abstract

Oxidative stress is a major contributor of disease aetiology, progression and outcomes. Host systems and parasite infectivity play critical roles in the generation and manipulation of oxidative stress in malaria. Host systems involve the immunological and inflammatory responses that generate free radical species as host signalling processes as well as parasite combating and destructive entities. Parasites trigger molecules with inherent free radical generation in the host. Without the ability to synthesize amino acids, the parasite depends on the breakdown of haemoglobin to salvage the same within the food vacuole. This creates a highly oxidative stress environment from the Fenton reaction through the central ferrous moiety. Elimination of host oxidative stress process, from haemoglobin degradation product haeme, has critically evolved to protect parasites from the hostile intracellular compartment where it is an obligatory inhabitant. Parasites produce antioxidant species from both enzymatic and non-enzymatic molecules which cushions the parasites proteins from oxidation. The parasite also converts haeme, through biocrystallization, to haemozoin, a seemingly biologically inert molecule. In the presence of parasite DNA, haemozoin induces oxidative and inflammatory mediators (cytokines, chemokines, inducible nitric oxide synthase, nitric oxide, oxygen free radical, nitrogen free radicals, per-

G. A. Mavondo (✉) · C. Mhlanga · R. Dangarembizi · N. Mukonowenzou · M. L. Mzingwane
Departments of Pathology and Physiology, Faculty of Medicine, Mpilo Hospital NUST Complex, National University of Science and Technology (NUST), Bulawayo, Zimbabwe
e-mail: Greanious.mavondo@nust.ac.zw

B. N. Mkhwanazi
Dietetics and Human Nutrition, School of Agriculture, Earth and Environment Science, College of Agricultural, Engineering and Science, University of KwaZulu Natal, Durban, South Africa

J. Mavondo
Imagegate Diagnostics c/o Dalony Distributors (PVT) Ltd, Bulawayo, Zimbabwe

S. Chakraborti et al. (eds.), *Oxidative Stress in Microbial Diseases*,
https://doi.org/10.1007/978-981-13-8763-0_23

oxynitrite etc.) with a high propensity toward oxidative stress able to override host antioxidant defence systems. Other parasite proponents, e.g. glycosylphosphotidylinositol, are instrumental in negatively modulating host oxidative stress. Without intervention, the disease machineries of oxidative stress go into a vicious cycle of self-propagation mode that leads to host debilitation, cachexia and death. Current drugs are mainly antiparasitic and relieve the "disease" aspect of malaria sparingly. Phytotherapeutics and phytochemicals (asiatic acid, maslinic acid, oleanolic acid), which display anti-oxidant and pro-oxidant properties, have shown both "antiparasite" and "anti-disease" effects promising efficacy in combating malaria. These pleiotropic properties are displayed in different environments with potential to buffer malarial disease syndrome.

Keywords

Malaria oxidative stress · Phytochemical · Phytotherapeutic · Asiatic acid · *Plasmodium berghei* · Anti-parastic · Anti-disease · Maslinic acid · Oleanolic acid

23.1 Introduction

Malaria is a major tropic and subtropic parasitic disease that has been in existence for centuries where it has determined the course of history and population distributions [1, 2], shaped blood group genotypes [3], caused over two million deaths and resulted in over a billion infections a year worldwide [4]. Over 100 countries have malaria as a significant parasitic disease with more deaths (90%) occurring in sub-Saharan Africa where a child dies from the disease every half a minute. In Asian countries and South American rural areas, the disease causes substantial morbidity as well as in sub-Saharan Africa, Southern Africa and Zimbabwe in particular.

The disease manifests more strongly with higher fatalities in individuals from non-endemic areas, pregnant women and children under 5 years [5] with immigrants and travellers being the main "importers" of the disease into countries where elimination of the disease has been reported. These groups have a common factor of having lower immunity to the disease, i.e., individuals from semi- to none-endemic areas have little to no parasitic genetic priming, while pregnant women and children <5 years have lowered immune protection against the parasite. Also, in pregnant women, pathologies concomitant with placental malaria arise from the effect of a change in immune distinction from antibody-mediated immune reaction to cell-mediated immune response with subsequent excessive inflammation, oxidative stress (OS), apoptosis and diminished heat shock protein manifestation [6], inducing a high risk to the mother and the unborn child or newborn.

Compounding the malarial disease is the fact that drug resistance to *Plasmodium falciparum* has odiously crept into the cache of current drugs like chloroquine [7] and artemisinin [8, 9] with factors leading to this resistance still opaque to research and other investigations. The pathophysiology of malaria is also a hazy arena making other impinging aspects to treatment obfuscated.

Clinical complications of maternal malarial infection in pregnant women consequences include severe malaria anaemia, pulmonary oedema, hypoglycaemia, cerebral malaria, puerperal sepsis and death. Outcomes of maternal malaria in foetus may include abortion, still birth, intrauterine growth retardation (IUGR), premature delivery and low birth weight (LBW) [10, 11]. LBW of the infant has been suspected for a causal relationship between poor cognitive and neurosensory development of the child [5, 12].

In neonates, there is a protective paradigm against malaria provided by the maternally acquired immunity making the disease a considerable rare occurrence. The foetus acquires immunoglobulin G (IgG) in utero from the foetal circulation [13], albeit little evidence does exist to this assertion [14]. Lactoferrin, an iron-binding molecule, and secretory IgA, both found in breast milk and in infant sera, sequester iron from the parasite and protect against malaria [15] and reduce OS potential. Haemoglobin F (HbF) and IgG impede parasite antigen *P. falciparum* erythrocyte membrane protein-1 (PfEMP-1) cytoadherence, while low levels of metabolic substrate para-aminobenzoic acid (pABA) inhibit parasite growth in infants < 6 weeks. However, susceptibility to the disease increases in infants >6 weeks [16]. Subsequent HbF-red blood cell destruction may also avail iron into the circulation and increase in oxidative stress. Continuous exposure to malaria infections experienced by people in malaria holoendemic areas induces development of semi-immunity, which is subsequently lost as one moves from these areas [17], due to possible antioxidant defence. Oxidative stress emanating from molecular derangements, current malaria management and phytotherapeutics play critical role in malaria disease outlays as we know them.

23.2 Molecular Derangements in Malaria as Generators of Oxidative Stress

There are a number of pathophysiological changes that influence particular malarial presentations. Among these, there are the vascular bed alterations with macro- and micro-occlusions that modify blood circulation creating hypoxic conditions. Inflammation and immunological perturbations yielded from activation of leukocytes also result in increased energy demands depleting metabolic intermediates that leads to glucose homeostasis derangements. Intravascular haemolysis introduces haemoglobin-related pathologies and generation of highly oxidant materials that offsets tissues and redox balance. The malaria parasite also unleashes its own repertoire of molecular mediators within the red blood cells that alter cell membrane topography tagging them for destruction by the endothelial system. Ultimately, the vascular system changes with reduced tissue perfusion, haemolysis that reduces RBC mass and oxygen carrying capacity, inflammation that yield oxygen reactive species and increased glycolytic metabolic process that give rises to an overall hypoxic environment that drives and is driven by oxidative stress. Therefore, the different disease processes evident in malaria have a strong bearing on unique tissue- and organ-specific changes revolving around how molecular oxygen is eventually handled by the host and the parasite in the micro-environment.

23.3 Malaria Management and Oxidative Stress

The *Plasmodium* parasite is sensitive redox balance such that it has an inbuilt system of antioxidant processes that protects it from damage by the highly oxidative environment obtaining in the RBC in malaria. Treatment of malaria has, thus, been aimed at overriding the antioxidant capacity of the parasite to enable the oxidative capacity to eradicate the food vacuole where the parasite has created growing space with the cell. One of the methods the parasite controls the highly oxidant haeme created from broken down haemoglobin is through its conversion to haemozoin through a biocrystallization process. Drugs like chloroquine and quinine interfere with this process and increase toxicity in the food vacuole and this way destroy the parasite during the blood stage. Artemisinin and its derivatives elicit oxygen reactive species that destroy the parasite, albeit in a non-selective fashion. Therefore, oxidative stress is used in the management of malaria even though both the parasite and the host also create more of the same that eventually cause disease through redox imbalance.

23.4 Emerging Malaria Treatment and Oxidative Stress

Triterpenes, second metabolites of certain plants, are emerging as treatment options due to the pleiotropic nature displaying both oxidant and antioxidant capacities depending on the prevailing environmental factors. They are thought to balance the redox environment in the food vacuole precipitating an unfavourable condition that destroys the parasite while reversing the effects and presence of pro-oxidant mediators of inflammation, immunological disturbances, hypoxia and vascular pathologies.

This chapter presents the interplay of malaria, oxidative stress-induced pathology and management of oxidative stress using current and upcoming remedies. An overall interplay of the disease process and how the disease environment has a feed-forward mechanism that, without intervention, will bring debilitation, morbidity and mortality is presented. How the intervention eventually manipulates the disease process, through both host and parasite manipulations, towards wellness is enunciated. How pathophysiological manifestations driven by OS during or after successful treatment of infection, which are major causes of high morbidity and mortality associated with malaria, may be ameliorated is also explored.

23.5 Biological Sources of Oxidative Stress and Its Handling in Malaria

The natural host defence mechanism when assaulted by *Plasmodium* malarial infection responds through the recruitment of phagocytic first-line immune cells that combat invasion [18]. Macrophages and neutrophils generate large quantities of both reactive oxygen species (ROS) and reactive nitrogenous species (RNS) that disrupt the equilibrium naturally existing between pro-oxidants and antioxidants in the host systems. The disequilibrium so created results in OS. The OS generated is meant to cause the creation of the parasite-killing apparatus for the host but more often than not it viciously turns its insults more on the host than the parasite. However, OS has been shown to kill or promote killing of the parasite in vitro. *P. yoelii* species were incubated in the presence of glucose and glucose oxidase, and H_2O_2 was produced, which enabled OS capable of killing the parasite. Incubating the same parasite species in the presence of xanthine oxidase yielded superoxide free radical ($O_2^{\bullet -}$), which resulted in more oxidative bursts consequently destroying the parasites [19]. Malaria-infected hosts (both humans and rats) have, compared to uninfected controls, displayed higher oxidative markers in general [20–23]. Increased free radicals tend to increase OS in such incidences when compared to decreasing antioxidants shown by increased concentrations of malondialdehyde (MDA), an important lipid peroxidation marker [24].

Thrombocytopaenia-associated anaemia pathogenesis has been suggested to be associated with OS formation as shown by reduced platelets count, superoxide dismutase (SOD) and glutathione peroxidase (GSH-Px) activities, while platelet lipid peroxidation was elevated in *vivax* malaria patients. Platelet lipid peroxidation and platelet count display a negative correlation, suggesting the thrombocytopaenia pathogenesis, which presents in malaria patients, could arise from the loss of membrane elasticity, increasing brittleness and receptor dysfunction that bring about thrombocytes functional impairment [18].

Free radicals are critical in the pathogenesis of rheological changes in malaria patients. This understanding may influence negatively iron replacement therapy in malaria infected even in the worst degree of anaemia as shown by RBC count [25, 26].

23.5.1 *Plasmodium* Infection-Induced Oxidative Alteration in the Host

While generations of free radical species are part and parcel of the organism, an antioxidant defence system has been created as well in response to the increasing oxidative stress pressures. Antioxidant defence systems evolved naturally to become physiological aspect of the organism protecting against damage caused by free radicals and depend on the utilization of cellular and systemic antioxidant reserves.

Synthesis of endogenous antioxidant is made up of interlinked systems of enzymatic, small molecules and metal chelators, which forestalls or prevents oxidative damage of biomolecules. Oxidative species generation or creation is avoided through the antioxidant ability to scavenge or reduce oxidants as they self-oxidize to form less reactive compounds [27].

By acting directly on some free radicals, the most important enzymatic antioxidant defence systems of glutathione peroxidase (GSH-Px), catalase and superoxide dismutase (SOD) reduce their reactivity although they are impotent against the highly reactive free radicals chiefly responsible for oxidative pathology processes in malaria. Examples of antioxidant-resistant free radicals include hydroxyl and perhydryl radicals and peroxynitrite. For an ancillary antioxidant defence system to combat the highly reactive free radicals, the body utilizes small molecules.

The group of small antioxidant molecule is large in number to include molecules like vitamins A, C and E, β-carotene, uric acid and reduced glutathione (GSH). Certain proteins that bind transition metals to prevent them from driving the Fenton and Haber-Weiss reactions are essential sources of reactive species. Ferritin, transferrin, the iron chelator lactoferrin, ceruloplasmin and albumin, which are copper chelators, and metallothioneins, armed with thiol groups that bind heavy metals, comprise the group of metallo-directed antioxidant [28].

The GSH molecule appears as the most prominent and powerful antioxidant protecting the eukaryotic cell in the human defence against OS utilizing many different mechanisms of action [27]. On the other hand, tumour necrosis factor-α (TNF-α) seems to drive OS through the manipulation of GSH homeostasis, which plays a crucial role in malaria pathophysiology. The behaviours of TNF-α when injected into rats revealed a reciprocal relationship of reduced GSH concentration as compared to the oxidized (GSSG) form in CD4+ and CD8+ spleen T cells [29] which point at increased OS whichever way this is observed from. Decreased GSH has been reported as well in malaria [27, 30]. In *P. vivax* infection, there is a generalized decrease in several antioxidants to include antioxidant enzymes including glutathione S-transferase (GST) [20, 31]; GSH-Px; catalase; SOD [30, 32]; NADPH-methaemoglobin reductase [32]; chelators of heavy metals (desferrioxamine, salicylaldehyde isonicotinoyl) [33] and ferritin [30]; vitamins A, E and C, the provitamins [34, 35]; α- and β-carotene; lycopene; lutein; and zeaxanthin [36], among many.

Lower levels of antioxidant ascorbate correlate with disease severity as compared to elevated levels of urate and ceruloplasmin [37]. Ischaemia-reperfusion syndrome (IRS) associated with free radical production in ischaemia or hypoxia may be indicated by high concentrations of urate resulting from parasite-induced haemolysis and cytoadherence. The severity of parasitaemia is also directly related to decreased activity of glutathione S-transferase (GST), once known as ligandin, as the enzyme is involved in the reduction of malaria complications and occurrence of severe malarial disease [20, 31]. Invariably, disease severity may be indicated by estimation concentrations of GST and lipid peroxidation (MDA) and catalase activity.

Acute-phase proteins (APP) are useful markers of severity and the type of inflammation in malaria. With the exception of α1-antichymotrypsin, most APP are significantly correlated with splenomegaly. C-reactive protein (CRP) [38–40] and α1-glycoprotein (AGP) are key markers of chronic inflammation in malaria [41]. When concentrations of albumin, apolipoprotein A1 (ApoA1), transferrin, zinc, vitamin A, immunoglobulins G and M, interleukin 10 (IL-10), TNF-α and interferon-gamma (INF-γ) were estimated, children with malaria had decreased levels of ApoA1 and albumin while IL-10 concentration.

In *P. berghei* murine malaria infection, significantly increased lactate [42] and alanine have been observed in the final stage of severe malaria syndrome as well as slightly elevated essential amino acids in the brain [43].

Intriguingly, *P. vinkei* infection in mice seems to confer protection against ROS through the action of the enzymes SOD, catalase, GSH-Px, glutathione reductase, NADPH and NADPH-methaemoglobin reductase during the erythrocytic stage of malaria [44].

Oxidative stress damage, mediated by polymorphonuclear leukocytes, may be reversed by haem-oxygenase 1 (HO-1) and ferritin mRNA induction as well as administration of antioxidants catalase, GSH and SOD [30]. Iron chelators (e.g. desferrioxamine, DFO) administration in malaria may reverse OS damage through inhibition of *P. falciparum* growth [33].

When mice are infected with *P. berghei* murine malaria parasite, they develop hepatocellular injury and display increased mRNA expression of interleukin-12 (IL-12), protein 40 (p40), IFN-γ, interleukin-4 (IL-4) and IL-10 with resultantly increase in NO synthesis. Anti-IL-12 treatment of such infected animals reduces free radical generation indirectly. This prolongs survival and reduces liver damage and weight loss but does not reduce parasitaemia, making malaria infection tolerable [45].

In a study of children with uncomplicated *P. falciparum* malaria, antioxidant activity was related to the disease pathogenesis. Acute malaria was linked to low plasma antioxidant activity. Fast parasitaemia clearance was found in children with highest lycopene concentrations [36].

Riboflavin and other antioxidant substances are adjuncts in drug therapy, as is lycopene, and provide reducing power in the parasite food vacuole and against haemozoin formation [46]. Circumsporozoite protein (CSP), an essential component of host cell invasion by the parasite, is inhibited by allicin, a cysteine protease inhibitor present in garlic which shares the same antioxidant mode of action with riboflavin against malaria [47].

23.5.2 Sources of Oxidative Stress in Malaria

There are five sources of OS commonly observed during disease pathophysiology: (1) inflammatory reactions initiated as host defence against infection, (2) catalysis of transitional metals such as free iron released from haemoglobin after parasitic feeding, (3) infection triggered cytoadherence and anaemia that cause ischaemia-reperfusion syndrome, (4) reactive species generated by the parasite directly, and (5) antimalarial drug activity during treatment.

Essentially, sources of redox-active materials originate from the human host as well as the *Plasmodium* parasite activities.

During the metabolic processes of the parasite, within the parasitophorous (food) vacuole (PV or FV), haemoglobin is broken down to yield amino acids from the globin fraction while releasing the haem component with its attending Fe(II) which are highly pro-oxidant. However, the preceding steps by the parasite are the detoxification of the haemoglobin through biocrystallization of the haem to haemozoin which is an inert redox-reaction component although a very biologically active one. As a result, to disrupt the redox balance, three groups of compounds are generally noted to include (1) molecules that bring about the de novo synthesis of ROS and thus lead to parasite demise, (2) molecules which impede the activities of redox balancing enzymes, and (3) inhibition of formation of molecules that restrict in the foraging of pro-oxidant metabolic products like haemozoin.

Oxidative stress depends mostly on the concentrations and functional balance between oxidant and antioxidants in the body of which the former may be grouped as organic or inorganic (metals) in origin. Human biology is critically anchored on metals and their homeostasis which is maintained within very narrow ranges and strict regulatory mechanisms of their uptake, how and where they are stored and their secretion or excretion from the body [48]. Metal ion transport mechanisms are required to keep ions within body compartments engendering life preservations by their roles in cellular functions of synthesis, immunology, respiratory, regulatory and reproduction such that derangements of metal-ion homeostasis will invariably affect some if not all of these processes. Metal-ion homeostasis dysregulation invariably results in decompartmentalization of metal ions with random metal bindings to unnatural binding sites which alters protein function triggering oxidative deterioration of micro- and macromolecules [49]. A plethora of diseases have been associated with metal-ion homeostasis dysregulation and oxidative stress-related tissue damage being the underlying cause [50–52].

Iron is one of the most abundant ions in the body where it is found within the red blood cell involved in oxygen transport and respiration. Abnormalities of iron metabolism are found in either iron deficiencies or overload. Reduced iron availability will result in impaired iron-containing proteins production, e.g. haemoglobin and cytochromes. Cell growth is inhibited with subsequent immature cell death and anaemia of iron deficiency [53]. Haemochromatosis, a hereditary disease associated with iron overload, results in tissue damage from free radical toxicity [54]. Modulation of gene expression may be regulated by metals interfering with signal transduction [55] bringing about deregulation of cell proliferation by activating transcription factors involved in cell cycle progression and apoptosis [56].

Redox-active metals maintaining the cellular redox state such as iron, copper, chromium and cobalt are linked to the generation of free radicals and are maintained within strict physiological reference intervals [52, 57]. Redox homeostasis is a phenomenon closely linked to redox-active metals where the latter is involved in metal cycling reactions and participate in the transfer of electrons to substrates [58].

Therefore, metal homeostasis disruption may lead to uncontrolled metal-induced construction of deleterious free radicals contributing in modifications to DNA, increased lipid peroxidation, altered calcium metabolism and sulfhydryl homeostasis [59, 60].

Redox-active metals rejuvenate through redox-cycling reactions; however, redox-inert elements have their primary route for detoxification depending on the concentration of oxidized-reduced glutathione balance, binding to sulfhydryl groups on proteins and other mechanism [61–63].

23.5.3 Host Defence Mechanism Against *Plasmodium* Infection Utilizes Oxidative Stress

Oxidative stress has a bearing in the protection against malaria in infected persons through the destruction of the *Plasmodium* parasite. H_2O_2 and $O_2^{\bullet -}$ are capable of operating autonomously to become cytotoxic agents or be converted to other toxic compounds that comprises hydroxyl radicals ($OH^{\bullet}$), hypochlorous acid (HOCl) and peroxynitrite ($ONOO^-$) [21]. Increased ROS production by phagocytes is part of the non-specific effectors in the primary host defence arsenal. There is a potential to contribute to oxidative damage in the parasite as well as in the host pRBCs, once ROS breaches pRBC's membrane [64]. In the same manner, proteolytic enzymes and ROS produced by neutrophils tend to trigger apoptosis of endothelial cells at low concentrations and at high concentrations they cause cell necrosis [65]. Intraerythrocytic destruction of the parasite by ROS ($O_2^{\bullet -}$, $ONOO^-$) occurs [23]. Endothelial injury is associated with elevated concentrations of cytokines emanating from activated neutrophils and monocytes in malarial infection. Organ failure and severe malaria, therefore, are a result of activation of neutrophils and their secreted products that cause both antiparasitic activity and endothelial damage in the process [65].

Immune response to malaria is usually insufficient in elimination of some species of the *Plasmodium* parasite and may act in ways that harm the host cell to some extent. Inflammatory response is usually necessary as the homeostasis of the host is altered to accommodate infection. The cytokine profile proven effective in combating malarial infection does so through a mechanism that involves T helper lymphocytes (Th1) with prevalence of TNF-α, IFN and IL-12 being secreted. On the opposite side, Th2 profile which consists of IL-4, IL-5, IL-10 and transforming growth factor-β (TGF-β) is seen to aggravate severe conditions of the malarial disease. Consensus on the correctness of these profiles is rather absent [64, 65] as different cellular responses have their effectiveness in different stages of parasitic elimination. The Th1 profile is essential at the initial process of infection and allows the host to mount an adaptive response against the parasite which turn to favour severe malaria anaemia process as the disease becomes chronic or prolonged. The Th2 promotes the clearance of the Th1 response giving way to the adaptive action response. Proper activation of these processes is essential to enable parasitic succumbing of the parasite to the host defences [66].

Most anti-malaria treatment is based on the participation of ROS/RNS in the parasitaemia elimination process and their regulation of the immune system response through stimulating or inhibiting production of particular cytokines, or transcription factors or even regulate cell death or apoptosis in the manner haeme does [67]. When cell death occurs from an internal process without releasing the intracellular contents to the extracellular environment, the process is referred to as apoptosis. In this process, intracellular contents do not cause damage to the extracellular matrix in a manner different to necrosis. This process is regulated by a mitochondrial protein, cytochrome c, among other substances. Under normal conditions, cytochrome c is involved in electron transfer when linked to cardiolipin, a phospholipid [68]. The apoptosis process may be activated under stressful conditions through the oxidation of cardiolipin which releases cytochrome c to form complexes with activating proteins of the apoptosis process. Cardiolipin oxidation by the hydroxyl radical appears to trigger lipid peroxidation in the apoptosis process of the hepatocytes in malaria [69].

Besides lipid peroxidation, NO may play a significant role in eradicating the parasite. Despite NO contribution as a free radical in *Plasmodium* destruction, increased concentrations of NO induce immune suppression that may lead to cerebral malaria development [70]. Haemozoin, a malarial parasite pigment formed from biocrystallization of haem, is well recognized for inducing NO synthesis directly by the enzyme nitric oxide synthase (iNOS) in macrophages. Extracellularly signal regulation kinase (ERK) and nuclear factor κβ (NF-κβ)-dependent pathways are involved in the generation of NO. At liver stage of parasite development, host defence mechanism is closely connected to the production of INF by natural killer (NK) cells and subsequently synthesized NO [71].

Haemozoin, apart from inducing NO generation, also activates macrophages through NO-partially dependent mechanisms [72] and on other ROS/RNS such as $O_2^{\bullet-}$ and $OH^{\bullet}$. Ironically, increasing activities of iNOS in human monocytes are not accompanying worsening malaria in *P. falciparum*-infected individuals [73].

Mice resistant to *P. berghei* cerebral malaria tend to have enhanced expression of cytokines, excluding IL-4 and RANTES, and NO in comparison to susceptible animals [74]. Increased production of TNF-α by malaria-resistant mice suggests that macrophage activation is significantly higher in these animals, which corresponds to increased elicit of toxic reactive oxygen intermediates (ROI) and parasite elimination [75]. Furthermore, free haem stimulates migration of neutrophils and production of ROS/RNS by a G-protein-coupled receptor especially the Gα receptor. Activation of protein kinase C follows with increasing inflammatory response [76] and delay apoptosis [67] and possibly contributes to malaria-induced immune suppression [77].

Oxidative stress and reduction of parasitaemia are correlated with an increase in granulocyte-macrophage colony-stimulating factor (GM-CSF). The stimulatory activity of the cytokine GM-CSF on granulocytes and macrophages promotes an increase in the number and activity of these cells to provide a cellular immunity against malaria. When administered alone or in combination with other cytokines, GM-CSF has been shown to protect against a number of parasitic infection [78, 79].

When animals lack or have low activity of GM-CSF, they tend to have impaired immune response to blood stage malaria parasitic infections [80]. GM-CSF and IL-4 are susceptible to modification by lipid peroxidation from products originating from haemozoin released by pRBC rapture, e.g. 4-hydroxynonenal (4-HNE), linking the cytokines to oxidative stress. What more, the two cytokines stimulate the differentiation of monocytes that are laden with haemozoin into dendritic cells. The monocyte-dendritic cellular differentiation, however, may be inhibited by 4-HNE, thus playing an important immune suppression role seen in malaria [81]. Additionally, the mechanism seems to be related to the expression of peroxisome proliferator-activated receptor gamma (PPAR-γ) in differentiation and maturation of dendritic cells laden with haemozoin [72, 82]. Kaur and others have also shown another relationship of GM-CSF and NO. Pre-treatment (prophylaxis) of malaria parasite-administered mice with GM-CSF and methionine encephalin (TGG) protects them from developing the disease [79]. Nevertheless, iNOS inhibition increases the mortality significantly alluding to the fact that NO conferred protection against malaria to some extent and also that the protection is exerted through the GM-CSF/TGG mechanism with OS a major player.

23.5.4 Nitric Oxide Mediation of Oxidative Stress in Malaria

The characteristics of NO present as a small and simply biosynthesized molecule [83], inorganic, colourless, free radical gas, comprising of 7-nitrogen and 8-oxygen electrons [84]. Naturally, NO is sometimes beneficial and sometimes harmful [85]. New dimensions of interest in NO have demonstrated it as an endogenous molecule with essential biological effects. NO reacts with guanylate cyclase in the smooth muscle cell to produce vasodilation. The functionally reported aspects of NO include blood pressure regulation and maintenance, inflammation and non-specific immunity activity, neuronal plasticity, peripheral nervous system and neurotransmission potential. This molecule is also capable of lysing tumour cells [86]. The main substrate for NOS, to produce NO, is l-arginine. Inflammatory response by macrophages and other immune cells expressing iNOS produces significant quantities of NO which tend to have lethal oxidative stress capabilities [87]. Endothelial NO synthase (eNOS), in the vascular endothelial cells, synthesizes NO that at times diffuses into the blood vessel lumen to facilitate red blood cell functions. Hb also facilitates the transportation of NO to the vasculature [88].

Notwithstanding being an essential cytotoxic mediator of activated immune cells with the capacity to destroy pathogens and tumour cells, NO has a high toxic potential during OS leading to reactive oxygen intermediates generation leading to antioxidant deficient in certain systems [85].

While the role of NO in erythrocytic stage malaria remains unknown, it plays essential role in physiological functions to such an extent that various diseases processes are associated with either elevated or reduced NO activity in the body. After comparing serum NO concentrations in *Plasmodium*-infected mice and humans, many researchers have concluded that nitrite and nitrate concentration differ insignificantly from controls throughout the blood stage of malaria. These findings invariably indicate an inadequate parasite-killing capability of NO as parasitaemia differed not in untreated controls to treated individuals [23, 89–92]. The protective role of haemoglobin against ROS toxicity has been cited as the reason why NO is ineffective in inhibition of the progression of malaria parasitaemia [23]. Furthermore, for NO to be able to effectively combat the parasite, concentrations of the gas need to reach near saturation point first. In iNOS-/- knockout mice, parasitaemia is observed to be similar to that of controls [91] and that NO is not necessary for parasite eradication in murine malaria [93].

However, other researchers suggest that NO has a protective function against erythrocytic stage of malaria as elevated concentrations of NO are seen as essential in *P. falciparum* malaria resolution [66, 94].

The seemingly contradictory findings in NO's efficacy in malaria is attributable to that the physiological gas acts directly and indirectly in malaria resolution. The direct antiparasitic effect of NO emanates from the action of peroxynitrite ($ONOO^-$) created by the reaction of NO with $O_2^{\bullet-}$. Indirectly, NO increases immune activity through the association of NO and GM-CSH/PPAR-γ and also the interaction of NO and increased ROS/RNS. The gas may also influence the inflammatory process directly while indirectly enabling cytokines to activate immune response to malaria.

In the NOS isoenzymes, the dimeric flavoproteins contain tetrahydrobiopterin, homologous to cytochrome P450, in an arrangement that distinguishes enzymes according to their physiological characteristics. Thus, they are named accordingly as inducible NOS (iNOS), endothelial NOS (eNOS) and constitutive NOS (cNOS). Location of and cell type NOS is used to identify them namely as NOS1 (cytoplasm of neurons) [95], NOS2 (cells of the immune system) [96] and NOS3 (endothelial cells) [97]. An important role of NO produced by iNOS through induction by haemozoin has been observed in the killing of *Plasmodium* parasites [98]. Endothelial NOS acts as a blood flow regulator inducing vasodilation, preventing platelet aggregation and inhibiting lymphocytes and monocytes adhesion to the endothelium [86] and prevents local ischaemia. Cerebral malaria is prevented in this manner affirming the use of exogenous NO as an adjuvant treatment for malaria [99]. Here it improves microcirculation, reduces brain inflammation, protects the blood-brain barrier [100, 101] and decreases oxidative stress. Administration of NO tends to have a prophylactic effect against lung injury in malaria [102].

23.5.5 Redox Reactions of Nitrite and Oxidative Stress in Malaria

Nitric oxide synthase is involved in the synthesis of NO which regulates a number of cellular activities. The molecule's strong pro-oxidant capacity is enhanced when it reacts with superoxide to produce peroxynitrite ($ONOO^-$). In the presence of O_2, NO is very unstable and easily forms N_2O_3 and nitrite which reaction contributes to the 500 nM of nitrite in plasma [103]. The concentration of NO entering the RBC is limited in this manner. The little amount of NO entering the RBC either reacts with oxyhaemoglobin of which nitrite is the product or irreversibly binds to deoxyhaemoglobin. This makes the RBC essentially the sink for endothelially generated NO.

Low concentrations of NO on the β-93 cysteine residue of haemoglobin bring about the assumption that the RBC may be involved in the transportation of NO to the vasculature [88]. Labile bioactive NO may be produced in the RBC from a redox reaction involving plasma produced nitrite and deoxygenated haemoglobin at low partial pressure of O_2. The reaction between nitrite and Hb Fe^{2+} releases hydroxide (OH) ions from the protonated nitrite yielding a nitrosonium cation [104]. Ferric iron (Fe(III) and NO are generated when electron transfer occurs from ferrous iron to the nitrosonium ion. Affinity for the formed Fe(III) by NO is much lower as compared to the Fe(II); therefore, labile NO is formed from this reaction. Analogous to this reaction is the autoxidation process of the redox reaction between ferrous haem and O_2. However, the former is a more efficient reaction since the nitrosonium cation is a much stronger oxidizing agent than O_2.

By altering the concentration of nitrite, NO production may be regulated by this reaction and in a way lessen oxidative stress. Possible oxidative damage may occur from the pro-oxidant NO. The labile form is able to leak from the RBC into the intravascular compartment where it will provide a source for NO regulation of blood flow other cellular functions.

23.6 Haemoglobin, Haemolysis and Oxidative Stress in Malaria

The erythrocytic phase of malaria is epitomized by parasitized red blood cell (pRBC) and non-parasitized red blood cell (npRBC) breakdown and release of haemozoin (ferriprotoporphyrin IX dimers and monomers [FP]), methaemoglobin and parasite proteins. Monocytes and macrophages release cytokines (TNF-α, IL-1β) when activated by haemozoin. Free haem yields free radicals with a high potential for oxidative damage to both parasites and host cells. Accumulation of free haem is one mechanism that chloroquine uses to disrupt plasma membrane structure and increase oxidative stress (OS) in *Plasmodium*. Haemolysis occurs when structural and functional changes to the fluid lipid membranes of RBCs have been peroxidized by OS. High concentration of thiobarbituric acid reactive substances (TBARS), markers of lipid peroxidation, has been observed in *P. falciparum*, *P. vinckei*, *P. berghei* and *P. chabaudi* pRBCs [105].

Inflammation and OS arise from haemolysis which increases free haeme. Haeme is capable of activating neutrophils through chemoattractant signalling. Molecules that are able to inhibit this haeme induced migration may be important therapeutics in inflammatory arising from either bleeding or haemolytic disorders. Mesoporphyrins, which lack vinyl groups in their rings, are such molecules that experimentally inhibited haem-induced neutrophil migration selectively [76]. Antimalarials that work to promote selective haem-induced ROS tend to find scope of development, e.g. a[(aryl)arysulfanylmethyl] Pyridine (AASMP)[106].

23.6.1 *Plasmodium*-Infected Red Blood Cell and Oxidative Stress in Malaria

The viscosity of blood increases at the epischizogeny developmental stage where *P. falciparum* trophozoites cause changes to the surface of the pRBCs. Consequently, there is a development of endothelial cell adhesion antigens as a parasite defence against spleen passage, sequestration and destruction [107]. Ironically, the increased viscosity and cell adhesion tendencies are the major causes of capillary blockage in the kidney, pulmonary and brain, making cerebral malaria the most common cause of coma and death in infected children [108]. Ischaemia and hypoxia result from the capillary occlusion with more ROS creation, forming an autonomous and self-perpetuating vicious cycle.

Changes occurring on the RBC surface include the lipid peroxidation phenomenon. Parasitized red blood cells encompass huge amounts of monohydroxy derivative of polyenoic fatty acids (OH-PUFA) in their lipids, indicating possible lipid peroxidation from released haem iron [109]. The toxic 12-and 15-hydroxy-arachidonic acid (HETE) is a well-known OH-PUFA whose concentration increases according to the evolutionary parasite stages.

Moreover, *P. falciparum*-pRBC connected with premature and accelerated aging contribute substantially to the severe malaria anaemia (SMA) [110]. The RBC senescence-driven anaemia alters circulatory physiology leading alternate existence moments of tissue hypoxia and tissue oxygenation at basal level. In this there is a preponderance for ischaemia-reperfusion syndrome (IRS) which generates free radical production [28]. Undeniably, amplified lipid peroxidation and oxidative stress described in malaria can disturb the membrane of pRBCs and npRBCs and also reduce cells deformity, which is linked to increased mortality in malaria. Increased cell rigidity has deleterious outcomes such as microcirculation blockage, which worsens tissue hypoperfusion, and cell stiffness that leads to removal by the spleen fuelling SMA [111]. In *P. falciparum* infection, however, pRBCs with mature trophozoites and schizonts avoid clearance by the spleen through sequestration in various organs capillaries. The large number of antigenic variants encoded on the surface of the parasite binds to the vascular endothelial receptors in what is called the cytoadherence phenomenon [112].

Stimulation of the var gene creates the cytoadherence phenomenon parasite proteins expression on the surface of pRBCs, e.g. *P. falciparum* erythrocyte membrane protein 1 (PfEMP1). The expression of the proteins allows nRBCs to attach to various host molecules located in the vascular endothelium which include intercellular adhesion molecule type 1 (ICAM-1), platelet endothelial cell adhesion molecule (PECAM), vascular cell adhesion molecule (VECAM), hyaluronic acid and heparin sulphate bringing about blood flow interruption and damage to the tissues flooded by the congested vessels [112, 113]. There are other proteins associated with membrane rigidity of pRBCs such as the histidine-rich protein associated with deformity (KAHRP) and erythrocyte membrane protein of *P. falciparum* 3 (PfEMP3). When these parasite proteins are absent, membranes display less altered mechanical properties [114].

23.7 Ischemia, Reperfusion Syndrome (IRS) and Oxidative Stress in Malaria

Sequestration of pRBCs in certain tissues during malarial paroxysm and their cytoadherence to blood vessels is the bed rock of IRS conception. Restriction of blood flow invariably leads to lower O_2 concentration to affected tissues which uncouples mitochondrial oxidative phosphorylation with reduced ATP generation. Stored cellular ATP consumption follows thereafter with creation of adenosine diphosphate (ADP) and adenosine monophosphate (AMP) and finally adenosine. Reperfusion may restore ATP from ADP through phosphorylation. Persistence of hypoxia leads to irreversible metabolism of adenosine to inosine and hypoxanthine. Sodium-potassium ATPase pump fails in hypoxia due to shortage of ATP which consequently causes accumulation of Na^+ and K^+ intracellularly and extracellularly, respectively. Increased osmotic pressure and cytoplasmic decompartmentalization will result from the occurrence of cellular oedema.

Cytoplasmic decompartmentalization is a cellular distress condition in which ions kept in watertight compartments flood the cytoplasm with resultant activation of enzymatic systems that should be inactive. Communication processes are thus triggered or free radical production is promoted. Calcium ions and Fe^{2+} become important in this situation. Free Ca^{2+} in the cytoplasm activates the protease calpain followed by the breaking of the peptide bond for xanthine dehydrogenase (XD) to alter its activity to xanthine oxidase ((XO). Conversion of hypoxanthine to xanthine by XO, unlike XD, required O_2. In ischaemia, three events take place, mainly: hypoxanthine production from ATP, conversion of XD to XO and cytoplasmic decompartmentalization. Iron is subsequently released from ferritin into cytoplasm. During reperfusion, on the other hand, hypoxanthine metabolism by XO is catalysed by the renewed O_2 supply with resultant products of xanthine, uric acid, $O_2^{\bullet-}$ and H_2O_2.

Production of ROS/RNS mainly occurs during reperfusion, where $O_2^{\bullet-}$ and H_2O_2 react together in the presence of iron (Haber-Weiss reaction) to yield $OH^{\bullet}$ radical that initiates magnification and propagation of oxidative damage.

23.8 Molecular Haemoglobin (Hb) and Oxidative Stress in Malaria

There is a continuous exposure of haemoglobin (Hb) to redox reactions which reactions with oxygen (O_2) produce superoxide and hydrogen peroxide (H_2O_2). Ferryl haemoglobin is produced when haemoglobin reacts with H_2O_2 which may degrade the haem. Nitrite found in the circulation may react with deoxygenated haemoglobin to produce NO which in turn produces $ONOO^-$ when it reacts with superoxide. The reactive species produced have both detrimental outcomes (OS) and beneficial effects such as of NO endothelial relaxation.

When malaria parasite merozoites invade the erythrocyte, they consume the red blood cell cytoplasm, haemoglobin included, to provide source of amino acids, toxic reactive oxygen species and free haem. The parasite evades the toxicity by the formation of relatively inert haemozoin from toxic free haem. Intrinsic red blood cell redox reactions and methods to increase or inhibit them provide potential techniques that may be utilized to increase parasite toxicity and/or minimize oxidative stress experience by the human host in malaria.

23.8.1 Auto-oxidation Redox Reactions and Oxidative Stress in Malaria

Reversible binding of O_2 to Fe(II)Hb is required during its transportation which is accompanied by a slow rate of haemoglobin autoxidation at 3% per 24 hours and subsequent superoxide release [115]. There is a quadratic increase in autoxidation at reduced O_2 pressure as in hypoxia of malaria when Hb is partially oxygenated. During reduced O_2 pressure, there is a tendency of Hb binding to the red blood cell membrane that makes the ROS generated from autoxidation to be in close proximity to sites they are inaccessible from antioxidant activity of the cell. This situation invariably exposes the red blood cell membrane to ROS damage and/or to their leakage out of the red blood cell thus increasing systemic oxidative especially in the microcirculation where O_2 partial pressure is naturally low [115]. The same phenomenon is amplified in malarial parasite pRBC.

The superoxide from autoxidation undergoes dismutation to H_2O_2 which initiates a secondary cascade of oxidative reactions (Fig. 23.1). In this reaction cascade, H_2O_2 reacts with Fe(II) Hb resulting in the formation of a highly reactive Fe(IV)

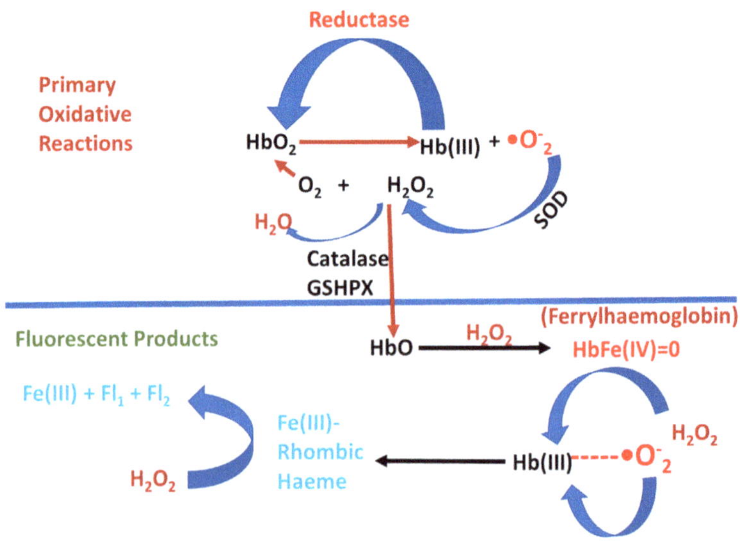

Fig. 23.1 The cascade of primary and secondary haemoglobin oxidative reactions triggered by autoxidation [115]

ferrylHb. A second H_2O_2 reacts with ferrylHb to yield superoxide radicals which attack the porphyrin before it can escape from the haem pocket [116]. A high-spin rhombic haem complex [117] with a g = 4.3 electron paramagnetic resonance signal is generated as the initially damaged tetragonal geometry haem rearranges. Further reaction with H_2O_2 of the complex causes haem degradation which releases iron and the formation of several fluorescent degradation products [116].

Interestingly, red blood cell antioxidant systems do not react with either the rhombic haem complex or the fluorescent degradation products which tend to accumulate in vivo initiating the additional oxidative processes that reflect the oxidative stress experienced by the RBC [118] which can also occur in malaria. The malaria parasite, on the other hand, does not have the antioxidant systems to combat rising rhombic haem and uses the conversion of haem to haemozoin as the detoxification method.

23.8.2 Haemoglobin Regulation of Redox Reaction in Hypoxia of Malaria

When fully oxygenated, Hb is quite stable and the autoxidation rate in negligible. However, nitrite may still react with oxyhaemoglobin in a complicated multi-step reaction which gives rise to oxidation of both Hb and nitrite. Methaemoglobin and nitrate are formed in the process without a direct electro-transfer from haem iron to nitrite, which normally requires reduced O_2 partial pressure and dissociation of some O_2 from Hb. While O_2 removal is necessary for a direct reaction of nitrite with haem iron, distal histidine is essential for both autoxidation and reduction of nitrite. The anion nitrite will normally not be able to react with ferrous haem. However, the distal histidine is required in the protonation of nitrite as it binds bringing about the cleavage of a OH ion and the nitrosonium cation formation that becomes bound to Fe(II) haem. In the malarial parasite parasitophorous vacuole, this process is avoided by the formation of haemozoin that sequesters the haem molecule away from the NO and other oxidizing agents. Coupled by the antioxidant systems of the parasite and generation of excess glutathione exported to the host cell cytoplasm, the parasite is usually safe from oxidative stress until the haemozoin formation mechanism is "busted" by antimalarials that infringes on the haem biocrystallization system.

Overall, Hb redox reactions produce reactive species under hypoxic conditions. These $O_2^{\bullet-}$, ferrylhaemoglobin, haem degradation products, free iron, NO and $ONOO^-$ are some of the products of the redox reactions. Oxidative stress increase may be experienced by subjects with malaria due to the reactive species. However, treatment for malaria may harness this phenomenon of haemoglobin reactions to destroy the parasite preferentially. On the other hand, Hb can be able to attenuate the hydrogen peroxide-induced oxidative stress through antioxidative peroxidase activity.

23.9 Haemoglobin Attenuation of Hydrogen Peroxide-Induced Oxidative Stress Through Antioxidative Peroxidase Activity

Oxidative stress and antioxidant defences are in built with the physiology of life in general. The protective role of free and bound Hb is an interesting phenomenon against a backdrop of most oxidative damage having been explored. Intravascular haemolysis, extravasation of red blood cells and micro- or macroscopic tissue injury lead to creation of free Hb. Free Hb consumes NO through the dioxygenase reaction and may be involved in vasoconstriction leading to hypertension [119, 120]. Hb may react with physiologic oxidant like H_2O_2 and cause oxidative cell damage through its peroxidative properties. In malaria, there is a large intravascular Hb exposures which may cause lipid peroxidation and eventual renal lipid peroxidation protein modification [121]. Evidence to Hb involvement in kidney parenchymal damage is given by the oxidatively modified Hb found in urine excreted by the exposed animals [122]. Propagation of inflammatory reactions that enhance cellular susceptibility to OS has been reported for free haem, Hb oxidation products or both [123, 124].

The reaction that yields oxidative toxicity by Hb and myoglobin (Mb) involves Fe(III) and H_2O_2. Consumption of H_2O_2 forms radicals that can abstract environmental hydrogen and initiate oxidative damage. Haem-protein reaction in a H_2O_2 environment provides an antioxidative protective and oxidative damage which needs to be balanced. Extensive haem degradation, α-globin cross-linking and specific β-globin amino acid oxidation resulting from Fe(II) and Hb reaction have been reported [125–127]. In malaria, the extensive Hb liberation from infected and non-infected red blood cells haemolysis may have its oxidative effects mitigated through the cross-linking reactions that may absorb a significant amount of radical-derived oxidative impact protecting the physiological environment. Compatible with the protective activity of Hb involvement in the neutralization of haem-associated oxidative processes is the extensive oxidation of Hb, globulin-chain amino acid and protein precipitation [125]. In an experiment to measure the effects of glucose-oxidase generated H_2O_2, extensive Hb amino acid oxidation and precipitation suggested that globin acted as scavenger for haem-generated radicals [125]. Hydrogen peroxide-Hb-generated radicals oxidized primarily Hb β-globin amino acids Cys93, Trp15 and Cs112 with the H_2O_2 originating from haeme redox cycling process.

The structural consequences of the Hb β-globin amino acid oxidation and formation of altered haem-protein products find potential substrates in highly redox cycling and associated radical-generating vulnerable groups. Lipoproteins and fatty acids, e.g. arachidonic acid and fatty acids, which are ubiquitously present in the circulation and as part of cell membranes, may be oxidized to generate biological active compounds with inflammatory and vasoactive properties. Specific oxidative-stress markers, e.g. 4-hydroxynonenol (4-HNE)-modified proteins, are exceedingly increased in tissues when animals are exposed to huge quantities of cell-free intravascular Hb. This gives ample evidence that oxidative reactions do take place [121]. In vitro, ferric myoglobin (Mb) has been shown to induce toxicity [128].

In prolonged OS or during extracellular Hb accumulation, small molecular antioxidants, e.g. ascorbic acid and uric acid, become depleted [129], as in malaria, making toxicity resulting from reaction of haem-proteins with H_2O_2 particularly essential in dissipating the highly pro-oxidant molecule. Hb and Mb have been known to possess pseudoperoxidase activity, and this could introduce a protective role of the proteins. The ability of Hb and Mb to degrade H_2O_2 provides an antioxidant and protective physiological function. The fascinating and multifaceted organizational variations detected when Hb is exposed to H_2O_2, and the apparent cytoprotective activity of Hb during H_2O_2 exposure, support a novel protective apparatus by which Hb seems to limit cellular insult, contrary to the previously reported data with ferric Mb and certain chemically modified Hb [130]. This understanding brings to the fore the fact that malarial oxidative stress is a gradual but insidious process formatted by a complicity of parasite and human host responses.

Certain mammalian and plant peroxidases have a haem covalently bound within the active site of the enzyme which prevents damage to the active site by reactive products of the peroxide enzymatic degradation [131–133]. There is a very high possibility that Hb uses a mechanism to defend the external cellular environment by redirecting the radical burden towards haem adduct development, internalization of haem within the protein, haem cross-linking or a combination [125]. In the same manner, a mechanism including irreversible oxidative alterations of globin amino acids could shield non-Hb substrates and living cells from OS. In certain experiments, shown by spectrophotometric and electron microscopy, data suggested that the cyclic redox reactions involving H_2O_2 and Hb resulted ultimately with destruction of the molecule. In this perspective, Hb may act as an antioxidant and cytoprotective peroxidase.

Typical clinical situations in which large quantities of extracellular Hb could be exposed to H_2O_2 within an inflammation-prone tissue environment might be found in atherosclerotic plaques [134] as well as in malaria. In the atherosclerotic plaque, macrophages confined to regions of the intraplaque haemorrhage exhibited expressively less oxidative impairment as equated to the cells within the lipid-core area of the same plaques. The antioxidant phenotype of the macrophages itself was invoked to elucidate the unexpected result. The Hb-rich environment may be a further protective antioxidant influence.

While the antioxidant effect of Hb in malaria has not been deduced, there is plausible hypothesis that the protein may have considerable effect in the amelioration of oxidative stress in the disease through its ability to decompose H_2O_2. This assertion intones the view of Hb as being only prooxidant as too simplistic. There is a plethora of interactions starting with the baseline Hb oxidative status, possible structural Hb modifications, presence or absence of specific Hb or haem scavengers, e.g. haptoglobin, hemopexin and their respective receptors that determine Hb impact in oxidative conditions [135–137] like in malaria. There being many cell types and functions, different cell types may have different responses to Hb or its oxidative products and adducts. Haem has been recognized as a ligand for Toll-like receptor 4 (TLR4) and may encourage or drive an inflammatory response in some particular cell types [138] like malaria. Consequently, certain pattern recognition receptors display specificity for adducted haem, oxidized Hb, oxidized globin chains or cross-linked Hb species. In view of all these factors, the role of Hb in tissue homeostasis during inflammation, tissue injury, haemolytic diseases (e.g. malaria) and vascular disease are of paramount importance [125].

23.10 Antioxidant Defence Mechanisms in the Malaria Parasite

23.10.1 Parasite Sources of Reactive Species

The parasite has ample capabilities of producing free radical just as the host generates ROS/RNS. These in turn interfere with pRBC's biochemical and physiological functions. Parasites use ROS/RNS as a means for re-invasion of other hepatocytes and npRBCs through the weakening of lipid peroxidized membranes.

Aerobic membrane transport systems are a major source of free reactive species generation by the *Plasmodium*. The absence of NADPH-oxidase expression, a free radical generation source in macrophages, did not affect the progression of parasitaemia in knockout mice for most murine malaria species as compared to controls. Therefore, free radical generation could be a result of infection as opposed to respiratory phagocytic burst or both. There is a high likelihood that the parasite is the source of OS [139]. To combat its own created oxidative stress, the parasite has a plethora of antioxidant defence mechanism.

23.10.2 Antioxidant Defence Mechanisms in the Parasite

In malaria, redox-active metals are readily available from the breakdown of RBCs that release free iron-containing haemoglobin which will increase redox reactions and free radical generation processes. Oxidative stress from generated ROS or reactive oxygen intermediates (ROI) feeds from the redox-cycling reactions as well as from the redox-inert compounds consumption of reduced glutathione and other natural antioxidative molecules. Little to reasonable quantities of ROS have advantageous effects on several physiological processes including killing of assaulting pathogens, wound curative effects and tissue repair processes [140], and the parasite has developed systems for its protection. The early blood stage of the *Plasmodium* is characterized by a continuous cascade gene expression of at least five different antioxidant proteins showing the importance of this process in malaria [141]. The parasite has also developed mechanism to reduce its own production of ROS/RNS while protecting against host-derived oxidative damage. The apicoplast, a symbiotic intracellular apparatus near the mitochondria, synthesizes lipoic acid for parasite antioxidant defence [142].

The natural redox homeostasis of the *Plasmodium* depends on the synthesis of reduced glutathione (GSH) and the thioredoxin system (Trx)/thioredoxin reductase (TrxR) minus the catalase system. Reduction of GSSH is supported by high proportion of Trx/TrxR system in cells that have no glutathione reductase which proves to be important in particular parasite stages [143].

The glutathione complex controls the redox status through the glutathione-glutathione disulphide (GSH-GSSG) combination which gives the major redox couple determining the antioxidant capacity of the cell and works together with the active site dithiol-disulphide of thioredoxins (Trx_{red}-Trx_{ox}) [144] limiting, to some extent, the deleterious effect malaria generated oxidative stress. Furthermore, there is the primary enzymatic antioxidants reacting directly with the pro-oxidants (catalase, superoxide dismutase) and the secondary antioxidant which regenerate low molecular weight antioxidant species (GSSG-Trx) [145]. Glutathione reductase (GSR) and Trx reductase restore the reduced states of GSSG and the Trx-enzymes using NADPH-sourced electrons. Intracellular disulphide reduction is also obtained by glutaredoxins utilizing GSH [146]. The Trx system comprised of the Trx, TrxR and the Trx peroxidase (peroxiredoxin, Prx) together with the GSH system (enzymes for GSH synthesis and cycling) protects against ROS-induced damage and other detoxification processes in malaria.

Glutathione reductase will emerge as an important antioxidant molecule which seems enhanced in malaria patients with elevated pRBCs. This may suggest that a significant portion of the increased GSH may be produced by the parasite itself from particular cellular compartmentalized sites of the regenerating enzyme [147].

However, export of GSH-conjugated metabolites and xenobiotics from the host (and parasite) is necessarily carried out by the GSH efflux transporters and pumps [148]. Additionally, GSH is linked to the degradation of unpolymerized ferriprotoporphyrin IX (FP) which is intended to increase OS in the parasite food vacuole. Increase in GSH generation by the parasite may be implicated in increasing chloroquine resistance [149]. Resistance of *P. vinckei vinckei* to artemisinin appears to be GSH action-mediated [150]. Also, the *Plasmodium falciparum* histidine-rich protein-2 (*Pf*HRP2) is connected to FP and has antioxidant characteristics that benefit the parasite[151].

Vitamin B6 is described as *P. falciparum* antioxidant molecule with a functional uptake system in place. Biosynthesis in the parasite depends on Pdx1 and Pdx2 which act simultaneously with glutaminase activity. Activation of vitamin B6 requires phosphorylation by pyridoxine kinase [152] making inhibition of the vitamin's synthesis mechanism a potential hit target in malaria.

23.10.3 Handling of ROS by Malaria Parasite (Fig. 23.2)

Disproportionate production of ROS in malaria from free haemoglobin poses problematic disruption of bodily homeostasis and may cause oxidative tissue damage. Accumulative inflammatory processes of untreated malaria, parasitic infection, blood flow occlusion-induced ischemia-reperfusion (I/R) injury and ingestion of non-steroidal anti-inflammatory drugs (NSAIDs), which is a common practice in malaria, result in elevated ROS generation. Oxidative stress components generated in malaria include, but not limited to, hydroxyl radicals (OH), ROS, $ONOO^-$ and

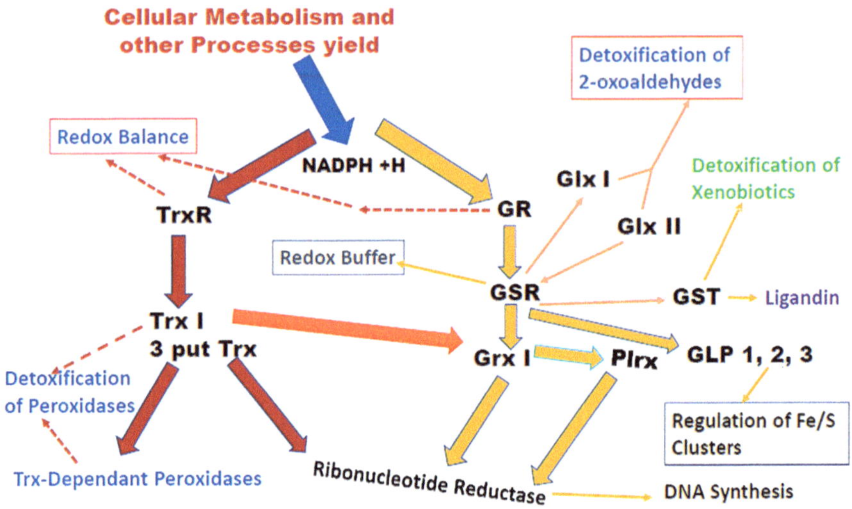

Fig. 23.2 The role of thioredoxins and glutaredoxins in the redox network of *P. falciparum*. Shown are only proteins/pathways that have been verified to exist in *P. falciparum*. The thioredoxin system is shown in dark red and the glutathione/glutaredoxin system in amber shades. Wide arrows indicate redox reactions, solid arrows indicate GSH-dependent reactions, and dotted arrows indicate functions of the systems. NADPH represents the major source of reducing equivalents cellular metabolism. Both thioredoxin reductase (TrxR) and glutathione reductase (GR) reduce their respective substrates, thioredoxin (Trx) and glutathione disulphide. Trx reduces Trx-dependent peroxidases as well as ribonucleotide reductase (RiboR). GR reduces glutathione which serves as a substrate for glutathione *S*-transferase (GST) or glyoxalases (Glx); when used by the glyoxalase system, GSH is recycled. GSH also reduces glutaredoxin, which in turn is able to provide RiboR with electrons. Also plasmoredoxin (Plrx), which can be reduced by glutaredoxin, is able to reduce RiboR [167, 170]

RNS. It is essential to note that ROS pervades all cells, tissues and organs alike in malaria.

Aerobic respiration invariably produces O_2 metabolites which comprise of superoxide anion $O_2^{\bullet-}$ and H_2O_2. In the process of forming the mentioned metabolites, one and two electrons reductions of molecular oxygen (O_2) or the extremely reactive hydroxyl radical (·OH) is experienced via the Fenton and/or Haber-Weiss reaction in the presence of iron abundant in malaria [153]. Naturally, ROS can function as a redox-signalling second messenger where H_2O_2 provides specificity of interactions with signalling process effectors [154]. To combat ROS, *P. falciparum* mostly uses two interacting systems as mentioned above, the GSH and TRX systems [155–159], which give maximum protection depending on the cellular context, ROS concentration and the redox state of the cell [160].

In the malaria parasite and the human host, harmonizing the creation and elimination of ROS maintains the proper function of redox-sensitive signalling proteins. Severe increases of ROS, as seen in malaria, overrun the GSH-TRX systems which are linked by redox protein plasmoredoxin [161]. The rapid multiplication of the *P. falciparum* and the consequent increased metabolic rate with attending large quantities of Hb digestion yield amounts of toxic haem iron. Of necessity, the antioxidant defence systems are necessary in the parasite where they play a critical role in the *P. falciparum* parasite and offer themselves as good targets for antimalarial interventions that use redox homeostasis [162]. The two *P. falciparum* NADPH-dependent redox systems (thioredoxin reductase [TrxR]-thioredoxin [Trx] and glutathione reductase [GR]-glutathione [GSH]) play an antioxidant role against hydroxyl radicals and other sources of ROS [163–165]. The four (thioredoxin-dependent) peroxidases [156, 166, 167] , one glutaredoxin [168], three glutaredoxin-like proteins and one glutathione *S*-transferase [169] are small redox-active proteins with functional and structural similarities whose failure to reduce H_2O_2 may contribute to ROS in the unbalanced redox homeostasis of malaria. Figure 23.2 [167] shows the interactions of these molecules and functions in malaria and possible areas of ROS build-up [170].

The different players of the antioxidant processes are given in the following sections.

23.10.3.1 *Plasmodium falciparum* Glutaredoxins (*Pf*Grxl)

*Pf*Grxl is a classical glutaredoxin which possesses an active CPYC12 and has been identified in *P. falciparum* [168, 171]. More so, recombinant *Pf*Grxl has been shown to display glutathione-dependent glutaredoxin activity in an assay that uses β-hydroxy-ethyldisulphide (BEDS) as reagent. Importantly, *Pf*Grx1 serves as a hydrogen donor for the first step in DNA synthesis catalysed by the enzyme ribonucleotide reductase [172]. In vitro *P. falciparum* trophozoite extracts from eight different chloroquine-resistant and chloroquine-sensitive strains displayed the same Grx1 activity pointing to a well-preserved protein homology in the malaria parasites.

23.10.3.2 *P. falciparum* Glutaredoxin-Like Proteins

P. falciparum glutaredoxin-like proteins designated *Pf*GLP1 are a second class of glutaredoxins, possessing one conserved cysteine residue at their active site [173]. An extension of the Grx's importance has been the detection of 1-Cys Grxs which interacts with protein kinase C that has been termed PICOT proteins (protein kinase C interacting cousin of thioredoxins) [174]. These proteins possess additional thioredoxin or glutaredoxin domains together with a PICOT homology domain. The PICOT proteins have a function in the synthesis of Fe/S-cluster protein [175] and protein deglutathionylation process in the yeast *Saccharomyces cerevisiae* [176]. There are other GLPs together with PfGLP1 that have been identified, mainly PfGLP2 and PfGLP3 which can be found in gametocyte stage [177], merozoites stage [168] and trophozoite stage. There also exist Trx-like proteins with the same function as thioredoxins but are yet to be verified.

23.10.3.3 *P. falciparum* Thioredoxin (*Pf*Trx)

The *P. falciparum* thioredoxin (*Pf*Trx1) provides reducing equivalents to peroxidases and ribonucleotide reductase and is in turn reduced by *Pf*TrxR. Notably, the parasite thioredoxin is a better substrate for the human TrxR with a K_m of 2mM, k_{cat}

of 3300 min^{-1} compared to the isofunctional parasite enzyme which has a K_m of 10.4; k_{cat} of 3100 min^{-1} making it a suitable alternative in malaria where the parasite extrudes excess redox balance materials into the cell cytoplasm. Although found in other organisms, the non-enzymatic reduction of glutathione disulphide (GSSG) by *Pf*Trx, which has a k_2 of 650 $M^{-1}S^{-1}$, is particularly efficient in the malarial parasite [178] to allow for excess GSH to be exported out of the parasitophorous vacuole to the cytoplasm.

The *Pf*Trx1is involved in the non-enzymatic buffering of ROS by influencing sensitivity to H_2O_2 as has been described for *Escherichia coli*-related molecule [179]. The hydroxyl radical scavenging is one of the functions of the human and *E. coli* thioredoxin (the former being 13 more potent than the latter) which may be extrapolated to the malarial parasite one as well. The human Trx is more likely to provide important endogenous antioxidant capacity [180]. When coupled with NADPH and thioredoxin reductase, Trx reduces H_2O_2 slowly [181]. The peroxide buffering capacity of *Pf*Trx has been demonstrated and is very relevant in the high oxidant potential seen in the parasitophorous vacuole. Therefore, the antioxidant capacity may contribute to the overall antioxidant activity of the parasite which does not possess glutathione peroxidase and catalase [172].

23.10.3.4 Plasmoredoxin (Plrx) and Oxidative Stress in Malaria

A 22-kDa redox-active thioredoxin-like protein has been identified in the *P. falciparum* besides *Pf*Trx named plasmoredoxin (Plrx) which displays high conservativeness and exclusivity to the malarial parasite [161]. As a member of the thioredoxin superfamily but clustering separately from other members in the phylogenetic tree, Plrx is reduced in the second-order rate kinetics, by GSH at pH 7.4, 25°C, k of 1.6 $M^{-1}S^{-1}$, at pH 6.9, k of 0.5 $M^{-1}S^{-1}$. Also, reduced Plrx reduces GSSG at pH 6.9, k of 640 $M^{-1}S^{-1}$, and provides electrons for ribonucleotide reductase [172]. Due to that, Plrx protein has been demonstrated in blood stage malarial parasite making the protein a good candidate for immunological and PCR-based diagnostics techniques.

23.11 Malaria Infection in the Mosquito and Oxidative Stress

The *Anopheles* mosquito is the vector responsible for transmitting the *Plasmodium* parasite sporozoites. There is a correlation existing between oxidative stress and antioxidant defences that allows the survival of the parasite passaging in the vector.

Vector susceptibility to certain *Plasmodium* species allows the development and maintenance in such a way that immunological barriers are trespassed during the evolution of the parasite to sporozoite stage. In some species of the *Anopheles gambiae*, there is resistance to parasitic evolution driven by OS. In this process, tyrosine is oxidatively converted to melanin [182] unlike in others where the melanin source is obtained for the covering of the mosquito ova, thereby exposing the parasite to possible sterility [183]. These strains also lack the antioxidant defence mechanism in such a way that even when the catalase mRNA is steadily expressed, there is no influence on H_2O_2 production.

The molecule NO exerts a protective role in the mosquito with parasitaemia reduction effects. Glycophosphatidylinositol (GPI), a parasite molecule, induces NOS expression in the same way as haemozoin.

Haemozoin changes malaria vector function as there is an inordinate amount of blood in acute phases of infection. The molecule induces NOS gene expression in *A. stephensi* and *A. gambiae* in vitro and in vivo in *A. stephensi* [184]. In the midgut of the mosquito, NO induction is mediated via GPI. Of note, while exerting insulin-like signalling, GPI is not insulinomimetic as it requires AKT/PTB activation [185].

In the *A. gambiae* mosquito, the GSH gene expression is missing which is compensated through the synthesis of TrxR-1 which has glutathione-like activity [186]. Glutathione transferase synthesizes GSH in *A. cracens*, a Thailand malaria vector to introduce antioxidant actions [187]. Different redox responses expressed in certain mosquitos have the possibility of determining resistance to parasite infection.

23.12 Oxidative Stress and Malaria Treatment

The antimalarial quinine is an active ingredient extracted from the bark of the tropical tree of the genus Cinchona. Although having been used for over 100 years, quinine principle of actions is not well known. However, DNA synthesis inhibition is thought to be the most plausible. Use of quinine is now assigned to acute cases; otherwise, synthetic drugs and derivatives of acridine and quinoline structures like chloroquine and mefloquine have become first-line drugs in current use as combination therapy. These inhibit haem polymerase preventing haem polymerization to haemozoin and subsequently increasing oxidative stress for the parasite through generation of free radicals from iron catalysed reactions [187]. Another similar antimalarial primaquine destroys the malaria parasite gametocytes.

Basically, the antimalarial therapeutics pharmacological mechanism against *plasmodium* depends on the oxidative susceptibility and inhibition of metabolic pathways synthesizing essential molecule(s) for parasite growth. Essentially, most molecules used as antimalarials have pro-oxidant characteristics, e.g. chloroquine, primaquine and artemisinin. Some drugs produce free radicals directly (artemisinin) [188], while others inhibit function of molecule with antioxidant capacities, e.g. GSH reductase by quinolones [189].

The most ancient antimalarial plant, *Artemisia annua* which has been used for >2000 years in China, contains artemisinin a parasiticidal agent against blood-stage parasite species resistant to chloroquine. When in contact with iron, which is abundant in pRBCs, the drug releases free radicals. Free radical produced by artemisinin is effective against the parasite with minimum host damages. The rapid action, low half-life of artemisinin ($t_{1/2}$ = 2.6hrs), quick patient recovery and fast decreases in plasma concentration favour parasite recrudesces [190] as some patients may stop treatment prematurely. Many have testified that artemisinin depends on oxidative stress for its antimalarial action [191–194].

Elevated lipid peroxidation markers and reduced ascorbate/glutathione have been observed in malaria-infected pregnant women as compared to non-infected pregnant women. Higher concentrations of lipid peroxidation with more intense reduction in GSH and ascorbate in women administered with antimalarials in comparison to those not treated have been reported [195]. Of note is the synergism noted between metalloporphyrins and artemisinin against chloroquine-resistant *P. falciparum* [196].

Increased antioxidant enzymes and NO expression with induction of lipid peroxidation and protein carbonylation have been observed in *P. yoelii*-infected mice treated with pyrimethamine [197]. Antioxidant protein expression by parasites seems to be the basis of antimalarial resistance by certain species as early antioxidant gene transcription is associated with adaptive capacity to chloroquine [198]. Inhibition of haemo-peroxidase by antifungal drug clotrimazole increases oxidative stress in malaria providing a pro-oxidant alternative therapy strategy [199].

23.13 Antimalarial Phytotherapeutics and Oxidative Stress in Malaria

Natural products continue to form basis of most modern medicine whether modified or not. Different classes of medicinal plants are currently used to manage malaria fever, and the mechanism has speculated their ability to be pro-oxidants and antioxidants. Medicinal plants are considered cost-effective and less toxic to human biological systems. Entry of medicinal plants through the system is mainly through fruits and vegetables. After all, the lead compounds used in the current antimalarials are from plant materials or natural compounds.

One of the mechanism for antimalarials is the creation or increase of OS; however, certain plants and their extract have been shown to have a high antioxidant content which is associated with interruption of malarial disease processes through modulation of cellular signalling pathways as well as inducing parasite death [200]. What is exciting about this "anti-disease" and anti-parasitic approach exerted by the phytotherapeutics is the schizonticidal effect with minor challenges to host redox balance. *Piper betle* L. leaves [201], *Anogeissus leiocarpus* [202], *Nigella sativa* seeds [203], *Artemisia annua* [204], asiatic acid (AC) [205], maslinic acid (MA), oleanolic acid (OA) and ursolic acid (UA) have been reported to have efficacy against malaria through redox balance mechanism.

A correlation of reduction in *P. berghei* parasitaemia with increase in total antioxidant capacity, decrease in lipid peroxidation and NO markers has been observed in malaria-infected animals administered with *Agaricus sylvaticus*, a mushroom with high antioxidant capacity [206]. Also, the damage caused by antimalarial treatment to the host may be reversed or minimized by administration of antioxidant supplements like curcumin [207].

23.13.1 Terpenes and Oxidative Stress in Malaria

Terpenes are one of the largest groups of compounds of plant-based natural products comprising of approximately 30 000 species. Terpenes are divided into monoterpene, sesquiterpene, diterpene, triterpene and tetraterpene [208]. A terpene is defined as a hydrocarbon molecule and terpenoid is a derivative of a terpene. For simplicity, these terms are used interchangeably. Naming these molecules is not short of complexity; the isoprene unit is a five-carbon molecule. The single isoprene unit, therefore, represents the most basic class of terpenes [209]. Monoterpenes (C10) are two-bonded isoprene units. Sesquiterpenes contain three isoprene units (C15); diterpenes (C20) and triterpenes (C30) contain two and three terpene units, respectively. Tetraterpenes consist of four terpene units and polyterpenes contain more than four terpene unit [209]. Of interest in this chapter are the antimalarial effects of diterpenes and triterpenes as they have recently attracted attention due to their widespread biological activities. These compounds exhibit hydroxyl, carboxyl, indole, acetoxy and thiol functional groups capable of redox reactions tending to alter or influence redox metabolism, homeostasis and/or environment with fatal outcomes for the parasite while ameliorating disease aspects of malaria [205, 210, 211].

23.13.2 Diterpenes

Diterpenes contain essential oils that is made up of four isoprene units (molecular formula: $C_{20}H_{32}$), which display antioxidants, pro-oxidant, antimicrobial and antiprotozoal effects. Several antimalarial diterpenes distributed in various medicinal plants provide promising lead/crude compounds for development of new derivatives [212]. *Caesalpinia crista* LINN. (Fabaceae) has previously been used to extract 44 cassane and norcassane-type diterpenes. Of the diterpenes extracted, norcaesalpin E and C-17 norcassane type with acetoxy group at carbon-1 and hydroxyl-group at carbon 17 have been shown to have three times more potency than chloroquine against malaria parasite in vitro [213]. Introduction of acetoxy group in carbon 1 and hydroxy group in carbon 17 plays a major role in inhibiting malarial growth presumably due to the redox reaction of the anion groups. A recent study by Nondo et al. suggests that plant-derived norcaesalpin D exhibits antiplasmodial activity against chloroquine-sensitive, chloroquine (CHQ)-resistant and artemisinin-resistant *P. falciparum* in vitro with less potency (IC 50: 0.98 μm) [214] compared to norcaesalpin E (IC 50 0.09 μM) [213]. Sesquiterpenes derived from *Acanthella klethra* can inhibit plasmodial activity and shows no cytotoxic effects on human cell lines [215]. Derivatives of natural diterpenes have developed through addition of more potent free radical generators which ultimately increase oxidative stress to the *Plasmodium* parasite.

Table 23.1 Chemical properties of the different triterpenes [209]

Compound	Hydrogen bond donor	Hydrogen bond acceptor
Ursolic acid	2	1
Oleanolic acid	2	3 [218]
Betulinic acid	2	3
Maslinic acid	3	4 [225]
Asiatic acid	4	7 [226]

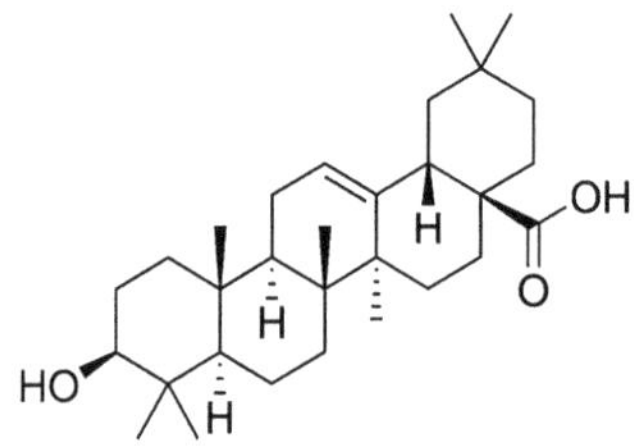

Fig. 23.3 Chemical structure of oleanolic acid. Molecular weight, 456.7 g/mol; hydrogen bond donor, 2; hydrogen acceptor 3

23.13.3 Triterpenes and Oxidative Stress in Malaria

Pentacyclic triterpenes (PT) are widely distributed throughout the plant kingdom and have been widely used as a natural product against infectious disease such as malaria. There are six main types of PTs, and these include hopane, taraxastane and friedelane, oleanane, ursane and lupane groups [216, 217]. Oleanane, ursane and lupane groups are thought to have more biological activities; thus more research focuses on these compounds [216]. Oleanane and derivatives demonstrated high to low antimalarial activity which is based on the molecule's oxidative functional groups. Triterpenes display hydrogen donor and hydrogen acceptor characteristics (Table 23.1) that allow them potential to participate in redox reactions.

23.13.3.1 Oleanolic Acid (OA) and Oxidative Stress in Malaria (Fig. 23.3)

Viola verecunda-derived epi-OA exhibits an IC_{50} of 0.018 μgmL^{-1} against chloroquine-sensitive *D10 strain of Plasmodium falciparum* which is comparable to that of the artemisinin (0.015 μgmL^{-1}) [218]. These interesting in vitro results still warranted verification in vivo. Oral administration of *Syzygium aromatic*-derived OA (160 mg/kg) eliminates malaria parasite after 6 days of infection and chloroquine-attenuated malaria after 5 days of infection in male Sprague-Dawley rats showing the adjuvant characteristics of the molecule derived from redox capacities. Animals administered with this compound displayed lower MDA, indicating reduced oxidative stress, while increased activity of the antioxidant enzymes SOD and GPx in renal and hepatic tissues was observed as compared to untreated controls or those treated with CHQ [219].

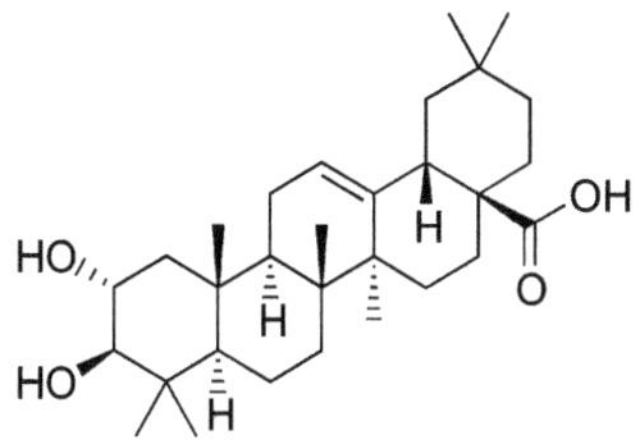

Fig. 23.4 Chemical structure of maslinic acid (MA), molecular weight 472.7 g/mol, donate 3 hydrogen and accept 4 hydrogen

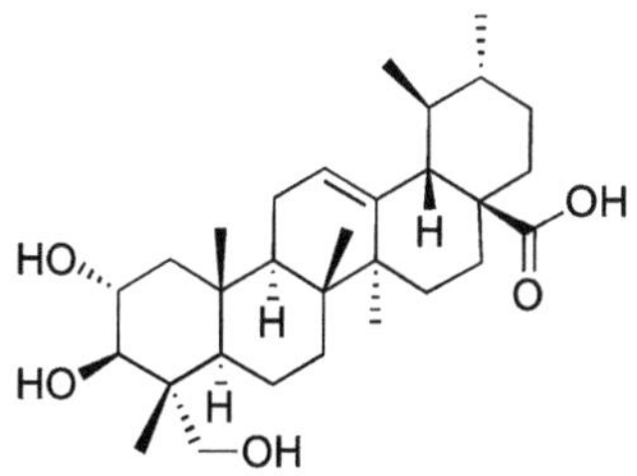

Fig. 23.5 Chemical structure of asiatic acid (AA). Molecular weight 488.7 g/mol, can donate 4 hydrogens and accept 7 hydrogens

23.13.3.2 Maslinic Acid and Oxidative Stress in Malaria (Fig. 23.4)

Antimalarial properties of MA (Fig. 23.1) have previously been determined on *P. falciparum* cell cultures [220]. MA inhibits proliferation of *P. falciparum* in infected red-blood cells, and these results were future validated by use of this triterpene in mice models [221]. Studies indicate that proteolytic enzymes as well as the oxidative stress play a role in the development of inflammation [222]. MA exhibits antioxidants properties by decreasing lipid peroxidation and enhancing production of antioxidants on diabetic animals [223]. MA also lowered lipid peroxidation and increased SOD and GPx in renal and liver tissues of the *Plasmodium berghei*-infected rats [224]. Structure relationship studies offer insight regarding the mechanism of action of this triterpene. The ability of this triterpene to donate 3 hydrogens and accept 4 hydrogens gives it an ability to have pro and anti-oxidant properties. Additionally, the presence of OH group in carbon 2 of MA is thought to enhance its properties when compared to OA [225]. Therefore, the postulated mechanism action is as follows: the lipid soluble MA is able to freely diffuse through the infected red-blood cell where it acts as an oxidant since the parasite is well known to thrive in an antioxidant environment. Notably, MA acts as an antioxidant (anti-disease) in other tissues (liver, kidney and hearts) possibly due to pH differences compared to the infected RBCs.

23.13.3.3 Asiatic Acid and Oxidative Stress in Malaria (Fig. 23.5)

Table 23.1 indicates that AA is a stronger oxidant and antioxidant compound than OA and MA as shown by the higher number of hydrogens that can be donated and accepted. Indeed, it has been indicated that treatment of *P. berghei* with AA (5 mg, 10 mg and 20mg/kg) significantly reduces parasitaemia [226]. The efficacy of AA may be attributed to its strong ability to reduce antioxidant capacity of the parasite within the pRBCs. The presence of an additional OH group in carbon 4 gives this triterpene more efficacy in comparison to its counterparts shown in Table 23.1.

Research also indicate oral administration of AA in streptozotocin-induced (STZ) diabetic rats lowers lipid peroxidation to levels close to normal [227], indicating its ability to prevent oxidative stress in the liver and heart tissues. These results were also accompanied by increased levels of SOD, CAT and GPx which are often low in diabetic rats [227]. Therefore, AA indeed offers the antiparasite and anti-disease effects which may be beneficial in malaria patients.

23.14 Conclusion

The host-parasite relationship is a complex system driven by a plethora of activities of which oxidative stress plays a critical role as it initiates and exacerbates malarial pathophysiology consequently expunging human resistance to the disease. The origins of ROS/RNS generation implicated in the malarial disease include (i) inflammatory host responses to parasitic infection, (ii) activation of Haber-Weiss and Fenton reaction from the abundant haem and free iron, (iii) alternating ischaemia-reperfusion syndrome cycles which enable oxygen transportation during malarial paroxysm and cytoadherence, (vi) parasitic sources of ROS and antioxidant defences and (v) effects of pro-oxidant antimalarials. Oxidative stress also drives processes in the mosquito affecting malarial transmission. Ultimately, the use of antioxidant phytotherapeutics such as triterpenes constitutes an effective adjuvant and/or therapeutic antimalarial strategy that protects the host from both malarial parasite and treatment oxidative damage.

Conflict of Interests Authors declare no conflict of interests.

Funding The authors would like to acknowledge the National University of Science and Technology and Durban University of Technology for providing an enabling environment for this work.

References

1. Hay SI, Guerra CA, Tatem AJ, Noor AM, Snow RW (2004) The global distribution and population at risk of malaria: past, present, and future. Lancet Infect Dis 4(6):327–336
2. Hay SI, Okiro EA, Gething PW, Patil AP, Tatem AJ, Guerra CA et al (2010) Estimating the global clinical burden of Plasmodium falciparum malaria in 2007. PLoS Med 7(6):e1000290. https://doi.org/10.1371/journal.pmed.1000290
3. de Carvalho GB, de Carvalho GB (2011) Duffy Blood Group System and the malaria adaptation process in humans. Revista Brasileira de Hematologia e Hemoterapia 33(1):55–64
4. WHO (2016) World malaria report 2016. In Malaria report, Organization WH (ed). World Health Organization, Geneva
5. Guyatt HL, Snow RW (2004) Impact of malaria during pregnancy on low birth weight in sub-Saharan Africa. Clin Microbiol Rev 17(4):760–769
6. Sharma L, Shukla G (2017) Placental malaria: a new insight into the pathophysiology. Front Med (Lausanne). 4:117. https://doi.org/10.3389/fmed.2017.00117

7. Talisuna AO, Bloland P, D'Alessandro U (2004) History, dynamics and public health importance of malaria parasite resistance. Clin Microbiol Rev 17:235–254
8. Walker DJ, Pitsch JL, Peng MM, Robinson BL, Peters W, Bhisutthibhan J et al (2000) Mechanisms of artemisinin resistance in the rodent malaria pathogen *Plasmodium yoelii*. Antimicrob Agents Chemother 44:344–347
9. Afonso A, Hunt P, Cheesman S, Alves AC, Cunha CV, Rosário V et al (2006) Malaria parasites can develop stable resistance to artemisinin but lack mutations in candidate genes atp6 (encoding the sarcoplasmic and endoplasmic reticulum Ca^{2+} ATPase), tctp, mdr1, and cg10 Antimicrob Agents Chemother 50:480–489
10. Menendez C, Ordi J, Ismail MR, Ventura PJ, Aponte JJ, Kahigwa E (2000) The impact of placental malaria on gestational age and birth weight. J Infect Dis 181:1740–1745. https://doi.org/10.1086/315449
11. Watkinson M, Rushton DI (1983) Plasmodial pigmentation of placenta and outcome of pregnancy in West African mothers. Br Med J 287:251–254. https://doi.org/10.1136/bmj.287.6387.251
12. Murphy SC, Breman JG (2001) Gaps in the childhood malaria burden in Africa: cerebral malaria, neurological sequelae, anemia, respiratory distress, hypoglycemia, and complications of pregnancy. Am J Trop Med Hyg 64:57–67. https://doi.org/10.4269/ajtmh.2001.64.57
13. Riley EM, Wagner GE, Akanmori BD, Koram KA (2001) Do maternally acquired antibodies protect infants from malaria infection? Parasite Immunol 23:51–59
14. Riley EM, Wagner GE, Ofori MF, Wheeler JG, Akanmori BD, Tetteh K et al (2000) Lack of association between maternal antibody and protection of African infants from malaria infection. Infect Immun 68:5856–5863
15. Doolan DL, Dobano C, Baird JK (2009) Acquired immunity to malaria. Clin Microbiol Rev 22:13–36
16. Kassim OO, Ako-Anai KA, Torimiro SE, Hollowell GP, Okoye VC, Martin SK (2000) Inhibitory factors in breastmilk, maternal and infant sera against in vitro growth of *Plasmodium falciparum* malaria parasite. J Trop Pediatr 46:92–96
17. Doolan DL, Dobaño C, Baird JK (2009) Acquired immunity to malaria. Clin Microbiol Rev 22(1):13–36. https://doi.org/10.1128/CMR.00025-08
18. Percário S, Moreira DR, Gomes BAQ, Ferreira MES, Gonçalves ACM, Laurindo PSO et al (2012) Oxidative stress in malaria. Int J Mol Sci. 13(12):16346-16372. https://doi.org/10.3390/ijms131216346
19. Dockrell HM, Playfair JH (1984) Killing of *Plasmodium yoelii* by enzyme-induced products of the oxidative burst. Infect Immun 43:451–456
20. Sohail M, Kaul A, Raziuddin M, Adak T (2007) Decreased glutathione-S-transferase activity: Diagnostic and protective role in vivax malaria. Clin Biochem 40:377–382
21. Guha M, Kumar S, Choubey V, Maity P (2006) Apoptosis in liver during malaria: Role of oxidative stress and implication of mitochondrial pathway. FASEB J 20:E439–E449
22. Clark IA, Hunt NH (1983) Evidence for reactive oxygen intermediates causing hemolysis and parasite death in malaria. Infect Immun 39:1–6
23. Sobolewski P, Gramaglia I, Frangos JA, Intaglietta M, Heyde HVD (2005) *Plasmodium berghei* resists killing by reactive oxygen species. Infect Immun 73:6704–6710
24. Cabrales P, Zanini GM, Meays D, Frangos JA, Carvalho LJM (2011) Nitric Oxide protection against murine cerebral malaria is associated with improved cerebral microcirculatory physiology. J Infect Dis 203:1454–1463
25. Erel O, Vural H, Aksoy N, Aslan G, Ulukanligil M (2001) Oxidative stress of platelets and thrombocytopenia in patients with *vivax* malaria. Clin Biochem 34:341–344
26. Griffiths MJ, Ndungu F, Baird KL, Muller DP, Marsh K, Newton CR (2001) Oxidative stress and erythrocyte damage in Kenyan children with severe *Plasmodium falciparum* malaria. Br J Haematol 113:486–491
27. Dumaswala UJ, Zhuo L, Mahajan S, Nair PN, Shertzer HG, Dibello P et al (2001) Glutathione protects chemokine-scavenging and antioxidative defense functions in human RBCs. Am J Physiol 208:C867–C873

28. Halliwell B, Gutteridge JMC (2007) Free radicals in biology and medicine, 3rd edn. Oxford University Press, New York
29. Glosli H, Tronstad KJ, Wergedal H, Müller F, Svardal A, Aukrust P et al (2002) Human TNF-alpha in transgenic mice induces differential changes in redox status and glutathione-regulating enzymes. FASEB J 16:1450–1452
30. Balla G, Jacob HS, Balla J, Rosenberg M, Nath K, Apple F et al (1992) Ferritin: A cytoprotective antioxidant stratagem of endothelium. J Biol Chem 267:18148–18153
31. Kavishe RA, Koenderink JB, McCall MB, Peter WH, Mulder B, Hermsen CC et al (2006) Short report: Severe *Plasmodium falciparum* malaria in Cameroon associated with the glutathione-S-transferase M1 null genotype. Am J Trop Med Hyg 75:827–829
32. Becker K, Tilley L, Vennerstrom JL, Roberts D, Rogerson S, Ginsburg H (2004) Oxidative stress in malaria parasite-infected erythrocytes: Host-parasite interactions. Int J Parasitol 34:163–189
33. Golensen J, Domb A, Teomim D, Tsafack A, Nisim O, Ponka P et al (1997) The treatment of animal models of malaria with iron chelators by use of a novel polymeric device for slow drug release. J Pharmacol Exp Ther 281:1127–1135
34. Caufield LE, Richard SA, Black RE (2004) Undernutrition as an underlying cause of malaria morbidity and mortality in children less than five years old. Am J Trop Med Hyg 71:55–63
35. Hassan G I, Gregory U, Maryam H (2004) Serum ascorbic acid concentration in patients with acute *falciparum* malaria infection: Possible significance.Braz J Infect Dis 8:378–381
36. Metzger A, Mukasa G, Shankar AH, Ndeezi G, Melikian G, Semba RD (2001) Antioxidant status and acute malaria in children in Kampala, Uganda. Am J Trop Med Hyg 65:115–119
37. Das BS, Patnaik JK, Mohanty S, Mishra SK, Mohanty D, Satpathy SK et al (1993) Plasma antioxidants and lipid peroxidation products in *falciparum* malaria. Am J Trop Med Hyg 49:720–725
38. Mavondo GA, Kasvosve I (2017) Antimalarial phytochemicals: delineation of the triterpene asiatic acid malarial anti-disease and pathophysiological remedial activities – Part I. J Infect Disease Pathol 1:104
39. Mavondo GA, Kasvosve I (2017) Antimalarial phytochemicals: delineation of the triterpene asiatic acid malarial anti-disease and pathophysiological remedial activities – Part II. J Infect Dis Pathol 1:103
40. Mavondo GA, Musabayane CT (2016) Asiatic acid-pectin hydrogel matrix patch transdermal delivery system influences parasitaemia suppression and inflammation reduction in *P. berghei* murine malaria infected Sprague-Dawley rats. Asian Pac J Trop Med 9(12):1172–1180. https://doi.org/10.1016/j.apjtm.2016.10.008
41. Rosales FJ, Topping JD, Smith JE, Shankar AH, A.C. R. (2000) Relation of serum retinol to acute phase proteins and malarial morbidity in Papua New Guinea children. Am J Clin Nutr 71:1582–1588
42. Mavondo GA, Mkhwananzi BN, Mabandla MV, Musabayane CT (2016) Asiatic acid influences glucose homeostasis in P. berghei murine malaria infected Sprague Dawley rats. Afr J Tradit Complement Altern Med 13(5):91–101. https://doi.org/10.21010/ajtcam.v13i5.13
43. Sanni LA, Rae C, Maitland A, Stocker R, Hunt NH (2001) Is ischemia involved in the pathogenesis of murine cerebral malaria? Am J Pathol 159:1105–1112
44. Stocker R, Hunt NH, Buffinton GD, Weidemann MJ, Lewis-Hughes PH, Clark IA (1985) Oxidative stress and protective mechanisms in relation to Plas*modium vinckei* load. Proc Natl Acad Sci USA 82:548–551
45. Yoshimoto T, Takahama Y, Wang CR, Yoneto T, Waki S, Nariuchi H (eds) (1998) A pathogenic role of IL-12 in blood-stage murine malaria lethal strain *Plasmodium berghei* NK65 infection. J Immunol 160:5500–5505
46. Akompong T, Ghori N, Haldar K (2000) In vitro activity of riboflavin against the human malaria parasite *Plasmodium falciparum.* Antimicrob Agents Chemother 44:88–96
47. Coppi A, Cabinian M, Mirelman D, Sinnis P (2006) Antimalarial activity of Allicin, a biologically active compound from garlic cloves. Antimicrob Agents Chemother 50:1731–1737

48. Bertini I, Cavallaro G (2008) Metals in the "omics" world: copper homeostasis and cytochrome c oxidase assembly in a new light. J Biol Inorg Chem 13:3–14
49. Nelson N (1999) Metal ion transporters and homeostasis. EMBO J 18:4361–4371
50. Halliwell B, Gutteridge JMC (2007) Free radicals in biology and medicine, 4th edn. Oxford University Press, Oxford
51. Valko M, Morris H, Cronin MTD (2005) Metals, toxicity and oxidative stress. Curr Med Chem 12:1161–1208
52. Matés JM, Segura JA, Alonso FJ, Márquez J (2008) Intracellular redox status and oxidative stress: implications for cell proliferation, apoptosis, and carcinogenesis. Arch Toxicol 82:273–299
53. Jomova K, Valko M (2011) Advances in metal-induced oxidative stress and human disease. Toxicology 283:65–87
54. Toyokuni S (1996) Iron-induced carcinogenesis: the role of redox regulation. Free Radic Biol Med 20:553–566
55. Valko M, Rhodes CJ, Moncol J, Izakovic M, Mazur M (2006) Free radicals, metals and antioxidants in oxidative stress-induced cancer. Chem Biol Interact 160:1–40
56. Evan GI, Vousden KH (2001) Proliferation, cell cycle and apoptosis in cancer. Nature 411:342–348
57. Rahman K (2007) Studies on free radicals, antioxidants, and co-factors. Clin Interv Aging 2:219–236
58. Lindeque JZ, Levanets O, Louw R, van der Westhuizen FH (2010) The involvement of metallothioneins in mitochondrial function and disease. Curr Protein Peptide Sci 11:292–309
59. Gutteridge JMC (1995) Lipid peroxidation and antioxidants as biomarkers of tissue damage. Clin Chem 41:1819–1828
60. Valko M, Leibfritz D, Moncol J, Cronin MTD, Mazur M, Telser J (2007) Free radicals and antioxidants in normal physiological functions and human disease. Int J Biochem Cell Biol 39:44–84
61. Speisky H, Gómez M, Burgos-Bravo F, López-Alarcón C, Jullian C, Olea-Azar C et al (2009) Generation of superoxide radicals by copper-glutathione complexes: redox-consequences associated with their interaction with reduced glutathione. Bioorg Med Chem 17:1803–1810
62. Sinicropi MS, Amantea D, Caruso A, Saturnino C (2010) Chemical and biological properties of toxic metals and use of chelating agents for the pharmacological treatment of metal poisoning. Arch Toxicol 84:501–520
63. Peralta-Videa JR, Lopez ML, Narayan M, Saupe GJ (2009) The biochemistry of environmental heavy metal uptake by plants: implications for the food chain. Int J Biochem Cell Biol 41:1665–1677
64. Das BS, Nanda NK (1999) Evidence for erythrocyte lipid peroxidation in acute *falciparum* malaria. Trans R Soc Trop Med Hyg 93:8–62
65. Hemmer CJ, Lehr HA, Westphal K, Unverricht M, Kratzius M, Reisinger EC (2005) *Plasmodium falciparum* malaria: Reduction of endothelial cell apoptosis in vitro. Infect Immun 73:1764–1770
66. Stevenson MM, Tam MF, Wolf SF, Sher A (1995) IL-12-induced protection against blood-stage *Plasmodium chabaudi* requires IFN-gamma and TNF-alpha and occurs via a nitric oxide-dependent mechanism. J Immunol 155:2545–2556
67. Arruda MA, Rossi AG, Freitas MS, Barja-Fidalgo C, Graça-Souza AV (2004) Heme inhibits human neutrophil apoptosis: Involvement of phosphoinositide 3-Kinase, MAPK, and NF-κB. J Immunol 173:2023–2030
68. Fariss MW, Chan CB, Patel M, Houten BV, Orrenius S (2005) Role of mitochondria in toxic oxidative stress. Mol Intervent 5:94–111
69. Dey S, Guha M, Alam A, Goyal M, Bindu S, Pal C et al (2009) Malarial infection develops mitochondrial pathology and mitochondrial oxidative stress to promote hepatocyte apoptosis. Free Radic Biol Med 46:271–281

70. Jaramillo M, Gowda DC, Radzioch D, Oliver M (2003) Hemozoin increase IFN-gamma-inducible macrophage nitric oxide generation through extracellular signal-regulated kinase- and NF-Kappa B-dependent pathways. J Immunol 171:4243–4253
71. Saeftel M, Krueger A, Arriens S, Heussler V, Racz P, Fleischer B et al (2004) Mice deficient interleukin-4 (IL-4) or IL-4 receptor α have higher resistance to sporozoite infection with *Plasmodium berghei* (ANKA) than do naive wild-type mice. Infect Immun 72:322–331
72. Jaramillo M, Godbout M, Olivier M (2005) Hemozoin induces macrophage chemokine expression through oxidative stress-dependent and independent mechanisms. J Immunol 174:475–484
73. Chiwakata CB, Hemmer CJ, Dietrich M (2000) High levels of inducible nitric oxide synthase mRNA are associated with increased monocyte counts in blood and have a beneficial role in *Plasmodium falciparum* malaria. Infect Immun 68:394–399
74. Syarifah HP, Masashi H, Somei K (2002) Cytokine and chemokine reponses in a cerebral malaria-susceptible or -resistant strain of mice to *Plasmodium berghei* ANKA infection: Early chemokine expression in the brain. Int Immun 15:633–640
75. Kumaratilake LM, Ferrante A, Rzepczyk CM (1990) Tumor necrosis factor enhances neutrophil-mediated killing of *Plasmodium falciparum*. Infect Immun 58:788–793
76. Porto BN, Alves LS, Fernández PL, Dutra TP, Figueiredo RT, Graça-Souza AV et al (2007) Heme induces neutrophil migration and reactive oxygen species generation through signaling pathways characteristic of chemotactic receptors. J Biol Chem 282:24430–24436
77. Taramelli D, Recalcati S, Basilico N, Olliaro P, Cairo G (2000) Macrophage preconditioning with synthetic malaria pigment reduces cytokine production via heme iron-dependent oxidative stress. Lab Invest 80:1781–1788
78. Singh PP, Singh S (2001) Protection of mice from malaria after co-administration of recombinant mouse granulocyte-macrophages colony- stimulating factor and methionine-enkephalin. Eur Cytokine Netw 12:528–536
79. Kaur A, Kinhikar AG, Singh PP (2004) Bioimmunotherapy of rodent malaria: Co-treatment with recombinant mouse granulocyte-macrophage colony-stimulating factor and an enkephalin fragment peptide Tyr-Gly-Gly. Acta Trop 91:27–41
80. Riopel J, Tam M, Mohan K, Marino MW, Stevenson MM (2001) Granulocyte-macrophage colony-stimulating factor-deficient mice have impaired resistance to blood-stage malaria. Infect Immun 69:129–136
81. Skorokhod OA, Schwarzer E, Grune T, Arese P (2005) Role of 4-hydroxynonenal in the hemozoin-mediated inhibition of differentiation of human monocytes to dendritic cells induced by GM-CSF/IL-4. Biofactors. 24:283–289
82. Skorokhod, O.A., , Alessio M, Mordmuller B, Arese P, and Schwarzer E. (2004)Hemozoin (malarial pigment) inhibits differentiation and maturation of human monocyte-derived dendritic cells: A peroxisome proliferator-activated receptor-gamma-mediated effect. J Immunol 173:4066-4074
83. Morris SM, Jr B, T.R. (1994) New insights into the regulation of inducible nitric oxide synthesis. Am J Physiol 266:E829–E889
84. Beckman JS, Koppenol WH (1996) Nitric oxide, superoxide, and peroxynitrite: The good, the bad, and ugly. Am J Physiol. 271:C1424–C1437
85. Dusse LMS, Vieira LM, Carvalho MG (2003) Revisão sobre óxido nítrico. J Bras Patol Med Lab 39:343–350
86. Förstermann U, W.C. S. (2012) Nitric oxide synthases: Regulation and function. Eur Heart J 33:829–837
87. Pacher P, Beckman JS, Liaudet L (2007) Nitric oxide and peroxynitrite in health and disease. Physiol Rev 87:315–424
88. Jia L, Bonaventura C, Bonaventura J, and Stamler J S. (1996)S-Nitrosohaemoglobin: a dynamic activity of blood involved in vascular control. Nature 380:221-226
89. Favre N, Ryffel B, Bordmann G, Rudin W (1997) The course of *Plasmodium chabaudi* infections in interferon-gamma receptor deficient mice. Parasite Immunol 19:375–383

90. Agbenyega T, Angus B, Bedu-Addo G, Baffoe-Bonnie B, Griffin G, Vallance P et al (1997) Plasma nitrogen oxides and blood lactate concentrations in Ghanaian children with malaria. Trans R Soc Trop Med Hyg 91:298–302
91. Van der Heyde HC, Gu Y, Zhang Q, Sun G, Grisham MBG (2000) Nitric oxide is neither necessary nor sufficient for resolution of *Plasmodium chabaudi* malaria in mice. J Immunol 165:3317–3323
92. Yoneto T, Yoshimoto T, Wang C R, Takahama Y, Tsuji M, Waki S, et al (1999) Gamma interferon production is critical for protective immunity to infection with blood-stage *Plasmodium berghei* XAT but neither NO production nor NK cell activation is critical. Infect Immun 67:2349–2356
93. Favre N, Ryffel B, Rudin W (1999) Parasite killing in murine malaria does not require nitric oxide production. Parasitology 118:139–143
94. Böhlke M (1999) Imunossupressão induzida pela malária: Existe um papel para o óxido nítrico? Rev Bras Alerg Imunopatol 22:173–178
95. Bredt DS, Snyder SH (1990) Isolation of nitric oxide synthetase, a calmodulin-requiring enzyme. Proc Natl Aca Sci USA 87:682–685
96. Green SJ, Scheller LF, Marletta MA, Seguin MC, Klotz FW, Slayter M et al (1994) Nitric oxide: Cytokine-regulation of nitric oxide in host resistance to intracellular pathogens. Immunol Lett 43:87–94
97. Pollock JS, Forstermann U, Mitchell JA, Warner TD, Schmidt HHHW, Nakane M et al (1991) Purification and characterization of particulate endothelium-derived relaxing factor synthase from cultured and native bovine aortic endothelial cells. Proc Natl Acad Sci USA 88:10480–10484
98. Keller CC, Kremsner PG, Hittner JB, Misukonis MA, Weinberg JB, Perkins DJ (2004) Elevated nitric oxide production in children with malarial anemia: Hemozoin-induced nitric oxide synthase type 2 transcripts and nitric oxide in blood mononuclear cells. Infect Immun 72:4868–4873
99. Zanini GM, Cabrales P, Barkho W, Frangos JA, E C L J M. (2011) Exogenous nitric oxide decreases brain vascular inflammation, leakage and venular resistance during *Plasmodium berghei* ANKA infection in mice. J Neuroinflam 8:66
100. Hawkes M, Opoka RO, Namasopo S, Miller C, Conroy AL, Serghides L et al (2011) Nitric oxide for the adjunctive treatment of severe malaria: Hypothesis and rationale. Med Hypothesis 77:437–444
101. Martins YC, Zanini GM, Frangos JA, Carvalho LJM (2012) Efficacy of different nitric oxide-based strategies in preventing experimental cerebral malaria by *Plasmodium berghei* ANKA. PLoS One. 7:e32048
102. Speyer CL, Neff TA, Warner RI, Guo RF, Sarma JV, Riedemann NC et al (2003) Regulatory effects of iNOS on acute lung inflammatory responses in mice. Am J Pathol 163:2319–2328
103. Preik-Steinhoff H, Kelm M (1996) Determination of nitrite in human blood by combination of a specific sample preparation with high-performance anion-exchange chromatography and electrochemical detection. J Chromatogr B Biomed Appl 685:348–352
104. Ranghino G, Scorza E, Sjogren T, Williams PA, Ricci M, Hajdu J (2000) Quantum mechanical interpretation of nitrite reduction by cytochrome cd1 nitrite reductase from *Paracoccus pantotrophus*. Biochemistry 39:10958–10966
105. Das BS, Nanda NK (1999) Evidence for erythrocyte lipid peroxidation in acute falciparum malaria. Trans R Soc Trop Med Hyg. 93:58–62
106. Kumar S, Das SK, Dey S, Maity P, Guha M, Choubey V et al (2008) Antiplasmodial activity of [(aryl)arylsulfanylmetyl]pyridine. Antimicrob Agents Chemother 52:705–715
107. Luse SA, Miller LH (1971) *Plasmodium falciparum* malaria: ultrastructure of parasitized erythrocytes in cardiac vessels. Am J Trop Med Hyg 20:655–660
108. Phiri H, Montgomery J, Molyneux M, Craig A (2009) Competitive endothelial adhesion between *Plasmodium falciparum* isolates under physiological flow conditions. Malar J 8:214

109. Schwarzer E, Kuhn H, Valente E, Arese P (2003) Malaria-parasitized erythrocytes and hemozoin nonenzymatically generate large amounts of hydroxy fatty acids that inhibit monocyte functions. Blood 101:722728
110. Omodeo-Salè F, Motti A, Basilico N, Parapini S, Olliaro P, Taramelli D (2003) Accelerated senescence of human erythrocytes cultured with *Plasmodium falciparum*. Blood. 102:705–711
111. Becker K, Tilley L, Vennerstrom J L, Roberts D, Rogersone S, and Ginsburg H. (2004) Oxidative stress in malaria parasite-infected erythrocytes: host–parasite interactions. Int J Parasitol. 34 163–189
112. Ferreira MU, Nunes MS, Wunderlich G (2004) Antigenic diversity and immune evasion by malaria parasites. Clin Diag Lab Immunol 11:987–995
113. Pettersson F, Vogt AM, Jonsson C, Mok BW, Shamaei-Tousi A, Bergstro S et al (2005) Whole-body imaging of sequestration of *Plasmodium falciparum* in the Rat. Infect Immun 73:7736–7746
114. Glenister FK, Coppel RL, Cowman AF, Mohandas N, Cooke BMC (2002) Contribution of parasite proteins to altered mechanical properties of malaria-infected red blood cells. Blood. 99:1060–1060
115. Rifkind JM, Nagababu E, Ramasamy S, Babu L (2003) Hemoglobin redox reactions and oxidative stress. Redox Report 8(5):234–237. https://doi.org/10.1179/135100003225002817
116. Nagababu E, Rifkind JM (2000) Reaction of hydrogen peroxide with ferrylhemoglobin: superoxide production and heme degradation. Biochemistry 39:12503–12511
117. Nagababu E, Ramasamy S, Rifkind JM, Jia Y, Alayash AI (2002) Site specific cross-linking of human and bovine hemoglobins differentially alters oxygen binding and redox side reactions producing rhombic heme and heme degradation. Biochemistry 41:7407–7415
118. Nagababu E, Chrest FJ, Rifkind JM (2003) Hydrogen-peroxide induced heme degradation in red blood cells: the respective roles of catalase and glutathione peroxidase. Biochim Biophys Acta 1620:211–217
119. Reiter CD, Wang X, Tanus-Santos JE, Hogg N, Cannon RO 3rd, Schechter AN et al (2002) Cell-free hemoglobin limits nitric oxide bioavailability in sickle-cell disease. Nat Med 8:1383–1289
120. Moore EG, Gibson QH (1976) Cooperativity in the dissociation of nitric oxide from hemoglobin. J Biol Chem 251:2788–2794
121. Boretti FS, Buehler PW, D'Agnillo F, Kluge K, Glaus T, Butt OI et al (2009) Sequestration of extracellular hemoglobin within a haptoglobin complex decreases its hypertensive and oxidative effects in dogs and guinea pigs. J Clin Invest 119:2271–2280
122. Buehler PW, Abraham B, Vallelian F, Linnemayr C, Pereira CP, Cipollo JF et al (2009) Haptoglobin preserves the CD163 hemoglobin scavenger pathway by shielding hemoglobin from peroxidative modification. Blood 113:2578–2586
123. Graca-Souza AV, Arruda MAB, de Freitas MS, Barja-Fidalgo C, Oliveira PL (2002) Neutrophil activation by heme: implications for inflammatory processes. Blood 99:4160–4165
124. Liu X, Spolarics Z (2003) Methemoglobin is a potent activator of endothelial cells by stimulating IL-6 and IL-8 production and E-selectin membrane expression. Am J Physiol Cell Physiol 285:C1036–C1046
125. Widmer CC, Pereira CP, Gehrig P, Vallelian F, Schoedon G, Buehler PW et al (2010) Hemoglobin can attenuate hydrogen peroxide-induced oxidative stress by acting as an antioxidative peroxidase. Antioxid Redox Signal 12(2):185–198. https://doi.org/10.1089/ars.2009.2826
126. Jia Y, Buehler PW, Boykins RA, Venable RM, Alayash AI (2007) Structural basis of peroxide-mediated changes in human hemoglobin: a novel oxidative pathway. J Biol Chem 282:4894–4907
127. Vallelian F, Pimenova T, Pereira CP, Abraham B, Mikolajczyk MG, Schoedon G et al (2008) The reaction of hydrogen peroxide with hemoglobin induces extensive alpha-globin cross-linking and impairs the interaction of hemoglobin with endogenous scavenger pathways. Free Radic Biol Med 45:1150–1158

128. D'Agnillo F, Alayash AI (2002) A role for the myoglobin redox cycle in the induction of endothelial cell apoptosis. Free Radic Biol Med 33:1153–1164
129. Cooper CE, Silaghi-Dumitrescu R, Rukengwa M, Alayash AI, Buehler PW (2008) Peroxidase activity of hemoglobin towards ascorbate and urate: a synergistic protective strategy against toxicity of hemoglobin-based oxygen carriers (HBOCs). Biochim Biophys Acta 1784:1415–1420
130. D'Agnillo F, Alayash AI (2001) Redox cycling of diaspirin cross-linked hemoglobin induces G2=M arrest and apoptosis in cultured endothelial cells. Blood 98:3315–3323
131. Huang L, Wojciechowski G, Ortiz de Montellano PR (2005) Prosthetic heme modification during halide ion oxidation: demonstration of chloride oxidation by horseradish peroxidase. J Am Chem Soc 127:5345–5353
132. Huang L, Wojciechowski G, Ortiz de Montellano PR (2006) Role of heme-protein covalent bonds in mammalian peroxidases: protection of the heme by a single engineered hemeprotein link in horseradish peroxidase. J Biol Chem 281:18983–18938
133. Wojciechowski G, Huang L, Ortiz de Montellano PR (2005) Autocatalytic modification of the prosthetic heme of horseradish but not lactoperoxidase by thiocyanate oxidation products: a role for heme-protein covalent cross-linking. J Am Chem Soc 127:15871–15879
134. Boyle JJ, Harrington HA, Piper E, Elderfield K, Stark J, Landis RC et al (2009) Coronary intraplaque hemorrhage evokes a novel atheroprotective macrophage phenotype. Am J Pathol 174:1097–1108
135. Schaer DJ, Alayash AI, Buehler PW (2007) Gating the radical hemoglobin to macrophages: the anti-inflammatory role of CD163, a scavenger receptor. Antioxid Redox Signal 9:991–999
136. Schaer DJ, Schaer CA, Buehler PW, Boykins RA, Schoedon G, Alayash AI, et al (2006) CD163 is the macrophage scavenger receptor for native and chemically modified hemoglobins in the absence of haptoglobin. Blood 107:373–380
137. Kapralov A, Vlasova II, Feng W, Maeda A, Walson K, Tyurin VA et al (2008) Peroxidase activity of hemoglobin haptoglobin complexes- covalent aggregation and oxidative stress in plasma macrophages. J Biol Chem 284(44):30395–30407
138. Figueiredo RT, Fernandez PL, Mourao-Sa DS, Porto BN, Dutra FF, Alves LS et al (2007) Characterization of heme as activator of Toll-like receptor 4. J Biol Chem 282:20221–20229
139. Potter SM, Mitchell AJ, Cowden WB, Sanni LA, Dinauer M, Haan JB et al (2005) Phagocyte-derived reactive oxygen species do not influence the progression of murine blood-stage malaria infections. Infect Immun 73:4941–4947
140. Bhattacharyya A, Chattopadhyay R, Mitra S, Crowe SE (2014) Oxidative stress as an essential factor in the pathogenesis of gastrointestinal mucosal diseases. Physiol Rev 94:329–354. https://doi.org/10.1152/physrev.00040.2012
141. Foth BJ, Zhang N, Chaal BK, Sze SK, Preiser PR, Bozdech Z (2011) Quantitative time-course profiling of parasite and host cell proteins in the human malaria parasite *Plasmodium falciparum*. Mol Cell Proteom 10:1–16
142. Toler S (2005) The plasmodial apicoplast was retained under evolutionary selective pressure to assuage blood stage oxidative stress. Med Hypotheses 65:683–690
143. Kanzok SM, Schirmer RH, Turbachova I, Iozef R, Becker K (2000) The thioredoxin system of the malaria parasite *Plasmodium falciparum*. Glutathione reduction revisited. J Biol Chem 275:40180–40186
144. Jones DP, Go YM (2010) Redox compartmentalization and cellular stress. Diabetes Obes Metab 12:116–125
145. Halliwell B (1999) Antioxidant defense mechanisms: from the beginning to the end (of the beginning). Free Radic Res 31:261–272
146. Fernandes AP, Holmgren A (2004) Glutaredoxins: Glutathione-dependent redox enzymes with functions far beyond a simple thioredoxin backup system. Antioxid Redox Signal 6:63–74
147. Richard D, Bartfai R, Volz J, Ralph SA, Muller S, Stunnenberg HG et al (2011) A genome-wide chromatin-associated nuclear peroxiredoxin from the malaria parasite *Plasmodium falciparum*. J Biol Chem 286:11746–11755

148. Sies H (1999) Glutathione and its role in cellular functions. Free Radic Biol Med 27:916–921
149. Meierjohann S, Walter RD, Müller S (2002) Regulation of intracellular glutathione levels in erythrocytes infected with chloroquine-sensitive and chloroquine-resistant *Plasmodium falciparum*. Biochem J 368:761–768
150. Chandra R, Tripathi LM, Saxena JK, Puri SK (2011) Implication of intracellular glutathione and its related enzymes on resistance of malaria parasites to the antimalarial drug arteether. Parasitol Int 60:97–100
151. Mashima R, Tilley L, Siomos MA, Papalexis V, Raftery MJ, Stocker R (2002) *Plasmodium falciparum* histidine-rich protein-2 (PfIHRP2) modulates the redox activity of ferri-protoporphyrin IX (FePPIX): Peroxidase-like activity of the PfIHRP2-FePPIX complex. J Biol Chem 277:14514–14520
152. Wrenger C, Eschbach ML, Müller IB, Warnecke D, Walter RD (2005) Analysis of the vitamin B6 biosynthesis pathway in the human malaria parasite *Plasmodium falciparum*. J Biol Chem 280:5242–5248
153. Massimine KM, McIntosh MT, Doan LT, Atreya CE, Gromer S, Sirawaraporn W et al (2006) Eosin B as a novel antimalarial agent for drug-resistant *Plasmodium falciparum*. Antimicrob Agents Chemother 50:3132–3141
154. Forman HJ, Maiorino M, Ursini F (2010) Signaling functions of reactive oxygen species. Biochemistry 49:835–842
155. Kanzok SM, Rahlfs S, Becker K, Schirmer RH (2002) Thioredoxin, thioredoxin reductase, and thioredoxin peroxidase of malaria parasite *Plasmodium falciparum*. Methods Enzymol. 347:370–381
156. Kawazu S, Komaki K, Tsuji N, Kawai S, Ikenoue N, Hatabu T et al (2001) Molecular characterization of a 2-Cys peroxiredoxin from the human malaria parasite *Plasmodium falciparum*. Mol Biochem Parasitol 116:73–79
157. Kehr S, Jortzik E, Delahunty C, Yates JR, Rahlfs S, Becker K (2011) Protein s-glutathionylation in malaria parasites. Antioxid Redox Signal 15:2855–2865
158. Kumar S, Das SK, Dey S, Maity P, Guha M, Choubey V et al (2008) Antiplasmodial activity of [(aryl)arylsulfanylmethyl]Pyridine. Antimicrob Agents Chemother 52:705–715
159. Krnajski Z, Gilberger TW, Walter RD, Cowman AF, Müller S (2002) Thioredoxin reductase is essential for the survival of *Plasmodium falciparum* erythrocytic stages. J Biol Chem 277:25970–25975
160. Wrenger C, Schettert I, Liebau E (2013) Oxidative stress in human infectious diseases-present and current knowledge about its druggability. In: InTechOpen (ed) Drug discovery. InTecOpen, London, pp 227–250
161. Becker K, Kanzok SM, Iozef R, Fischer M, Schirmer RH, Rahlfs S (2003) Plasmoredoxin, a novel redox-active protein unique for malarial parasites. Eur J Biochem. 270:1057–1064
162. Schirmer RH, Müller JG, Krauth-Siegel RL (1995) Disulfide-reductase inhibitors as chemotherapeutic agents: the design of drugs for trypanosomiasis and malaria. Angew Chem Int Edn 34:141–154
163. Kanzok SM, Schirmer RH, Türbachova II, Ozef R, K. B. (2000) The thioredoxin system of the malaria parasite *Plasmodium falciparum*. Glutathione reduction revisited. J Biol Chem 275:40180–40186
164. Krnajski Z, Gilberger T, Walter RD, Müller S (2001) The malaria parasite *Plasmodium falciparum* possesses a functional thioredoxin system. Mol Biochem Parasitol 112:219–228
165. Färber PM, Arscott LD, Williams CH Jr, Becker K, Schirmer RH (1998) Recombinant *Plasmodium falciparum* glutathione reductase is inhibited by the antimalarial dye methylene blue. FEBS Lett 422:311–314
166. Krnajski Z, Walter RD, Müller S (2001) Isolation and functional analysis of two thioredoxin peroxidases (peroxiredoxins) from *Plasmodium falciparum*. Mol Biochem Parasitol 113:303–308
167. Rahlfs S, Becker K (2001) Thioredoxin peroxidases of the malarial parasite *Plasmodium falciparum*. Eur J Biochem 268:1404–1409

168. Rahlfs S, Fischer M, Becker K (2001) *Plasmodium falciparum* possesses a classical glutaredoxin and a second, glutaredoxin-like protein with a PICOT homology domain. J Biol Chem 276:37133–37140
169. Harwaldt P, Rahlfs S, Becker K (2002) Glutathione S-transferase of the malarial parasite *Plasmodium falciparum*: characterization of a potential drug target. Biol Chem 383:821–830
170. Holmgren A (2000) Antioxidant function of thioredoxin and glutaredoxin systems. Antioxid Redox Signal 2:811–820
171. Becker K, Rahlfs S, Nickel C, Schirmer RH (2003) Glutathione-function and metabolism in the malarial parasite *Plasmodium falciparum*. Biol Chem. 384:551–566
172. Rahlfs S, Nickel C, Deponte M, Schirmer RH, Becker K (2003) *Plasmodium falciparum* thioredoxins and glutaredoxins as central players in redox metabolism. Redox Report 8(5):246–250. https://doi.org/10.1179/135100003225002844
173. Rodriguez-Manzaneque MT, Ros J, Cabiscol E, Sorribas A, Herrero E (1999) Grx5 glutaredoxin plays a central role in protection against protein oxidative damage in Saccharomyces cerevisiae. Mol Cell Biol 19:8180–8190
174. Isakov N, Witte S, Altman A (2000) PICOT-HD: a highly conserved protein domain that is often associated with thioredoxin and glutaredoxin modules. Trends Biochem Sci 25:537–539
175. Belli G, Polaina J, Tamarit J (2002) Structure-function analysis of yeast Grx5 monothiol glutaredoxin defines essential amino acids for the function of the protein. J Biol Chem 277:37590–37596
176. Shenton D, Perrone G, Quinn KA, Dawes IW, Grant CM (2002) Regulation of protein S-thiolation by glutaredoxin 5 in the yeast Saccharomyces cerevisiae. J Biol Chem 277:16853–16859
177. Lasonde E, Ishihama Y, Andersen JS (2002) Analysis of the *Plasmodium falciparum* proteome by high-accuracy mass spectrometry. Nature 419:537–542
178. Kanzok SM, Fechner A, Bauer H (2001) Substitution of the thioredoxin system for glutathione reductase in *Drosophila melanogaster*. Science 296:643–646
179. Takemoto T, Zhang QM, Yonei S (1998) Different mechanisms of thioredoxin in its reduced and oxidized forms in defense against hydrogen peroxide in *Escherichia coli*. Free Radic Biol Med 24:556–562
180. Das KC, Das CK (2000) Thioredoxin, a singlet oxygen quencher and hydroxyl radical scavenger: redox independent functions. Biochem Biophys Res Commun 277:443–447
181. Björnstedt M, Hamberg M, Kumar S, Xue J, Holmgren A (1995) Human thioredoxin reductase directly reduces lipid hydroperoxides by NADPH and selenocystine strongly stimulates the reaction via catalytically generated selenols. J Biol Chem 270:11761–11764
182. Kumar S, Christophides GK, Cantera R, Charles B, Han YS, Meister S et al (2003) The role of reactive oxygen species on *Plasmodium* melanotic encapsulation in *Anopheles gambiae*. Proc Natl Acad Sci USA 100:14139–14144
183. Crompton PD, Moebius J, Portugal S, Waisberg M, Hart G, Garver LS et al (2014) Malaria immunity in man and mosquito: insights into unsolved mysteries of a deadly infectious disease. Annu Rev Immunol 32:157–187. https://doi.org/10.1146/annurev-immunol-032713-120220
184. Akman-Anderson L, Olivier M, Luckhart S (2007) Induction of nitric oxide synthase and activation of signaling proteins in *Anopheles mosquitoes* by the malaria pigment, hemozoin. Infect Immun 75:4012–4019
185. Lim J, Gowda DC, Krishnegowda G, Luckhart S (2005) Induction of nitric oxide synthase in *Anopheles* stephensi by *Plasmodium falciparum*: Mechanism of signaling and the role of parasite glycosylphosphatidylinositols. Infect Immun 73:2778–2789
186. Bauer H, Gromer S, Urbani A, Schnölzer M, Schirmer RH, Müller HM (2003) Thioredoxin reductase from the malaria mosquito *Anopheles gambiae*. Eur J Biochem 270:4272–4281
187. Wongtrakul J, Pongjaroenkit S, Leelapat P, Nachaiwieng W, Prapanthadara LA, Ketterman AJ (2010) Expression and characterization of three new glutathione transferases, an epsilon (AcGSTE2-2), omega (AcGSTO1-1), and theta (AcGSTT1-1) from Anopheles cracens (Diptera: Culicidae), a major Thai malaria vector. J Med Entomol 47:162–171

188. Zhang S, Chen H and Gerhard G S. (2010)Heme synthesis increases artemisinin-induced radical formation and cytotoxicity that can be suppressed by superoxide scavengers. Chem Biol Interact 186::30-35
189. Grellier P, Maroziene A, Nivinskas H, Sarlauskas J, Aliverti A, Cenas N (2010) Antiplasmodial activity of quinones: Roles of aziridinyl substituents and the inhibition of *Plasmodium falciparum* glutathione reductase. Arch Biochem Biophys 494:32–39
190. Giao PT, Binh TQ, Kager PA, Long HP, van Thang N, van Nam N et al (2001) Artemisinin for treatment of uncomplicated *falciparum* malaria: Is there a place for monotherapy? Am J Trop Med Hyg 65:690–695
191. Krungkrai SR, Yuthavong Y (1987) The antimalarial action on *Plasmodium falciparum* of qinghaosu and artesunate in combination with agents which modulate oxidant stress. Trans R Soc Trop Med Hyg 81:710–714
192. Scott MD, Meshnick SR, Williams RA, Chiu DT, Pan HC, Lubin BH et al (1989) Qinghaosu-mediated oxidation in normal and abnormal erythrocytes. J Lab Clin Med 114:401–406
193. Hartwig CL, Rosenthal AS, D'Angelo J, Griffin CE, Posner GH, R.A. C. (2009) Accumulation of artemisinin trioxane derivatives within neutral lipids of *Plasmodium falciparum* malaria parasites is endoperoxide-dependent. Biochem Pharmacol 77:322–336
194. Klonis N, Crespo-Ortiz MP, Bottova I, Abu-Bakar N, Kenny S, Rosenthal PJ et al (2011) Artemisinin activity against *Plasmodium falciparum* requires hemoglobin uptake and digestion. Proc Natl Acad Sci USA 108:11405–11410
195. Akabi OM, Odaibo AB, Ademowo OG (2010) Effect of antimalarial drugs and malaria infection on oxidative stress in pregnant women. Afr J Reprod Health 14:209–212
196. Benoit-Vical F, Robert A, Meunier B (1999) Potentiation of artemisinin activity against chloroquine-resistant *Plasmodium falciparum* strains by using heme models. Antimicrob Agents Chemother 43:2555–2558
197. Legorreta-Herrera M, Retana-Ugalde R, Ventura-Gallegos JL, Narvaez VP (2010) Pyrimethamine induces oxidative stress in *Plasmodium yoelii* 17XL-infected mice: A novel immunomodulatory mechanism of action for an old antimalarial drug? Exp Parasitol 126:381–388
198. Nogueria F, Diez A, Radfar A, Perez-Benavente S, do Rosario VE, Puyet A et al (2010) Early transcriptional response to chloroquine of the *Plasmodium falciparum* antioxidant defence in sensitive and resistant clones. Acta Trop 114:109–115
199. Trivedi V, Chand P, Srivastava K, Puri S, Maulik PR, Bandyopadhyay U (2005) Clotrimazole inhibits hemoperoxidase of *Plasmodium falciparum* and induces oxidative stress. J Biol Chem 280:41129–41136
200. Mavondo GA (2017) Malaria disease perspective and an opinion: should malaria treatment target the parasite or the malarial pathophysiology generated by the parasite or both? EC Microbiology 7(5):149–154
201. Al-Adhroey AH, Nor ZM, Al-Mekhlafi HM, Amran AA, Mahmud R (2011) Antimalarial activity of methanolic leaf extract of *Piper betle* L. Molecules 16:107–118
202. Akanbi OM, Omonkhua AA, Cyril-Olutayo CM, Fasimoye RY (2012) The antiplasmodial activity of *Anogeissus leiocarpus* and its effect on oxidative stress and lipid profile in mice infected with *Plasmodium berghei*. Parasitol Res 110:219–226
203. Okeola VO, Adaramoye OA, Nneji CM, Falade CO, Farombi EO, Ademowo OG (2011) Antimalarial and antioxidant activities of methanolic extract of *Nigella sativa* seeds (black cumin) in mice infected with *Plasmodium yoelli* nigeriensis. Parasitol Res 108:1507–1512
204. Ferreira JF, Luthria DL, Sasaki T, Heyerick A (2010) Flavonoids from *Artemisia annua* L. as antioxidants and their potential synergism with artemisinin against malaria and cancer. Molecules 15:3135–3170
205. Mavondo G A, Mkhwananzi B N, Mabandla M V, and Musabayane C T. (2016)Asiatic acid influences parasitaemia reduction and ameliorates malaria anaemia in *P. berghei* infected Sprague-Dawley male rats. BMC Complement Altern Med. 16(1):357.10.1186/s12906-016-1338-z

206. Percario S, Naufal AS, Gennari JL, Gennari MS (2009) Antioxidant activity of edible blushing wood mushroom, Agaricus sylvaticus Schaeff. (Agaricomycetideae) in vitro. Int J Med Mushrooms 11:133–140
207. Dattani JJ, Rajput DK, Moid N, Highland HN, George LB, Desai KR (2010) Ameliorative effect of curcumin on hepatotoxicity induced by chloroquine phosphate. Environ Toxicol Pharmacol 30:103–109
208. Connolly JD, Hill RA (1991) Dictionary of terpenoids. Chapman and Hall, London
209. Zwenger S (2008) Plant terpenoids: applications and future potentials. Biotechnol Mol Biol Rev 3:001–007
210. Mavondo GA, Mzingwane MML (2017) Parasitic causes of hemolytic anemia: severe malaria anemia pathophysiology, current and experimental phytotherapeutics in malaria anemia. In Khan J (ed) Current topics on anaemia. Intechopen, Rijeka, Croatia-EU
211. Mavondo GA, Musabayane CT (2018) Transdermal drug delivery of asiatic acid influences renal function and electrolyte handling in *Plasmodium berghe*i-infected sprague-dawley male rats. J Dis Med Plants 4(1):18–29. https://doi.org/10.11648/j.jdmp.20180401.13
212. Batista R, García P, Castro M (2013) Synthesis, cytotoxicity and antiplasmodial activity of novel ent-kaurane derivatives. Eur J Med Chem 62:168–176
213. Kalauni SK, Awale S, Tezuka Y, Banskota AH, Linn TZ, Asih PB et al (2006) Antimalarial activity of cassane- and norcassane-type diterpenes from Caesalpinia crista and their structure-activity relationship. Biol Pharm Bull 29(5):1050–1052
214. Nondo RS, Moshi MJ, Erasto P, Masimba PJ, Machumi F, Kidukuli AW et al (2017) Antiplasmodial activity of Norcaesalpin D and extracts of four medicinal plants used traditionally for treatment of malaria. BMC Complement Altern Med 17(1):167. https://doi.org/10.1186/s12906-017-1673-8
215. Angerhofer CK, Pezzuto JM, Konig GM, Wright AD, Sticher O (1992) Antimalarial activity of sesquiterpenes from the marine sponge Acanthella klethra. J Nat Prod 55(12):1787–1789
216. Isah MB, Ibrahim MA, Mohammed A, Aliyu AB, Masola B, Coetzer TH (2016) A systematic review of pentacyclic triterpenes and their derivatives as chemotherapeutic agents against tropical parasitic diseases. Parasitology 143(10):1219–1231. https://doi.org/10.1017/s0031182016000718
217. Dzubak P, Hajduch M, Vydra D, Hustova A, Kvasnica M, Biedermann D, et al (2006) Pharmacological activities of natural triterpenoids and their therapeutic implications. Nat Prod Rep 23(3):394–411. https://doi.org/10.1039/b515312n
218. Moon HI, Jung JC, Lee J (2007) Antiplasmodial activity of triterpenoid isolated from whole plants of Viola genus from South Korea. Parasitol Res 100(3):641–644. https://doi.org/10.1007/s00436-006-0279-8
219. Mbatha B (2015) Evaluation of the efficacy of Syzygium Aromaticum-derived oleanolic acid on malaria parasited in Plasmodium BergheiI-infected male Sprague-Dawley rats-and effects on blood glucose and renal electrolyte handling. In Human physiology. University of KwaZulu-Natal: South Africa, Durban, p 90
220. Moneriz C, Mestres J, Bautista JM, Diez A, Puyet A (2011) Multi-targeted activity of maslinic acid as an antimalarial natural compound. Febs J 278(16):2951–2961. https://doi.org/10.1111/j.1742-4658.2011.08220.x
221. Moneriz C, Marin-Garcia P, Bautista JM, Diez A, Puyet A (2011) Parasitostatic effect of maslinic acid. II. Survival increase and immune protection in lethal Plasmodium yoelii-infected mice. Malar J 10:103. https://doi.org/10.1186/1475-2875-10-103
222. Ben Salem M, Affes H, Athmouni K, Ksouda K, Dhouibi R, Sahnoun Z, et al (2017) Chemicals compositions, antioxidant and anti-inflammatory activity of cynara scolymus leaves extracts, and analysis of major bioactive polyphenols by HPLC. Evid Based Complement Alternat Med 2017:4951937. https://doi.org/10.1155/2017/4951937
223. Mkhwanazi BN, Serumula MR, Myburg RB, Van Heerden FR, Musabayane CT (2014) Antioxidant effects of maslinic acid in livers, hearts and kidneys of streptozotocin-induced diabetic rats: effects on kidney function. Ren Fail 36(3):419–31. https://doi.org/10.3109/0886022x.2013.867799

224. Thaane T (2014) Evaluation of the efficacy of maslinic acid on malaria parasites in plasmodium Berghei-infected male sprague-dawley rats: effects on blood glucose and renal fluid and electrolyte handling. In Human physiology. University of KwaZulu-Natal, p 109
225. Marquez Martin A, de la Puerta VR, Fernandez-Arche A, Ruiz-Gutierrez V (2006) Supressive effect of maslinic acid from pomace olive oil on oxidative stress and cytokine production in stimulated murine macrophages. Free Radic Res 40(3):295–302. https://doi.org/10.1080/10715760500467935
226. Mavondo GA, Mkhwananzi BN, Mabandla MV, Musabayane CT (2016) Asiatic acid influences parasitaemia reduction and ameliorates malaria anaemia in P. berghei infected Sprague-Dawley male rats. BMC Complement Altern Med 16:357. https://doi.org/10.1186/s12906-016-1338-z
227. Ramachandran V, Saravanan R (2013) Asiatic acid prevents lipid peroxidation and improves antioxidant status in rats with streptozotocin-induced diabetes. J Funct Foods 5(3):1077–1087. https://doi.org/10.1016/j.jff.2013.03.003

Exploring Endoperoxides as Leishmanicidal Compounds

24

Sritama De Sarkar and Mitali Chatterjee

Abstract

With advances in genomics, proteomics, and bioinformatics, identification of unique parasite-specific metabolic pathways has facilitated development of antileishmanial chemotherapeutics. In view of *Leishmania* parasites having a compromised antioxidant defense mechanism, induction of oxidative stress is a universal strategy adopted by conventional antileishmanial drugs with mitochondrial dysfunction being the major source of free radicals. However, a limitation is that mammalian mitochondria too are inhibited. Therefore, an attractive therapeutic option would be compounds like endoperoxides which owing to their unusual peroxide bridge mediate parasiticidal activity primarily via generation of free radicals. Accordingly, exploring the leishmanicidal potential of endoperoxides like artemisinin and ascaridole having established antimalarial and antihelminthic properties respectively, was the focus of this study. *Leishmania* proved to be a susceptible target for these free radical generating endoperoxides, making this therapeutic modality worthy of future pharmacological consideration.

Keywords

Antileishmanial drugs · Drug repurposing · Endoperoxides · Free radicals · Mitochondria · Oxidative stress

S. De Sarkar · M. Chatterjee (✉)
Department of Pharmacology, Institute of Post Graduate Medical Education and Research, Kolkata, West Bengal, India
e-mail: ilatim@vsnl.net

S. Chakraborti et al. (eds.), *Oxidative Stress in Microbial Diseases*,
https://doi.org/10.1007/978-981-13-8763-0_24

24.1 Is there a Role for Drug Repurposing in Leishmaniasis?

Drug repurposing or reprofiling has been responsible for 30% of the new drugs and vaccines approved by the Food and Drug Administration [1], and as these repurposed drugs are typically approved for clinical uses, this process translates into significant savings in terms of time and cost. Thus, repositioning of approved drug(s) is a particularly attractive strategy for neglected tropical diseases like leishmaniasis, a group of diverse diseases that ranges from a life-threatening visceral form to multiple dermal manifestations caused by various species of *Leishmania.* An estimated 700,000 to 1 million new cases and 20,000–30,000 deaths occur due to leishmaniasis annually, and is endemic in 89 countries with 350 million people being susceptible [2, 3]. Owing to the limited R&D resources, the antileishmanial arsenal includes several repurposed drugs, e.g., amphotericin B, which was discovered in the 1950s, against threatening fungal and bacterial infections, along with miltefosine, which was established in the 1980s to have antitrypanosomal and antineoplastic activities [4–6]. The discovery and development of effective drugs against parasitic diseases are challenging due to differences in clinical forms of the disease, variable pharmacokinetics, and difficulty in genetic manipulation of some parasites. Accordingly, to maximize the chances of success against the *Leishmania* parasite, a practical approach would be to study established antiparasitic drugs, e.g., artemisinin, chloroquine etc. [7–10].

24.2 Modulation of the Redox Status of Macrophages

Macrophages upon encountering a pathogen conventionally respond by phagocytosis followed by phagosome-lysosome fusion and generate toxic reactive oxygen and nitrogen intermediates to eliminate the engulfed pathogen [11]. The first and foremost player among the prooxidants include the superoxide anion ($O_2^{\bullet -}$) produced by NADPH oxidases present in host cell membranes. This molecule marks the start of a rampage by free radicals aimed at creating a milieu where the parasite would find it difficult to survive (Table 24.1). Nonetheless, the parasite evolves devious strategies to bail themselves out, and therefore, the persistence and proliferation of parasites within the macrophages is determined by a balance between the host's ability

Table 24.1 Reactive oxygen and nitrogen species encountered by *Leishmania* within host macrophages

Free radicals	Host source	Stages of infection
Superoxide ($O_2^{\bullet -}$)	NADPHox	Phagocytosis
Nitric oxide ($NO^{\bullet}$)	iNOS	Amastigotes
Peroxynitrite ($ONOO^-$)	$O_2^{\bullet -}$ + $NO^{\bullet}$	Amastigotes
Hydrogen peroxide (H_2O_2)	SOD	Promastigotes and Amastigotes
Hypochlorous acid (HOCl)	Myeloperoxidase	Promastigotes
Hydroxyl radical ($^{\bullet}OH$)	H_2O_2	Promastigotes

to sufficiently activate *Leishmania*-infected macrophages and tenacity of the parasite to resist the host's cytotoxic mechanisms [12].

Oxidative burst is a potent signal against pathogens manifested by an enhanced production of reactive oxygen species (ROS) intermediates with NADPH oxidase and xanthine oxidase being the major sources. Activated intracellular NADPH oxidase reacts with the substrate NADPH, to reduce oxygen to superoxide ($O_2^{\bullet-}$). Through enzyme catalysis, superoxide dismutase spontaneously converts superoxide into hydrogen peroxide (H_2O_2), hydroxyl radical ($OH^{\bullet-}$), hypochlorous acid (HOCl), and peroxynitrite ($ONOO^-$). Xanthine oxidase is also responsible for the synthesis of $O_2^{\bullet-}$and nitric oxide (NO) by donating an electron from xanthine to oxygen and reducing nitrite to NO, respectively [13; Fig. 24.1].

24.2.1 Leishmania Possesses a Unique Antioxidant Defense

Leishmania lacks two major H_2O_2-metabolizing enzymes usually present in eukaryotes, namely catalase and classical selenocysteine-containing glutathione peroxidase, thus rendering the parasites more vulnerable to free radical toxicity [14]. Instead, the presence of ascorbate peroxidase (LmAPX) from *Leishmania major* is considered as a potential candidate for scavenging ROS and serves as the central redox defense system [15]. Another important difference in African trypanosomes and *Leishmania* is the unique presence of trypanothione, the glutathione-spermidine conjugate N1, N8-*bis*(L-γ-glutamyl-L-hemicystinylglycyl) spermidine [16, 17]. The biosynthesis of trypanothione comprises the synthesis of GSH and spermidine, followed by their conjugation. Taking into account all the enzymes involved in the synthesis and reduction of trypanothione, virtually all components of the *Leishmania* trypanothione pathway are potential drug targets [18–20]. However, drug development approaches have focused on two central enzymes, trypanothione reductase and trypanothione synthase [21].

24.2.2 Antileishmanials Cause Induction of Oxidative Stress

The parasiticidal activity of antimonials like sodium antimony gluconate (SAG) involves increased generation of ROS and NO as studied in murine macrophages and modulation of the mitogen-activated protein kinase (MAPK) cascade, resulting in enhanced production of TNF-α [22, 23]. Similarly, other antileishmanials such as amphotericin B, miltefosine, and pentamidine also enhance generation of free radicals within parasites, along with enhanced membrane permeability and collapse of the mitochondrial membrane potential [24–26]. The orally administrable repurposed anticancer drug miltefosine mediates its leishmanicidal activity by mitochondrial dysfunction including membrane depolarization, cytochrome c release, and DNA fragmentation leading to parasite apoptosis [27, 28]. Likewise, amphotericin B-triggered the generation of free radicals which led to lipid peroxidation, cell cycle arrest, and DNA damage of *L. donovani* promastigotes [29, 30].

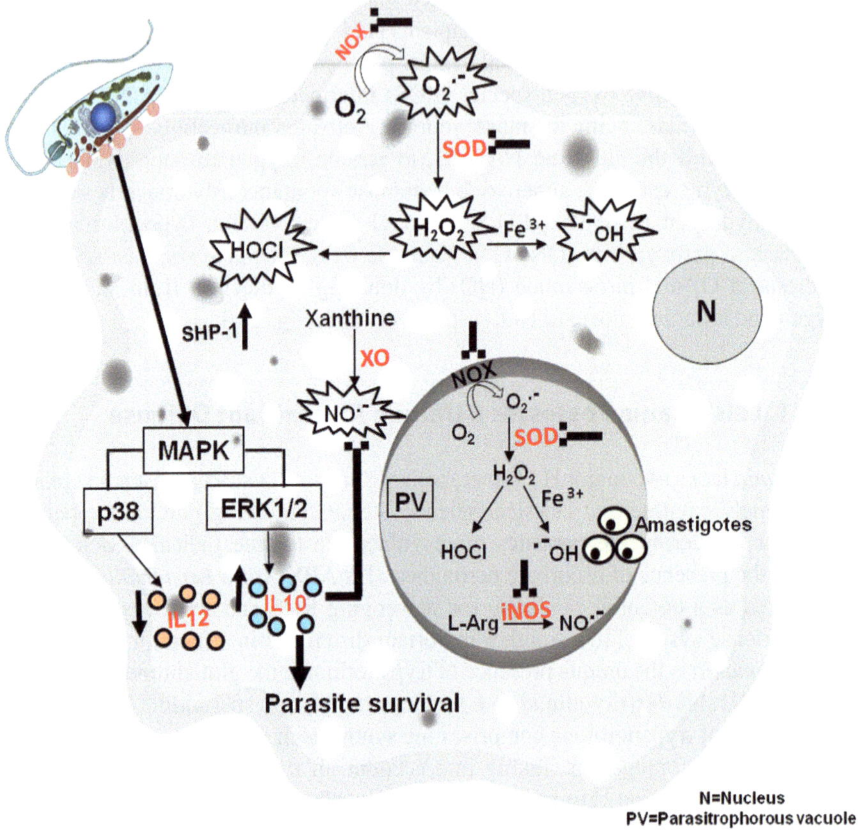

Fig. 24.1 **Modulation of the redox status of macrophages as a critical survival strategy of *Leishmania***
Leishmania encounter several free radicals inside the host macrophages. Within the macrophages, molecular oxygen (dioxygen) following addition of one electron is reduced to superoxide ($O_2^{\bullet -}$) in the presence of activated intracellular NADPH oxidase (NOX), which upon enzyme catalysis in the presence of superoxide dismutase (SOD) is converted into hydrogen peroxide (H_2O_2), which further gets converted into hydroxyl ($^{\bullet -}OH$) and hyp Fochlorous acid (HOCl). Nitric oxide (NO) is formed either by xanthine oxidase (XO) or by inducible nitric oxide synthase (iNOS) [13]. The parasite employs several sequestering mechanisms by inhibiting these enzymes (—¦) along with modulation of signalling pathways, e.g., MAPK pathway to ensure its survival within macrophages

24.2.3 Endoperoxides and Generation of Free Radicals Within Parasites

Organic peroxides generate a major group of ROS as a by-product of carbon-centered radicals with oxygen. More than 600 peroxides have been isolated and structurally characterized from natural sources, mainly as constituents of the family *Compositae* or *Asteraceae*. Among different peroxides studied, fatty acid derivatives, sesquiterpene endoperoxides, and Quinghaosu have been clinically approved as antimalarials. These endoperoxides, a class of heterocyclic compounds, contain a peroxide bridge that can partially escape the host's detoxification systems and target the parasite. Both synthesized and natural peroxides have been reported to exhibit biological activities [31]. Among naturally occurring peroxides, fatty acid derivatives represent a large group of compounds which have been shown to possess anticancer activities [32]. Naturally occurring cholesterols and terpenoids identified from marine and terrestrial plants exhibited cytotoxic activities against human lung carcinoma and colon carcinoma cells [31, 33].

Ascaridole, a bicyclic monoterpene with an unusual peroxide bridge has been isolated from the leaves of *Chenopodium ambrosioides*, and reported to exhibit anti-inflammatory, antifungal, antimalarial, as well as anticancer activities [34–37]. Dihydroascaridole, a derivative of ascaridole, showed antimalarial activity against *Plasmodium falciparum* and in mice against *Plasmodium berghei* [31]. Another sesquiterpene endoperoxide artemisinin and its derivatives are clinically used as antimalarials. Currently, new artemisinin analogues and/or derivatives are being developed, and studies of their structure-activity relationships, antimalarial mechanisms, their interaction with ferrous ions, and the DNA damage associated with these processes are being actively pursued. In addition, both ascaridole and artemisinin derivatives demonstrated antiparasitic activities against *Trypanosoma cruzi* [38] and *Leishmania amazonensis* [39] and anticancer activities against tumor cell lines *in vitro* (CCRF-CEM, HL60, MDA-MB-231) [40].

Endoperoxides affect parasites differently from other prooxidants and are proposed to mediate their mechanism of action by generation of carbon-centered free radicals causing damage to cellular biomolecules causing DNA, protein, and lipid damage secondary to oxidative stress. The endoperoxide bond present in artemisinin and ascaridole is proposed to be activated by reduced heme or ferrous iron (FeII), which following reductive scission of the peroxide bridge produced oxygen-centered radicals and following rearrangement, produces carbon-centered radicals [41, 42 and references therein]. These proved to be cytotoxic, alkylating agents and caused damage to cellular organelles [41–43]. The importance of the carbon-centered free radicals for mediating the antimalarial action of endoperoxides was confirmed by studies using artemisinin derivatives that lacked the endoperoxide bridge and failed to demonstrate antimalarial activity [44]. In *Plasmodium*, it has been postulated that artemisinin may target organelles such as the mitochondrion, endoplasmic reticulum, and the digestive vacuoles [45]. However, the precise target for the free radicals generated remains open-ended, as unlike conventional prooxidants, one molecule of artemisinin can only generate one free radical, and therefore the proportion of oxidant end products does not justify its parasiticidal activity [46].

The bioactivation of artemisinin as an endoperoxide has been proposed with the aid of different model systems where hemolytic or heterolytic cleavage of the pharmacophore leads to formation of primary and secondary radicals which directly cause damage to cellular components (Fig. 24.2). Heterolytic cleavage of the endoperoxide bridge in artemisinin with subsequent capture of water leads to the formation of unsaturated hydroperoxides and hydroxyl radical, which is capable of

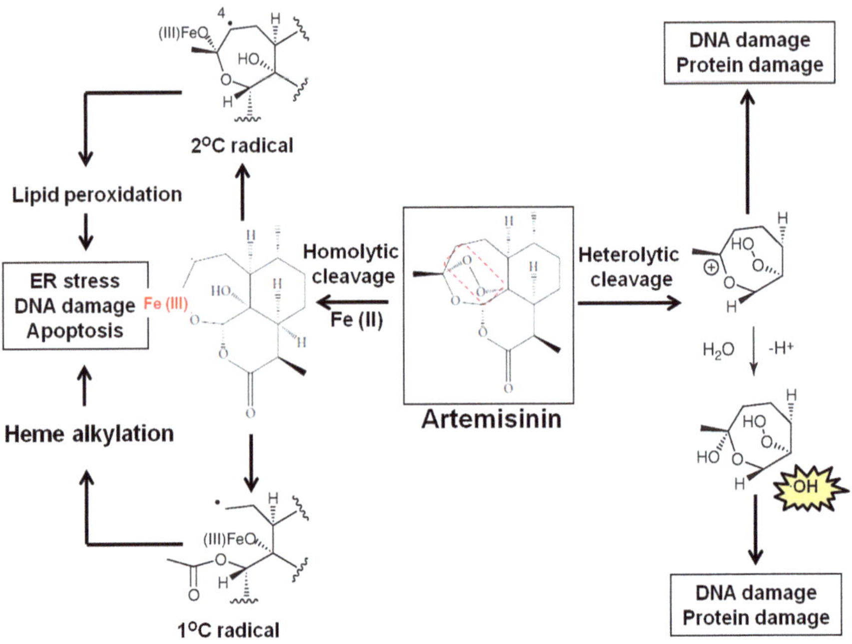

Fig. 24.2 Generation of carbon-centered radicals by artemisinin
Homolytic cleavage in the presence of Fe(II) and heterolytic cleavage with subsequent capture of water lead to the formation of 1°, 2°C radicals and unsaturated hydroperoxides, respectively, which are capable of irreversibly modifying protein residues by heme alkylation or lipid peroxidation causing protein/DNA damage leading to artemisinin-mediated apoptosis [47, 48]

irreversibly modifying protein residues by direct oxidation [47, 48]. On the other hand, due to the asymmetrical nature of the endoperoxide bridge, iron was found to interact with the peroxide in different ways to produce either a primary carbon-centered radical or a secondary carbon. Both primary and secondary radicals were efficiently spin trapped by electroparamagnetic resonance spin-trapping techniques after being activated by iron [44].

Similarly, the prooxidant action of ascaridole was delineated during the search for reasons behind the generation of contact allergy using the essential oils of *Chenopodium* and *Melaleuca* plants that contains the compound, a bicyclic monoterpene having an unusual bridging peroxide. Mechanistically, activation of the endoperoxide with Fe(II) was proposed to facilitate cleavage of the O-O bond to form oxygenated radicals (Fig. 24.3). Thin-layer chromatography studies have shown formation of isoascaridole along with many other diastereomers (DP1–DP5) which were further confirmed by NMR spectroscopy [35, 49]. These compounds give rise to Michael adducts along with other radical adducts which caused further radical-mediated damage [50]. This was also demonstrated in *Leishmania tarentolae* when spin adducts were detected in the presence of ascaridole incubated with increasing concentration of Fe^{2+} [51].

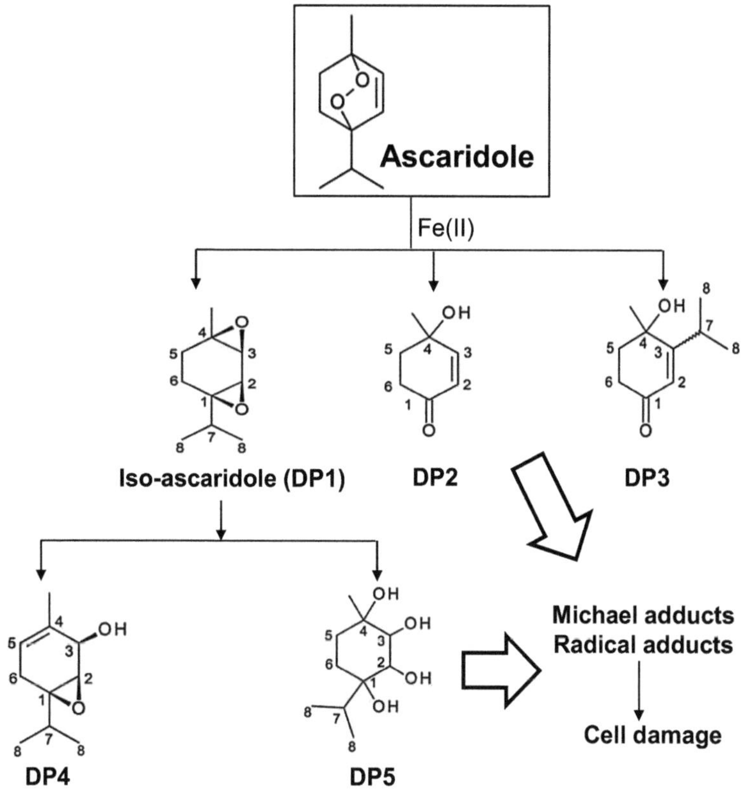

Fig. 24.3 Radical formation by ascaridole
Ascaridole is cleaved into carbon-centered radicals in the presence of Fe(II) forming isoascaridole and diastereomers which upon transformation into Michael adducts causes cell damage [50, 51]

In principle, endoperoxides as free radical generators can act against *Leishmania* akin to their mode of action as antimalarials [52, 53]. Indeed, artemisinin and ascaridole generated free radicals causing cell death in *Leishmania* promastigotes [7, 54]. Identification of the divalent cation transporter Nramp1 as a host susceptibility gene for infections with *L. donovani* and *L. infantum* was among the first indications that *Leishmania* is critically dependent on iron availability for intracellular replication [55]. It has been reported that *Leishmania* expresses two membrane proteins that play a central role in iron acquisition: (i) LFR1, a ferric reductase which reduces the insoluble Fe^{3+} to soluble Fe^{2+} and (ii) LIT1, a ferrous ion transporter which is a member of the ZIP family of metal transporters [56]. Endoperoxides require low molecular iron for initiation of cleavage of their peroxide bridge which is the hallmark of their mechanism of action, and as *Leishmania* can acquire iron from the host, it enhanced their vulnerability and selectivity to endoperoxides (Fig. 24.4). Exploiting the compromised antioxidant defense of *Leishmania*, this 'Achilles heel' enhances the potential of endoperoxides to create a redox imbalance (Table 24.2).

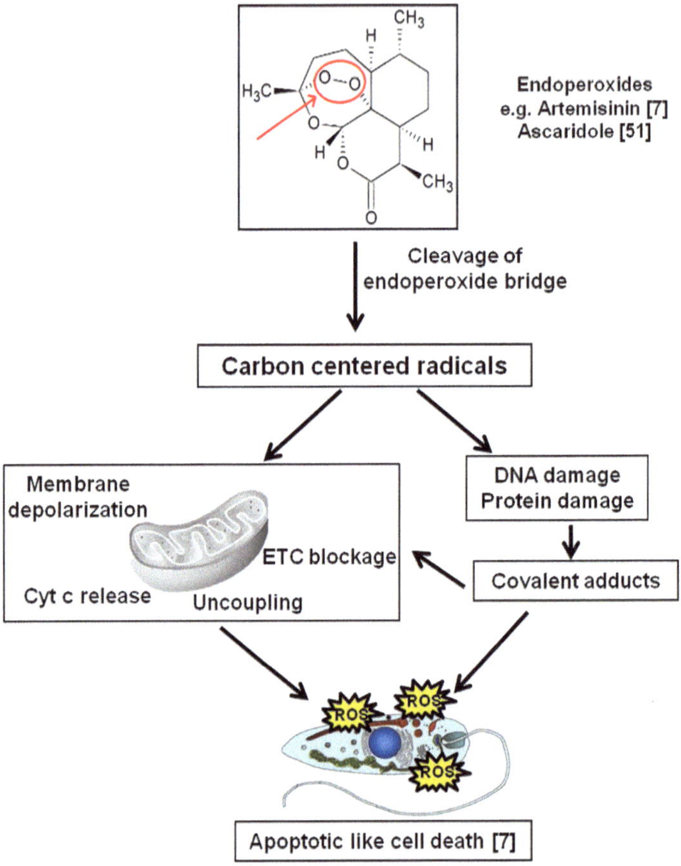

Fig. 24.4 **A proposed mechanism of action for endoperoxides**
Endoperoxides upon cleavage of their peroxide bridges generate carbon-centered radicals which either cause damage to biomolecules directly or via impacting upon mitochondria leading to membrane depolarization, cytochrome C release along with blockage of the electron transport chain leading to an apoptotic-like cell death in *Leishmania*

24.3 Mitochondria: A Source or a Target for Free Radicals?

Mitochondria are considered as the cell's powerhouse being the primary cellular source of ATP generated via oxidative phosphorylation [63]. In this process, electrons liberated from reducing substrates are delivered to O_2 so as to establish an electrochemical gradient needed for driving ATP synthesis. Following the monoelectronic reduction of O_2, the primary ROS generated by mitochondria is superoxide anion ($O_2^{\bullet-}$), which is the precursor of most ROS and a mediator of oxidative chain reactions. This formation of superoxide occurs on the outer mitochondrial membrane, in the matrix, and on both sides of the inner mitochondrial membrane [64].

Table 24.2 Peroxides as Antileishmanials

No.	Name	Source	Mode of action	References
1.	Ascaridole	Essential oil from *Chenopodium ambrosioides*	Generation of free radicals by cleavage of the endoperoxide bridge in presence of low molecular iron	[51, 57]
2.	Artemisinin	Leaves of *Artemisia annua*	Oxidative stress leading to caspase-independent apoptosis	[7]
3.	Analogues of artemisinin	Synthesized	Presence of peroxide bridge	[58, 59]
4.	Fluoro-artemisinin derivatives	Synthesized		[60]
5.	Isoascaridole	Essential oils from *C. ambrosioides*	–	[57]
6.	Artesunate	Artemisinin	–	[61]
7.	Artemisinin and its derivatives (dihydroartemisinin (DHA), anhydrodihydroartemisinin, and seven derivatives from DHA)	Artemisinin	–	[62]

The mammalian mitochondria are composed of four integral membrane enzyme complexes and are provided with a robust energy package of 100–1000 mitochondria. This backup is adequate as it ensures compensation of functionally impaired mitochondria. However, parasites being unicellular organisms do not enjoy such privileges and have to make do with a single mitochondria, whose functioning is imperative for its survival. Furthermore, mitochondria of mammalian vs. parasites have important morphological and functional differences (Table 24.3), the strongest differences being in the respiratory complexes e.g., complex I, which could potentially be a target for the development of antiparasitic drugs [71]. In parasites, the complex I is either absent or not very active based on the lack of impact of the classical inhibitor rotenone. However, irrespective of the mitochondrial source, complex III appears to be responsible for the maximal contribution of the $O_2^{-\bullet}$ [77–80]. Taken together, efforts are ongoing to validate mitochondria as a potential target [71, 81].

Established antileishmanials (e.g., miltefosine) as well as plant-derived products (e.g., chalcones) have been shown to inhibit specific mitochondrial respiratory complexes, enhance membrane permeability, and trigger collapse of the mitochondrial membrane potential which culminate in an apoptosis-like death in *Leishmania* parasites [75, 82; Table 24.4]. Experimental evidence about the exact target in mitochondria has been described for 2-hydroxybenzoic acid that acts on trypanosome alternative oxidase in *Trypanosoma brucei* [128], and chalcones that inhibited the fumarate reductase in *L. donovani* and *L. major* [75].

Table 24.3 Morphological and functional differences between mammalian and *Leishmania* mitochondria

No.	Features	Mammals	*Leishmania*	References
1.	Number of mitochondria per cell	>1000	1	[65]
2.	Kinetoplast containing mitochondrial DNA	Absent	Present	[66]
3.	Location of mitochondrial DNA	In mitochondrial matrix	In kinetoplast	[67, 68]
4.	% of cell volume occupied	20–40	12	[68]
5.	Classical complex I of electron transport chain	Present	Absent	[69]
6.	Number of NADH dehydrogenase subunits	Multiple	1	[70]
7.	Ability to use preformed pyrimidines	Present	Absent	[71]
8.	Coupling efficiency, respiratory control ratio, spare respiratory capacity	High	Low	[72, 73]
9.	Dihydroorotate:Ubiquinone oxidoreductase	Part of a multienzyme complex	Present as a single enzyme	[71]
10.	Acetate:Succinate CoA transferase	Absent	Present	[74]
11.	Fumarate reductase (FRD)	Absent	Present	[75]
12.	Import of tRNA for intramitochondrial protein synthesis	No	Yes	[76]

A common strand for triggering parasiticidal activity of conventional antileishmanial drugs like amphotericin B, miltefosine, and pentamidine is collapse of the MMP which occurs secondary to multiple factors that include (i) direct inhibition of the ETC, (ii) blockade of ATP synthase, (iii) stimulation of uncoupling proteins, and/or (iv) permeabilization of the inner membrane [68]. Plant-derived compounds like xanthohumol have been shown to exert their antileishmanial activity, secondary to inhibition of the mitochondrial electron transfer complexes II–III [73]. This strategy is adopted by several compounds that include albendazole, an alkaloid from *Peschiera australis* [127], azasterols [129], nifurtimox [130], and triazolopyrimidine derivatives [123], among others (Table 24.4).

During normal/unstressed conditions, mammalian cells require only a fraction of their mitochondrial bioenergetics capacity. However, when energy demands increase, supply is ensured as the difference between the maximal respiratory capacity and basal respiratory capacity, referred to as the "spare or reserve respiratory capacity," is increased. In mammalian cells, this uncoupler-accelerated respiration is enhanced by 200% above the basal respiration. However, parasites have a poor respiratory reserve; as demonstrated in *L. tarentolae*, the maximal uncoupled mitochondrial respiration was barely above normal mitochondrial respiration [73] and thereby parasite vulnerability is enhanced.

The generation of ATP is one of the major functions of mitochondria and depends on mitochondrial F_0F_1-ATP synthase and consequently is a chemotherapeutic target.

Table 24.4 Antileishmanial drugs that impact on parasite mitochondria

No.	Compounds	Type of compounds	Target	Effect on mitochondria	References
1	Quinoline derivatives	Natural	*Leishmania amazonensis*	Loss of mitochondrial membrane potential (MMP), swelling, ultrastructural changes	[83]
2	Flau-A [2-(2,3,4-tri-*O*-acetyl-6-deoxy-β-L-galactopyranosyloxy)-1,4-naphthoquinone], a naphthoquinone derivative	Natural	*L. infantum* *L. amazonensis*	Depolarization of mitochondrial membrane potential	[84]
3	Anthracene compounds	Synthetic	*L. tarentolae*	Generation of superoxide, decrease in oxygen consumption	[85]
4	Essential oil from *Chenopodium ambrosioides*	Natural	*L. tarentolae* *L. amazonensis* *L. infantum*	Increased mitochondrial superoxide and impaired coupling, mitochondrial membrane depolarization	[54, 57]
5	Kalsome TM10	Synthetic	*L. donovani*	Depolarization of MMP, disruption of mitochondrial integrity, depletion of ATP	[29]
6	Fruits of *Sapindus saponaria* L.	Natural	*L. amazonensis*	Ultrastructural changes, mitochondrial membrane depolarization	[86]
7	Phthalimido-triazole derivatives	Synthetic	*L. infantum*	Swelling, loss of MMP	[87]
8	Edelfosine	Synthetic	*L. panamensis*	Kinetoplast DNA cleavage, translocation of F_0F_1-ATP synthase	[88]
9	Perifosine	Synthetic	*L. amazonensis* *L. donovani*	Loss of MMP, decreased ATP levels	[89]
10	A3K2A3, a dibenzylideneacetone	Synthetic	*L. amazonensis*	Swelling, depletion of ATP	[90]

(continued)

Table 24.4 (continued)

No.	Compounds	Type of compounds	Target	Effect on mitochondria	References
11	Xanthohumol and resveratrol	Natural	*L. tarentolae*	Decrease in oxygen consumption, inhibition of ETC complexes, loss of MMP	[73]
12	Plumbagin, a plant-derived naphthoquinone metabolite	Natural	*L. donovani*	Generation of superoxide, loss of MMP, depletion of ATP	[91]
13	Sesquiterpene (−)- αbisabolol	Natural	*L. major* *L. tropica*	Loss of MMP, mitochondrial swelling	[92]
14	Phytol-rich hexane fraction (PRF) from the leaves of *Lacistema pubescens*	Natural	*L. amazonensis*	Mitochondrial membrane depolarization	[93]
15	Apigenin	Natural	*L. donovani*	Swelling, decrease in mitochondrial membrane potential	[94]
16	Strychnobiflavone from *Strychnos pseudoquina*	Natural	*L. infantum*	Depolarization of the mitochondrial membrane potential	[95]
17	Nemorosone and guttiferone A	Natural	*L. tarentolae*	Inhibition of ETC complexes, decrease in mitochondrial respiration, generation of superoxide	[96]
18	Fractions from *Arrabidaea chica*	Natural	*L. amazonensis* *L. infantum*	Swelling	[97]
19	Melatonin	Natural hormone	*L. infantum*	Inhibition of ETC complexes, increased mitochondrial nitrite levels	[98]
20	Clerodane diterpene (K-09)	Natural	*L. donovani*	Impairment of ATP production, depolarization of mitochondrial membrane, release of cytochrome c	[99]
21	5′-Trityl-substituted thymidine derivatives	Synthetic	*L. infantum*	Alterations in mitochondrial transmembrane potential, generation of superoxide	[100]

(continued)

Table 24.4 (continued)

No.	Compounds	Type of compounds	Target	Effect on mitochondria	References
22	2,2-dipyridyl, an iron chelator	Synthetic	*L. (Viannia) braziliensis*	Decrease in mitochondrial membrane potential	[101]
23	Nitazoxanide, a nitrothiazolyl-salicylamide derivative	Synthetic	*L. infantum*	Enlargement and progressive swelling of mitochondria, loss of MMP	[102]
24	LQB-118, a pterocarpanquinone	Synthetic	*L. amazonensis*	Decrease in MMP	[103]
25	Itraconazole and posaconazole	Synthetic	*L. amazonensis*	Collapse of MMP, ultrastructural alterations in the mitochondrion	[104]
26	Benzaldehyde thiosemicarbazone	Synthetic	*L. amazonensis*	Generation of superoxide, mitochondrial membrane depolarization	[105]
27	Geranylgeraniol	Natural	*L. amazonensis*	Generation of superoxide, swelling	[106]
28	Epigallocatechin-3-gallate (EGCG)	Natural	*L. amazonensis*	Mitochondrial membrane depolarization, swelling	[107]
29	Tomatidine	Synthetic	*L. amazonensis*	Swelling, loss of MMP, ATP depletion	[108]
30	Quercetin	Synthetic	*L. amazonensis*	Loss of MMP	[109]
31	Chromanol derivatives from *Cuban propolis* extracts	Synthetic	*L. amazonensis* *L. tarentolae*	Inhibition of ETC complex III, loss of MMP	[110]
32	Benzophenone-derived bisphosphonium salts	Synthetic	*L. donovani* *L. pifanoi*	Inhibition of ETC complex II, loss of MMP	[111]
33	Tafenoquine	Synthetic	*L. donovani*	Depletion of ATP levels, mitochondrial membrane depolarization, inhibition of ETC complex III	[112]
34	Berberine chloride	Natural	*L. donovani*	Mitochondrial membrane depolarization	[113]

(continued)

Table 24.4 (continued)

No.	Compounds	Type of compounds	Target	Effect on mitochondria	References
35	2,6-dimethyl-4-(3-*O*-benzyl-1,2-*O*-isopropylidene-beta-L-threo-pentofuranos-4-yl)-1-phenyl-1,4-dihydro-pyridine-3,5-dicarboxylic acid diethyl ester, a glycosyl dihydropyridine analogue	Synthetic	*L. donovani*	Depolarization of MMP	[114]
36	Nimodipine	Synthetic	*L. chagasi*	Enlargement of mitochondria	[115]
37	Ethanolic extract of leaves of *Piper betle* (Paan) Linn	Natural	*L. donovani*	Depolarization of mitochondrial membrane potential	[116]
38	3,3′-Diindolylmethane (DIM)	Natural	*L. donovani*	Inhibition of F_0F_1-ATP synthase, decrease in ATP levels, loss of MMP, inhibition of ETC complexes	[117]
39	Human antimicrobial peptide (Histatin 5)	Natural	*Leishmania*	Decrease of mitochondrial ATP synthesis	[118]
40	*Aloe vera* leaf exudate	Natural	*L. donovani*	Loss of MMP, release of cytochrome c	[119]
41	Artemisinin	Natural	*L. donovani*	Loss of MMP	[7]
42	Sterol methenyl transferase inhibitors	Synthetic	*L. amazonensis*	Ultrastructural changes in the inner mitochondrial membrane	[120]
43	Miltefosine	Synthetic	*L. donovani*	Loss of MMP and release of cytochrome c	[27]
44	Dinitroaniline sulfonamides	Synthetic	*L. donovani* *L. mexicana*	Loss of MMP	[121]
45	3-substituted quinolines	Synthetic	*L. chagasi*	Swelling	[122]
46	Pt/Ru-Triazole-pyrimidine complexes	Synthetic	*L. donovani*	Ultrastructural changes in the inner mitochondrial membranes	[123]
47	Polyamine derivatives	Synthetic	*L. infantum*	Loss of MMP	[124]
48	3-(biphenyl-4-yl)-3-hydroxyquinuclidine (BPQ-OH)	Synthetic	*L. amazonensis*	Swelling, ultrastructural changes	[125]

(continued)

Table 24.4 (continued)

No.	Compounds	Type of compounds	Target	Effect on mitochondria	References
49	Phenyl-phenalenone phytoalexins from the *Musa acuminata*	Natural	*L. donovani*	Inhibition of FRD	[126]
50	Indole alkaloid from *Peschiera australis*	Natural	*L. amazonensis*	Ultrastructural alterations	[127]
51	Chalcones	Synthetic	*L. donovani*	Inhibition of FRD	[75]
			L. major		

The F_0F_1-ATPase in *Leishmania* spp. consists of two oligomeric components F_0 and F_1, the former being an integral membrane protein that contains the proton channel, whereas the latter is a peripheral membrane protein [131]. Importantly, ATP synthesis requires both F_0 and F_1 components, and to generate ATP from this chemiosmotic gradient, presence of a proton impermeable inner mitochondrial membrane and a functional complex V (ATP synthase) is necessary. Therefore, compounds capable of inducing oxidative stress by generating free radicals can impact on the mitochondrial membrane potential and have a deleterious effect on the F_0F_1-ATPase. Indeed, in plant-derived antileishmanial compounds like camptothecin and 3,3′-diindolylmethane (DIM), the depolarization of MMP and inhibition of complex V, F_0F_1-ATP synthase decreased levels of ATP [68 and references therein].

24.4 Endoperoxides Cause Minimal Mitochondrial Dysfunction

With regard to ascaridole, there was minimal involvement of the mitochondria in mediating its leishmanicidal activity, as evidenced by its inability to act as an uncoupler, except upon prolonged incubation and at higher doses. Furthermore, it was unable to inhibit oxygen consumption, indicating its minimal impact on the mitochondrial ETC [51, 54]. Similarly, artemisinin demonstrated a minimal inhibitory effect on the mitochondrial ETC, as it did not increase oxygen consumption and showed no immediate uncoupling effect, collectively endorsing its minimal impact on mitochondria [132].

24.5 Conclusion

Antileishmanial therapy remains challenging largely because of the disease complexity coupled with the ongoing problem of resistance and drug toxicity emphasizing the need for identifying new compounds. One of the characteristic features in the process of drug development is target identification, and in view of trypanosomatids phylogenetically branching out quite early from higher eukaryotes, significant differences in their cell organization have been exploited. Indeed, this has translated into the identification and validation of potential targets in *Leishmania* that include enzymes of the thiol metabolism pathways, cell cycle, and DNA replication, as well as proteins and biomolecules critical for their survival and virulence [133, 134]. Naturally occurring and synthetic endoperoxides exhibited promising antiparasitic activities based on their potential to be potent free radical generators, without impacting majorly on mitochondrial function [135], and therefore, further studies to delineate the in vivo efficacy of endoperoxides are warranted.

References

1. Jin G, Wong ST (2014) Toward better drug repositioning: prioritizing and integrating existing methods into efficient pipelines. Drug Discov Today 19:637–644
2. Torres-Guerrero E, Quintanilla-Cedillo MR, Ruiz-Esmenjaud J, Arenas R (2017) Leishmaniasis: a review. F1000Res 6:750
3. World Health Organization. http://www.who.int/mediacentre/factsheets/fs375/en/. Accessed 19 Sept 2018
4. Croft SL, Seifert K, Duchêne M (2003) Antiprotozoal activities of phospholipid analogues. Mol Biochem Parasitol 126:165–172
5. Dorlo TP, Balasegaram M, Beijnen JH, de Vries PJ (2012) Miltefosine: a review of its pharmacology and therapeutic efficacy in the treatment of leishmaniasis. J Antimicrob Chemother 67:2576–2597
6. Andrade-Neto VV, Cunha-Junior EF, Dos Santos FV, Pereira TM, Silva RL, Leon LL, Torres-Santos EC (2018) Leishmaniasis treatment: update of possibilities for drug repurposing. Front Biosci (Landmark Ed) 23:967–996
7. Sen R, Bandyopadhyay S, Dutta A, Mandal G, Ganguly S, Saha P, Chatterjee M (2007) Artemisinin triggers induction of cell-cycle arrest and apoptosis in *Leishmania donovani* promastigotes. J Med Microbiol 56:1213–1218
8. Sen R, Ganguly S, Saha P, Chatterjee M (2010a) Efficacy of artemisinin in experimental visceral leishmaniasis. Int J Antimicrob Agents 36:43–49
9. Sen R, Saha P, Sarkar A, Ganguly S, Chatterjee M (2010b) Iron enhances generation of free radicals by artemisinin causing a caspase-independent, apoptotic death in *Leishmania donovani* promastigotes. Free Radic Res 44:1289–1295
10. Huang Z, Srinivasan S, Zhang J, Chen K, Li Y, Li W, Quiocho FA, Pan X (2012) Discovering thiamine transporters as targets of chloroquine using a novel functional genomics strategy. PLoS Genet 8:e1003083
11. van Assche T, Deschacht M, da Luz RA, Maes L, Cos P (2011) Leishmania-macrophage interactions: insights into the redox biology. Free Radic Biol Med 51:337–351
12. Mukbel RM, Patten C Jr, Gibson K, Ghosh M, Petersen C, Jones DE (2007) Macrophage killing of *Leishmania amazonensis* amastigotes requires both nitric oxide and superoxide. Am J Trop Med Hyg 76:669–675
13. Courret N, Fréhel C, Gouhier N, Pouchelet M, Prina E, Roux P, Antoine JC (2002) Biogenesis of *Leishmania* harbouring parasitophorous vacuoles following phagocytosis of the metacyclic promastigote or amastigotes stage of the parasites. J Cell Sci 115:2303–2316
14. Krauth-Siegel RL, Comini MA (2008) Redox control in trypanosomatids, parasitic protozoa with trypanothione-based thiol metabolism. Biochim Biophys Acta 1780:1236–1248
15. Dolai S, Yadav RK, Pal S, Adak S (2009) Overexpression of mitochondrial *Leishmania major* ascorbate peroxidase enhances tolerance to oxidative stress-induced programmed cell death and protein damage. Eukaryot Cell 8:1721–1731
16. Fairlamb AH, Blackburn P, Ulrich P, Chait BT, Cerami A (1985) Trypanothione: a novel *bis*(glutathionyl)spermidine cofactor for glutathione reductase in trypanosomatids. Science 227:1485–1487

17. Saudagar P, Dubey VK (2011) Cloning, expression, characterization and inhibition studies on trypanothione synthetase, a drug target enzyme, from *Leishmania donovani*. Biol Chem 392:1113–1122
18. Flohé L, Hecht HJ, Steinert P (1999) Glutathione and trypanothione in parasitic hydroperoxide metabolism. Free Radic Biol Med 27:966–984
19. Comini MA, Flohé L (2013) Trypanothione-based redox metabolism of trypanosomatids. In: Trypanosomatid diseases: molecular routes to drug discovery. Wiley-VCH Verlag GmbH & Co. KGaA, Weinheim, pp 167–199
20. Field MC, Horn D, Fairlamb AH, Ferguson MAJ, Gray DW, Read KD, De Rycker M, Torrie LS, Wyatt PG, Wyllie S, Gilbert IH (2017) Anti-trypanosomatid drug discovery: an ongoing challenge and a continuing need. Nat Rev Microbiol 15:217–231
21. Leroux AE, Krauth-Siegel RL (2016) Thiol redox biology of trypanosomatids and potential targets for chemotherapy. Mol Biochem Parasitol 206:67–74
22. Mookerjee Basu J, Mookerjee A, Sen P, Bhaumik S, Sen P, Banerjee S, Naskar K, Choudhuri SK, Saha B, Raha S, Roy S (2006) Sodium antimony gluconate induces generation of reactive oxygen species and nitric oxide via phosphoinositide 3-kinase and mitogen-activated protein kinase activation in *Leishmania donovani*-infected macrophages. Antimicrob Agents Chemother 50:1788–1797
23. Sundar S, Chatterjee M (2006) Visceral leishmaniasis—current therapeutic modalities. Indian J Med Res 123:345–352
24. Lee N, Bertholet S, Debrabant A, Muller J, Duncan R, Nakhasi HL (2002) Programmed cell death in the unicellular protozoan parasite *Leishmania*. Cell Death Differ 9:53–64
25. Mehta A, Shaha C (2004) Apoptotic death in *Leishmania donovani* promastigotes in response to respiratory chain inhibition: complex II inhibition results in increased pentamidine cytotoxicity. J Biol Chem 279:11798–11813
26. Shaha C (2006) Apoptosis in *Leishmania* species & its relevance to disease pathogenesis. Indian J Med Res 123:233–244
27. Verma NK, Singh G, Dey CS (2007) Miltefosine induces apoptosis in arsenite-resistant *Leishmania donovani* promastigotes through mitochondrial dysfunction. Exp Parasitol 116:1–13
28. Getachew F, Gedamu L (2012) *Leishmania donovani* mitochondrial iron superoxide dismutase A is released into the cytosol during miltefosine induced programmed cell death. Mol Biochem Parasitol 183:42–51
29. Shadab M, Jha B, Asad M, Deepthi M, Kamran M, Ali N (2017) Apoptosis-like cell death in *Leishmania donovani* treated with KalsomeTM10, a new liposomal amphotericin B. PLoS One 12:e0171306
30. Singh K, Ali V, Pratap Singh K, Gupta P, Suman SS, Ghosh AK, Bimal S, Pandey K, Das P (2017) Deciphering the interplay between cysteine synthase and thiol cascade proteins in modulating amphotericin B resistance and survival of *Leishmania donovani* under oxidative stress. Redox Biol 12:350–366
31. Dembitsky VM (2008) Bioactive peroxides as potential therapeutic agents. Eur J Med Chem 43:223–251
32. Dörsam B, Fahrer J (2016) The disulfide compound α-lipoic acid and its derivatives: a novel class of anticancer agents targeting mitochondria. Cancer Lett 371:12–19
33. Dembitsky VM, Gloriozova TA, Poroikov VV (2007) Natural peroxy anticancer agents. Mini-Rev Med Chem 7:571–589
34. Usman LA, Hamid AA, Muhammad NO, Olawore NO, Edewor TI, Saliu BK (2010) Chemical constituents and anti-inflammatory activity of leaf essential oil of Nigerian grown *Chenopodium album* L. EXCLI J 9:181–186
35. Jardim CM, Jham GN, Dhingra OD, Freire MM (2008) Composition and antifungal activity of the essential oil of the Brazilian *Chenopodium ambrosioides* L. J Chem Ecol 34:1213–1218
36. Pollack Y, Segal R, Golenser J (1990) The effect of ascaridole on the in vitro development of *Plasmodium falciparum*. Parasitol Res 76:570–572

37. Efferth T, Olbrich A, Sauerbrey A, Ross DD, Gebhart E, Neugebauer M (2002) Activity of ascaridole from the anthelmintic herb *Chenopodium anthelminticum* L. against sensitive and multidrug-resistant tumor cells. Anticancer Res 22:4221–4224
38. Kiuchi F, Itano Y, Uchiyama N, Honda G, Tsubouchi A, Nakajima-Shimada J, Aoki T (2002) Monoterpene hydroperoxides with trypanocidal activity from *Chenopodium ambrosioides*. J Nat Prod 65:509–512
39. Monzote L, Montalvo AM, Almanonni S, Scull R, Miranda M, Abreu J (2006) Activity of the essential oil from *Chenopodium ambrosioides* grown in Cuba against *Leishmania amazonensis*. Chemotherapy 52:130–136
40. Meshnick SR (2002) Artemisinin: mechanisms of action, resistance and toxicity. Int J Parasitol 32:1655–1660
41. Meshnick SR, Thomas A, Ranz A, Xu CM, Pan HZ (1991) Artemisinin (qinghaosu): the role of intracellular hemin in its mechanism of antimalarial action. Mol Biochem Parasitol 49:181–189
42. Posner GH, Wang D, Cumming JN, Oh CH, French AN, Bodley AL, Shapiro TA (1995) Further evidence supporting the importance of and the restrictions on a carbon-centered radical for high antimalarial activity of 1,2,4-trioxanes like artemisinin. J Med Chem 38:2273–2275
43. Olliaro P (2001) Mode of action and mechanisms of resistance for antimalarial drugs. Pharmacol Ther 89:207–219
44. Meshnick SR (1994) Free radicals and antioxidants. Lancet 344:1441–1442
45. O'Neill PM, Barton VE, Ward SA (2010) The molecular mechanism of action of artemisinin—the debate continues. Molecules 15:1705–1721
46. Dong Y, Vennerstrom JL (2003) Mechanisms of in situ activation for peroxidic antimalarials. Redox Rep 8:284–288
47. Biagini GA, O'Neill PM, Nzila A, Ward SA, Bray PG (2003) Antimalarial chemotherapy: young guns or back to the future? Trends Parasitol 19:479–487
48. O'Neill PM, Posner GH (2004) A medicinal chemistry perspective on artemisinin and related endoperoxides. J Med Chem 47:2945–2964
49. Cavalli JF, Tomi F, Bernardini AF, Casanova J (2004) Combined analysis of the essential oil of *Chenopodium ambrosioides* by GC, GC-MS and 13C-NMR spectroscopy: quantitative determination of ascaridole, a heat-sensitive compound. Phytochem Anal 15:275–279
50. Chittiboyina AG, Avonto C, Khan IA (2016) What happens after activation of ascaridole? Reactive compounds and their implications for skin sensitization. Chem Res Toxicol 29:1488–1492
51. Geroldinger G, Tonner M, Hettegger H, Bacher M, Monzote L, Walter M, Staniek K, Rosenau T, Gille L (2017) Mechanism of ascaridole activation in *Leishmania*. Biochem Pharmacol 132:48–62
52. Ruiz J, Mallet-Ladeira S, Maynadier M, Vial H, André-Barrès C (2014) Design, synthesis and evaluation of new tricyclic endoperoxides as potential antiplasmodial agents. Org Biomol Chem 12:5212–5221
53. Rudrapal M, Chetia D (2016) Endoperoxide antimalarials: development, structural diversity and pharmacodynamic aspects with reference to 1,2,4-trioxane-based structural scaffold. Drug Des Devel Ther 10:3575–3590
54. Monzote L, Geroldinger G, Tonner M, Scull R, De Sarkar S, Bergmann S, Bacher M, Staniek K, Chatterjee M, Rosenau T, Gille L (2018) Interaction of ascaridole, carvacrol, and caryophyllene oxide from essential oil of *Chenopodium ambrosioides* L. with mitochondria in *Leishmania* and other eukaryotes. Phytother Res 32:1729–1740
55. Blackwell JM, Goswami T, Evans CA, Sibthorpe D, Papo N, White JK, Searle S, Miller EN, Peacock CS, Mohammed H, Ibrahim M (2001) SLC11A1 (formerly NRAMP1) and disease resistance. Cell Microbiol 3:773–784
56. Mittra B, Cortez M, Haydock A, Ramasamy G, Myler PJ, Andrews NW (2013) Iron uptake controls the generation of *Leishmania* infective forms through regulation of ROS levels. J Exp Med 210:401–416

57. Monzote L, García M, Pastor J, Gil L, Scull R, Maes L, Cos P, Gille L (2014) Essential oil from *Chenopodium ambrosioides* and main components: activity against *Leishmania*, their mitochondria and other microorganisms. Exp Parasitol 136:20–26
58. Avery MA, Muraleedharan KM, Desai PV, Bandyopadhyaya AK, Furtado MM, Tekwani BL (2003) Structure-activity relationships of the antimalarial agent artemisinin. 8. Design, synthesis, and CoMFA studies toward the development of artemisinin-based drugs against leishmaniasis and malaria. J Med Chem 46:4244–4258
59. Menon RB, Kannoth MM, Tekwani BL, Gut J, Rosenthal PJ, Avery MA (2006) A new library of C-16 modified artemisinin analogs and evaluation of their anti-parasitic activities. Comb Chem High Throughput Screen 9:729–741
60. Chollet C, Crousse B, Bories C, Bonnet-Delpon D, Loiseau PM (2008) In vitro antileishmanial activity of fluoro-artemisinin derivatives against *Leishmania donovani*. Biomed Pharmacother 62:462–465
61. Raffetin A, Bruneel F, Roussel C, Thellier M, Buffet P, Caumes E, Jauréguiberry S (2018) Use of artesunate in non-malarial indications. Med Mal Infect 48:238–249
62. Slade D, Galal AM, Gul W, Radwan MM, Ahmed SA, Khan SI, Tekwani BL, Jacob MR, Ross SA, Elsohly MA (2009) Antiprotozoal, anticancer and antimicrobial activities of dihydroartemisinin acetal dimers and monomers. Bioorg Med Chem 17:7949–7957
63. Marchi S, Giorgi C, Suski JM, Agnoletto C, Bononi A, Bonora M, De Marchi E, Missiroli S, Patergnani S, Poletti F, Rimessi A, Duszynski J, Wieckowski MR, Pinton P (2012) Mitochondria-ROS crosstalk in the control of cell death and aging. J Signal Transduction 2012:329635
64. Turrens JF (2003) Mitochondrial formation of reactive oxygen species. J Physiol 552:335–344
65. Alvarez-Fortes E, Ruiz-Pérez LM, Bouillaud F, Rial E, Rivas L (1998) Expression and regulation of mitochondrial uncoupling protein 1 from brown adipose tissue in *Leishmania major* promastigotes. Mol Biochem Parasitol 93:191–202
66. Gull K (1999) The cytoskeleton of trypanosomatid parasites. Annu Rev Microbiol 53:629–655
67. Taanman JW (1999) The mitochondrial genome: structure, transcription, translation and replication. Biochim Biophys Acta 1410:103–123
68. Fidalgo LM, Gille L (2011) Mitochondria and trypanosomatids: targets and drugs. Pharm Res 28:2758–2770
69. Opperdoes F, Michels P (2008) The metabolic repertoire of *Leishmania* and implications for drug discovery. After The Genome, Leishmania
70. Santhamma KR, Bhaduri A (1995) Characterization of the respiratory chain of *Leishmania donovani* promastigotes. Mol Biochem Parasitol 75:43–53
71. Monzote L, Gille L (2010) Mitochondria as a promising antiparasitic target. Curr Clin Pharmacol 5:55–60
72. Brand MD, Nicholls DG (2011) Assessing mitochondrial dysfunction in cells. Biochem J 435:297–312
73. Monzote L, Lackova A, Staniek K, Steinbauer S, Pichler G, Jäger W, Gille L (2016) The antileishmanial activity of xanthohumol is mediated by mitochondrial inhibition. Parasitology 2016:1–13
74. van Hellemond JJ, Opperdoes FR, Tielens AG (1998) Trypanosomatidae produce acetate via a mitochondrial acetate:succinate CoA transferase. Proc Natl Acad Sci U S A 95:3036–3041
75. Chen M, Zhai L, Christensen SB, Theander TG, Kharazmi A (2001) Inhibition of fumarate reductase in *Leishmania major* and *L. donovani* by chalcones. Antimicrob Agents Chemother 45:2023–2029
76. Schneider A (2001) Unique aspects of mitochondrial biogenesis in trypanosomatids. Int J Parasitol 31:1403–1415
77. Turrens JF, Boveris A (1980) Generation of superoxide anion by the NADH dehydrogenase of bovine heart mitochondria. Biochem J 191:421–427
78. Turrens JF, Freeman BA, Levitt JG, Crapo JD (1982) The effect of hyperoxia on superoxide production by lung submitochondrial particles. Arch Biochem Biophys 217:401–410

79. Dröse S, Brandt U (2008) The mechanism of mitochondrial superoxide production by the cytochrome bc1 complex. J Biol Chem 283:21649–21654
80. Nohl H, Gille L, Kozlov A, Staniek K (2003) Are mitochondria a spontaneous and permanent source of reactive oxygen species? Redox Rep 8:135–141
81. Sen N, Majumder HK (2008) Mitochondrion of protozoan parasite emerges as potent therapeutic target: exciting drugs are on the horizon. Curr Pharm Des 14:839–846
82. Luque-Ortega JR, Rivas L (2007) Miltefosine (hexadecylphosphocholine) inhibits cytochrome c oxidase in *Leishmania donovani* promastigotes. Antimicrob Agents Chemother 51:1327–1332
83. Calixto SL, Glanzmann N, Xavier Silveira MM, da Trindade GJ, Gorza Scopel KK, Torres de Aguiar T, DaMatta RA, Macedo GC, da Silva AD, Coimbra ES (2018) Novel organic salts based on quinoline derivatives: the in vitro activity trigger apoptosis inhibiting autophagy in *Leishmania* spp. Chem Biol Interact 293:141–151
84. Mendonça DVC, Lage DP, Calixto SL, Ottoni FM, Tavares GSV, Ludolf F, Chávez-Fumagalli MA, Schneider MS, Duarte MC, Tavares CAP, Alves RJ, Coimbra ES, Coelho EAF (2018) Antileishmanial activity of a naphthoquinone derivate against promastigote and amastigote stages of *Leishmania infantum* and *Leishmania amazonensis* and its mechanism of action against *L. amazonensis* species. Parasitol Res 117:391–403
85. Geroldinger G, Tonner M, Fudickar W, De Sarkar S, Dighal A, Monzote L, Staniek K, Linker T, Chatterjee M, Gille L (2018 Jul 10) Activation of anthracene Endoperoxides in Leishmania and impairment of mitochondrial functions. Molecules 23(7). pii: E1680). https://doi.org/10.3390/molecules23071680
86. Moreira AL, Scariot DB, Pelegrini BL, Pessini GL, Ueda-Nakamura T, Nakamura CV, Ferreira ICP (2017) Acyclic sesquiterpenes from the fruit pericarp of *Sapindus saponaria* induce ultrastructural alterations and cell death in *Leishmania amazonensis*. Evid Based Complement Alternat Med 2017:5620693
87. Aliança ASDS, Oliveira AR, Feitosa APS, Ribeiro KRC, de Castro MCAB, Leite ACL, Alves LC, Brayner FA (2017) In vitro evaluation of cytotoxicity and leishmanicidal activity of phthalimido-thiazole derivatives. Eur J Pharm Sci 105:1–10
88. Villa-Pulgarín JA, Gajate C, Botet J, Jimenez A, Justies N, Varela-M RE, Cuesta-Marbán Á, Müller I, Modolell M, Revuelta JL, Mollinedo F (2017) Mitochondria and lipid raft-located F_0F_1-ATP synthase as major therapeutic targets in the antileishmanial and anticancer activities of ether lipid edelfosine. PLoS Negl Trop Dis 11:e0005805
89. López-Arencibia A, Martín-Navarro C, Sifaoui I, Reyes-Batlle M, Wagner C, Lorenzo-Morales J, Maciver SK, Piñero JE (2017) Perifosine mechanisms of action in *Leishmania* species. Antimicrob Agents Chemother 61(4):e02127-16
90. Garcia FP, Henrique da Silva Rodrigues J, Din ZU, Rodrigues-Filho E, Ueda-Nakamura T, Auzély-Velty R, Nakamura CV (2017) A3K2A3-induced apoptotic cell death of *Leishmania amazonensis* occurs through caspase- and ATP-dependent mitochondrial dysfunction. Apoptosis 22:57–71
91. Awasthi BP, Kathuria M, Pant G, Kumari N, Mitra K (2016) Plumbagin, a plant-derived naphthoquinone metabolite induces mitochondria mediated apoptosis-like cell death in *Leishmania donovani:* an ultrastructural and physiological study. Apoptosis 21:941–953
92. Corpas-López V, Merino-Espinosa G, Díaz-Sáez V, Morillas-Márquez F, Navarro-Moll MC, Martín-Sánchez J (2016) The sesquiterpene (−)-α-bisabolol is active against the causative agents of Old World cutaneous leishmaniasis through the induction of mitochondrial-dependent apoptosis. Apoptosis 21:1071–1081
93. da Silva JM, Antinarelli LM, Ribeiro A, Coimbra ES, Scio E (2015) The effect of the phytol-rich fraction from *Lacistema pubescens* against *Leishmania amazonensis* is mediated by mitochondrial dysfunction. Exp Parasitol 159:143–150
94. Fonseca-Silva F, Canto-Cavalheiro MM, Menna-Barreto RF, Almeida-Amaral EE (2015) Effect of apigenin on *Leishmania amazonensis* is associated with reactive oxygen species production followed by mitochondrial dysfunction. J Nat Prod 78:880–884

95. Lage PS, Chávez-Fumagalli MA, Mesquita JT, Mata LM, Fernandes SO, Cardoso VN, Soto M, Tavares CA, Leite JP, Tempone AG, Coelho EA (2015) Antileishmanial activity and evaluation of the mechanism of action of strychnobiflavone flavonoid isolated from *Strychnos pseudoquina* against *Leishmania infantum*. Parasitol Res 114:4625–4635
96. Monzote L, Lackova A, Staniek K, Cuesta-Rubio O, Gille L (2015) Role of mitochondria in the leishmanicidal effects and toxicity of acyl phloroglucinol derivatives: nemorosone and guttiferone A. Parasitology 142:1239–1248
97. Rodrigues IA, Azevedo MM, Chaves FC, Alviano CS, Alviano DS, Vermelho AB (2014) *Arrabidaea chica* hexanic extract induces mitochondrion damage and peptidase inhibition on *Leishmania* spp. Biomed Res Int 2014:985171
98. Elmahallawy EK, Jiménez-Aranda A, Martínez AS, Rodriguez-Granger J, Navarro-Alarcón M, Gutiérrez-Fernández J, Agil A Activity of melatonin against *Leishmania infantum* promastigotes by mitochondrial dependent pathway. Chem Biol Interact 220:84–93
99. Kathuria M, Bhattacharjee A, Sashidhara KV, Singh SP, Mitra K (2014) Induction of mitochondrial dysfunction and oxidative stress in *Leishmania donovani* by orally active clerodane diterpene. Antimicrob Agents Chemother 58:5916–5928
100. Casanova E, Moreno D, Gigante A, Rico E, Genes CM, Oliva C, Camarasa MJ, Gago F, Jiménez-Ruiz A, Pérez-Pérez MJ (2013) 5′-Trityl-substituted thymidine derivatives as a novel class of antileishmanial agents: *Leishmania infantum* EndoG as a potential target. ChemMedChem 8:1161–1174
101. Mesquita-Rodrigues C, Menna-Barreto RF, Sabóia-Vahia L, Da-Silva SA, de Souza EM, Waghabi MC, Cuervo P, De Jesus JB (2013) Cellular growth and mitochondrial ultrastructure of *Leishmania (Viannia) braziliensis* promastigotes are affected by the iron chelator 2,2-dipyridyl. PLoS Negl Trop Dis 7:e2481
102. Mesquita JT, Pinto EG, Taniwaki NN, Galisteo AJ Jr, Tempone AG (2013) Lethal action of the nitrothiazolyl-salicylamide derivative nitazoxanide via induction of oxidative stress in *Leishmania* (L.) *infantum*. Acta Trop 128:666–673
103. Ribeiro GA, Cunha-Júnior EF, Pinheiro RO, da-Silva SA, Canto-Cavalheiro MM, da Silva AJ, Costa PR, Netto CD, Melo RC, Almeida-Amaral EE, Torres-Santos EC (2013) LQB-118, an orally active pterocarpanquinone, induces selective oxidative stress and apoptosis in *Leishmania amazonensis*. J Antimicrob Chemother 68:789–799
104. de Macedo-Silva ST, Urbina JA, de Souza W, Rodrigues JC (2013) *In vitro* activity of the antifungal azoles itraconazole and posaconazole against *Leishmania amazonensis*. PLoS One 8:e83247
105. Britta EA, Silva AP, Ueda-Nakamura T, Dias-Filho BP, Silva CC, Sernaglia RL, Nakamura CV (2012) Benzaldehyde thiosemicarbazone derived from limonene complexed with copper induced mitochondrial dysfunction in *Leishmania amazonensis*. PLoS One 7:e41440
106. Lopes MV, Desoti VC, Caleare Ade O, Ueda-Nakamura T, Silva SO, Nakamura CV (2012) Mitochondria superoxide anion production contributes to geranylgeraniol-induced death in *Leishmania amazonensis*. Evid Based Complement Alternat Med 2012:298320
107. Inacio JD, Canto-Cavalheiro MM, Menna-Barreto RF, Almeida-Amaral EE (2012) Mitochondrial damage contribute to epigallocatechin-3-gallate induced death in *Leishmania amazonensis*. Exp Parasitol 132:151–155
108. Medina JM, Rodrigues JC, De Souza W, Atella GC, Barrabin H (2012) Tomatidine promotes the inhibition of 24-alkylated sterol biosynthesis and mitochondrial dysfunction in *Leishmania amazonensis* promastigotes. Parasitology 139:1253–1265
109. Fonseca-Silva F, Inacio JD, Canto-Cavalheiro MM, Almeida-Amaral EE (2011) Reactive oxygen species production and mitochondrial dysfunction contribute to quercetin induced death in *Leishmania amazonensis*. PLoS One 6:e14666
110. Monzote Fidalgo L, Sariego Ramos I, García Parra M, Cuesta-Rubio O, Márquez Hernández I, Campo Fernández M, Piccinelli AL, Rastrelli L (2011) Activity of Cuban propolis extracts on *Leishmania amazonensis* and *Trichomonas vaginalis*. Nat Prod Commun 6:973–976

111. Luque-Ortega JR, Reuther P, Rivas L, Dardonville C (2010) New benzophenone-derived bisphosphonium salts as leishmanicidal leads targeting mitochondria through inhibition of respiratory complex II. J Med Chem 53:1788–1798
112. Carvalho L, Luque-Ortega JR, Manzano JI, Castanys S, Rivas L, Gamarro F (2010) Tafenoquine, an antiplasmodial 8-aminoquinoline, targets *Leishmania* respiratory complex III and induces apoptosis. Antimicrob Agents Chemother 54:5344–5351
113. Saha P, Sen R, Hariharan C, Kumar D, Das P, Chatterjee M (2009) Berberine chloride causes a caspase-independent, apoptotic-like death in *Leishmania donovani* promastigotes. Free Radic Res 43:1101–1110
114. Kaur J, Singh BK, Tripathi RP, Singh P, Singh N (2009) *Leishmania donovani*: a glycosyl dihydropyridine analogue induces apoptosis like cell death via targeting pteridine reductase 1 in promastigotes. Exp Parasitol 123:258–264
115. Tempone AG, Taniwaki NN, Reimão JQ (2009) Antileishmanial activity and ultrastructural alterations of *Leishmania* (L.) *chagasi* treated with the calcium channel blocker nimodipine. Parasitol Res 105:499–505
116. Sarkar A, Sen R, Saha P, Ganguly S, Mandal G, Chatterjee M (2008) An ethanolic extract of leaves of *Piper betle* (Paan) Linn mediates its antileishmanial activity via apoptosis. Parasitol Res 102:1249–1255
117. Roy A, Ganguly A, BoseDasgupta S, Das BB, Pal C, Jaisankar P, Majumder HK (2008) Mitochondria-dependent reactive oxygen species-mediated programmed cell death induced by 3,3′-diindolylmethane through inhibition of F0F1-ATP synthase in unicellular protozoan parasite *Leishmania donovani*. Mol Pharmacol 74:1292–1307
118. Luque-Ortega JR, van't Hof W, Veerman EC, Saugar JM, Rivas L (2008) Human antimicrobial peptide histatin 5 is a cell-penetrating peptide targeting mitochondrial ATP synthesis in *Leishmania*. FASEB J 22:1817–1828
119. Dutta A, Bandyopadhyay S, Mandal C, Chatterjee M (2007) *Aloe vera* leaf exudate induces a caspase-independent cell death in *Leishmania donovani* promastigotes. J Med Microbiol 56:629–636
120. Rodrigues JC, Bernardes CF, Visbal G, Urbina JA, Vercesi AE, de Souza W (2007) Sterol methenyl transferase inhibitors alter the ultrastructure and function of the *Leishmania amazonensis* mitochondrion leading to potent growth inhibition. Protist 158:447–456
121. Delfín DA, Bhattacharjee AK, Yakovich AJ, Werbovetz KA (2006) Activity of and initial mechanistic studies on a novel antileishmanial agent identified through in silico pharmacophore development and database searching. J Med Chem 49:4196–4207
122. Tempone AG, da Silva AC, Brandt CA, Martinez FS, Borborema SE, da Silveira MA, de Andrade HF Jr (2005) Synthesis and antileishmanial activities of novel 3-substituted quinolines. Antimicrob Agents Chemother 49:1076–1080
123. Magán R, Marín C, Rosales MJ, Salas JM, Sánchez-Moreno M (2005) Therapeutic potential of new Pt(II) and Ru(III) triazole-pyrimidine complexes against *Leishmania donovani*. Pharmacology 73:41–48
124. Tavares J, Ouaissi A, Lin PK, Tomás A, Cordeiro-da-Silva A (2005) Differential effects of polyamine derivative compounds against *Leishmania infantum* promastigotes and axenic amastigotes. Int J Parasitol 35:637–646
125. Rodrigues JC, Urbina JA, de Souza W (2005) Antiproliferative and ultrastructural effects of BPQ-OH, a specific inhibitor of squalene synthase, on *Leishmania amazonensis*. Exp Parasitol 111:230–238
126. Luque-Ortega JR, Martínez S, Saugar JM, Izquierdo LR, Abad T, Luis JG, Piñero J, Valladares B, Rivas L (2004) Fungus-elicited metabolites from plants as an enriched source for new leishmanicidal agents: antifungal phenyl-phenalenone phytoalexins from the banana plant (*Musa acuminata*) target mitochondria of *Leishmania donovani* promastigotes. Antimicrob Agents Chemother 48:1534–1540
127. Delorenzi JC, Attias M, Gattass CR, Andrade M, Rezende C, da Cunha PA, Henriques AT, Bou-Habib DC, Saraiva EM (2001) Antileishmanial activity of an indole alkaloid from *Peschiera australis*. Antimicrob Agents Chemother 45:1349–1354

128. Ott R, Chibale K, Anderson S, Chipeleme A, Chaudhuri M, Guerrah A, Colowick N, Hill GC (2006) Novel inhibitors of the trypanosome alternative oxidase inhibit *Trypanosoma brucei brucei* growth and respiration. Acta Trop 100:172–184
129. Dantas-Leite L, Urbina JA, de Souza W, Vommaro RC (2004) Selective anti-*Toxoplasma gondii* activities of azasterols. Int J Antimicrob Agents 23:620–626
130. Muelas-Serrano S, Le-Senne A, Fernandez-Portillo C, Nogal JJ, Ochoa C, Gomez-Barrio A (2002) In vitro and in vivo anti-*Trypanosoma cruzi* activity of a novel nitro-derivative. Mem Inst Oswaldo Cruz 97:553–557
131. Faccenda D, Campanella M (2012) Molecular regulation of the mitochondrial F(1)F(o)-ATPsynthase: physiological and pathological significance of the inhibitory factor 1 (IF(1)). Int J Cell Biol 2012:367934
132. De Sarkar S, Sarkar D, Sarkar A, Dighal A, Chakrabarti S, Staniek K, Gille L, Chatterjee M (2019) The leishmanicidal activity of artemisinin is mediated by cleavage of the endoperoxide bridge and mitochondrial dysfunction. Parasitology 146:511–520
133. Sen R, Chatterjee M (2011) Plant derived therapeutics for the treatment of leishmaniasis. Phytomedicine 18:1056–1069
134. Jain V, Jain V (2018) Molecular targets and pathways for the treatment of visceral leishmaniasis. Drug Discov Today 23:161–170
135. Chatterjee M, Saha P, Sarkar A, Ghosh S, Mukherjee S, Roy S, Mukhopadhyay D (2012) In: Ray A, Gulati K (eds) Emerging druggable targets in leishmaniasis. Vidyanilyam Prakashan, Delhi. Chapter 18, pp 319–340

Photodynamic Therapy Against Bacterial Biofilm: Role of Reactive Oxygen Species

25

Lama Misba and Asad U. Khan

Abstract

Microbial infections remain to be one of the main causes of mortality and the culprits behind these infections are biofilm-forming multiple drug-resistant strains of bacteria. Biofilm are the surface attached, three-dimensional structure of heteromorphic microbial communities embedded in self-producing extracellular polymeric substances (EPSs). Biofilm-associated bacteria exhibit several antibiotic-resistance mechanisms; antibiotic penetration, efflux of the antibiotic, and EPS production are the main mechanisms of antibiotic resistance as they deny drug access to the cell interior. Due to the severity of biofilm-related infections, there is an urgent need to explore novel approaches like photodynamic therapy (PDT) to circumvent this increased resistance. PDT employs a nontoxic, light-sensitive dye called photosensitizer (PS), and harmless visible light of appropriate wavelength to match the absorption peak of the PS in presence of oxygen-rich environment produces a phototoxic response. PDT appears the most promising alternative methodology against multidrug resistance and biofilm-related infections.

Keywords

Antimicrobial resistance · Biofilm · Photodynamic therapy · Type I and Type II · Reactive oxygen species

L. Misba · A. U. Khan (✉)
Medical Microbiology and Molecular Biology Laboratory, Interdisciplinary Biotechnology Unit, Aligarh Muslim University, Aligarh, Uttar Pradesh, India
e-mail: asad.k@rediffmail.com; asadukhan72@gmail.com

S. Chakraborti et al. (eds.), *Oxidative Stress in Microbial Diseases*,
https://doi.org/10.1007/978-981-13-8763-0_25

25.1 Biofilm

Earlier bacteria have been regarded as unicellular organism growing in homogeneous planktonic free-floating populations. They are able to grow, divide, sense, and adapt to environmental changes. Despite their self-sufficiency, bacteria communicate with neighbors to carry out cooperative activities such as biofilm development, bioluminescence production, and exoenzyme secretion. The rediscovery of a microbiological phenomenon was first described by Anton van Leeuwenhoek, who scraped the plaque biofilm from his teeth and observed the "animalculi" that produced this microbial community with his primitive microscope [1]. However, the general theory of biofilm predominance was not appreciated until 1978. The existence of biofilm was well-known for a long time, but the comprehensive research began from the 1980s [2]. Microbial biofilm is an assemblage of sessile microbial communities that is irreversibly associated with substratum and embedded in the pool of self-produced extracellular polymeric matrix consisting of proteins, polysaccharides, and nucleic acids [3]. Biofilm-associated bacteria differ from their planktonic counterparts with respect to growth rate and gene transcription because embedded bacteria experience a different microenvironment with higher osmolarity, nutrient limitation, and higher cell density. The three-dimensional structure of biofilm formation is a dynamic process with heterogeneous bacterial communities. Biofilm is formed to capitalize on energy, communication, and special arrangement and to maintain the community of microorganisms. Bacteria living within the biofilms are protected from environmental stress, such as desiccation, antimicrobials attack by the immune system, and protozoa ingestion; hence this architecture proves to be superior over planktonic bacteria [4]. The fluidity of biofilm is due to the formation of void and water channels which help in the uptake and flow of nutrient and oxygen and elimination of harmful metabolic by-products [5]. Coordination within the biofilm bacteria occurs through a cell-to-cell communication called quorum sensing. Quorum sensing is an intracellular response of high population density due to the accumulation of signaling compound in the extracellular environment which regulates the expression of specific genes. Many bacterial species use QS to coordinate the disassembly of the biofilm community [6].

25.2 Photodynamic Therapy

Photodynamic therapy (PDT) is a light-based treatment for cancer and infectious diseases. PDT involves delivering visible light of appropriate wavelength to excite nontoxic photosensitizers (PS). This sensitization reaction occurs in an ambient environment which may be oxygen rich. The excitation occurs when the wavelength of the applied light overlaps with the absorption spectrum of the PS. These nontoxic photoactivatable dye or photosensitizer (PS) absorbs the photons of light of the correct wavelength to excite the PS. After excitation, PS from excited singlet state converts to a long-lived triplet excited state. Reactive triplet-state photosensitizers transfer this energy to surrounding oxygen molecules that results in the formation of reactive oxygen species (ROS). These reactive oxygen species are highly toxic and consequently damage the microbial cell by targeting membrane lipids, proteins, nucleic acids, and other cellular components [7].

25.2.1 History of Photodynamic Therapy

Light has been used as therapy for more than thousands of years. Ancient Egyptian, Indian, Greece, Rome, and Chinese civilizations used light to treat various diseases, including vitiligo, psoriasis, rickets, and skin cancer [8]. The utilization of sunlight as a therapeutic agent was introduced in ancient Greeks. Heliotherapy, the whole body sun bath or exposure for the treatment of disease, was well familiarized in the olden days. The famous Greek physician Herodotus was regarded as the father of heliotherapy and emphasized the importance of sun exposure for the restoration of health [9]. It disappeared for many centuries but reemerged recently; the therapeutic effects of sunlight in medicine were rediscovered by the Western civilization through Arnold Rikli, Oscar Raab, Niels Finsen, and Herman von Tappeiner [10]. In the eighteenth and nineteenth centuries, Danish physician Niels Finsen further developed "phototherapy" or the use of light to treat diseases. He successfully treated small pox using red light and cutaneous tuberculosis using ultraviolet light and developed the use of carbon arc phototherapy for which he was awarded a Nobel Prize in 1903 [11].

The PDT concept of using a dye as a PS in the photodynamic process was developed by Oscar Raab, a medical student working with Professor Herman von Tappeiner in Munich. He accidently discovered the combination light and acridine red that had the adverse effect on paramecium while performing the experiment during a thunderstorm. He identified the effect caused by fluorescent compound acridine in the presence of light and without light this compound showed no adverse effect on paramecium [12]. He hypothesized that acridine dye converts light into a form of active chemical energy, which was a finding that formed the basis of PDT. After this finding, von Tappeiner predicted the promising application of compounds of fluorescent type in the clinical use [13]. Later in 1905, von Tappeiner, in coordination with a dermatologist named Jesionek, treated skin tumors with topically applied eosin and white light. They described this phenomenon as "photodynamic action," and they discovered that oxygen is required in photosensitization reactions [14, 15]. In the current scenario, photodynamic therapy has been successful in trial and is also approved for clinical application (Fig. 25.1).

25.2.2 Mechanism of Photodynamic Therapy

Photodynamic therapy involves the use of a nontoxic light-sensitive dye called a photosensitizer (PS) combined with harmless visible light of appropriate wavelength to match the absorption spectrum of the PS. These are individually harmless components but when combined result in the formation of reactive oxygen species

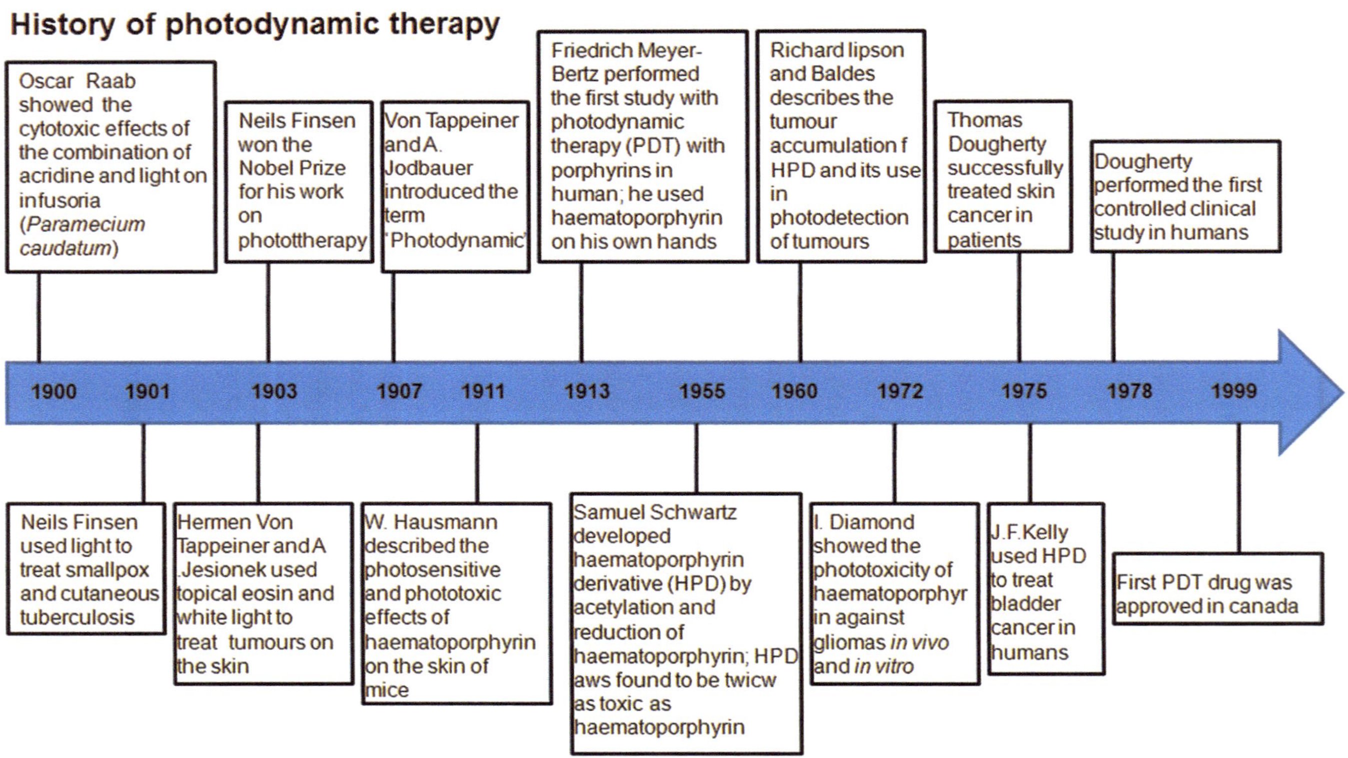

Fig. 25.1 History of photodynamic therapy. (Modified version from the source: Dolmans et al. [16]

(ROS). PS are usually aromatic molecules and have molecular structures that are typified by conjugated double bonds containing a delocalized system of π electrons [17]. In its ground (singlet) state, the electrons of the PS are spin paired in low-energy orbitals. When a photosensitizer absorbs a photon of light corresponding to the absorption peak of the PS, an electron in the highest occupied molecular orbital (HOMO) of the PS is excited to the lowest unoccupied molecular orbital (LUMO). The elevation of the photosensitizer from ground (singlet) state (S_0) to an unstable, very reactive, and short-lived excited singlet state (electron spins paired). In this state, several processes may rapidly occur. The excited photosensitizer molecule can release its energy via different electronic or physical processes (intersystem crossing) and return to the stable ground state. It can lose energy either by fluorescence or by phosphorescence. The excited electron from the excited singlet state (S_1) returns back to the ground state (S_0) by losing the absorbed energy via fluorescence ($S_1 \rightarrow S_0$) through internal conversion system. The lifetime of conversion of singlet state to ground state by fluorescence is 10^9–10^6 s. Another most important process is spin reversal; in this process reversal of the excited electron's spin takes place, known as intersystem crossing, to give the triplet state of the PS. This excited triplet state has a longer lifetime and is less reactive than the excited singlet state. Now the excited electron spins parallel to its previous paired electron, and it may not be able to return to its lower energy level according to Pauli exclusion principle. In alternative phosphorescence, the electron from triplet state may change its spin orientation and emits its energy in the form of phosphorescence; it is relatively a slow process. Apart from phosphorescence, excited triplet state of PS may interact with molecular oxygen. According to the selection rules, triplet state are spin allowed with triplet state while triplet state are spin-forbidden with that singlet state. So, the triplet state of PS can react with molecular oxygen because ground electronic state of oxygen is a triplet [18]. The two outermost orbital of oxygen are unpaired and spin parallel and this excited triplet photosensitiser can react further by one or both the pathway known as the type I and type II photo chemical reaction [19] (Fig. 25.2).

The type I reaction involves electron transfer or sometimes with proton donation from the triplet-state photosensitizer to the ambient substrate (unsaturated membrane phospholipids or aminolipids) to produce lipid-derived radicals or hydroxyl radicals (HO•) derived from water. These radicals can further react with oxygen to produce reactive oxygen species such as superoxide, hydroxyl radical, and hydrogen peroxide. These radical can cause lipid peroxidation leading to cellular damage and cell death [20]. Hydroxyl radical (•OH) is considered as the most reactive among the three, and it is a strong electrophile. H_2O_2 is less reactive, and O• − ^{2} is considered as least reactive because O• − ^{2} may be converted into H_2O_2 and O_2 by superoxide dismutase. H_2O_2 shows its reactivity when it reacts with ferrous ion, known as Fenton reaction [21].

$$H_2O_2 + Fe_2 + \rightarrow OH - + {}^{\bullet}OH + Fe^{3+}$$

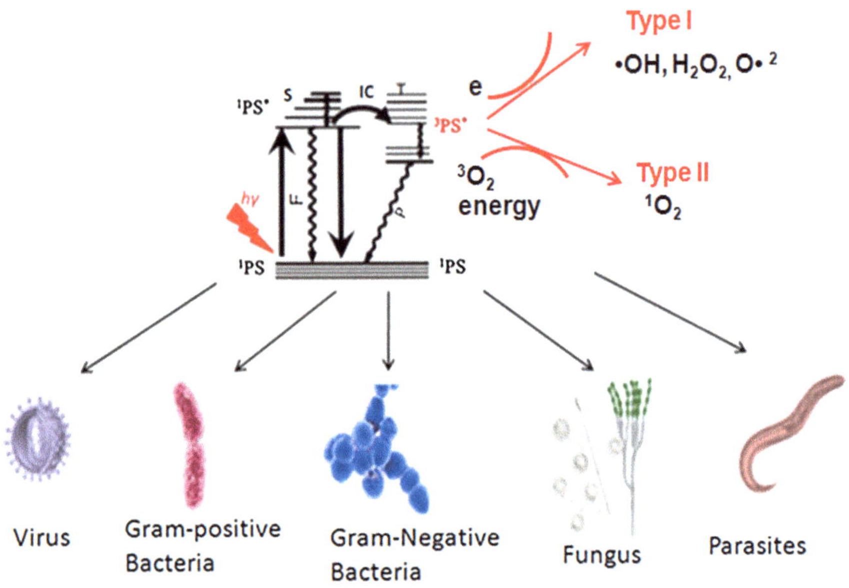

Fig. 25.2 Mechanism of photodynamic therapy

H_2O_2 is also converted into water and oxygen by catalase, while •OH is not degraded by any enzymatic pathway; it may be quenched by some antioxidants like glutathione and ascorbic acid.

Type II reaction involves energy transfer from the triplet-state PS to ground-state (triplet) molecular oxygen to produce excited singlet-state oxygen, which is responsible for the oxidation of various cellular constituents [17]. The Type II reaction involves in flipping the spin of the two outermost unpaired and parallel spin electron of oxygen and shifting it into one orbital, which in turn leaves one orbital entirely unoccupied known as singlet oxygen ($^1O_2$). This is extremely reactive, short-lived, and electronically unstable species and is not considered a radical as its electrons are spin paired which is against Hund's rule. The $^1O_2$ Σg form of singlet oxygen is too short-lived; it is the $^1O_2$ Δg form that is involved in photodynamic response [22]. $^1O_2$ can react with sulfur moieties, double bonds, and aromatic components of macromolecules. Singlet oxygen is known to modify guanine residue selectively and few stand break and site of base loss (AP site), the major product of base modification caused by singlet oxygen is 8-hydroxyguanine (7,8- dihydro-8-oxoguanine) [23]. The susceptibility of proteins is due to double bond or sulfur moieties; sulfides are generally oxidized to sulfoxides, while disulfides react to form thiolsulfinates [24, 25]. Singlet oxygen reacts with unsaturated lipids to form lipid hydroperoxides and with the aid of ferrous iron can play an important role in initiating free radical chain reactions [26].

Many biological molecules such as proteins, nucleic acids, and lipids contain these components and can easily be oxidized by $^1O_2$ [27, 28].

25.2.3 Effect of Photodynamic Therapy

Low levels of superoxides and hydrogen peroxides are normally produced by the cells as a consequence of aerobic respiration and metabolism; these are neutralized by the antioxidant defense (superoxide dismutase (SOD) and catalase) system without causing any damage. However, higher concentrations of these ROS may overcome these defenses. Small amounts of hydroxyl radical (•OH) and singlet oxygen species may prove lethal, especially to microorganisms because organisms do not have adequate defenses for these ROS. Membrane lipid is the major site of action by ROS by readily attacking the polyunsaturated fatty acids. The pathway of lipid peroxidation is illustrated in Fig. 25.3. Lipid peroxidation is a self-propagating chain reaction; if only few lipids are initially oxidized, it results into significantly large tissue damage [30]. These three (O^{-2}, •OH, and H_2O_2) species together with unstable intermediates from the peroxidation of lipids are responsible for many diseases such as Alzheimer's, Parkinson's, atherosclerosis, myocardial infarction, and autoimmune diseases during oxidative stress condition [31]. The major by-product of high-energy ionization of water molecules and from Fenton reaction cascade produces •OH radicals; this radical is an extremely aggressive oxidant, and it can degrade various biological molecules and impair their proper functioning.

$$\mathbf{H_2O \rightarrow \bullet OH + \bullet H + e^-{}_{aq} \rightarrow H_2O_2 \bullet}$$

Fenton reaction

$$\mathbf{Fe^{2+} + H_2O_2 \rightarrow Fe^{3+} + {}^-OH + \bullet OH}$$

Singlet-state oxygen is deemed to be one of the most damaging ROS. It is a very reactive species form of oxygen and can react with its environmental biomolecules in the cell, such as peptides, nucleic acids, and phospholipids causing the destruction of target cell.

25.2.4 Broad-Spectrum Antimicrobial Activity

Photodynamic therapy is used to treat various types of infection that include bacterial, fungal, viral, protozoan, or even parasitic. The advantage of the broad spectrum exhibited by PDT is that it could be used to treat localized infection before identifying the microbes. It was reported that BODIPY photosensitizers at nanomolar concentrations and short illumination times are sufficient to reduce clinically relevant microbes, including gram-positive, gram-negative, and drug-resistant bacteria, as well as pathogenic yeast and viruses [32]. It has been reported that neutral, anionic, and cationic photosensitizers effectively destroyed gram-positive bacteria, but only cationic photosensitizers were found to photoinactivate gram-negative bacteria [33]. The resistance of gram-negative bacteria to the action of photosensitizers can

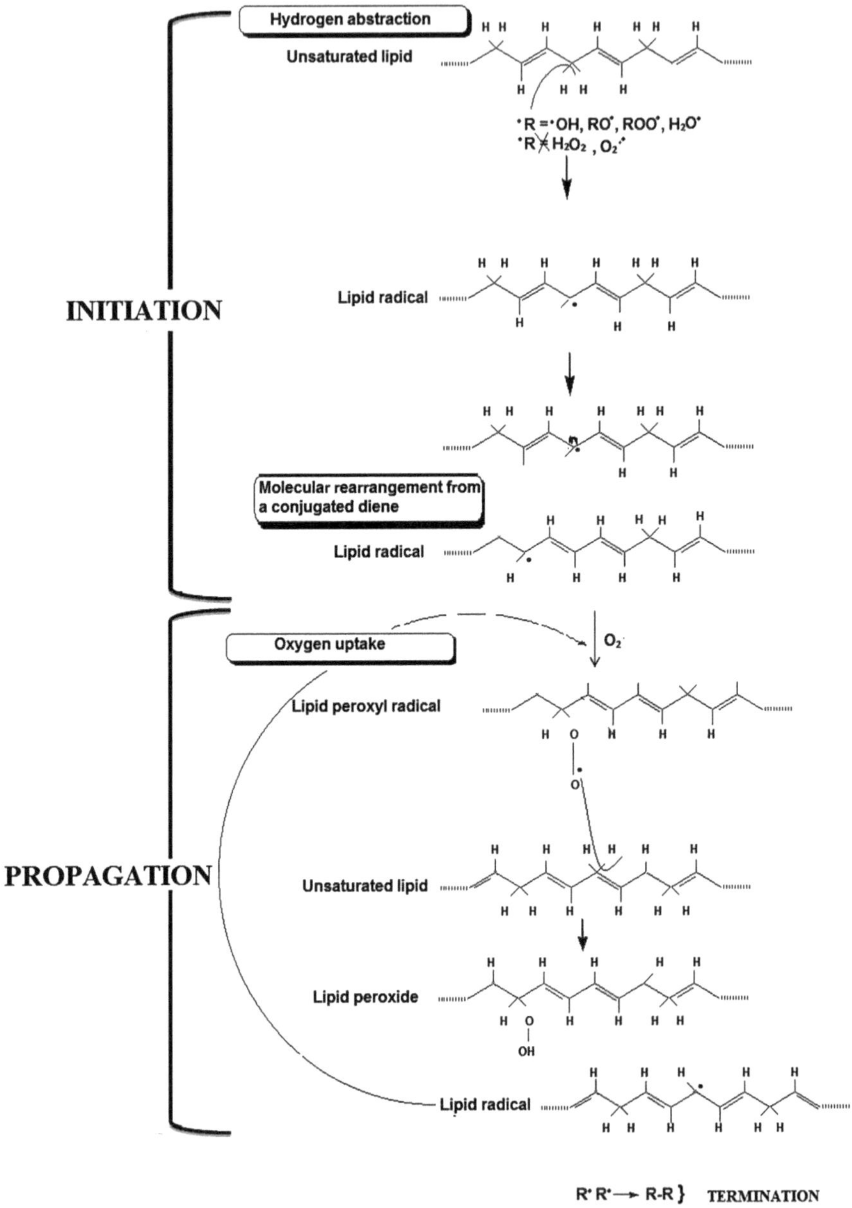

Fig. 25.3 Oxidative degradation of lipid. (Adapted from [29])

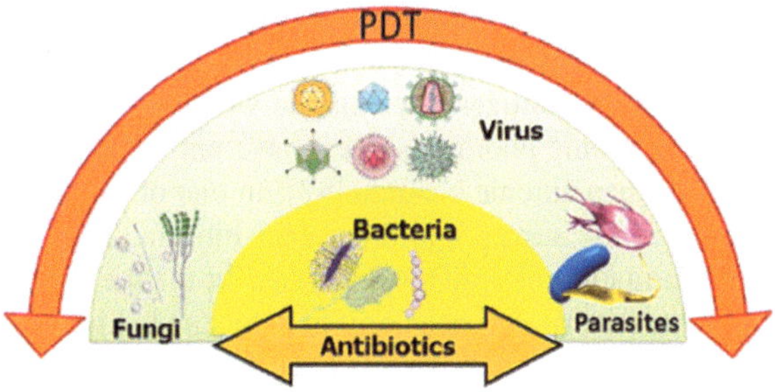

Fig. 25.4 Broad-spectrum effects of PDT. (Source: [39])

be explained by the physiology of outer membrane, which hinders the interaction of the photosensitizer with the cytoplasmic membrane. Gram-positive bacteria consist of a relatively simple envelope; its cytoplasmic membrane is surrounded by a relatively porous cell wall, while gram-negative bacterial cell envelope is more complex, and its relatively impermeable structure consists of an inner cytoplasmic membrane and an outer membrane which are separated by a peptidoglycan layer [34, 35]. It was also reported that positively charged PS binds to negatively charged bacterial cell membrane and in some cases penetrated to the microbial cells due to the porous nature of gram-positive bacterial cell wall [36]. Photosensitizing of gram-negative bacteria can be increased by the presence of cell membrane disorganizing substances that include EDTA, nitrilotriacetic acid, sodium hexametaphosphate, and polycationic agent polymyxin nonapeptide (PMNP) [37]. PDT has significant advantages over conventional treatment owing to its ability of selective binding to the membranes of pathogenic cells and the possibility for accurate delivery of light to the affected tissue, such that there is maximum damage to the microbes with minimal damage to the host tissue [38] (Fig. 25.4).

25.2.5 Photodynamic Therapy Against Microbial Biofilm

As described above, the major mode of action of PDT against microbes is mediated by the reactive oxygen species. Broad spectrum of antimicrobial action suggests that there is no or little selectivity for microbes, but the selectivity of the microbial cells over human cells is entirely dependent on the property of the photosensitizer. The effect of PDT on biofilm was demonstrated by Wainwright and Crossley [40]. Phenothiazinium photosensitiser new methylene blue was used to photoeliminate extracellular polymeric substance (EPS) in *Pseudomonas aeruginosa* biofilm. Breakdown of EPS leads to disruption of biofilm, and it may also inhibit plasmid exchange in drug-resistant bacteria. Electron microscopic study suggested that the photosensitizer (cationic zinc phthalocyanine) is taken up by the dental biofilm, and when treated with light significant reduction in the number of cells was observed due to reduced cell-to-cell and cell-to-matrix binding [41].

The outermost layer of dental biofilm is more susceptible than the innermost biofilm; this is due to the inability of the photosentizer to diffuse through the inner biofilm. Most of the photosensitizers are charged which are taken up by the outermost cells and EPS. Cationic photosensitizers have tenfold higher rate of binding efficiency by the cells than anionic charged [42]. In one of the report, it was found that young biofilm are less susceptible than older biofilm. They suggested that it may be due to the formation of voids and channel in mature biofilm which allow greater access of photosensitizer to cells or it may be due to higher metabolic activity and more effective repair mechanism in young biofilm [43].

Inactivation of *Staphylococcus* biofilm by toluidine blue O (TBO) is light dose dependent. Confocal laser scanning microscope confirmed that bacterial cell membrane is the major site of damage caused by photodynamically treated biofilm. Presence of dead cells throughout the biofilm implies that there is no hindrance in the penetration of photosensitizers through the biofilm [44].

25.2.6 Advantages of Photodynamic Therapy Biofilm

Some of the advantages of PDT are:

- PDT provides a broad spectrum and is able to kill a wide variety of microorganism, such as yeast, fungi, filamentous algae, virus, protozoa, and gram-positive and gram-negative bacteria.
- It has not yet been reported to develop any kind of photoresistance species even after multiple treatments.
- There is a low risk of inducing mutagenic effects.
- It is fast, and it can kill many folds of microbial cells over a minute, while conventional antibiotics can take days to work.
- PS and drug-light intervals can be designed such that there is maximum damage to the microbes with minimal damage to the host tissue.
- PS can also be topically delivered into infected areas in traumatic infections where the blood supply is compromised preventing antibiotics reaching the microbes.
- It is also reported that PDT can effectively reduce biofilm infection that are resistant to antibiotics.
- Antibiotic penetration is a major challenge in biofilm-based infection; PDT is not hindered by EPS production in biofilm, since ROS produced at the cell exterior are still effective in reducing bacterial biofilm.
- It is selective and painless.
- Last but not the least, it is inexpensive.

References

1. Costerton JW, Geesey GG, Cheng KJ (1978) How bacteria stick. Sci Am 238:86–95
2. Donlan RM, Costerton JW (2002) Biofilms: survival mechanisms of clinically relevant microorganisms. Clin Microbiol Rev 15:167–193
3. Hoiby N, Ciofu O, Johansen HK, Song ZJ, Moser C, Jensen PO, Molin S, Givskov M, Tolker-Nielsen T, Bjarnsholt T (2011) The clinical impact of bacterial biofilms. Int J Oral Sci 3:55
4. Wilkins M, Hall-Stoodley L, Allan RN, Faust SN (2014) New approaches to the treatment of biofilm-related infections. J Infect 69:S47–S52
5. Lang NP, Mombelli A, Attström R (2008) Oral biofilms and calculus. In: Clinical periodontology and implant dentistry. Blackwell Munksgaard, Oxford, UK, pp 197–205
6. Solano C, Echeverz M, Lasa I (2014) Biofilm dispersion and quorum sensing. Curr Opin Microbiol 18:96–104
7. Hamblin MR, Hasan T (2004) Photodynamic therapy: a new antimicrobial approach to infectious disease. Photochem Photobiol Sci 3:436–450
8. Spikes JD (1985) In: Berghausen RV, Jori G, Land EJ, Truscott TH (eds) Primary photoprocesses in biology and medicine. Plenum Press, New York, pp 209–227
9. Ackroyd R, Kelty C, Brown N, Reed M (2001) The history of photodetection and photodynamic therapy. Photochem Photobiol 74:656–669
10. Daniell MD, Hill JS (1991) A history of photodynamic therapy. ANZ J Surg 6:340–348
11. Finsen NR (1901) Phototherapy. Edward Arnold, London
12. Raab O (1900) Uber die Wirkung fluoreszierender Stoffe auf Infusorien. Z Biol 39:524–546
13. von Tappeiner H (1900) Uber die Wirkung fluoreszierender Stoffe auf Infusorien nach Versuchen von O. Raab. Muench Med Wochenschr 47:5
14. von Tappeiner H, Jesionek A (1903) Therapeutische versuche mit fluoreszierenden stoffen. Muench Med Wochenschr 47:2042–2044
15. von Tappeiner H, Jodlbauer A (1904) Uber wirkung der photodynamischen (fluorieszierenden) stoffe auf protozoan und enzyme. Dtsch Arch Klin Med 80:427–487
16. Dolmans DE, Fukumura D, Jain RK (2003) Photodynamic therapy for cancer. Nature Rev Cancer 3(5):380
17. Wainwright M (1998) Photodynamic antimicrobial chemotherapy (PACT). J Antimicrob Chemother 42:13–28
18. Tegos G, Dai T, Fuchs BB, Coleman JJ, Prates RA, Astrakas C, St Denis TG, Ribeiro MS, Mylonakis E, Hamblin MR (2012) Concepts and principles of photodynamic therapy as an alternative antifungal discovery platform. Front Microbiol 3:120
19. Ochsner M (1997) Photophysical and photobiological processes in the photodynamic therapy of tumours. J Photochem Photobiol B 39:1–18
20. Athar M, Mukhtar H, Bickers DR (1988) Differential role of reactive oxygen intermediates in photofrin-I- and photofrin-II-mediated photoenhancement of lipid peroxidation in epidermal microsomal membranes. J Invest Dermatol 90:652–657
21. Valko M, Morris H, Cronin MT (2005) Metals toxicity and oxidative stress. Curr Med Chem 12:1161–1208
22. Wilkinson F, Helman WP, Ross AB (1993) Quantum yields for the photosensitized formation of the lowest electronically excited singlet state of molecular oxygen in solution. J Phys Chem Ref Data 22(1):113–262
23. Epe B (1991) Genotoxicity of singlet oxygen. Chem Biol Interact 80:239–260
24. Bensasson RV, Land EJ, Pariente R (1988) Deactivation of singlet molecular oxygen by thiols and related compounds, possible protectors against skin photosensitivity. Photochem Photobiol 47(4):485–489
25. Buettner GR, Robert DH (1987) Superoxide, hydrogen peroxide and singlet oxygen in hematoporphyrin derivative-cysteine, -NADH and -light systems. Biochim Biophys Acta (BBA) Gen Subj 923(3):501–507

26. Bors W, Erben-Russ M, Michel C, Saran M (1990) Radical mechanisms in fatty acid and lipid peroxidation. Free radicals, lipoproteins, and membrane lipids. Springer, Boston, pp 1–16
27. Leach AG, Houk KN (2002) Diels-Alder and ene reactions of singlet oxygen, nitroso compounds and tri azolinediones: transition states and mechanisms from contemporary theory. Chem Commun (Camb) 21:1243–1255
28. Singleton DA, Hang C, Szymanski MJ, Meyer MP, Leach AG, Kuwata KT, Chen JS, Greer A, Foote CS, Houk KN (2003) Mechanism of ene reactions of singlet oxygen. A two-step no-intermediate mechanism. J Am Chem Soc 125:1319–1328
29. Vatansever F, de Melo WC, Avci P, Vecchio D, Sadasivam M, Gupta A, Chandran R, Karimi M, Parizotto NA, Yin R, Tegos GP (2013) Antimicrobial strategies centered around reactive oxygen species–bactericidal antibiotics, photodynamic therapy, and beyond. FEMS Microbiol Rev 37:955–989
30. Mylonas C, Kouretas D (1999) Lipid peroxidation and tissue damage. In vivo (Athens) 13:295–309
31. McCord JM, Fridovich I (1969) Superoxide dismutase. An enzymic function for erythrocuprein (hemocuprein). J Biol Chem 244:6049–6055
32. Carpenter BL, Situ X, Scholle F, Bartelmess J, Weare WW, Ghiladi RA (2015) Antiviral, antifungal and antibacterial activities of a BODIPY-based photosensitizer. Molecules 20:10604–10621
33. Huang L, Xuan Y, Koide Y, Zhiyentayev T, Tanaka M, Hamblin MR (2012) Type I and type II mechanisms of antimicrobial photodynamic therapy: an in vitro study on gram-negative and gram-positive bacteria. Lasers Surg Med 44:490–499
34. Minnock A, Vernon DI, Schofield J, Griffiths J, Parish JH, Brown SB (2000) Mechanism of uptake of a cationic water-soluble pyridinium zinc phthalocyanine across the outer membrane of *Escherichia coli*. Antimicrob Agents Chemother 44:522–527
35. Pereira MA, Faustino MAF, Tomé JPC, Neves MGPMS, Tome AC, Cavaleiro Â, Cunha JAS, Almeida A (2014) Influence of external bacterial structures on the efficiency of photodynamic inactivation by a cationic porphyrin. Photochem Photobiol 13:680–690
36. Huang L, Huang YY, Mroz P, Tegos GP, Zhiyentayev T, Sharma SK, Lu Z, Balasubramanian T, Krayer M, Ruzie C, Yang E, Kee HL, Kirmaier C, Diers JR, Bocian DF, Holten D, Lindsey JS, Hamblin MR (2010) Stable synthetic cationic bacteriochlorins as selective antimicrobial photosensitizers. Antimicrob Agents Chemother 54:3834–3841
37. Nitzan Y, Gutterman M, Malik Z, Ehrenberg B (1992) Inactivation of gram-negative bacteria by photosensitized porphyrins. Photochem Photobiol 55:89–96
38. Plaetzer K, Krammer B, Berlanda J, Berr F, Kiesslich T (2009) Photophysics and photochemistry of photodynamic therapy: fundamental aspects. Lasers Med Sci 24:259–268
39. Sharma K, Sulbha DT, Kharkwal GB, Huang YY, Huang L, Bil DAVJ, Tegos GP, Hamblin MR (2011) Drug discovery of antimicrobial photosensitizers using animal models. Curr Pharm Des 17:1303–1319
40. Wainwright M, Crossley KB (2002) Methylene blue—a therapeutic dye for all seasons. J Chemother 14:431–443
41. Wood S, Nattress B, Kirkham J, Shore R, Brookes S, Griffiths J, Robinson C (1999) An in vitro study of the use of photodynamic therapy for the treatment of natural oral plaque biofilms formed in vivo. J Photochem Photobiol B 50:1–7
42. Zanin ICJ, Goncalves RB, Junior AB, Hope CK, Pratten J (2005) Susceptibility of *Streptococcus mutans* biofilms to photodynamic therapy: an in vitro study. J Antimicrob Chemother 56:324–330
43. Wood TK, Barrios AF, Herzberg M, Lee J (2006) Motility influences biofilm architecture in *Escherichia coli*. Appl Microbiol Biotechnol 72:361–367
44. Sharma M, Visai L, Bragheri F, Cristiani I, Gupta PK, Speziale P (2008) Toluidine blue-mediated photodynamic effects on staphylococcal biofilms. Antimicrob Agents Chemother 52:299–305

Potentials of Phytopharmaceuticals for Treating Microbiological and Oxidative Stress-Induced Type 2 Diabetes

26

Saloni Khogta, V. Addepalli, Harpal S. Buttar, and Ginpreet Kaur

Abstract

Currently, type 2 diabetes is one of the most widespread noncommunicable maladies that inflict more than 400 million people globally. Diabetes-induced oxidative stress and distortion in gastrointestinal microbiota seem to play an important role in the occurrence of metabolic syndrome and related disorders. Oxidative stress causes imbalance between highly reactive oxygen species (ROS) and antioxidant systems of the body, resulting in metabolic syndrome, atherosclerosis, cardiovascular diseases, kidney problem, neuropathy, retinopathy, cancer risk, etc. Some studies show that imbalance in the gut microbiota also contributes in the development of type 2 diabetes. Microbial dysbiosis is considered to cause diabetic complications like cardiometabolic syndrome. Soluble dietary fibers have beneficial effects on gut microbiota and promote metabolic benefits on glucose control, but the underlying mechanisms remain unknown. Oral administration of purified and standardized phytopharmaceuticals containing bioactive compounds of different medicinal plants which help to normalize blood sugar level and promote healthy microbiota have proven useful in the prevention, mitigation, and management of diabetes as well as associated cardiometabolic syndrome. Recently, probiotics/prebiotics/synbiotics have gained importance in promoting healthy microbiota in diabetic patients. Probiotics consist of live bacteria present in foods like yogurt/cheese that confer health benefits on the host, while prebiotics are indigestible oligosaccharides that promote growth of beneficial bacteria in the gut. The focus of the review is to highlight the therapeutic

S. Khogta · V. Addepalli · G. Kaur (✉)
Department of Pharmacology, SPP School of Pharmacy & Technology Management, SVKM's NMIMS, Mumbai, India
e-mail: ginpreet.kaur@nmims.edu

H. S. Buttar
Department of Pathology & Laboratory Medicine, Faculty of Medicine, University of Ottawa, Ottawa, ON, Canada

S. Chakraborti et al. (eds.), *Oxidative Stress in Microbial Diseases*,
https://doi.org/10.1007/978-981-13-8763-0_26

potential of various types of phytomedicines and their known mechanistic actions involved in the management of type 2 diabetes. Results of clinical trials of herbal remedies and probiotics/prebiotics in humans and in vivo effects observed in animal models will also be described in this review. Randomized, placebo-controlled, and double-blind clinical studies with robust endpoints are needed for establishing the long-term safety, efficacy, and therapeutic relevance of phyto-pharmaceuticals and probiotics/ prebiotics/synbiotics in treating patients suffering from type 2 diabetes.

Keywords
Type 2 diabetes · Phytopharmaceuticals · Oxidative stress · Gut microbiota · Probiotics/prebiotics/synbiotics

26.1 Introduction

Type 2 diabetes mellitus or diabetes mellitus, also known as non-insulin-dependent diabetes, is a serious malady that afflicts approx. 8% of adults in the United States and over 400 million people globally. The classic symptoms of type 2 diabetes include increased thirst, weight loss, and frequent urination. One of the major complications resulting from type 2 diabetes is hyperglycemia, which can be treated with insulin injection and oral antidiabetic drugs (e.g., metformin) that normalize blood sugar. However, other diabetes-related risks such as metabolic syndrome, insulin resistance, hypercholesterolemia, atherosclerosis, heart attack and stroke, kidney problems, obesity, neuropathy, and cancer cannot be managed by antidiabetic drugs. Lifestyle changes, dietary interventions, exercise, and alternative therapies help to reduce the incidence of type 2 diabetes. It has been reported that lifestyle interventions are far more effective than metformin in high-risk persons [1]. Administration of aqueous extracts of *Cassia fistula* stem bark have shown promising results in the treatment of hyperglycemia in streptozotocin-induced diabetes in rats [2].

Oxygen is required for the survival of each and every aerobic living organism and caters to the need for high energy requirement. While reactive oxygen radicals are needed for cell signaling, excessive production of ROS can be toxic due to the generation of highly reactive intermediate species. Cellular membrane and DNA damage is caused by intermediate reactive species that include superoxide radical, hydrogen peroxide, and hydroxyl radical. Collectively, these three species are referred to as reactive oxygen species (ROS). In order to prevent or reduce the damage caused by ROS, nature has developed numerous enzymatic and nonenzymatic systems for cell survival. The defense mechanism includes enzymes like superoxide dismutase (SOD), catalase, and glutathione peroxidase, as well as macromolecules like ceruloplasmin, albumin, and vitamins E and C [3].

Recently, a hypothesis stated that oxidative stress causes superoxide production that is considered to be the most common pathogenic factor that leads to β-cell dysfunction, insulin resistance, weakened glucose tolerance, ultimately leading to type 2 diabetes mellitus. It also plays a part in the development of long-term diabetes complications, such as macrovascular and microvascular dysfunction [4]. Oxidative stress can be defined as a disruption in the level of reactive oxygen species/free radicals and antioxidant defenses [5]. In this condition, a disturbance in the antioxidant and prooxidant balance occurs. Superoxide dismutases (SOD) catalyze the disproportionation of superoxide into oxygen and hydrogen peroxide. Corresponding to oxidative stress, SOD levels show a significant increase. Due to the presence of superoxide, polyunsaturated fatty acids and other macromolecules in membrane lipids undergo alteration leading to the formation of malondialdehyde (MDA). MDA, a naturally occurring organic compound is considered a reactive species as well as a marker of oxidative stress. In patients suffering from type 2 diabetes, the levels of both MDA and SOD are increased due to increased oxidative stress [6]. Protein malnutrition can result due to an increased level of oxidative stress.

Type 2 diabetes is caused due to hyperglycemia, along with peripheral and hepatic insulin resistance. According to the recent studies, the gut microbiota plays a crucial role in the development of diabetes. Gut microbiota alters fatty acid and glucose metabolism. Thus, microbial dysbiosis is common in patients suffering from type 2 diabetes. Patients diagnosed with type 2 diabetes possess less amounts of butyrate-producing bacteria and increased number of potential pathogens in comparison with metabolically healthy individuals [7]. Vrieze et al. showed a relationship between altered microbial diversity and insulin resistance. Metagenomics also showed that microbial dysbiosis is commonly observed in patients with T2D. A reduction in the number of butyrate-producing bacteria and an increase of pathogens such as Gram-negative bacteria like *Bacteroides caccae* and *Escherichia coli* as well as Gram-positive bacteria like *Clostridia* are observed in diabetes patients when compared to healthy individuals [8]. Both Qin et al. [9] and Karlsson et al. [10] reported a decrease in the number of butyrate-producing bacteria such as *Faecalibacterium prausnitzii* and *Roseburia, both* are Gram-positive bacteria, in the microbiota of patients compared with healthy people. The gut microbiota of 345 Chinese participants were examined in the study of Qin et al. Moderate dysbiosis with a difference of 3% of the microbial genes was reported in patients suffering from the disorder [9]. A total of 145 Scandinavian women in postmenopausal stage, some having normal glucose metabolism, and some with impaired glucose tolerance were examined by Karlsson et al. An increase in the number of *Streptococcus mutans, Escherichia coli*, and *Lactobacillus grasseri* was considered to be a precursor for insulin resistance in obese, postmenopausal females in Sweden [10]. Wu et al. reported that metformin influences the microbial composition in individuals recently diagnosed with type 2 diabetes. The composition of microbiota showed changes due to metformin, along with these changes, improvement in fasting blood glucose and HbA1c concentrations were also observed. These changes were even found to be transferable to mice after colonization with microbiota of donors treated with metformin. Thus, it was concluded that metformin's antidiabetic activity is due to an alteration in the composition of microbiota [11].

The gut microbiota influences the progression of insulin resistance, a prediabetes condition. Clinical studies have suggested that people with insulin resistance and obesity showed an altered composition of gut microbiota, where an elevated *Firmicutes/Bacteroidetes* ratio was observed in comparison with healthy people [12, 13]. It is also suggested that changes in the microbiota due to obesity increases the metabolic endotoxin secretion by modulating intestinal permeability, leading to extreme low-level inflammation, insulin resistance, and finally T2DM [14, 15]. *Akkermansia muciniphila, Escherichia coli, Bacteroidetes thetaiotaomicron*, and other commensal bacterial species have varying influence on the intestinal mucus as well as glycocalyx layer, thus effecting intestinal permeability [16]. Microbiota-dependent changes in the intestinal alkaline phosphatase activity, endocannabinoid system, and gut tight-junction proteins can be responsible for changed intestinal permeability. This can result into insulin resistance [15].

Moreover, the association between triggered low-level inflammation and changed microbiota has also been considered as a reason for the development of type 2 diabetes. Gut-derived bacterial inflammatory molecules (e.g., flagellin and peptidoglycans) in intestine are responsible for the acceleration in the inflammation in type 2 diabetes [17, 18].

Apart from the gut microbiota, the oral microbiota also plays a significant role in the diabetic conditions. Studies suggest that there is a direct connection between periodontal pathogenic bacteria (such as *Porphyromonas gingivalis* and *Aggregatibacter actinomycetemcomitans*) and diabetes risk and glycemic control [19–22].

Diabetes is one such malady that has affected many and is known to be the fifth leading cause of death [23]. Effective treatment is highly required as this disease has high prevalence, progressive process, variable pathogenesis, and complications of diabetes. Treatment options like pharmacotherapy, insulin therapy, as well as diet therapy can be used in the management of diabetes. Pharmacotherapy involves the usage of many types of drugs that help by lowering glucose levels. Antidiabetic drugs like biguanides and thiazolidinediones act by increasing the peripheral absorption of glucose, sulfonylurea, and meglitinides drugs act by stimulating the insulin secretion [24], alpha-glucosidase act by delaying the absorption of carbohydrates from the intestine, biguanides act by reducing the hepatic gluconeogenesis [25]. However, these treatments possess certain disadvantages like resistance toward drug, adverse effects, and even toxicity. If we consider an example of sulfonylureas, sulfonylureas lose their effectiveness in 44% of patients after 6 years of treatment. In addition, the side effects of medicines and their interactions with other drugs are numerous. Thus, numerous treatments that make use of medicinal plants are recommended [26]. Carotenoids, flavonoids, terpenoids, alkaloids, glycosides are present in most of the plants; they often have antidiabetic effects [27]. Plants improve the performance of pancreatic tissue, by enhancing insulin secretions or reducing the intestinal absorption of glucose, thus demonstrating antihyperglycemic activity.

26.1.1 Oxidative Stress-Induced Type 2 Diabetes

26.1.1.1 Oxidative Stress and Antioxidants

Oxidative stress is generally defined as a disturbance in the balance of antioxidants and oxidants or prooxidants due to different factors such as diabetes-induced hyperglycemia, lesser amount of glutathione, environmental toxicants, drug-induced toxicity, blood vessel obstruction and reperfusion, and addiction due to substances of abuse [28]. Basically, oxidative stress results from the excessive production and/or insufficient removal of highly reactive oxygen/nitrogen/chlorine species [29]. It is well recognized that oxidative stress and hypoxia promote inflammation in the vascular wall and initiate coronary heart disease and stroke. Antioxidants reduce oxidative stress by mopping free radicals.

Free radicals have extremely short half-life. They are highly reactive chemical entities that contain a single or more unpaired electrons. The free radicals can cause damage to the cell membrane components and DNA. They also adversely affect the cell signaling mechanisms [30].

26.1.1.2 Classification of Free Radicals

Free radicals can be classified as:

- Reactive Nitrogen species (RNS).
- Reactive oxygen species (ROS).
- Reactive chlorine species (RCS).

26.1.1.3 Oxidative Stress and Diabetes Mellitus

Oxidative stress plays a crucial role in the development of cardiovascular complications in diabetic patients, especially type 2 diabetics [31]. The decreased mopping of ROS observed in diabetes is usually due to the lowered production of superoxide dismutase (SOD), glutathione peroxidase (GSH-Px), and catalase (CAT—enzymatic/nonenzymatic) antioxidants. Tissues become vulnerable to oxidative stress due to the reduced amounts of these enzymes, leading to the development of metabolic syndrome and other complications associated with diabetes mellitus [32].

In diabetes, free radical production occurs due to enhanced lipid peroxidation and glucose oxidation as well as nonenzymatic glycation of proteins, which leads to the dysfunction of cellular machinery and greater insulin resistance because of oxidative stress [33]. As per latest research, due to hydroxyl radical-induced cross-linkage occurring between apo-B monomers, lipids as well as LDL's apolipoprotein component form insoluble aggregates that are responsible for oxidative damage in complications related to diabetes [31]. Mitochondria act as the main sources of oxidative stress in diabetes mellitus. During oxidative metabolism in mitochondria, a reduction of a component of the utilized oxygen occurs to water, and the remaining oxygen gets converted to oxygen free radical (O·) that is further converted to other reactive species such as ONOO−, OH, and H_2O_2 [34]. ROS/RNS modulates

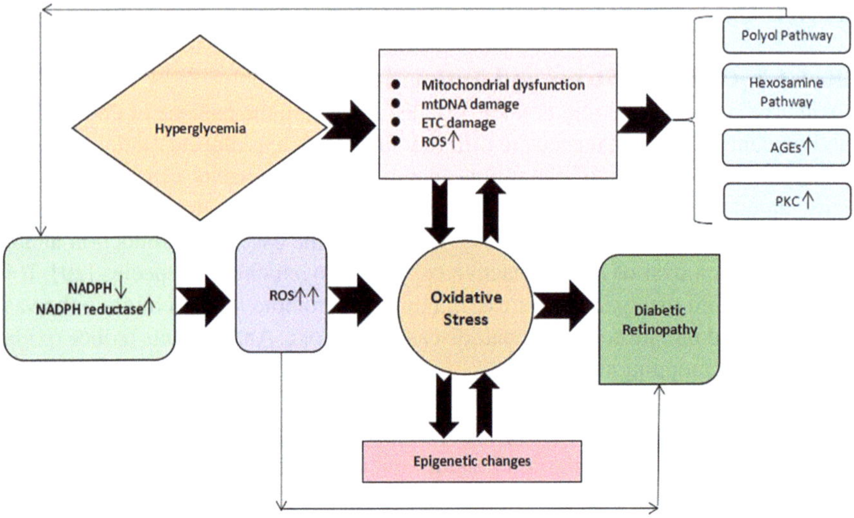

Fig. 26.1 Diagrammatic representation of pathogenesis of oxidative stress-induced type 2 diabetes

insulin signaling by two ways. As per the first way, in response to insulin, the ROS/RNS are produced to exert full physiological function. As per the second way, both of them have got negative regulation on insulin signaling, leading them to develop insulin resistance, a risk factor for type 2 diabetes [35] (Fig. 26.1).

26.1.2 Impact of Gut Microbiota in Type 2 Diabetes

Recently, several studies have suggested that the gut microbiota play a crucial role in the development of type 2 diabetes. Differentiation of the gut microbiota between patients suffering from diabetes and healthy individuals can help in obtaining further knowledge about the pathogenic mechanisms. It may also help in providing preventive incentives for prediabetic patients about the modification of the gut microbes for prevention of diabetes.

Larsen et al. made a comparison between the gut microbiota of 18 adult males suffering from type 2 diabetes, who had a high range of age and body-mass indices with 18 healthy subjects with similar range of BMI and age. The compositional changes seen in the gut microbiota were linked with type 2 diabetes at both class and phylum levels [36]. In a recent study, three groups of healthy individuals were investigated: those who were (a) recently diagnosed with type 2 diabetes (b) prediabetes and who had (c) normal glucose tolerance; the gut microbiota was found to be different in the three groups. The authors made use of a high-throughput sequencing based on 16S rRN, by which a type 2 diabetes-related dysbiosis was observed. A total of 28 taxonomic units associated with type 2 diabetes was observed, whereas the normal glucose tolerance group showed a higher amount of

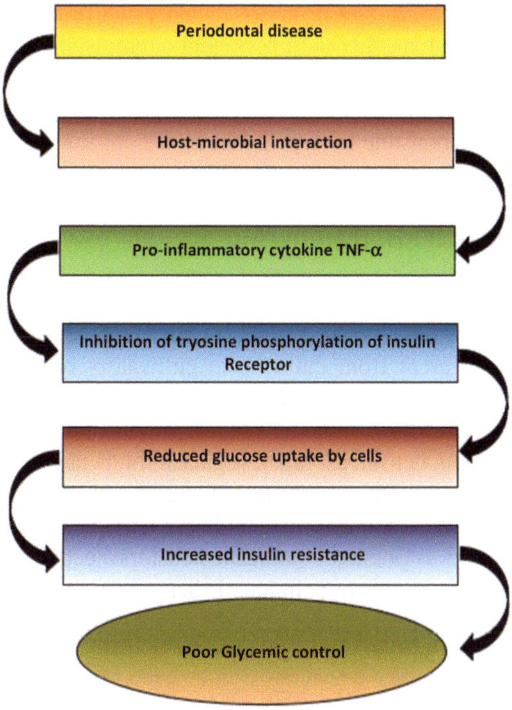

Fig. 26.2 Pathogenesis of periodontal disease and type 2 disease

butyrate-producing *Faecalibacterium prausnitzii L2–6* and A. *Muciniphila ATCCBAA-835* than the group with prediabetes. Both the prediabetes and type 2 diabetes groups had reduced amount of *Verrucomicrobiae* in comparison to the normal glucose tolerance group [37]. These results conclude that not only the gut microbiota of type 2 diabetes sufferers is different from healthy control individuals, but also the alterations in gut microbiota are linked with the progress of glucose intolerance (Fig. 26.2).

26.2 Anti-inflammatory and Antioxidant Effects of Botanical Products in Oxidative Stress-Induced Type 2 Diabetes

Oxidative stress has a crucial role in β-cell dysfunction and insulin resistance [38]; hence, a hypothesis based on loads of data has been made, according to which, a viscous connection exists between free radicals and hyperinsulinemia. These free radicals may be responsible for degeneration of insulin action [39], mainly by downregulating insulin-mediated glucose uptake [40].

Antioxidants are mostly used as complementary agents to reduce complications related to diabetes in patients [41–44]; persistent attempts are going on to discover antioxidants that can be used as drugs to combat complications related to diabetes.

Herbal drugs or phytopharmaceuticals have significance in oxidative stress, a major factor that is responsible for type 2 diabetes.

26.2.1 *Ficus carica* (Figs) [Family: Moraceae]

Chemical composition: Figs comprise of many phytochemicals, polyphenols like chlorogenic acid, gallic acid, syringic acid, (−)-epicatechin, rutin, and (+)-catechin. Its color may be different from cultivars to cultivators because of the varying amounts of anthocyanins, with particularly high content of cyanidin-3-O-rutinoside. Leaves comprise of flavonoids, mucilages, vitamins, nicotinic acid, enzymes, and tyrosin. Bergaptene, ficusin, psoralen, stigmasterol, beta-sitosterol, taraxasterol, sapogenin, rutin, lepeolacetate, calotropenyl acetate, and oleanolic acid sistosterol are other bioactive components present in the leaves of this plant.

Uses: It is reported to possess significant hepatoprotective activity, anticancer activity, hypoglycemic activity, hypolipidemic activity, antifungal activity, antibacterial activity, antipyretic activity, and antioxidant activity.

Ficus carica exhibit antioxidant properties by restoring levels of vitamin E and fatty acids [45].

26.2.2 *Azadirachta indica* (Neem) [Family: Meliaceae]

Chemical composition: Neem tree possess manifold medicinal properties because of its chemical composition. Azadirachtin, in highest concentration is present in the seeds of neem tree. Apart from azadirachtin, gedunin, salannin, azadirone, nimbidine, nimbin, nimbinol, nimbicidine, etc. are other liminoids of neem

Uses: Neem has numerous medicinal uses such as antiarthritic, anti-inflammatory, antipyretic, antifungal, diuretic, antimalarial, spermicidal, antibacterial, and hypoglycemic properties. Leaves, flowers, seeds, and bark of neem are used in home remedies. Bark of neem has antipyretic properties; thus it helps in relieving fever. Flowers find their use in intestinal disorders. Skin diseases, obesity, and wounds can be treated by the juice from fresh leaves. Neem helps in treating skin diseases due to its antifungal, antibiotic, as well as blood-purifying properties. Ayurveda principles state that skin diseases occur due to increased pitta and kapha. Neem controls vitiated pitta and kapha, thus relieving skin ailments. Wound healing is achieved, as it is antibacterial and astringent in nature. It reduces irritation, itching, and roughness of the skin and heals the psoriatic patches. Similarly, it heals eczema too. It also treats acne by reducing infection and inflammation. Neem helps to maintain the condition of scalp thus preventing dandruff

Arthritis, muscular sprains, and skin diseases can be treated by oil obtained from neem seeds. Neem is very useful in the treatment of gum diseases

Azadirachta indica extract acts by lowering the glucose level in blood, inhibiting lipid peroxides production, reactivating the antioxidant enzymes, and restoring metal and GSH levels [46]

26.2.3 *Allium sativum* (Garlic) [Family: Liliaceae]

Chemical composition: The most active ingredient is Allicin. It also comprises of allin, mucilage, albumin, alpha-glutamyl peptides, volatile oils, and vitamin C amino acids such as methionine, lucine, and cysteine.

Uses: It has significant anticancer activity, cardioprotective activity, antidiabetic activity, and antifungal activity. It is used as a flavoring agent. It is also used as an expectorant and a stimulant.

Allium sativum also acts by lowering the glucose level in blood, inhibiting lipid peroxides production, reactivating the antioxidant enzymes, and restoring metal and GSH levels [46]. Antidiabetic action is exhibited by its antioxidant nature as well as its capacity to increase insulin secretion [63].

26.2.4 *Momordica charantia (*Bitter Melon or Bitter Gourd) [Family: Cucurbitaceae]

Chemical composition: It mainly comprises of curcubitane, momordenol, momordicin I, and momordicinin. It also comprises of numerous vitamins and minerals.

Uses: Traditionally, bitter melon is famous for its medicinal uses against diseases like diabetes and cancer. It also possesses antivirus, anti-inflammatory, and cholesterol-lowering effects. Antimutagen and antioxidant properties are mainly due to the presence of many phenolic compounds.

Extracts of *Momordica charantia* act by lowering the glucose level in blood, inhibiting lipid peroxides production, reactivating the antioxidant enzymes, and restoring metal and GSH levels [46]

26.2.5 *Ocimum sanctum* (Tulsi) [Family: Lamiaceae]

Chemical composition: It comprises of eguenol, sitosterol, urosolic acid, euginal, carvacrol, limatrol, linalool, methyl carvicol, caryophyllene, and fatty acids.

Uses: It has antidiabetic activity, cardioprotective activity, wound healing activity, radio-protective effect, genotoxicity, and antioxidant activity.

Ocimum sanctum extracts act by lowering the blood glucose level, inhibiting lipid peroxides production, reactivating the antioxidant enzymes, and restoring metal and GSH levels [46]

26.2.6 *Momordica grosvenori* (*Momordica* Fruit, Luohanguo) [Family: Cucurbitaceae]

Chemical composition: Luohanguo is a round green fruit that upon drying turns brown in color. Mogrosides are a group of terpene glycosides that are present at the level of about 1% of the fleshy part of the fruit and are mainly responsible for the sweet taste of the fruit. A powder that is 80% or more mogrosides can be extracted from both the dried and fresh fruits. The main component of mogrosides is known as mogroside-5, previously known as esgoside. Neomogroside and siamenoside are also present. The mixed mogrosides are considered to be about 300 times as sweet as sugar in terms of weight; thus 80% extracts are nearly 250 times sweeter than sugar; pure mogrosides, i.e., 4 and 5, may be 400 times as sweet as sugar in terms of weight

Uses: Antioxidative, anticarcinogenic, and anti-inflammatory. It is used to relieve sore throat and in the removal of phlegm

In alloxan-induced diabetic mice, *Momordica grosvenori*, a Chinese medicinal herb, used as a substitute for sugar for patients with obesity and diabetes, was tested. A correction in the altered glucose level was observed. Also, the immune balance was effectively regulated in diabetic mice. These effects were linked to the upregulation of heme oxygenase-1 (HO-1) protein, which has antioxidant and anti-inflammatory properties [47]

26.2.7 *Scutellaria baicalensis* (*Scutellaria* Root, Scute, Skullcap, Baical Skullcap Root) [Family: Lamiaceae]

Chemical composition: It comprises of baicalin, wogonin, oroxylin A, and baicalein as the main constituents. Flavones like wogonoside, baicalin, and their aglycones, wogonin, and baicalein are the most important bioactive compounds extracted from the root of *S. baicalensis.*

Uses: It finds use in the treatment of dysentery, diarrhea, hemorrhaging, hypertension, insomnia, respiratory infections, and inflammation. Flavones have various pharmacological functions like hepatoprotective, anticancer, antiviral, antibacterial, anticonvulsant, and antioxidant as well as neuroprotective effects.

The ethanolic extract of *Scutellaria baicalensis* has antioxidant role in rats that had STZ-induced diabetes; reports have also stated that it enhances the antidiabetic effect of metformin [48].

26.2.8 *Acanthopanax senticosus* (Siberian Ginseng) [Family: Araliaceae]

Chemical composition: Lignans, triterpenoid, coumarins, saponins, and flavones are present. Eleutheroside E and syringing that are phenolic in nature were considered to be the most active components.

Uses: It has hepatoprotective, antiulcer, anti-stress, anticancer, anti-irradiation as well as anti-inflammatory properties.

Based on a study, in most of the assays, except the assay for inhibiting formation of early glycation products, *Aralia taibaiensis* proved to be better in comparison to other extracts. *Acanthopanax senticosus* extract showed maximum inhibition of the early glycation product formation. High saponin content mainly was responsible for its antiglycation and antioxidant activities [49].

26.2.9 *Albizia lebbeck* (Siris, Shiris) [Family: Fabaceae]

Chemical composition: The chemical constituents include D-catechin, melacacidin, β-sitosterol, albiziahexoside, betulinic acid, and echinocystic acid glycosides.

Uses: The traditional uses include antidiarrheal, anti-inflammatory, antiasthmatic, antifertility, antiseptic, antidysenteric, as well as antitubercular activities.

In alloxan-induced diabetic rats, the antioxidant property of aqueous extract of *Albizia lebbeck* was checked [50].

26.2.10 *Lycium barbarum* (Chinese Boxthorn, Chinese Wolfberry, Tibetan Goji, Himalayan Goji) [Family: Solanaceae]

Chemical composition: The main chemical constituents include palmitic acid, linoleic acid, and myristic acid.

Uses: It is used in diabetes, dry eyes, cancer, blood circulation problems, fever, dizziness, hypertension, ringing in the ears (tinnitus), malaria, as well as sexual problems (impotency).

Purified polysaccharide fractions (LBP-X) and crude polysaccharides (LBP) extract of *Lycium barbarum* fruit showed hypolipidemic as well as hypoglycemic effects that were shown in diabetic rabbits induced by alloxan [51]. Although the blood glucose-lowering effect of LBP-X was better than the rest of the extracts, the latter showed better antioxidant activity because of the presence of antioxidants like riboflavin, carotenoids, thiamine, nicotinic acid, and ascorbic acid in the crude extracts. Li isolated polysaccharides of *Lycium barbarum*, which were considered to be the chief ingredients of the fruits, and studied its ability to fight the oxidative insult using a STZ-induced hyperglycemic model [52]. Wu et al. also studied these polysaccharides and the antidiabetic effects associated with them, using rats with NIDDM. The study concluded that *Lycium barbarum* polysaccharides can control glucose levels in blood by modulating glucose metabolism; this further helps in improving oxidative stress markers (SOD, MDA) and decreasing DNA damage by creating a balance between antioxidants and oxidants [53]. It was concluded by these findings that *Lycium barbarum* polysaccharides can fight against oxidative stress; hence, they prevent the progress of complications related to diabetes.

26.2.11 *Strobilanthes crispus* (Pokok Pecah Kaca or Pokok Pecah Beling) [Family: Acanthaceae]

Chemical composition: Rutin, (+)-catechin, myricetin, (−)-epicatechin, luteolin, naringenin, kaempferol, and apigenin are the active compounds present in *Strobilanthes crispus*.

Uses: It is used to treat constipation as it has potential laxative effect. It helps in the management of liver maladies and gallbladder stones. Hemorrhoids can also be treated. *S. crispus*'s leaves have diuretic properties and thus are used extensively to treat bladder and kidney stones. Baby corns along with the decoction of the leaves can expel kidney stone within a short period of time. It also finds use in the relief from poisonous bites like snakebites. Skin diseases like eczema can also be treated.

Strobilanthes crispus, when tested in rats suffering from diabetes, induced by STZ, has shown antilipidemic and antihyperglycemic properties. The antioxidant effect of both fermented and unfermented herbal hot water extract is responsible for these properties [54].

26.2.12 *Silybum marianum* (Milk Thistle) [Family: Asteraceae]

Chemical composition: Extract of milk thistle prepared from the seeds contains 4–6% silymarin approximately. The extract consists of about 20–35% fatty acids, including linoleic acid, and the rest is all silymarin [8]. Silymarin comprises of many polyphenolic molecules that also include seven minutely related flavonolignans (isosilybin A, silybin A, isosilybin B, silybin B, silydianin, isosilychristin, and silychristin) along with taxifolin, a flavonoid. A semipurified fraction of silymarin known as silibinin is primarily a mixture of silybin A and B, in a 1:1 ratio roughly.

Uses: Milk thistle helps in treating disorders related to the liver and bile duct. Silymarin prevents kidney and liver cells from adverse reactions of drugs. It has also lowered enzyme levels in the liver and demonstrated T-cell modulation and anti-inflammatory effects. A major preclinical evidence exists for silymarin's anticarcinogenic and hepatoprotective effects. It is used in the treatment of skin, prostrate, cervical, and breast cancer. In case of gastrointestinal upset, milk thistle acts as a mild laxative.

In a double-blind, randomized, placebo-controlled, clinical study of 51 patients suffering from type 2 diabetes, it was concluded that *Silybum marianum* seed extract (silymarin) has potential antioxidant activity, and this extract showed a tremendous improvement in the blood glucose profile of these individuals [55].

26.2.13 *Pueraria lobata* (Kudzu Root) [Family: Fabaceae]

Chemical constituents: It comprises of beta-sitosterol, beta-sitosterol palmitate, lupeol, pueraro, lupeone, bis(2-ethylhexyl) phthalate, diisobutyl phthalate, sophoracoumestan A, allantoin, coumestrol, diadzein, 3′-methoxy diadzein, formononetin, 3′-hydroxy diadzein, ononin, genistin, 8-methoxy ononin, diadzein, sucrose, (6S,9R)-roseoside, (−)-puerol B 2-O-glucopyranoside, and sissotorin.

Uses: It has been used as a medicinal drug for various ailments such as cancer, alcoholism, angina, migraine, and hypertension. Furthermore, it has been used traditionally as a remedy for psoriasis, diarrhea, muscle pain, some symptoms related to menopause, and infections in the upper respiratory tract such as sinus infection, the common cold, and hay fever.

Xiong et al. studied the preventive effect of puerarin, an isoflavone purified from *Pueraria lobate* on rat pancreatic islets damage induced by hydrogen peroxide. It was concluded by this study that puerarin is capable of preserving pancreatic islet cells from the damage caused by reactive oxygen species [56].

26.2.14 *Plantago depressa* (Plantains) [Family: Plantaginaceae]

Chemical constituents: It comprises of ursolic acid, plantamajoside, acteoside, plantainoside D, majoroside, 10-hydroxymajoroside, and geniposidic acid.

Uses: It is commonly used as a bulk laxative as it soothes membranes that are irritated. Sometimes, only seed husks are used.

Similarly, the extract of *Plantago depressa* var. montata showed effects due to its free radical scavenging capability as well as antioxidant property. It was found that the herb was capable of correcting homeostasis of lipid and glucose; it also helps in restoring redox status in mice suffering from diabetes induced by alloxan [57].

26.2.15 *Morinda officinalis* (Indian Mulberry) [Family: Rubiaceae]

Chemical constituents: It mainly comprises of 2-methyl-anthraquinone, beta-sitosterol, 24-ethylcholesterol, and rubiadin-1-methyl ether.

Uses: It helps in providing nourishment to the kidney and strength to the bone. Conditions of impotence, depression, osteoporosis, and inflammatory diseases such as dermatitis and rheumatoid arthritis can be treated by enhancing immunofunction.

The dried roots of *Morinda officinalis* is also tested for its antioxidant and hypoglycemic activities, A test was conducted in STZ-treated rats and results showed a reduction in lipid peroxide and fasting glucose levels, along with the resurrection of the imbalanced redox indices. Thus, antidiabetic and antioxidant potentials of *Morinda officinalis* were concluded by the study [58].

26.2.16 *Amaranthus esculentus* (Blood Amaranth, Purple Amaranth) [Family: Amaranthaceae]

Chemical constituents: It comprises of thiamine, riboflavin, niacin ascorbic acid, amino acids like lysine, leucine, alanine, cysteine, serine, proline, and arginine.

Uses: It finds use against cancer and cardiovascular diseases. It has high antioxidant properties.

Oil and grain fraction of *Amaranthus esculentus* have both antidiabetic and antioxidant activity, suggesting their usefulness in combatting against hyperglycemia and preventing complications related to diabetes [59].

26.2.17 *Gentiana olivieri* (Gentian) [Family: Gentianaceae]

Chemical constituents: It comprises of isoorientin along with secoiridoidal and iridoid glycosides such as xanthones, gentiopicroside, polyphenol, flavones, and monoterpene alkaloid.

Uses: It acts as an anodyne, antibilious, anthelmintic, antioxidant, anti-inflammatory, astringent, antipyretic, cholagogue, bitter, digestive stimulant, diaphoretic, tonic, hepatic, stomachic, and sialagogue. The herb encourages the gallbladder and liver to function more efficiently by acting as a stimulant. It also provides nourishment to the pancreas, spleen, kidneys, and stomach. It was used as a cure for liver and stomach ailments by both Greeks and Arabs. It is useful in conditions like splenitis, edema, hepatitis, stomatitis, jaundice, and cirrhosis. Stimulation of digestion and appetite occurs through stimulation of digestive juices, which in turn increases the production of saliva, bile, and gastric juices. It also helps in quick gastric emptying. It is suggested in the remedy of dyspepsia and flatulence. It is also advised in conditions like anorexia, exhaustion, and debility.

Due to its potent antioxidant properties, *Gentiana olivieri* Griseb. (Gentianaceae), a Turkish folkloric medicine, is utilized as a hypoglycemic plant [60]. Isoorientin contributed to the hypoglycemic activity as it saves β-cells from oxidative damage [61].

26.2.18 *Panax ginseng* (Korean Ginseng) [Family:Araliaceae]

Chemical constituents: Ginseng consists of several saponin glycosides like panaxosides, chikusetsusaponin, and ginsenosides. Aglycone dammarol is present in ginsenosides; panaxosides contain aglycone, i.e., oleanolic acid. Panaxosides give panaxadiol, panaxatriol, and oleanolic acid on decomposition.

Uses: It acts as an immunomodulator by increasing natural resistance. It also increases the ability to overcome illness and exhaustion. It acts as a stimulating agent, sedative, and aphrodisiac. It is useful for thyroid dysfunctioning and adrenal dysfunctioning. It cures giddiness and prolongs life. It is used as demulcent. It is used to treat gastritis and anemia. Ginseng extracts are used externally in cosmetics.

Ginseng is capable of scavenging free radicals, and thus this adds to its antidiabetic mechanisms [62]. Ginseng acts by decreasing the rate of auto-oxidation of monosaccharides, elevates the activity of enzymes such as SOD, and directly eliminates the free radicals.

26.2.19 *Phyllanthus amarus* (Gale of Wind) [Family: Euphorbiaceae]

Chemical constituents: It comprises of lignans (hypophyllanthin, phyllanthin, nirtetralin, niranthin, phyltetralin), ellagitannins (amariinic acid, phyllanthusiin D, elaeocarpusin, geraninic acid B, and repandusinic acid A), tannins, and flavonoids (quercetin-3-O-glucosides).

Uses: It helps fight problems related to the genitourinary system, stomach, liver, spleen, and kidney. It is astringent, bitter, diuretic, stomachic, antiseptic, and febrifuge. It also finds its use in menorrhagia, gonorrhea, and other genital affections. It helps in diarrhea, gastropathy, intermittent fevers, dysentery, scabies, ophthalmopathy, wounds, and ulcers. Pharmacological activities include antibacterial, antiviral, anti-inflammatory, antimalarial, antiplasmodial, anticancer, antimicrobial, hypolipidemic, antidiabetic, antioxidant, nephroprotective, diuretic, and hepatoprotective properties.

Phyllanthus amarus's methanolic extract depicted a potent antioxidant activity and antihyperglycemic efficacy tested in diabetes induced due to alloxan in rats [64].

Punica granatum, *Capparis deciduas*, *Emblica officinalis*, *Camellia sinensis*, *Ficus bengalensis*, and *Musa sapientum* are also known to have significant antidiabetic property [65]. Also, *Vaccinium arctostaphylos* L. (Ericaceae) fruit was checked for its antidiabetic property. It shows notable increase in the amount of catalase (19%), glutathione peroxidase (35%), and erythrocyte superoxide dismutase (57%) activities of the alloxan-treated rats [66].

Activation of enzyme aldose reductase due to hyperglycemia leads to the reduction of NADPH that is necessary for GSH reductase; thus, alteration of the endogenous defense system occurs. Therefore, components that inhibit aldose reductase can be beneficial for the management of diabetes. Feng et al. showed that flavonoid compounds like puerarin, quercetin, silymarin, and others and their derivatives are capable of inhibiting the activity of aldose reductase [67]. Also, mangiferin, an active constituent derived from the extract of *S. reticulata* stems demonstrated aldose reductase inhibitory activity [68].

26.3 Phytopharmaceuticals in Microbiological-Induced Type 2 Diabetes

Oral efficacies of herbs like *amlaki* and *triphala* for stress modulation and other bioactivities increase with increasing numbers, and that is mostly because of their modulating effects on microbiological ecology inside the gastrointestinal tract [69]. Bacterial diarrhea was treated with berberine, active ingredient of a Chinese herb *Coptis chinensis* (Huang-Lian) [70–72], found to be clinically useful for the treatment of diabetes in a randomized, multicentered, placebo-controlled, and double-blinded clinical trial [73]. It was found out in a recent study that berberine helped in preventing obesity caused due to high-fat diet. It also combatted insulin resistance by enriching short-chain fatty acid-producing bacteria, reducing the numbers of beneficial pathogens, and reliving inflammation in Wistar rats [74]. As both type 2 diabetes and bacterial diarrhea have common dysbiotic gut microbiota, drugs like berberine that can treat bacterial diarrhea might also be useful for T2D treatment. Gegen Qinlian Decoction (GQD), a Chinese herbal formula, can relieve T2D. In order to understand the mechanism behind it, 187 patients suffering from T2D, in a double-blinded trial, were randomly decided to receive placebo ($n = 41$), low dose GQD ($n = 50$), moderate dose ($n = 52$), or high dose ($n = 44$) for 12 weeks. Patients administered with high or moderate dose showed remarkable reductions in adjusted mean changes from baseline of glycated hemoglobin (HbA1c) and fasting blood glucose in comparison with the placebo and low-dose groups. Gut microbiota showed a dose-dependent deviation in response to GQD treatment due to pyrosequencing of the V3 regions of 16S rRNA genes. This deviation preceded improvement in symptoms of T2D. 47 GQD-enriched species level phylotypes were discovered, out of which 17 were negatively correlated with FBG and 9 with HbA1c by redundancy analysis. It was confirmed with real-time quantitative PCR that *Faecalibacterium prausnitzii* were enriched by GQD significantly, a negative correlation with FBG, 2-h postprandial blood glucose levels and HbA1c was made. A positive correlation was made with homeostasis model assessment of β-cell function. Therefore, as per this data, it was indicated that Chinese herbal formula GQD made structural changes of gut microbiota. Also, the quantities of beneficial bacteria like *Faecalibacterium* spp. were increased due to treatment by GQD. Thus, it can be concluded that a correlation exists between antidiabetic effect of GQD and alterations in the gut microbiota [75]. In order to understand the effects on gut microbiota due to type 2 diabetes, rats suffering from diabetes were fed a bitter melon formulation as a lyophilized superfine powder (BLSP). BLSP treatment significantly reduced serum insulin levels ($p < 0.05$) along with glucose levels in blood under fasting conditions ($p < 0.05$) of the diabetic rats. Pyrosequencing of 16S rRNA genes (V3–V9 region) and PCR amplification of the gut microbiota of control and treated rats were done. The ratio of *Firmicutes* to *Bacteroidetes* in diabetic rats was reduced due to BLSP administration; however, the relative quantity of *Bacteroides*, *Ruminococcus*, and *Ruminococcaceae* showed severe reduction in rats treated with BLSP in comparison to diabetic rats. BLSP also showed a significant suppression of the activation of MAPK (p38 and JNK). As per the results, BLSP can significantly alter the amount of a specific gut microbiota in rats with diabetes, without disturbing diversity in the normal population. Thus, BSLP acts against type 2 diabetes by suppression of MAPK signaling pathway.

Phytopharmaceuticals have been considered beneficial against type 2 diabetes induced due to both oxidative stress and/or microbiota. However, caution must be taken when preclinical data are interpreted that evaluate the effects of natural functional compounds on antioxidation [76].

26.4 Conclusions

Oxidative stress was found to be one of the many factors that help in progression of diabetes as per numerous studies. This involves disruption of insulin action and enhancement of incidences of complications related to the disease. Many studies have also proved the efficacy of antioxidants against diabetes of both type 1 and type 2. Nonenzymatic glycation of proteins, oxidation of glucose, and lipid peroxidation are all related to rise in the levels of nitrogen and oxygen free radicals. These conditions contribute to diabetes and its complications. Relationship between oxidative stress and diabetes along with their complications related to the liver, kidney, eye, and heart has also been established. Thus, oxidative stress is a grave danger to metabolic disorders specially type 2 diabetes. Similarly, Microbiota contributes not only during onset of type 2 diabetes by causing low-level inflammation, but also in the development of the malady due to faulted inflammatory components. It also is responsible for the rise in diabetes-related complication like kidney toxicity, diabetic retinopathy, hypertension, atherosclerosis, diabetic foot ulcers, Alzheimer's disease, and cystic fibrosis. These studies thus support the important role of microbiota in the maintenance of the integrity of intestinal barrier. Microbiota also helps in keeping metabolic homeostasis as per normal levels, protecting infection to host by pathogens, and thus improving defense system of host and even impacting the nervous system in type 2 diabetes. However, the potential mechanisms that can link type 2 diabetes with the microbiota have not been fully cleared, and extensive research efforts are required. Phytopharmaceuticals have been considered beneficial against type 2 diabetes induced due to both oxidative stress and/or microbiota.

Conflict of Interest The authors declare no conflict of interest.

References

1. Diabetes Prevention Program Research Group (2002) Reduction in the incidence of type 2 diabetes with lifestyle intervention or metformin. N Engl J Med 346(6):393–403
2. Daisy P, Saipriya K (2012) Biochemical analysis of *Cassia fistula* aqueous extract and phytochemically synthesized gold nanoparticles as hypoglycemic treatment for diabetes mellitus. Int J Nanomedicine 7:1189
3. Piconi L, Quagliaro L, Ceriello A (2003) Oxidative stress in diabetes. Clin Chem Lab Med 41(9):1144–1149

4. Wright E, Scism-Bacon JL, Glass LC (2006) Oxidative stress in type 2 diabetes: the role of fasting and postprandial glycaemia. Int J Clin Pract 60(3):308–314
5. Betteridge DJ (2000) What is oxidative stress? Metabolism 49(2 Suppl 1):3–8
6. Morteza A, Jenab Y, Nargesi AA, Ghazizadeh Z, Salabati M (2012) Urea and oxidative stress in type 2 diabetes. J Metabolic Synd 1(105):2167–0943
7. Crommen S, Simon MC (2017) Microbial regulation of glucose metabolism and insulin resistance. Genes 9(1):10
8. Vrieze A, Van Nood E, Holleman F, Salojärvi J, Kootte RS, Bartelsman JF et al (2012) Transfer of intestinal microbiota from lean donors increases insulin sensitivity in individuals with metabolic syndrome. Gastroenterology 143(4):913–916
9. Qin J, Li Y, Cai Z., Li S, Zhu J, Zhang F, … Peng Y (2012) A metagenome-wide association study of gut microbiota in type 2 diabetes. Nature 490(7418):55–last page
10. Karlsson FH, Tremaroli V, Nookaew I, Bergström G, Behre CJ, Fagerberg B et al (2013) Gut metagenome in European women with normal, impaired and diabetic glucose control. Nature 498(7452):99
11. Wu H, Esteve E, Tremaroli V, Khan MT, Caesar R, Mannerås-Holm L et al (2017) Metformin alters the gut microbiome of individuals with treatment-naive type 2 diabetes, contributing to the therapeutic effects of the drug. Nat Med 23(7):850
12. Tilg H, Kaser A (2011) Gut microbiome, obesity, and metabolic dysfunction. J Clin Invest 121(6):2126–2132
13. Tremaroli V, Bäckhed F (2012) Functional interactions between the gut microbiota and host metabolism. Nature 489(7415):242
14. Diamant M, Blaak EE, De Vos WM (2011) Do nutrient–gut–microbiota interactions play a role in human obesity, insulin resistance and type 2 diabetes? Obes Rev 12(4):272–281
15. Everard A, Cani PD (2013) Diabetes, obesity and gut microbiota. Best Pract Res Clin Gastroenterol 27(1):73–83
16. Devaraj S, Hemarajata P, Versalovic J (2013) The human gut microbiome and body metabolism: implications for obesity and diabetes. Clin Chem 59(4):617–628
17. Kootte RS, Vrieze A, Holleman F, Dallinga-Thie GM, Zoetendal EG, de Vos WM et al (2012) The therapeutic potential of manipulating gut microbiota in obesity and type 2 diabetes mellitus. Diabetes Obes Metab 14(2):112–120
18. Mingqian H, Shi B (2017) Gut microbiota as a potential target of metabolic syndrome: the role of probiotics and prebiotics. Cell Biosci 7(1):54
19. Delzenne NM, Cani PD (2011) Gut microbiota and the pathogenesis of insulin resistance. Curr Diab Rep 11(3):154
20. Makiura N, Ojima M, Kou Y, Furuta N, Okahashi N, Shizukuishi S, Amano A (2008) Relationship of Porphyromonas gingivalis with glycemic level in patients with type 2 diabetes following periodontal treatment. Mol Oral Microbiol 23(4):348–351
21. Aemaimanan P, Amimanan P, Taweechaisupapong S (2013) Quantification of key periodontal pathogens in insulin-dependent type 2 diabetic and non-diabetic patients with generalized chronic periodontitis. Anaerobe 22:64–68
22. Thorstensson H, Dahlen G, Hugoson A (1995) Some suspected periodontopathogens and serum antibody response in adult long-duration insulin-dependent diabetics. J Clin Periodontol 22(6):449–458
23. Kazi S (2014) Use of traditional plants in diabetes mellitus. Intl J Pharmaceut 4(4):283–289
24. Bathaie SZ, Mokarizade N, Shirali S (2012) An overview of the mechanisms of plant ingredients in the treatment of diabetes mellitus. J Med Plants 4(44):1–24
25. Hui H, Zhao X, Perfetti R (2005) Structure and function studies of glucagon-like peptide-1 (GLP-1): the designing of a novel pharmacological agent for the treatment of diabetes. Diabetes Metab Res Rev 21(4):313–331
26. Kooti W, Moradi MT, Ali-Akbari S, Sharafi-Ahvazi N, Asadi-Samani M, Ashtary-Larky D (2014) Therapeutic and pharmacological potential of Foeniculum vulgare Mill: a review. J Herb Med Pharmacol:4

27. Afrisham R, Aberomand M, Ghaffari MA, Siahpoosh A, Jamalan M (2015) Inhibitory effect of Heracleum persicum and *Ziziphus jujuba* on activity of alpha-amylase. J Bot 2015:1
28. Sies H, Cadenas E (1985) Oxidative stress: damage to intact cells and organs. Phil Trans R Soc Lond B 311(1152):617–631
29. Johansen JS, Harris AK, Rychly DJ, Ergul A (2005) Oxidative stress and the use of antioxidants in diabetes: linking basic science to clinical practice. Cardiovasc Diabetol 4(1):5
30. Bansal AK, Bilaspuri GS (2011) Impacts of oxidative stress and antioxidants on semen functions. Vet Med Intl 2011:1–7
31. Pham-Huy LA, He H, Pham-Huy C (2008) Free radicals, antioxidants in disease and health. Intl J Biomed Sci IJBS 4(2):89
32. Lipinski B (2001) Pathophysiology of oxidative stress in diabetes mellitus. J Diabetes Complicat 15(4):203–210
33. Maritim AC, Sanders A, Watkins J (2003) Diabetes, oxidative stress, and antioxidants: a review. J Biochem Mol Toxicol 17(1):24–38
34. Moussa SA (2008) Oxidative stress in diabetes mellitus. Romanian J Biophys 18(3):225–236
35. Erejuwa OO (2012) Oxidative stress in diabetes mellitus: is there a role for hypoglycemic drugs and/or antioxidants. Oxidative Stress Diseases 217:246
36. Larsen N, Vogensen FK, van den Berg FW, Nielsen DS, Andreasen AS, Pedersen BK et al (2010) Gut microbiota in human adults with type 2 diabetes differs from non-diabetic adults. PLoS One 5(2):e9085
37. Zhang X, Shen D, Fang Z, Jie Z, Qiu X, Zhang C, Ji L (2013) Human gut microbiota changes reveal the progression of glucose intolerance. PLoS One 8(8):e71108
38. Ceolotto G, Bevilacqua M, Papparella I, Baritono E, Franco L, Corvaja C, Avogaro A (2004) Insulin generates free radicals by an NAD (P) H, phosphatidylinositol 3′-kinase-dependent mechanism in human skin fibroblasts ex vivo. Diabetes 53(5):1344–1351
39. Levy J, Gavin JR, Sowers JR (1994) Diabetes mellitus: a disease of abnormal cellular calcium metabolism? Am J Med 96(3):260–273
40. Wohaieb SA, Godin DV (1987) Alterations in free radical tissue-defense mechanisms in streptozocin-induced diabetes in rat: effects of insulin treatment. Diabetes 36(9):1014–1018
41. Sardesai VM (1995) Role of antioxidants in health maintenance. Nutr Clin Pract 10(1):19–25
42. Gibaldi M (1996) Antioxidant vitamins and health. J Clin Pharmacol 36(12):1093–1099
43. Gökkus C, Palanduz S, Ademoğlu E, Tamer S (2001) Oxidant and antioxidant systems in NIDDM patients: influence of vitamin E supplementation. Endocr Res 27(3):377–386
44. Garcia-Medina JJ, Pinazo-Duran MD, Garcia-Medina M, Zanon-Moreno V, Pons-Vazquez SA (2011) A 5-year follow-up of antioxidant supplementation in type 2 diabetic retinopathy. Eur J Ophthalmol 21(5):637–643
45. Perez C, Canal JR, Torres MD (2003) Experimental diabetes treated with ficus carica extract: effect on oxidative stress parameters. Acta Diabetol 40(1):3–8
46. Welihinda J, Karunanayake EH (1986) Extra-pancreatic effects of Momordica charantia in rats. J Ethnopharmacol 17(3):247–255
47. Song F, Chen W, Jia W, Yao P, Nussler AK, Sun X, Liu L (2006) A natural sweetener, Momordica grosvenori, attenuates the imbalance of cellular immune functions in alloxan-induced diabetic mice. Phytother Res 20(7):552–560
48. Waisundara VY, Hsu A, Huang D, Tan BKH (2008) Scutellaria baicalensis enhances the antidiabetic activity of metformin in streptozotocin-induced diabetic Wistar rats. Am J Chin Med 36(03):517–540
49. Xi M, Hai C, Tang H, Chen M, Fang K, Liang X (2008) Antioxidant and antiglycation properties of total saponins extracted from traditional Chinese medicine used to treat diabetes mellitus. Phytother Res 22(2):228–237
50. Resmi CR, Venukumar MR, Latha MS (2006) Antioxidant activity of Albizzia lebbeck (Linn.) Benth. in alloxan diabetic rats. Indian J Physiol Pharmacol 50(3):297
51. Luo Q, Cai Y, Yan J, Sun M, Corke H (2004) Hypoglycemic and hypolipidemic effects and antioxidant activity of fruit extracts from *Lycium barbarum*. Life Sci 76(2):137–149

52. Li XM (2007) Protective effect of *Lycium barbarum* polysaccharides on streptozotocin-induced oxidative stress in rats. Int J Biol Macromol 40(5):461–465
53. Wu H, Guo H, Zhao R (2006) Effect of *Lycium barbarum* polysaccharide on the improvement of antioxidant ability and DNA damage in NIDDM rats. Yakugaku Zasshi 126(5):365–371
54. Fadzelly AM, Asmah R, Fauziah O (2006) Effects of Strobilanthes crispus tea aqueous extracts on glucose and lipid profile in normal and streptozotocin-induced hyperglycemic rats. Plant Foods Hum Nutr 61(1):6–11
55. Huseini HF, Larijani B, Heshmat RA, Fakhrzadeh H, Radjabipour B, Toliat T, Raza M (2006) The efficacy of *Silybum marianum* (L.) Gaertn. (silymarin) in the treatment of type 2 diabetes: a randomized, double-blind, placebo-controlled, clinical trial. Phytother Res 20(12):1036–1039
56. Xiong S, Melton LD, Easteal AJ, Siew D (2006) Stability and antioxidant activity of black currant anthocyanins in solution and encapsulated in glucan gel. J Agric Food Chem 54(17):6201–6208
57. Wu FH, Liang JY, Yu P, Cai SF (2005) Studies on the hypoglycemia and lipids regulating effects of Plantago depressa var. montata. Zhongguo Zhong yao za zhi= Zhongguo zhongyao zazhi China J Chinese materia medica 30(15):1179–1183
58. Soon YY, Tan BKH (2002) Evaluation of the hypoglycemic and anti-oxidant activities of Morinda officinalis in streptozotocin-induced diabetic rats. Singap Med J 43(2):077–085
59. Kim HK, Kim MJ, Cho HY, Kim EK, Shin DH (2006) Antioxidative and anti-diabetic effects of amaranth (Amaranthus esculentus) in streptozotocin-induced diabetic rats. Cell Biochem Funct 24(3):195–199
60. Sezik E, Aslan M, Yesilada E, Ito S (2005) Hypoglycaemic activity of Gentiana olivieri and isolation of the active constituent through bioassay-directed fractionation techniques. Life Sci 76(11):1223–1238
61. Orhan DD, Aslan M, Aktay G, Ergun E, Yesilada E, Ergun F (2003) Evaluation of hepatoprotective effect of Gentiana olivieri herbs on subacute administration and isolation of active principle. Life Sci 72(20):2273–2283
62. Xie ZC, Qian ZK, Liu ZW (1993) Effect of ginseng on antiperoxidate injury in myocardium and erythrocytes in streptozocin-induced diabetic rats. Zhongguo Zhong xi yi jie he za zhi Zhongguo Zhongxiyi jiehe zazhi Chinese J Integr Trad Western Med 13(5):289–290
63. Anwar MM, Meki ARM (2003) Oxidative stress in streptozotocin-induced diabetic rats: effects of garlic oil and melatonin. Comp Biochem Physiol A Mol Integr Physiol 135(4):539–547
64. Raphael KR, Sabu MC, Kuttan R (2002) Hypoglycemic effect of methanol extract of Phyllanthus amarus Schum & Thonn on alloxan induced diabetes mellitus in rats and its relationship with antioxidant potential. Indian J Exp Biol 40(8):905–909
65. Patel DK, Prasad SK, Kumar R, Hemalatha S (2012) An overview on antidiabetic medicinal plants having insulin mimetic property. Asian Pac J Trop Biomed 2(4):320–330
66. Feshani AM, Kouhsari SM, Mohammadi S (2011) Vaccinium arctostaphylos, a common herbal medicine in Iran: molecular and biochemical study of its antidiabetic effects on alloxan-diabetic Wistar rats. J Ethnopharmacol 133(1):67–74
67. Feng CG, Zhang LX, Liu X (2005) Progress in research of aldose reductase inhibitors in traditional medicinal herbs. Zhongguo Zhong yao za zhi Zhongguo zhongyao zazhi China J Chin Materia Medica 30(19):1496–1500
68. Yoshikawa M, Nishida N, Shimoda H, Takada M, Kawahara Y, Matsuda H (2001) Polyphenol constituents from Salacia species: quantitative analysis of mangiferin with alpha-glucosidase and aldose reductase inhibitory activities. Yakugaku zasshi J Pharmaceut Soc Japan 121(5):371–378
69. Thakur AK, Shakya A, Husain GM, Emerald M, Kumar V (2014) Gut-microbiota and mental health: current and future perspectives. J Pharmacol Clin Toxicol 2(1):1016
70. Khin-Maung-U MK, Nyunt-Nyunt-Wai AK (1985) Clinical trial of berberine in acute watery diarrhoea. Br Med J (Clin Res Ed) 291(6509):1601
71. Rabbani GH, Butler T, Knight J, Sanyal SC, Alam K (1987) Randomized controlled trial of berberine sulfate therapy for diarrhea due to enterotoxigenic Escherichia coli and Vibrio cholerae. J Infect Dis 155(5):979–984

72. Tang J, Feng Y, Tsao S, Wang N, Curtain R, Wang Y (2009) Berberine and Coptidis rhizoma as novel antineoplastic agents: a review of traditional use and biomedical investigations. J Ethnopharmacol 126(1):5–17
73. Zhang Y, Li X, Zou D, Liu W, Yang J, Zhu N, Ren G (2008) Treatment of type 2 diabetes and dyslipidemia with the natural plant alkaloid berberine. J Clin Endocrinol Metabol 93(7):2559–2565
74. Zhang X, Zhao Y, Zhang M, Pang X, Xu J, Kang C, Li X (2012) Structural changes of gut microbiota during berberine-mediated prevention of obesity and insulin resistance in high-fat diet-fed rats. PLoS One 7(8):e42529
75. Xu J, Lian F, Zhao L, Zhao Y, Chen X, Zhang X, Pang X (2015) Structural modulation of gut microbiota during alleviation of type 2 diabetes with a Chinese herbal formula. ISME J 9(3):552
76. Peluso I, Villaño Valencia D, Chen CYO, Palmery M (2018) Antioxidant, anti-inflammatory, and microbial-modulating activities of nutraceuticals and functional foods. Oxidative Med Cell Longev 2018:1

Modulation of the Host-Parasite Redox Metabolism to Potentiate Antimalarial Drug Efficiency

27

Federica Turrini, Raffaella Boggia, Paola Zunin, and Francesco Michelangelo Turrini

Abstract

Artemisinin-based combination therapy (ACT) is nowadays the most effective treatment for *P. falciparum* malaria: artemisinin is the most active drug able to rapidly kill all erythrocyte stages of the malaria parasite. However, due to its short half-life, it requires the association with other long-acting drugs. Even if the exact mechanism of action of most antimalarial drugs is still unknown, many of these compounds are able to interact directly or indirectly with the redox metabolism of the parasite and/or the host, enhancing the effectiveness of the antimalarial therapy. This review focuses on many natural compounds, isolated mainly from plants, and used as traditional antimalarial treatments, known to possess a potent antimalarial activity (IC50 lower than 1 ug/mL). These compounds belong to some specific chemical family, mainly alkaloids, terpenoids, quassinoids, limonoids, and polyphenols, sharing some common chemical features. These natural molecules could offer new possibilities of combination therapies development as antimalarials when associated with artemisinin.

Keywords

Antimalarial drugs · Artemisinin-based combination therapy · Natural antimalarial compounds · Host-parasite redox metabolism

F. Turrini (✉) · R. Boggia · P. Zunin
Department of Pharmacy, University of Genoa, Genoa, Italy
e-mail: turrini@difar.unige.it

F. M. Turrini
Department of Genetics, Biology and Biochemistry, University of Turin, Turin, Italy

S. Chakraborti et al. (eds.), *Oxidative Stress in Microbial Diseases*,
https://doi.org/10.1007/978-981-13-8763-0_27

27.1 Introduction

27.1.1 Current Antimalarial Therapy

The natural antimalarial product artemisinin and its semisynthetic derivatives represent the frontline treatment of *P. falciparum* malaria, as they are the most active antimalarials available, rapidly killing all blood stages of the malaria parasite. Artemisinins contain an endoperoxide bridge which plays a key role in the antimalarial activity with a mode of action starting from radical transient species initiated by the cleavage of this bridge. Few other natural compounds with such a peroxide bridge are known. On the other hand, the oxygen-oxygen bridge, being chemically unstable, determines a very short plasma half-life, constituting a major limiting factor to the use of artemisinin as a single drug [1]. To solve this problem, artemisinin was used early on in combination with partner drugs characterized by much longer half-life. Artemisinin-based combination therapy (ACT) is the most effective treatment for *P. falciparum* malaria. Artemisinin derivatives such as dihydroartemisinin, artesunate, and artemether are combined with a partner drug such as lumefantrine, mefloquine, amodiaquine, and piperaquine. Since the introduction of artemisinin-combination therapies (ACTs), the overall number of malaria cases displayed a marked decline, but since the last few years, the rate of decline has stalled or even reversed in some regions [2].

The reasons of the recent increase of the number of malaria cases are plausibly multifactorial, including insufficient investments for treatment and prevention, insecticide resistance, and antimalarial drug resistance. The relative role of each factor is undefined.

In the Greater Mekong subregion, artemisinin resistance raised concern and is currently defined as "partial artemisinin resistance" in patients showing a delayed parasite clearance following treatment with an ACT. Notably, in the same region, resistance to the partner drugs is present. To rule out between artemisinin and partner drugs resistance in the development, treatment failure is obviously very difficult. Currently, no evidence of artemisinin resistance has been observed in African countries accounting for about 90% of malaria cases and deaths worldwide.

27.2 Interactions of Antimalarial Drugs with Host-Parasite Redox Homeostasis

27.2.1 Antimalarial Drugs Showing Redox Activity

Although the precise mechanism of action of most of the antimalarial drugs is still unknown, most of the antimalarial drugs have the potential of interacting directly or indirectly with redox metabolism of the parasite and/or of the host.

A direct redox effect exerted by some antimalarial drugs on the host cells is clearly evidenced as hemolysis (oxidative damage and rapid destruction of erythrocytes leading to variable degrees of anemia) in G6PD-deficient individuals. Powerful antimalarial drugs such as primaquine, methylene blue, and sulfonamides cause acute and severe hemolytic anemia in G6PD-deficient subjects [3, 4]. In addition, popular antimalarial drugs and their combinations such as halofantrine, quinine, chloroquine, and chlorproguanil-dapsone have been associated with variable degrees of hemolytic anemia, generation of ROS, and depletion of erythrocyte GSH [5, 6]. Artemisinin and its derivatives also cause delayed hemolysis. The central role of the endoperoxide bridge of artemisinin and the generation of free radicals following its cleavage has been clearly established. Artemisinin activation needs iron provided by the host cells, resulting in the rapid generation of free radicals and the formation of heme-artemisinin adducts. To explain its high activity (IC50 ≈ 2 nM), specific molecular targets are expected to play a role in its mechanism of action.

27.2.2 Antimalarial Drugs Causing the Accumulation of Free Heme Through the Inhibition of Hemozoin Synthesis

Many antimalarial drugs including chloroquine, lumefantrine, mefloquine, amodiaquine, and piperaquine show the capability to inhibit heme polymerization to form hemozoin, an inert crystal, in a specialized digestive vacuole of *P. falciparum* [7, 8]. Heme detoxification and its polymerization constitute a central step of the parasite metabolism, and its inhibition leads to parasite death. On the other hand, free iron released during hemoglobin digestion and heme constitute a powerful source of free radicals needing to be neutralized by both parasite and erythrocyte enzymes [9, 10]. Although the chemical and metabolic interactions occurring between artemisinin derivatives and heme polymerization inhibitory drugs are scarcely understood, it is interesting to notice that the drugs that inhibit hemoglobin metabolism are the best candidates to be utilized in combination with artemisinin in ACTs [11] suggesting a combined mechanism of action [12, 13].

27.3 Response of the Host Cell to Redox Changes Exerted by Parasite Growth and/or Antimalarial Drugs

Redox metabolism of the parasitized erythrocyte depends on the equilibrium between the antioxidant defenses of both erythrocyte and parasite and on the free radicals produced by the parasite and by the erythrocyte [14]. Iron plays a central role in free radical production in the parasite through hemoglobin digestion and in the parasitized erythrocyte which accumulates large amounts of denatured hemoglobin generated by the oxidative stress exerted by the parasite [15, 16].

As an evidence of this unstable equilibrium, a large number of mutations have been selected by malaria such as G6PD deficiency (defect of production of NADPH), sickle cell anemia, thalassemias, HbC, HbE, and many other hemoglobin disorders triggering oxidative stress through hemoglobin denaturation and plausibly amplifying the redox stress exerted by the intracellular parasite. More than 500 million people living in malaria endemic areas are affected by one or more of those mutations conferring variable levels of protection to severe malaria [17, 18].

It has been demonstrated that mild oxidative stress exerted by malaria parasite activates a specific redox signaling pathway in the host erythrocyte inducing the Tyr phosphorylation of some membrane proteins [19] which, in turn, appear to be essential for the remodeling of the parasitized erythrocyte membrane needed to activate new functions essential for the parasite growth such as import of nutrients and insertion of parasite proteins binding to endothelium [20]. The activation of Syk kinase, constituting a key element of such pathway, has been directly implicated both in the egress of merozoites and in the infection of new erythrocytes occurring after the 48-hour life cycle of the parasite. A new class of drugs has been developed to interfere with this pathway revealing good antiplasmodial efficacy [21].

Targeting the redox metabolism of *P. falciparum* by antimalarial drugs is believed to create an overload of oxidative stress leading to parasite death [22]. Anyway, much more complex mechanism of action appears to be involved in the interactions occurring between parasite and host cell metabolism and in the interference exerted by a large number of redox active antimalarial drugs (Fig. 27.1).

27.4 Natural Antimalarial Compounds

In addition to artemisinin, a large number of natural compounds present in plants, are known to possess a potent antimalarial activity. In some cases, their clinical efficiency and activity have been demonstrated by their use as traditional antimalarial treatments.

A comprehensive list of plant compounds with antimalarial activity, quantified as in vitro IC50, is reported in Table 27.1. Only the compounds whose IC50 is lower than 1 ug/mL have been selected.

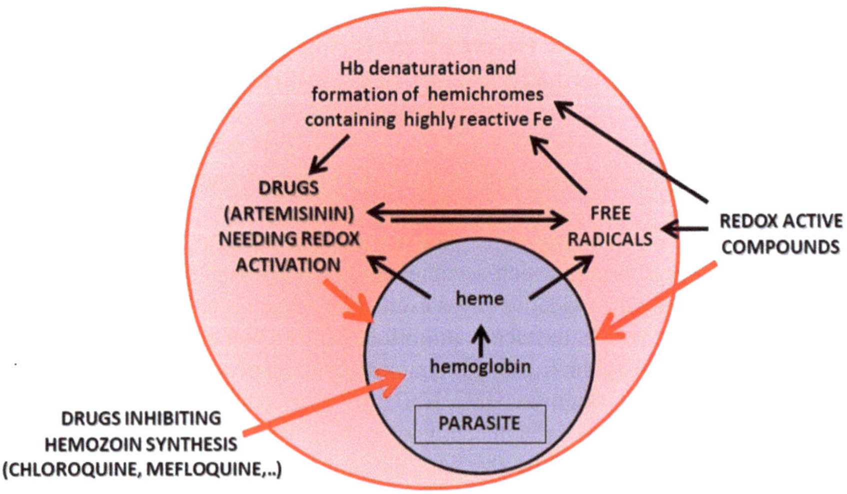

Fig. 27.1 Redox homeostasis perturbation between host and parasite

Table 27.1 Summary of antimalarial natural compounds selected for their activity "in vitro" (IC50 < 1 ug/mL)

Natural compound	CAS number	IUPAC name	Chemical structure	Chemical group	Botanical source	Anti-malarial activity	IC50 (μg/mL) and strains used	Potential mechanism of action	Other antiparasitic activities
10 -Hydroxyusambarine	n.a	(2S,3R,12bS)-3-ethenyl-2-[[(1S)-2-methyl-1,3,4,9-tetrahydropyrido[3,4-b]indol -1-yl]methyl]-1,2,3,4,6,7,12, 12b-octahydroindolo[2,3-a] quinolizin-9-ol	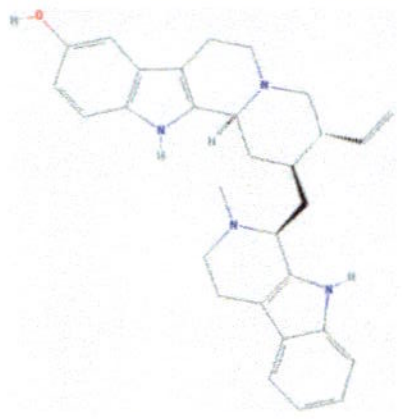	Indole alkaloids	*Strychnos usambarensis*	In vitro [32]	0.48 (FCA 20) 0.16 (W2)	Not discussed	Activities against *Entamoeba histolytica* In vitro [33]
Akuammicine	639-43-0	2,16,19,20-tetradehydrocuran -17-oic acid methyl ester	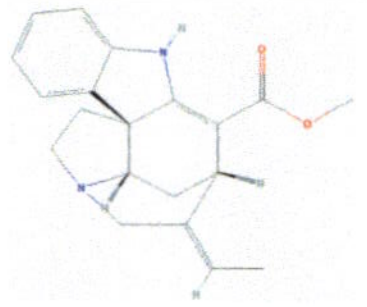	Indole alkalcids	*Picralima nitida*	In vitro [34]	0.45 (D6) 0.73 (W2)	Not discussed	
Akuammine	3512-87-6	Methyl (1S,9S,14Z,15R)-14-ethylidene-6-hydroxy-2-methyl-18-oxa-2,12-diazahexacy clo[13.3.2.01,9.03,8.09,16.012, 19] icosa-3,5,7-triene-16-carboxylate	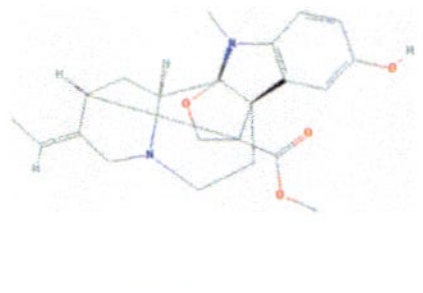	Indole alkaloids	*Picralima nitida*	In vitro [34]	0.95 (D6) 0.66 (W2)	Not discussed	
Alstonine	642-18-2	(19α,20α)-16-(Methoxycarbonyl)-19-methyl-3,4,5,6,16,17-hexadehydro-18-oxayohimban-4-ium	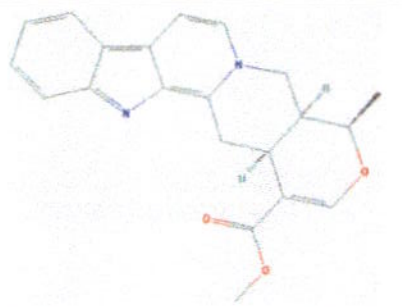	Indole alkaloids	*Picralima nitida* *Alstonia scholaris*	In vitro [34] In vivo [35]	0.017 (D6) 0.038 (W2)	Not discussed	

(continued)

Table 27.1 (continued)

Natural compound	CAS number	IUPAC name	Chemical structure	Chemical group	Botanical source	Anti-malarial activity	IC50 (μg/mL) and strains used	Potential mechanism of action	Other antiparasitic activities
Berberine	633-66-9	5,6-dihydro-9,10-dimethoxybenzo[g]-1,3-benzodioxolo[5,6-a] quinolizinium	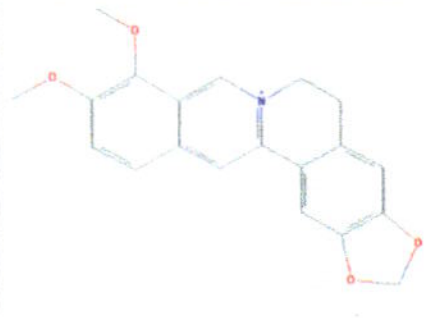	Indole alkaloids	*Enantia Chlorantha*	In vitro [36] In vitro and in vivo [37]	0.14 (D6) 0.15 (W2)	Inhibitor of both nucleic acid and protein synthesis in human malarial parasite, *P. falciparum*, FCR-3	Antileishmanial activity [38, 39]
Cryptolepine	480-26-2	5-methylindolo[3,2-b] quinoline	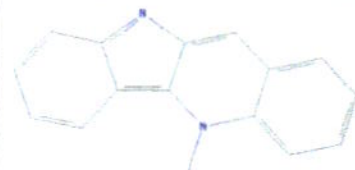	Indole alkaloids	*Quassia indica*	[23]	0.19 (K1) 0.05 (T996)	Not discussed	
Fagaronine	52259-65-1	3,8,9-trimethoxy-5-methy lbenzo[c]phenanthridin-5 -ium-2-ol	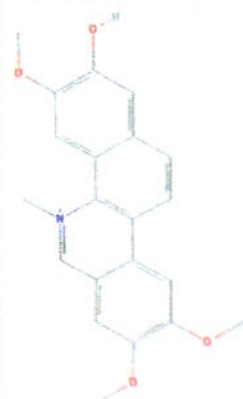	Indole alkaloids	*Fagara zanthoxyloides*	In vitro [40]	0.018 (3D7)	Not discussed	
Jatrorrhizine	3621-38-3	2,9,10-trimethoxy-5,6-dihydroisoquinolino[2,1-b] isoquinolin-7-ium-3-ol		Indole alkaloids	*Enantia Chlorantha Penianthus longifolius*	In vitro [41] In vitro and in vivo [42]	0.42 (D6) 1.61 (W2)	Not discussed	

Palmatine	3486-67-7	2,3,9,10-tetramethoxy-5,6-dihydroisoquinolino[2,1-b] isoquinolin-7-ium	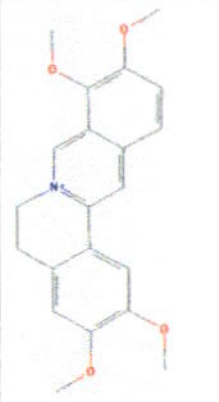	Indole alkaloids	*Enantia Chlorantha* *Penianthus longifolius*	In vitro [41;43] In vitro and in vivo [42]	0.28 (D6) 0.16 (W2)	Not discussed	Antibacterial activity against *Clostridium perfringens* [44] Antiflavivirus activity In vitro [43]
Picraline	2671-32-1	Methyl (1R,9S,11S,14Z,15S,17S,19S)-19-[(acetyloxy)methyl]-14-ethylidene-18-oxa-2,12-diazahexa cyclo[9.6.1.1^{9,15}0.0^{1,9}0.0 ^{3,8}0.0^{12,17}] nonadeca-3,5,7-triene-19-carboxylate	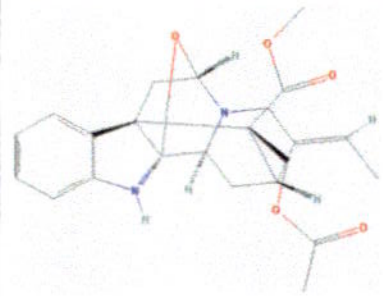	Indole alkaloids	*Picralima nitida*	[45]	0.44 (D6) 0.53 (W2)	Not discussed	
Ψ-Akuammigine	642-17-1	(7aS,8S,11aS,12aR)-8-Methyl-5,6,7a,8,11a,12,12a,13-octahydro-7H-9-oxa-6a,13-diaza-indeno[2,1-a] anthracene-11-carboxylic acid methyl ester	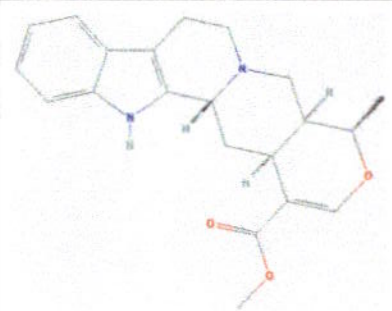	Indole alkaloids	*Picralima nitida*	[45]	0.42 (D6) 0.10 (W2)	Not discussed	

(continued)

Table 27.1 (continued)

Natural compound	CAS number	IUPAC name	Chemical structure	Chemical group	Botanical source	Anti-malarial activity	IC50 (μg/mL) and strains used	Potential mechanism of action	Other antiparasitic activities
2,6-Dihydroxyfissinolide	n.a	Methyl 2-[(1R,2S,5R,6R,13R,14S,16S)-14-(acetyloxy)-6-(furan-3-yl)-13-hydroxy-1,5,15,15-tetramethyl-8,17-dioxo-7-oxatetracy clo[11.3.1.0^{2,11}0.0^{5,10}] heptadec-10-en-16-yl]-2-hydroxyacetate	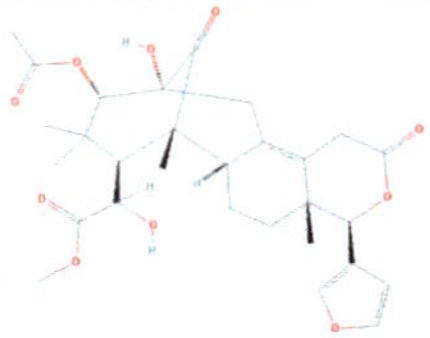	Limonoids	*Khaya senegalensis*	In vitro [46]	0.12 (chloroquine-sensitive (3D7) *P. Falciparum* strain)	Not discussed	IC50 > 0.20 mM against *Leishmania major* In vitro [46]
Gedunin	2753-30-2	(4aR,6R,6aS,6bR,7aS,10S,10aS,12aR,12bR)-10-(3-Furyl)-4,4,6a,10a,12b-pentamethyl-3,8-dioxo-3,4,4a,5,6,6a,7a,8,10,10a,11,12,12a,12b-tetradecahydronaphtho[2,1-f] oxireno [d]isochromen-6-yl acetate	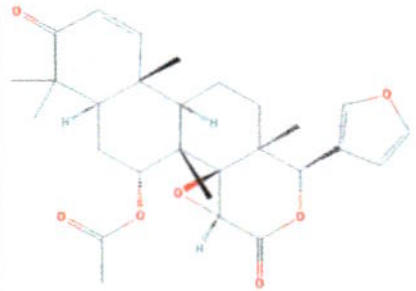	Limonoids	*Azadirachta indica* *Cedrela odorata* *Khaya grandifoliola*	In vitro [47, 48] In vivo [49]	0.03 (D6) 0.02 (W2) 1.25 (W2-Indochina clone)	Not discussed	Insect antifeedant activity In vivo [50]

Ancistrocladidine	52659-52-6	2-[(3S)-6,8-dimethoxy-1,3-dimethyl-3,4-dihydroisoquinolin-7-yl]-8-methoxy-3-methylnaphthalen-1-ol	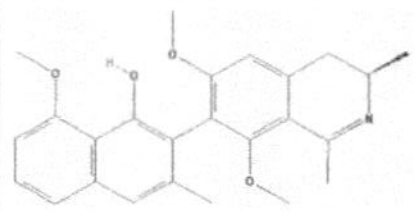	Naphthoisoquinolines	*Ancistrocladustanzaniensis*	In vitro [51]	0.3 (K1) 1.9 (3D7)	Not discussed	Antileishmanial activity against *Leishmania donovani* (IC50 = 2.9 μg/mL) Antitrypanosomal activity against *Trypanosoma brucei rhodesiense* (IC50 = 2.0 μg/mL) and *T. cruzi* (IC50 = 23.4 μg/mL) In vitro [51]
Ancistrotanzanine C	692755-31-0	(1R,3S)-7-(1-hydroxy-8-methoxy-3-methylnaphthalen-2-yl)-8-methoxy-1,2,3-trimethyl-3,4-dihydro-1H-isoquinolin-6-ol	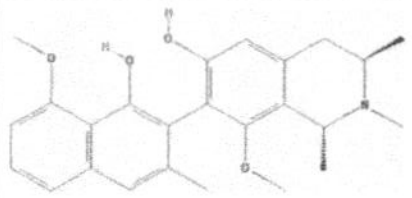	Naphthoisoquinolines	*Ancistrocladustanzaniensis*	In vitro [51]	0.1 (K1) 4.2 (3D7)	Not discussed	Antitrypanosomal activity against *Trypanosoma brucei rhodesiense* (IC50 = 1.3 μg/mL) and *T. cruzi* (IC50 = 14 μg/mL) In vitro [51]

(continued)

Table 27.1 (continued)

Natural compound	CAS number	IUPAC name	Chemical structure	Chemical group	Botanical source	Anti-malarial activity	IC50 (μg/mL) and strains used	Potential mechanism of action	Other antiparasitic activities
Ancistrotectorine	98985-59-2	2-[(1R,3S)-6,8-dimethoxy-1,2,3-trimethyl-3,4-dihydro-1H-isoquinolin-7-yl]-8-methoxy-3-methylnaphthalen-1-ol	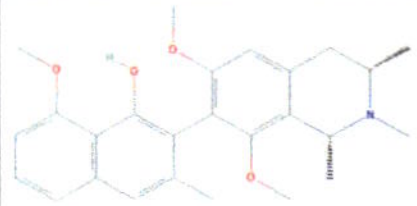	Naphthoisoquinolines	*Ancistroc ladustan zaniensis*	In vitro [51]	0.7 (K1) 9.1 (3D7)	Not discussed	Antitrypanosomal activity against *Trypanosoma brucei rhodesiense* (IC50 = 4.3 μg/mL) And T. cruzi (IC50 = 4.3 μg/mL) In vitro [51]
Dioncopeltine A	n.a	(1R,3R)-7-[5-hydroxy-2-(hydroxymethyl)-4-methoxynaphthalen-1-yl]-1,3-dimethyl-1,2,3,4-tetrahydroisoquinolin-8-ol	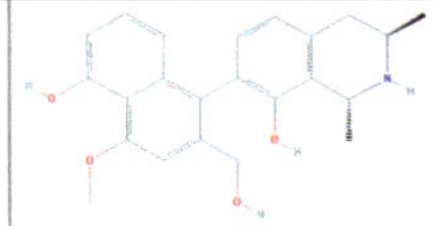	Naphthoisoquinolines	*Triphyoph yllumpe ltatum*	In vitro and in vivo [52, 53]	0.0048 (K1) 0.0033 (NF54)	Not discussed	Antileishmanial activity against *Leishmania major* [54]
Dioncophylline C	n.a	(1R,3R)-5-(5-hydroxy-4-methoxy-2-methylnaphthalen-1-yl)-1,3-dimethyl-1,2,3,4-tetrahydroisoquinolin-8-ol	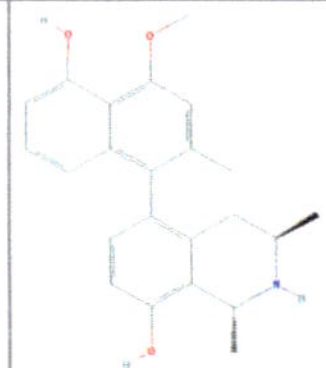	Naphthoisoquinolines	*Triphyoph yllumpe ltatum*	In vitro and in vivo [52, 53]	0.014 (blood form of *P. falciparum*) 0.015 (blood form of *P. berghei*)	Not discussed	Cytotoxicity against cloned virulent *Leishmania major* promastigotes [54]
Habropetaline A	n.a	(1R,3R)-7-[2-(hydroxymethyl)-4,5-dimethoxynaphthalen-1-yl]-1,3-dimethyl-1,2,3,4-tetrahydroisoquinolin-8-ol	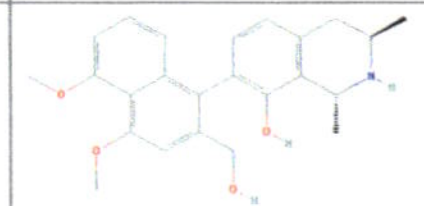	Naphthoisoquinolines	*Triphyoph yllumpe ltatum*	In vitro [55]	0.005 (K1) 0.0023 (NF54)	Not discussed	

Ellagic acid	476-66-4	2,3,7,8-Tetrahydroxy chromeno[5,4,3-cde] chromene-5,10-dione	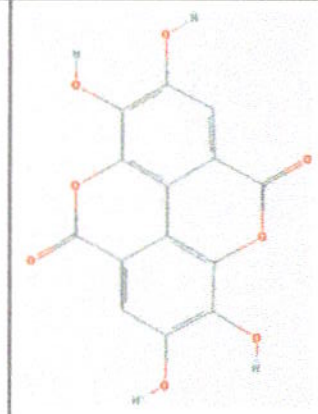	Polyphenoliccompounds	*Alchornea cordifolia*	In vitro [56]	0.10 (F32) 0.03(Dd2) 0.09 (FcB1) 0.10 (W2) 0.05(FcM29)	Inhibition of the β-hematin formation [31]	
Ailanthone	981-15-7	(1β,11β,12α)-1,11,12-Trihydroxy -11,20-epoxypicrasa-3,13(21)-diene-2,16-dione	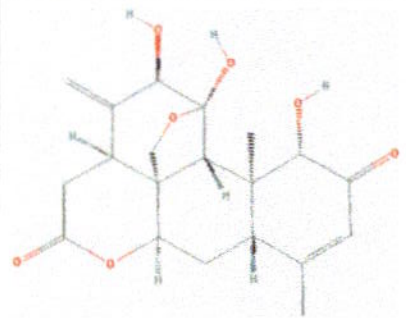	Quassinoids	*Ailanthus altissima*	In vitro [57]	0.003	Not discussed	
6alpha-Tigloyloxychaparrinone	n.a.		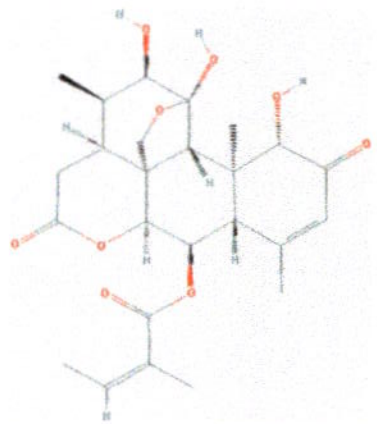	Quassinoids	*Ailanthus altissima*	In vitro [57]	0.061	Not discussed	

(continued)

Table 27.1 (continued)

Natural compound	CAS number	IUPAC name	Chemical structure	Chemical group	Botanical source	Anti-malarial activity	IC50 (μg/mL) and strains used	Potential mechanism of action	Other antiparasitic activities
Simalikalactone D	35321-80-3	(1β,11β,12α,15βb)-1,11,12-trihydroxy-2,16-dioxo-13,20-epoxypicrasa-3-en-15-yl (2R)-2-methylbutanoate	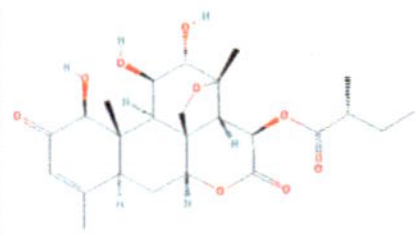	Quassinoids	*Simaba orinocensis*	In vitro [58]	0.003 (D6) 0.00367 (W2)	Inhibitor of the protein biosynthesis in vitro in translation system from the Krebs cell	Antileishmanial activity against *Leishmania donovani* [58]
Tagitinin C	59979-56-5	[(3aR,4R,6R,7E,10Z,11aR)-6-hydroxy-6,10-dimethyl-3-methylidene-2,9-dioxo-3a,4,5,11a-tetrahydrocyclodeca[b] furan-4-yl] 2-methylpropanoate	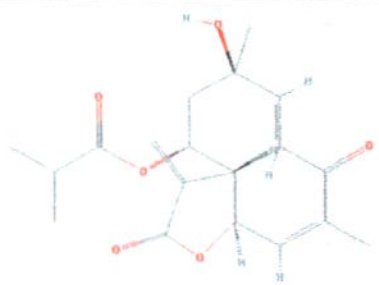	Sesquiterpenes	*Tithonia diversifolia*	In vitro [59]	0.33 (FCA) Cytoxicity IC50 = 0.71 (HTC-116)	Not discussed	
Urospermal A-15-O-acetate	n.a	[(3aR,4R,6E,10Z,11aR)-6-formyl-4-hydroxy-3-methylidene-2-oxo-3a,4,5,8,9,11a-hexahydrocyclodeca[b] furan-10-yl]methyl acetate	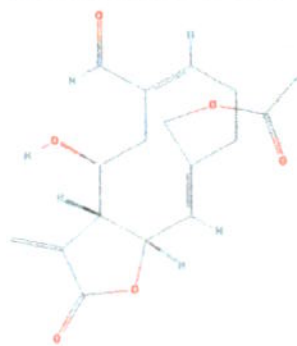	Sesquiterpenes	*Dicoma tomentosa*	In vitro [60]	0.92 (3D7) 0.77 (W2)	Not discussed	
Artemisinin	63968-64-9	(4S,5R,8S,9R,12S,13R)-1,5,9-Trimethyl-11,14,15,16-tetraoxatet racyclo[10.3.1.04,13.08,13] hexadecan-10-one	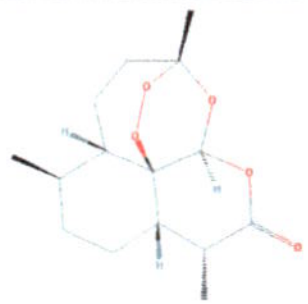	Sesquiterpenes	*Artemisia annua*	In vitro [61]	0.006 (D6 e W2)	Heme-mediated decomposition of the endoperoxide bridge to produce carbon-centered free radicals	In vitro anti-promastigote activity in visceral leishmaniasis In vitro [62]

Plakortin	n.a.	Methyl 2-[(3R,4R,6S)-4-ethyl-6-[(E,2R)-2-ethylhex-3-enyl]-6-methyldioxan-3-yl]acetate	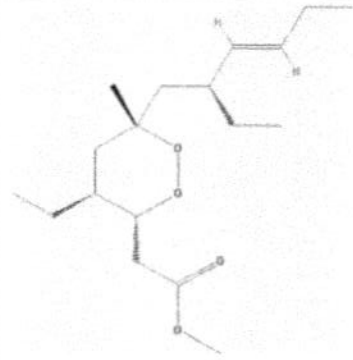	Endoperoxyketal polyketides	*Plakortiscfr. Simplex*	In vitro [63, 64]	0.12 (W2) 0.27 (D10)	ROS generation, lipid peroxidation, and generation of 4-HNE-protein conjugate with essential parasite proteins	Anti-promastigote activity in *Leishmania Chagasi* (IC50 = 6.0 μg/mL) [65]

n.a not available

It shall be noticed that most of these compounds belong to the families of alkaloids, in particular, indole and naphthoisoquinoline alkaloids, with few exceptions including sequiterpenes, quassinoids, limonoids, and a polyphenolic compound (ellagic acid).

27.4.1 Alkaloids

Alkaloids are one of the most important antimalarial compound classes known from ancient time. Quinine, which represents the first antimalarial natural drug used, belongs to this group. It is a quinidine alkaloid isolated from the bark of *Cinchona succirubra* and it has been used for more than three centuries for the treatment of malaria [23]. Numerous studies reported in literature described the significant antimalarial activity of over 100 natural alkaloids, which have been considered more potent than chloroquine [24]. Some of the active reported alkaloids can be also grouped according to their structural chemical classes. In particular, indole alkaloids which from a chemical point of view contain an important indole group in their structure and derive from the amino acid tryptophan, represent one of the largest class of alkaloids grouping together many antimalarial compounds [25]. The naphthoisoquinoline alkaloids represent as well potential antimalarial compounds. These compounds are secondary metabolites naturally present mainly in some plants of the Ancistrocladaceae and Dioncophyllaceae families. Structurally, they are chiral molecules characterized by the presence of C-C and C-N bonds typical of the naphthalene and isoquinoline structures contained in them. In addition to the antimalarial activity, these alkaloids have shown many biological activities including anti-leishmanial, anti-trypanosomal, fungicidal, insecticidal, and larvicidal [26].

27.4.2 Sesquiterpenes

An important medicinal plant used from ancient time because of its content in antimalarial artemisinins is the *Artemisia annua*, an herb used in Chinese traditional medicine. As already mentioned, artemisinin-combination therapies (ACTs) are nowadays the standard treatment in the world against *P. falciparum* [1, 27]. Chemically, artemisinin belongs to sesquiterpenes, namely, terpenes characterized by three isoprene units ($C_{15}H_{24}$), which represent a very important class of secondary metabolites obtained from plants. In addition to artemisinin, several other sesquiterpene compounds have shown a potentially antimalarial activity [25]. Structurally, as already mentioned, artemisinin presents a characteristic endoperoxidic group which is considered responsible for its activity. However, the other sesquiterpene compounds reported in Table 27.1, do not contain the typical peroxide group of artemisinin. Interestingly, this chemical group is found in another antimalarial compound, named Plakortin, recently isolated from a tropical sponge instead of from plants. It is a polyketide endoperoxide isolated from the sponge *Plakortis simplex*, and it has shown antiparasitic activity against *P. falciparum*. It has been shown that the activity of this compound also depends on the functionality of the peroxide [28].

27.4.3 Quassinoids

Quassinoids are related to terpenes group too. They are a class of degraded triterpenes, and most of them are characterized by a C-20 and δ-lactone skeleton [22]. Concerning the quassinoids, in particular, three compounds have reported significant antimalarial activity with very low IC50 values as a result of our free research: Ailanthone and 6-alpha-tigloyloxychaparrinone isolated from *Ailanthus altissima* and simalikalactone D obtained from the roots of *Simabaorinocensis*. The presence of the ester function seems to be important for in vitro antiplasmodial activity of these compounds [29].

27.4.4 Limonoids

Limonoids present a similar structure to quassinoids. These compounds, classed as tetranortriterpenes, present different variations of the furanolactone core structure. *Meliaceae*, *Cucurbitaceae*, and *Rutaceae* are the plant families richer of these phytochemical compounds, some of which have demonstrated different biological properties such as antimalarial activity [23].

27.4.5 Polyphenolic Compounds

As regards the polyphenolic compounds, only the ellagic acid (EA) has shown a very interesting IC50 value. This is an antioxidant molecule which chemically belongs to the hydrolysable tannins class, known for its antimalarial activity. EA is derived from the hydrolysis of ellagitannins (ETs). ETs are hydrolyzed, chemically by acids or bases or enzymatically, into hexahydroxy diphenic acid, which spontaneously tends to EA [30]. EA explains its antiplasmodial activity by interfering with hemoglobin metabolism and, in particular, inhibiting the β-hematin formation [31]. For these reasons, ellagic acid could be a potential candidate to be utilized in combination with artemisinin in ACTs.

27.5 Conclusions

This work has revealed a wide range of natural products with potent antimalarial activity, belonging to some specific chemical group, mainly alkaloids, terpenoids, quassinoids, limonoids, and polyphenols.

Selected compounds from a physicochemical point of view are characterized for larger molecular mass and higher hydrophobicity index (logP). These properties may be considered necessary for the achievement of the intracellular target [66]. Furthermore, most of these compounds are hydrogen bond acceptors. Structurally, there are recurrent chemical groups that share these molecules. Most compounds present a rigid scaffold of at least 3 cycles which confers high rigidity to the molecule. However, further structure-activity relationship studies are necessary for the possible identification of a pattern of functional groups (pharmacophore) responsible for the antimalarial activity.

Many of the selected compounds, including artemisinin, are likely expected to exert redox activity in biological systems, but, in most instances, a direct evidence is missing.

At this regard, it should be considered that the erythrocyte-plasmodium system represents a unique environment combining two interconnected oxidative defense systems, extremely high concentration of heme iron contained in the host compartment and intense hemoglobin digestion and heme detoxification in the parasite [22]. Experiments to test the interference with the host-parasite redox state of any potential antimalarial compounds should be, therefore, tested in infected erythrocytes at different development stages.

It should be also noticed that, the antimalarial mechanism of action of redox compounds is complex involving very sensitive host targets [67] that could be considered for the screening of antimalarial drug candidates.

References

1. Cui L, Su X (2007) Discovery, mechanisms of action and combination therapy of artemisinin. Expert Rev Anti-infective Therapy 7(8):999–1013
2. World Health Organization (2017) World malaria report. WHO, Geneva
3. Gassner C (2012) Aging and clearance of erythrocytes. Transfus Med Hemother 39(5):297–298
4. Luzzatto L, Arese P (2018) Favism and Glucose-6-phosphate dehydrogenase deficiency. N Engl J Med 378(1):1067–1069
5. AnabaF C, Ahiante BO, Pepple DJ (2012) In vitro hemolytic effect of sulfadoxine/pyrimethamine and artemether/lumefantrine on malaria parasitized erythrocytes of female patients. Pak J Pharm Sci 25(4):851–855
6. Chikezie PC, Monago CC, Uwakwe A (2009) Studies of methemoglobin concentrations of three human erythrocyte genotypes (Hb AA, HbAS, and Hb SS) in the presence of five antimalarial drugs. Iranian J Blood Cancer 4:151–157
7. Niles JC, DeRisi JL, Marletta MA (2009) Inhibiting Plasmodium falciparum growth and heme detoxification pathway using heme-binding DNA aptamers. Proc Natl Acad Sci U S A 106(32):13266–13271
8. Abshire JR, Rowlands CJ, Ganesan SM, So PTC, Niles JC (2017) Quantification of labile heme in live malaria parasites using a genetically encoded biosensor. Proc Natl Acad Sci U S A 114(11):E2068–E2076
9. Amri I, Mancini E, De Martino L, Marandino A, Lamia H, Bassem J, Scognamiglio M, Reverchon E, De Feo V (2012) Chemical composition and biological activities of the essential oils from three Melaleuca species grown in Tunisia. Int J Mol Sci 13(12):16346–16372
10. Francis SE, Sullivan DJ, Goldberg DE (1997) Hemoglobin metabolism in the malaria parasite Plasmodium falciparum. Annu Rev Microbiol 51:97–123
11. World Health Organization (2015) Guidelines for treatment of malaria. WHO, Geneva

12. Gupta S, Thapar MM, Wernsdorfer WH, Björkman A (2002) In vitro interactions of artemisinin with Atovaquone, quinine, and Mefloquine against Plasmodium falciparum. Antimicrob Agents Chemother 46(5):1510–1515
13. Ekong R, Warhurst DC (1990) Synergism between artemether and mefloquine or quinine in a multidrug-resistant strain of Plasmodium falciparum in vitro. Trans R Soc Trop Med Hyg 84(6):757–758
14. Becker K, Tilley L, Vennerstrom JL, Roberts D, Rogerson S, Ginsburg H (2004) Oxidative stress in malaria parasite-infected erythrocytes: host–parasite interactions. Int J Parasitol 34(2):163–189
15. Giribaldi G, Ulliers D, Mannu F, Arese P, Turrini F (2001) Growth of *Plasmodium falciparum* induces stage-dependent haemichrome formation, oxidative aggregation of band 3, membrane deposition of complement and antibodies, and phagocytosis of parasitized erythrocytes. Br J Haematol 113(2):492–499
16. Parker PD, Tilley L, Klonis N (2004) *Plasmodium falciparum* induces reorganization of host membrane proteinsduring intraerythrocytic growth. Blood 103(6):2404–2406
17. Lelliott PM, McMorran BJ, Foote SJ, Burgio G (2015) The influence of host geneticson erythrocytes and malaria infection: is there therapeutic potential? Malar J 14:289
18. Williams TN (2011) How do Hemoglobins S and C result in malaria protection? J Infect Dis 204(11):1651–1653
19. Mohandas N, An X (2012) Malaria and human red blood cells. Med Microbiol Immunol 201(4):593–598
20. Udeinya IJ, Schmidt JA, Aikawa M, Miller LH, GreenI (1981) Falciparum malaria-infected erythrocytes specifically bind to cultured human endothelial cells. Science 213(4507): 555–557
21. Pantaleo A, Kesely KR, Pau MC, Tsamesidis I, Schwarzer E, Skorokhod OA, Chien HD, Ponzi M, Bertuccini L, Low PS, Turrini FM (2017) Syk inhibitors interfere with erythrocyte membrane modification during P falciparum growth and suppress parasite egress. Blood 130(8):1031–1040
22. Nepveu F, Turrini F (2013) Targeting the redox metabolism of Plasmodium falciparum. Future Med Chem 5(16):1993–2006
23. Kirandeep K, Meenakshi J, Tarandeep K, Rahul J (2009) Antimalarials from nature. Bioorg Med Chem 17:3229–3256
24. Saxena S, Pant N, Jaic DC, Bhakuni RS (2003) Antimalarial agents from plant sources. Curr Sci 85:1314–1329
25. Onguéné A (2013) The potential of anti-malarial compounds derived from African medicinal plants, part I: a pharmacological evaluation of alkaloids and terpenoids. Malar J 12:449
26. Ibrahim SRM, Mohamed GA (2015) Naphthylisoquinoline alkaloids potential drug leads. Fitoterapia 106:194–225
27. World Health Organization (2018) Status report on artemisinin resistance and ACT efficacy. WHO, Geneva
28. Skorokhoda OA, Davalos-Schaflera D, Gallo V, Valente E, Ulliersa D, Notarpietro A, Mandili G, Novelli F, Persico M, Taglialatela-Scafati O, Arese P, Schwarzer E (2015) Oxidative stress-mediated antimalarial activity of plakortin, a natural endoperoxide from the tropical sponge Plakortis simplex. Free Radic Biol Med 89:624–637
29. Mojab F (2012) Antimalarial natural products: a review. Avicenna journal of Phytomedicine 2(2):52–62
30. Landete JM (2011) Ellagitannins, ellagic acid and their derived metabolites: a review about source, metabolism, functions and health. Food Res Intl 44(5):1150–1160
31. Dell'AgliM PS, Basilico N, Verotta L, Taramelli D, Berry C, Bosisio E (2003) In vitro studies on the mechanism of action of two compounds with antiplasmodial activity: ellagic acid and 3,4,5-trimethoxyphenyl(6'-O-aalloyl)-beta-D-glucopyranoside. Planta Med 69(2):162–164
32. Frederich M, Hayette MP, Tits M, De Mol P, Angenot L (1999) In vitro activities of strychnos alkaloids and extracts against Plasmodium falciparum. Antimicrob Agents Chemother 43(9):2328–2331

33. Wright CW, Bray DH, O'Neill MJ, Warhurst DC, Phillipson JD, Quetin-Leclercq J, Angenot L (1991) Antiamoebic and antiplasmodial activities of alkaloids isolated from Strychnos usambarensis. Planta Med 57(4):337–340
34. Adebayo JO, Krettli AU (2011) Potential antimalarials from Nigerian plants: a review. J Ethnopharmacol 133(2):289–302
35. Gandhi M, Vinayak VK (1990) Preliminary evaluation of extracts of Alstoniascholaris bark for in vivo antimalarial activity in mice. J Ethnopharmacol 29(1):51–57
36. Elford BC (1986) L-glutamine influx in malaria-infected erythrocytes: a target for antimalarials? Parasitol Today 2(11):309–312
37. Vennerstrom JL, Klayman DL (1988) Protoberberine alkaloids as antimalarials. J Med Chem 31(6):1084–1087
38. Ghosh AK, Bhattacharyya FK, Ghosh DK (1985) Leishmania donovani: amastigote inhibition and mode of action of berberine. Experim Parasitol 60(3):404–413
39. Fata J, Rakhshandeh H, Berenji F, Jalalianfard A (2006) Treatment of cutaneous Leishmaniasis in murine model by alcoholic extract of Berberis vulgaris. Iran J Parasitol 1(1):39–42
40. Kassim OO, Loyevsky M, Elliott B, Geall A, Amonoo H, Gordeuk VR (2005) Effects of root extracts of Fagarazanthoxyloides on the in vitro growth and stage distribution of Plasmodium falciparum. Antimicrob Agents Chemother 49:264–268
41. Bidla G, VPK T, Joko B, El-Ghazali G, Bolad A, Berzins K (2004) Antiplasmodial activity of seven plants used in African folk medicine. Indian J Pharmacol 36:245–246
42. Vennerstrom JL, Klayman DL (1988) Protoberberine alkaloids as antimalarials. J Med Chem 31:1084–1087
43. Jia F, Zou G, Fan J, Yuan Z (2010) Identification of palmatine as an inhibitor of West Nile virus. Arch Virol 155(8):1325–1329
44. Kim JH, Ryu YB, Lee WS, Kim YH (2014) Neuraminidase inhibitory activities of quaternary isoquinoline alkaloids from Corydalis turtschaninovii rhizome. Bioorg Med Chem 22:6047–6052
45. Adebayo JO, Krettli AU (2011) Potential antimalarials from Nigerian plants: a review. J Ethnopharmacol 133:289–302
46. Khalid SA, Friedrichsen GM, Kharazmi A, Theander TG, Olsen CE, Christensen SB (1998) Limonoids from Khaya senegalensis. Phytochemistry 49:1769–1772
47. MacKinnon S, Durst T, Arnason JT (1997) Antimalarial activity of tropical Meliaceae extracts and Gedunin derivatives. J Nat Prod 60:336–341
48. Bickii J, NjifutieN FJA, Basco LK, Ringwald P (2000) In vitro antimalarial activity of limonoids from Khaya grandifoliola C.D.C. (Meliaceae). J Ethnopharmacol 69:27–33
49. Omar S, Godard K, Ingham A, Hussain H, Wongpanich V, Pezzuto J, Durst T, Eklu C, Gbeassor M, Sanchez-Vindas P, Poveda L, Philogene BJR, Arnason JT (2005) Antimalarial activities of gedunin and 7-methoxygedunin and synergistic activity with dillapiol. Ann Appl Biol 143(2):135–141
50. Suresh G, Gopalakrishnan G, Wesley SD, Singh NDP, Malathi R, Rajan SS (2002) Insect Antifeedant activity of Tetranortriterpenoids from the Rutales. A perusal of structural relations. J Agric Food Chem 50:4484–4490
51. Bringmann G, Dreyer M, Faber JH, Dalsgaard PW, Stærk D, Jaroszewski JW, Ndangalasi H, Mbago F, Brun R, Christensen SB (2004) Ancistrotanzanine C and related 5,1′- and 7,3′-coupled Naphthylisoquinoline alkaloids from Ancistrocladustan zaniensis. J Nat Prod 67(5):743–748
52. François G, Bringmann G, Dochez C, Schneider C, Timperman G, Assi LA (1995) Activities of extracts and naphthylisoquinoline alkaloids from Triphyophyllumpeltatum, Ancistrocladus abbreviatus and Ancistrocladusbarteri against Plasmodium berghei (Anka strain) in vitro. J Ethnopharmacol 46(2):115–120
53. François G, Timperman G, Eling W, Assi LA, Holenz J, Bringmann G (1997) Naphthylisoquinoline alkaloids against malaria: evaluation of the curative potentials of dioncophylline C and dioncopeltine a against Plasmodium berghei in vivo. Antimicrob Agents Chemother 41(11):2533–2539

54. Ponte-Sucre A, Faber JH, Gulder T, Kajahn I, Pedersen SEH, Schultheis M, Bringmann G, Moll H (2007) Activities of Naphthylisoquinoline alkaloids and synthetic analogs against Leishmania major. Antimicrob Agents Chemother 51(1):188–194
55. Bringmann G, Messer K, Schwobel B, Brun R, Assi LA (2003) Habropetaline A, an antimalarial naphthylisoquinoline alkaloid from Triphyophyllum peltatum. Phytochemistry 62:345–349
56. Soh PN, Witkowski B, Olagnier D, Nicolau ML, Garcia-Alvarez MC, Berry A, Benoit-Vical F (2009) In vitro and in vivo properties of Ellagic acid in malaria treatment. Antimicrob Agents Chemother 53(3):1100–1106
57. Okunade AL, Bikoff RE, Casper SJ, Oksman A, Goldberg DE, Lewis WH (2003) Antiplasmodial activity of extracts and quassinoids isolated from seedlings of Ailanthus altissima (Simaroubaceae). Phytother Res 17(6):675–677
58. Muhammad I, Bedir E, Khan SI, Tekwani BL, Khan IA, Takamatsu S, Pelletier J, Walker LA (2004) A new antimalarial quassinoid from Simabaorinocensis. J Nat Prod 67(5):772–777
59. GoffinE ZE, De Mol P, Madureira M, Martins AP, Proença da Cunha A, Philippe G, Tits M, Angenot L, Frederich M (2002) In vitro Antiplasmodial activity of Tithoniadiversifolia and identification of its Main active constituent: Tagitinin C. Planta Med 68(6):543–545
60. Jansen O, Tits M, Angenot L, Nicolas JP, De Mol P, PNikiema JB, Frédérich M (2012) Antiplasmodial activity of Dicomatomentosa (Asteraceae) and identification of urospermal A-15-O-acetate as the main active compound. Malar J 11:289
61. Labadie GR, Choi SR, Avery MA (2004) Diamine derivatives with antiparasitic activities. Bioorg Med Chem Lett 14:615–619
62. Hassan GM, Ali HZ (2017) In-vitro effect of artemisinin on L. tropica promastigotes. J Pharm Biol Sci 12(6):82–86
63. Fattorusso C, Persico M, Calcinai B, Cerrano C, Parapini S, Taramelli D, Novellino E, Romano A, Scala F, Fattorusso E, Taglialatela-Scafati O (2010) Manadoperoxides A-D from the Indonesian sponge Plakortis cfr. simplex. Further insights on the structure-activity relationships of simple 1,2-Dioxane Antimalarials. J Nat Prod 73:1138–1145
64. Skorokhod OA, Davalos-Schafler D, Gallo V, Valente E, Ulliers D, Notarpietro A, Mandili G, Novelli F, Persico M, Taglialatela-Scafati O, Arese P, Schwarzer E (2015) Oxidative stress-mediated antimalarial activity of plakortin, a natural endoperoxide from the tropical sponge plakortis simplex. Free Rad Biol Med 89:624–637
65. Kossuga MH, Nascimento AM, Reimão JQ, Tempone AG, Nosomi Taniwaki N, Veloso K, Ferreira AG, Cavalcanti BC, Pessoa C, Moraes MO, Mayer AMS, Hajdu E, Berlinck RGS (2008) Antiparasitic, antineuroinflammatory, and cytotoxic polyketides from the marine sponge Plakortisangulospiculatus collected in Brazil. J Nat Prod 71(3):334–339
66. Gamo FJ, Sanz LM, Vidal J, de Cozar C, Alvarez E, Lavandera JL, Vanderwall DE, Green DVS, Kumar V, Hasan S, Brown JR, Peishoff CE, Cardon LR, Garcia-Bustos JF (2010) Thousands of chemical starting points for antimalarial lead identification. Nature 465:305–312
67. Pantaleo A, Ferru E, Vono R, Giribaldi G, Lobina O, Nepveu F, Ibrahim H, Nallete JP, Carta F, Mannu F, Pippia P, Campanella E, Low PS, Turrini F (2012) New antimalarial indolone-N-oxides, generating radical species, destabilize the host cell membrane at early stages of Plasmodium falciparum growth: role of band 3 tyrosine phosphorylation. Free Rad Biol Med 52(2):527–536

Mechanistic and Structural Insights into Oxidative Stress in Malaria and Anti-malarial Drug Metabolism

28

Vijeta Sharma and Shailja Singh

Abstract

Malaria is a devastating infectious disease affecting mostly tropical and subtropical regions. Owning to the emergence of resistance to the existing chemotherapy, the development of anti-malarial drugs as novel chemotherapeutics remains unavoidable. Malaria parasite, *Plasmodium*, experiences oxidative stress throughout its life cycle upon infection, and underlying redox metabolism is quite complex. Alterations in the redox homeostasis occur during host-pathogen interactions. Parasite is highly vulnerable to such alterations in redox homeostasis. To circumvent this, the parasites engage in an efficient redox system having protective roles towards the turbulence faced by the parasite. Targeting the redox system of malaria parasite is tempting in developing novel antimalarial drugs. On the other hand, oxidative stress, generated during anti-malarial drug metabolism, acts as a source of inhibition against progression of this outrageous parasite. This review aims to provide updated knowledge on redox networks of parasite and structural insights of redox system enzymes, underpinning the balance between antioxidant and pro-oxidant candidates throughout the host-parasite interactions. Furthermore, it also highlights the importance of reactive oxygen species generation during anti-malarial drug metabolism. This review summarises on the vulnerabilities of the malaria parasite due to oxidative stress and the potential cues towards development of the novel antimalarial drugs.

Keywords

Anti-malarials · Oxidative stress · Plasmodium · Reactive oxygen species (ROS) · Drug resistance · Host-Pathogen interaction

V. Sharma · S. Singh (✉)
Special Centre for Molecular Medicine, Jawaharlal Nehru University, New Delhi, India
e-mail: shailjasingh@mail.jnu.ac.in

S. Chakraborti et al. (eds.), *Oxidative Stress in Microbial Diseases*,
https://doi.org/10.1007/978-981-13-8763-0_28

28.1 Introduction

Malaria is still a major health threat affecting millions of people worldwide. 212 million cases of malaria had been estimated with 429,000 deaths globally (WHO malaria report 2016) [1]. Apicomplexan parasite of genus *Plasmodium* causes malaria by completing its developmental life stages in human host and female anopheles vector [2]. Cure to malaria is artemisinin chemotherapy which is under threat due to the emergence of resistance. Thus, immediate measures are required to develop novel antimalarial drugs [3, 4].

28.1.1 Oxidative Stress Faced by Malaria Parasite

During the developmental stages of malaria parasite in human host, most clinical symptoms of malaria occur during the erythrocytic stage of asexual development of the parasite. During infection in human host, the parasite encounters oxidative stress leading to the enhanced production of reactive oxygen species (ROS, e.g. superoxide anion, hydroxyl ions, hydrogen peroxide) [5]. Inside the host cells, ROS is majorly produced by activated neutrophils, due to increased metabolic rate as the parasite enters trophozoite stage and degradation of haemoglobin during the erythrocytic stage. Malaria parasite is exposed to the oxidative stress. Evidently, the parasite has to cope up with this stress as susceptibility to the oxidative stress is a widely known phenomenon in the case of pathogenic parasites [6].

28.1.2 Redox Systems

Aerobic microorganisms have evolved a defence mechanism like redox and antioxidant systems, which are efficient in protecting the oxidative effects of ROS and keeping a reduced environment. These systems regulate the antioxidant and redox homeostasis and comprise of antioxidant enzymes and molecules. The major players of these systems are cellular antioxidant enzymes and molecules, such as glutathione peroxidase, superoxide dismutase, catalase and peroxiredoxin, NADPH, and thiols like thioredoxin and glutathione (Fig. 28.1).

In malaria parasite, compartmentalisation of the redox network within the organelles like cytosol, mitochondria and apicoplast is done into sub-cellular systems. The parasites engage in an efficient redox system comprising NADPH-dependent thioredoxin system and glutathione system, to impede the oxidative stress and maintain a redox homeostasis under normal physiological conditions [7]. These systems detoxify and maintain the redox state by thiol-sulphide exchange reactions. This is achieved by oxidised and reduced glutathione, thioredoxins and their disulphides with NADP+/NADPH. However, the intracellular reduced state is maintained by various enzymatic and non-enzymatic antioxidants as well as peroxiredoxins and superoxide dismutases (Figs. 28.1 and 28.2).

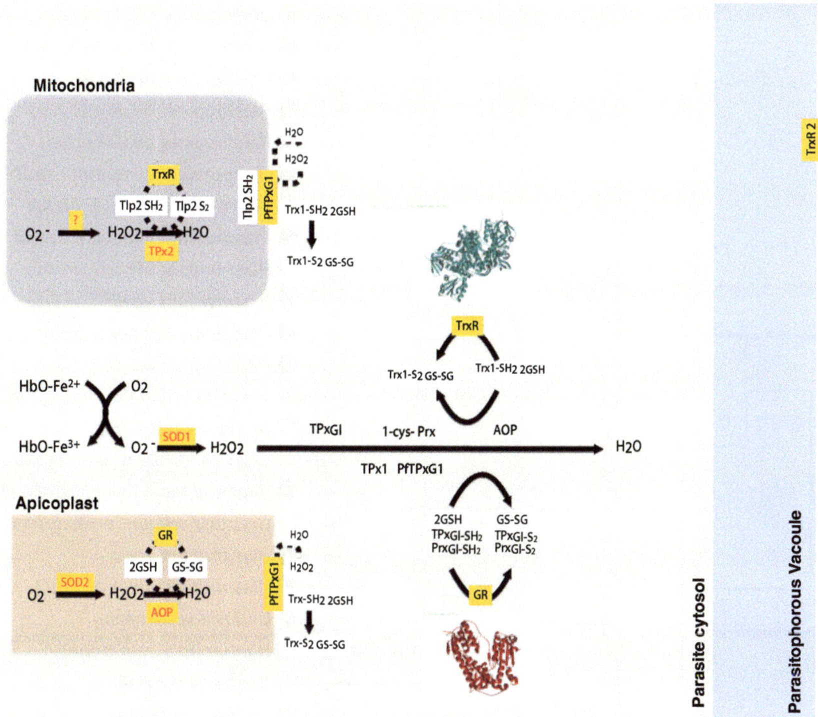

Fig. 28.1 Model showing schematic of redox networks of malaria parasite *Plasmodium*. Proteins of glutathione system and thioredoxin system involved in the redox homeostasis of parasite are shown in different compartments like mitochondria, apicoplast and parasite cytosol. The proteins are shown in their reduced and oxidised forms. Abbreviations: TPx_{GI}, glutathione peroxidase-like thioredoxin peroxidase, *1-cys-Prx* 1-cysteine peroxiredoxin, *AOP* antioxidant protein, *TPx1, 2* thioredoxin peroxidase 1,2, *Trx1,2,3* thioredoxin 1,2,3, *Tlp1,2* thioredoxin-like protein 1,2, *TrxR* thioredoxin reductase, *GR* glutathione reductase, *GSH* reduced glutathione, *GS-SG* oxidised glutathione, $Trx\text{-}S_2$, oxidised thioredoxin, $Trx\text{-}SH_2$, reduced thioredoxin, $HbO\text{-}Fe^{3+}$ oxyhaemoglobin with ferriprotoporphyrin, $HbO\text{-}Fe^{3+}$ oxyhaemoglobin containing ferriprotoporphyrin, *SOD1* superoxide dismutase 1, *SOD2* superoxide dismutase 2

Malaria parasites *Plasmodium* possess ubiquitous peroxidases as peroxiredoxins (Prxs), instead of catalase or genuine peroxidase. The basal cellular peroxide flux is maintained by Prxs by using cysteine residue in their active site to reduce reactive oxygen species [8, 9]. Prxs belong to the distinct classes of protein family: typical 2-Cys peroxiredoxins, atypical 2-Cys peroxiredoxins and 1-Cys peroxiredoxins. These are localised mainly in cytosol and mitochondria. *Plasmodium* parasites also have a glutathione peroxidase-like protein which uses thioredoxin as a reducing factor, which might be localised in apicoplast, not experimentally verified though [10–14].

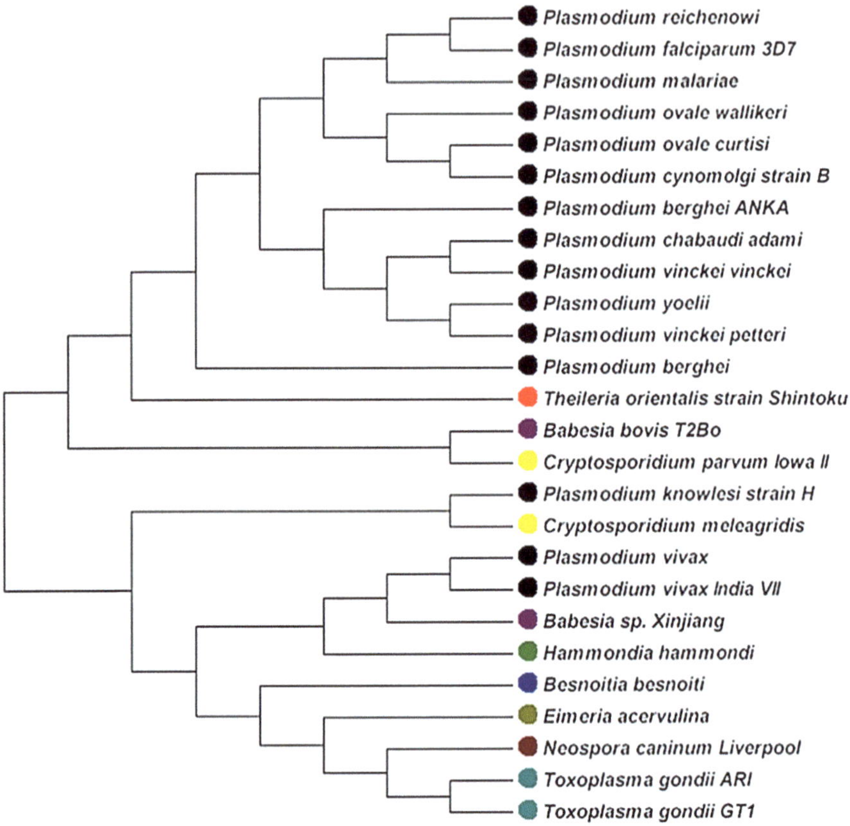

Fig. 28.2 Phylogenetic relationships of thioredoxin reductase of parasites belonging to different taxonomic groups. Phylogenetic analysis involved 26 amino acid sequences was done by Maximum Likelihood method conducted in MEGA 7

28.1.3 Glutathione System

In malaria parasite, the glutathione system is composed of glutathione and the low molecular weight thiol which is present in abundance and regulates the redox balance primarily. The other thiol-disulphide exchange enzymes like oxidoreductases are glutathione reductase, glutathione S-transferase, thioredoxin peroxidase-like glutathione peroxidase, glutathione redoxins and glutathione-like proteins. In this system, the thiol-sulphide exchange (GSH-GSSG) acts as the major redox regulator in malaria parasite [15, 16].

28.1.4 Thioredoxin System

The thioredoxin system in malaria parasite comprises small proteins like thioredoxins, thioredoxin-like proteins, thioredoxin reductase, thioredoxin peroxidase and peroxiredoxins, which are localised in endoplasmic reticulum, cytosol and parasitophorous vacuole [17, 18]. The antioxidant enzyme thioredoxin reductase is the most important enzyme of this system which helps the *Plamsodium* parasites to maintain a reduced intracellular environment during intraerythrocytic development. Hence, this enzyme is crucial for the survival of the parasites in the intraerythrocytic stage of malaria parasite life cycle.

28.1.5 The Unique Feature of Enzyme Thioredoxin Reductase

There are two low molecular weight isoforms of thioredoxin reductase (TrxR) which are expressed by *Plasmodium falciparum*. These two forms of TrxR are located in mitochondria and cytosol, act by reducing its substrate thioredoxin [19]. The catalytic site of *P. falciparum* TrxR is composed of two active sites in which NADPH-dependent electron transfer occurs to its disulphide substrate. The active centre of the malaria parasite is unique from its counterpart in human and insects in having the two cysteine redox pair (535 and 540) separated by four amino acid linker in contrast to cysteine-selenocysteine redox pair at the C-terminal redox centre. The active site C-VNVG-C is conserved among the species of *Plasmodium* [20–23]. The N-terminal of parasite TrxR has an extended coil region which replaced by a helix and strand in human TrxR. This suggests the difference in the electrostatic surface properties of the enzymes in human and malaria parasite which is responsible for the different modes of interactions with other protein candidates. It has been previously reported that *P. falciparum* TrxR shows almost 40–42% sequence identity to the human TrxR. This distinctive feature of malaria parasite TrxR makes it unique from human vis-a-vis making it a fascinating druggable target [24–26].

28.1.5.1 Known Inhibitors Against *Pf*TRxR

As the enzymes of redox system help the parasite to survive and dodge the deadly oxidative effects of its metabolism or host immune response, these enzymes present to be attractive targets for the design of repertoire of anti-parasitic compounds. The unique specificity of the enzyme makes it convenient for the design of rational drug candidates. There had been a few inhibitors which are reported till date as inhibitors of TrxR. Some of the inhibitors are illustrated in Table 28.1. Andricopulo et al. have discovered plausible inhibitor of parasite TrxR, which was believed to interact with the interactive cavity of the enzyme [33]. One of the inhibitors reported was aurothiomalate which showed inhibition of enzyme TrxR by inducing eryptosis. Eryptosis-like condition is triggered when red blood cells experience oxide stress [34, 35]. Similarly, the compounds 1,4-napthoquinone (1,4-NQ) and 4-nitrobenzothiadiazole (4-NBT) were reported to be active against *Plasmodium falciparum. O*ther inhibitors reported are bis-(2,4-dinitrophenyl)sulphide (2,4-DNPS) and menadione (MD) which showed moderate activity against malaria parasite [36]. However, the active compounds 1,4-NQ, 2,4- DNPS, 4-NBT and MD showed more toxicity than 3-DAP against the mammalian cells.

Table 28.1 Table showing the known inhibitors of *Plasmodium falciparum* thioredoxin reductase (PfTrx) and glutathione reductase (PfGR)

	Inhibitors	References
PfTrx	Aurothiomalate (ATM)	Sharma et al. [21]
PfTrx	1,4-napthoquinone (1,4-NQ), 3-dimethylaminopropiophenone (3-DAP)	Munigunti et al. [36]
PfTrx	Bis-(2,4-dinitrophenyl)sulfide (2,4-DNPS),	Munigunti and Calderón [27]
PfTrx	Demethoxycurcumin (DMC)	Munigunti et al. [28]
PfTrx	4-nitrobenzothiadiazole (4-NBT),	Andricopulo et al. [33]
PfTrx	Menadione (MD)	Tiwari et al. [29]
Pf GR	BCNU(1,3-bis(2-chloroethyl)-1-nitrosourea) and HeCNU(1-(2-chloroethyl)-3-(2-hydroxythyl)-1-nitrosourea),	Zhang et al. [30]
Pf GR	Quinoline-based alcohols 4-Anilinoquinoline	Davioud-Charvet et al. [31]
Pf GR	Methylene blue	Färber et al. [32]

28.1.5.2 Pro-oxidant and Antioxidant Property of Redox Enzymes

In malaria parasites, the administration of drugs like artemisinin and its derivatives acts as pro-oxidants and inhibits the growth of parasite both in vivo and in vitro during intra-erythrocytic stage. Likewise, in the case of cerebral malaria, the ROS production is responsible for the pathology of the host tissue causing vascular endothelial dysfunction. When pro-oxidants act on vascular tissue, it causes the damage potentially in contrast to supporting the host defence while acting near the infected cell. The paradox of ROS production remains, as it could be both beneficial and pathological which highly depends on the site and amount of production.

References

1. WHO malaria report 2016
2. Ball P (2001) Roll up for the revolution. Nature 414:42–144. (2008) Roll back malaria: the global malaria action plan: for a malaria-free world. Roll back Malaria Partnership, Geneva
3. Visser BJ, Vugt M, Grobusch MP (2014) Expert Opin Pharmacother 15:2219–2254
4. Dondorp AM, Nosten F, Yi P et al (2009) N Engl J Med 361:455–467
5. Becker K, Tilley L, Vennerstrom JL, Roberts D, Rogerson S, Ginsburg H (2004) Oxidative stress in malaria parasite-infected erythrocytes: host–parasite inter- actions. Int J Parasitol 34:163–189
6. Francis SE, Sullivan DJ Jr, Goldberg DE (1997) Hemoglobin metabolism in the malaria parasite Plasmodium falciparum. Annu Rev Microbiol 51:97–123
7. Muller S (2004) Redox and antioxidant systems of the malaria parasite Plasmodium falciparum. Mol Microbiol 53:1291–1305
8. Gretes MC, Poole LB, Karplus PA (2012) Peroxiredoxins in parasites. Antioxid Redox Signal https://doi.org/10.1089/ars.2011.4404
9. Sztajer H, Gamain B, Aumann K-D, Slomianny C, Becker K, Brigelius-Flohé R, Flohé L (2001) The putative glutathione peroxidase gene of Plasmodium falciparum codes for a thioredoxin peroxidase. J Biol Chem 276(10):7397–7403
10. Kehr S, Sturm N, Rahlfs S, Przyborski JM, Becker K (2010) Compartmentation of redox metabolism in malaria parasites. PLoS Pathog 6:e1001242
11. Rahlfs S, Becker K (2001) Thioredoxin peroxidases of the malarial parasite Plas- modium falciparum. Eur J Biochem 268:1404–1409
12. Nickel C, Trujillo M, Rahlfs S, Deponte M, Radi R, Becker K (2005) Plasmodium falciparum 2-Cys peroxiredoxin reacts with plasmoredoxin and peroxynitrite. Biol Chem 386(11):1129–1136
13. Yano K, Komaki-Yasuda K, Tsuboi T, Torii M, Kano S, Kawazu S (2006) 2-Cys peroxiredoxin TPx-1 is involved in gametocyte development in Plasmodium berghei. Mol Biochem Parasitol 148:44–51
14. Yano K, Otsuki H, Arai M, Komaki-Yasuda K, Tsuboi T, Torii M, Kano S, Kawazu S (2008) Disruption of the Plasmodium berghei 2-Cys peroxiredoxin TPx-1 gene hinders the sporozoite development in the vector mosquito. Mol Biochem Parasitol 159:142–145
15. Becker K, Rahlfs S, Nickel C, Schirmer RH (2003b) Glutathione – functions and metabolism in the malarial parasite Plasmodium falciparum. Biol Chem 384:551–566
16. Sarma GN, Savvides SN, Becker K, Schirmer M, Schirmer RH, Karplus PA (2003) Glutathione reductase of the malarial parasite Plasmodium falciparum: crystal structure and inhibitor development. J Mol Biol 328:893–907
17. Nickel C, Rahlfs S, Deponte M, Koncarevic S, Becker K (2006) Thioredoxin networks in the malarial parasite Plasmodium falciparum. Antioxid Redox Signal 8(7–8):1227–1239

18. Kanzok SM, Schirmer RH, Turbachova I, Iozef R, Becker K (2000) The thioredoxin system of the malaria parasite Plasmodium falciparum. Glutathione reduction revisited. J Biol Chem 275:40180–40186
19. Tamura T, Stadtman TC (2002) Mammalian thioredoxin reductases. Methods Enzymol 347:297–306
20. Gilberger T-W, Bergmann B, Walter RD, Müller S (1998) The role of the C-terminus for catalysis of the large thioredoxin reductase from Plasmodium falciparum. FEBS Lett 425(3):407–410
21. Kanzok SM, Rahlfs S, Becker K, Schirmer RH (2002) Thioredoxin, thioredoxin reductase, and thioredoxin peroxidase of malaria parasite *Plasmodium falciparum*. Methods Enzymol 347:370–381
22. Fritz-Wolf K, Jortzik E, Stumpf M, Preuss J, Iozef R, Rahlfs S, Becker K (2013) Crystal structure of the *Plasmodium falciparum* Thioredoxin reductase–Thioredoxin complex. J Mol Biol 425(18):3446–3460
23. Gladyshev VN, Jeang K-T, Stadtman TC (1996) Selenocysteine, identified as the penultimate C-terminal residue in human T-cell thioredoxin reductase, corresponds to TGA in the human placental gene. Proc Natl Acad Sci 93(12):6146–6151
24. Gilberger T-W, Walter RD, Müller S (1997) Identification and characterization of the functional amino acids at the active site of the large Thioredoxin reductase from Plasmodium falciparum. J Biol Chem 272(47):29584–29589
25. Kanzok SM, Schirmer RH, Türbachova I, Iozef R, Becker K (2000) The thioredoxin system of the malaria parasite Plasmodium falciparum glutathione reduction revisited. J Biol Chem 275(51):40180–40186
26. Krnajski Z, Gilberger T-W, Walter RD, Cowman AF, Müller S (2002) Thioredoxin reductase is essential for the survival of Plasmodium falciparum erythrocytic stages. J Biol Chem 277(29):25970–25975
27. Munigunti R, Calderón AI (2012) Development of liquid chromatography/mass spectrometry based screening assay for PfTrxR inhibitors using relative quantitation of intact thioredoxin. Rapid Commun Mass Spectrom 26(17):2051–2056
28. Munigunti R, Gathiaka S, Acevedo O, Sahu R, Tekwani B, Calderón AI (2014) Determination of antiplasmodial activity and binding affinity of curcumin and demethoxycurcumin towards PfTrxR. Nat Prod Res 28(6):359–364
29. Tiwari NK, Reynolds PJ, Calderón AI (2016) Preliminary LC-MS based screening for inhibitors of *Plasmodium falciparum* thioredoxin Reductase (PfTrxR) among a set of antimalarials from the Malaria Box. Molecules 21(4):424
30. Zhang YA, Hempelmann E, Schirmer RH (1988) Glutathione reductase inhibitors as potential antimalarial drugs. Effects of nitrosoureas on *Plasmodium falciparum* in vitro. Biochem Pharmacol 37(5):855–860
31. Davioud-Charvet E, Delarue S, Biot C, Schwöbel B, Boehme CC, Müssigbrodt A, Maes L, Sergheraert C, Grellier P, Schirmer RH, Becker K (2001) A prodrug form of a *Plasmodium falciparum* glutathione reductase inhibitor conjugated with a 4-anilinoquinoline. J Med Chem 44(24):4268–4276
32. Färber PM, Arscott LD, Williams CH Jr, Becker K, Schirmer RH (1998) Recombinant *Plasmodium falciparum* glutathione reductase is inhibited by the antimalarial dye methylene blue. FEBS Lett 422(3):311–314
33. Andricopulo AD, Akoachere MB, Krogh R, Nickel C, McLeish MJ, Kenyon GL, Arscott LD, Williams CH Jr, Davioud-Charvet E, Becker K (2006) Specific inhibitors of Plasmodium falciparum thioredoxin reductase as potential antimalarial agents. Bioorg Med Chem Lett 16(8):2283–2292
34. Kuntz AN, Davioud-Charvet E, Sayed AA, Califf LL, Dessolin J, Arnér ES, Williams DL (2007) Thioredoxin glutathione reductase from Schistosoma mansoni: an essential parasite enzyme and a key drug target. PLoS Med 4(6):e206
35. Sharma A, Sharma A, Dixit S, Sharma A (2011) Structural insights into thioredoxin-2: a component of malaria parasite protein secretion machinery. Sci Rep 1:179
36. Munigunti R, Gathiaka S, Acevedo O, Sahu R, Tekwani B, Calderón AI (2013) Characterization of Pf TrxR inhibitors using antimalarial assays and in silico techniques. Chem Cent J 7(1):175

Oxidative Stress Inducers as Potential Anti-Leishmanial Agents

29

Bhanu Priya Awasthi, Swetapadma Majhi,
and Kalyan Mitra

Abstract

Targeting redox metabolism of the parasite may prove to be a rewarding strategy in leishmaniasis since it plays a crucial role in the survival of *Leishmania* parasites inside the host cells where they face intense oxidative stress. Enzymes involved in redox metabolism have been widely studied establishing their significance in survival and virulence of *Leishmania*. Moreover, the selective presence of some of these enzymes makes them validated drug targets in *Leishmania*. Kinetoplastids like *Leishmania* are one of the most primitive eukaryotes possessing a single large branched mitochondrion spanning the cell body of the parasite, and their survival depends on the proper functioning of this single mitochondrion, which makes them vulnerable to mitochondrial stress. Mitochondria are primary sites where intracellular reactive oxygen species (ROS) are generated by the electron transport chain composed of multi-protein complexes present in the inner mitochondrial membrane. Several leishmanicidal agents have been identified to induce oxidative stress and disrupt the structural as well as functional integrity of the mitochondrion, thereby triggering an apoptosis-like cell death in *Leishmania*. Here, we provide an overview of the role of the antioxidant defense mechanism in *Leishmania* and highlight the potential of compounds inducing oxidative stress for their development as new anti-leishmanial agents.

Keywords

Leishmaniasis · Redox metabolism · Oxidative stress · Anti-leishmanial agents · Apoptosis

B. P. Awasthi · S. Majhi · K. Mitra (✉)
Electron Microscopy Unit, Sophisticated Analytical Instrument Facility, CSIR-Central Drug Research Institute, Lucknow, India

Academy of Scientific and Innovative Research, Chennai, India
e-mail: k_mitra@cdri.res.in

S. Chakraborti et al. (eds.), *Oxidative Stress in Microbial Diseases*,
https://doi.org/10.1007/978-981-13-8763-0_29

29.1 Introduction

Leishmaniasis is a neglected vector-borne parasitic disease, which is prevalent mainly in tropical and sub-tropical regions of the world. This disease is caused by more than 20 species of parasitic protozoa of the genus *Leishmania* and is transmitted by phlebotomine sand flies. The life cycle of *Leishmania* includes two forms – extracellular motile promastigotes residing in the gut of the sand fly, while intracellular non-motile amastigotes reside inside parasitophorous vacuoles of mammalian host cells. There are three manifestations of leishmaniasis – cutaneous leishmaniasis (CL), mucocutaneous leishmaniasis (MCL), and visceral leishmaniasis (VL), among which VL, caused by *Leishmania donovani*, is the most fatal one if left untreated. According to the recent report by World Health Organization, 0.2–0.4 million new cases and 20,000–30,000 deaths per annum occur due to VL commonly in developing countries. The standard drugs currently used to cure leishmaniasis patients are antimonials, miltefosine, amphotericin B, and its formulations, whereas paromomycin and pentamidine are also effectively used. However, the limitations of these drugs like resistance, high toxicity, high cost, and painful routes of drug administration restrict their uses and necessitate the development of new safer drugs [1–4]. To achieve this goal, researchers have been targeting unique and essential metabolic pathways of the parasite, which differ from their mammalian counterpart. Here, we discuss the redox metabolic pathways in *Leishmania* and highlight how oxidative stress-inducing agents could have a potential for developing new chemotherapeutics against leishmaniasis.

29.2 The Unique Mitochondrion of *Leishmania*

The unicellular *Leishmania* parasites have several organelles that are unique to Kinetoplastids out of which the mitochondrion deserves special mention. A single ramified mitochondrion with disk-shaped DNA containing kinetoplast is present in these organisms which constitutes 12% of body weight differing from mammalian cells containing thousands of mitochondria (Fig. 29.1). Kinetoplast DNA is made up of parallel DNA fibers and constitutes ~30% of whole genome that encodes rRNAs and subunits of respiratory chains and guide RNA. However, most of the mitochondrial proteins are encoded by nuclear DNA [5–7].

The inner mitochondrial membrane of trypanosomatids comprises various complexes to generate energy – NADH-ubiquinone oxidoreductase (complex I), succinate-ubiquinone oxidoreductase (complex II), ubiquinol-cytochrome c oxidoreductase (complex III or cytochrome bc1), and cytochrome c oxidase (complex IV) as in higher eukaryotes. These complexes generate proper proton gradient involving ubiquinone and cytochrome c as electron carriers across the inner mitochondrial membrane. This proton gradient helps in ATP synthesis by ATP synthase (complex V) which is called coupling of respiration with oxidative phosphorylation. Despite these similarities, there are growing evidences of differences in ETCs (electron transport chains) in *Leishmania* parasites from their mammalian host. First, there is

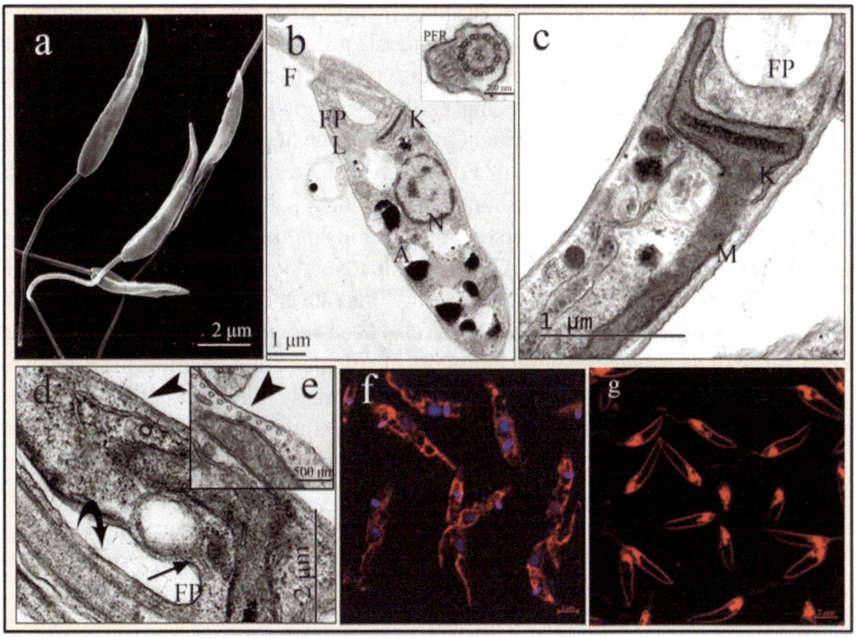

Fig. 29.1 Unique features of the *Leishmania* parasite – (**a**) SEM (scanning electron microscope) micrograph depicting elongated cell body with smooth cell membrane and a long flagellum. TEM (transmission electron microscope) micrographs of thin sections displaying ultrastructural characteristics – (**b**) well-defined nucleus, flagellum [inset revealing 9 + 2 microtubule arrangement and ribbed paraflagellar rod (PFR)], flagellar pocket, acidocalcisomes, and lipid droplets. (**c**) Single mitochondrion extended along the body length with electron-dense kinetoplast DNA. (**d**) TEM image showing the presence of subpellicular microtubules beneath the cell membrane. (**e**) TS of microtubule (arrowhead) which is absent in the membranes of flagella and flagellar pocket (arrow and curved arrow). (**f**) Confocal micrograph showing single-branched mitochondrion with nucleus and kinetoplast (dually stained with MitoTracker Red CMXROS and Hoechst). (**g**) Confocal micrograph exhibiting uniform cell membrane with flagellar pocket stained with FM 4–64 dye. *F* flagellum, *FP* flagellar pocket, *K* kinetoplast, *M* mitochondria, *N* nucleus, *A* acidocalcisomes, *L* lipid droplets, *TS* transverse section

controversy over presence of complex I in trypanosomatids as it has been reported that oxygen consumption is independent of rotenone, a complex I inhibitor, and also no effect in oxygen consumption after deletion of its subunits (*ND4*, *ND5*, and *ND7*) in case of *T. cruzi* has been reported [8, 9]. Interestingly, some reports have shown that KCN, inhibitor of complex IV, cannot slow down the respiratory process of trypanosomatids suggesting the presence of other alternative oxidase (AOX) [10]. Though the presence of AOX and its inhibition by salicylhydroxamic acid (SHA) have been confirmed in trypanosomes, SHA does not affect AOX of *Leishmania* parasites that are cyanide-insensitive. However, cyanide-insensitive respiration is impaired by another cytochrome c inhibitor orthohydroxydiphenyl (OHD) explaining the need of further study related to this enzyme in trypanosomatids. Succinate, substrate of complex II, has been reported to participate actively in oxidative phosphorylation in these parasites. Besides, in these parasites, there is an alternative electron path to produce ATP in which fumarate accepts electrons from NADH and produces succinate using fumarate reductase (absent in mammals) [11]. The single mitochondrion of these parasites makes them especially vulnerable since proper functioning of this organelle is critical for the survival of the parasites in the absence of any compensatory mechanism. For this reason, researchers are targeting this organelle for developing new leishmanicidal agents.

Since amastigotes live in hypoxic condition inside the parasitophorous vacuole (PV), they do not synthesize ATP via oxidative phosphorylation, but through substrate-level phosphorylation (SLP) using fumarate and pyruvate as substrates. In addition, lack of complex I, II, and IV has also been shown in amastigotes validating occurrence of SLP. Nevertheless, inhibitors of complex I and III affect the growth of amastigotes, suggesting the presence of substantially different ETC complexes which need further investigation [12, 13]. Interestingly, fumarate reductase and pyruvate reductase, which participate in SLP, may prove to be potential drug targets for amastigotes besides ETC complexes.

29.3 Oxidative Stress in *Leishmania*

Oxidative stress is a condition when there is an imbalance between the generation of reactive oxygen species (ROS) and their regulation by radical scavenging systems. ROS molecules include peroxides, superoxides, hydroxyl radicals, and singlet oxygen, which are chemically reactive molecules and can react with proteins, lipids, and DNA of the cells. Under normal conditions, ROS molecules perform physiological functions and actively participate in signaling pathways, but excessive ROS levels have detrimental effects and sometimes may lead to cell death. Therefore, it is essential to regulate ROS generation by reducing its level both with the help of antioxidant enzymes and non-enzymatic molecules [14]. Like other higher organisms, *Leishmania* parasites also possess redox pathways including antioxidants, enzymes, and molecules to control ROS levels. Nevertheless, it appears that the parasites are very sensitive to oxidative stress whether they face it inside the macrophages in the form of amastigotes or in the presence of anti-leishmanial compounds. Several reports state that oxidative stress leads to peroxidation of membrane lipids, mitochondrial dysfunction, and ATP depletion in these parasites resulting in hampering of other physiological events or organelles and finally causing death of the parasites [15].

29.3.1 Oxidative Stress Encountered by *Leishmania* Parasites Due to Immunological Response During Infection

Macrophages are the major host cells in *Leishmania* infection, whereas other host cells may also be infected like dendritic cells, eosinophils, neutrophils, and fibroblast. The first type of immune response that occurs post-infection is innate immune response which includes activation of cells like natural killer cells, macrophages, neutrophils, monocytes, and dendritic cells through recognition receptor mechanisms (e.g., toll-like receptors) producing soluble products (complement, released cytokines including interleukin 12 and TNF) [16]. TLR 2, TLR 3, and TLR 4 have been shown to display leishmanicidal activity by inducing the production of TNF-α (tumor necrosis factor alpha), NO (induction of iNOS, inducible nitric oxide synthase), and ROS in various species of *Leishmania* parasites [17]. The adaptive immune response during infection includes the participation of T-cells, i.e., CD4 T cells. CD4+ T cells comprise of Th (T helper) 1 and Th2 population [18]. Th1 cells produce pro-inflammatory cytokines like interleukin 12 (IL-12), gamma interferon (IFN-γ), and TNF causing induction of iNOS and suppression of arginase-1 consequently leading to potent anti-leishmanial activities [19, 20]. On the other hand, Th2 response promotes disease progression by producing anti-inflammatory cytokines like IL-4, IL-13, and IL-10 and by enhancing expression of arginase-1 [21]. Infection by *L. donovani* has been shown to suppress TLR 2 action and increase IL-10 production which both help in persistence of infection. Furthermore, the parasites develop various strategies like the presence of thick layer of lipophosphoglycan (LPG) [22] and generation of arginase causing the formation of L-ornithine from arginine that is utilized by the parasites for their survival. Metalloprotease gp63 of *Leishmania* also inhibits the NOX2 (phagocytic NADPH-oxidase) production and iNOS signaling in macrophage [23]. Hence, oxidative stress is one of the important defense mechanisms employed by macrophage against the parasite. However, the parasite adopts strategies to combat the oxidative stress in the host for its survival. The fine line balance between defense mechanism of the host and the efficiency of the parasite to combat it decides the progression of infection in this disease.

29.3.2 Oxidative Stress-Mediated Programmed Cell Death in *Leishmania*

Several studies have illustrated the importance of programmed cell death (PCD) or apoptosis in multicellular organisms. It is mostly beneficial over necrosis where cellular lysis generates inflammatory responses [24]. Earlier, apoptosis was believed to be restricted only to metazoans. However, there are increasing evidences affirming the existence of apoptosis-like cell death in unicellular organisms such as *Leishmania* induced by stress like heat [25], oxidative stress [26], starvation, and serum complement or stimulated by treatment with various drugs [27]. The significance of apoptosis in these parasites is understood to maintain cell number in culture to get enough resources as well as to maintain proper cell density or clonality of population to evade host immune response [28, 29]. Some of the morphological and physiological features of apoptosis in *Leishmania* are common to multicellular organisms like cell shrinkage, blebbing on the membrane (in some cases, hence controversial), nuclear condensation, DNA fragmentation, PS externalization, and mitochondrial membrane depolarization. However, there are dissimilarities with typical metazoan apoptosis like the absence of classical regulatory or effector molecules, for example, TNF-related family receptors, Bcl-2 family members, and typical caspases in these unicellular parasites. The molecular machinery involved in cell death is not clear, and further research in this area may aid in determining the importance of these molecules in the life cycle of these organisms and help in designing new drugs against leishmaniasis.

In trypanosomatids, interplay of ROS (reactive oxygen species) generation, mitochondrial membrane depolarization, and rise in cytosolic calcium (Ca^{2+}) have been widely studied with respect to cell death [27, 30]. Rise in ROS level causes peroxidation in mitochondrial membrane lipids resulting in mitochondrial depolarization and opening of mitochondrial pore with initiation of PCD. Mitochondrial dysfunction leads to disruption in cytosolic Ca^{2+} homeostasis and to activation of various proteases (particularly caspases) and nucleases. A number of reports related to apoptosis in trypanosomatids have confirmed the presence of metacaspases containing homologues of caspases that are arginine/lysine specific instead of cysteine specific (caspases of mammals) [31]. Activation of metacaspases may result in the activation of nucleases causing DNA fragmentation, which is observed in apoptosis-like cell death in trypanosomatids. Pro-apoptotic endonuclease G (Endo G) is released from the mitochondrion causing DNA fragmentation in *Leishmania* under oxidative stress [32, 33]. The study of novel molecules involved in apoptosis in these organisms may help in development of new drugs against leishmaniasis (Fig. 29.2).

29.4 Redox Metabolic Pathways

Regulation of redox metabolism plays a vital role in the survival of *Leishmania* parasites which experience oxidative stress during their life cycle. Here, we focus on the oxidative stress generation and the defense mechanisms of the parasites.

29.4.1 Redox Metabolism in Mitochondria

Since *Leishmania* parasites are aerobic, ROS generation is a normal process in mitochondria. In agreement with other eukaryotes, the transfer of electrons from one complex to another in electron transport chain (ETC) may lead to escape of some electrons from the chain and cause reduction of molecular oxygen to superoxide radicals ($O_2^{\bullet-}$) (precursor for hydrogen peroxide and other ROS agents). In higher eukaryotes, complex I of ETC actively produces ROS while the presence of this complex in trypanosomatids is debatable. Nevertheless, in *L. donovani*, production of superoxide radicals is reported in response to rotenone (inhibitor of complex I) [34, 35]. In spite of presence of NADH dehydrogenase of complex I having low activity, the existence of type II NADH dehydrogenase has also been reported in *Leishmania* genome (LinJ.36.5620; http://tritrypdb.org/tritrypdb), but its role in ROS production has not been identified in the case of *Leishmania* species though having ROS-producing activity in trypanosomes [9]. Although ROS generation by

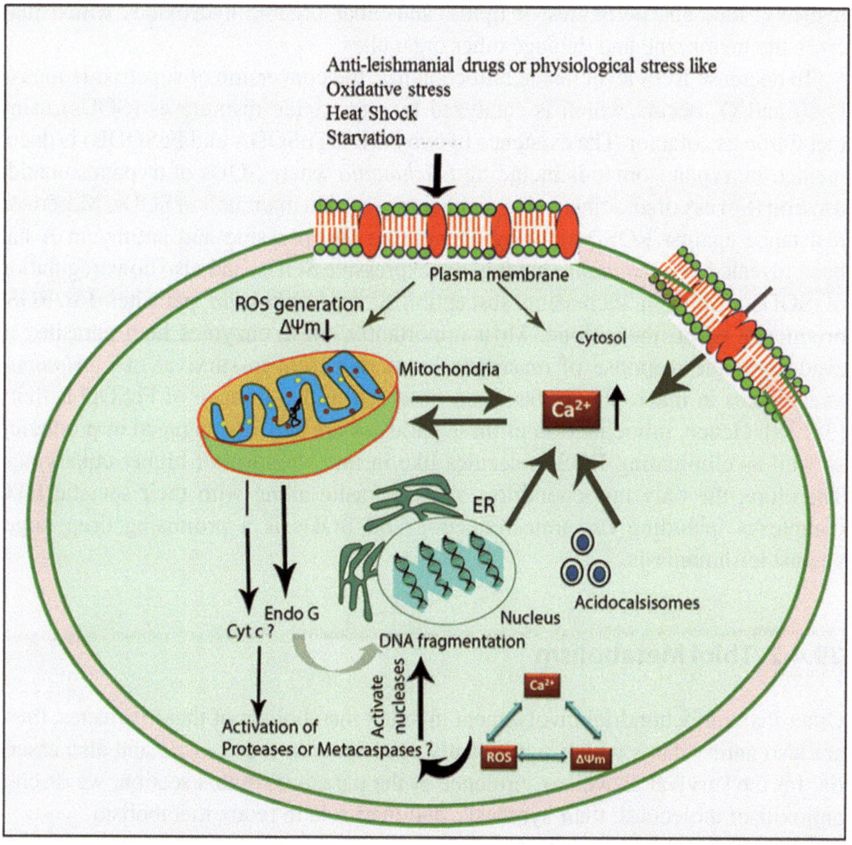

Fig. 29.2 Schematic diagram describing events leading to apoptosis-like cell death in *Leishmania* parasites induced by either physiological stress or chemicals. Various types of stimuli provoke different stress such as oxidative stress (ROS generation), depolarization of mitochondrial membrane (Delta Psi m indicates mitochondrial membrane potential), or disruption in calcium homeostasis. These three events are interrelated to one another and ultimately lead to cell death involving activation of caspase-like proteases and nucleases causing DNA fragmentation

complex II is rarely documented in mammalian cells, ROS is significantly produced by complex II inhibitor (TTFA, thenoyltrifluoroacetone) in *L. donovani* [35]. ROS generation by complex III has been observed in *Leishmania* with antimycin A, inhibitor of complex III, and tafenoquine treatments [35, 36]. In general, complex IV is not involved in ROS production but some obstruction of electron flow even in complex IV can lead to generation of ROS from molecular oxygen (O_2). In spite of having typical respiratory chain complexes, trypanosomatids including *Leishmania* parasites contain NADH-fumarate reductase subunit that is also responsible for ROS production [37]. Superoxide radicals ($O_2^{\bullet-}$) that are primarily produced in mitochondria may generate other ROS molecules like hydrogen peroxide (H_2O_2), hydroxyl radicals, peroxides of lipids, and other organic hydroxides which may cross the membrane and damage other organelles.

To decrease ROS level inside mitochondria, the conversion of superoxide ions to H_2O_2 and O_2 occurs which is catalyzed by superoxide dismutases (SODs) using metal iron as cofactor. The existence of two SODs (FeSODA and FeSODB) is documented in trypanosomatids including *Leishmania* where SODs of trypanosomatids use iron (Fe) as cofactor instead of manganese (Mn) in mammalian SODs. Moreover, resistance against ROS-producing agents like nitroprusside and antimycin A has been revealed in *Leishmania* having overexpressive SODs, and also downregulation of SODs results in increasing susceptibility of *Leishmania* parasites for ROS-producing agent, menadione. Most importantly, these enzymes help parasites to evade immune response of macrophages as reduction in survival of *Leishmania* amastigotes in macrophages has been observed in the absence of FeSOD activity [38, 39]. Hence, mitochondria of these parasites are actively engaged in producing as well as eliminating ROS molecules like in mitochondria of higher eukaryotes. Therefore, the only mitochondrion of the parasite along with their specific ETC complexes including fumarate reductase and SODs is a promising drug target against leishmaniasis.

29.4.2 Thiol Metabolism

Other than mitochondrial involvement in redox metabolism of these parasites, there are also antioxidants which significantly participate in redox cycle and also essential for the survival as well as virulence of the parasites. In this section, we discuss antioxidant molecules, their synthesis, and their role in redox metabolism.

29.4.2.1 Trypanothione Synthesis

The occurrence of specific molecule (N^1,N^8-bis(*L*-γ-glutamyl-*L*-hemicystinyl-glycyl)spermidine) of redox reaction named trypanothione (Try) with a molecular weight of 721 has been validated in trypanosomatids [40]. Try in these parasites is predominantly used to maintain reducing environment inside the cells while this function is performed by glutathione in higher eukaryotes. Try is synthesized using glutathione and spermidine as substrates and using ATP as energy source (Fig. 29.3). Previous reports have confirmed the presence of TRYS gene encoding trypanothione synthetase (TS) which catalyzes the reaction of glutathione (GSH) and spermidine/glutathionylspermidine to form Try [41, 42]. TS in *L. major* displays substrate inhibition at different concentrations of GSH. TS from *L. donovani* has also been isolated which represents 95% homology with TS of *L. major* [42]. Interestingly, these parasites need TS for their survival in oxidative stress, and inhibition of this enzyme suppresses the growth of *Leishmania* parasites as well [43]. Moreover, the abovementioned enzymes of polyamine biosynthesis have been well-studied as promising drug targets against leishmaniasis since several polyamine analogues are documented to affect the growth of parasites [44–46]. Transport of

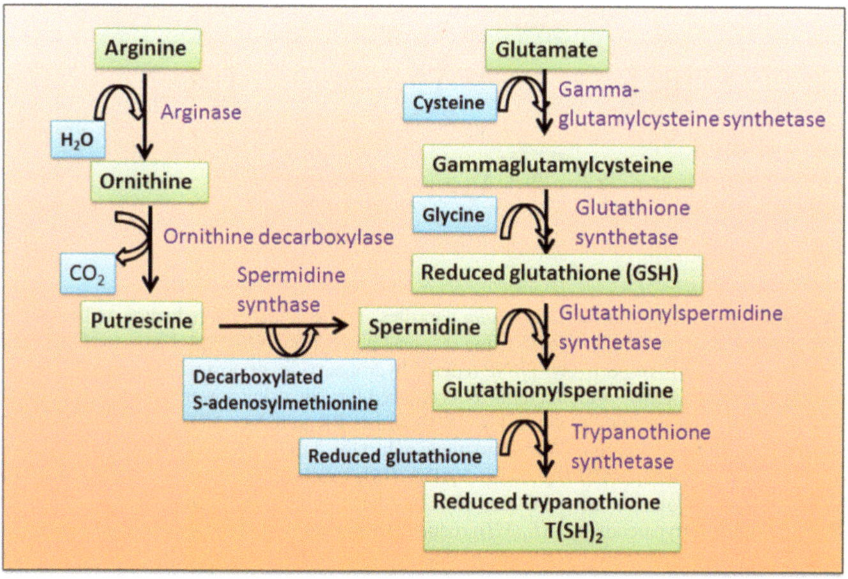

Fig. 29.3 Synthesis of trypanothione. Reduced trypanothione is synthesized using reduced glutathione and glutathionylspermidine as substrates in a reaction catalyzed by trypanothione synthetase enzyme. Glutathionylspermidine is produced by the reaction of reduced glutathione and spermidine. Spermidine is formed by the reaction of decarboxylated S-adenosylmethionine with putrescine which is produced by decarboxylation of ornithine. Arginine reacts with water molecule to produce ornithine catalyzed by arginase (only exists in *Leishmania*, not in other trypanosomatids). Reduced form of glutathione is produced by the reaction of glycine with gamma-glutamylcysteine which is formed by reaction of cysteine with glutamate

polyamines is equally important for the growth of parasites since molecules targeting polyamine transport exhibit potent activity against *Leishmania* parasites [47].

29.4.2.2 Trypanothione Metabolism

$T(SH)_2$ and $T(S)_2$ are representative forms of reduced trypanothione and oxidized trypanothione, respectively. Using NADPH, $T(S)_2$ is enzymatically reduced to $T(SH)_2$ by the enzyme trypanothione reductase (TR), a nicotinamide adenine dinucleotide phosphate ($NADP^+$)-dependent enzyme (Fig. 29.4). $T(SH)_2$ then reacts with GSSG (oxidized glutathione) and reduces it to GSH by oxidizing itself to $T(S)_2$. The sum of both the above reactions leads to final reaction explaining how trypanothione acts as cofactor for the reduction of GSSG [40]. Interestingly, $T(SH)_2$ is also involved in detoxification of metals as well as different drugs since more $T(SH)_2$ has been found in metal-resistant parasites [48]. TR is the member of FAD (flavin adenine dinucleotide) disulfide oxidoreductase protein family and is the key enzyme of thiol metabolism of trypanosomatids including *Leishmania* which is not present in mammals where similar function is performed by glutathione reductase (GR). Though TR and GR share 40% homology in sequence, there are some differences in their active sites. The mechanism of reduction of $T(S)_2$ by TR begins with

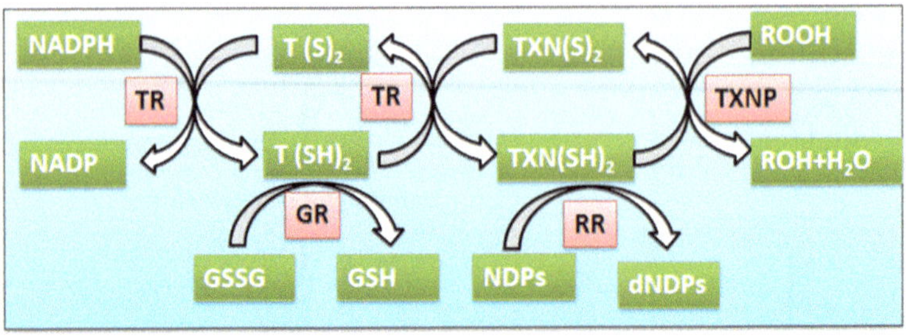

Fig. 29.4 Thiol metabolism in *Leishmania* parasites. TR participates in reduction and oxidation of trypanothione from $T(S)_2$ to $T(SH)_2$ using reducing power from NADPH (change to NADP) and from $T(SH)_2$ to $T(S)_2$, respectively. TXNP reduces hydroperoxides or organic peroxides by oxidizing $TXN(SH)_2$ to $TXN(S)_2$ which returns to its reduced form by using power from $T(SH)_2$. $T(SH)_2$ also helps GR to reduce GSSG while $TXN(SH)_2$ assists RR to reduce NDPs to generate dNDPs (precursor for DNA synthesis). *TR* trypanothione reductase, *NADP* oxidized nicotinamide adenine dinucleotide phosphate, *NADPH* reduced NADP, *$T(S)_2$* oxidized trypanothione, *$T(SH)_2$* reduced trypanothione, *GR* glutathione reductase, *GSSG* oxidized glutathione, *GSH* reduced glutathione, *$TXN(S)_2$* oxidized tryparedoxin, *$TXN\ (SH)_2$* reduced tryparedoxin, *TXNP* tryparedoxin peroxidase, *RR* ribonucleotide reductase, *NDPs* nucleoside diphosphates, *dNDPs* deoxynucleoside diphosphates, *ROOH* peroxides (R-H atom or any other)

donation of electrons from NADPH to FAD producing a transient $FADH^-$ anion, which attacks the enzyme's disulfide bond (Cys52– Cys57) at the active site rendering redox-active Cys52 and Cys57 in the dithiol form of the enzyme. Cys57 reacts with flavin and Cys52 breaks the disulfide group of trypanothione disulfide, forming a covalent enzyme-substrate complex. Next, Cys57 interacts with nucleophile and breaks the sulfide of Cys52, displacing Try from active site of TR and generating protein disulfide group. The histidine (His461′) residue of TR participates as an acid and base catalyst. The same reaction is followed in reduction of GSSG by GR with differences in the active sites of TR and GR. The active site of TR contains hydrophobic and negatively charged region (Glu19, Trp22, Ser110, Met114, and Ala343) which helps in binding of spermidine of trypanothione disulfide, while active site of GR possesses hydrophilic and positively charged region (Ala34, Arg37, Ile113, Asn117, and Arg347) to bind glycine carboxylates of glutathione disulfide [49]. Furthermore, the requirement of TR for survival as well as virulence has been evidently demonstrated in these parasites particularly inside the host cells. *Leishmania* cells having only one wild-type allele for TR show weakened activity of TR for redox metabolism, thereby with reduced ability to survive in activated macrophages [50, 51]. TR is a well-studied enzyme as a drug target against leishmaniasis, and we will be describing some inhibitors of TR later.

29.4.2.3 Tryparedoxins and Tryparedoxin Peroxidase

The other antioxidant enzymes found in higher organisms, like catalase, typical selenium-containing glutathione peroxidases, and classical thioredoxins which reduce hydroperoxides, are absent in trypanosomatids. Thus, to perform functions of the above enzymes, there are proteins like tryparedoxins (TXNs) instead of thioredoxins (TXNs) and tryparedoxin peroxidase (TXNP) and also ascorbate peroxidases (APXs) instead of typical peroxidases in these parasites. TXNs are dithiol proteins that act as oxidoreductases and share only 15% homology with thioredoxins and possess an active site (Trp-Cys-Pro-Pro-Cys-Arg) different from the active site of thioredoxins (Trp-Cys-(Gly/Ala)-Pro-Cys-Lys). TXNs use $T(SH)_2$ for their reduction during electron transfer between TR and flavoprotein and hence are termed as tryparedoxins. TXNs are used as electron donor to reduce peroxides by TXNPs [52, 53]. TXNP of *L. donovani* (LdTXNPx) is a typical 2-cysperoxiredoxin (Prx) or peroxidoxin that shares common features with other trypanosomatids and with mammals present in cytosol as well as in mitochondria. Basically, catalytic reaction of TXNPs involves reduction of peroxides using proximal thiol group of Cys (Cys52) that is oxidized to cysteine sulfenic acid (Cys-SOH). The other c-terminal thiol group of Cys (Cys173) of TXNPs attacks Cys-SOH and forms a stable intersubunit disulfide bond. This disulfide bond is then broken by TXN attacking on cys173 of TXNP and reducing it being oxidized itself. Since conformation of TXNP in its reduced form makes cys52 highly activated and accessible, only cys52 is attacked by peroxides, not cys173, whereas oxidized conformation of TXNP exposes cys173 and provides space for bulky molecules so that TXN can bind and reduce it. Oxidized TXN is reduced by trypanothione as mentioned previously, indicating Try-TXN-TXNP collectively perform redox metabolism and reduce toxic peroxides in these parasites [54]. TXNP in *L. donovani* is very essential for the survival, drug response, as well as virulence of the parasites [55]. Moreover, elevated level of TXNPs in *L. donovani* cells resistant to antimonials as well as to amphotericin B has been revealed, suggesting importance of TXNP [56, 57]. Furthermore, overexpression of mTXNP in *L. donovani* results in protection of cells from PCD due to oxidative stress [58]. Cytosolic TXNPs of *L. infantum* (*LicTXNPx1* and *LicTXNPx2*) are reduced by *LiTXN1* involving reduction of hydroperoxides, alkyl peroxides, and also peroxynitrite and therefore help parasites to live inside host cells. On the other hand, *LiTXN2* has shown specificity for mitochondrial TXNP (*LimTXNPx*) which prevents parasites from internal peroxides [59]. In addition to detoxifying peroxides and other ROS molecules, these peroxidoxins are also involved in detoxifying reactive nitrogen species (RNS) challenged by *L. chagasi* parasites harbored inside the host cells [60]. Thus, both cytosolic and mitochondrial TXNPs complement one another to maintain reducing environment of the parasites. Though mTXNP is abundant in promastigotes, their important role in establishing infection by amastigotes in host has shown to be based on the chaperone-like activity of this enzyme during heat stress inside host cells [61]. Other than assisting in redox metabolism, these enzymes also help in DNA synthesis by donating electrons to ribonucleotide reductase for the synthesis of deoxyribonucleotides. Further, mitochondrial TXNs regulate mitochondrial DNA replication by reacting with Universal Minicircle Sequence Binding Protein essential for kDNA replication [62]. Recently, purified cTXNP (c, cytosolic) of *L. donovani* has been shown to supress immune response of the host cells against parasites and hence help in enhancing growth of the parasites [63]. Interestingly, in the structure of TXN-TXNP pair in *L. major*, the location of the epitopes has been recognized which makes the protein antigenic and therefore possibly fit to become anti-leishmanial polyprotein vaccine [64].

29.4.2.4 Other Peroxidases

Another well-studied peroxidase enzyme of *Leishmania* is ascorbate peroxidase (APX) which is actually a class I heme-containing peroxidase enzymes present in photosynthetic organisms. It participates in oxidation of ascorbate in H_2O_2-dependent manner and is extensively studied in *L. major*. Previous studies have reported that APXs of trypanosomatids are different from plant-APXs in using cytochrome c as substrate instead of ascorbate [65, 66]. The significance of APX in *L. major* is affirmed as its overexpression protects cells from ROS-mediated oxidation of cardiolipins as well as ROS-mediated PCD [67, 68]. Also, the deletion of gene for APX makes parasites more susceptible toward ROS molecules, indicating this enzyme is essential for virulence and survival of parasites inside macrophages [69].

29.4.2.5 Other Antioxidants Present in Small Amount

More than 1 mM Ovothiol A has also been reported to be present in five species of *Leishmania* promastigotes while less than 0.25 mM in other trypanosomatids. Though ovothiol is present equal to or more than trypanothione in *Leishmania* promastigotes, its non-enzymatic activity to eliminate H_2O_2 is less efficient than trypanothione. Also, ovothiols have been shown to be absent in amastigotes of *L. donovani* but present in amastigotes of *L. major* in lesser amount. Varied presence and functions in trypanosomatids suggest further investigation of this molecule in these parasites [70]. 3-Mercaptopyruvate sulfurtransferase possessing unusual C-terminal extension has been revealed to be present in *L. major* and *L. mexicana* and participates in thioredoxin and antioxidant metabolism. Being itself oxidized, this enzyme can reduce TXN using thiosulfate as substrate. This enzyme generally catalyzes transfer of sulfur to cyanide to become thiocyanate [71]. Because of its selectivity, this enzyme could be an attractive target for *Leishmania* parasites. In *L. major*, pteridine reductase 1 (PTR1) is generally involved in reduction of biopterin to dihydrobiopterin and tetrabiopterin and also participates in balancing oxidative stress as its deletion makes parasites more susceptible toward oxidative stress [72].

Trypanothione is reported to mediate not only reduction of GSSG but also reduction of ovothiol, thioredoxin, and NDP with the help of tryparedoxin/ribonucleotide reductase and hence is a validated drug target in leishmaniasis [43].

29.5 Anti-Leishmanial Molecules

Our earlier sections have described how oxidative stress is deleterious for the parasites and how important it is to control this stress for parasite survival. Now, we discuss the different anti-leishmanial molecules killing the parasites by either inhibiting enzymes of redox metabolism or causing ROS-mediated apoptosis-like cell death.

29.5.1 Anti-Leishmanial Molecules Targeting TR

The structural differences together with significance of TR in survival of *Leishmania* inside macrophages make TR a validated drug target against leishmaniasis. Consequently, a number of natural as well as synthetic molecules have been studied to inhibit TR but here we focus on TR inhibitors with potent anti-leishmanial activity. To begin with, antimonials which have been a first choice to treat leishmaniasis inhibit TR to kill parasites. Sb(III) (bioactive trimonial antimony) is coordinately attached to TR using the two redox-active catalytic cysteine residues (Cys52 and Cys57), one threonine residue (Thr335), and one histidine residue (His461′) from the second subunit of the dimer [73]. Furthermore, several tricyclic compounds have been screened for TR-inhibiting activity in trypanosomatids, but phenothiazine is the most studied in case of *Leishmania* parasites. ED_{50} (maximum dose at which 50% of population is affected) of chlorpromazine, one of the phenothiazine in *L. donovani*, has been calculated to be 4.9 μM and also chlorpromazine has shown strong in vivo activity for the parasites in hamster infected with *L. donovani* [74, 75]. Besides, other tricyclic antidepressant molecules also inhibit TR significantly along with promising anti-leishmanial activity among which clomipramine displays the most potent activity in amastigotes with LD_{50} (dose at which 50% of population is killed) of 1.18 μM [76]. Based on the previous study of quinacrine (also called mepacrine) showing TR inhibition, quinacrine derivatives have been synthesized using sulfonamides and urea derivatives and have been examined to improve TR activity as well as antiproliferative activity in *Leishmania* parasites. Among them, the derivatives which are found more potent and selective inhibitor of TR also present promising anti-leishmanial activity, suggesting strong relation between TR inhibition and anti-leishmanial activity [77]. Quaternary arylalkylammonium 2-chlorophenyl phenylsulfides have been shown to enhance TR inhibitory activity along with better anti-parasitic activity [78]. In spite of tricyclic TR inhibitors, there are also reports of bicyclic and heterocyclic compounds showing strong inhibition of TR. However, they have been mostly found toxic toward mammalian cells explaining why there are only few reports of these derivatives against *Leishmania* parasites [79]. Considering different substrates used by TR of parasites and GR of mammals, several polyamine (spermine and spermidine) derivatives have been designed and investigated for their TR-inhibiting activity, and their potency against trypanosomes has been evaluated. Conjugates of quinones and natural polyamines have been observed to possess potent activity against *Leishmania* parasites with mitochondrial dysfunction and ATP depletion [80]. Moreover, several plant-derived 1,4-naphthoquinones (like plumbagin, menadione, and lapachol) have been demonstrated as subversive substrates of TR which convert antioxidant TR enzyme to pro-oxidant form. Basically, TR reacts with subversive substrates like quinones and reduces them into semihydroquinones in a single-electron step which subsequently react with molecular oxygen to form superoxide radicals leading to parasite death. In addition, these naphthoquinones are also known to inhibit TR in non-competitive manner. Therefore, using dual effects on TR making harsh oxidative stress status, these naphthoquinones present potent anti-leishmanial activity

[81, 82]. In addition to naphthoquinones, doxorubicin and mytomycin are also reported to act as subversive substrates and to produce ROS as well as to inhibit TR in *L. donovani* [83]. Furthermore, some nitrofurans like nifurtimox and furazolidone also act as subversive substrates for TR with promising anti-leishmanial activity; however, later studies point out their similar activity toward GR [84]. Some other nitro heterocycles like megazole and their derivatives have revealed TR-inhibiting activity together with anti-leishmanial activity [85]. Among them, dinitrodiphenylthioethers have been studied to kill intracellular amastigotes effectively through generation of ROS. Modifications in the structure of dinitrodiphenylthioethers result in more efficient molecules exhibiting good in vivo activity against intracellular *Leishmania* parasites [86]. Quinazolinone molecules have also been investigated to design TR inhibitors which have presented promising leishmanicidal activity [87]. Among organometallic compounds inhibiting TR, auranofin (a gold-containing drug) is found to be very effective against *Leishmania* parasites specifically in in vivo studies [88]. Using bioinformatics tools, several natural compounds have been screened for inhibiting TR of *Leishmania* along with anti-leishmanial activity [89]. Recently, some peptides have been designed for promising TR inhibition as well as anti-leishmanial activity [90]. Several TR inhibitors possessing anti-leishmanial activity have been discussed above, suggesting that TR activity is very essential for parasite survival.

29.5.2 Leishmanicidal Compounds Inhibiting TS

Being uniquely present in trypanosomatids, TS (trypanothione synthetase) is also a potent drug target against leishmaniasis which has been widely studied after TR for its importance. However, there are fewer studies on TS and its inhibitors in *Leishmania* parasites. In view of this essentiality, various analogues of GSH have been designed to inhibit TS activity in trypanosomatids but have not got successful results as their peptidic nature interferes with their entry into the cells without compromising membrane [91]. In spite of that, some paullones like N5-substituted paullone in *L. infantum* also have been tested against TS, displaying both anti-leishmanial and TS inhibition activity [92]. In addition, some naturally occurring molecules are also reported to possess TS-inhibiting and leishmanicidal activity like conessine, tomatine, uvaol, and betulin [93]. Large-scale screening of synthetic compounds using bioinformatics tools has been carried out, presenting oxabicyclo[3.3.1]nonanones as both TR and TS inhibitors with promising leishmanicidal activity both in promastigotes and amastigotes without affecting host [94]. This study suggests a validated approach for targeting both the important enzymes of redox metabolism simultaneously TR and TS, not allowing the parasites to compensate for one enzyme.

29.5.3 Compounds Inhibiting Other Enzymes of Redox Reactions

As described previously, there are several enzymes that participate in redox metabolism of parasites other than TS and TR. Though other enzymes of redox reaction have not been studied in detail yet, some successful results in trypanosomatids suggest other ways to target these parasites. Along these lines, N,N-disubstituted 3-aminomethyl quinolone derivatives have been reported recently that target tryparedoxin peroxidase of *L. major* [95]. Recently, a series of ascorbate peroxidase inhibitors have also been designed to find promising anti-leishmanial hits among several thousand molecules, and three hits have been identified showing potent anti-leishmanial activity without affecting host [96]. Moreover, hypericin, an inhibitor of spermidine synthase of *L. donovani* (LdSS), has been identified to impair parasitic growth effectively causing necrotic cell death. Although LdSS is an enzyme of redox metabolism, the mode of action of hypericin is not based on ROS generation but on other functions of spermidine that are essential for parasites survival [97]. Thus, inhibiting other enzymes of redox metabolism also halts parasitic growth, whereas function of these enzymes sometimes may be performed by TR in their absence.

29.5.4 ROS-Generating Compounds Causing Apoptosis-Like Cell Death

The significance and occurrence of apoptosis-like cell death in these parasites has already been described. Here, we discuss anti-leishmanial molecules that generate ROS and remarkably alter the physiology of the parasites causing cell death.

29.5.4.1 Apoptosis-Like Cell Death Induced by Standard Drugs

Starting with standard anti-leishmanial drugs, antimonials have been reported to increase ROS production in *Leishmania*-infected macrophages through phosphorylation of protein kinase C (PKC) Ras, phosphoinositide 3-kinase (PI3K), and extracellular-signal regulated kinase (ERK) [98]. In antimonial-treated promastigotes, ROS is generated which is preceded by both increase in cytosolic calcium level and mitochondrial depolarization leading to apoptotic-like cell death of parasites [15]. Involvement of ROS generation in cell death induced by antimonial has also affirmed where the presence of antioxidants like N-acetyl cysteine reduces the effects of drugs and also by high level of thiols in antimony-resistant parasites [15, 99]. Recently, it has been reported that liposomal formulation of amphotericin B (KalsomeTM10) causes apoptosis-like cell death in *Leishmania* promastigotes including generation of H_2O_2 and superoxide radicals [100]. Miltefosine-induced cell death has also been demonstrated to be dependent on oxidative stress since overexpression of superoxide dismutase of *L. donovani* results in inhibition of PCD induced by miltefosine [101]. Also, a recent report has described how clinical isolates of *L. donovani* have not shown susceptibility toward miltefosine which are resistant to oxidative stress [102].

29.5.4.2 Natural Compounds Causing ROS-Mediated Apoptosis-Like Cell Death

In addition to standard anti-leishmanial drugs, various naturally occurring plant-derived metabolites have been identified to induce ROS-mediated apoptosis in these parasites. Curcumin, a well-studied natural molecule, is known to produce ROS and disturb calcium homeostasis to kill *Leishmania* parasites via PCD involving mitochondrial dysfunction [103]. Plumbagin is a plant-derived naphthoquinone that was previously reported to inhibit TR and to generate ROS. In our studies on plumbagin, the role of ROS in triggering PCD in both promastigotes and amastigotes is evidenced with mitochondrial dysfunction. Pretreatment of *N*-acetyl cysteine (NAC) prior to plumbagin causes restoration of cell viability, suggesting cell death principally depends on disturbing redox metabolism [82, 104]. Another plant-derived natural molecule quercetin exhibits promising anti-leishmanial activity in both forms of the parasites with significant production of ROS mediating apoptotic-like cell death. Also, in the presence of reduced glutathione and NAC, cell death caused by quercetin is prevented [105, 106]. The other potent anti-leishmanial agent, baicalein, induces caspase-independent PCD through ROS generation. Though baicalein mainly targets endonuclease G of *L. donovani* to cause PCD, reduced level of ROS has been revealed in endonuclease G-downregulated cells, confirming relation of cell death with oxidative stress [33]. The most potent pterocarpanquinone, LQB-118, is demonstrated to stimulate oxidative stress-mediated apoptosis-like cell death, including mitochondrial dysfunction in *L. amazonensis* [107]. Zerumbone isolated from *Zingiber zerumbet* presents promising anti-leishmanial activity in both forms of *L. donovani* parasites and induces apoptosis-like cell death which is mediated by ROS generation [108]. One of the broad-spectrum plant-derived flavonoids, apigenin, produces oxidative stress in *L. amazonensis*, resulting in apoptosis-like cell death along with significant mitochondria damage. By providing reducing environment, the effect of apigenin can be decreased [109]. (−)-Epigallocatechin 3-O-gallate shows promising leishmanicidal activity in *L. braziliensis* caused by ROS generation together with loss of mitochondrial membrane potential and ATP depletion [110]. A plant-derived metabolite from *Nyctanthes arbor-tristis* termed iridoid glucosides inhibits TR and exhibits a rise in ROS level and hence disturbs cellular redox homeostasis in *Leishmania* parasite finally leading to apoptosis-like cell death [111]. Other natural molecules which have been shown to imbalance redox homeostasis by inhibiting TR to kill *L. donovani* are b-carboline-quinazolinone hybrids, while one of them has displayed features of apoptosis-like cell death [87]. Our studies have shown that K-09, an orally active clerodane diterpene, isolated from the leaves of *Polyalthia longifolia*, induces prominent damage in mitochondrial structure with abnormal function which mediates apoptosis involving ROS production and disruption in calcium homeostasis [112] (Fig. 29.5). Recently, mahanine, isolated from an Indian medicinal plant *Murraya koenigii*, reveals strong anti-leishmanial activity both in vitro and in vivo with apoptosis-like cell death triggers via oxidative stress in *L. donovani*. Additionally, mahanine activates host macrophages to kill parasites by increasing ROS and NO (nitric oxide) levels. Also, mahanine has shown to decrease ascorbate peroxidase level due to excess ROS

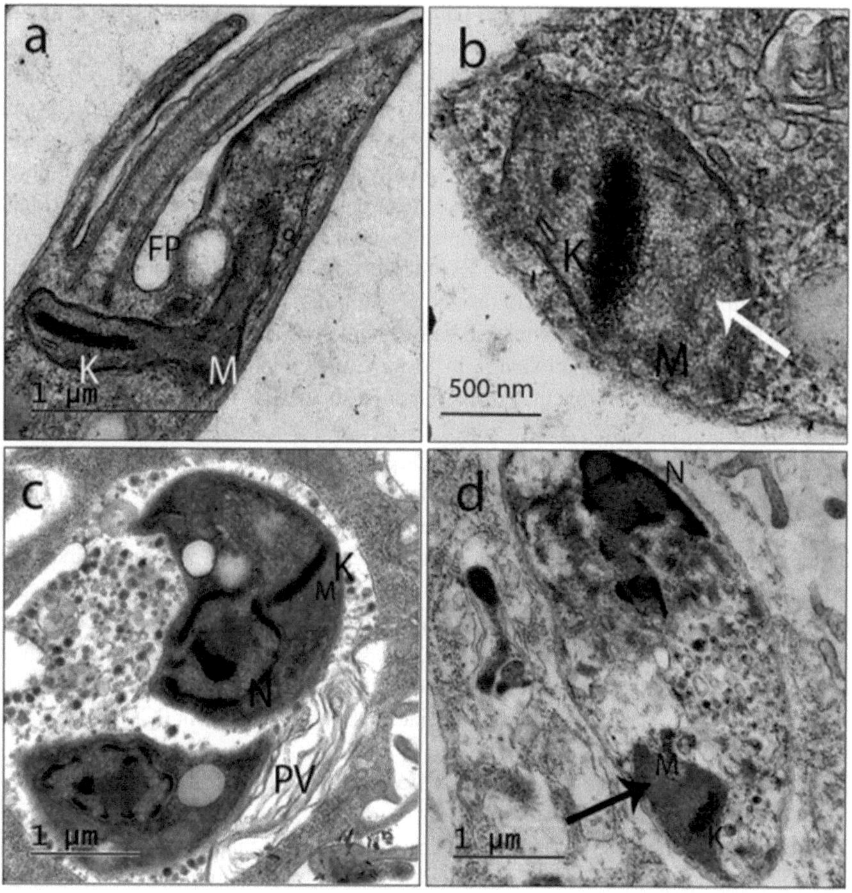

Fig. 29.5 Ultrastructural changes in the mitochondrion of *Leishmania* parasites under oxidative stress – (**a**) healthy mitochondrion of untreated promastigotes. (**b**) Mitochondrial damage with decondensation of kDNA after treatment with plumbagin (white arrow). (**c**) Normal mitochondrion of untreated amastigotes. (**d**) Swollen mitochondrion in amastigote post K-09 treatment (black arrow). *K* kinetoplast, *FP* flagellar pocket, *M* mitochondria, *PV* parasitophorous vacuole, *N* nucleus

[113]. Betulin, previously discussed to inhibit TS, has been reported to cause apoptosis-like cell death mediated through ROS [114]. Also, the derivative of betulin, disuccinyl betulin, has been shown to trigger caspase-dependent cell death in *L. donovani* involving ROS generation [115].

29.5.4.3 PCD Caused by Inhibitors of Mitochondrial Complexes Involving ROS Production

Importantly, inhibitors of mitochondrial ETC complexes are also known to induce ROS-mediated apoptosis-like cell death. Tafenoquine inhibits mitochondrial complex III which is linked to loss of mitochondrial membrane potential, fall in ATP level, and rise in ROS level leading to PCD. 3,3-Diindolylmethane impedes F0F1-ATP synthase (complex V) functioning and causes depletion in ATP level, and increase in ROS level which is followed by collapse of mitochondria ultimately results in cell death [116]. 8-Aminoquinoline analogue, sitamaquine, inhibits mitochondrial complex II and induces depletion in ATP level and depolarization of mitochondrial membrane in association with ROS production which consequently lead to cell death [117]. Furthermore, the classical inhibitors of ETC complexes also display apoptosis-like cell death features in *Leishmania* parasites, for example, rotenone, thenoyltrifluoroacetone (TTFA), and antimycin A as complex I, II, and III inhibitors, respectively. Inhibition of complex I induces superoxide generation followed by hyperpolarization of mitochondrial membrane. TTFA which inhibits complex II causes high production of hydrogen peroxides and low production of superoxide with loss of mitochondrial membrane potential. Parasites treated with antimycin A result in rise in superoxide level only [35].

29.5.4.4 PCD Induced by Nutrition Deprivation

Moreover, as we have mentioned earlier, apoptosis may occur due to deprivation of nutrition in these parasites; the parasitic cells deprived of L-arginine exhibit ROS-mediated apoptosis-like cell death. Lack of L-arginine in culture interferes with the synthesis of polyamine and hence trypanothione, which therefore results in increase of ROS level leading to cell death [118]. Additionally, in a way to find the importance of zinc (Zn) for the growth and survival of *Leishmania* parasites, it has been observed that depletion of Zn in these parasites causes ROS generation, promoting mitochondrial depolarization and DNA fragmentation and finally causing ROS-mediated caspase-independent PCD [119].

29.5.4.5 Some Other Anti-Leishmanial Agents Related to ROS-Mediated Cell Death

Furthermore, among membrane-acting anti-fungal imidazoles, econazole has been identified to increase ROS level significantly with disruption in membrane integrity of *Leishmania* parasites [120]. In our study, another anti-fungal Natamycin has also shown to induce apoptosis-like cell death via increase in cytosolic calcium level which is tightly linked with ROS generation and mitochondrial dysfunction [121]. Camptothecin, a topoisomerase inhibitor, induces ROS generation in *Leishmania* parasites along with typical features of PCD [122]. Another report of ROS-mediated apoptosis-like cell death is interestingly shown in 15d-prostaglandin J2, a lipid mediator-treated parasite [123].

Collectively, we have described how ROS generation significantly causes apoptosis-like cell death in these parasites, giving examples of various ROS-producing molecules (Table 29.1).

29.6 Concluding Remarks

Redox systems play a crucial role in the survival of *Leishmania* parasites inside the host cells where they are challenged by various ROS and RNS molecules. Trypanothione reductase (TR) and trypanothione synthetase (TS) enzymes of redox

Table 29.1 Anti-leishmanial compounds showing diverse mode of cell death due to oxidative stress

Sl. No.	Compound	Target/Mode of action	References
1.	Sb(III)	Inhibits TR, ROS-mediated PCD	[15, 73]
2	KalsomeTM10 (formulation)	ROS-mediated PCD	[100]
3.	Miltefosine	PCD and cell death dependent on SOD	[101]
4.	Chlorpromazine	Inhibits TR	[74, 75]
5.	Clomipramine	Inhibits TR	[76]
6.	Quinacrine	Inhibits TR	[77]
7.	Quaternary arylalkyl ammonium 2-Chlorophenyl phenyl sulfides	Inhibits TR	[78]
8.	Quinone conjugates	Inhibits TR	[80]
9.	Napthoquinones: Plumbagin, Menadione	Inhibits TR, ROS-mediated apoptosis	[81, 82, 104]
10.	Doxorubicin, mitomycin	Inhibits TR, subversive substrate	[83]
11.	Nifurtimox, furazolidone	Inhibits TR	[84]
12.	Megazole	Inhibits TR	[85]
13.	Dinitrodiphenylthioethers	Inhibits TR	[86]
14.	Quinazoline molecules: b-carboline-quinazolinone	Inhibits TR, Apoptotic cell death	[87]
15.	Organometallic compounds: auranofin	Inhibits TR	[88]
16.	N5-substituted paullone	Inhibits TS	[91]
17.	Naturally occurring molecules: conessine, tomatine, uvaol, and betulin	Inhibits TS Betulin, apoptotic cell death	[92, 114]
18.	Oxabicyclo[3.3.1]nonanones	Inhibits TS	[93]
19.	N,N-disubstituted 3-aminomethyl quinolone	Inhibits TXNP	[95]
20.	Asinex and ZINC compound databases	Inhibits APX	[96]
21.	Hypericin	Spermidine synthase inhibitor	[97]
22.	Curcumin	ROS-mediated PCD	103]
23.	Quercetin	ROS-mediated PCD	[105, 106]
24.	Baicalein	ROS-mediated PCD	[33]
25.	Pterocarpanquinones, LQB-118	Oxidative stress-mediated apoptosis	[107]
26.	Zerumbone	Apoptosis-like cell death	[108]
27.	Apigenin	Oxidative stress-mediated apoptosis	[109]
28.	(−)-epigallocatechin 3-O-gallate	ROS-mediated cell death.	[110]
29.	Iridoid glucosides	ROS-mediated apoptosis	[111]
30.	Mahanine	Oxidative stress-mediated apoptosis	[113]

(continued)

Table 29.1 (continued)

Sl. No.	Compound	Target/Mode of action	References
31.	K-09	Mitochondrial dysfunction, PCD	[112]
32.	Disuccinyl betulin	Apoptotic cell death	[115]
33.	Tafenoquine	Complex III inhibitor, ROS-generated cell death	[36]
34.	3,3-Diindolylmethane	F0F1-ATP synthase inhibitor	[116]
35.	8-Aminoquinoline analogue, sitamaquine	Complex II inhibitor, PCD	[117]
36.	Rotenone	Complex I inhibitor	[35]
37.	Thenoyltrifluoroacetone (TTFA)	Complex II inhibitor	[35]
38.	Antimycin A	Complex III inhibitors; apoptosis-like cell death	[35]
39.	Econazole	ROS-mediated cell death with disruption of membrane integrity	[120]
40.	Camptothecin	Topoisomerase inhibitor producing ROS	[122]
41.	Natamycin	Disruption in Ca^{2+} homeostasis with ROS generation	[121]
42.	15d-prostaglandin J2	ROS-mediated apoptosis	[123]

PCD programmed cell death, *TR* trypanothione reductase, *ROS* reactive oxygen species, *APX* ascorbate peroxidase, *TXNP* tryparedoxin peroxidase

system have been confirmed as validated drug targets involved in defending the parasites against oxidative stress. In addition, TXNs, TXNPs, and ascorbate peroxidases are also involved in maintaining a reduced environment in the cells as well as for the virulence and intracellular survival of parasites. Therefore, in the search for molecules inhibiting redox enzymes, several compounds have been designed and screened for anti-leishmanial activity using virtual, high throughput screening and biochemical studies. A few molecules with potent leishmanicidal activity have been identified demonstrating TR-inhibiting activity such as antimonials; tricyclic, bicyclic, and heterocyclic compounds; polyamines; and various natural products such as naphthoquinones, subversive substrates of TR. TS inhibitors include analogues of GSH, paullones, oxabicyclo[3.3.1]nonanones, conessine, tomatine, uvaol, and betulin. However, as compared to TR, fewer studies have been performed in case of TS. Also, inhibitors of TXNs, TXNPs, and ascorbate peroxidases exhibit promising anti-leishmanial activity like *N*, *N*-disubstituted 3-aminomethyl quinolone. We have cited several studies related to ROS-mediated apoptosis-like cell death and have mentioned the importance of apoptosis-like regulated cell death in these unicellular parasites. Interestingly, some molecules induce ROS-mediated cell death of these parasites along with inhibition of redox enzymes. In addition, our studies have also demonstrated the effectiveness of oxidative stress inducers in killing both forms of the parasite. We have provided evidences that oxidative stress results in severe structural and functional disruption of the single parasite mitochondrion, which results in apoptosis-like cell death. Unfortunately, among several inhibitors of redox enzymes, only a few hold drug-like properties such as selectivity over human counterpart enzymes, low toxicity and side effects, high bioavailability, and other drug-like characteristics. Several redox inhibitor molecules do not possess potent anti-leishmanial activity since these parasites are able to survive with as low as 10% of TR activity or 5–10% $T(SH)_2$ levels [94].

The complications of leishmaniasis and the limitations of currently available drugs necessitate the discovery of novel and validated drug targets along with new safer drug molecules. Probably, a combination of drugs could be used to target both TR and TS simultaneously thereby not allowing parasites to use alternatives for $T(SH)_2$. Moreover, molecules which are subversive substrates of TR along with TR inhibitors may prove to be good approach for drug development because of their dual role in killing parasites. Repurposing of available drugs, especially those involved in redox metabolism, could be investigated for anti-leishmanial activity. Further studies should be undertaken to design new leishmanicidal molecules with drug-like properties and to improve drug-like properties of existing redox inhibitors. Likewise, the molecules triggering ROS-mediated apoptosis-like cell death along with disrupting redox homeostasis can open another way to target these parasites. In addition, other molecules, which have shown potent activity against redox enzymes of other trypanosomatids and protozoan parasites, should also be evaluated for leishmanicidal activity. Therefore, targeting the redox system of these parasites could be a very attractive strategy for new drug discovery in leishmaniasis.

Acknowledgments This work was supported by CSIR network project BSC0114. Fellowships from ICMR and CSIR to BPA and SPM are acknowledged. This is CSIR-CDRI communication no. 9858.

References

1. Singh N, Kumar M, Singh RK (2012) Leishmaniasis : current status of available drugs and new potential drug targets. Asian Pac J Trop Med 5(6):485–497. https://doi.org/10.1016/S1995-7645(12)60084-4
2. Kamhawi S (2006) Phlebotomine sand flies and Leishmania parasites : friends or foes ? Trends Parasitol 22(9):439–445. https://doi.org/10.1016/j.pt.2006.06.012
3. Singh N, Mishra BB, Bajpai S et al (2014) Natural product based leads to fight against leishmaniasis. Bioorg Med Chem 22(1):18–45. https://doi.org/10.1016/j.bmc.2013.11.048
4. Singh OP, Singh B, Chakravarty J, Sundar S (2016) Current challenges in treatment options for visceral leishmaniasis in India: a public health perspective. Infect Dis Poverty 5:19. https://doi.org/10.1186/s40249-016-0112-2
5. De Souza W, Attias M, Rodrigues JC (2009) Particularities of mitochondrial structure in parasitic protists (Apicomplexa and Kinetoplastida). Int J Biochem Cell Biol 41(10):2069–2080. https://doi.org/10.1016/j.biocel.2009.04.007
6. Liu B, Liu Y, Motyka SA, Agbo EEC, Englund PT (2005) Fellowship of the rings: the replication of kinetoplast DNA. Trends Parasitol 21(8):363–369

7. De Souza W (2008) An introduction to the structural organization of parasitic protozoa. Curr Pharm Des 14(9):822–838
8. Menna-Barreto RF, de Castro SL (2014) The double-edged sword in pathogenic trypanosomatids: the pivotal role of mitochondria in oxidative stress and bioenergetics. Biomed Res Int 2014:614014. https://doi.org/10.1155/2014/614014
9. Carranza JC, Kowaltowski AJ, Mendonça MA, de Oliveira TC, Gadelha FR, Zingales B (2009) Mitochondrial bioenergetics and redox state are unaltered in Trypanosoma cruzi isolates with compromised mitochondrial complex I subunit genes. J Bioenerg Biomembr 41(3):299–308
10. Chaudhuri M, Ott RD, Hill GC (2006) Trypanosome alternative oxidase: from molecule to function. Trends Parasitol 22(10):484–491
11. Santhamma KR, Bhaduri A (1995) Characterization of the respiratory chain of Leishmania donovani promastigotes. Mol Biochem Parasitol 75(1):43–53
12. Chakraborty B, Biswas S, Mondal S, Bera T (2012) Stage specific developmental changes in the mitochondrial and surface membrane associated redox systems of Leishmania donovani promastigote and amastigote. Biochemistry (Mosc) 75(4):494–518
13. Mondal S, Roy JJ, Bera T (2014) Characterization of mitochondrial bioenergetic functions between two forms of Leishmania donovani – a comparative analysis. J Bioenerg Biomembr 46(5):395–402. https://doi.org/10.1007/s10863-014-9569-5
14. Dennery PA (2010) Oxidative stress in development: nature or nurture? Free Radic Biol Med 49(7):1147–1151. https://doi.org/10.1016/j.freeradbiomed.2010.07.011
15. Mukherjee SB, Das M, Sudhandiran G, Shaha C (2002) Increase in cytosolic Ca2+ levels through the activation of non-selective cation channels induced by oxidative stress causes mitochondrial depolarization leading to apoptosis-like death in Leishmania donovani promastigotes. J Biol Chem 277(27):24717–24727
16. Sacks D, Sher A (2002) Evasion of innate immunity by parasitic protozoa. Nat Immunol 3(11):1041–1047
17. Faria MS, Reis FC, Lima AP (2012) Toll-like receptors in Leishmania infections: guardians or promoters? J Parasitol Res 2012:930257. https://doi.org/10.1155/2012/930257
18. McMahon-Pratt and Alexander J. (2004) Does the Leishmania major paradigm of pathogenesis and protection hold for New World cutaneous leishmaniases or the visceral disease? Immunol Rev 201:206–224
19. Liew FY, Wei XQ, Proudfoot L (1997) Cytokines and nitric oxide as effector molecules against parasitic infections. Philos Trans R Soc Lond Ser B Biol Sci 352(1359):1311–1315
20. Das P, Lahiri A, Lahiri A, Chakravortty D (2010) Modulation of the arginase pathway in the context of microbial pathogenesis: a metabolic enzyme moonlighting as an immune modulator. PLoS Pathog 6(6):e1000899. https://doi.org/10.1371/journal.ppat.1000899
21. Chandra D, Naik S (2008) Leishmania donovani infection down-regulates TLR2-stimulated IL-12p40 and activates IL- 10 in cells of macrophage/monocytic lineage by modulating MAPK pathways through a contact-dependent mechanism. Clin Exp Immunol 154(2):224–234. https://doi.org/10.1111/j.1365-2249.2008.03741.x
22. Spath GF, Garraway LA, Turco SJ, Beverley SM (2003) The role(s) of lipophosphoglycan (LPG) in the establishment of Leishmania major infections in mammalian hosts. Proc Natl Acad Sci U S A 100(16):9536–9541
23. Olivier M, Atayde VD, Isnard A, Hassani K, Shio MT (2012) Leishmania virulence factors: focus on the metalloprotease GP63. Microbes Infect 14(15):1377–1389. https://doi.org/10.1016/j.micinf.2012.05.014
24. Meier P, Finch A, Evan G (2000) Apoptosis in development. Nature 407:796–801
25. Moreira ME, Del Portillo HA, Milder RV, Balanco JM, Barcinski MA (1996) Heat shock induction of apoptosis in promastigotes of the unicellular organism Leishmania (Leishmania) amazonensis. J Cell Physiol 167(2):305–313
26. Das M, Mukherjee SB, Shaha C (2001) Hydrogen peroxide induces apoptosis-like death in Leishmania donovani promastigotes. J Cell Sci 114(Pt 13):2461–2469

27. Smirlis D, Duszenko M, Ruiz AJ, Scoulica E, Bastien P, Fasel N, Soteriadou K (2010) Targeting essential pathways in trypanosomatids gives insights into protozoan mechanisms of cell death. Parasit Vectors 3:107. https://doi.org/10.1186/1756-3305-3-107
28. Kaczanowski S, Sajid M, Reece SE (2011) Evolution of apoptosis-like programmed cell death in unicellular protozoan parasites. Parasit Vectors 4:44. https://doi.org/10.1186/1756-3305-4-44
29. Lee N, Bertholet S, Debrabant A, Muller J, Duncan R, Nakhasi HL (2002) Programmed cell death in the unicellular protozoan parasite Leishmania. Cell Death Differ 9(1):53–64
30. Brookes PS, Yoon Y, Robotham JL, Anders MW, Sheu SS (2004) Calcium, ATP, and ROS: a mitochondrial love-hate triangle. Am J Physiol Cell Physiol 287(4):C817–CC33
31. Casanova M, Gonzalez IJ, Sprissler C, Zalila H, Dacher M, Basmaciyan L, Späth GF, Azas N, Fasel N (2015) Implication of different domains of the Leishmania major metacaspase in cell death and autophagy. Cell Death Dis 6:e1933. https://doi.org/10.1038/cddis.2015.288
32. Gannavaram S, Vedvyas C, Debrabant A (2008) Conservation of the proapoptotic nuclease activity of endonuclease G in unicellular trypanosomatid parasites. J Cell Sci 121(Pt 1):99–109
33. BoseDasgupta S, Das BB, Sengupta S, Ganguly A, Roy A, Dey S, Tripathi G, Dinda B, Majumder HK (2008) The caspase independent algorithm of programmed cell death in Leishmania induced by baicalein: the role of LdEndoG, LdFEN-1 and LdTatD as a DNA 'degradosome'. Cell Death Differ 15(10):1629–1640. https://doi.org/10.1038/cdd.2008.85
34. Kowaltowski AJ, de Souza-Pinto NC, Castilho RF, Vercesi AE (2009) Mitochondria and reactive oxygen species. Free Radic Biol Med 47(4):333–343. https://doi.org/10.1016/j.freeradbiomed.2009.05.004
35. Mehta A, Shaha C (2004) Apoptotic death in Leishmania donovani promastigotes in response to respiratory chain inhibition: complex II inhibition results in increased pentamidine cytotoxicity. J Biol Chem 279(12):11798–11813
36. Carvalho L, Luque-Ortega JR, Manzano JI, Castanys S, Rivas L, Gamarro F (2010) Tafenoquine, an antiplasmodial 8-aminoquinoline, targets Leishmania respiratory complex III and induces apoptosis. Antimicrob Agents Chemother 54(12):5344–5351. https://doi.org/10.1128/AAC.00790-10
37. Turrens JF (1987) Possible role of the NADH-fumarate reductase in superoxide anion and hydrogen peroxide production in Trypanosoma brucei. Mol Biochem Parasitol 25(1):55–60
38. Paramchuk WJ, Ismail SO, Bhatia A, Gedamu L (1997) Cloning, characterization and overexpression of two iron superoxide dismutase cDNAs from Leishmania chagasi: role in pathogenesis. Mol Biochem Parasitol 90(1):203–221
39. Ghosh S, Goswami S, Adhya S (2003) Role of superoxide dismutase in survival of Leishmania within the macrophage. Biochem J 369(Pt 3):447–452
40. Fairlamb AH, Blackburn P, Ulrich P, Chait BT, Cerami A (1985) Trypanothione: a novel bis(glutathionyl)spermidine cofactor for glutathione reductase in trypanosomatids. Science 227(4693):1485–1487
41. Krauth-Siegel RL, Leroux AE (2012) Low-molecular-mass antioxidants in parasites, Antioxid. Redox Signal 17(4):583–607. https://doi.org/10.1089/ars.2011.4392
42. Oza SL, Shaw MP, Wyllie S, Fairlamb AH (2005) Trypanothione biosynthesis in Leishmania major. Mol Biochem Parasitol 139(1):107–116
43. Steenkamp DJ (2002) Thiol metabolism of the trypanosomatids as potential drug targets. IUBMB Life 53(4–5):243–248
44. Heby O, Persson L, Rentala M (2007) Targeting the polyamine biosynthetic enzymes: a promising approach to therapy of African sleeping sickness, Chagas' disease, and leishmaniasis. Amino Acids 33(2):359–366
45. Vannier-Santos MA, Menezes D, Oliveira MF, de Mello FG (2008) The putrescine analogue 1,4-diamino-2-butanone affects polyamine synthesis, transport, ultrastructure and intracellular survival in Leishmania amazonensis. Microbiology 154(Pt 10):3104–3111. https://doi.org/10.1099/mic.0.2007/013896-0

46. Roberts SC, Jiang Y, Gasteier J, Frydman B, Marton LJ, Heby O, Ullman B (2007) Leishmania donovani polyamine biosynthetic enzyme overproducers as tools to investigate the mode of action of cytotoxic polyamine analogs. Antimicrob Agents Chemother 51(2):438–445
47. Reguera RM, Tekwani BL, Balaña-Fouce R (2005) Polyamine transport in parasites: a potential target for new antiparasitic drug development. Comp Biochem Physiol C Toxicol Pharmacol 140(2):151–164
48. Mukhopadhyay R, Dey S, Xu N, Gage D, Lightbody J, Ouellette M, Rosen BP (1996) Trypanothione overproduction and resistance to antimonials and arsenicals in Leishmania. Proc Natl Acad Sci U S A 93(19):10383–10387
49. Krauth-Siegel RL, Bauer H, Schirmer RH (2005) Dithiol proteins as guardians of the intracellular redox milieu in parasites: old and new drug targets in trypanosomes and malaria-causing plasmodia. Angew Chem Int Ed Engl 44(5):690–715
50. Tovar J, Cunningham ML, Smith AC, Croft SL, Fairlamb AH (1998) Down-regulation of Leishmania donovani trypanothione reductase by heterologous expression of a trans-dominant mutant homologue: effect on parasite intracellular survival. Proc Natl Acad Sci U S A 95(9):5311–5316
51. Krieger S, Schwarz W, Ariyanayagam MR, Fairlamb AH, Krauth-Siegel RL, Clayton C (2000) Trypanosomes lacking trypanothione reductase are avirulent and show increased sensitivity to oxidative stress. Mol Microbiol 35(3):542–552
52. Castro H, Tomás AM (2008) Peroxidases of trypanosomatids. Antioxid Redox Signal 10(9):1593–1606. https://doi.org/10.1089/ars.2008.2050
53. Krauth-Siegel LR, Comini MA, Schlecker T (2007) The trypanothione system. In Flohé L, Harris JR (eds) Peroxiredoxin systems. Subcellular biochemistry, vol 44. Springer, New York, pp 231–251
54. Flohé L, Budde H, Bruns K, Castro H, Clos J, Hofmann B, Kansal-Kalavar S, Krumme D, Menge U, Plank-Schumacher K, Sztajer H, Wissing J, Wylegalla C, Hecht HJ (2002) Tryparedoxin peroxidase of Leishmania donovani: molecular cloning, heterologous expression, specificity, and catalytic mechanism. Arch Biochem Biophys 397(2):324–335
55. Iyer JP, Kaprakkaden A, Choudhary ML, Shaha C (2008) Crucial role of cytosolic tryparedoxin peroxidase in Leishmania donovani survival, drug response and virulence. Mol Microbiol 68(2):372–391. https://doi.org/10.1111/j.1365-2958.2008.06154.x
56. Wyllie S, Mandal G, Singh N, Sundar S, Fairlamb AH, Chatterjee M (2010) Elevated levels of tryparedoxin peroxidase in antimony unresponsive Leishmania donovani field isolates. Mol Biochem Parasitol 173(2):162–164. https://doi.org/10.1016/j.molbiopara.2010.05.015
57. Suman SS, Equbal A, Zaidi A, Ansari MY, Singh KP, Singh K, Purkait B, Sahoo GC, Bimal S, Das P, Ali V (2016) Up-regulation of cytosolic tryparedoxin in Amp B resistant isolates of Leishmania donovani and its interaction with cytosolic tryparedoxin peroxidase. Biochimie 121:312–325. https://doi.org/10.1016/j.biochi.2015.12.017
58. Harder S, Bente M, Isermann K, Bruchhaus I (2006) Expression of a mitochondrial peroxiredoxin prevents programmed cell death in Leishmania donovani. Eukaryot Cell 5(5):861–870
59. Castro H, Sousa C, Santos M, Cordeiro-da-Silva A, Flohé L, Tomás AM (2002) Complementary antioxidant defense by cytoplasmic and mitochondrial peroxiredoxins in Leishmania infantum. Free Radic Biol Med 33(11):1552–1562
60. Barr SD, Gedamu L (2003) Role of peroxidoxins in Leishmania chagasi survival. Evidence of an enzymatic defense against nitrosative stress. J Biol Chem 278(12):10816–10823
61. Castro H, Teixeira F, Romao S, Santos M, Cruz T, Flórido M, Appelberg R, Oliveira P, Ferreira-da-Silva F, Tomás AM (2011) Leishmania mitochondrial peroxiredoxin plays a crucial peroxidase-unrelated role during infection: insight into its novel chaperone activity. PLoS Pathog 7(10):e1002325. https://doi.org/10.1371/journal.ppat.1002325
62. Onn I, Milman-Shtepel N, Shlomai J (2004) Redox potential regulates binding of universal Minicircle sequence binding protein at the Kinetoplast DNA replication origin. Eukaryot Cell 3(2):277–287
63. Fiorillo A, Colotti G, Boffi A, Baiocco P, Ilari A (2012) The crystal structures of the tryparedoxin-tryparedoxin peroxidase couple unveil the structural determinants of

Leishmania detoxification pathway. PLoS Negl Trop Dis 6(8):e1781. https://doi.org/10.1371/journal.pntd.0001781
64. Suman SS, Amit A, Singh KP, Gupta P, Equbal A, Kumari A, Topno RK, Ravidas V, Pandey K, Bimal S, Das P, Ali V (2018) Cytosolic tryparedoxin of Leishmania donovani modulates host immune response in visceral leishmaniasis. Cytokine 108:1–8. https://doi.org/10.1016/j.cyto.2018.03.010
65. Adak S, Datta AK (2005) Leishmania major encodes an unusual peroxidase that is a close homologue of plant ascorbate peroxidase: a novel role of the transmembrane domain. Biochem J 390(Pt 2):465–474
66. Jasion VS, Poulos TL (2012) Leishmania major peroxidase is a cytochrome c peroxidase. Biochemistry 51(12):2453–2460. https://doi.org/10.1021/bi300169x
67. Dolai S, Yadav RK, Pal S, Adak S (2008) Leishmania major ascorbate peroxidase overexpression protects cells against reactive oxygen species-mediated cardiolipin oxidation. Free Radic Biol Med 45(11):1520–1529. https://doi.org/10.1016/j.freeradbiomed.2008.08.029
68. Dolai S, Yadav RK, Pal S, Adak S (2009) Overexpression of mitochondrial Leishmania major ascorbate peroxidase enhances tolerance to oxidative stress-induced programmed cell death and protein damage. Eukaryot Cell 8(11):1721–1731. https://doi.org/10.1128/EC.00198-09
69. Pal S, Dolai S, Yadav RK, Adak S (2010) Ascorbate peroxidase from Leishmania major controls the virulence of infective stage of promastigotes by regulating oxidative stress. PLoS One 5(6):e11271. https://doi.org/10.1371/journal.pone.0011271
70. Ariyanayagam MR, Fairlamb AH (2001) Ovothiol and trypanothione as antioxidants in trypanosomatids. Mol Biochem Parasitol 115(2):189–198
71. Williams RA, Kelly SM, Mottram JC, Coombs GH (2003) 3-Mercaptopyruvate sulfurtransferase of Leishmania contains an unusual C-terminal extension and is involved in thioredoxin and antioxidant metabolism. J Biol Chem 278(3):1480–1486
72. Cunningham ML, Titus RG, Turco SJ, Beverley SM (2001) Regulation of differentiation to the infective stage of the protozoan parasite Leishmania major by tetrahydrobiopterin. Science 292(5515):285–287
73. Baiocco P, Colotti G, Franceschini S, Ilari A (2009) Molecular basis of antimony treatment in Leishmaniasis. J Med Chem 52(8):2603–2612. https://doi.org/10.1021/jm900185q
74. Pearson RD, Manian AA, Hall D, Harcus JL, Hewlett EL (1984) Antileishmanial activity of chlorpromazine. Antimicrob Agents Chemother 25(5):571–574
75. Chan C, Yin H, Garforth J, McKie JH, Jaouhari R, Speers P, Douglas KT, Rock PJ, Yardley V, Croft SL, Fairlamb AH (1998) Phenothiazine inhibitors of trypanothione reductase as potential antitrypanosomal and antileishmanial drugs. J Med Chem 41(2):148–156
76. Zilberstein D, Dwyer DM (1984) Antidepressants cause lethal disruption of membrane function in the human protozoan parasite Leishmania. Science 226(4677):977–979
77. Chibale K, Haupt H, Kendrick H, Yardley V, Saravanamuthu A, Fairlamb AH, Croft SL (2001) Antiprotozoal and cytotoxicity evaluation of sulfonamide and urea analogues of quinacrine. Bioorg Med Chem Lett 11(19):2655–2657
78. Parveen S, Khan MO, Austin SE, Croft SL, Yardley V, Rock P, Douglas KT (2005) Antitrypanosomal, antileishmanial, and antimalarial activities of quaternary arylalkylammonium 2-amino-4-chlorophenyl phenyl sulfides, a new class of trypanothione reductase inhibitor, and of N-acyl derivatives of 2-amino-4-chlorophenyl phenyl sulfide. J Med Chem 48(25):8087–8097
79. Bernardes LS, Zani CL, Carvalho I (2013) Trypanosomatidae diseases: from the current therapy to the efficacious role of trypanothione reductase in drug discovery. Curr Med Chem 20(21):2673–2696
80. Lizzi F, Veronesi G, Belluti F, Bergamini C, López-Sánchez A, Kaiser M, Brun R, Krauth-Siegel RL, Hall DG, Rivas L, Bolognesi ML (2012) Conjugation of quinones with natural polyamines: toward an expanded antitrypanosomatid profile. J Med Chem 55(23):10490–10500. https://doi.org/10.1021/jm301112z

81. Croft SL, Evans AT, Neal RA (1985) The activity of plumbagin and other electron carriers against Leishmania donovani and Leishmania mexicana amazonensis. Ann Trop Med Parasitol 79(6):651–653
82. Sharma N, Shukla AK, Das M, Dubey VK (2012) Evaluation of plumbagin and its derivative as potential modulators of redox thiol metabolism of Leishmania parasite. Parasitol Res 110(1):341–348. https://doi.org/10.1007/s00436-011-2498-x
83. Shukla AK, Patra S, Dubey VK (2011) Evaluation of selected antitumor agents as subversive substrate and potential inhibitor of trypanothione reductase: an alternative approach for chemotherapy of Leishmaniasis. Mol Cell Biochem 352(1–2):261–270. https://doi.org/10.1007/s11010-011-0762-0
84. Berman JD, Lee LS (1983) Activity of oral drugs against Leishmania tropica in human macrophages in vitro. Am J Trop Med Hyg 32(5):947–951
85. Riente RR, Souza VP, Carvalho SA, Kaiser M, Brun R, da Silva EF (2009) Leishmanicidal activity of new megazol derivatives. Med Chem 5(4):392–397
86. Delfín DA, Morgan RE, Zhu X, Werbovetz KA (2009) Redox-active dinitrodiphenylthioethers against Leishmania: synthesis, structure-activity relationships and mechanism of action studies. Bioorg Med Chem 17(2):820–829. https://doi.org/10.1016/j.bmc.2008.11.031
87. Ramu D, Garg S, Ayana R, Keerthana AK, Sharma V, Saini CP, Sen S, Pati S, Singh S (2016) Novel β-carboline-quinazolinone hybrids disrupt Leishmania donovani redox homeostasis and show promising antileishmanial activity. Biochem Pharmacol 129:26–42. https://doi.org/10.1016/j.bcp.2016.12.012
88. Ilari A, Baiocco P, Messori L, Fiorillo A, Boffi A, Gramiccia M, Di Muccio T, Colotti G (2012) A gold-containing drug against parasitic polyamine metabolism: the X-ray structure of trypanothione reductase from Leishmania infantum in complex with auranofin reveals a dual mechanism of enzyme inhibition. Amino Acids 42(2–3):803–811. https://doi.org/10.1007/s00726-011-0997-9
89. Venkatesan SK, Saudagar P, Shukla AK, Dubey VK (2011) Screening natural products database for identification of potential antileishmanial chemotherapeutic agents. Interdiscip Sci 3(3):217–231. https://doi.org/10.1007/s12539-011-0101-x
90. Ruiz-Santaquiteria M, Sánchez-Murcia PA, Toro MA, de Lucio H, Gutiérrez KJ, de Castro S, Carneiro FAC, Gago F, Jiménez-Ruiz A, Camarasa MJ, Velázquez S (2017) First example of peptides targeting the dimer interface of Leishmania infantum trypanothione reductase with potent in vitro antileishmanial activity. Eur J Med Chem 135:49–59. https://doi.org/10.1016/j.ejmech.2017.04.020
91. Sousa AF, Gomes-Alves AG, Benítez D, Comini MA, Flohé L, Jaeger T, Passos J, Stuhlmann F, Tomás AM, Castro H (2014) Genetic and chemical analyses reveal that trypanothione synthetase but not glutathionylspermidine synthetase is essential for Leishmania infantum. Free Radic Biol Med 73:229–238. https://doi.org/10.1016/j.freeradbiomed.2014.05.007
92. Saudagar P, Dubey VK (2011) Cloning, expression, characterization and inhibition studies on trypanothione synthetase, a drug target enzyme, from Leishmania donovani. Biol Chem 392(12):1113–1122. https://doi.org/10.1515/BC.2011.222
93. Saudagar P, Saha P, Saikia AK, Dubey VK (2013) Molecular mechanism underlying antileishmanial effect of oxabicyclo[3.3.1]nonanones: inhibition of key redox enzymes of the pathogen. Eur J Pharm Biopharm 85(3 Pt A):569–577. https://doi.org/10.1016/j.ejpb.2013.08.014
94. Leroux AE, Krauth-Siegel RL (2016) Thiol redox biology of trypanosomatids and potential targets for chemotherapy. Mol Biochem Parasitol 206(1–2):67–74. https://doi.org/10.1016/j.molbiopara.2015.11.003
95. Brindisi M, Brogi S, Relitti N, Vallone A, Butini S, Gemma S, Novellino E, Colotti G, Angiulli G, Di Chiaro F, Fiorillo A, Ilari A, Campiani G (2015) Structure-based discovery of the first non-covalent inhibitors of Leishmania major tryparedoxin peroxidase by high throughput docking. Sci Rep 5:9705. https://doi.org/10.1038/srep09705
96. Mansuri R, Kumar A, Rana S, Panthi B, Ansari MY, Das S, Dikhit MR, Sahoo GC, Das P (2017) In vitro evaluation of Antileishmanial activity of computationally screened com-

pounds against ascorbate peroxidase to combat amphotericin B drug resistance. Antimicrob Agents Chemother 61(7):pii:e02429-16. https://doi.org/10.1128/AAC.02429-16
97. Singh S, Sarma S, Katiyar SP, Das M, Bhardwaj R, Sundar D, Dubey VK (2015) Probing the molecular mechanism of hypericin-induced parasite death provides insight into the role of spermidine beyond redox metabolism in Leishmania donovani. Antimicrob Agents Chemother 59(1):15–24. https://doi.org/10.1128/AAC.04169-14
98. Mookerjee Basu J, Mookerjee A, Sen P, Bhaumik S, Sen P, Banerjee S, Naskar K, Choudhuri SK, Saha B, Raha S, Roy S (2006) Sodium antimony gluconate induces generation of reactive oxygen species and nitric oxide via phosphoinositide 3-kinase and mitogen-activated protein kinase activation in Leishmania donovani-infected macrophages. Antimicrob Agents Chemother 50(5):1788–1797
99. Mandal G, Wyllie S, Singh N, Sundar S, Fairlamb AH, Chatterjee M (2007) Increased levels of thiols protect antimony unresponsive Leishmania donovani field isolates against reactive oxygen species generated by trivalent antimony. Parasitology 134(Pt 12):1679–1687
100. Shadab M, Jha B, Asad M, Deepthi M, Kamran M, Ali N (2017) Apoptosis-like cell death in Leishmania donovani treated with KalsomeTM10, a new liposomal amphotericin B. PLoS One 12(2):e0171306. https://doi.org/10.1371/journal.pone.0171306
101. Getachew F, Gedamu L (2012) Leishmania donovani mitochondrial iron superoxide dismutase A is released into the cytosol during miltefosine induced programmed cell death. Mol Biochem Parasitol 183(1):42–51
102. Deep DK, Singh R, Bhandari V, Verma A, Sharma V, Wajid S, Sundar S, Ramesh V, Dujardin JC, Salotra P (2017) Increased miltefosine tolerance in clinical isolates of Leishmania donovani is associated with reduced drug accumulation, increased infectivity and resistance to oxidative stress. PLoS Negl Trop Dis 11(6):e0005641. https://doi.org/10.1371/journal.pntd.0005641
103. Das R, Roy A, Dutta N, Majumder HK (2008) Reactive oxygen species and imbalance of calcium homeostasis contributes to curcumin induced programmed cell death in Leishmania donovani. Apoptosis 13(7):867–882. https://doi.org/10.1007/s10495-008-0224-7
104. Awasthi BP, Kathuria M, Pant G, Kumari N, Mitra K (2016) Plumbagin, a plant-derived naphthoquinone metabolite induces mitochondria mediated apoptosis-like cell death in Leishmania donovani: an ultrastructural and physiological study. Apoptosis 21(8):941–953. https://doi.org/10.1007/s10495-016-1259-9
105. Fonseca-Silva F, Inacio JD, Canto-Cavalheiro MM, Almeida-Amaral EE (2011) Reactive oxygen species production and mitochondrial dysfunction contribute to quercetin induced death in Leishmania amazonensis. PLoS One 6(2):e14666. https://doi.org/10.1371/journal.pone.0014666
106. Fonseca-Silva F, Inacio JD, Canto-Cavalheiro MM, Almeida-Amaral EE (2013) Reactive oxygen species production by quercetin causes the death of Leishmania amazonensis intracellular amastigotes. J Nat Prod 76(8):1505–1508. https://doi.org/10.1021/np400193m
107. Ribeiro GA, Cunha-Júnior EF, Pinheiro RO, da Silva SA, Canto-Cavalheiro MM, da Silva AJ, Costa PR, Netto CD, Melo RC, Almeida-Amaral EE, Torres-Santos EC (2013) LQB-118, an orally active pterocarpanquinone, induces selective oxidative stress and apoptosis in Leishmania amazonensis. J Antimicrob Chemother 68(4):789–799. https://doi.org/10.1093/jac/dks498
108. Mukherjee D, Singh CB, Dey S, Mandal S, Ghosh J, Mallick S, Hussain A, Swapana N, Ross SA, Pal C (2016) Induction of apoptosis by zerumbone isolated from Zingiber zerumbet (L.) smith in protozoan parasite Leishmania donovani due to oxidative stress. Braz J Infect Dis 20(1):48–55. https://doi.org/10.1016/j.bjid.2015.10.002
109. Fonseca-Silva F, Canto-Cavalheiro MM, Menna-Barreto RF, Almeida-Amaral EE (2015) Effect of Apigenin on Leishmania amazonensis is associated with reactive oxygen species production followed by mitochondrial dysfunction. J Nat Prod 78(4):880–884. https://doi.org/10.1021/acs.jnatprod.5b00011
110. Inacio JD, Gervazoni L, Canto-Cavalheiro MM, Almeida-Amaral EE (2014) The effect of (-)-epigallocatechin 3-O--gallate in vitro and in vivo in Leishmania braziliensis: involvement

of reactive oxygen species as a mechanism of action. PLoS Negl Trop Dis 8(8):e3093. https://doi.org/10.1371/journal.pntd.0003093
111. Shukla AK, Patra S, Dubey VK (2012) Iridoid glucosides from Nyctanthes arbortristis result in increased reactive oxygen species and cellular redox homeostasis imbalance in Leishmania parasite. Eur J Med Chem 54:49–58. https://doi.org/10.1016/j.ejmech.2012.04.034
112. Kathuria M, Bhattacharjee A, Sashidhara KV, Singh SP, Mitra K (2014) Induction of mitochondrial dysfunction and oxidative stress in Leishmania donovani by orally active clerodane diterpene. Antimicrob Agents Chemother 58(10):5916–5928. https://doi.org/10.1128/AAC.02459-14
113. Roy S, Dutta D, Satyavarapu EM, Yadav PK, Mandal C, Kar S, Mandal C (2017) Mahanine exerts in vitro and in vivo antileishmanial activity by modulation of redox homeostasis. Sci Rep 7(1):4141. https://doi.org/10.1038/s41598-017-03943-y
114. Saudagar P, Dubey VK (2014) Molecular mechanisms of in vitro betulin-induced apoptosis of Leishmania donovani. Am J Trop Med Hyg 90(2):354–360. https://doi.org/10.4269/ajtmh.13-0320
115. Chowdhury S, Mukherjee T, Chowdhury SR, Sengupta S, Mukhopadhyay S, Jaisankar P, Majumder HK (2014) Disuccinyl betulin triggers metacaspase-dependent endonuclease G-mediated cell death in unicellular protozoan parasite Leishmania donovani. Antimicrob Agents Chemother 58(4):2186–2201. https://doi.org/10.1128/AAC.02193-13
116. Roy A, Ganguly A, BoseDasgupta S, Das BB, Pal C, Jaisankar P, Majumder HK (2008) Mitochondria-dependent reactive oxygen species-mediated programmed cell death induced by 3,3′-diindolylmethane through inhibition of F0F1-ATP synthase in unicellular protozoan parasite Leishmania donovani. Mol Pharmacol 74(5):1292–1307. https://doi.org/10.1124/mol.108.050161
117. Carvalho L, Luque-Ortega JR, López-Martín C, Castanys S, Rivas L, Gamarro F (2011) The 8-aminoquinoline analogue sitamaquine causes oxidative stress in Leishmania donovani promastigotes by targeting succinate dehydrogenase. Antimicrob Agents Chemother 55(9):4204–4210. https://doi.org/10.1128/AAC.00520-11
118. Mandal A, Das S, Roy S, Ghosh AK, Sardar AH, Verma S, Saini S, Singh R, Abhishek K, Kumar A, Mandal C, Das P (2016) Deprivation of L-arginine induces oxidative stress mediated apoptosis in Leishmania donovani promastigotes: contribution of the polyamine pathway. PLoS Negl Trop Dis 10(1):e0004373. https://doi.org/10.1371/journal.pntd.0004373. eCollection 2016 Jan
119. Kumari A, Singh KP, Mandal A, Paswan RK, Sinha P, Das P, Ali V, Bimal S, Lal CS (2017) Intracellular zinc flux causes reactive oxygen species mediated mitochondrial dysfunction leading to cell death in Leishmania donovani. PLoS One 12(6):e0178800. https://doi.org/10.1371/journal.pone.0178800
120. Mesquita JT, da Costa-Silva TA, Borborema SE, Tempone AG (2014) Activity of imidazole compounds on Leishmania (L.) infantum chagasi: reactive oxygen species induced by econazole. Mol Cell Biochem 389(1–2):293–300. https://doi.org/10.1007/s11010-013-1954-6
121. Awasthi BP, Mitra K (2018) In vitro leishmanicidal effects of the anti-fungal drug natamycin are mediated through disruption of calcium homeostasis and mitochondrial dysfunction. Apoptosis 23(7–8):420–435. https://doi.org/10.1007/s10495-018-1468-5
122. Sen N, Das BB, Ganguly A, Mukherjee T, Tripathi G, Bandyopadhyay S, Rakshit S, Sen T, Majumder HK (2004) Camptothecin induced mitochondrial dysfunction leading to programmed cell death in unicellular hemoflagellate Leishmania donovani. Cell Death Differ 11(8):924–936
123. Vishwakarma P, Parmar N, Yadav PK, Chandrakar P, Kar S (2016) 15d-prostaglandin J2 induced reactive oxygen species-mediated apoptosis during experimental visceral Leishmaniasis. J Mol Med (Berl) 94(6):695–710. https://doi.org/10.1007/s00109-016-1384-5

MIX
Papier aus verantwortungsvollen Quellen
Paper from responsible sources
FSC® C105338

If you have any concerns about our products,
you can contact us on
ProductSafety@springernature.com

In case Publisher is established outside the EU,
the EU authorized representative is:
Springer Nature Customer Service Center GmbH
Europaplatz 3, 69115 Heidelberg, Germany

Printed by Libri Plureos GmbH
in Hamburg, Germany